7th International Conference on Renewable Energy and Materials Technology (ICOREMT 2021)

IOP Conference Series: Earth and Environmental Science
Volume 877

Erbil, Iraq
1 - 2 September 2021

Editors:

Dhiya Al-Jumeily
Rafid Al Khaddar
Talib Al-Khaseraji
Sabeeh Lafta Farhan
Mohammed Al-Jumaili
Laith Abdul-Rahaim
Jasim Salman
Khalaf Gaeid
Anas Al-Hayawi
Abbas Al-Hdabi
Khalid Hashim
Monower Sadique
Patryk Kot
Salah L. Zubaidi
Michaela Gkantou
Magomed Muradov
Alaa Abd Ali
Shatha Al-Samarrai
Saleem Ethaib Mohammad
Jareer Jaber Mohammed

ISBN: 978-1-7138-4859-2
ISSN: 1755-1307

Printed from e-media with permission by:

Curran Associates, Inc.
57 Morehouse Lane
Red Hook, NY 12571

CURRAN ASSOCIATES INC. proceedings .com

Some format issues inherent in the e-media version may also appear in this print version.

Institute of Physics
Dirac House, Temple Back
Bristol BS1 6BE UK

Phone: 44 1 17 929 7481
Fax: 44 1 17 920 0979

techtracking@iop.org

Additional copies of this publication are available from:

Curran Associates, Inc.
57 Morehouse Lane
Red Hook, NY 12571 USA
Phone: 845-758-0400
Fax: 845-758-2633
Email: curran@proceedings.com
Web: www.proceedings.com

TABLE OF CONTENTS

IOP Conf. Series: Earth and Environmental Science **877** (2021) 011001 doi:10.1088/1755-1315/877/1/011001

Preface

It is our great pleasure to introduce the Proceedings of the 7th International Conference on Renewable Energy and Materials Technology (ICOREMT 2021), held at Erbil, Iraq, from the 2nd to the 3rd of August 2021.

The core aim of ICOREMT-2021 has always been to bring together early-career researchers, scientists, academics, engineers, and postgraduate students to exchange and share their experiences and research results on all aspects of science, technology, and engineering. Accordingly, the professional keynote speakers and researchers have presented various perspectives on research.

All submitted papers were have gone through an initial assessment by the editors of ICOREMT-2021 before sending them for reviewers. The latter were carefully selected from many countries to ensure reliable outcomes. The committee of STEPS-2020 followed an accurate and professional double-blind peer-reviewing process that involved 223 reviewers from all over the world. In total, 272 papers were submitted to ICOREMT-2021; 22 were rejected during the initial assessment process, and 56 papers were accepted (acceptance rate of 20.59%). The accepted papers demonstrated novel ideas and impressive effort in engineering science and technology.

We wish to express our sincere gratitude to all individuals, authors, editors, reviewers, and committees of ICOREMT-2021 who have contributed to ICOREMT-2021. Without their support, it would not have been possible to make ICOREMT-2021 a successful event in this challenging time.

We would also like to express our sincere gratitude to our partner, Tikrit University, Iraq. Without their support, it would not have been possible to host such a successful international event.

Warmest regards,

STEPS Team

Editors

- Prof Dhiya Al-Jumeily
- Prof Rafid Al Khaddar
- Prof Talib Al-Khaseraji
- Prof Sabeeh Lafta Farhan
- Prof Mohammed Al-Jumaili
- Prof Laith Abdul-Rahaim
- Prof Jasim Salman
- Assistant Prof Khalaf Gaeid
- Assistant Prof Anas Al-Hayawi
- Assistant Prof Abbas Al-Hdabi
- Dr Khalid Hashim
- Dr Monower Sadique
- Dr Patryk Kot
- Dr Salah L. Zubaidi
- Dr Michaela Gkantou
- Dr Magomed Muradov
- Dr Alaa Abd Ali
- Dr Shatha Al-Samarrai
- Dr Saleem Ethaib Mohammad
- Dr Jareer Jaber Mohammed

Scientific Committee

- Prof Rafid Al Khaddar, United Kingdom
- Prof Sabeeh Lafta Farhan, Iraq
- Prof Dhiya Al-Jumeily, United Kingdom
- Prof. Abdullah Al-Ani, Canada
- Prof. Aura Conc, Brazil
- Prof. Adel Al- Al-Baghdadi, Iraq
- Prof. Chani Park, Korea
- Prof. Alexander Ivanov, Bulgaria
- Prof. Fouad A. Al-Saady, Iraq
- Prof. Hussain Ahmed, UAE
- Prof. Ivan Yrmolov, United Kingdom
- Prof. Kahtan Hadi Hussein, Iraq
- Prof Karkaz M. Thalij, Iraq
- Prof. Laith A. AbdulRahaim, Iraq
- Prof. Jasim Salman, UAE
- Prof. Marley Maria, France
- Prof. Mohamed Hamdan, United Kingdom
- Prof. Raouf Naguib, United Kingdom
- Prof. Abdulkhaleq Alwan, Iraq
- Prof. Ahmed Ghazali, Indonesia
- Assist Prof. Anas Al-Hayawi, Iraq
- Assist Prof. Abbas Al-Hdabi, Iraq
- Assist Prof. Khalid Safaa Hashim, Iraq
- Assist Prof. Asad Al Defae, Iraq
- Assist Prof. Thaar S. Al- Gasham, Iraq
- Dr Jakub S. Markiewicz, Poland
- Dr Ariel Edesess, United Kingdom
- Dr Ibijoka Idowu, United Kingdom

- Dr Khalid Hashim, United Kingdom
- Dr Patryk Kot, United Kingdom
- Dr Michaela Gkantou, United Kingdom
- Dr Magomed Muradov, United Kingdom
- Dr Daneil Anton, Spain

IOP Conf. Series: Earth and Environmental Science **877** (2021) 011002 doi:10.1088/1755-1315/877/1/011002

Peer review declaration

All papers published in this volume of IOP Conference Series: Earth and Environmental Science have been peer reviewed through processes administered by the Editors. Reviews were conducted by expert referees to the professional and scientific standards expected of a proceedings journal published by IOP Publishing.

- **Type of peer review:** Double-blind
- **Describe criteria used by Reviewers when accepting/declining papers. Was there the opportunity to resubmit articles after revisions?**

  Reviewers of ICOREMT-2021 considered the following vital points in the assessment process:

A- Technical Criteria (such as scientific merits and adequacy of discussion)
B- Quality Criteria (such as the scientific interest of the results and the ratio of length and importance of the idea)
C- Presentation Criteria (quality of figures/tables and conclusions)

  One opportunity was given to the authors to address reviewers' comments and corrections.

- **Conference submission management system:** Direct submission via official email address.

- **Number of submissions received:** 272 papers

- **Number of submissions sent for review:** 250 papers

- **Number of submissions accepted:** 56 papers

- **Acceptance Rate (Number of Submissions Accepted / Number of Submissions Received X 100):** 20.59%

- **The average number of reviews per paper:** 2 reviewers

- **Total number of reviewers involved:** 203 reviewers

- **Any additional info on the review process (i.e. plagiarism check system):**
  All papers were checked using Turnitin software; papers with a more than 20% similarity percentage were rejected without review. The reviewers checked the quality of the submitted papers, including the contribution to the knowledge, the importance of the idea, adequacy of discussion, quality of figures and tables, and the conclusions. After the reviewing process, one opportunity was given to the authors to address reviewers' comments and corrections.

- **Contact person for queries:**

Dr Khalid Hashim,
Liverpool John Moores University, UK

IOP Conf. Series: Earth and Environmental Science **877** (2021) 011002 doi:10.1088/1755-1315/877/1/011002

Email: k.s.hashim@ljmu.ac.uk

IOP Conf. Series: Earth and Environmental Science **877** (2021) 012001 doi:10.1088/1755-1315/877/1/012001

Simulation of a perovskite sandwich solar cell with the p-CZTS / p-$CH_3NH_3PbCI_3$ / p-CZTS absorber layers

Marwah S Mahmood and N K Hassan
Dept. of Physics, College of Education for Pure Sciences, Tikrit University, Iraq

E-mail: mmiru7761@st.tu.edu.iq

Abstract. Perovskite solar cells attract the attention because of their unique properties in photovoltaic cells. Numerical simulation to the structure of Perovskite on p-CZTS/p-CH3NH3PbCI3/p-CZTS absorber layers is performed by using a program solar cell capacitance simulator (SCAPS-1D), with changing absorber layer thickness. The effect of thickness p-CZTS/p-CH3NH3PbCI3/p-CZTS, layers at (3.2μm, 1.8 μm, 1.1 μm) respectively are studied. The obtained results are short circuit current density (J_{sc}), open circuit voltage (V_{oc}), fill factor ($F.F$) and power conversion efficiency (PCE) equal to (28 *mA/cm²*, 0.83 *v*, 60.58 % and 14.25 %) respectively at 1.1 μm thickness. Our findings revealed that the dependence of current - voltage characteristics on the thickness of the absorbing layers, an increase in the amount of short circuit current density with an increase in the thickness of the absorption layers and thus led to an increase in the conversion efficiency and improvement of the cell by increasing the thickness of the absorption layers.

1. Introduction

In the past decades. Several articles have been published in the development of perovskite solar cell PSC, using different techniques under multiple parameters,in addition to practical fabrication, there are many numerical simulations to simulate proposed structure. Numerical simulation have been mad using SCAPS by Kojima et.al 2009 [1] the peroveskite solar cells was two organolead halide, power conversion efficiency PCE was 3.8 %. Tin Halide peroveskite solar cells studied by B.J.Babu et.al using SCAPS-1D 2018 [2], the power conversion efficiency PCE was 25.91 %.The absorber layers in the previous studies was highly toxic.

Our study aimed to improve such kind of solar cell by changing the absorber layers thickness with low toxicity. (p-CZTS/p-$CH_3NH_3PbCl_3$/p-CZTS) absorber layers with different thickness ($\mathbf{3.2\ \mu m, 1.8\ \mu m}$ **and** $\mathbf{1.1\ \mu m}$) respectively. For peroveskite solar have been simulated using SCAPS Method. Perovskite solar cell sandwiched (PSCs) advantages are high absorption coefficient and easy fabrications. These advantages make them very attractive for future solar cell technologies. The PCE obtained was 14.25 %.The advantages of changing absorber layer thickness leads to efficient solar cells.

2. Numerical simulation

Basic semiconductor equations simulation by SCAPS-1D Computer-based window application, Possion and continuity equation [3]. Thus, is used in this study to simulate PSCs and study properties (I–V) [4], as given in equation (1).

$$Voc = \frac{kt}{q} In\left(\frac{Jsc}{J_\circ} + 1\right) \quad (1)$$

Where k= Boltzmann constant, t = Operating Temperature, $J_\circ$ = Reverse saturation current, q = Electron charge. The parameters set for simulation are selected from the previous experimental work Ismaila.B, et.al to study the hole transporting material effect on perovskite based solar cells using SCAPS – 1D [5-6]. Table 1 summarizes all the parameters in the simulation process.

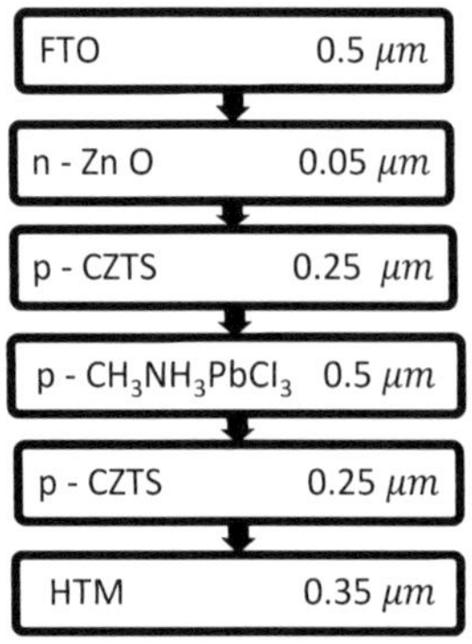

Figure 1.Architecture of simulated model.

Table 1.The parameters set based at *(300°k) and AM1.5sun.*

Layers	FTO	Zn O	CZTS	$CH_3NH_3PbCl_3$	HTM
Thickness, x (μm)	0.5	0.05	0.25	0.5	0.35
Band gab, Eg (eV)	3.5	3.35	1.55	1.55	2.45
Electron affinity, x (eV)	4	4.21	4.5	3.9	3
Dielectric permittivity, $\varepsilon/\varepsilon_o$	9	9E 0	10	6.5	3
$N_C\ (cm^{-3})$	2.2E+18	2.2E+18	2.2E+18	2.2E+18	1E+15
$N_V\ (cm^{-3})$	1.8E+19	1.8E+19	1.8E+19	1.8E+19	1E+19
Electron thermal velocity (cm/s)	2E+1	2.5E+1	1E+2	1.6E 0	2E-4
Hole thermal velocity (cm/s)	1E 0	1E+2	2E+1	2E-1	2E-4
Electron mobility $(cm^2/v.s)$	1E+7	1E+7	1E+2	3E+7	1E+7
Hole mobility $(cm^2/v.s)$	1E+7	1E+7	1E+7	3E+7	1E+7
$N_D\ (cm^{-3})$	2E+19	1E+18	0 E 0	0 E 0	0 E 0
$N_A\ (cm^{-3})$	0 E 0	0 E 0	8.22E+18	6E+18	2E+18
Reference	[7]	[8]	[9]	[10,11,12]	[13]

3. Results and discussion

The study is done on the structure of the perovskite solar cell sandwiched FTO/n-Zn O/p-CZTS/p-CH3NH3PbCI3/p-CZTS/ HTM,these layers include **HTM** hole transport material ,**Zn O** buffer, and **FTO** window, as shown in figure 1. The incident photons will interact with the absorber layer resulting in photo-exited carriers. Based on the fact the depending of photo generated carriers on the thickness of the absorber layer, the absorber layer thickness has been changed to an optimum thickness to obtain more efficient solar cell. Modification of absorber layer thickness clearly affected the

IOP Conf. Series: Earth and Environmental Science **877** (2021) 012001 doi:10.1088/1755-1315/877/1/012001

diffusion length of carriers, very thin absorber layer will decrease the absorption rate. The optical absorption in addition of photo generated carriers depending on the length of diffusion which are directly proportional with the width of the absorber layer .The dependence of solar cell parameters such as PCE, F.F, J_{sc} , V_{oc} with varying thickness are shown in figure 2.

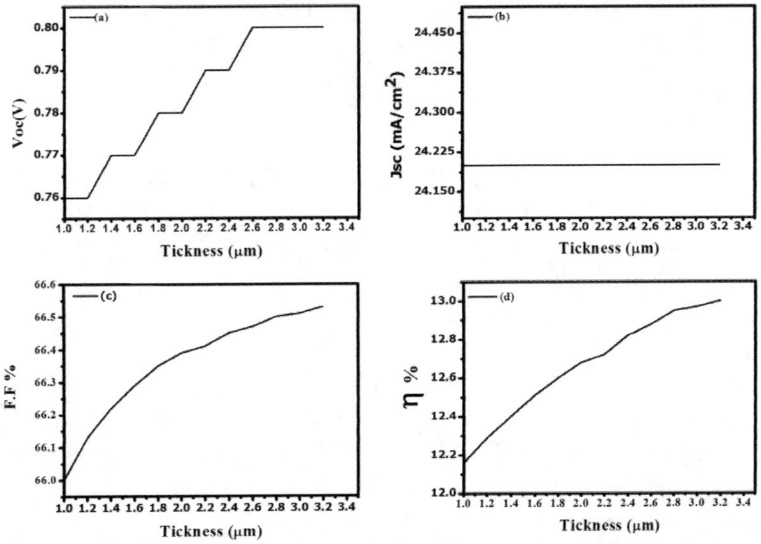

Figure 2. The first absorber layer (p-CZTS).

Based on figure 2, the V_{oc} is increasing slightly with the increasing of thickness. In this figure, the first absorber layer (p-CZTS) PCE of 13% with, J_{sc}=24.58 *mA/cm²*, V_{oc}= 0.8 *v*, and *F.F*= 66.53 %. This is achieved when the thickness reaches (3.2 μm).When the thickness is increasing from(1 μm) to (3.2 μm),both of PCE and, J_{sc} increase up to(3.2 μm).This represents the increase in the generation of the electron and hole pairs in the absorber layer. But when it increases from (1 μm) to higher amount, the PCE and, J_{sc} both undergo saturation. This indicates the recombination of charges inside the absorber layer before reaching the contacts.

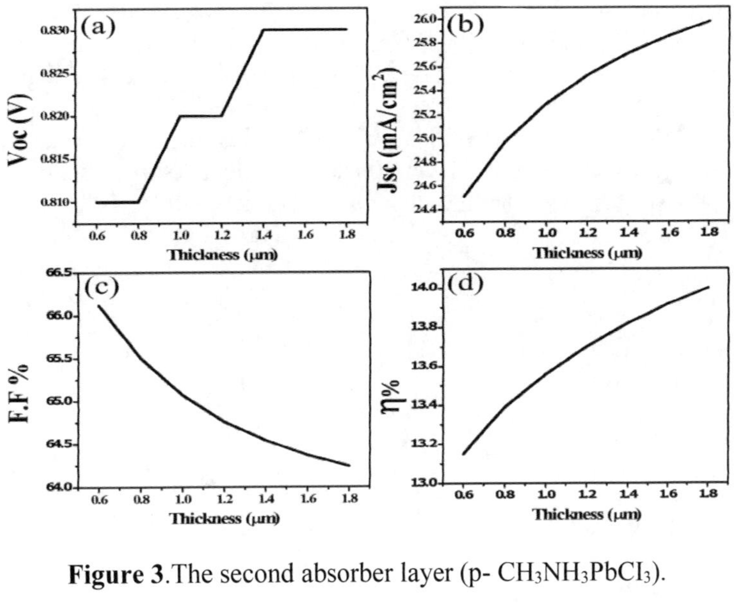

Figure 3.The second absorber layer (p- $CH_3NH_3PbCl_3$).

In this layer, PCE is(14%),with, J_{sc} = 25.98 *mA/cm²*, V_{oc} = 0.83 *v*, and *F.F*= 64.24 %.This is achieved when the thickness reaches (1.8 μm).In this case .When the thickness increases from(0.6 μm) to (1.8 μm), both of PCE and, J_{sc} increase up to(1.8 μm) .

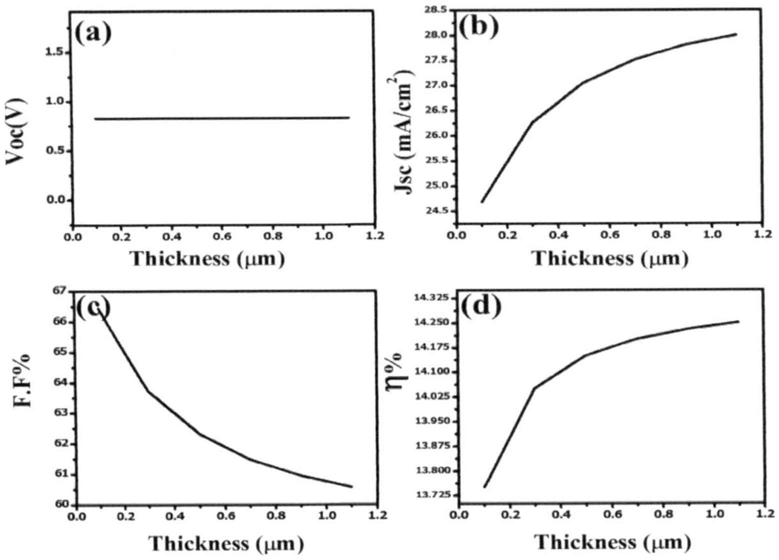

Figure 4.The third absorber layer (p-CZTS)

In this layer, PCE is 14.25% with, J_{sc} =28 *mA/cm²*,V_{oc} =0.83 *v*, and F.F = 60.58 %.When the thickness increases from (0.1 μm) to (1.1 μm),the PCE and, J_{sc} also increase up to (1.1 μm).The thickness of first absorber layer (p- CZTS)was changed from(1 μm)to(3.2μm);the second layer(p-CH3NH3PbCI3) from(0.6 μm)to(1.8 μm)and the third layer (p-CZTS)from(0.1 μm)to(1.1 μm).To reach a higher performance level, the architecture of perovskite solar cell should have significant characteristics like use less harmful materials and easy fabrication, and low cost .

4. Conclusion

Using simulation program SCAPS-1D.The changes done thickness layers of the three absorption for the first (p-CZTS), second (p-CH3NH3PbCI3)and third(p-CZTS)layers as follows (1- 3.2) μm,(0.6 - 1.8) μm, (0.1-1.1) μm, respectively. Hence, the high performance of a solar cell is achieved at the thickness of (1.1 μm). When the absorption layer defects are reduced, the performance of the solar cell improves,and we find through simulation that a layer CH3NH3PbCI3 layer attained (14%) for PCE . By simulating the model, the absorption layers thickness influences perovskite solar cell to achieve higher efficiency. The results show that the maximum rates attained for PCE , J_{sc} , F.F and V_{oc} were (14.25 %,28 *mA/cm²*,60.58 %,and 0.83 *v*),respectively .

Acknowledgment

We to thank Marc .B, of University Gent, Belgium (the Electronic information system) to get a SCAPS-1D simulation software.

5. Reference

[1] Kojima A, Teshima K, Shirai Y and Miyasaka T 2009 Organometal Halide Perovskite as Visible-Light Sensitizers for Photovoltaic Cells *Journal of American Chemical Society* **131** 6050–6051

[2] Mandadapu U, Vedanayakam S V, Thyagarajan K and Babu B J 2018 Optimisation of high efficiency tin halide perovskite solar cells using SCAPS-1D *International Journal of Simulation and Process Modelling* **13** 221-227
[3] Wang K, Gunawan O, Todorov T, Shin B, Chey S J, Bojarczuk N A, Mitzi D and Guhaa S Thermally Evaporated Cu2ZnSnS4 Solar Cell *Applied Physics Letters* **97**
[4] Lin P, Lin L, Yu J, Cheng S, Lu P and Zheng Q 2013 Numerical simulation of Cu2ZuSnS4 based solar cells with In2S3 buffer layers by SCAPS-1D *Journal of Applied Science and Engineering* **17** 383-390
[5] Ismaila T B 2019 *Hole Transporting Material Effect on sandwiched perovskite Based solar cells using SCAPS – 1D*
[6] Du H-J, Wang W-C and Zhu J- Z 2016 Device simulation of lead-free CH3NH3SnI3 perovskite solar cells with high-efficiency *China Physics B* **25**
[7] Chandrasekharan R 2012 *Numerical modeling of tin – based absorber devices for cost effective solar photovoltaic* Ph. D. thesis, the Pennsylvania states university, united states of America
[8] Olopade M A, Oyabola O O and Adeleke B S 2012 Investigation of some materials as buffer layer in copper zinc tin sulphade (Cu2ZnSnS4) solar cell by SCAPS – 1D pelagia Research library *Advanced in Applied Science research* **3** 3396 – 3400
[9] Mebarkia C, Dib D, Zerfaoui H and Belght R 2016 Energy efficiency of photovoltaic cell based thin films CZTS by SCAPS *Journal of foundamental and Applied science* **8** 363 – 371
[10] Lee M M, Teuscher J, Miyasaka T, Murakami T N and Snaith H J 2012 Efficient hybrid solar cells based on meso-super structure d'organo metal halide perovskites *Science* **338** 643-647
[11] Noh J H, Im S H, Heo J H, Mandal T N, Seok S2013 Chemical management for colorful, efficient, and stable inorganic – organic hybrid nanostructured solar cells", *Nano letters* **13**
[12] Hirasawa M, Ishihara T, Goto T, Uchida K, Miura N 1994 Magneto absorption of the lowest excition in perovskite – type compound (CG3NH3)PbcI3 *Physica B:condensed Matter* **201**
[13] Soni R, Eva R, Barea M, and Santiago F. F 2019 Analysis of the origin of Open Circuit Voltage in Dye Solar Cells *Journal of Physical Chemical Letters* **3** 1629-1634

IOP Conf. Series: Earth and Environmental Science **877** (2021) 012002 doi:10.1088/1755-1315/877/1/012002

The operation characteristics of air cathode Microbial Desalination Cell to treat oil refinery wastewater

S S Jaroo[1,3], G F Jumaah[1] and T R Abbas[2]

[1] Civil Engineering Department, University of Technology, Baghdad, Iraq
[2] Environment and Water Directorate, Ministry of Science and Technology, Baghdad, Iraq
[3] Corresponding author E-mail: 42102@student.uotechnology.edu.iq

Abstract. This system [microbial desalination cell (MDC)] is considered an excellent sustainable process to treat wastewater by biological anaerobic oxidation of the organic material by electroactive bacteria, desalinate saltwater, and electrical power generation. In the present work, MDC was used for treating oil refinery wastewater in the anode chamber by anaerobic bacteria. Simultaneously, an air pump was used to provide the oxygen to the cathode chamber as an electron acceptor to generate bioelectricity power. The power density generated by this air cathode MDC with 1KΩ external resistance at the 1st experiment was 71.11 $\mu W/m^2$. It increased to a peak value of 570.86 $\mu W/m^2$ at the last experiment. The maximum chemical oxygen demand (COD) removal percent of oily wastewater was 96%. The higher salinity removal rate 150.39 ppm/h with a first salt concentration in a desalinating chamber of 35000 ppm.

1. Introduction

Oil refineries usually generate a significant quantity of wastewater from different refining processes, for example, percolation, desalting, distillation, thermal cracking, and other treatment processes used in oil refining to produce beneficial products [1,2]. The quantity of the generated wastewater is estimated to be as high as 1.6 times the crude oil volume remedied [3]. Discharging this type of wastewater into surface water bodies without treatment causes significant damages to environmental and human health due to the poisonous contaminants such as (hydrocarbons, phenol, and dissolved minerals) [4,5]. Therefore, this wastewater most has sufficiently remedied to satisfy the environmental regulations standards [6]. Different physical and chemical oily wastewater treatment processes are implemented utilizing different electrocoagulation methods, electrochemical oxidation, and dissolved air flotation [7]. However, these processes are costly due to the enormous quantities of chemicals are used and expensive pieces of equipment, and extravagant sludge production amounts. So, biological methods are an excellent alternative to the chemicals and physical processes due to modest and uncostly with friend-environment operating conditions [2-8]. Biological processes use aerobic, anaerobic, or semi aerobic microorganisms to transform complex organic material into simple materials (CO_2, H_2O, and CH_4) [9]; a bioelectrochemical system one of the current biological treatment methods to treat many types of industrial wastewater.

Bioelectrochemical systems can produce the excellent value of chemical energy products or clean electricity from different wastewater sources. Organic or inorganic wastes can avail to use by electroactive bacteria as fuel [10,11]. The microbial desalination cell (MDC) combines wastewater and saltwater treatment to benefit from electrical power recapture without any exterior electrical power chip or mechanistic activity or pressure implementation [12]. MDC is progressed from the microbial fuel cell

(MFC) by adding the desalination compartment in the middle of the anode and cathode compartments [13,14]. The anode chamber used many types of wastewater as a substrate, synthetic wastewater, or real wastewater. The anode chamber is considered a powerhouse for the MDC; microbes donate their electrons to the chamber's electrode. The electrons later pass to the electrode of the cathode chamber through an outer circuit. In the cathode chamber, the definitive response happens at the cathode compartment's electrode in the oxygen existence. Air cathode is utilized in MDC; this supplies a more suitable and efficient system to reduce the synoptic cost [15,16].

Moreover, many MDC designs are suggested by many researchers to promote the process's execution, and one of these designs air cathode MDC. This design has two ways to supply the oxygen in the cathode chamber: exposing the cathode's electrode to the air directly. The second way was used in this research; an air pump provides the oxygen to the cathode chamber. While widely utilizing synthetic wastewater as a substrate in MDCs [17,18], little works focus on treating actual wastewater. For example, Surajbhan et al., [15] studied seawater desalination with simultaneous bioelectricity generation from MDC treatment. In this work, MDC was used to treat petroleum refinery wastewater and study the saltwater desalination and catholyte solution's effect on the treating process. Atieh et al., [19] compared the execution of an MDC under batch and continuous municipal wastewater feeding conditions. The results of this study presented relatively improve MDC performance by anolyte enrichment under controlled continuous feeding conditions. Long-term execution and description of this system in domestic wastewater remediation studied by Haiping et al., [20]. This work described MDC's long-term execution pending wastewater treatment and determined the critical factors of execution decrease. The 8-month was the operation period for this work, and during this period, the MDC execution declined by time. The dropping in current density was 47%, the decline in Columbic efficiency was 46%, and the decline in desalination rate was 27%. Haiping et al., [17] studied MDCs for improving the execution in municipal wastewater treating, desalination, and electric power generation.

This work investigates air cathode MDC's performance to treat oil refinery wastewater, desalination of salty water, and electric power production by recording COD removal percentage, desalination rate, and amount of electrical power produced.

2. Materials and method

2.1. MDC consistency

In this study, the air cathode MDC was constructed of plexiglass chambers (anode, desalination, and cathode chamber). The anode and the cathode chambers each have dimensions of 10x10x7 cm, while the desalination chamber has a size of 10x10x3 cm, as described in figure 1-a [21]. Two types of ion-exchange membranes supplied from Membrane International Ltd. were used, anion exchange membrane (AEM, AMI 7001) and cation exchange membrane (CEM, CMI 7000). The AEM disconnected the anode and desalination chambers; the CEM disconnected the cathode and desalination chambers. Before use, the membranes were preconditioned for 24h by flooding in a 5% salt (NaCl) solution and swilled with deionized water as recommended by the supplier to allow membrane hydration and expansion. The anode and cathode were made of graphite sheets; each has a 9x9 cm (length x width) dimension and 2 mm (thickness). The distance between each electrode and the adjacent membrane was 1 cm. Copper filaments with sealed contact were utilized to connect the external resistance to the electrodes.

All chambers were provided with inlet and outlet ports of 1 cm in diameter, except the cathode chamber has a 3 cm diameter inlet port suitable for dissolved oxygen (DO) meter. The anode chamber was sealed from ambient air to attain the required anaerobic conditions. The anolyte and catholyte volume was 650 mL, while the desalination chamber solution volume was 300 mL. The air cathode MDC is shown in figure 1-b.

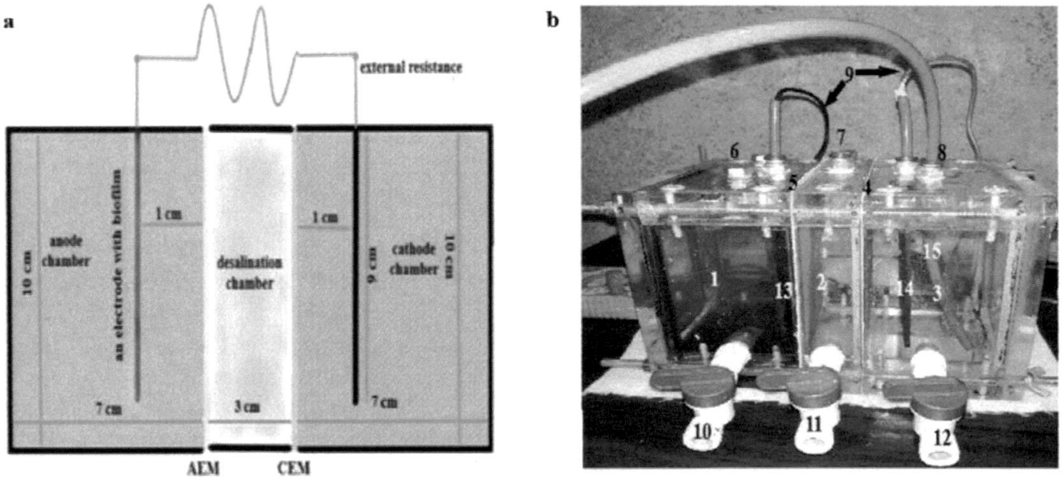

Figure 1. a- Graphic diagram of air cathode MDC and **b-** Photograph of air cathode MDC (1. Anode chamber, 2. Desalination chamber, 3. Cathode chamber, 4,5. CEM and AEM membranes, respectively, 6,7,8. Inlets, 9. Filaments, 10,11,12. Outlets, 13,14. The electrodes, and 15. The tube of air pump).

2.2. The materials

In this study, the experimental work consists of three experiments in the same conditions to treat the same wastewater. The three chambers of air cathode MDC have the following:

1. Microbial consortium and anolyte solution:

A mixed culture microbial consortium was used; it was obtained from the conventional activated sludge of the Al-Doura refinery wastewater treatment plant in Baghdad city. Acclimation of the sludge was done in anaerobic conditions for one week. For the biofilm development on the anode, the acclimation process is done inside the MDC's anode chamber. After that, the oily wastewater treatment was started by feeding the anode chamber with oily wastewater of COD 75 mg/L (oil refinery wastewater after pretreated with dissolved air flotation). The characteristics of oily wastewater are illustrated in table 1.

Table 1. The characteristics of oily wastewater.

COD (mg/L)	Oil (mg/L)	Ph (mg/L)	S.S (mg/L)	S^{-2} (mg/L)
75	6	1.02	124	0.09

2. Catholyte solution:

The buffer solution was used as a catholyte solution composed of 2.25 g/L KH_2PO_4, 2 g/L $K_2HPO_4.3H_2O$ dissolved in 1L of DI water [22].

3. Desalination chamber solution:

The NaCl solution with 35000 ppm was used in the desalination chamber to simulate seawater.

2.3. Analyses and calculations

A digital multimeter (Aswar, DT860D) was used to record the voltage with a 1KΩ external resistance was used in closed circuit tests to connect the anode and the cathode. Ohm's law determined the electric stream, I = E/R, where E is the voltage and R is the external resistance. The power density was studied as per the electrode surface area. The coulombic efficiency (CE), a portion of the electrons recovered as an electric stream from the substrate, was estimated using eq. (1) [23]:

$$CE=\frac{M.\int_0^t I\,dt}{n.F.V_a.(COD_0-COD_t)}\times 100\% \quad (1)$$

Where (M) is the oxygen molecular weight, (I) is the electric stream, (F) is Faraday's constant, n = 4 is the electrons numbering mutual per mole of oxygen, and V_a is the anolyte volume. COD_o appears to influence anolyte chemical oxygen demand, and COD_t is the COD value after the time (t).

The MDC was operated on for several days to reach a steady state. The polarization curves were determined with the aid of variable external resistance ranged from 10 kΩ to 100 Ω in steps (leaving about 20 min between successive steps to attain a steady-state). The COD was measured by utilizing a standard method. The multimeter (Lovi Bond Senso Direct 150) was used to measure TDS, Electrical conductivity, DO in the cathode chamber, and pH of the samples. The TDS removal was calculated as follows:

$$TDS_{\text{removal rate}}\ (ppm/h) = \left(\frac{C_0-C_t}{t}\right) \quad (2)$$

Where C_o is the initial TDS, and C_t is the final TDS of NaCl solution in the desalination chamber over a batch cycle of time t.

3. Outcomes and discussion

3.1. Power generation of the MDC

Figure 2-a shows the voltage generated across the MDC, while figure 2-b shows the power density during three successive experiments. At the beginning of each experiment, the three chambers fed with the solutions were illustrated in section 2.2. These two figures show that the maximum voltage difference between the electrodes in the anode and cathode chamber for a fixed value of the external electrical resistance at 1 KΩ was 24 mV, 56 mV, and 68 mV for experiments 1, 2, and 3, respectively; for three days of each experiment. Similarly, the maximum power density was 71.11, 387.2, and 570.86 $\mu W/m^2$ for experiments 1,2, and 3, respectively. The voltage and power density values increased at the beginning over time until they reach the maximum value. After that, the values dropped over time due to the substrate's depletion in the anode chamber.

It was clear that the electricity production effectiveness has developed between the 1st and the 2nd to the 3rd experiment, which may be the reason is due to the biofilm layer thickness increasing on the anode with time, so the number of ions migrating between chambers increased, and electrical power production efficiency improved.

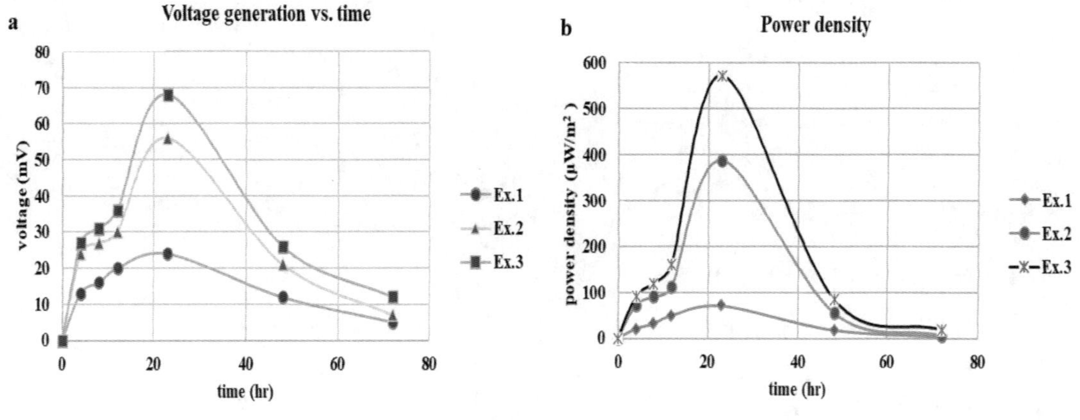

Figure 2. a- The voltage generation, and **b-** The power density for all experiments vs. time.

The previous study by Surajbhan et al., [15] treated the petroleum refinery wastewater in the MDC with the maximal value of power density was 243.4 mW/m^3 more than the power density in this study might be attributed to the difference of the initial value of substrate COD of the anolyte (the COD was 806 mg/L of the prior study and 75 mg/L of the present study).

After the reach of each experiment's steady-state operation, the polarization curve was drawn using the way presented in section (2.3), and it was illustrated in figure 4. While the voltage values vs. multi resistance for all experiments as shown in figure 3. The maximal values of power density were 100, 707, and 436 μW/m^2 for the experiment 1,2, and 3, respectively.

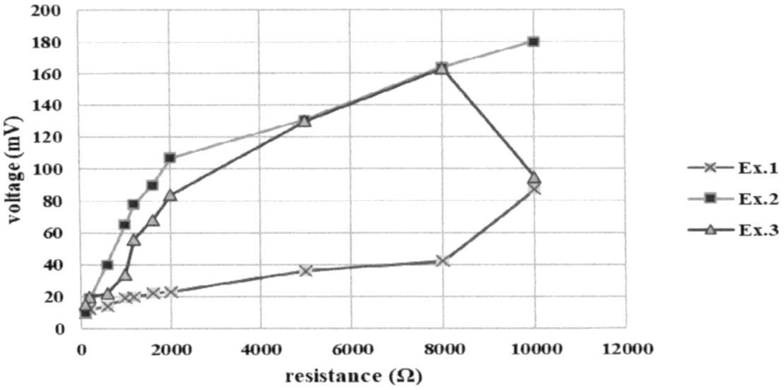

Figure 3. The voltage values of the three experiments vs. multi-resistance values.

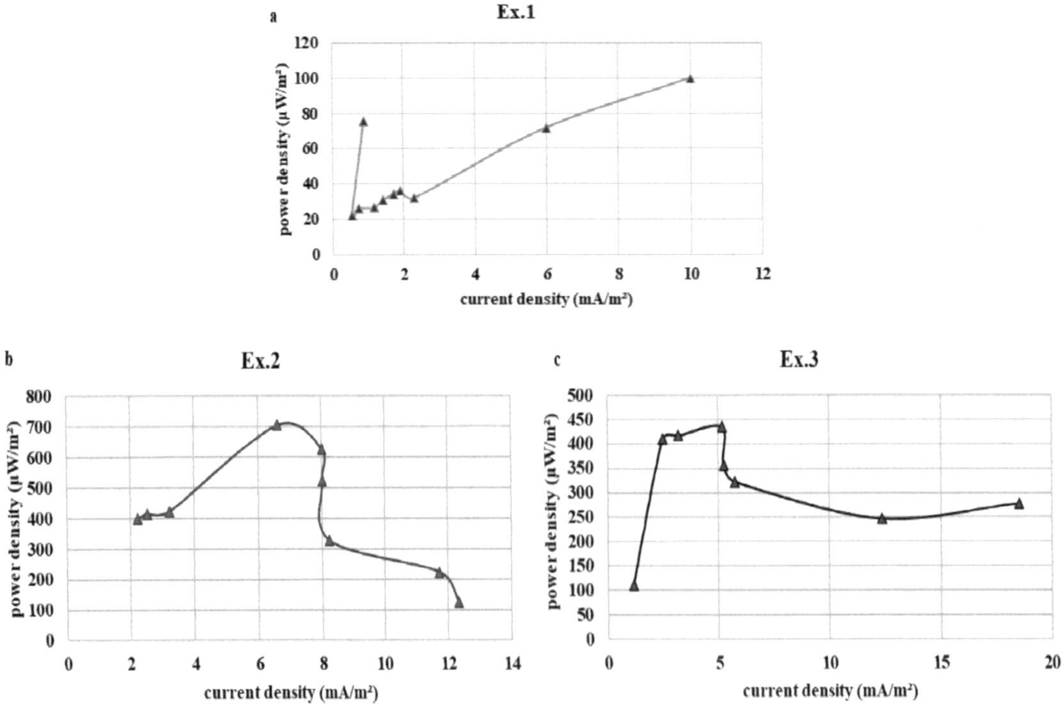

Figure 4. the polarization curves of all experiments, **a-** experiment 1, **b-** experiment 2, and **c-** experiment 3.

3.2. Salinity removal rate of the MDC

The salinity removal rate accomplished was different for each experiment. The salt removal values were 150.39, 149.25, and 147.92 ppm/h for the 1st, 2nd, and 3rd experiment, respectively, with a first salt concentration, which was 35000 ppm for all experiments as illustrated in figure 5-a. The results appear little difference in the salinity removal rate between the three experiments due to the fouling and scaling on the AEM and CEM, respectively. The desalination process affected the cathode and anode chamber's pH values, and the pH changed in all experiments.

During the 3rd experiment, the anolyte solution's pH value decreased from the first value of approximately 7 to 6.13 and increased the catholyte solution from 7 to 8.2 figure 5-b. Accumulation of H^+ ions in the anode chamber was produced from biodegradation of the substrate by bacteria caused the pH to decline. The AEM prevents H^+ ions from migrating to the desalination chamber; conserve charge neutrality allows chloride ions to transfer from saltwater in the desalination compartment to the anolyte solution. On the other hand, the depletion of H^+ ions in the cathode chamber due to stimulating them with oxygen caused the pH to increase. The CEM prevents OH^- ion from migrating to the desalination chamber; conserve charge neutrality allows sodium ions to transfer from the saltwater in the desalination compartment to the catholyte.

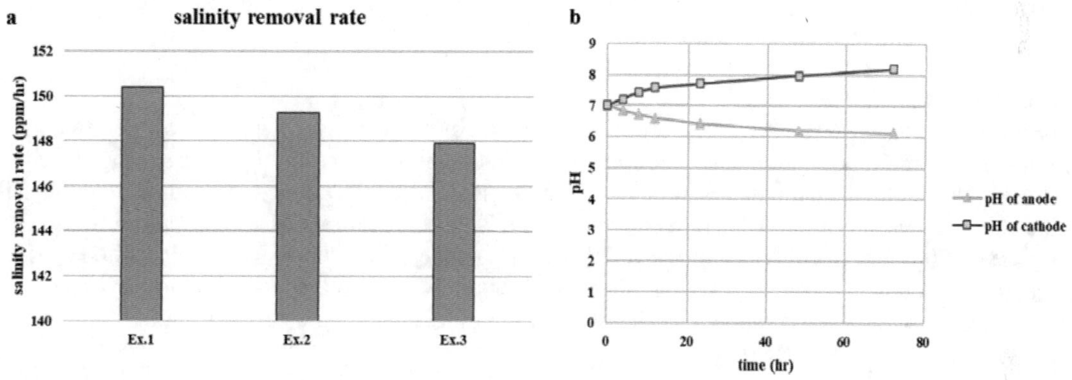

Figure 5. a- Salinity removal rate from the desalination chamber for the three experiments, and **b-** The pH changes of anolyte and catholyte for the 3rd experiment.

3.3. COD removal efficiency of the MDC

The air cathode MDC performance to treat oil refinery was monitored by recording the COD removal percentage of anolyte. The COD removal was 53.33%, 74%, and 96% for the 1st, 2nd, and 3rd experiments, respectively, each with three days cycle as illustrated in figure 6-a. The COD removal activity in this study developed by the time, the COD removal percentage of the 2nd experiment better than the 1st experiment, and the 3rd experiment were the best. The increasing of the biofilm thickness and biofouling on the walls and membrane of the anode chamber and the improving bacteria's ability to consume this kind of wastewater over time are the reasons for this development in the COD removal efficiency. Surajbhan et al., [15] used MDC to treat petroleum refinery wastewater, and the maximal COD removal percentage was 64%. Haiping et al., [20] investigated the MDC performance to treat domestic wastewater in long-term operation, and the percentage of COD removal was 55%. On the other hand, Haiping et al., [17] investigated the MDC performance to remedy municipal wastewater, and the COD removal activity was 53%. So, the COD removal efficiency in this work better than other works due to the difference in the initial COD; in this work, the initial concentration of COD lower than in other works.

The values of COD reduction in the anode chamber and the achieved columbic efficiency (CE) values versus time for the 3rd experiment are illustrated in figure 6-b. The highest value of CE was 13.6%. This value was less than the value achieved by Haiping et al., [17] (25.4%). That may be due to the significant variation between COD's first value in both works (for the sitting work was 75 mg/L while for the prior work was more than 2700 mg/L).

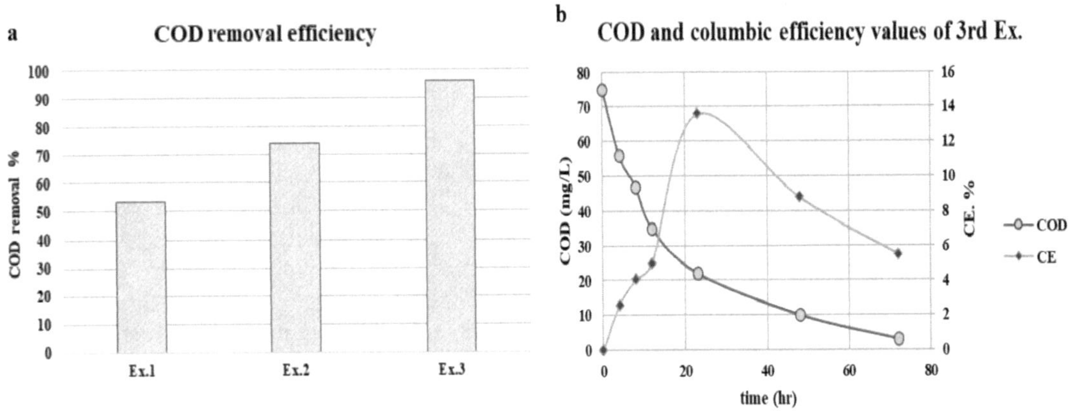

Figure 6. a- The three experiments' COD removal efficiency and **b-** COD with columbic efficiency values of the 3rd experiment.

4. Conclusions

This work investigated the air cathode MDC's performance in oil refinery wastewater treatment, simulated seawater desalination, and electrical power generation. The peak of power density generated by this MDC with 1KΩ resistance was 570.86 μW/m^2. The COD percentage removal of the oily wastewater was 96%. The MDC achieved a maximum TDS removal rate of 150.39 ppm/h for the first TDS was 35000 ppm of saltwater in the desalination chamber. This work proved air cathode MDC's conceptual capability to remedy actual oil refinery wastewater by the anode chamber's microbial community. MDC is an environmental-friendly method to treat actual wastewater, desalination sea, or brackish water and power generation.

Acknowledgment

The University of Technology, Iraq, supported this work.

References

[1] Metcalf & Eddy 2003 *Wastewater Engineering: Treatment and Reuse* vol 158 (McGraw-Hill Inc., New York).
[2] Josiane M, Heloisa B, Antonio S, Adriano S, Selene A and Guelli S 2010 Biodegradation of BTEX compounds in a biofilm reactor – modeling and simulation *Journal of Petroleum Science and Engineering* **70** 131-39.
[3] Alessandra C, Antonio C, Marcia D and G.L.Sant A 2006 Treatment of petroleum refinery wastewater by advanced oxidation processes *Journal of Hazardous Materials* **137** 178-84.
[4] Mi-Seon J, Eldon R, Soo-Hong K and Hung-Suck P 2008 An analysis of synergistic and antagonistic behavior during BTEX removal in a batch system using response surface methodology *Journal of Hazardous Materials* **152** 1276-84.
[5] Basheer D, Wan D and A.R. Abdul Aziz 2011 Treatment technologies for petroleum refineries effluents: A review *Process Safety and Environmental Protection* **89** 95-105.
[6] *Technical support document for the 2004 effluent guidelines* 2008 [United States Environmental Protection Agency (US EPA), USA].

[7] Malik H, M.A. Al-Hashimi and M.M. Al-Doori 2007 Effect of activated carbon on BOD and COD removal in a dissolved air flotation unit treating refinery wastewater *Desalination* **216** 116-22.
[8] Dania M, Carlos L, Dejanira A and Maria M 2010 BTEX biodegradation by bacteria from effluents of a petroleum refinery *Science of The Total Environment* **408** 4334-40.
[9] Fakhru'l-Razi A, Alireza P, Luqman A, Dayang A, Sayed M and Zurina A 2009 Review of technologies for oil and gas produced water treatment *Journal of Hazardous Materials* **170** 530-51.
[10] Veera G 2016 Wastewater treatment in microbial fuel cells–an overview *Journal of Cleaner Production* **122** 287–307.
[11] Shailesh K, Yi T, Min L and How N 2017 Development and long- term stability of a novel microbial fuel cell BOD sensor with MnO_2 catalyst *Int. Journal of Molecular Sciences* **18** 276.
[12] Bahareh K and Veera G 2015 Sustainable photosynthetic biocathode in microbial desalination cells *Chemical Engineering Journal* **262** 958–65.
[13] Henna S, Ghaleb H, Sharifeh Y, Jawaria S, Sameer A, Abdullah A.F, Sara A, Rehab K and Ahmed A 2015 Microbial desalination cell technology: a review and a case study *Desalination* **359** 1–13.
[14] Surajbhan S, Heyang Y, Zhen H and Ibrahim A.R 2015 Microbial desalination cells as a versatile technology: functions, optimization, and prospective *Desalination* **371** 9–17.
[15] Surajbhan S, Ibrahim A.R, Heyang Y and Zhen H 2017 Bioelectricity generation from the treatment of petroleum refinery wastewater with simultaneous seawater desalination in microbial desalination cells *Energy Conversion and Management* **141** 101–7.
[16] Surajbhan S and Ibrahim M. A.R 2017 Energy production in microbial desalination cells and its effects on desalination *Journal of Energy and Environmental Sustainability* **3** 53–8.
[17] Haiping L, Pei X, Timberley R, Peter J and Zhiyong R 2012 Microbial desalination cells for improved performance in wastewater treatment, electricity production, and desalination *Bioresource Technology* **105** 60–6.
[18] Kristen B and Zhen H 2013 Water softening using microbial desalination cell technology *Desalination* **309** 32–7.
[19] Atieh E, Ghasem N and Daryoush K 2016 Effect of batch vs. continuous mode of operation on microbial desalination cell performance treating municipal wastewater *Iranian Journal of Hydrogen & Fuel Cell* **3** 281-90.
[20] Haiping L, Pei X and Zhiyong R 2012 Long-term performance and characterization of microbial desalination cells in treating domestic wastewater *Bioresource Technology* **120** 187–93.
[21] Suhad J, Ghufran J and Talib A 2019 Photosynthetic Microbial Desalination Cell to Treat Oily Wastewater Using Microalgae Chlorella Vulgaris *Civil Engineering Journal* **5** 2686-99.
[22] Xiaoxin C, Xia H, Peng L, Kang X, Yingjun Z, Xiaoyuan Z and Bruce L 2009 A new method for water desalination using microbial desalination cells *Environ. Sci. Technol.* **43** 7148-52.
[23] Bruce L, Bert H, Rene R, Uwe S, Jurg K, Stefano F, Peter A, Willy V and Korneel R 2006 Microbial fuel cells: methodology and technology *Environ. Sci. Technol.* **40** 5181–92.

Contamination methods of milk with pesticides residues and veterinary drugs

Walaa S Raheem[1] and Alaa Niamah[2]
[1]Department of Genetic Engineering, college of Biotechnology, Al- Qasim Green University Iraq
[2]Department of Food Science, College of Agriculture, University of Basrah, Iraq

E-mail: wala.s@biotech.uoqasim.edu.iq

Abstract. The use of agricultural pesticides is very important in improving production, but the residue of these pesticides on crops have a serious health effect on humans. On the other hand, there is another type of pollutants resulting from the use of veterinary medicines, as well as heavy metals and mycotoxins that also result in health risks to humans due to their consumption of milk and dairy products. Milk is an important and necessary food for the body as it is consumed daily by many people because its contain fat, protein, vitamin, and minerals. In this work, we present a review of most scientific articles from 1964 until 2020, related to evaluating milk contamination with various chemicals, especially pesticides and veterinary drugs. We observe from many research papers at different country in the world Lingering presence detectable values of chemical contaminate The maximum remainder, as a final result of this review paper we need to establish monitor program system for all chemical contaminate residue in milk and many food type to improve safety and reduce the risk for consumers in Iraq.

Keywords: Chemical contaminate, pesticide, veterinary drugs, milk.

1. Introduction

The essential need for the human body is a food because it's very important for human body growth and also perform the repair and reproduction process for all human activities [1]. Milk it is an integrated food because it contains all the important nutrients like , carbohydrates ,fats, proteins. and minerals Although the milk have high nutritional values, but it is an good Medium and suitable for the growth of microorganisms

So it is vulnerable to contamination and damage. Milk is a complex chemical compound. It is a mixture of fats, vitamins, proteins, carbohydrates, minerals and other compounds in small quantities [2]. For this reason, milk was considered a staple of a balanced healthy diet due to its formula rich in minerals like calcium and sodium and also rich in amino acids and free fatty acids [3]. Milk has a high chemical composition, conjugated linoleic acid, total elements and microelements see table 1 [4].

In order to protect agricultural crops from harmful and destructive pests and enhance food production, agricultural pesticides are using to control pests. The residues of these pesticides on agricultural yields and in food are contamination that poses a risk to the consumer's life. [5]. Consumption of food containing pesticide residues is not safe for humans. So pesticide residues in foods are considering dangerous for human health [6].

However, the ways in which milk is treated such as sterilization, pasteurization and fermentation cannot eliminate the remaining quantities of pesticides in milk and dairy products [4]. Spraying pesticides on field crops for the purpose of increasing production. However, the residues of these pesticides in the environment, such as soil, water, and various vegetable parts, such as fruits and vegetables, are harmful to human health [7]. Pesticides are a major source of toxic and carcinogenic compounds that pose a threat to both animals and humans, as well as cause environmental pollution.

Veterinary medicines that are given to animals are part of the problem of contamination for animal products, the European Union has determined the maximum residual values (MRL) for a drugs such as 100 mg/kg for tetracycline and sulfonamide 50 mg/kg for macrolides and quinolones and also 100 mg/kg for pesticide. Several methods have also been developed to monitor and analyze the remaining substances in dairy products [8]. Milk is a complex food complex containing protein, fats, and other substances. Such a complex composition causes interference between its compounds during the analysis. During laboratory analysis, the sample must be prepared in practice by performing the extraction and purification process [9].

In Iraq, a few research deals with contamination of cow's milk with chemical materials resulting from agricultural pesticide residues that used to improve agricultural production, and also from the remnants of veterinary medicines used to treat animals.

Table 1. Milk composition [3].

Type of milk	Protein %	Lactose %	Fat %	Total solid %	Ash %	Moisture %
Human	1.25	6.95	3.20	11.8	0.21	88.25
Goat	3.30	4.40	3.90	12	0.7	88
Camel	2.90	4.30	3.60	11.7	0.75	88.35
Cow	3.40	4.80	3.75	12.8	0.71	87.23
Sheep	6.35	5.0	6.90	19.3	0.85	80.7
Buffalo	4.30	4.7	7.7	17.5	0.8	82.7

2. Origin of Milk Contaminants

2.1. Physical contaminants in milk
Milk is expose during the manufacturing process and also due to, handling, marketing, and storage to many physical contaminants such as animal hair, bush parts, dust, metal parts, udder peelings, dirt from the body of the cow, and unclean pots. Another source of contamination result from workers supervising , the milking and handling process, such as chewing tobacco, all previous sources physical contamination [2].

2.2. Chemical contaminants in milk
The milk and dairy products may be pollutant by chemical materials, the main source of which is from the pesticide residue, more than raw materials that are plant-based. Another source of chemical pollutants in milk and its product come from veterinary medicines such as antibiotics, sulfotamides, hormones, anti-worms, bush pesticides, and insects [10]. The major common chemical contaminants in milk and other products are worms antiseptics, insecticides, chlorine, nitrite, somatotropin and mycotoxins [11]. The chemical pollutants to which milk is exposed come from two main sources: the environment and the chemicals materials that are residue in milk. Where they are dangerous or cause other side effects that harm the health of the consumer [2].

2.3. Aflatoxins
Many molds can produce toxic metabolic products under appropriate conditions of temperature and humidity. These products are called aflatoxins, which cause a risk to human health and are found in animal milk that feeds on a bush containing fungi toxins of type B1. The toxin produced in milk M1 produced from metabolism of ruminants produces These molds toxins in wet feed that is not well dried or stored in a moist environment [2, 12 ,13].

2.4. Veterinary drugs (Antimicrobial Antibiotic residue drugs residue)
The common use of veterinary medicines by livestock breeders used to enhance the production of milk and its products. And also for the purpose of treating diseases such as mastitis and parasites and failure to observe the

adequate time period for disposal of these medicines by the animal, which leads to the exit of these medicines with milk and thus causes health risks to the consumer [14].

The human consumption of the remnants of veterinary medicines in milk is followed by many side effects that affect the health of the individual. The most important risks that accompany the consumption of milk containing antibiotics are the cancerous and genetic toxic response that often relates to chloramphenicol and sulfamethasin and others, see table 2.

It can also cause the creation of bacterial strains resistant to antibiotics as well as possible that the resistance move to other sensitive microorganisms, and the effect is not limited to the health aspect, but there are industrial effects, such as killing the starting bacteria, so fermentation does not occur [15,16] .

Table 2. Represent the MRL of different drugs [2].

No.	Drugs	MRL Concentration µg/kg	Ref.
1	Amoxicillin	4	Codex
2	Ampicillin	4	EU
3	Cloxacillin	30	EU
4	Benzylpenicillin	4	MERCOSUR
5	Nafciline	30	EU
6	Tetracycline	100-200	EU
7	Oxytetracycline	100	EU
8	Chlortetracycline	100	EU
9	Trimethoprim	50	EU
10	Ceftiofur	100	EU
11	Striptomycine	200-1000	EU
12	Oxfendazole	10	EU
13	Sulphonamiduse	100	EU
14	Cephalexin	20	EU

2.5. Heavy metals residue

The definition of heavy metals are the materials that has density in standard state more than 5 g/cm3 .The presence of heavy metals in milk and dairy products results from consumption the animals to feed and water contain environmental pollution , and also the raw milk may be contaminates during manufacturing process. The existing of many toxic from heavy metals like cadmium , lead with low concentration produce disturbance in metabolic and health problems [17]. Other researchers also mentioned that the reason for the presence of heavy metals in dairy products is to eat lactating cows contaminated animal feed or the cause of pollution is contaminated manufacturing processes [18].

2.6. Hormone residue in milk

The reason for the contamination of dairy products with hormones is to give these hormones and veterinary medicines to animals in order to help expedite the production process and its quality, or it may give for a therapeutic purpose. During the incorrect treatment of animals, they cause pollution, part of which remains in the animal tissues, in undesirable proportions and not suitable for human consumption. An example of these materials is a natural and synthetic hormones [19, 20].

From an epidemiological point of view the contaminated milk can cause symptoms of allergic diseases for humans as well as there are some hormone dependent diseases and constitute a risk factor that stimulates many diseases for humans such as cancer and other diseases [21]. There should not be any hormone residues in the animal tissues. The highest permissible concentration according to European laboratories is 1 micrograms per kilogram for muscles and tissues to control hormone residues.

3. Pesticides and insecticides residue

In the agricultural sector, fruits, vegetables, and grains are sprayed with artificial chemicals that are known as pesticides, including rodenticides, herbicides, fungicides and pesticides. These pesticides are used before harvest and during the storage period and these pesticides have the ability to transfer through the food chain from simple plants and animals to high-end animals which leads to the bioaccumulation of pesticides in animal products such as meat, milk, fats and eggs. The accumulation is caused by continuous exposure to the pesticide [11]. Chlorinated hydro carbonate is a highly effective and disturbing material for endocrine and bioaccumulative chemical toxins that spreads widely and enters the food chain by controlling animal and environmental pests. Contamination of fodder occurs in warehouses in the field due to the use of chlorinated insecticides such as Dichlorodiphenyltrichloroethane (DDT), Dichloro-dimethyl dichloroethane (DDD) and others. Dairy products are contaminated due to consumption of the contaminated cow feed.

The change that occurs in chlorine compounds in the environment is relatively slow over the years. It can be detected in meat, eggs, fish and dairy products, which are the main source of chlorine compounds reaching the human body [8]. The percentage of chlorinated hydro-carbon residues of milk is about 20%, which is mainly concentrated in the milk and butter fats which contain large quantities of pesticides [22]. Table 3 shows the residues of some chlorinated compounds in milk and its products. (DDT) is concentrated in animal fat tissue that is transferred to milk. That causes pollution by moving through the food chain.

Table 3. Average percentages of pesticides in raw milk and its products [11].

Pesticide Residue	Product type	Maximum residue limit mg per kg
HCB	milk	0.06
	Powder milk	0.0656
	Ras cheese	0.0037
	Damietta cheese	0.0045
	Spanish pasteurized milk	0.007
Aldrin-dieldrin	Cheese	0.2 ng/g
	milk	0.0074-0.0271
	Milk powder	0.0038
Dieldrin	Ras cheese	0.0039-0.0068
	Damietta cheese	0.0025
HCH	milk	0.094
	Milk pasteurized	0.015
	Milk powder	0.0149
	Ras cheese	0.00865
	cheese	0.0115
	Butter	0.093
DDT	Milk	0.159
	Butter	0.049
	Milk powder	0.0546

HCB: Hexachlorobenzene; HCH: Hexachlorocyclohexane; DDT: Dichlorodiphenyltrichloroethane

4. Pesticide Residues (PR) In Milk

The study of milk contamination and the ratios of pesticide residues in raw milk are very important because of their relationship to human health [4]. To study the residues of agricultural pesticides with raw milk, it is necessary to know the chemical composition of the milk components. Here two samples of raw milk were taken, one for bovine milk and the other for buffalo milk. And their chemical composition was studied to determine the percentages of agricultural pesticide residues in milk.Where the study showed that the buffalo milk sample showed higher levels of fats, total protein, total solid materials and whey than the cow's milk sample, while the lactose levels of the cow's milk sample were higher if compared to buffalo milk see table 4.

The presence of chlorine insecticides (OC) such as (HCB, Lindane, Aldrin and DDT) has been verified with values exceeding the MRL, according to the World Health Organization (WHO). Note that some pesticide residues that disappeared with skim milk and buttermilk during the production process. The yogurt has low level of pesticides due to heat treatment.

Table 4. pesticide residue of (mg/kg/fat basis) in milk samples % (+) ve = percentage of positive sample [21].

Pesticide type	% (+) ve	Buffaloes milk Range of (+) ve	% (+) ve	Cow milk Range of (+) ve	MRL (WHO,FAO) 2008
HCB	41.7	0.081- 0.186	36.7	0.091-0.180	-
Lindane	50	0.021-0.082	43.3	0.012-0.046	0.01
Aldrin	33.3	0.040-0.080	26.7	0.030-0.072	0.006
Heptachlor	8.3	0.012-0.042	10	0.016-0.030	0.006
H epoxide	25	0.044-0.078	20	0.066-0.082	-
chlordane	16.7	0.010-0.036	20	0.010- 0.022	0.002
Endrin	20	0.016-0.020	16.7	0.010-0.016	0.0008

Commercial milk contamination is a very important topic, so many research and studies in this field has been reviewed, which relate to milk contamination with agricultural pesticide residues, including chlorine pesticides. This work provides us with an evaluation study of milk contamination of various types of organic pesticides in different parts of the world. From this work, note the existence of the chlorine pesticide in milk samples although it was banned since1980s. Where in some samples exceeded the limits allowed by the World Health Organization, where it poses cause a large risk on human health due to its biological accumulation, causing many diseases [23].

The collected samples were examined by gas chromatography GC technique where used electron capture detector for detect the organchloride pesticide OCP residuesusing detector such as pulsated flame photometric. The dicofol at concentrations between 0.07+-0.0007 remain at fodder sample while the organophosphorus like dimetheoate exist in all milk samples at concentrations between 0.13 all the samples collected from Musi river The results of test showed the fodder samples with residues of dicofol at concentration of 0.07±0.0007 (0.071-0.077) [11].

The milk contaminates by different organochlorine pesticides (OCPs) like (α HCH,β HCH,ϵHCH) heptachlor, aldrin, heptachlor epoxide, γchlordane, endosulfane, dieldrin, Endrin, p,p' -DDE, p,p' -DDD, and p,p' –DDT) was investigated and also the ,fat samples was extracted from milk to calculate pesticide residues. All the analysed samples contained detectable residues of organ chloride pesticide . Also a high level of contamination with metabolites of DDT and HCH, p, p'- DDE, also another contaminant by p, p'-DDD. All the test samples showed acceptable values according to WHO [24].

Table 5. Average ratio of Recovery and RT for raw milk [24].

Pesticides	% *Recovery± SE	RT (min)	Limit of detection (ng $g^{-1)}$	Fortification levels (ng g^{-1})
$\alpha\, HCH$	833.35 ± 0.2998	6.02	0.03	0.5
$\beta\, HCH$	81.446 ± 0.8119	7.57	0.04	0.4
ϵHCH	86.779 ± 1.2288	8.36	0.06	0.5
Heptachlor	84.91 ± 1.4453	8.92	0.05	0.1

Previous work provided a practical study to analyze level of organic chlorine pesticide residues and chemical composition in buffalo milk samples. Monitoring of the analyzed samples gives the research team information about the quantities and type of pesticides in the environment. The research team analyzed three types of pesticide residue like organochlorine pesticides, namely Hexachlorocyclohexane (HCH), Dichlorodiphenyltri- chloroethane (DDT) and Endosulfan while ,the Lindane Residues exceeded MRL values in 50% and cause the serious concern ,the detected valused of p,p'-DDT showed exist in 70% of the samples with p,p'-DDE (dichlorodiphenyldichloroethylene) in 80% from all tested milk samples. The DDD (Di-chlorodiphenyldichloroethane) was detected in 65% of the milk samples. The results of study show that DDT is the major contaminants in different parts of Delhi state, due to statistical analysis shows no significant correlation between chemical compositions of the samples [25].

The study and analysed of 92 samples from raw milk taken from India's Ludhina province during the three years 1999, 2000 and 2001 and the results were 6 samples with a percentage of DDT of 4% and in two samples exceeding the maximum waste level (MRL). The remainder was fixed for DDT 0.05 mg / kg (MRL) For pesticide (HCH) was 0.01 mg / kg in whole milk, the collection of raw milk samples amounting to 49 samples exceeded the specified values, the study indicated that the milk was contaminated with pesticide residues despite the urban imposed by the agriculture and health programs on the use of pesticide (DDT) and (HCH) analysis of 40 samples of butter showed the presence of quantities of pesticide (DDT) in 28 samples.

And the presence of the pesticide (HCH) in 8 samples, however, none of the samples exceeded the value of the remaining (MRL) of (DDT), (HCH) .DDT residues from the P-DDE, P, P-TDE. HCH residues were present in 6 samples that are HCH and 2 samples revealed the presence of b-HCH the amount of MRL exceeded the acceptable value during daily consumption For tainted milk [26].

The study and analysis of milk samples was presented in [1] to clarify the ratios of pesticide residue in pasteurized and sterilized milk in Tehran province (Iran) such this pesticide OCP pesticides (such as α-HCH- . γ -HCH, HCB, dieldrin, o. p- DDE, β-HCH p,p- DDE. p,p- DDD, p,p- DDT) and polychlorinated biphenyl (PCBs) congeners (28, 52, 101, 138, 153, 180) in containing 1.5%, 2.5%, and 3% fat from . The results clarify and indicated that is no direct correlation between fat percent and the concentration of PCBs and OC pesticides [1].

Experimental study from two different locations in Poland, one agriculture and the other industrial. were the levels of OPC evaluate such asDDT ,the metabolites in cow's milk also study . All the samples, collected from different areas agricultural (n D 25) and industrial (n D 25) , the evaluation study was achieved by GC technique, DDT residue detected in all milk samples specially from agriculture sections where the concentration of DDT about 0.336 µg/ Liter. While in the industrial samples found the concentration lower, at 0.131 µg/ Liter approach, 4, 4-DDT was the main metabolite, constituting 83% of total DDT metabolites.

The collected samples in some site exceeded the accepted ratios will cause health risk on human body [17]. The researcher presented a study to determine the quantities of pesticides such as organic phosphate pesticides in raw milk and also to detect residues of maloxone and malathion in cow's milk and buffalo consumed in Turkey. Therefore, we collected 75 samples of buffalo milk and 75 samples of cow's milk to determine the remaining concentration by using HPLC technology [21].

As review of past research introduce another study to determine the concentration of OPPpesticide and carbamates pesticide using GC technology and found the reminder ratios of pesticide in cow milk and also in feedstuff provided to diary cattle [27], the research team used HPLC method with .NH2 column and also 97% acetonitrile elute to determine the insecticide cyromazine and metabolite melamine residues in pork and milk. The cyromazine and melamine was separated at 8 and 12 min retention time. We can get the linear calibration curves for cyromazine and melamine in range of concentration beyween 0.01–1.0 µg/mL at with correlation coefficients of 0.9999 and 0.9997. The study presents a limit of detection for two compounds 0.2 µg/ kg, and the limit of quantitation was 0.02 mg/kg.

The classic method based on HPLC was introduced to measurements the residue of twenty six veterinary and 187 pesticide residue samples in milk [28]. Moreover, many analytical methods used to study the pesticide residue in milk and fish. All or main of these results prove that the milk sample has own lower level of organochlorine pesticides than of fish sample [29].

Two fast methods were studied and evaluated. These methods used to prepare and analyze samples for fatty foods such as milk, eggs and avocado. The evaluation and comparison study was achieved for 32 samples to determine the agricultural pesticide residues because its represent a wide range of physical and chemical properties. One of them the method called the fast (quick), and easy to use also cheap and effective abbreviated as (Q,u,E,Ch,ER,S) method for pesticide residue analysis. Many researchers study the pesticide residue in milk and other dairy product but this work we take different five OCP pesticide such as hexchlorobenzene, methoxychlor alachlor, dieldrin, lindane and also other three OPP pesticide parathion-methyl chlorpyrifos, and malathion, all the analysis perform on milk samples [30].

We note from the results 44% from the samples have exist the lindane (HCH) pesticide and malathion residues with different concentrations exceeded not all time the levels set by the European union according code 2008. As a results of the work, the concentrations of residues for the methoxychlor, chlorpyrifos, and hexachlorobenzene exceeded the limit set by EU 2008 in 33%, 66%, and 88% of the examined samples. Also the other levels of residues for parathion-methyl alachlor, and dieldrin. Below the EU maximum residue limit in Egypt [31]. A develop reverse phase HPLC method, this method used to measure the melamine in milk with different process such as pasteurized and UHT milk and dairy products [32]. Using the fast method

QuEChERS (quick, easy, cheap, effective, rugged and safe) to extract the tetracyclines from the milk sample [33].

The main idea behind the work in reference [34] is to identify all the parameters affecting the quality of food as a result of pesticide residues. Due to direct impact of pesticides on the growth of microorganisms necessary for events of fermentation process such as bacteria and yeasts, as well as biological transformations of pesticides and their residues that may negatively affect the quality of sensory products (chemical and physical). Production of compounds responsible for flavoring fermented foods [34]. Present a experimental study to measure the residues of organic phosphorous pesticides, which included 13 types of pesticides that are frequently used to combat the external parasites of milk cows or in the feed used for feeding them.

Homogeneous samples were taken from homogenized pasteurized Mexican milk. 6% of the samples are on op residue. There were 8 samples that exceeded the permissible limit (MRL) of the pesticide. 5 samples contained dichlorvos, folate and one sample that contained chlorpyrifos, chlorine phenvinus. The remaining pesticide rate for 13 samples was less than the minimum permissible value ranging from 0.0015 to 0.0203 [35].

An experimental study was achieved to monitor the food treatments and the effect on reducing pesticide residue and their degradation of organic chlorine like HCH in milk samples. We note the concentration of residue for pesticide in milk will decay after heat process. The heat treatment will effect highly on the residue in milk sample and also the sterilization and boiling provide effective degradation of HCH isomers. Many chemical contaminants like heavy metals, veterinary drugs, pesticide residues, and, mycotoxins exist in milk and milk products [36, 37].

The analysis of pesticide residues in milk is a very important process to know the extent of their impact on human health, as pesticide residues are a major source of infection with many cancerous diseases and diseases of the nervous system. The MRLlimit proposed by EC for pesticides residue acceptable in different products. This work used Multiresidue LC-mass spectrometry methods (LC-MS) for testing food products [38].

Samples of buffalo milk were collected from different regions in Egypt to know the ratios of the pesticide residues in it. Then, the effect of the different process such as pasteurization, sterilization, fermentation, coagulation and temperature was studied on the pesticide residues in the milk. The main results of this study conclude all the treatment on milk will reduce the concentration of residue of pesticide [39].

Table 6. Effect of different milk process on pesticides [38].

Process	Concentration (mg/100 g)		Degradation (%)	
	dimethoate	malathion	dimethoate	malathion
Un treatment	3.852	1.695	-	-
Pasteurization	1.024	0.000	73.4	100
Sterilization	0.935	0.068	75.72	95.99
Fermentation	0.520	0.048	86.5	97.17
Coagulation curd	0.618	0.193	50.6	85.02
Whey	1.27	0.06	-	-

The use of chlorinated hydrocarbons as an insecticide remains in water, soil and biological materials and enters the food chain in many ways [40]. Here we note the effect of different heat process on pesticide residue in milk and diary products. The organochlorin pesticide reminder that detected in all test milk samples like DDE the found values in yoghurt reach to 2.7298±0.1126 ppb, with a mean value of 1.9161±0.0992 ppb., the residues detected in cottage cheese samples approach to zero. The main conclude results ,that the sterilization process of raw milk is the most effective method for reduction the pesticide residues to minimum value [41].

Milk samples were collected from different places in Romania, as well as studying the percentages of residual pesticides and knowing the extent of their impact on humans when consuming dairy products. Fifty-four samples of raw milk were collected, where proportions of the remaining pesticides were found of a type HCH while DDT not found all analysis achieved by using liquid chromatography.In this work, residual ratios of lindane were found ranging from 0.0042 to 0.2124 (mg / kg fat) in cow's milk samples, as well as ratios ranging from 0.0028 to 0.1408 (mg / kg fat) for pasteurized cow's milk. Also, the values ranged from 0.0042 to 0.2682 (mg / kg fat) for sour cream. The maximum values of detected OCP residues exceeded the maximum residue levels stipulated [42].

Table 7. Reminder ratio of concentration (mg kg-1) of 7 OPP pesticides in milk. Cultured @ Lactobacillus spp. For 42 C [41].

Pesticide	Strain inoculated	Residual concentration at different times (h)				
		8	12	16	20	24
Dimethoate	L. bulgaricus	1.100 ± 0.025	0.987 ± 0.047	0.909 ± 0.046	0.876 ± 0.013	0.808 ± 0.039
	L. paracasei	1.178 ± 0.042	1.092 ± 0.047	0.986 ± 0.031	0.957 ± 0.013	0.897 ± 0.024
	L. plantarum	0.990 ± 0.017	0.861 ± 0.029	0.841 ± 0.048	0.798 ± 0.026	0.735 ± 0.032
	Control	1.050 ± 0.021	0.985 ± 0.042	0.908 ± 0.035	0.860 ± 0.022	0.808 ± 0.037
Malathion	L. bulgaricus	0.987 ± 0.026	0.907 ± 0.022	0.853 ± 0.042	0.760 ± 0.043	0.677 ± 0.046
	L. paracasei	0.991 ± 0.034	0.847 ± 0.025	0.728 ± 0.032	0.699 ± 0.027	0.614 ± 0.041
	L. plantarum	0.912 ± 0.044	0.854 ± 0.027	0.663 ± 0.036	0.565 ± 0.014	0.484 ± 0.033
	Control	0.986 ± 0.046	0.900 ± 0.055	0.821 ± 0.034	0.750 ± 0.033	0.699 ± 0.056
Trichlorphon	L. bulgaricus	1.210 ± 0.041	1.190 ± 0.024	0.988 ± 0.033	0.907 ± 0.012	0.877 ± 0.053
	L. paracasei	1.136 ± 0.025	0.987 ± 0.033	0.900 ± 0.011	0.830 ± 0.046	0.778 ± 0.019
	L. plantarum	1.230 ± 0.033	1.160 ± 0.019	0.990 ± 0.014	0.930 ± 0.021	0.847 ± 0.012
	Control	1.036 ± 0.030	0.923 ± 0.044	0.892 ± 0.034	0.869 ± 0.054	Not detected

Continue to review the research that deals with calculating the remaining agricultural pesticides with milk and its products in this work it was studied adding seven types of pesticides such as trichlorphon methyl parathion fenthion monocrotophos dimethoate malathion phorate and were added to skimmed milk. We use one strain of. Lactobacillus. spp. including Lb. bulgaricus, Lb. paracasei and Lb. plantarum, to inoculated the milk at 42oC for 24 hours to check its impact on decay the OPPs . All the residue of pesticide in test samples were extracted by using extraction solvent And measured by GC technique after purification. Decay rate constant and half live periods of the OPPs were measured from a first-order reaction kinetics model. The main results indicate that the Lactobacillus spp. exhibited some speed up on Organo phosphorus degradation totally see table 7 [43].

Etect organic phosphate residues in milk based on acetylcholine esterase inhibition. The study revealed that raw milk samples are stable for eight days if they are stored at a temperature of four degrees. Milk samples were extracted with acetonitrile prior to enzyme assay [44] .

The effect of ultraviolet (UV) light on degradation of Chlorinated Hydrocarbon Pesticides in milk. The light was degrade organochlorine insecticides contaminating fluid whole milk or butter oil. The percentage of decomposition and decay depends on the thickness of the film and the depth of introduction of the rays into the sample when operating conditions are constant.

The effect of ultraviolet light on the degradation of chlorinated hydrocarbons in milk has been studied. Where the light analyzes the organic pesticides contaminated with chlorine in whole milk or butter oil. The degradation rate depends on thickness of the sample and also penetration depth. For the sample into the sample where the work was done under constant operating conditions. Decay of methoxychlor in butter oil was 96% after exposure for first time to UV light, little percentages insecticides were degraded also [45].

Thirty-five commercial samples taken from market for cow milk from the Iranian city of al-Ahwaz. Where seven types of insecticides of organic chlorine were discovered. All the identification and measurement process of the toxins were performed by using HPLC technique with an Ultra violet detector. The collected results noted the presence of lindane in ratios ranging from (0.042 mg/Kg) and DDT (0.28 mg/Kg) and the results above maximum limit ratio.by FAO and WHO [46].

The effect of fermentation, break down pesticide residues in milk. Seven samples of cow's milk were collected and added to them pesticides from organic phosphorous like, i.e., malathion, fenthion, methyl, parathion, dimethoate phorate, trichlorphon, monocrotophos and then fermented with 42°C. All the samples undergo to further process to extract pesticides by using organic solvent and tested by gas chromatography GC after purification process. The researchers noticed a significant degradation of the pesticide residues after supplementing the yoghurt with one of the initiators with the exception of the pesticide malathion [47].

The measurements of the OCPs pesticide in goat milk determined for seven OCPs. The researcher found the values in fat approximately 2.7%. If milk samples left at a temperature of 4°C for one week this will deceasing and cause decay the reminder of OCPs. OCPs were calculated and found that most were present with a mean of Heptachlor0.091695 while Dieldrin had a value of 0.034547 [46]. The effect of pesticides as Ivermectin and deltamethrin has been investigated. Because it used by farmers as an anti-parasite, and are very widely used for animals. The pesticides were detected by using liquid chromatography at wavelength of 233 nm.The QuEchERS method was used to extract pesticides from milk samples. All results confirmed that the recovery rate from bovine milk samples was 92.22-96.7% for ivermectin.and for deltamethrin average ratio recoveries from 92.56- 97.78%.

The minimum detection limits of ivermectin from 2 to 6 micrograms / kg and for deltamethrin from from 3 to 9 micrograms/kg. All results indicated that the ratios were 18.18, 28.57, 40, 100 and 60% of the samples for bovine milk containing residues of the type Ivermectin less than the maximum, while the rest of the samples contained 81.82, 71.43, 60, 0 and 40% of Ivermectin residues. All of them are above the maximum residue limit. While the deltamethrin cannot exist in 33 samples. The method indicated as good and reliable method for identification pesticide from type avermectin and deltamethrin in milk samples [48]. The veterinary medicines in milk from 2009 to 2011 were studied in Brazil and their effect on animals and what are the residual proportions of cow's milk for these drugs, where it was found that the proportions did not exceed the maximum permissible residue. The drugs were of the type of ivermectin that are given to dairy cows, especially quinolones and tetracyclines.

Table 8. Summary of literate review.

Researchers	Studied Samples	Country	Elements	Test methods	Year
J. E. Beam and D. J. tankinson	Milk	Italy	organophosphate	Acetyl cholinesterase inhibition	1964
B.J. Liska	Milk	China	chlorinated hydrocarbons	GC	1967
C, F. Li and R. L. Bradley	milk	USA	methoxychlor	UV	1968
Salwa M. Dogheim	Milk	Egypt	Pesticide	GC	1990
J. H. Salas	Milk	India	organophosphorus	gas chromatography	2003
R.S. Battu et al	Milk	India	DDT, HCH	Gc	2004
S.J. Lehotay	milk,	Egypt	Pesticide	QuEChERS, LC	2005

	eggs				
Nevein S. Ahmed and Eman M.S. Zaki	Milk	Egypt	$\alpha\ HCH, \beta\ HCH, \epsilon HCH)$	HPLC	2009
Ashnagar et al	Cow Milk	Iran	organochlorine	HPLC	2009
Ruicheng Wei1 et al	milk and pork	China	cyromazine and melamine	HPLC	2009
S. K. Nag	Milk	India	veterinary drugs, Pesticide	GC	2010
Li-Ying Bo et al	Bovine milk	china	dimethoate, fenthion, malathion	gas chromatograph (GC)	2010
A.M. Abo. Donia et al	Cow milk, Buffalo milk	Egypt	HCB, lindane, aldrin and DDT	GC	2010
Sara Bayat , Abbas Esmaili Sari	Milk	Australia	α-HCH, β-HCH, γ -HCH, HCB, dieldrin	GC	2010
Rafael Fagnani et al	Milk	Brazil	Organophosphorus, carbamate (CB)	GC	2011
Xin-Huai Zhao , Jing Wang	Milk	China	organophosphorus	gas chromatography (GC)	2011
Ayoub, M. M et al	Milk	Egypt	organochlorine	GC	2012
Filazi et al	Milk	Turkey	melamine	Reverse phase HPLC	2012
Julijana et al	Goat milk	India	organochlorine	gas chromatograph (GC)	2013
Jorge Regueiro	Milk	United kingdom	organochlorine	HPLC	2013
Mohd Aslam, Sumbul Rais, Masood Alam	buffalo milk	Egypt	Organochlorine (Hexachlorocyclohexane (HCH), Dichlorodiphenyltri-chloroethane (DDT) and Endosulfan)	HPLC	2013
Eman M. Shaker, Eman E. Elsharkawy	buffalo milk	Egypt	alachlor, dield-rin, hexachlorobenzene, lindane , ,methoxychlor , organophosphorus pesticideschlorpyrifos, malathion, , parathion-methyl	gas chromatography (GC)	2015
Anna Stachniuk and Emilia Fornal	Milk	Poland	Pesticide	Multiresidue liquid chromatography-mass spectrometry methods (LC-MS)	2015
J. Kuba et al	Milk	India	dichlorodiphenyltrichloroethane (DDT)	GC	2015
Korrapati et al	Milk	India	organochlorine	GC	2015
Husniye Imamoglu and Elmas Oktem Olgun	Milk	Turkey	veterinary drugs, pesticide	LC-MS/MS	2016
Fawzia Abd-Rabo et al	buffalo milk	Egypt	organophosphorus	GC	2016
L. RUSU et al	Milk	Romania	α-HCH, β-HCH and γ-	gas chromatography	2016

			HCH	(GC)	
Hanan et al	Milk	Egypt	Deltamethrin Ivermectin	HPLC	2016
Stefani Faro et al	Cow milk	Brazil	tetracyclines	GC	2017
Sujatha Singh and Krishnaiah Nelapati	Milk	India	α, β, ɣ and δ hexachlorocyclohexane isomers (HCH))	HPLC	2017
E. Marinou et al	Milk	Greece	tetracyclines	QuEChERS	2019
Isabella C. Vasconcelos et al	Milk	Brazil	Organochlorine	-	2019

These drugs have been found in proportions, such as doxycycline (9%) and abamectin (1.6%), noting that these substances are not intended for use in milk-producing animals due to their effect on human consumption. Also the researcher note Norfloxacin drug at ratio 15% not exceed the allowed maximum ratio limit there is no residues for the treptomycin, chlortamphenicol, β-lactams were confirmed. As a final result, milk in Brazil contains low levels of veterinary medication so that it is not considered a risk as well as it does not contain toxicity risks related to milk consumption and does not raise public health concerns. All the previous research can be summarize as in table 8.whrer this table express about the briefly of review literatures.

5. Conclusions

We note from the review of previous research papers, the chemical contaminate of the milk especially pesticide residues remain the effective parameters in all the research papers because its relate directly to human health and dairy industry. There are many methods that lead to reducing the percentage of pesticide residues remaining in dairy products and these methods, such as thermal treatments, fermentation and radiation treatment. All fermentative microbiota in the reduction of the pesticide residues, during the fermentation time. A way in metabolomics to understand the common effects of pesticide residues on fermenting microorganisms (yeasts and bacteria) would be of essential, fundamental importance, together with on accounts in extraction methods and analytical mechanisms with the utilize of increasingly precise exposition tools mechanisms to estimate pesticide residues.

ACKNOWLEDGMENTS
The authors wish to thank Professor Ibrahim Murdas for valuable comments to the article and advice in the model formulation.

6. References

[1] Bayat S, Bahramifar A and Younesi N 2011 Survey of organochlorine pesticides and polychlorinated biphenyls in commercial pasteurized milk in Iran *J. of Envi. moni. and assess.* **175** 469-474
[2] Cserháti T and Szőgyi M 2012 Chromatographic determination of pesticides in foods and food products *European Chemical Bulletin* **1** 58-68
[3] Fletouris D, Botsoglou N, Psomas I and Mantis A 1997 Determination of the marker residue of albendazole in cheese by ion-pair liquid chromatography and fluorescence detection *J.of dairy. scie.* **80**(11) 2695-2700
[4] Donia M, Abou-Arab A, Enb A, El-Senaity M and Abd-Rabou N 2010 Chemical composition of raw milk and the accumulation of pesticide residues in milk products *J. of Glob. veter.* **4**(1) 6-14
[5] Kovacova J, Kocourek V, Kohoutkova J, Lansky M and Hajslova J 2014 Production of apple-based baby food: changes in pesticide residues. *Food Additives and Contaminants: Part A* **31**(6) 1089-1099
[6] Lozowicka B 2015 Health risk for children and adults consuming apples with pesticide residue *Science of the total environment* **502** 184-198
[7] Wasi S,Tabrez S, Ahmad M 2013 Toxicological effects of major environmental pollutants: an overview *J. of Envi. Moni. and asse.* **185**(3) 2585-2593

[8] Imamoglu H and Oktem O 2016 Analysis of veterinary drug and pesticide residues using the ethyl acetate multiclass/multiresidue method in milk by liquid chromatography-tandem mass spectrometry. *J.of analy. Meth. in chem.* **3**(5) 123-126
[9] Raikos V 2010 Effect of heat treatment on milk protein functionality at emulsion interfaces A review. *J. of Food Hydrocolloids* **24**(4) 259-265
[10] Cohick W 1988 Role of the insulin-like growth factors and their binding proteins in lactation *J. of Dairy Scie.* **81**(6) 1769-1777
[11] Kotinagu K and Krishnaiah N 2015 Organochlorine and organophosphorus pesticide residues in fodder and milk samples along Musi River belt, India *J. of Veterinary worl.* **8**(4)
[12] Ghanem I and Orfi M 2009 Aflatoxin M_1 in raw, pasteurized and powdered milk available in the Syrian market *J. of Food Cont.* **20**(6) 603-605
[13] Daou R, Afif C, Joubrane K, Khabbaz L, Maroun R, Ismail A andEl-Khoury A 2020 Occurrence of aflatoxin M1 in raw, pasteurized, UHT cow's milk, and dairy products in Lebanon *J. of Food Cont.* **11** (10)
[14] Solomon D, Shpige N, Salamon H and Goshen T 2020 Epidemiology and Risk Factors of Pyelonephritis in Israeli Dairy Cattle. Israel *J. of Veter. Medi.* **3**(5) 3-8
[15] Klaassen C and Amdur M 2013 *Casarett and Doull's toxicology: the basic science of poisons* New York: McGraw-Hill
[16] Ben Y, Fu C, Hu M, Liu L, Wong M and Zheng C 2019 Human health risk assessment of antibiotic resistance associated with antibiotic residues in the environment: a review *Environmental research* **169** 483-493
[17] Kuba J,Tomza M, Pilarczyk B, Tarasewicz N, Pilarczyk R and Ligocki M 2015 Comparison of DDT and its metabolites concentrations in cow milk from agricultural and industrial areas. *J. of Environmental Science and Health Part B* **50**(1) 1-7
[18] Alimentarius C 2004 *Code of hygienic practice for milk and milk products* CAC/RCP, 57
[19] Moussa F, Mokh S, Doumiati S, Barboni B, Bernabò N and Al Iskandarani M 2020 LC-MS/MS method for the determination of hormones: Validation, application and health risk assessment in various bovine matrices *Food and Chemical Toxicology* **11**(12)
[20] Hess C and Boehmer A 2020" Toxicology of natural and synthetic aphrodisiacs. Rechtsmedizin" 1-15.
[21] Kara R.,Ince S. 2016 Evaluation of malathion and malaoxon contamination in buffalo and cow milk from Afyonkarahisar Region, Turkey, using liquid chromatography/tandem mass spectrometry–a short report Polish *J. of Food and Nutr. Scie.* **66**(1) 57-60
[22] Ray P and Sen C 2019 Biochemical Residues in milk and milk products-A review *Indian J. Anim. Hlth.* **58**(2) 145-152.
[23] Rêgo I, Santos G N, Santos G, Ribeiro S, Lopes B, Santos S and Vasconcelos A 2019 Organochlorine pesticides residues in commercial milk: a systematic review *Acta Agronómica* **68**(2) 99-107
[24] Ahmed N and Zaki E 2009 Detection of Some Organochlorine Pesticides in Raw Milk in Giza Governorate *J. of App. Scie. Resea.* **5** (12) 2520-2523
[25] Aslam M, Rais S and Alam M 2013 Quantification of organochlorine pesticide residues in the buffalo milk samples of Delhi City, India *Journal of Environmental Protection* **6**(12) 63-68
[26] Battu R, Singh B and Kang B 2004 Contamination of liquid milk and butter with pesticide residues in the Ludhiana district of Punjab state, India *Ecotoxicology and environmental safety* **59**(3) 324-331
[27] Fagnani R, Beloti V, Battaglini A, Dunga K and Tamanini R 2011 Organophosphorus and carbamates residues in milk and feedstuff supplied to dairy cattle *Pesquisa Veterinária Brasileira*, **31**(7) 598-602
[28] Wei R, Wang R, Zeng Q, Chen M and Liu T 2009 High-performance liquid chromatographic method for the determination of cyromazine and melamine residues in milk and pork *J. of chroma. science* **47**(7) 581-584.
[29] Aerts R, Van Overmeire I, Colles I and Covaci A 2019 Determinants of persistent organic pollutant (POP) concentrations in human breast milk of a cross-sectional sample of primiparous mothers in Belgium *Enviro. inter.* **131**
[30] Lehotay S J, Maštovská K, Yun S J 2005 Evaluation of two fast and easy methods for pesticide residue analysis in fatty food matrixes *J. of AOAC Inter* **88**(2) 630-638

[31] Shaker E M and Elsharkawy E 2015 Organochlorine and organophosphorus pesticide residues in raw buffalo milk from agro industrial areas in Assiut, Egypt *Envir. Toxi. and pharm* **39**(1) 433-440
[32] Filazi A, Sireli U, Ekici H, Can Y, Karagoz A 2012 Determination of melamine in milk and dairy products by high performance liquid chromatography *J. of dairy science* **95**(2) 602-608
[33] Marinou E, Samanidou V and Papadoyannis I 2019 Development of a High Pressure Liquid Chromatography with Diode Array Detection Method for the Determination of Four Tetracycline Residues in Milk by Using QuEChERS Dispersive Extraction *Separations* **6**(2) 21
[34] Regueiro J, Lopez F, Rial O, Cancho B and Gandara J 2015 A review on the fermentation of foods and the residues of pesticides—biotransformation of pesticides and effects on fermentation and food quality *Critical reviews in food science and nutrition* **55**(6) 839-863
[35] Salas J H, González M, Noa M, Pérez N, Díaz G and Osuna I 2003 Organophosphorus pesticide residues in Mexican commercial pasteurized milk *J. of agricul. and food chem.* **51**(15) 4468-4471
[36] Singh S and Nelapati K 2017 Effect of food processing on degradation of hexachlorocyclohexane and its isomers in milk *Veterinary world* **10**(3) 270
[37] Nag S K 2010 *Pesticides, veterinary residues and other contaminants in milk. In Improving the safety and quality of milk* Woodhead Publishing
[38] Stachniuk A, Fornal E 2016 Liquid chromatography-mass spectrometry in the analysis of pesticide residues in food *Food Analytical Methods* **9**(6) 1654-1665
[39] Abd-Rabo F, Elsalamony H and Sakr S 2016 Reduction of pesticide residues in Egyptian buffalo milk by some processing treatments *Int J. Dairy Sci* **11** 75-80
[40] Liaska B J 1968 Effects of processing on pesticide residues in milk *Journal of animal science* **27**(3) 827-830
[41] Ayoub M, Desoki M, Hassanin A, Thabet M, Mansour M, Loutfy N M and Raslan A 2012 Detection of pesticide residues in milk and some dairy products *J. Plant Prot. and Path. Mansoura Univ* **3**(8) 865-880
[42] Rusu L, Harja M, Suteu D, Dabija A and Favier L 2016 Pesticide residues contamination of milk and dairy products. A case study: Bacau district area Romania *Journal of Environmental Protection and Ecology* **17**(3) 1229-1241
[43] Zhao X H and Wang J 2012 A brief study on the degradation kinetics of seven organophosphorus pesticides in skimmed milk cultured with Lactobacillus spp. at 42 C *Food chemistry* **131**(1) 300-304
[44] Beam J E and Hankinson D 1964 Application of the acetylcholinesterase inhibition method for detecting organophosphate residues and related compounds in milk *J. of Dairy Science* **47**(12) 1297-1305
[45] Li C F and Bradley R L 1969 Degradation of Chlorinated Hydrocarbon Pesticides in Milk and Butteroil by Ultraviolet Energy *J. of Dairy Science* **52**(1) 27-30
[46] Ashnagar A, Naseri N and Farmad M C 2009 Determination of organochlorine pesticide residues in cow's milk marketed in Ahwaz city of Iran *Inter. J. of Pharm. Resea.* **1**(2) 247-251
[47] Bo L Y, Zhang Y H and Zhao X H 2011 Degradation kinetics of seven organophosphorus pesticides in milk during yoghurt processing *J. of the Serbian Chem. Soci.* **76**(3) 353-362
[48] Henan A, Fayza A and Fayza A 2016 Detection of ivermectin and deltamethrin in the bulk milk tank of some dairy herd *Advances in Environmental Biology* **10**(11) 1-10

Investigation of pollution level of groundwater in the ALjazaier Area, Babil

N S Hadi [1*], Z S Jabbur [1]

[1] Department of Environmental Engineering, College of Engineering, Babylon University, Iraq
* Corresponding author: nabaa.hadi@uobabylon.edu.iq

Abstract

A total with 15 samples was collected from 3 wells in the aljazaier distreet in Babil governorate for the period from January to April 2020, to study the microbial contamination represented by fecal bacteria, while chemical contamination includes a physochemical and the heavy element's properties. The goal of this study: Study the water of some wells from some microbes and chemical aspects and evaluate this according to Iraqi standards and specifications. Results showed microbial pollution by fecal Coliform bacteria in well water when using Using the agar media. Results showed the difference in the mean of (MPN) of well water. It was found that the EC value (1061) µs/cm was exceeded the permissible limit in W1, Cl values (325, 430)mg/l, pb values (0.43, 0.14) µg/l and Ni values (0.11, 0.11) µg/l and the permissible limit were exceeded in W1, W2 and SO4 values (681, 653, 584) mg / l also the permissible limit was exceeded in W1, W2 and W3 as well as Ag value (0.12) µg/l is in W1 only. The total risk index (HIT) in the groundwater for W1, W2, W3 was acceptable for adult values ΣHI(0.23, 0.97, 3.8E-4) and children values ΣHI (2.6E-2, 1.6E-2, 1.3E-3) respectively.

Keywords: Ground water, heavy metal , Microbial, Pollution.

1. Introduction

Groundwater is the water that has sunk through the layers of rain, rivers and fresh lake water on the earth at various depths and images, and since ancient times the amount of water in the globe has been fixed and does not change [1]. Groundwater, although often free from any bacteriological contaminations, require groundwater study and complete analyzes in the past allowing their use, And according to the geological composition of the soil, they reside in various dimensions of the surface area of the Earth to another. The importance of groundwater is increasing on an ongoing basis in particular, due to the growing demand for water for both agriculture purpose and industrial drink, the volume of groundwater exceeds the water of rivers and lakes significantly [2]. [3] Studied the Groundwater pollution in some areas between the Al-Kifl stream and the Shatt al-Hindi. The study relied on the collection and analysis of five models for selected areas of groundwater. The result was Hg [0.031]mg per l, Ni[0.01]mg per l, pb [0.068]mg/l , Cu [0.390mg/later Zn [0.171], Cd [0.002]mg per l, When compared with the Iraqi limitations for drinking water, it was found that it does not exceed the permissible limit. [4] studied trace element of groundwater contamination was measured in middle-class neighbourhoods of Lagos. Using atomic absorption spectrophotometers, approximately forty-nine well and borehole water samples were tested for their aluminum, cadmium and lead content

IOP Conf. Series: Earth and Environmental Science **877** (2021) 012004 doi:10.1088/1755-1315/877/1/012004

and their levels relative to the WHO maximum amount of pollutants. The results are [0.2, 0.003 and 0.01mg/L], respectively, for Al, Cd, pb. Given the toxicity of these metals, the findings obtained from this study indicate a major risk to this population and the fact that for many, hand drilled wells and bore hole are really the only sources of their water supply in this area.

2. Methods

The methodology of this study divides into three main sections counting data set, data pre-processing, and hybrid machine learning model with the optimisation algorithm.

2.1 Well water samples

Three wells were chosen in the Aljzaier distreet of the Babel Governorate to study the microbial and chemical contamination in the water of these wells. 15 water samples were collected. Five samples from each well. The samples were placed in sterile glass bottles with a sealed seal and brought to the laboratory after being placed in a plastic box containing ice to preserve the samples and reduce microbial activity. And bottles of polyethylene for chemical trace elements characterized by [pb, Ag, Cr, Cd, Ni] were estimated by(Atomic absorption spectroscopy) device, type shimadzu Japanese origin, model AA-700.properties, as samples were collected for the period from January to April [5].

2.2 Screening for coliform bacteria

2.2.1 Most Probable Numbers (MPN)

For middle user (FC Agar). A filter paper (sterile) is placed in the designated place in the filter unit. Transfer 100 ml of sample of the filter unit container. After finished from the filtered the filter paper is transported with sterile forceps and placed on the hardened agar surface (-M) FC agar pre-prepared without any voids. Leave the plate for a short period until the paper adheres well to the food medium, and this appears in the color Red filter paper. Incubate the dish up to 44.5 ° C for 24 hours and record the final result after 48 hours , according to the equation [6].

$$\frac{NPM}{100\,ml} = (\frac{NPM}{ml}\ (from\ table * dilution\ factor\ of\ P2) * 100 \qquad (1)$$

2.2.2 Physiochemical properties of well water were estimated, according to the methods given:

1. ion exchang of water (pH), using meter-pH: It was measured directly in the field according to [7] by using (3540-pH-meter model).The instrument was calibrated before each sampling with standard buffer solution (pH: 4, 7 and 9).
2. Electrical conductivity (EC) using meter-EC: it measured directly in the field by using EC-meter(3540-pH-meter model). The instrument was supplied with an automatic temperature corrector. The results were expressed as µs.cm -1 according to [8].
3. Sulfate corrected with Na2-EDTA: It measured in the laboratory using the gravity burning method. Where 50 ml of sample were taken, adding dilute hydrochloric acid with a concentration of 0.1N, the mixture was heated to a boiling stage. Added to form 10 ml of barium chloride (BaCl2). The model is placed in a water bath at a temperature of 100 ° C for 15 minutes, then it is lifted, leaving for two hours to cool down, add to the form 3 ml of buffer ammonium and a small amount of aero chrome dye where the color becomes purple to blue and thus represents HARD2. Take 25 ml of the raw form and add to it 1 ml of ammonium puffer with a little dye of aero chrome and then wipe it with EDTA solution until the color changes to blue and thus represents HARD1[9]. And also using UV-spectrophotometer at wave length 420nm [8].
4. Chlorides with silver nitrate correction method: It measured according to the (Mohr) method, which is the approved method for determining the chloride ion in contaminated water by the American health association. 50 ml of sample , and 1 ml of potassium chromate solution(0.0141N) as an indicator was added, the sample titrated against silver nitrate solution until convert the color from yellow to brown color [8].
5. Total suspended solid (TSS) by using the filtration method: The method approved by [9] was counted to estimate the total suspended solid, that's by drying filter papers openings diameter of

0.45 micron, then the papers placed in the oven at temperature between 103-105 °C for one hour, cooled to room temperature in a glass bell, weighed by sensitive balance, Filter 100ml of sample.

6. Nitrates using the UV device: It measured using ultraviolet radiation with wavelengths of 220nm and 275nm with a spectrometer whereas the reagents were prepared. Reagents that were used (anhydrous potassium nitrate and hydrochloric acid).Standard solutions were prepared at concentrations of (0-350) micrograms from the nitrogen sample by diluting several volumes to 50ml with distilled water, which was (0, 1,..., 35) ml and adding 1 ml of 1N HCL, 220nm wavelength absorbance calculated.
7. Heavy metals characterized by [pb, Ag, Cr, Cd, Ni] were estimated by(Atomic absorption spectroscopy) device, type shimadzu Japanese origin, model AA-700.

3. Theory

3.1 Risk assessment

Risk assessment is the process for assessing risks related to environmental presence, fate as well as travel of chemicals. More specifically, risk assessment of environmental is an assessment of the risk for adverse effects from a chemical that is of concern to a site in order to identify the need for remedial action or to develop targeted levels where remedial action is needed. This includes the analysis of sources of release, mechanisms for chemical transport, and possible health risks to receptors. The Environmental Risk Assessment (ERA) includes an assessment of the risks associated with natural events (floods, extreme weather, etc.), technologies, practices, processes, products, factors (chemical, biological, radiological, etc.) and industrial activities that may pose a threat to ecosystems [10].

3.1.1 Chemical risks

Exposure is a contact with infected substance, such as persons or the endangered species. An exposure assessment is an estimate the exposure size, duration in addition to route of exposure frequency. The purpose of the exposure assessment is to estimate pollutant concentrations and dosages for at-risk populations [11].

The general chemical water intake equation is:

1.1. Performance Assessment of the ANN Model

After integrating the ANN model by selecting optimum (Lr, N1 and N2), the ANN technique was run a couple of times to get the optimum architecture that can simulate accurately stochastic signal of water demand. Various statistical criteria were applied to assess the capability of a prediction model. The correlation between the observed and predicted stochastic signal of water data was tested via utilising R^2 as shown in Figure 3. It reveals that R^2=0.99 and the data are scattered match to slop line 1:1.

$$MDI_i = \frac{C_w \times IR \times EF \times ED}{BW \times AT} \qquad (2)$$

Where the :

MDI: Maximum Daily Intakes (mg per kg. d),

C_(w) : chemical concentration (mg/L) in water.

IR: rate (L/day) of inhalation.

EF : frequency of exposure(day/years).

ED ;duration of exposure (year).

BW : the mean mass of the body over the exposure period (kg).

AT :average time; the time the exposure is averaged over (d).

3.1.2 Risk Assessment and Risk characterization

Risks for non-carcinogens are measured for all receptors that could be exposed to hazardous waste. Some general criteria include the measurement of risks for non-carcinogens for all types of exposure to hazardous chemicals (ingestion, inhalation and skin). A risk index (HI), which is the reference dose intake ratio, is measured as a non-cancer risk (RfD) [10]

$$HI = \frac{MDI}{RFD} \qquad (3)$$

Which:
HI = index of hazards (dimensionless).
MDI = Total intake per day (mg/kg.day).
RfD = (mg/kg.day) comparison dose.
In the strict sense of the term, this numerical value of HI, a risk and the measurement here is qualitative: a ratio of just under 1 means that a population exposed outside theory is hazardous, whereas a score greater than 1 means that a toxic effect will occur without estimating its likelihood [12]. In this assessment, three levels were considered to interpret the outcomes.
HI less than 1
HI equal 1
HI more than 1
The EPA sums the risk for each contaminant in a medium to account for multiple substances in one route:

$$Risk_T = \Sigma\, Risk\ i \quad (4)$$

The risk $_T$ = total risk
The hazard index multiple substances in one route is estimated as:

$$HI_T = \Sigma\, HI\ i \quad (5)$$

3.1.3 Microbial Risk Assessment

Risk assessment for FC in the ground water Calculated using the following equations [13] [14].

$$PI(\frac{inf iction}{day}) = 1 - [1 + 1\frac{d}{N50}]^{-a} \quad (6)$$

PI (infection/y) = (PI (infection/ d)) ×365 (7)

N50= Average dose of infection equivalent to 8.60×10^7
α= Parameter of probability distribution function equivalent to 0.1778
$d(Ingestion\ dose) = V \times C$
V= Drinking water intake (2L/day).
C= Drinking water exposure, species per liter .
Infection is conditional on sickness, and the risk of being sick can be written as: [15].

P (ill/y) = PI (infection/y) × S (ill per infection) (8)
Where:
S (ill/infection) is the germ's degree of infectivity.
A value of S (ill per infection) given for diarrheal disease infection was assumed to be equal to 0.255 [16].
S (ill/infection) = 0.25.

4. Results and discussion

4.1 Microbial contamination of well water

The results showed the presence of microbial contamination, including coliform fecal, which is believed to be a cause contamination with these bacteria as a result of the different locations of these wells in which animal waste is abundant and as a result of this water being contaminated with animal waste or sewage water, because the group coliform bacteria, including the fecal coliform fecal, are naturally present in the intestine of humans and animals through which these microbes reach Waste to groundwater, and then the increase in numbers of these bacterial in waters of wells, and their presence in the water indicates the arrival of feces in any form to the water and its contamination.

4.2 Calculation of the most probable number of well water: (MPN)

The results showed a variation in the rate of the most likely number of well water for fecal coliform bacteria Table -1, as the rate of bacterial numbers was the water in well No. 1 was the most, amounting to 40*103 NPP/ 100 ml, while the lowest rate was in well No. 3, which was 22*103 NPP/100ml . It is believed that the reason for the difference in number of bacterial in well water such as the result of a distance or proximity of the well to the sources of pollution and on the other hand it could be The difference in pH rates or different well water temperatures, the water temperature affects the viability of the water and promotes the growth of Pathological bacteria in it. As shown in Figure (1).

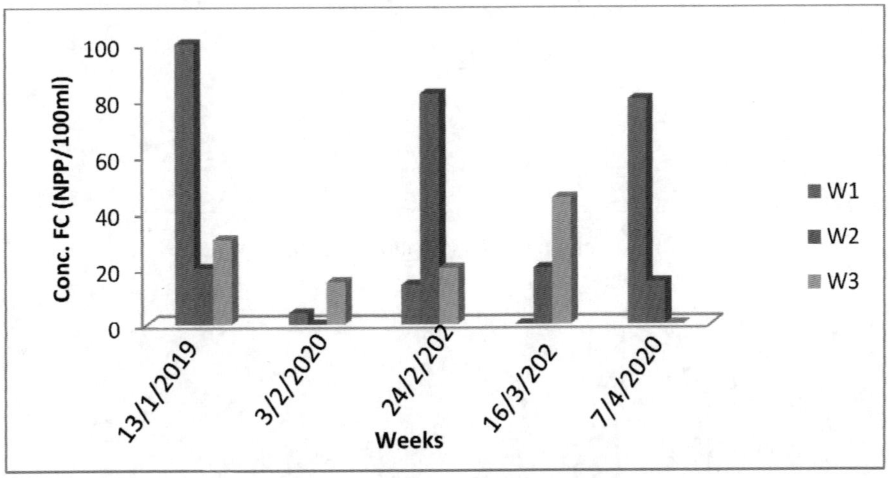

Figure 1. Fecal coliform (FC) variations of the ground water.

4.3 Physical and chemical and bacteriological groundwater analysis

The results shown in a table 1. Showed that physiochemical properties represented (pH, EC, TSS, Cl, SO4, NO3, BOD5, COD) in addition to five types of heavy elements (Pb, Cr, Ni, Cd, Ag) of the groundwater of the three wells. Through experimental work and its comparison with the Iraqi limits for drinking water, it was found that the EC value (1061) μs/cm had exceeded the permissible limit in W1 as shown in figure 3, Cl values (325, 430)mg/l, Pb values (0.43, 0.14) μg/l and Ni values (0.11, 0.11) μg/l were and the permissible limit were exceeded in W1, W2 as shown in figures (5, 10, 11)and SO4 values (681, 653, 584) mg / l as well as the permissible limit was exceeded in W1 and W2 And W3 as shown in figure 6 as well as Ag value (0.12) μg/l in W1 only as shown in figure14. The rest of the properties did not exceed the permissible limit. As shown in figures (2, 4, 7, 8, 9, 12,13).

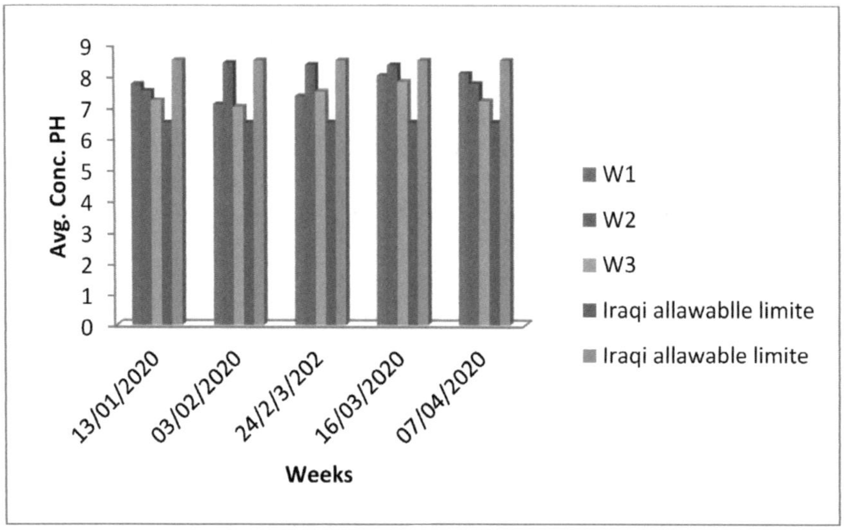

Figure 2. Ion exchange (pH) variations of the ground water.

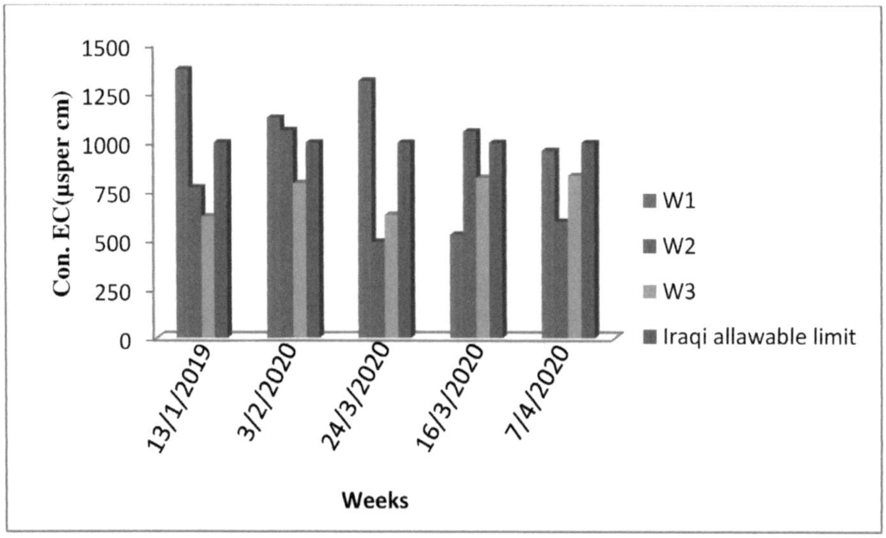

Figure 3. Electric conductivity (EC) variations of the ground water.

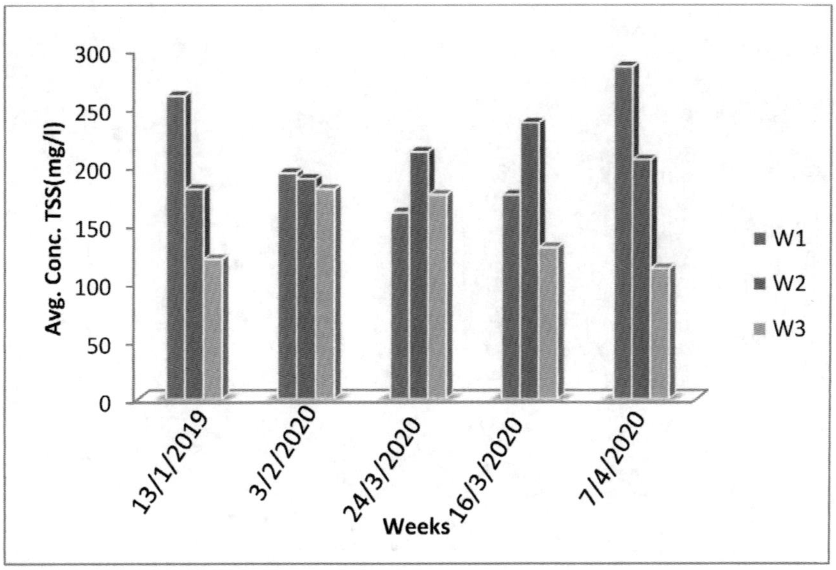

Figure 4. Total suspended solid (TSS) variations of the ground water.

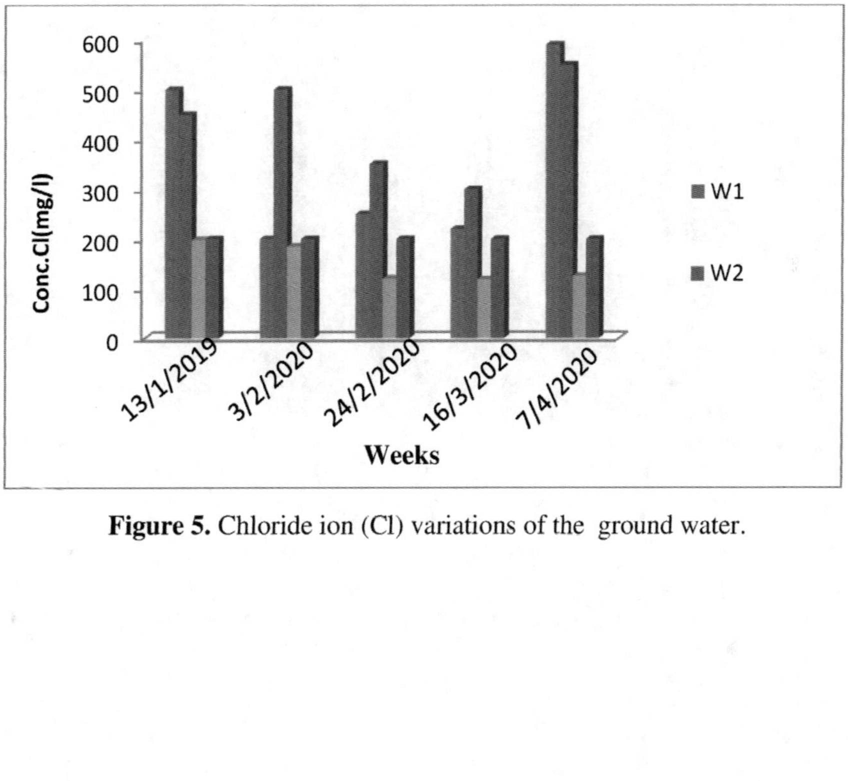

Figure 5. Chloride ion (Cl) variations of the ground water.

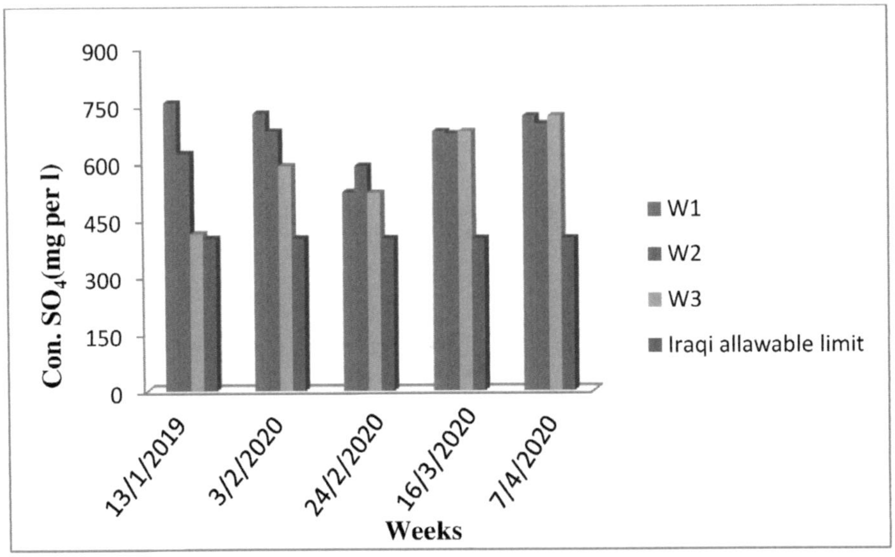

Figure 6. Sulfate ion (SO_4) variations of the ground water.

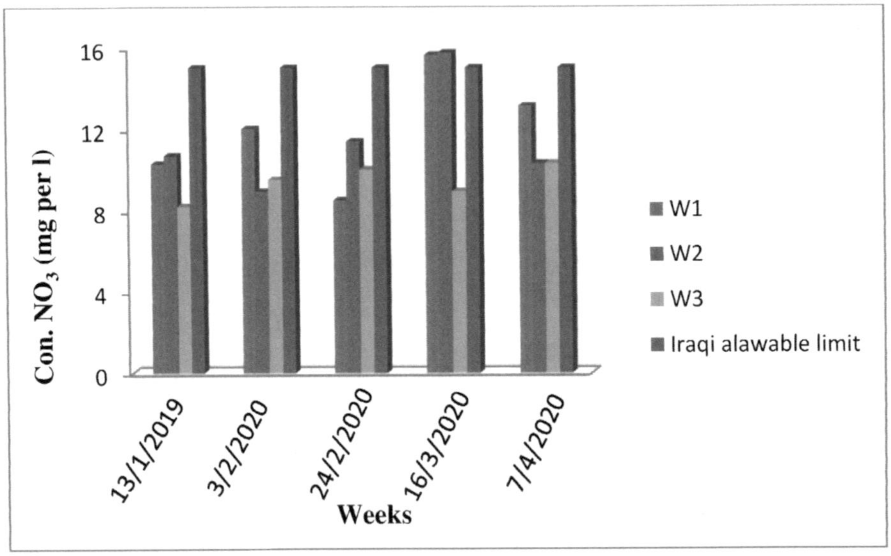

Figure 7. Nitrite ion (NO_3) variations of the ground water.

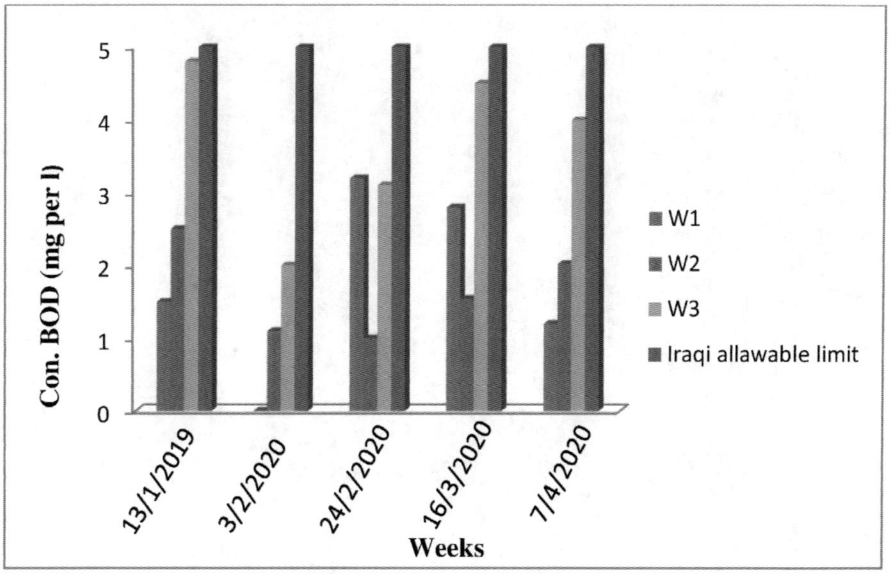

Figure 8. (BOD) variations of the ground water.

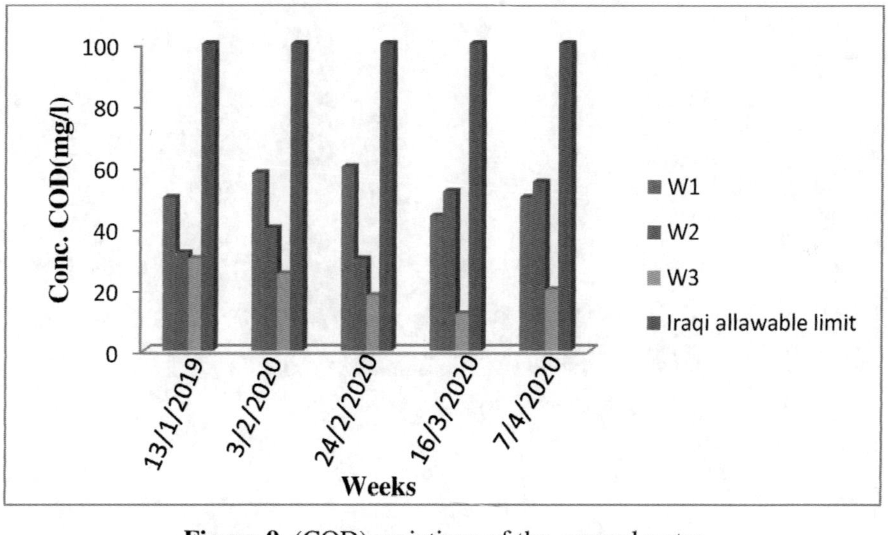

Figure 9. (COD) variations of the ground water.

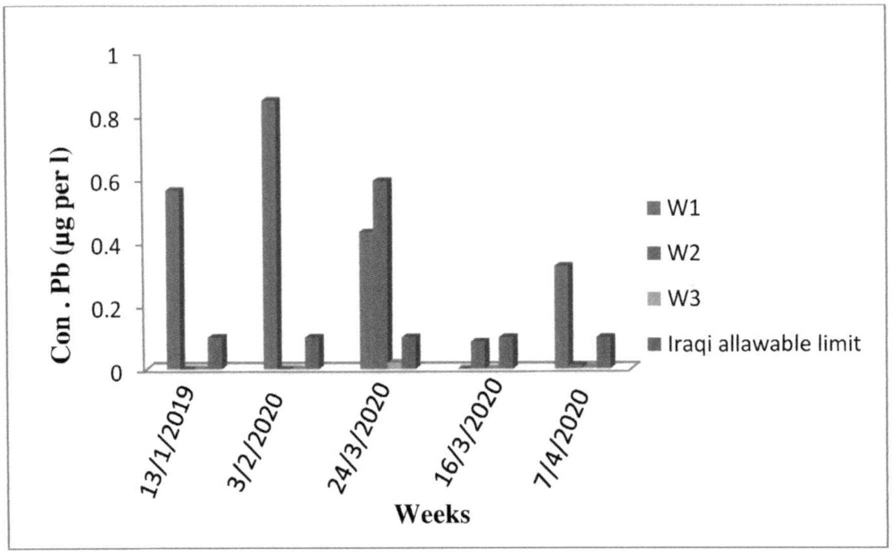

Figure 10. Lead (pb) variations of the ground water.

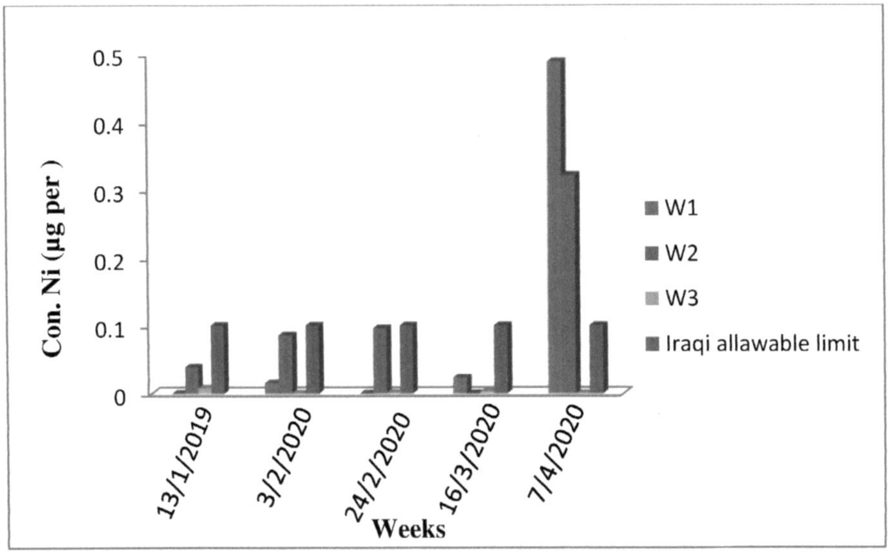

Figure 11. Nickel (Ni) variations of the ground water.

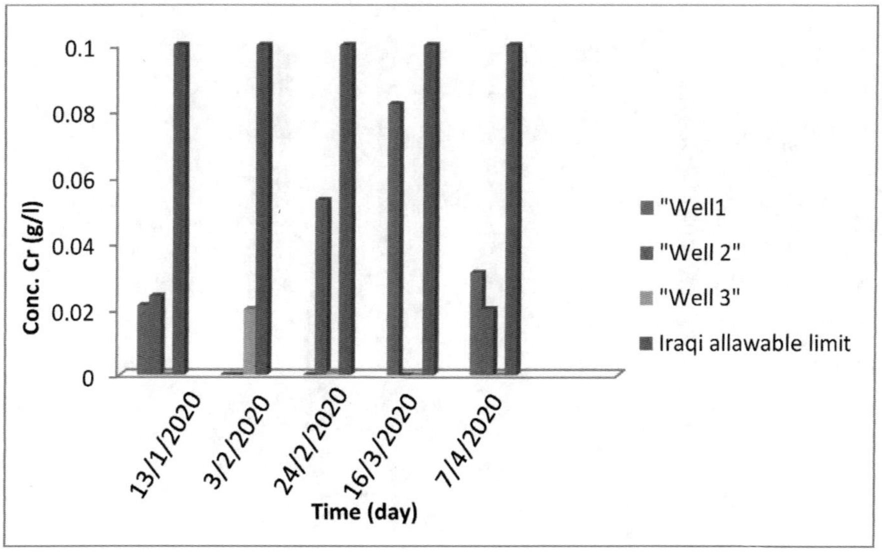

Figure 12. Chromium (Cr) variations of the ground water

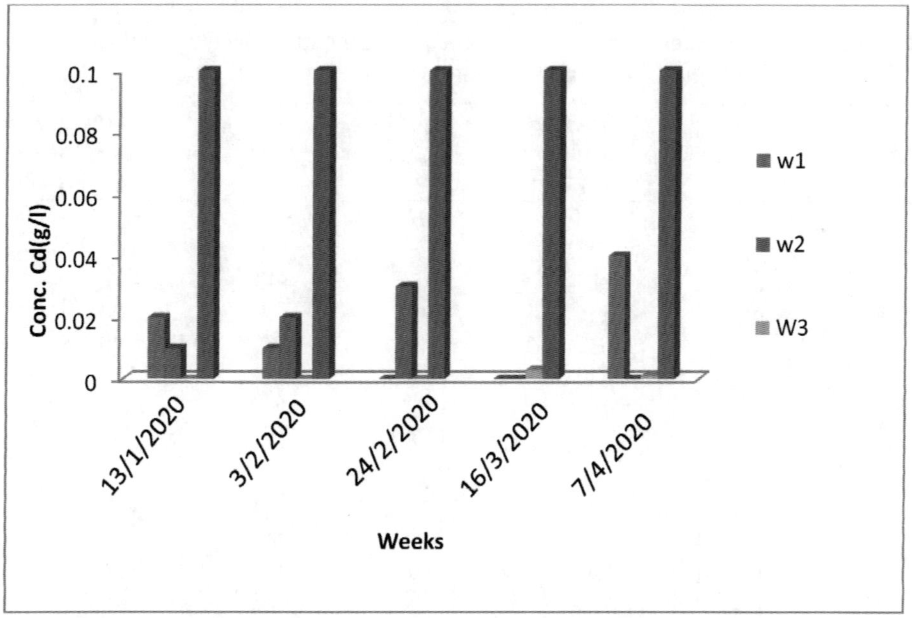

Figure 13. Cadmium (Cd) variation of the ground water.

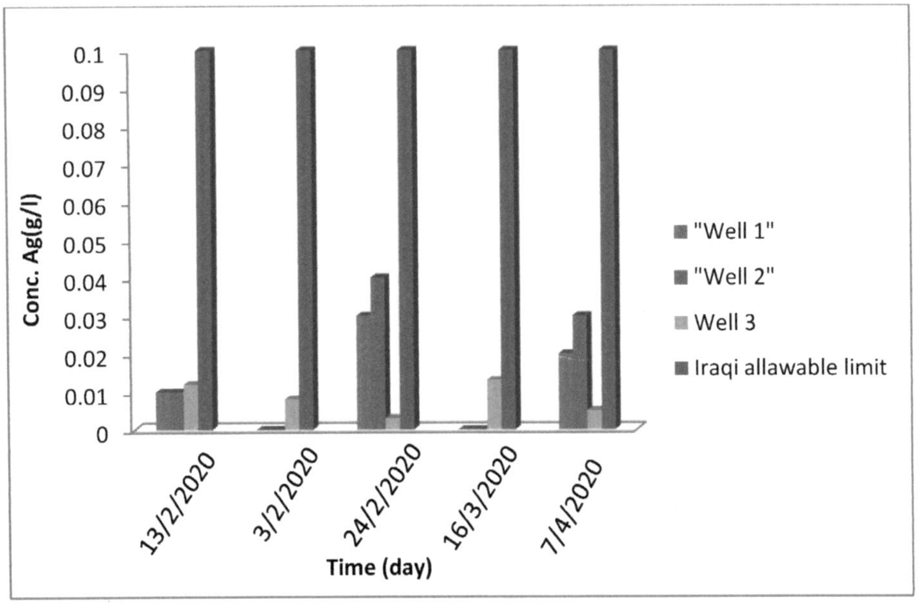

Figure 14. Silver (Ag) variations of the ground water.

Table 1. Results of the average concentration of the physiochemical, heavy metal and bacteriological characterization of the ground water in different locations.

Parameters	Unit	Average Conc. For W1	Average Conc. For W2	Average Conc. For W3	Iraqi allowable limit for drinking water 1967[17]
pH	Unit less	7.65	8.07	7.35	6.5-8.5
EC	µS/cm	1061	796	739	1000
TSS	mg/l	215	205	143	-
Chlorides	mg/L	352	430	150	200
SO_4	mg/L	681	653	584	400
NO_3	mg/L	12	11	9.4	15
BOD_5	mg/L	2	1.63	4	<5
COD	mg/L	53	41.8	21	<100
Pb	µg per L	0.43	0.14	0.005	0.1
Ni	µg per L	0.11	0.11	0.002	0.1
Cr	µg per L	0.03	0.02	0.004	0.1
Cd	µg per L	0.01	0.01	0.001	0.1
Ag	µg per L	0.12	0.02	0.001	0.1
Fecal c.	NPP per 100 ml	$40*10^3$	$28*10^3$	$22*10^3$	0

4.4 Chemical Risk Assessment

Table (2) shows the standard values that were used in an equation of max. daily intake of water. Exposure to heavy metal contaminants in Using the equation, ground water was measured by (2).While HI for non carcinogenic chemicals in Using the equation, ground water was measured by (3).And for the hazard index multiple substances used the equation (5).

IOP Conf. Series: Earth and Environmental Science **877** (2021) 012004 doi:10.1088/1755-1315/877/1/012004

Table 2. EPA Suggested values for intake estimation.

Limit	**The Standard value**	
	Adult	$\boldsymbol{Child (< 1.5\ year)}$
weight of the Average body (kg)	70	10
Ingestion rate (L/day)	2	1
Exposure duration (carcinogenic effect)(year)	70	70
Exposure duration (non carcinogenic) effect	30	30
exposure frequency (day/year)	365	365

Table (3) shows the results of the ingestion intake (MDI) and the hazard index (HI) for the heavy metal [pb, Ni, Cr, Cd, Ag] in the ground water. Where it's found that the total non-carcinogenic risk index (HI_T) in the groundwater for W1, W2, W3 was acceptable for adults and children.

Table 3. Calculated the ingestion intake of substances in the ground water and risk of threshold-effect substances.

	Substance	**Max. Conc. (mg/L)**	**MDI ($mg/kg.d^{-1}$) Adult**	**MDI ($mg/kg.d^{-1}$) Child**	**Oral RfD (mg/kg. day)**	**HI Adult**	**HI Child**
W1	**Pb**	0.84E-3	0.024	8.4E-5	0.006	4E-3	1.4E-2
	Ni	0.48E-3	4.8E-4	4.8E-5	0.02	6.85E-4	2.4E-3
	Cr	0.08E-3	8E-5	8E-6	0.005	3.8E-4	1.6E-3
	Cd	0.04E-3	1.14E-3	4E-7	0.0005	0.228	8E-3
	Ag	0.03E-3	8.57E-4	3E-6	0.005	1.71E-2	6E-2
						ΣHI=0.23	ΣHI=2.6E-2
W2	**Pb**	0.59E-3	1.68E-5	5.9E-7	0.006	2.8E-3	9.83E-5
	Ni	0.32E-3	9.14E-3	3.2E-5	0.02	0.457	1.6E-3
	Cr	0.05E-3	1.42E-3	5E-6	0.005	0.285	1E-3
	Cd	0.04E-3	8.57E-4	3E-3	0.0005	1.71E-3	6E-3
	Ag	0.04E-3	1.14E-3	4E-6	0.005	0.228	8E-3
						ΣHI=0.97	ΣHI=1.6E-2
W3	**Pb**	0.002E-3	5.71E-8	2E-7	0.006	9.52E-6	33E-5
	Ni	0.008E-3	2.28E-7	8E-7	0.02	1.14E-5	4E-5
	Cr	0.02E-3	5.71E-7	2E-6	0.005	1.14E-4	1E-4
	Cd	0.004E-3	1.14E-7	4E-7	0.0005	2.28E-4	8E-4
	Ag	0.003E-3			1.71E-5	0.005	6E-5
						ΣHI=3.8E-4	ΣHI=1.3E-3

-EPA allowable limits: If HI ≥1: risk unacceptable else the risk acceptable.

4.5 Microbial Risk Assessment

Risk assessment for FC in the ground water, was calculated by using equation (6) and (7), and for the probability of becoming ill was calculated by the equation (8).

Table (4) shows the results of risk assessment of FC in ground water. average value PI (infection/day) and P (ill/dose) does not exceed the EPA allowable limits

Table 4. Risk assessment of FC in the ground water.

Location	FC cell/L	PI (infection/day)	PI (infection/year)	S (ill/infection)	P (ill/dose)
W1	40E3	2E-3	0.73	0.25	0.18
W2	28E3	1.15E-4	4.22E-2	0.25	1E-1
W3	22E3	9.09E-5	3.31E-2	0.25	8.29E-3

** EPA allowable limits: If PI (infection/day) and P (ill/dose) < 1×10^{-6} : the risk acceptable else the risk unacceptable .

5. Conclusion

1. The result indicated that most of the physical, chemical, parameters did not exceed the Iraqi determinants for drinking water. In all wells chosen for study except for the EC value (1061) µs/cm only exceeded the normative parameters in W1 while chlorine values (325, 430)mg/l exceeded the Iraqi determinants of water in W1 and W2, and also SO4 values (681, 653, 584) mg / l exceeded the determinants in all wells studied. And for the examining the heavy elements of groundwater showed that the pb values (0.43, 0.14) µg/l and Ni element values (0.11, 0.11) µg/l were in the W1 and W2 exceeded the permissible limits, while Ag value (0.12) µg/l exceeded the permissible limit inW1 only, the other elements did not exceed the permissible limits.

2. The results showed that the total risk index in the groundwater for W1, W2, W3 was acceptable for adult values ΣHI(0.23, 0.97, 3.8E-4) and children values ΣHI (2.6E-2, 1.6E-2, 1.3E-3) respectively.

3. For the microbial risk assessment, the mean annual incidence of PI infection (infection / year) caused by fecal coliform bacteria in groundwater and the risk of diarrhea due to infection caused by fecal coliform in groundwater (mean P (ill / dose)) Until it becomes spatial in the groundwater used for drinking. Where it was found that the ratio of PI(infiction/day) values(2E-3, 1.15E-4, 9.09E-5) and P (ill / dose) values (0.18, 1E-1, 8.29E-3) had exceeded the environmental determination allowable for EPA .

Reference

[1] Al-Madhlom Q, Al-Ansari N and Hussain H M 2016 Assessment of Groundwater Vulnerability in Northern Babylon Governorate, Iraq *Engineering* **8** 883–902
[2] Modood T 1988 'Black', racial equality and Asian identity *J. Ethn. Migr. Stud.* **14** 397–404

[3] Hassan K 2016 Groundwater pollution in some areas between the Al-Kifl stream and the Shatt al-Hindi *J. Hum. Sci. / Coll. Educ. Humanit.* **23**
[4] Momodu M A and Anyakora C A 2010 Heavy metal contamination of ground water: The Surulere case study *Res. J. Environ. earth Sci.* **2** 39–43
[5] Devi R, Singh V and Kumar A 2008 COD and BOD reduction from coffee processing wastewater using Avacado peel carbon *Bioresour. Technol.* **99** 1853–1860
[6] Amal Alghamdi 2016 Study, isolation and definition of microorganisms in drinking water samples. 492.
[7] Lee J, Durst R W, Wrolstad R E and JD C E T G M M H J H H K S K D A K S M S K M B K M T C P F R A S G T U W 2005 Determination of total monomeric anthocyanin pigment content of fruit juices, beverages, natural colorants, and wines by the pH differential method: collaborative study *J. AOAC Int.* **88** 1269–1278
[8] APHA. (American Public Health Association) 2005 Standard Methods for the Examination of Water and Wast Water, Twentieth First Edition.

[9] APHA, AWWA AND WEF, 22nd ed. Washington: American public Health Association 2012 Standard Method for the Examination of water and wastewater
[10] Mohammad A.M. Al-Tufaily and Sabreen Lateef Kareem 2010 Risk Assessment Analysis Procedure to Control Pollution Problems in Euphrates Region in Iraq
[11] Richard J.Watts, 1997 Hazardous Wastes department of Civil and Environmental engineering, Washington state university, copy by John Wiley & Sons.
[12] Evens Emmanuel, Marie Gisèle Pierre , Yves Perrodin ,2009 Groundwater contamination by microbiological and chemical substances released from hospital wastewater: Health risk assessment for drinking water consumers *journal home page* : www. elsevier. com / locat e / envint.
[13] Fewtrell, L. and Bartram, J., 2001 water Quality: Guidelines, Standards and Health Published by IWA Publishing, London, UK. ISBN: 1 900222 28 0.
[14] Bruce A. Macler a and Stig Regli 1992 Use of microbial risk assessment in setting US drinking water standards. International Journal of Food Microbiology, **18** 245-256.
[15] Navier, D., Laroche, L. and Hartemann, P 2006 Waterborne Microbial Risk Assessment, Grenoble University Hospital, 38700 La Trochee, France, http://www.biomedcentral.com/1471-2458/6/122.
[16] Howard G and Pedley S 2003 Quantitative Microbial Risk Assessment for piped Water Supplies in Uganda Robens Center for Public and Environmental Health (University of Surrey_ https://www.researchgate.net/publication/221911472.
[17] WHO 1996 Water quality monitoring: a practical guide to the design and implementation of freshwater quality studies and monitoring programmes. CRC Press.

Surface Heat Budget estimation for Laurance Lake, US

Dahlia S. Abed-Zaid[1] and Hussein A. M. Al-Zubaidi[1*]

[1]Department of Environmental Engineering, Faculty of Engineering, University of Babylon, Babylon, Iraq
* Corresponding author: hussein.alzubaidi@uobabylon.edu.iq

Abstract
Estimating heat budget factors are important to understand the many physical processes of large lakes and their reaction to the atmosphere. Some of these components are affected by water temperature, while the other depends on atmospheric conditions. This paper estimates the total heat flux for Lawrence lake via a code developed in MATLAB environment. The code can deal with different time resolutions if the lake water surface temperature data were at different time resolutions from the meteorological data. Results showed that solar energy peaks at 842 Watt/m^2 at 540 Julian day, which is very normal for a sunny summer day, while the longwave radiation has 204 Watt/m^2 as a min value. The back radiation did not make any reaction for the variation, but it revealed a small gradient. Furthermore, evaporation recorded -67 Watt/m^2 as a minimum value at 659 Julian day and 360 Watt/m^2 as a maximum value at 578.43 Julian day close to the maximum water surface temperature event.

Keywords
Evaporation, heat exchange, heat transfer, heat flux, longwave radiation, lakes and reservoirs, shortwave radiation, solar radiation.

1. Introduction

Lakes have a rapid response to climate changes. Some of the most pervasive and concerning physical effects of climate change on lakes are changes in water heat budgets, evaporation [1,2] and surface water temperature [3]. As climate variations affect water temperature. Water temperature plays a big role in many physical, chemical, and aquatic biological processes. With the increasing water temperatures, the solubility of dissolved oxygen decreases. This leads to reducing the available-oxygen for aquatic fauna and effects on fish [4,5]. In addition, temperature has been changing because of weather, shading of stream bank, discharge of cooling water and others [6]. As a result, surface heat exchange keeps varying surface heat fluxes over time.

Surface heat fluxes calculation techniques have been reported in many references. These fluxes include five components. Some of them depend on water temperature, and others depend on external conditions [7,8]. Monika (2019) worked on problems related to the spread of thermal pollution in rivers and tried to summary it [9]. Edinger's equation was used for calculating the net heat rate, Stephan-law for calculating long wave, and Stefan Boltzmann law for back long radiation. Also, short-wave radiation was determined and then corrected by using Magnusson's equation. Shanahan (1984) described different styles for developing required data [10]. Shanahan took the average reflectance of the water surface to calculate short-wave radiation and adopted the EPA-Model [11]. For evaporation and conduction, Dalton's equation and Bowen's ratio were used [9], or different equations may be applied to be recalibrated for getting the best match with field data [10].

The purpose of this work is to calculate heat fluxes and perform a comparison between the obtained fluxes and typical values. Several plots were made to show the behaviour of each flux during different periods.

2. Materials and Methods

2.1.Study Area

Laurance Lake is a reservoir at the base of the Mountain Hood in Oregon, drainage into the Middle Fork of the Hood River. The reservoir was constructed in 1968 for irrigation storage, with capacity 3564 acres- feet at full pool [12, 7], see Figure 1.

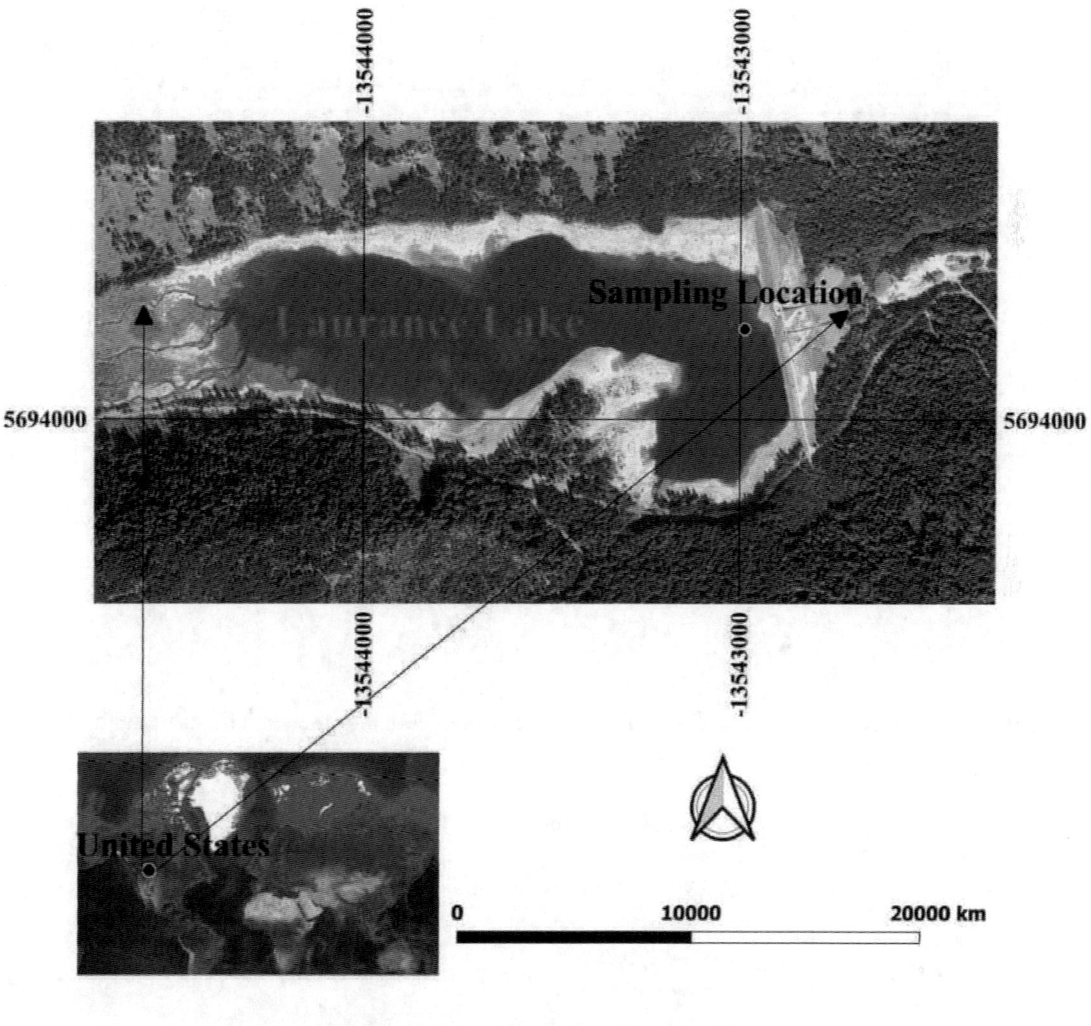

Figure 1. Laurance Lake, Oregon, US.

2.2.Data and coding

The MATLAB code was summarized in a schematic flowchart as shown in Figure 2. The required meteorological input data was measured at the dam (For more details, see [12]) for the period from 480 to 680 Julian day (Jday). In addition, the water surface temperature was demanded by the model.

The available water temperature was in two-time resolution from meteorological data. Therefore, linear interpolation was taken into account to determine water surface temperature at any required time. For atmospheric long radiation calculation, the required data of clouds cover and air temperatures were available. Dew point temperature was also used to determine the saturated vapor pressure. Wind profile was calculated based on the corrected wind speed data, and then it was entered along with saturated vapor pressure and air vapour pressure to get evaporation. Conduction radiation was calculated based on Bowen's ratio, and the calculation of short-wave radiation based on the EPA model [11]. Figure 3 and Table 1 show the total heat flux components, and the typical values, respectively.

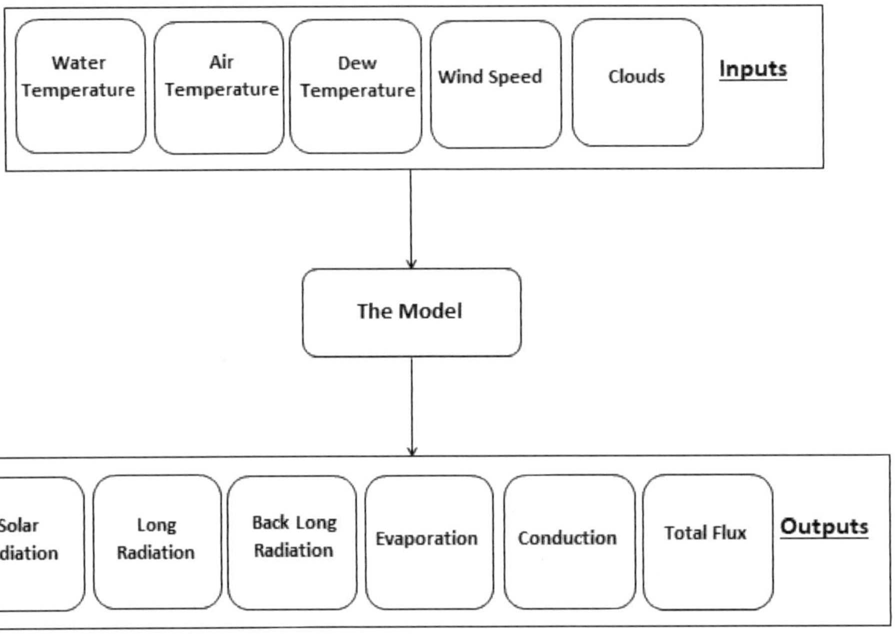

Figure 2 MATLAB code summarization.

Finally, the Laurance Lake heat fluxes (Watt/m^2) were obtained using the term by term method [7] as follows:

$$H_n = H_{sn} + H_{an} - H_{Br} - H_e - H_c$$

IOP Conf. Series: Earth and Environmental Science **877** (2021) 012005 doi:10.1088/1755-1315/877/1/012005

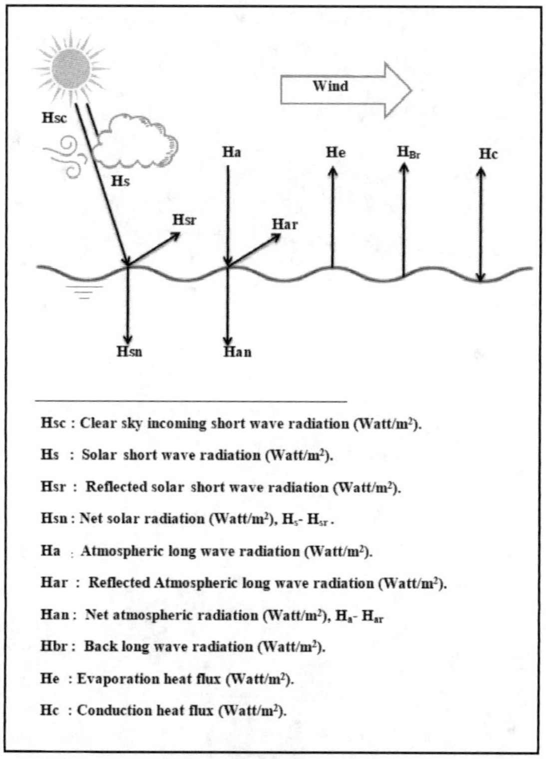

Figure 3 Surface heat flux components.

Table 1. Approximate values of the heat flux components (Watt/m^2) [9, 7]:

Fluxes	Range
Solar radiation	(50 – 350) For a sunny day, (800 – 1000) For a very cloudy day, (100 – 300)
Atmospheric radiation	(200 – 400) (30 – 450)
Back radiation	(250 – 500) (300 – 500)
Evaporation	(0 - 35)
Conduction	(-70 – 200)

Solar short wave radiation comes from the sun, and atmospheric radiation is produced from the atmosphere [13]. The value of short-wave radiation depends on various factors such as solar altitude, scattering and absorption, reflection and shading [14]. There are several models for calculating shortwave radiation. In this paper, the EPA model was adopted. This model was used in the water quality model CE-QUAL- W2 [8]. For more details about the heat budget determination approach, see [7].

3. Results and Discussion

The data covered the period from 480 to 680 Julian day. Solar energy becomes maximum in summer days [15]. It recorded 842 Watts/m^2 as a maximum value at 540 Julian day and of 606 Watt/m^2 as a minimum value. For a summer sunny day, this is typical. Atmospheric radiation recorded 204.1 watts/m^2 as the min value. These high values is returning to air temperature interacting with the high strength of short-wave radiation during summer days (Figure 4 and 5). While Conduction heat flux curve simulates the water temperature curve, back long radiation goes down to its minimum value of 340 Watts/m^2 at 487 Julian day (Figure 6) when the water surface temperature has the minimum value.

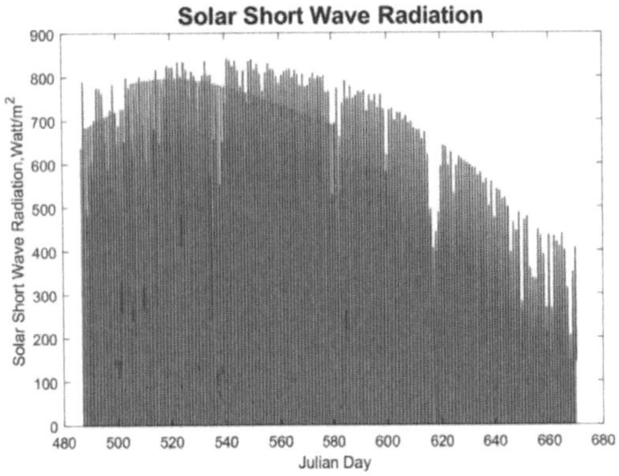

Figure 4. Solar radiation.

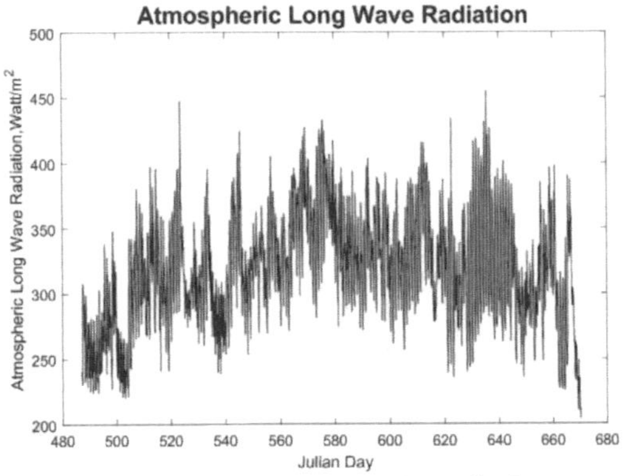

Figure 5. Atmospheric radiation.

Evaporation has a value of -67 Watt/m^2 as a minimum value at 659 Julian day and a maximum value of 360 Watt/m^2 at 578.43 Julian day when the maximum water surface temperature was at 576 Julian day. Conduction and evaporation are in inverse proportion by a ratio called Bowen's ratio. Figure 7 and 8 clearly illustrate this relationship.

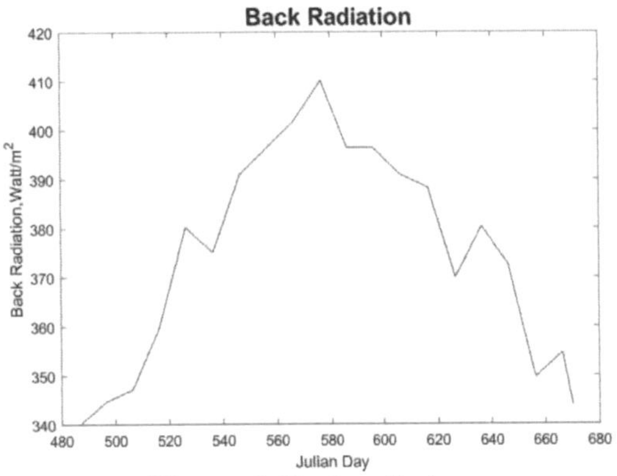

Figure 6. Back radiation.

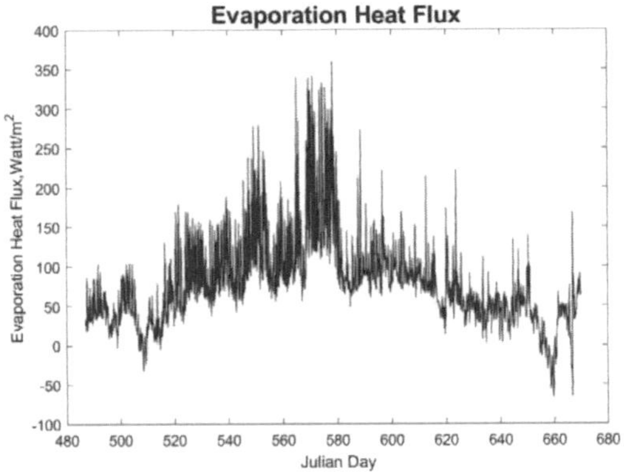

Figure 7. Evaporation radiation.

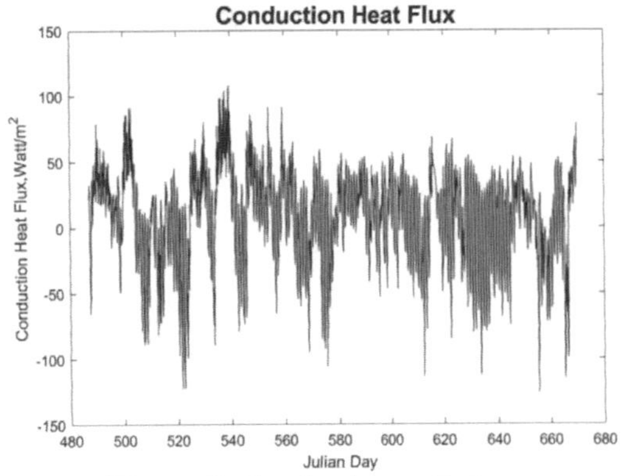

Figure 8. Conduction radiation.

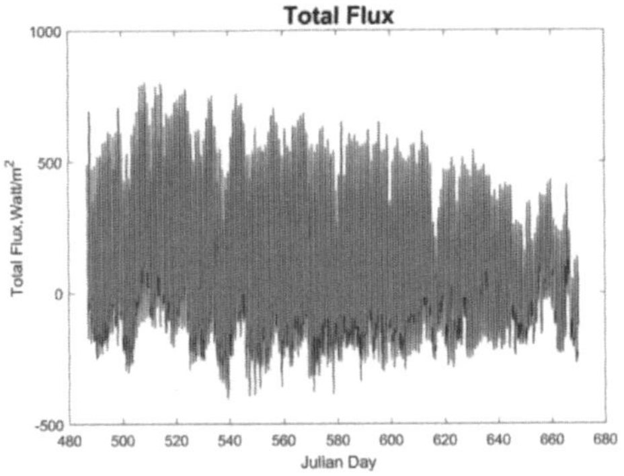

Figure 9. Total flux.

Figure 10 shows the fluxes during the period (500- 530) Julian day, where the variation of the fluxes become more obvious. The fluctuation of short-wave radiation is due to the natural phenomenon of sunrise and sunset. When short-wave radiation recorded the maximum value during day hours (particularly in summer days), an expectable increase in air temperature and water temperature. As a result, the related heat fluxes for each temperature will experience an increase too. The increases vary differently based on the heat fluxes dynamics. Initially, the solar flux transmitted from the sun, and it takes the highest amount in the daytime and literally in the middle of the day with a maximum value of 833 Watt/m^2. Atmospheric radiation appears rather calm with fluctuations at the start and end of the day. These fluctuations are getting between (447 to 220) Watt/m^2 with a mean value of 319 Watt/m2. Fluxes in Figure 10 present a clear inverse relationship between conduction and evaporation. In contrast, the back radiation seems unaffected by any variable other than water temperature. It looks like that the flux takes the shape of a straight line, but in fact, it is presenting by a weak gradient of 34.65. The total flux nearly copies the solar radiation shape, showing the role of solar radiation.

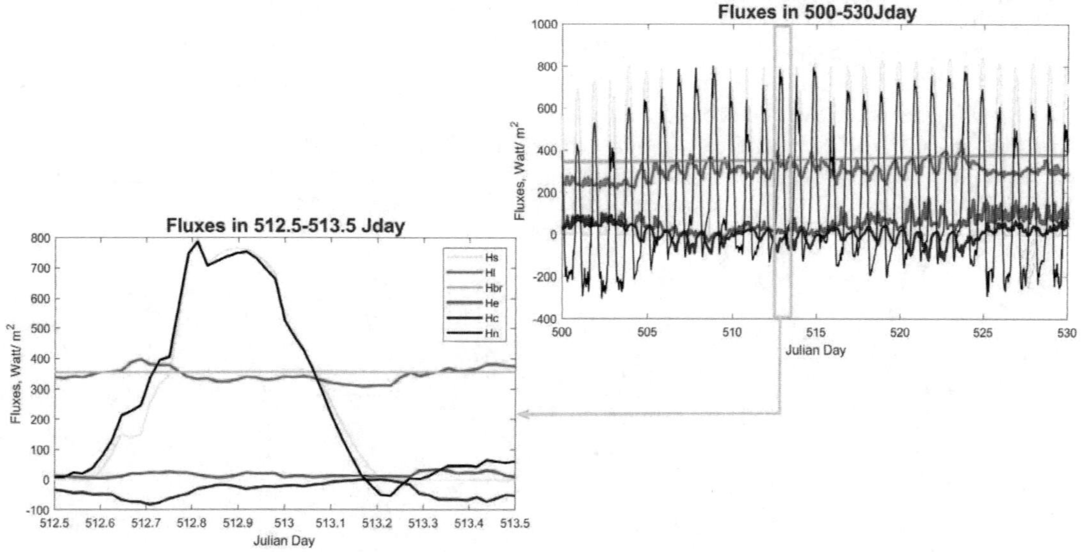

Figure 10. Fluxes during the period (500- 530) Julian day.

4. Conclusions.

In this paper, a development for a MATLAB model was made to calculate surface heat exchange Water-bodies. To calculate solar and atmospheric radiation, conduction, evaporation, long-wave back radiation and total net heat flux, the data from Laurance Lake, US was used. During the period of the study, the fluxes Results were highlighted for daily and seasonal. This was obvious in summer, where the total heat flux has positive values and negative during winter. Each flux has its role to play in the total heat flux, but solar radiation has the dominator role. After the important role of solar radiation, the influence of longwave back radiation comes.

References

[1] Rodell M, Famiglietti J S, Wiese D N, Reager J T, Beaudoing H K and Landerer F W 1240 and Lo, M.-H.: Emerging trends in global freshwater availability *Nature* **557**

[2] evaporation accelerated by changes in surface energy allocation in a warmer climate *Nat. Geosci.* **11** 410–414

[3] Lenters J D, McIntyre P B and Kraemer B M 2015 Rapid and highly variable warming of lake surface waters around the globe *Geophys. Res. Lett.* **42** 10–773

[4] Wanders N, van Vliet M T H, Wada Y, Bierkens M F P and van Beek L P H 2019 High-resolution global water temperature modeling *Water Resour. Res.* **55** 2760– 2778

[5] Bhateria R and Jain D 2016 Water quality assessment of lake water: a review. Sustain Water Resour Manag 2: 161–173

[6] Spellman F R and Drinan J E , "Drinking water regulations," in *The drinking water handbook, second edition*, 2nd ed., CRC Press, 2012, pp. 27–58.

[7] Al-Zubaidi H A M 2018 3D Hydrodynamic, Temperature, and Water Quality Numerical Model

for Surface Waterbodies: Development, Verification, and Field Case Studies
[8] Cole T M and Wells S A, "CE-QUAL-W2: A two-dimensional, laterally averaged, hydrodynamic and water quality model, version 3.5," 2006. [Online]. Available: https://pdxscholar.library.pdx.edu/cengin_fac/157/.
[9] Kalinowska M B 2019 Effect of water–air heat transfer on the spread of thermal pollution in rivers *Acta Geophys.* **67** 597–619
[10] Shanahan P, "Water temperature modeling: a practical guide.," in *Environmental Protection Agency*, 1984, vol. U.S. EPA R, no. 1, pp. 1–25.
[11] Annear R L and Wells S A 2007 A comparison of five models for estimating clear-sky solar radiation *Water Resour. Res.* **43**
[12] Berger C, Wells S A and Annear R L 2005 Laurance Lake Temperature Model
[13] Grossnickle S C 2000 *Ecophysiology of northern spruce species: the performance of planted seedlings* (NRC Research Press)
[14] Chapra S C 2008 *Surface water-quality modeling* (Waveland press)
[15] Lofgren B M and Zhu Y 2000 Surface energy fluxes on the Great Lakes based on satellite-observed surface temperatures 1992 to 1995 *J. Great Lakes Res.* **26** 305–314

The effect of saturation Methods on natural gypsum rocks

Dhafar abdulrahman [1*], Falah Hasan [1], Amjad Ibrahim [2]

[1] Civil Engineering Department, University of Technology, Baghdad, Iraq
[2] Department of Production Engineering and Metallurgy, University of Technology, Baghdad, Iraq
* Corresponding author: eng.mahmood.aa@uobabylon.edu.iq

Abstract

When the gypsum rock is exposed to the relative humidity in a closed environment for a continuous period, the result is an increase in the weight of these rocks after the passage of the first 24 hours, and there is no effect on the percentage of gypsum that the rocks contain, as the weight increase occurred at the same time for the different gypsum ratios. For rocks submerged in potable water and located within a humid environment, most of these rocks witnessed weight gain, with some exhibiting dissolving behavior. As for the gypsum rocks submerged with water saturated with aqueous calcium sulfate salt, they witnessed weight gain without any solubility. It is worth noting the important effect that relative humidity plays in the atmosphere at high levels, as it stimulated the saturation property to cover and overcome the famous property of gypsum rocks, which is the melting at high humidity.

Keywords: Natural gypsum rocks; saturation; humidity; exposure.

1. Introduction

Iraq includes several geological formations, including Al-Fatha formation, which is revealed on the surface of the earth in different governorates, such as Erbil, Kirkuk, Mosul, Diyala, Wasit, Sulaymaniyah, and Anbar, and it includes several deposits, including the primary gypsum

Taking the engineering view, we find that the presence of gypsum rocks within the soil layers of a specific project negatively affects the effectiveness of the project[1], and the reason is due to the high sensitivity of the gypsum rock to the existence of water [2, 5]. This sensitivity is represented in the saturation process that rocks are subjected to when in contact with water and the relative humidity for continuous periods depending on the surrounding conditions that accelerate or slow the saturation process. Exploring the effect of saturation is important to find out the highest percentage of water that can fill the pores of the gypsum rock and the effect that it can have on the weight of the rock and its resistance to the stresses, and later the stability of the structures built on layers of gypsum rocks saturated with water, relative humidity or both, so the researchers addressed this effect under consideration.

Sadeghiamirshahidi and Vitton [3] examined the effectiveness of different types of saturation methods during the test (freshwater, gypsum-saturated water, vacuum saturation, and improved vacuum saturation) they concluded that 30 h is a convenient saturation time for gypsum cores using the first three methods and 24 h was recoded as the required time to saturate gypsum cores using the improved vacuum saturation with additional 24 h time needed for sample preparation. Mohammadi et al.[4] investigated the saturation of gypsum rocks in terms of sensitivity for dry vacuum saturation, wet vacuum saturation, and time of applying each one of them their result appointed 5 h was enough time to achieve saturation using a dry vacuum method under initial dry vacuum pressure P vac(dry) = -0.5 atm, and Pvac(wet) = -0.5 recorded 95% weight gain at 96 h. Hoxha et al. [5] studied the deformation that occurs in the gypsum rock due to the existence of relative humidity in the underground quarries they used three different percentages of relative humidity RH = 30%, RH = 60%, RH = 90% kept in under-controlled conditions. relative humidity was achieved using the salt solution method which maintains placing rock samples in closed-cell with atmosphere responding to a source of an over-saturated solution. For a constant relative humidity, a specific type of salt should be added to the solution placing a rock sampling in that cell accompanied with mass change (absorbing \ releasing of water) because of the present equilibrium process between vapour pressure and water pressure in the rock sample pores. When equilibrium complete no mass change happen in the rock sample mass in this study equilibrium took place between 15 to 30 days.

In this paper we present a new perspective of saturation using a new method for imposing the relative humidity, studying the effect of the relative humidity environment on the gaining \ losing process and the time required for achieving the maximum weight gain.

2.Methods and Materials

Geology of the studied areas

The rock samples have been taken from three different districts all over Iraq Figure (2) Zurbatiya in Wasit governorate(E:589795 N:3693005 UTM), Bazyan in AL-Sulymania governorate (E:512004 N:3940373 UTM), and hit in AL-Anbar governorate (E:307715 N:3705260 UTM) containing three different percentages of gypsum content (42.28%,73.91%, and 72.48%) respectively. The total number of the samples 36 sets in three groups each group represent a district then each group was subdivided into 4 samples to resemble each saturation case the samples have an irregular shape, dissimilar in weight, their colors included white, dark grey mixed with white, and pale yellow.

Figure 1. cool-steam humidifier.

Figure 2. relative humidity devise.

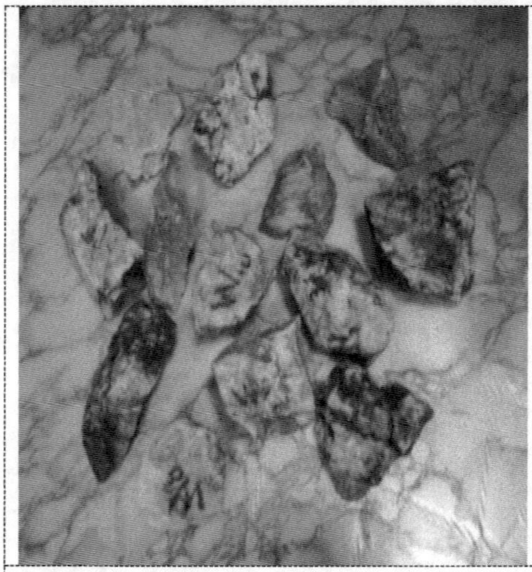

Figure 3. rock samles from Zurbatiya, Bazyan, and Hit.

Figure 4. gypsum salt ($CaSO_4.2H_2O$) powder.

Figure 5. (a) left samples submerged in a potable (b) right samples submerged in a gypsum-saturated wate.

Cool-steam Humidifier to stimulate the high relative humidity environments 'Figure 1'.

Container to maintain the relative humidity in a specific place and keeping the water level in the vessels under control from the evaporation process 'Figure 6'.

vessels for submerging samples in potable water and gypsum-saturated water 'Figure 5'.

Hygrometer measures the relative humidity in a closed container 'Figure 3'.

A sensitive balance that is accurate 0.01 in grams 'Figure 4'.

IOP Conf. Series: Earth and Environmental Science **877** (2021) 012006 doi:10.1088/1755-1315/877/1/012006

Figure 6. Container where all the saturation methods kept.

3.Results

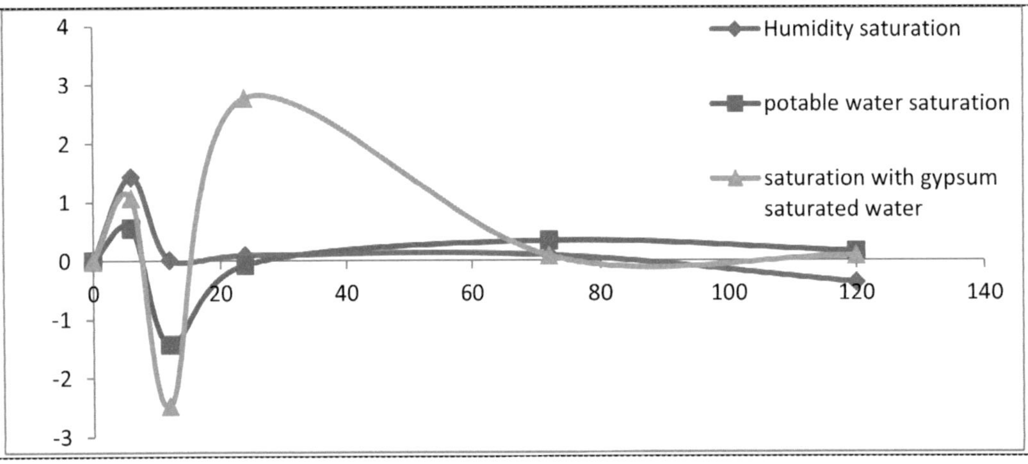

Figure 7. Saturation methods effect on Hit district samples.

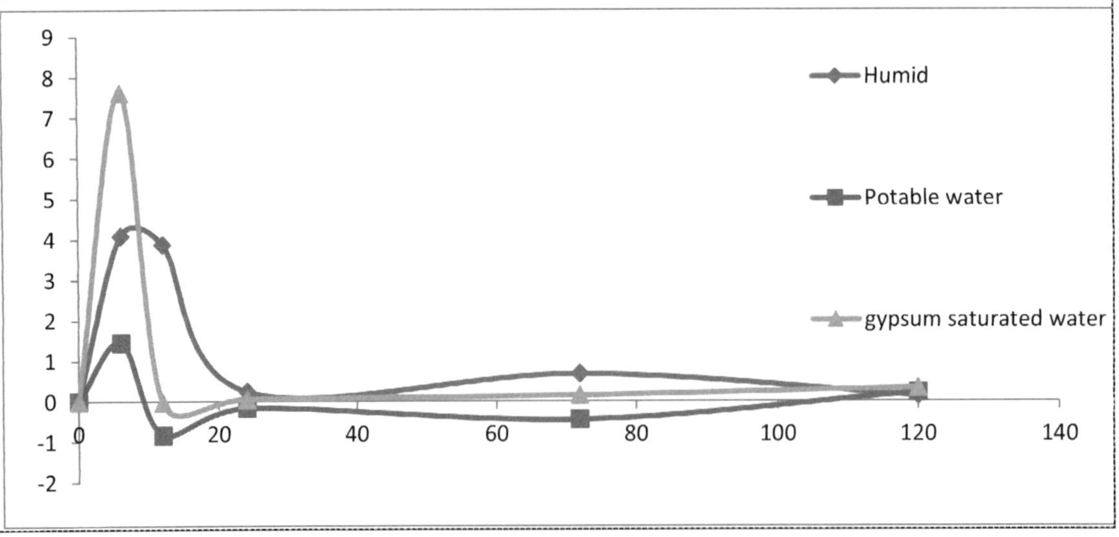

Figure 8. Saturation methods effect on Bazyan district samples.

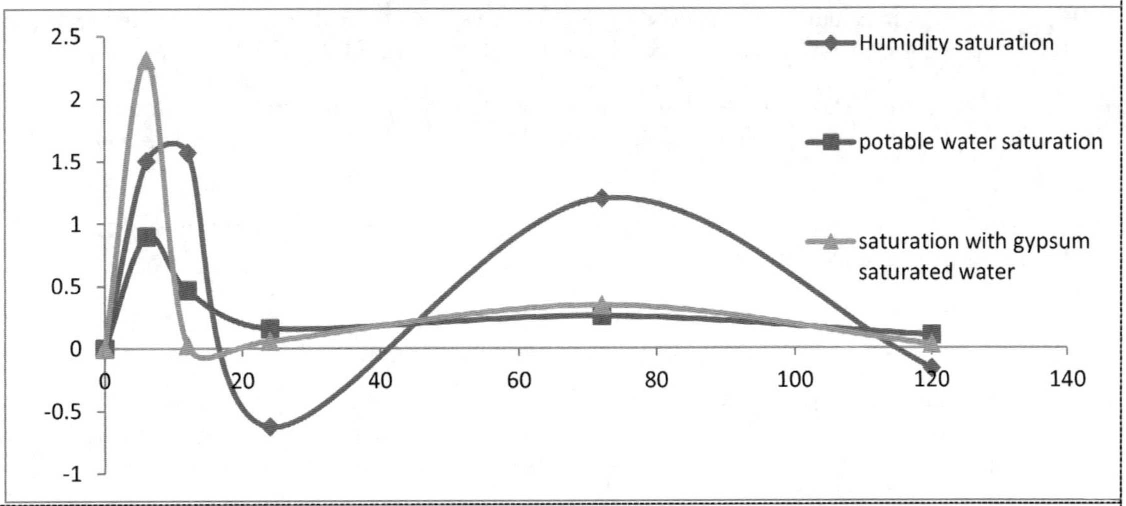

Figure 9. Saturation methods effect on Zurbatiya district samples.

Gypsum is considered a salt with high solubility and has a solubility in liquids with different properties[5], and in this paper, the results showed that gypsum samples gain weight upon the passage of the first 24 hours and maintain this gained weight, but the gypsum shows its lack of stability throughout the subsequent period as it shows signs Gain and loss But these changes, if any, are small, and they are a phenomenon that expresses the tendency of gypsum to interact with the influencing effect, Therefore, there was no need to continue the experiment after 200 hours.
The use of relative humidity with its direct effect and simulating its presence in nature to discover its effect on samples is considered unprecedented and the mechanism of its effect is still unknown there is a lot to know about relative humidity especially since there are regions in the world known for their high humidity and may be associated with low temperature. Many researchers considered the solubility of gypsum in water[6, 7]and with different influences to find out the reaction of rocks with variable gypsum content[8]. Because this solubility is the source of the problems that threaten the stability of structures built on layers with gypsum content, and in the worst case, it is the cause of damage to the structure and rendering it unusable.
In addition to the reported human losses but the main results from this paper cleared weight gain might lead to a different kind of failure in the gypseous strata and the samples to show different types of failure criteria. For samples saturated using gypsum- saturated water also experienced a dissolution behavior due to the gypsum ability to form an over-saturated solution[3].

4.Conclusion

•There was no effect for gypsum content percentage on the saturation (dissolution \ absorption) behavior.
•Further investigation and studies need to be made about the relative humidity effect on the mechanical behavior of gypsum rock, especially with the low-temperature present.
•All three methods effective and can be used for saturation 24 h is enough time to achieve a weight gain in the three methods inside a high humid environment.

References

[1] Wang J-A, Shang X C and Ma H T 2008 Investigation of catastrophic ground collapse in Xingtai gypsum mines in China *Int. J. Rock Mech. Min. Sci.* **45** 1480–99

[2] Shafiei A, Dusseault M B and Baghdardokht Z 2008 Geotechnical properties of soluble rocks from a dam site in Iran *The 42nd US Rock Mechanics Symposium (USRMS)* (American Rock

Mechanics Association)

[3] Sadeghiamirshahidi M and Vitton S J 2019 Analysis of drying and saturating natural gypsum samples for mechanical testing *J. Rock Mech. Geotech. Eng.* **11** 219–27

[4] Mohammadi S D, Rahimi M R and Taleb Beydokhti A 2020 Saturation Procedures of Soluble Rocks (Emphasizing on Gypsum-Anhydrite Rocks) *Sci. Q. J. Iran. Assoc. Eng. Geol.* **13** 97–111

[5] Hoxha D, Homand F and Auvray C 2006 Deformation of natural gypsum rock: Mechanisms and questions *Eng. Geol.* **86** 1–17

[6] Hong D, Fan M, Yu L and Cao J 2018 An experimental study simulating the dissolution of gypsum rock *Energy Explor. Exploit.* **36** 942–54

[7] Klimchouk A B and Klimchouk A *Dissolution and conversions of gypsum and anhydrite Porocarste View project Focused Reactive Fluid Flow in the Upper Crust: Processes and Manifestations of Hypogene Karstification View project 4.1.6. Dissolution and Conversions of Gypsum and Anhydrite*

[8] Nihad [, Salih B, Collins P E F and Kershaw S *[Dissolution Influences on Gypsum Rock under Short and Long-Term Loading: Implications for Dams]*

[9] Hashemi M, Saki A and Nikudel M R 2020 Effects of chemical composition on the solubility and geomechanical characteristics of gypsum rocks from Gachsaran Formation, Iran *Bull. Eng. Geol. Environ.* **79** 3785–96

IOP Conf. Series: Earth and Environmental Science **877** (2021) 012007 doi:10.1088/1755-1315/877/1/012007

Impact of Wastewater on Ditches of Rainwater Drain in Al-Hilla City

M D M Ali[1], M Husseini[2*] and N A Khudhair[3]

[1] Babylon sewage directorate, Babylon, Iraq.
[2]University of Babylon, College of Engineering, department of chemical engineering, Babylon, Iraq.
[3]Gilgamesh Ahliya University, Faculty of Health & Medical Techniques, Baghdad, Iraq.

* Corresponding author: eng.marwa.dawood@uobabylon.edu.iq

Abstract

In this research, the environmental reality of the ditches to rainwater drain located in the residential neighborhoods of Al-Hilla city in Iraq was studied. The study was conducted on three ditches near schools and health center within three residential neighborhoods. Chemical laboratory tests (BOD, COD, TSS, TN, TP, NO2-, NO3-, SO42-, Cl-, CN-, pH, T, DO) were proceeded on this water and it was found that the water is high polluted in sewage water compared to the sewage entering the treatment plant in the city and comparing with samples of water taken from ditches outside the city (as a standard sample). The pollution of this ditches with sewage water is considerable hazard on people health in this urban area, especially on children, students and the people come in to the health center in this areas.

1.Introduction

Water is considered one of the most important natural resources at all, as it is a basic factor on which human life and all the resources on the globe depend [1,2].

Wastewater is used water, comes from daily using for water from many activities such as human personally use, commercial, industrial and agricultural activities. This water contains metals, organic solvents and material, toxic sludge and microorganism. Wastewater also has many pollutants from rainfall carries road salts, clay, oil, grease, inorganic and organic chemical materials [3-6].

Sewage water, whether domestic, industrial or rainwater, it causes numerous hazards that directly effect on the environment and human, especially when it is in direct contact with the environment and people, this sewage water have greater harm in terms of odors, unwanted looks, diseases. Microorganisms and insects (especially mosquitoes and flies) that live in it cause many diseases, and these microorganisms may cause pollution of drinking water in this area [7-10]. In many countries water pollution is a major problem lead to death, according to the World Health Organization more than 1 in 4 deaths of children under 5 years of age are unsafe water, lack of sanitation, and inadequate hygiene [11,12].

Given the seriousness of wastewater and the health and environmental problems that result from it, it is necessary to assess the extent of pollution it causes in some residential neighborhoods in Al-Hilla city, where this research conducted to study the possibility of contamination of rivers water in these neighborhoods with sewage water and the extent of potential harm to the people of these areas and the surrounding environment.

2. Experimental section

2.1. Materials and instruments

The chemical were purchased from CDH and Fluka companies, chemical tests were made by Hach spectrophotometer (DR 2800) with DRB200 reactor, pH value was determined by using Hach HQD portable meter, dissolved oxygen (DO) was determined by Hach HQ30D.

2.2 Estimation of chemical oxygen demand (COD)

COD was measured by using high-range ampoules (Hach chemical kit) with spectrophotometer (Hach DR2800) and reactor (Hach DRB200) according to specific method of Hach company.

2.3 Estimation of biochemical oxygen demand (BOD) [13]

First a known amount of wastewater sample was placed on the airtight BOD bottle the (volume = 300 mL), then bottle was filled with dilution water contains the nutrients required for organisms growth and saturated with oxygen. The dissolved oxygen was measured for solution by Hach HQD dissolved oxygen meter, the bottle was incubated for five days at 20 °C, after five days the dissolved oxygen concentration was measured again, the BOD value of sample is the difference between the initial value of dissolved oxygen before incubation and dissolved oxygen value after incubation divided by decimal fraction of sample used.

2.4 Estimation of total nitrogen (TN) and total phosphor (TP)

Total nitrogen and total phosphor were measured by using ampoules of (Hach chemical kit) with spectrophotometer (Hach DR2800) and reactor (Hach DRB200) according to specific methods of Hach Company.

2.5 Estimation of dissolved oxygen (DO) and temperature (T)

Dissolved oxygen (DO) and temperature (T) were measured by using Hach HQD dissolved oxygen meter immediately after bringing samples.

2.6 Estimation of nitrite (NO_2^{2-}), nitrate (NO_3^{2-}), cyanide (CN^-) and total suspended solid (TSS)

The nitrite, nitrate and cyanide ions were measured by using ampoules of (Hach chemical kit) with spectrophotometer (Hach DR2800) according to specific methods of Hach company.

2.7 Estimation of chloride ion (Cl^-) [14]

Determination of chloride concentration in waste water done by using Mohr method, first step is taking 25 mL from collected sample into conical flask, then added 2-3 drops from indicator (potassium chromate), the color of solution was turn into light yellow. The solution was titrated against silver nitrate solution until reach the end point when light yellow color was in turn to permanent brownish-red color. The volume of silver nitrate added was noted then concentration of chloride was calculated.

2.8 Estimation of sulfate ion (SO_4^{2-}) [15]

Sulfate ion was measuring by using gravimetric method, 100 mL of sample was placed into beaker then added 100 mL distilled water then added 4 mL of 6 M HCl, the solution was heated nearly to boiling. For the warm solution 10 mL of barium chloride was added with string. The solution was lifted to 24 h for digestion, and then the precipitate was filter with ashless filter paper and rinsed with distilled water.

The precipitate with filter paper putted into crucible (the crucible was rinsed and dried in furnace then weighted after cooling in desiccator), then the crucible with sample was heated on Bunsen burner to burning the filter paper. After that the crucible was dried at 800 C in a furnace for 1 h then cooled, weighed and calculations were done.

3. Result and discussion

Pollution is one of the serious problems and has a direct impact on human life, animals and plants. One of great importance pollution problems is water pollution, especially its contamination with sewage water. Therefore, we notice in all developed countries of the world attaching great importance to this aspect and wastewater treatment is one of the priorities for them to reduce the possibility of contamination of surface or ground water with wastewater. But in developing countries and third world countries that suffer from poverty and economic problems, we notice that there is a great lack of interest in this aspect due to ignorance and lack of awareness of the hazards generated by this issue.

In the city of Al-Hilla, as an example of this side, there is a great insufficiency of sewage networks and the lack of areas served by these networks. In addition, the flood of most neighborhoods during the winter season has induced the local government and the people to replace the sewage networks by drilling a ditch to rainwater drain. These ditches have become a focal point of pollution instead of being a solution to this problem. In this research, the pollution of accumulation rainwater in these ditches with sewage water was studied in addition to studying the presence of chemical pollutants. This study was conducted in three areas in the city center of Al-Hilla, and samples were collected from the ditches in them.

the results of the chemical analysis of the samples were taken from these areas are shown in Table (1) these results were compared with samples taken from ditch located outside the city and far from residential areas (as a standard sample), as well as compared with heavy water in water treatment plant in the city.

During studied the results in table (1),the research found ditches inside the city have bad quality comparing with another ditches outside city (that is usually used for rainwater drainage, surface water drain, land reclamation, and soil salinity reduction) in this water the research found that the percentage of salts is high, especially sulfates and chlorides, and in contrast, a decrease in the value of the chemical oxygen demand and the biological oxygen demand, or even its nonappearance, due to the absence of organic pollutants in the water, as well as for biological pollutants. On the contrary, the research found that the water in the drains under study has a high percentage of pollution with organic materials and chemicals, where we notice the high percentage of the chemical requirement for oxygen and the biological requirement and this is a clear indication of its great pollution of organic materials and its values are close to the values of wastewater offered from the city to the station Sewage.

In addition to increase in the value of total nitrogen and total phosphorous, which is one of the important elements in the process of food enrichment, and this causes a problem in the growth of suspended organisms and filamentous algae as noted in the microscopic examination of the water samples and as seen in the images (1-3). The surface algae cover and its decomposition, as well as the emission of unpleasant odors that are not desirable, and this was clearly noticeable when collecting samples and a decrease in the value of the oxygen absorbed in the water in comparison with the natural state found in rivers and non-polluted waters.

Also, by examining these samples under the microscope, it was found that these samples contain a large number of microorganisms from parasites, as well as eggs of worms, filamentous algae and many other types as shown in images (4).

By comparing three samples found the first location is more pollution and has high odor and more turbidity and dark color where the third location is less pollution and there is an empty space and a pond of water with the reed plant on the site also, the soil present high salinity.

The results appeared this type of drain is contaminated directly with sewage water, and this constitutes an environmental disaster, especially in the rainy season, because the capacity of these rivers is little and the speed of their drainage depends on the slope of these rivers and sometimes on water pumps installed on these rivers and this is not sufficient for the purpose drainage of water in it and rainfall leads to flooding of these rivers on the neighboring streets, and in the event of heavy rain, the entire areas will be pools of water contaminated with sewage water, and this constitutes an environmental disaster, especially most of these areas do not contain public sewage networks or contain a rainwater network that will take this water is transferred to the main rivers in the city, and this means pollution of water sources and thus pollution of drinking water as well. In normal times, most of the time the water of these drainages reaches the street, and here the pollution will be transmitted through the cars that drive on it, and the pollution will spread in a larger way, and it can also be transmitted through the dust after the streets dry up and the traffic of cars spread this dust, which is contaminated with secondary pollutants that They exist in these waters.

Table 1. chemical tests of samples, standard sample and sewage water.

Test	Sample 1	Sample 2	Sample 3	Standard*	Sewage water average for one month
BOD	70	70	18	Zero	181
COD	198	135	117	Zero	328
TSS	345	116	48	9	344
TN	42	34	27	0.39	17.04
TP	9.67	8.69	3.79	0.9	7.46
NO_3^-	0.9	1.6	1.3	1.2	2.45
NO_2^-	0.005	0.01	0.001	0.006	-
Cl^-	360	470	1159	90	434
CN^-	Zero	Zero	0.032	Zero	0.034
SO_4^{2-}	820	977	1760	91	1039
pH	7.15	7.38	7.14	8.49	7.48
T	15	15	17	14	19
DO	0.16	0.37	4.41	9.22	-

*standard; the sample from river outside the city.

Figure 1. The position of sample (1).

Figure 2. The position of sample (2).

Figure 3. The position of sample (3)

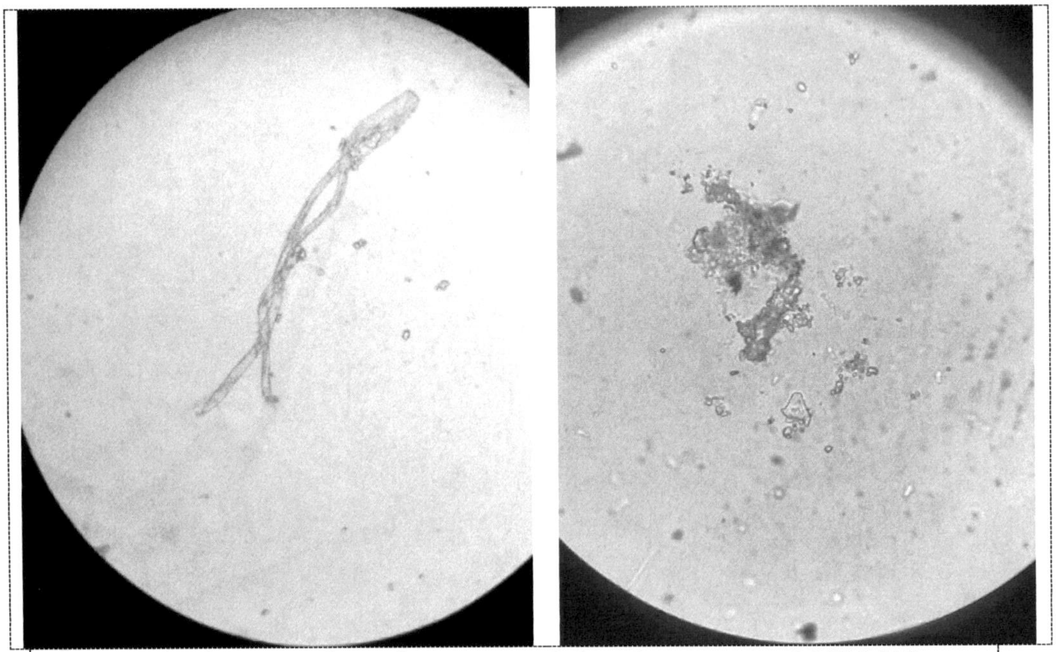

Figure 4. microscope picture for water sample by microscope model BM-120

4. Conclusion

- Through the results of the tests, it was found that the ditches waters polluted with high percentage of sewage water , which represents a major environmental problem that affects the residents of the areas where these ditches pass.
- the pollution increase in the rainy season because of the flood of polluted ditches waters to streets and near places.
-The banks of these ditches have become a dumping ground, and even inside these ditches noticed that there are a lot of waste which causes the clogging of the drainage water and lead to the crossing of this water on the streets and increasing pollution. In addition these ditches have become a shelter for many animals and rodents.
- It is possible to treat this phenomenon by burying these ditches and replacing them with an underground network of pipes, as well as by establishing a sewage network for the areas insulated from drainage networks for rainwater to reduce the pollution in rivers where the ditches pour.

References

[1] Hossain M 2015 *global J. of advanced reseach* **2** 1436
[2] Martina Z and Gabriela H 2020 *Rainwater Infiltration in Urban Areas* (Switzerland: springer Nature)
[3] Fernando S G E 2011 *Waste Water–Evaluation and Management* (Croatia: InTech)
[4] Tebbutt T H Y 1979 Principles of water Quality Control (, New York: Pergamon press)
[5] Glavan M 2018 *Water Challenges of an Urbanizing World* (InTech)
[6] Owa F W 2014 *International Letters of Natural Sciences* **3** 1
[7] Akpor O B 2011 *3rd International Conference on Chemical, Biological and Environmental Engineering IPCBEE* **20** 85
[8] Nesaratnam S T 2014 *Water pollution control* (Wiley) 1st edition
[9] Dwivedi A K 2017 *International Research Journal of Natural and Applied Sciences* **4** 118
[10] Haseena M, Malik M F, Javed A, Arshad S, Asif N, Zulfiqar S and Hanif J 2017 *Environ Risk Assess Remediat* **1** 16
[11] Amoatey P and Bani R 2011 *Wastewater Management. In Waste Water - Evaluation and Management* (InTech)

[12] Sawyer C N, McCarty P L and Parkin 2000 *Chemistry for Environmental Engineering and Science* 4th Edition (New York : McGraw-Hill, Inc.)
[13] Tchobanoglous G, Burton F and Stensel H D 2003 *Wastewater Engineering treatment and Reuse* (McGraw-Hill)
[14] Skoog D M, West D M and Holler F J 1996 Fundamentals of Analytical Chemistry 7th Edition (USA: Thomson Learning, Inc.)
[15] Rice E W, Baird R B, Eaton A D and Clesceri L S 2011 *Standard Methods for the Examination of Water and Wastewater* (American Public Health Association) 22nd Edition

Acknowledgment
The research team presents to Mr. Ali Najah with thanks and appreciation for the assistance in completing this research. In addition, appreciation to the Babil Sewerage Directorate, Laboratory and Environment department.

Treatability influence of municipal sewage effluent on surface water quality assessment based on Nemerow pollution index using an artificial neural network

R Mohammed [1], B Al-Obaidi[2]

[1] M.Sc. Civil Engineering Department, University of Baghdad, Iraq
[2] Assistant Professor, University of Baghdad, Civil Engineering Department
* Corresponding author: r.mohsin1901m@coeng.uobaghdad.edu.iq

Abstract

Assessing water quality provides a scientific foundation for the development and management of water resources. The objective of the research is to evaluate the impact treated effluent from North Rustumiyia wastewater treatment plant (WWTP) on the quality of Diyala river. The model of the artificial neural network (ANN) and factor analysis (FA) based on Nemerow pollution index (NPI). To define important water quality parameters for North Al-Rustumiyia for the line(F2), the Nemerow Pollution Index was introduced. The most important parameters of assessment of water variation quality of wastewater were the parameter used in the model: biochemical oxygen demand (BOD), chemical oxygen demand (COD), suspension solids (SS), chloride, cl, hydrogen ion concentration, pH, sulfate, SO4-2, nitrate, NO3- and phosphate, PO4-3. Taking these criteria into account, samples of water from the sampling sites were graded as C, indicating the pollutant of the waste treatment. Then the water quality map using neural network model was based on the results of water quality assessment. The results showed that the model North Al-Rustumiyia for line F2 was more efficient and R2 was 0.965 with the impotence parameter was chloride (CL).

1.Introduction

The most valuable freshwater resources have always been rivers and most emerging activities still rely on them[1], So, one of the main factors and a scientific foundation for enhanced water resources growth and management plans is the river water quality assessment. In order to summarize vast volumes of water quality data into a single number and detect its suitability (for example, great, good, poor, etc.) for various purposes, water quality indices (WQIs, suggested by comparison of real water quality parameters with their corresponding regulatory standards) [2]. This assessment tools are used to evaluate the water quality of the Beiyun River in a systematic manner using 17 parameters of water quality including temperature, pH, conductivity, dissolved oxygen (DO), the chemical oxygen requirement (COD), biochemical demand for oxygen (BOD5), ammonia nitrogen, phosphorus (TP) totals and sulphid. The spatial distribution of WQI values on the Beiyun River was substantially different and the quality of the water upstream and downstream was higher than midstream quality[3]. The principal cause of the degradation of water quality in the Sitnica River is anthropogenic activities, including wastewater discharge, an assessment of the water quality in the Sitnica River by applying the Canadian Water Quality Index (WQI) and results from the WQI show that it decreases quality as a result of its pollutant impact. This reinforces the urgent need for serious steps to track and control the river properly [4].

US scholarly N. In its 1974 establishment L. Nemerow introduced the Pollution Index (NPI), which has many advantages in the evaluation of water quality: simple mathematical method, easy procedure, integrated effect satisfying, high concentration pollutant effect manifestation and NPI, as an effective approach to assess the water quality of an environment of fresh water. In recent years, many academics and government officials in the field of environment have been very interested in the application of the NPI in water quality assessment[5]. The Zhangze Reservoir water quality was evaluated using the integrated NPI system for the water quality identification. The various assessment methods were analysed and compared and each method's characteristics determined. Based on these results, we assessed and analyzed the suitability of the water quality evaluation approaches. In every section of water quality monitoring the results produced by the standard Nemerow index method fluctuated considerably [6]. The tool used for the assessment of the Cimanuk water from the years 2013 to 2018 is the Nemerow Pollution Index (NPI). The deterioration of the quality of water in the river Cimanuk showed that the NPI was rising from 1.04 to 7.51. The Cimanuk River has improved from lightly to moderately polluted waters and is not ideal as a potable water source[7]. Moreover, some physicochemical water quality parameters selected were measured and the findings were compared to BIS standards whereby key cations and anionic values were observed below NPI limit according to pollutant status standards. NPI values show that the water of the river is in good condition throughout the year [8]. The assessment of water quality for drinking and irrigation in some marshlands in the Basrah Province to classify major surface water contaminants using NPIs based on the physical and chemical parameters reported in the study region by five monitoring stations during 2014. The findings are compared to Iraqi and WHO requirements and have shown unwanted values for closer purposes [9].

The rising environmental problems have led to an emphasis on the proper operation and regulation of wastewater treatment plants (WWTPs). This could have an effect on the plant at operational risk. The WWTP can result in serious environmental and public health issues because the introduction of polluted effluent to a receiving water bodyshell can result in different pathogens and transfer them to humans and to the marine community if they are introduced into the water before treatment. Environmental laws also set constraints on effluent consistency that any WWTP must follow[10]. From the performance studies Bangalore WWTP revealed that in controst to removal/reduction in other parameters, the efficiency of both treatment plants was low in the removal of total dissolved solids. The performance studies have shown that the device is optimistic and that aeration tank performance and secondary clarifier in both cases were practically matched. Fresh sludge with higher population of micro-organisms must be recycled and the aerators worked continuously in order to securely discharge the handled effluent into streams, and rivers can even be reutilized to irrigate groundwater [11]. The efficiency of the wastewater treatment facility in Boujaad Area, focused on physical and chemical control, both spatially and temporarily, of raw and refined wastewater in order to diagnose and assess pollution mitigation in this wastewater treatment plant. The study findings show that the efficiency of Boujaad's wastewater treatment plant complies with national and international standards and that its operation was sufficient[12].

The major reasons for lack of state of wastewater treatment facilities in the individual countries are the inadequate coverage of treatment systems in urban and rural areas, poor functionality of wastewater networks, the low architecture, expertise, inability, insufficient support for waste water treatment, the potential for overload of existing facilities and inefficiency[13]. The quality index for wastewater for the calculation of wastewater treatment effluent. Primary consideration of the existence of plant effluents included elevated standards of such requirements limiting the reuse of effluents for human health and the agriculture environment. WWQI is an important method to measure and assess wastewater productivity for decision makers and planners[14].

In order to decide how each of those variables impacts on the BOD regular inlet is used as a basis for the Artificial Neural Network used as a standard prediction model of inputs to waste water biochemical treatment plants with various compositions of daily water quality results[15]. ANNs have been widely previewed and predicted in many areas, including environmental science and water management, economics, pharmacies, power generation. The findings show that an ANN was an economic model of water quality, and the results show that repetitive neural networks had better results than radial neural networks with multiple components[16]. Due to its accurate, acceptable and promising use of advanced technology, ANN can be used for modeling certain WWTP processes to assist the process efficiency forecasting [17]. ANN's ability to learn basic data generations led to an extensive application of its data in measurement, estimation, practical approximation, classification, and data analysis, given sufficient data samples[18].

The concentration of dissolved oxygen (DO) was estimated based on the relationship between the dissolving oxygen and hydrological parameters using the MLR technique and Back-Propagation Neural Network (BPNN). The findings suggested that the neural network could reliably predict the DO concentration [19]. The ANN can provide a trustworthy forecast model for water quality with land use attributes in accordance with the methods of monitoring water quality, and can find a low-end solution and ensure sustainable development which mainly has an impact on water quality [20]. As water quality deteriorates, serious water quality control activities have been conducted to analyze the ANN methodology that can identify water quality automatically. The findings therefore showed low errors and a high correlation between the parameters measured and predicted and showed a strong ANN ability to forecast water quality parameters [21] Correlation was tested in three separate Korapuzha river stations at Kozhikode in Kerala with Biochemical Oxygen Demand (BOD) and Chemical Oxygen Demand (COD), using Artificial Neural Networks in the research area. The achievement was the high correlation coefficient (R) and low mean square errors of the neural network model provided by BOD and COD. It may also be inferred that ANN is an important modeling method [22]. The ANN-COA hybrid evaluation model developed to forecast a WWTP accurately. This model was developed to optimize model efficiency by using the Artificial Neural Network (ANN). The study showed a greater degree of precision in WWTP prediction and regulation by the hybrid ANN-COA model[23]. Both water amounts and efficiency parameters are agreed by the comparative findings of the optimized urban water model and ANN. The input-output data produced by an integrated urban wastewater model were formed for a feedback-propagation network. CSO dumps and treatment plant effluent are supplied inputs including plumbing and dry-weather and the quantity and consistency are considered. Comparisons between the integrated model and ANN have been shown to be effective with both water quantity and quality parameters [24]. Applications of ANNs are studied in the area of prediction of wastewater treatment. so this study will also discuss a case study documenting detailed simulation work on designing nonlinear neural network prediction models for the Gold Bar WWTP, Edmonton, Alberta [25] . Konya WWTP modeling was explored by using an artificial neural network in Matlab applications with various architectures. Plant efficiency was calculated by using the pH, temperature, COD, TSS and BOD input values with TSS output. The performance and correlation coefficient of the model were compared using Mean Squared Error (MSE) parameters (R). ANN can predict the plant performances of the observed and expected output variable up to 0.96 with correlation coefficient (r)[26]

Furthermore, Fajr Industrial WWTP, located in Mahshahr—Iran has been used to prepare, calibrate, and validate its neural model with qualitative and quantitative characteristics in its units. Primary component analysis (PCA) has also been introduced to enhance the performance of neural network models developed. The unit L-TDS was good model accuracy in estimating the qualitative waste water profile but it did not have enough accuracy to use the biological unit [27] Method has been developed to estimate the odor content with ANN in the WWTP.

ANN calculations of the odor concentration of a WWTP based on data of water quality including biological oxygen needs, dissolved oxygen and pH were used. The WWTP water quality and odor data were seasonally calculated in the spring, summer, and autumn as variations on feedback to the ANN model. In comparison with the calculated results was the odor expected by the ANN model and the forecast accuracy was evaluated. Suggestions are provided to improve prediction accuracy [28] Artificial neural network also had to forecast the effluent Chemical Oxygen Demand output of Touggourt WWTP. As an input variable for the neural networks, influential variables like pH, temperature, solid suspended, biochemical oxygen requirements and chemical oxygen requirements

were used. In the phases of learning, evaluation, and checking, the ANN model could predict experimental outcomes at 0.89, 0.96, and 0.87 high correlation coefficient. The overall results showed that the ANN modeling methodology can provide an efficient method to simulate, monitor and forecast WWTP output [29]. Models for artificial neural networks (ANN) to predict the efficiency of the WWTP have been developed based on past data obtained at an average flow rate of 1 million m3/day from an effective traditional treatment plant located in Greater Cairo,Egypt. There are two ANN-based models for plant effluent BOD and SS estimation. The ANN models have been found to be an accurate and stable method to forecast WWTP efficiency[30].
The investigation of the overall performance and evaluation of the effluent quality from Rustamiya sewage treatment plant (STP), Baghdad, Iraq by evaluating the effluent quality index (EQI) on the basis of an artificial neural network (ANN) model which was developed to forecast an effluent quality index based on selected water qualities, it was indicated that the treating sewage effluent quality was within Iraqi quality standards (IQS) The overall performance suggested positive efficiencies and the findings were more successful in the EQI model than relative parameters of 47,3 % in the EQI data sets with a high degree of determination coefficient of R2 = 99,8 % [31]. This paper aims to find the impact of sewage treatment on the quality of surface water based on Nemerow pollution index using Artificial neural network.

2.Case Study Disruption

Al-Rustamiya WWTP is Iraq's largest treatment plant for wastewater. It is situated on the south side of al-Rusafa in Bagdad, approximately 500 meters from the river Diyala with 400 000 m2 area as shown in Fig. 1 [32]. The inflow line of raw wastewater to plant is Zeplin line. That collect wastewater from Rusafa at design capacity and actual flow 300,000, and 450,000 m3/day respectively. It consists of separate two output lines (F1 and F2). The sewage passes through several stages within the project before being put in the Diyala River. Diyala River is one of the most important rivers in Iraq with a flow rate ranging between 25-650 m3 /s [31]. The stage of treatment of this plant are including; screens, grit chambers, aeration tanks, primary sedimentation tanks, secondary sedimentation tanks, chlorination and sludge treatment which treated the raw water physically, chemically and biologically to make it adequate with Iraqi standards for effluent to river [33].

Figure 1. North Al-Rustumiyia plant.

3.Data Collection and Analysis

Data from Al-Rustamiya WWTP is extracted as average monthly reports in five years of laboratory experimentation (2015-2019), biochemical oxygen demand (BOD), chemical oxygen demand (COD), suspension solids (SS), chloride, cl-, hydrogen ion concentration, pH, sulfate, SO4-2, nitrate, NO3- and phosphate, PO4-3. This is data mining to find Nemerow emission index by building models and artificial neural networks with the use of SPSS software Excel-Microsoft Office

4.Mathematical model and methodology

4.1.Nemerow pollution index (NPI)

After obtaining a water quality index in various surface-water bodies a significant role can be played to minimize pollution problems. In this analysis the use of the Water Quality Index (WQI) was considered useful to determine the total water quality and to exclude the water quality decision. This approach tends to be more standardized and compares water quality assessments of sampling stations [34]. In water quality assessment, although some pollutants' concentration exceeded standard and induce harm to environment, while the average value of the index is not exceeding the standard. Considering this effect, assessment method of Nemerow pollution index, which combine average value of pollutants with that of maximum value, was used to evaluate the water quality of River [35]. The physical, chemical and biological tests measured by the North Rustumiyia plant for two effluents line (F1, F2) were used. The parameter used in calculation are biochemical oxygen demand (BOD), chemical oxygen demand (COD), suspension solids (SS), chloride (CL-), ph., SO4-2, NO3- and PO4-3and through these data the monthly Nemerow pollution index rate was calculated for 5 years (2015-2019) depending on Iraqi standard for disposals for rivers. Equation of Nemerow pollution index is specified, each specific use index PIj is represented by a function of the relative values, where [36]:

$$Relative\ value = \ Ci\ (Lij\ )^{-1} \quad (1)$$

For the cases the contaminant pH:

$$Ci\ (Lij\ )^{-1} = \frac{\mathrm{Ci} - [\frac{\mathrm{Lij + Lij\ max}}{2}]}{\mathrm{Lij\ min\ or\ Lij\ max; whichever\ i\ closer\ to\ Ci} - [\frac{\mathrm{Lij\ min + Lij\ max}}{2}]} \quad (2)$$

And below equation to obtain the value of PI:

$$\mathrm{PI} = [\frac{\left(\frac{Ci}{Lij}\right)^2 maximum + \left(\frac{Ci}{Lij}\right)^2 mean}{2}]^{1/2} \quad (3)$$

The general water-quality index is computed as the weighted sum of the three specific use indices:

$$\mathrm{PI} = \sum_{\mathrm{j}=1}^{3} \mathrm{wj\ PIj} \quad (4)$$

Where:
Ci= The parameter level of water quality i
Lij= The allowable parameter value standard I at a water-use position j.
Wj= the weight coefficient
PI= Nemerow pollution index

Then, results that obtained after complete calculation for Nemerow Pollution Index and classified as shown in Table 2:

Table 2. NPI Nemerow Pollution Index classes [37].

NPI	Classes
$0 \leq \mathrm{NPI} \leq 1$	Standard/Good Quality
$1 \leq \mathrm{NPI} \leq 5$	Lightly Polluted
$5 \leq \mathrm{NPI} \leq 10$	Moderately Polluted
NPI > 10	Heavily Polluted

4.2.Artificial neural network (ANN)

The aim of a neural network is to measure output values. Neurons (or cells) are processed components conducting basic input value measurements. A neuron transforms the weighted sum of incoming neutron inputs in a nonlinear way to generate the neuron output [38]. ANN consists of layers of artificial neurons or nodes which is in a layer are not linked in the same layer with other nodes. This indicates that only relations between layers and not inside a layer exist. The first layer which receives input information is called the input layer, as shown in figure 2. The last layer that generates information on the output is called an output layer. There are hidden layers between the input and output layers. One or more hidden layers can be found in an ANN. Knowledge is conveyed through links between nodes in various layers; hidden layers are used to convert the non-linear input space for computational purposes [17]. Based on the results from Nemerow Pollution Index calculations, a neural network was drawn where the input layers represented the properties of treated sewage and 9 layers of input, while the output was one output layer and represented Nemerow Pollution Index.

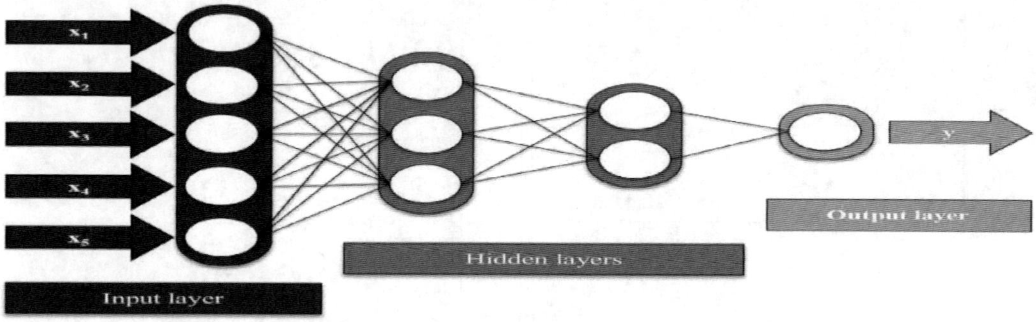

Figure 2. Structure of artificial neural network.

5.Result and discussion

5.1.Assessment of treated wastewater effluent quality

Figure 3 represents the biochemical oxygen demand (BOD) of treated sewage effluent characteristics over twelve months for period five years (2015-2019). The maximum value was 32.52 mg/L in April and minimum value was 22.368 mg/L in December. The values were within the recommended limits of the Iraqi standard (IQS) for disposal into the rivers. April values observed for five years in this month was maximum value that may be back to be that month in spring season that mean warm weather and this is critical condition because the bacterial utilization rate are high and the saturation concentration is low.in addition the lower flows of river [39]. In other hand, comparing with another standard of BOD5 is out of standard is more than 20 mg/L [40], also, the chemical oxygen demand (COD) of treated sewage effluent characteristics over eleven months for period five years (2015-2019) were with in recommended limits of Iraqi standards (IQS) for disposal into the rivers over months of five years, the maximum value was 33.08 mg/L in August and minimum value was 24.412 mg/L in September as showed below in (Figure 4).

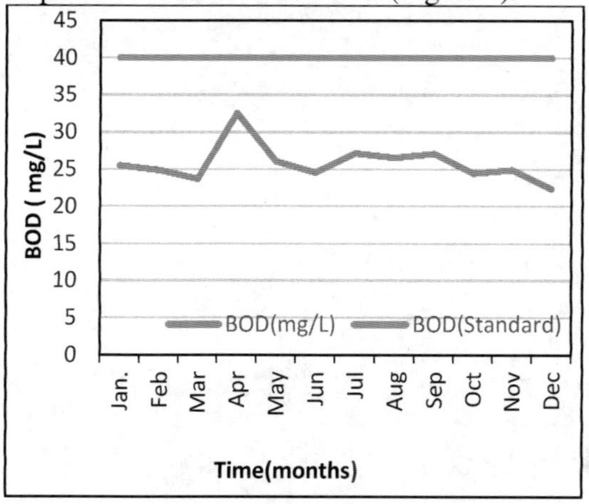

Figure 3. change biochemical oxygen demand (BOD) monthly over five years (2015-2019).

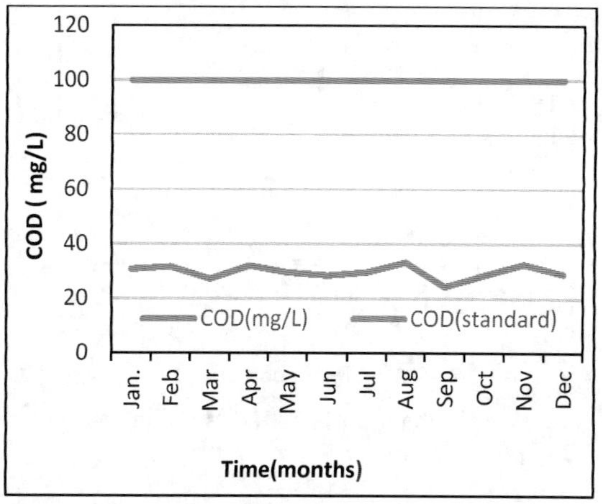

Figure 4. change chemical oxygen demand (COD) monthly over five years (2015-2019).

The suspension solids of effluent disposal monthly for five years was with standards of Iraqi effluent disposals in rivers. The maximum value observed was 27.89 mg/L in September and minimum value was 21.5 mg/L in June. This variation back to seasonal variation in Iraq; the suspended solids increase in winter season that back to density of rainfall (Figure 5). The quantity of chloride in effluent in

IOP Conf. Series: Earth and Environmental Science **877** (2021) 012008 doi:10.1088/1755-1315/877/1/012008

disposals with comparing Iraqi standards was adequate The maximum value 345.95 mg/L in December and minimum value was 275.04 mg/L in July (Figure 6).

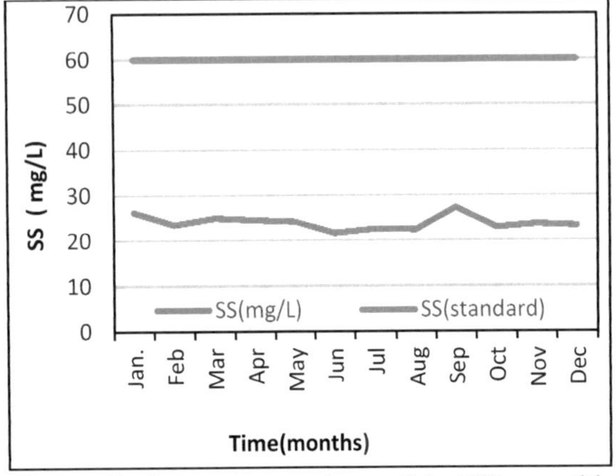

Figure 5. change suspension solids (SS) monthly over five years (2015-2019).

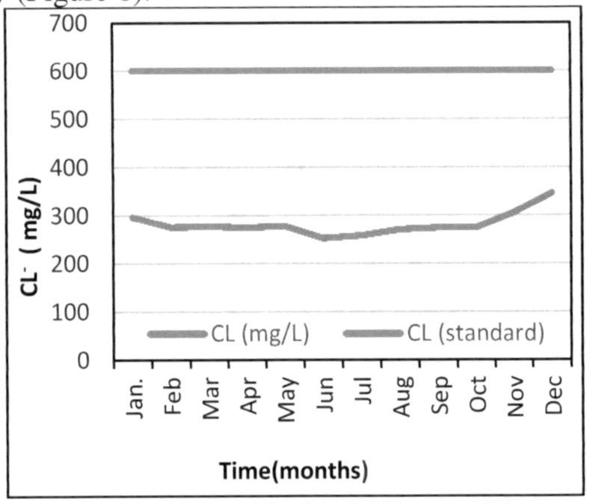

Figure 6. change chloride (Cl^-) monthly over five years (2015-2019).

The parameter pH referring to acidy or the alkalinity of water. The maximum monthly values for the 5-year evolution of PH were recorded between the period (April and September) that back to activity of factories and industrial processes that associated with activity of people the residual period observed minimum values of pH but surely all records were compatible with Iraqi standards (IQS).in (Figure 7) all value that calculate monthly for five years are with the limits of Iraqi standards for disposals in rivers but in other opinion almost all values is more than 7.3 so it doesn't adequate with the standard and this indicative the raw wastewater influent to plants include industrial wastewater. changing Sulfate(SO4)) monthly over five years (2015-2019) was out the Iraqi limits these lead to that the raw water have a lot of sulphurous organic matter which passes a lot of processes to produces SO4 which also the quantity of SO4 over 200 mg/L cause sever hindered for sludge digesters (Figure 8).

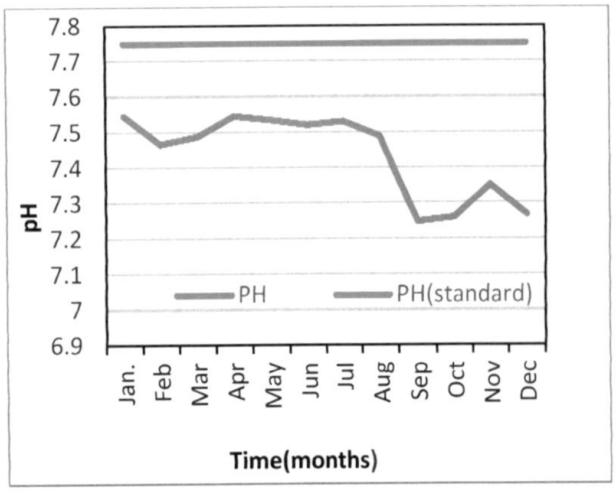

Figure 7. change pH monthly over five years (2015-2019).

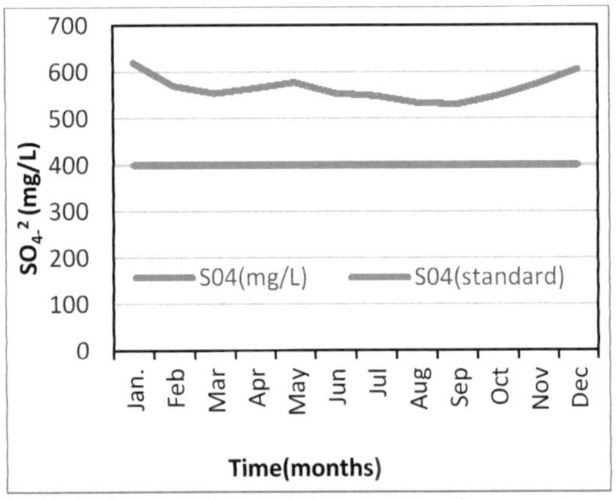

Figure 8. change Sulphate(SO_4^{-2}) monthly over five years (2015-2019)

changing Nitrate (NO3) monthly over five years (2015-2019) was very good comparing with limits of Iraqi standards for effluent disposals in rivers in opposite the value of phosphate were increasing in value in period (October –December) also these value don't adequate with Iraqi standard (IQS) for effluent disposal in river (figure 9, and figure 10).

IOP Conf. Series: Earth and Environmental Science **877** (2021) 012008 doi:10.1088/1755-1315/877/1/012008

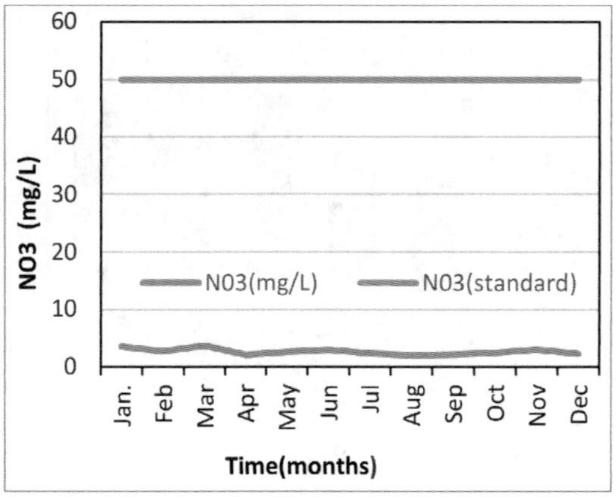

Figure 9. change Nitrate (NO_3^-) monthly over five years (2015-2019).

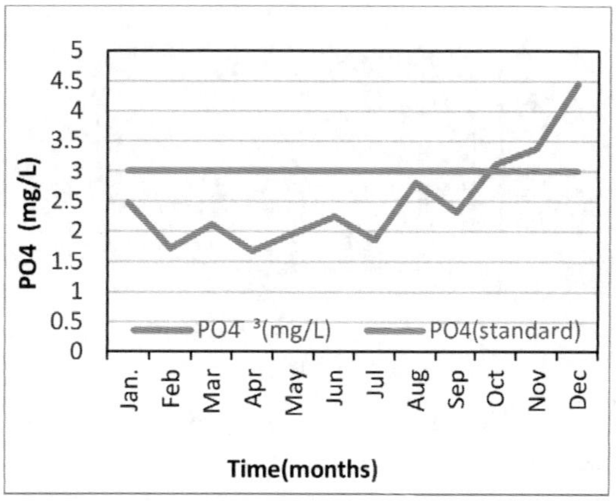

Figure 10 change Phosphate (PO_4^{-3}) monthly over five years (2015-2019)

By analyzing data of laboratory tests of treated wastewater that disposal in the Diyala River statistically and over five years from 2015 to the end of 2019 it was found to be the highest level of BOD, CL and SO4 recorded was in 2017 and were with Iraqi standard (IQS) for BOD, CL and out for SO4 (figure 11, figure 14 ,and figure 16).in other hand, in 2015 was recorded maximum value in SS, PH. NO3 and PO4-3 also were with Iraqi standard (IQS) expect PO4-3 was out the standard(figure 13, figure 15, figure 17 ,and figure 18) .in addition, in figure 12 appeared the maximum value was in 2016 and was adequate when comparing with (IQS).

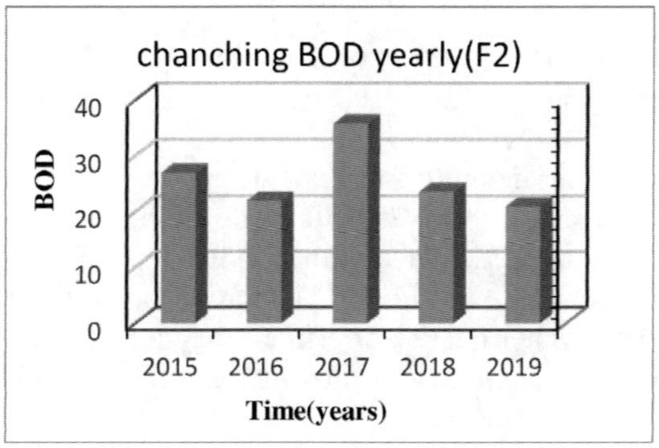

Figure 11. change biochemical oxygen demand (BOD) yearly over five years (2015-2019)

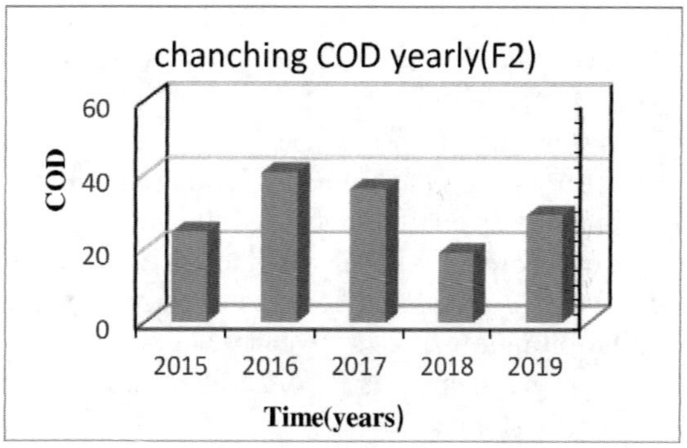

Figure 12. change chemical oxygen demand (COD) yearly over five years (2015-2019)

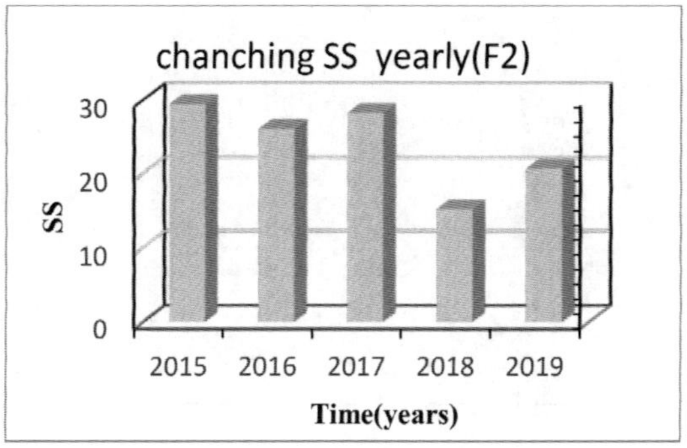

Figure 13. change suspension solids (SS) yearly over five years (2015-2019)

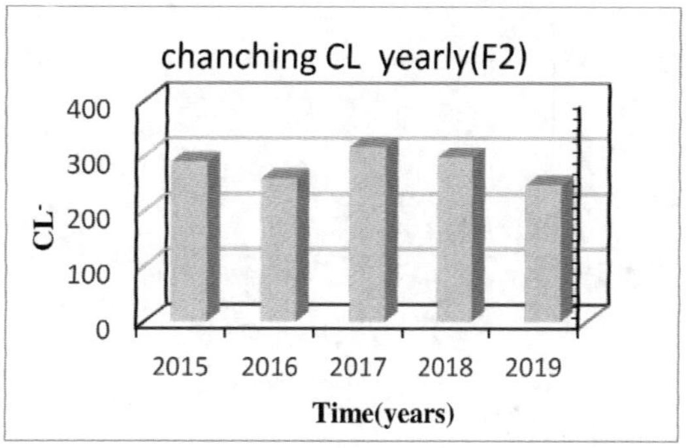

Figure 14. change Chloride (CL^-) yearly over five years (2015-2019)

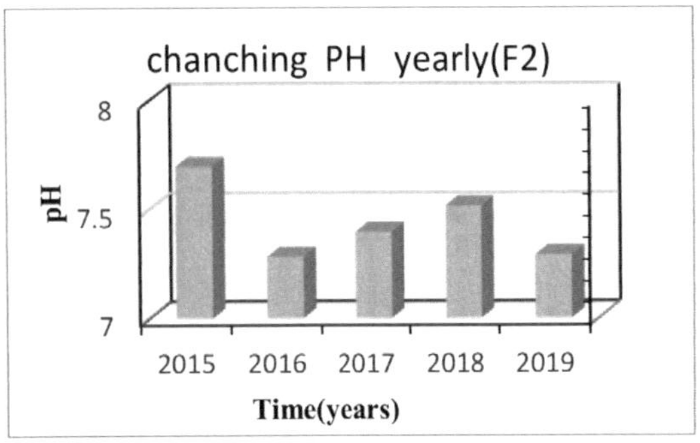

Figure 15. change pH yearly over five years (2015-2019)

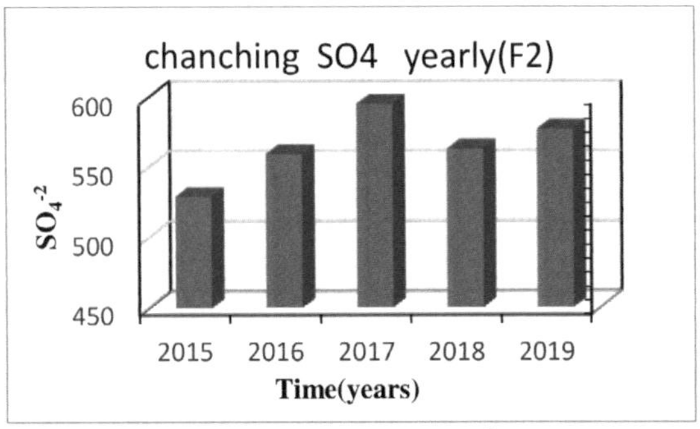

Figure 16. change Sulphate(SO_4^{-2}) yearly over five years (2015-2019)

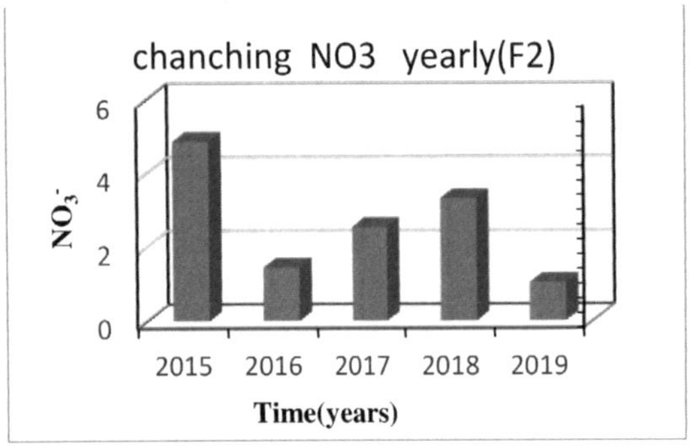

Figure 17. change (NO_3^-) yearly over five years (2015-2019).

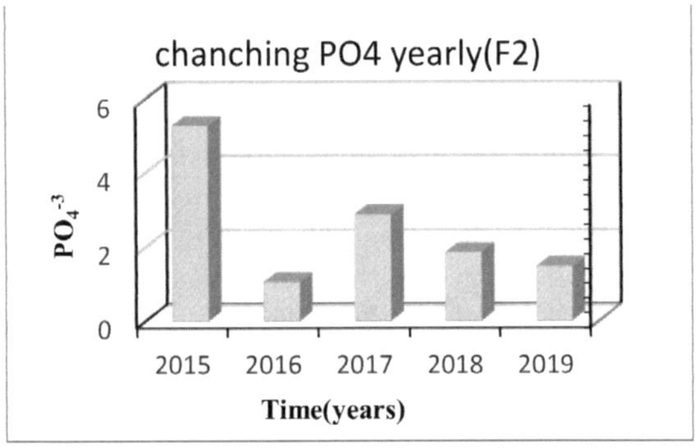

Figure 18. change Phosphate (PO_4^{-3}) yearly over five years (2015-2019)

5.2.Nemerow pollution index NPI

Figure 19 shows the amount of contamination of treated water monthly and years where the maximum pollution was in December while the highest year was 2019 which give indicate of pollution of effluent increase with increase temperature of weather and this associated with activity of people also decrease in flow make influent parameters more concentrate. In figure 20 represents Nemerow pollution index changing yearly and as showed above the water quality of effluent be more polluted with progress in years because increasing population in this area serviced that increase the flow that can able to treated it.

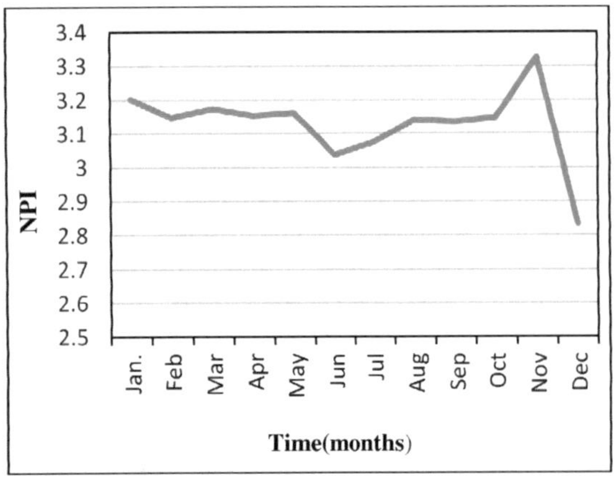

Figure 19. change Nemerow Pollution Index (NPI) monthly over five years (2015-2019)

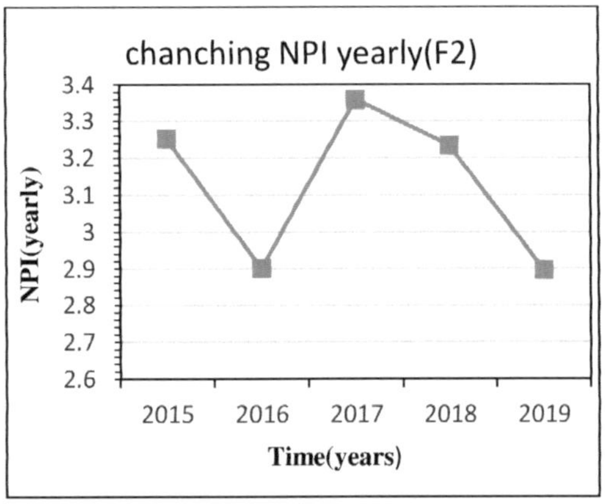

Figure 20. change Nemerow Pollution Index (NPI) yearly over five years (2015-2019)

5.3.Artificial Neural Network ANN

The neural network model was built using a program SPSS where the input layer was eight parameters of treated sewage, the number of hidden layers was two as show in figure 21. The model was successful and the R2 0. 965 (figure 22). The most important and influential factor in the model was chloride (CL) and less important parameter that effect in the built model is pH (figure 23).

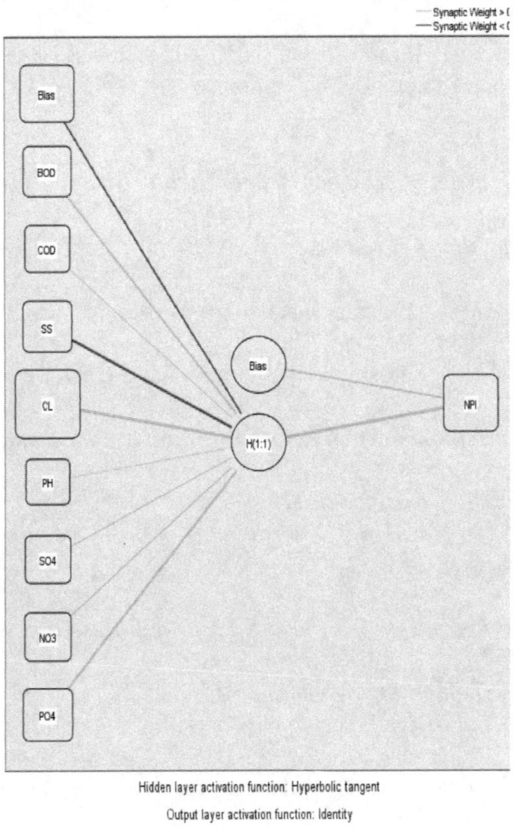

Figure 21. the model of artificial neural network ANN with layers.

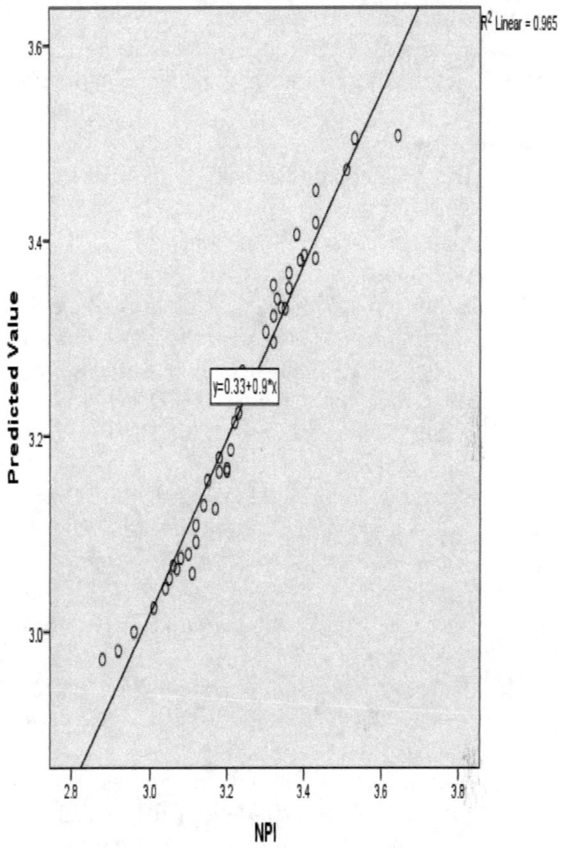

Figure 22. predicted value of Nemerow Pollution Index (NPI).

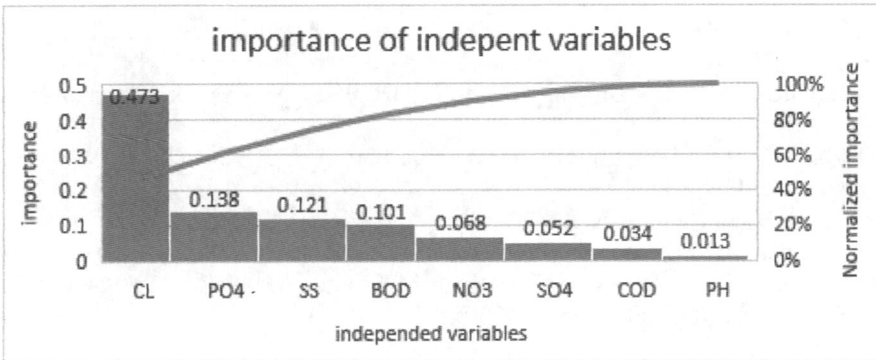

Figure 23. importance of parameters in create the model.

6.Conclusion:

The major results achieve from this research can be explained below:

1- The monthly and yearly changes were within the limits of the Iraqi specification (IQS) except SO4-2 and PO4 -3 which record some values not adequate to the Iraqi specification (IQS).

2-The Nemerow Pollution Index (NPI) showed that the deterioration over the years which need maintenance and development.

3- A successful ANN model was built based on Nemerow Pollution Index (NPI) where the R2 was 0.965 and the most important influence in the construction of the model was chlorine.

Reference:

[1] Donald A E and Blessing U A 2019 *International Journal of Fisheries and Aquatic Studies* **7(1)** 153-159

[2] Chu H B, Lu W X and Zhang L 2013 *International Journal of Fisheries and Aquatic Studies* **15** 343-356

[3] Huihui W, Wenjie Y, Ruihua Y, Yue Z, Yunqiang Z, Yuhang Z, Qianhui Y, Aijun L 2020 *Environmental Science and Pollution Research* **27** 35449–35458

[4] Abazi A M S, Durmishi B H, Sallaku F S, Çadraku H S, Fetoshi O B, Ymeri P H and Bytyçi P S 2020 *Rasayan J. Chem.*, **13(1)** 146-159

[5] Ming Z, Yong L and L Jia 2010 *IEEE 4th International Conference on Bioinformatics and Biomedical Engineering*

[6] Zhang Q, Feng M, Hao X 2018 IOP *Conf. Ser.: Earth Environ. Sci.* **128** 012160

[7] Sulthonuddin I, Hartono D M, Said C A A 2019 *EAI 1st International Conference on Environmental Science and Sustainable Development*

[8] Jain CK, Malik DS, Tomar G 2019 *International Journal of Advanced Science and Research* **4** 01-03

[9] Dawood A S 2017 *Journal of Babylon University/Engineering Sciences* **25** 1708-1720

[10] Dhande E S and Mane S J 2014 *International Journal of Engineering Research & Technology* **3** 1508-1511

[11] Kumar R, P., Pinto L B, Somashekar, R.K. 2010 *kathmandu university journal of science engineering and technology* **6** 115-125

[12] Ouhamchich K A, Arioua A, Lbayny M, Karaoui I, Habbari K E, Idrissi A E, Hssaisoune M, Boulli A, Elhamdouni D and Elfiraoui F E, 2018 *European Scientific Journal* **14** 1857 – 7881

[13] Edokpayi J N, Odiyo J O and Durowojuumar O S 2017 *Impact of Wastewater on Surface Water Quality in Developing Countries: A Case Study of South Africa water Quality*

[14] Jamshidzadeh Z, Barzi M T 2020 *Environmental Science and Pollution Research* **27** 7366–7378

[15] Dogan E, Ates A, Yilmaz E C and Eren B 2008 *American Institute of Chemical Engineers* **27** 439-446

[16] Al-obaidi B H K, Ali S K and Jassim D T 2020 *Journal of Engineering Science and Technology* **15** 2610 - 2623

[17] Neelakantan T R, Brion G M and Lingireddy S 2001 *Water Science and Technology* **43** *125–132*

[18] Delgrange N, Cabassuda C, Cabassudb M, Bourlierc L D, Laine J M 1998 *Journal of Membrane Science* **150** 111±123

[19] Van T P, Dao M P and Han P N N 2017 *Organic & Medicinal Chem IJ* **4(4)** *001-008*

[20] Gorash F, Abdullah A 2012 *Journal of Engineering Science and Technology* **7(4)** *447 – 461*

[21] Sulaiman K, Ismail L H, Razi M A M, Adnan M S and Ghazali R 2019 IOP Conf. Series: Materials Science and Engineering **601** 012005

[22] Nidhisha C, Kuruvila E C 2019 *International Research Journal of Engineering and Technology* **06** 4741-4747

[23] Khademikia S, Haghizadeh A, Godini H, Khorramabadi G S 2016 *Civil Engineering Journal* **2** 555-567

[24] Fu G, Makropoulos C and Butler D 2010 *Journal of Hydroinformatics* **12** 140-149
[25] El-Din A G, Smith D W and El-Din M G 2004 *J. Environ. Eng. Sci.* **3** S81–S95
[26] TÜMER A E, EDEBALİ S 2015 *International Journal of Intelligent Systems and Applications in Engineering* **3(4)** 131-135
[27] Mehrdadi N, Hasanlou H, Jafarzadeh M T, Hasanlou H and Abdolabadi H 2012 *Journal of Water Resource and Protection* **4** 370-376
[28] Kang J H, Song J, Yoo S S, Lee B J and Ji H W *2020 Atmosphere* **11** 784
[29] Bekkari N and Zeddouri A 2018 *MEQ* **30** 593-608
[30] Hamed M M, Khalafallah M G, Hassanien E A 2004 *Environmental Modelling & Software* **19** 919–928
[31] Al-obaidi B H K 2020 *Journal of Engineering Science and Technology* **15** 3571 – 3587
[32] Al-sakini 2013 *Journal of KerbalaUniversity* **11** 195-203
[33] Ismail H A 2016 *Journal of Basra* **76** 243-267
[34] Jena V, Dixit S and Gupt S 2013 *International Journal of ChemTech Research* **5** 278-283
[35] Gummadi S, latha S, Vijayakumar, Rao P B, Venkatarathnamma V 2015 *Int. Journal of Applied Sciences and Engineering Research* **4** 500-506
[36] ABBASI T and ABBASI S A 2012 WATER *QUALITY INDICES (*LONDON: Elsevier)
[37] Ajani G E, Popoola S O and Oyatola O O 2020 International *Journal of Environment and Climate Change*
[38] Nasr M S, Moustafa M A E, Seif H A E and El Kobrosy G 2012 *Alexandria Engineering Journal* **51** 37-43
[39] Steel, E. W. and Mcghee, T. J. 1991 *Water Supply and Sewerage. Sixth edition.*
[40] Punmia B C,Jain A K, Jain A K 1998 Wastewater Engineering(New Delhi: Golden House)

Acknowledgment
The authors wish to thank the Mayoralty of Baghdad for the opportunity to obtain data and to be given the facilities to be in touch with the station manager. The authors would like to thank the staff of the sanitary engineering laboratory and the Civil Engineering Department/Engineering College-University of Baghdad for their valuable support to complete this work.

Effect of Fullerenes Additions on Physical - Mechanical Properties of Hot-Forged Iron-Based Powder Materials

Mohammed Qasim Kareem[1*], **Vladimir Dorofeyev** [2]

[1] Hydraulic Structures Department, Factuality of Water Resources Engineering, Al-Qasim Green University, Hilla, Al-Qasim, Iraq
[2] Mechanical Engineering Department, Platov South Russian- State Polytechnic University, 346428 Novocherkassk, Rostov Region, Russian Federation
* Corresponding author: mohammed.qasim@wrec.uoqasim.edu.iq

Abstract

It is possible to expand the applications ranges of powder material products by enhancing the performance properties of these products in addition to their manufacturability and reliability together, it's possible by materials structures modification. In this paper, the effect of fullerene (C60) additives to iron-based powder material has been studied. All samples produced by Hot-Forging (HF) powder materials technology. Green and HF density of the obtained samples calculated by volume / weight and Archimede's principle, respectively. The effect of technological parameters on the microstructure of carbon steels' samples was done by an ALTAMI MET-1M metallographic microscope. Tensile test executed by using of a universal testing machine UMM –5 and the microhardness (HV10) was measured by REICHERT hardness test machine. The results showed that the HF C60 steels' samples had higher density and strength of 0.81 and 25%, respectively, with a good plasticity in comparison with graphite steels' samples.

Keywords

hot-forging technology, fullerene (C60), iron powder, mechanical properties.

1.Introduction

Powder metallurgy methods applied to the production of parts for various purposes occur in tough competition with traditional technologies of stamping, casting, machining, rolled products. Further expansion of the range of powder metallurgy products depends on the development of effective and competitive methods to improve their operational properties. Among such techniques may be mentioned especially Hot –Forging (HF) of porous preforms which has worked well as a simple and reliable method for obtaining high density powder products [1–3].

Providing the necessary combination of manufacturability and reliability of the product together with ensuring the required properties is carried out by improving (changing) the structure of materials. Iron-based materials and parts of them are one of the main types of powder metallurgy products. In powder metallurgy, iron-carbon materials are produced by doping carbon into iron powder in the form of graphite, carbon black or cast iron powder. In [4,5], carbon powder in the form of fullerene C60 was used.

Previously it was thought that carbon had three allotropic forms: diamond, graphite and carbon. At present, the fourth allotropic form of carbon is known, the so-called fullerene, named after the American architect Richard Buckminster Fuller, who designed hemispherical architectural structures consisting of hexagons and pentagons. The C60 molecule is the most common and most stable of all fullerenes [6-8].
In metallurgy, of interest is the synthesis of fullerenes and their compounds in the structure of iron-carbon alloys. In [9], fullerene and iron powder was sintered at high temperature and pressure. As a result, a composite material was obtained, where microparticles of the superhard carbon phase were formed in the iron matrix. This material had a high wear resistance. It was shown [10] that fullerenes formed in the steel structure are synthesized during sharp quenching into diamond-like compounds. This makes it possible to increase the hardness and wear resistance of ordinary carbon steels.
The possibility of expanding the field of application of powder metallurgy methods by creating powder materials (PM) with enhanced performance properties is of paramount importance. So, the goal of this paper was studying the effect of fullerenes additives to iron-based powder materials by HF technology.

2.Experimental Procedure:

2.1 Materials:

For the preparation of samples according to the procedure described in [11], the following materials were used:

- Reduced iron powder NC100.24 (from Höganäs AB);
- Fullerene C_{60} (purity 99.5%; particle size < 10 nm; from Nano product company, Saint-Petersburg, Russia);
- Graphite powder (GK-3, composition, wt.%: ≥ 94 C; ≤ 5 ash; ≤ 1 volatile; ≤ 1 humidity; D50 = 43 μm).

2.2Technology of Obtaining Samples:

Powder charges were carried out in a conical blender for 2 hours with ethyl alcohol as a wetting medium. Green samples were obtained by filling the die cavity with 44 ± 0.01 g of powder charges, which consisted of holder, two semi-matrices, and a lower and upper punch [12] then ,powder charges had compacted by cold-pressing (CP),which was performed on a hydraulic press HPM60 model (maximum force 1000 kN) down to the abutment. The working parts of the die were made of KhVG (composition, wt.%: *0.9 – 1.05 C; 0.1 – 0.4 Si; 0.8 – 1.1 Mn; < 0.35 Ni; < 0.03 S; <0.03 P; 0.9 – 1.2 Cr; <0.3 Mo; 1.2 – 1.6 W; <0.3 Cu; bal. Fe*) and Kh12M (*1.45 – 1.65 C; 0.15 – 0.35 Si; 0.15 – 0.4 Mn; < 0.03 S; <0.03 P; 11.0 – 12.5 Cr; 0.4 – 0.6 Mo; 0.15 – 0.3 V; bal. Fe*) steels, hardened to HRC 56-62.
Hot-forging green samples was executed by using the method developed at the Platov South Russian State Polytechnic University(PSRSPU) [13].So, the forging applied work which was used here, is approximately, equal to 342 MJ/m^3, which had calculated by using the following equation:

$$work_{HFS} = w_f\,[kg] * 9.8\,[m/s^2] * H_f\,[m] * \rho_k\,[g/cm^3] / w_{HFS}\,[g],\ MJ/m^3;$$

Where:
$work_{HFS}$: the work done by forging the preheated samples; w_f: weight of falling-mass, it was 100 kg; H_e: the height of the falling-mass dropped from, here, was 1.95 m; ρ_k : theoretical density of carbon steel, which is approximately, 7.88 g/cm^3; w_{HFS}: dry weight of hot-forged sample.

The die for HF was similar to the die for CP of samples (Fig.1.) But has a slightly larger size. Working parts of the HF die were made of steel 5KhNM (composition, wt.%: *0.5 – 0.6 C; 0.1 – 0.4 Si; 0.5 – 0.8 Mn; 1.4 – 1.8 Ni; < 0.03 S; <0.03 P; 0.5 – 0.8 Cr; 0.15 – 0.3 Mo; <0.3 Cu; bal. Fe*) with hardness HRC 45 - 48. Heating of green samples before theHF was carried out in a chamber electricalfurnace with silicate heaters. Green samples were laid in a container of heat-resistant steel 10Kh23N18 (composition, wt.%: < 0.1 *C; < 1.0 Si; < 2.0 Mn; 17.0 – 20.0. Ni; < 0.02 S; <0.035 P; 22.0 – 25.0 Cr; < 0.2 Ti; < 0.3 Cu; bal. Fe*), placed in the working space of the furnace and purged with protective gas - dissociated ammonia. Extraction of heated sample from the furnace was after 10 minutes. Then, the heated sample was placed in a HF die then a seal was applied and leaved to cool in air. Thus, HF technology (Table 1) of powder material includes: preparation of powder charges; formation of green samples after CP with porosity of 14-18%

(400 - 490 MPa); hot-forging with 342 MJ/m^3 at a heating temperature of 1100 °C to a residual porosity of 0.55-2.7%. Finally, prepare samples for studying physical-mechanical properties.

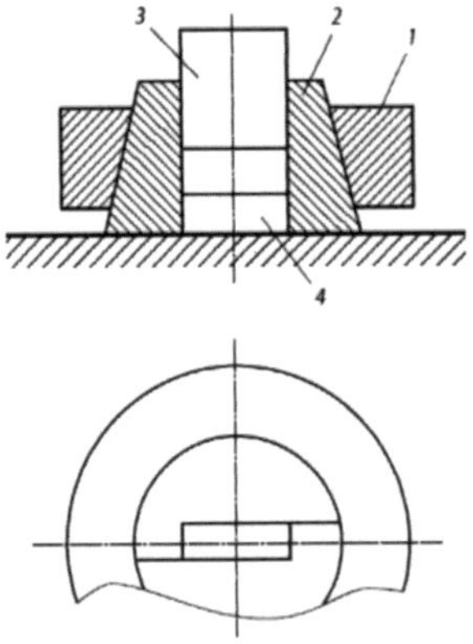

Figure 1. Die of green and forged samples 1-holder , 2-semi-matrix, 3- upper punch, 4-insert (lower punch).

Table 1. Composition of powder blends.

Powder Charge	1	2	3	4	5	6	7	8	9	10
NC100.24,wt.%	99.8	99.6	99.4	99.2	99.0	99.8	99.6	99.4	99.2	99.0
C_{60},wt.%	0.2	0.4	0.6	0.8	1.0	0	0	0	0	0
GK-3, wt. %	0	0	0	0	0	0.2	0.4	0.6	0.8	1.0

3.Methodology of Investigation of Physical-Mechanical Properties:

Green and HF density of steel powder samples were determined by volume / weight and Archimedes'principle, respectively. Weighing was carried out on an analytical balance of the type WA - 35 with an accuracy of up to 0.0001 g. The error in determining the density was 0.8%.

A metallographic microscope was used by ALTAMI MET-1M to study the effect of technological parameters on the microstructure of carbon steels of samples obtained after HF .For this, micro sections were made with the test surface d = 5 mm (Fig.2 a). The etching modes of microsections were chosen experimentally. Etching was carried out in a 3% solution of HNO3 in ethanol, as well as Marble reagent.

IOP Conf. Series: Earth and Environmental Science **877** (2021) 012009 doi:10.1088/1755-1315/877/1/012009

Testing of powder materials produced to determine the strength properties and hardness. The main criterion for evaluating the strength properties of powder materials is tensile strength. To study the mechanical properties of the steels, Gagarin tensile specimens of forged samples (Fig.2 a and b) were machined with the working part diameter d0 = 5mm and the gage length l0 = 25mm. The ultimate tensile strength (UTS, MPa) as determined by using a universal testing machine UMM - 5 according to GOST 1497-84.

The microhardness (HV10) of the structural components of the HF samples was determined according to GOST 9450-76 with the use hardness test machine (REICHERT, Austria).

a

b

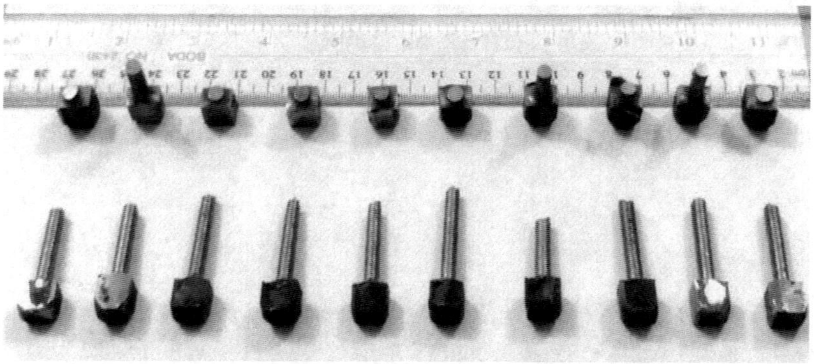

Figure 2. Represents (a) microstructures' studying sections which prepared from; b) hot-forged Gagarin tensile test samples after fracture.

4.Results and Discussion:

4.1.Density

The density of produced carbon steels materials decreased with increasing carbon content (Fig.3a and b). After CP, density of green samples with graphite carbon type were higher than those with fullerene (C60) due to bigger density of graphite, 2.2670 g.cm-3 than for C60 ,1.5651 g cm -3 [14]. Conversely, the porosity of cold pressed samples decreased with increasing density of carbonaseous additive. The porosity decreased from 16 % and 16.9 % (highest porosity values after CP) for samples with graphite and fullerene, respectively to 1.3 % with graphite, 0.6 % with fullerene (lowest porosity values after HF). So, samples of fullerene steels after HF had higher densities (lower porosities) about 0.81% than HF graphite samples. The higher density of the samples with fullerene is due to the smaller size of the initial fullerene particles, as well as the greater diffusion mobility of carbon atoms in the fullerene composition [14]. Also, with increasing C60 additives (decreasing of relative density) the volume change increased significantly in comparison with graphite powder content of HF steels samples because of relatively low density of C60.

One of the important factors influencing on the strength of iron-based powder materials is the density. Higher densities (7.7 - 7.8 g/cm3) can be obtained by powder forging technology among many others techniques. Higher densities lead to high physical-mechanical properties of iron-based powder parts [15].

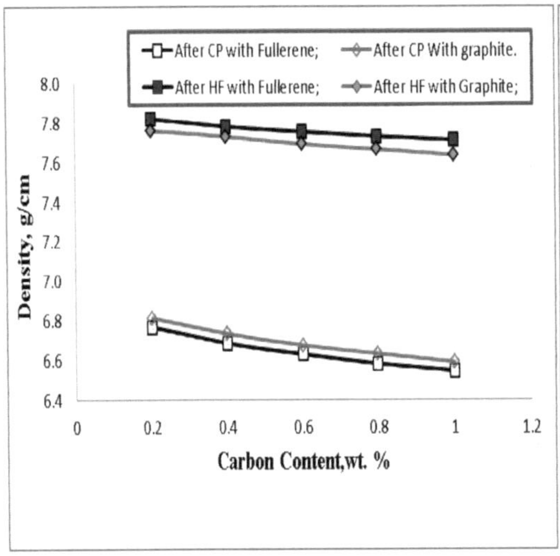

a

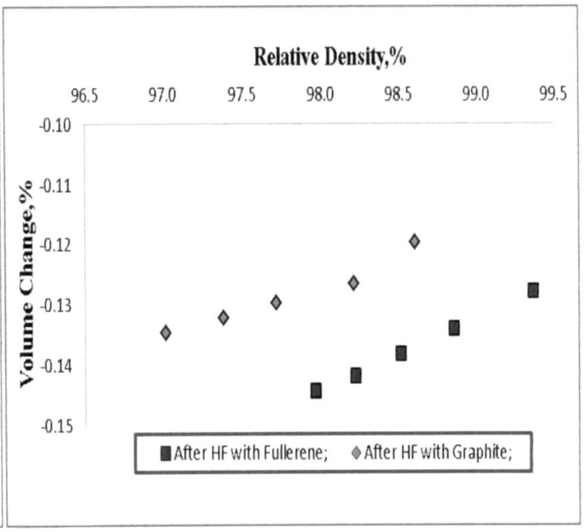

b

Figure 3. represents the effect of a) carbon additives' type and content on green and HF density ; b) relative density on volume change of hot-forged steels' samples.

4.2.Mechanical Properties

In the case after HF (Fig.4) the strength of samples with fullerene 25% more than samples with graphite. Relative elongation of samples with fullerene higher than with graphite. The reduction in area is proportional to elongation.

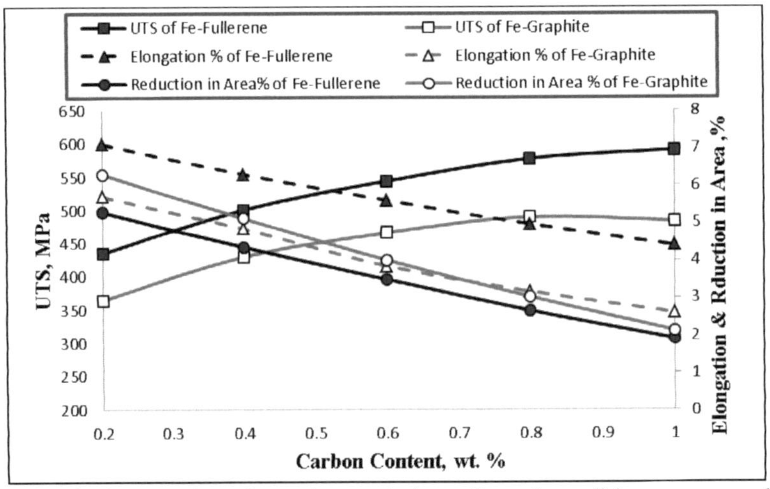

Figure 4. Mechanical properties of powder steels in as-forged condition versus carbon content.

As this study work compared with other works, table 2 showed that the mechanical properties of hot-forged powder carbon steels have been demonstrated and the effective rule of fullerene (C_{60}) additions in improving the UTS of steel samples. In work [16] it was found significant transformation of ferrite to martensite and austenite of compacted samples with fullerene and nanotubes additions. So, the higher strength of steel samples with fullerene additions almost came from these transformations.

IOP Conf. Series: Earth and Environmental Science **877** (2021) 012009 doi:10.1088/1755-1315/877/1/012009

Table 2. represents the comparison of mechanical properties of different production processes.

Materials	Producing Conditions	Produced Samples: Density, g/cm^3	UTS, MPa	Elongation, %	Reference
Astaloy Mo+0·5%G	Samples were pressed at 600 MPa and sintered at temperatures of 1140 °C, and cooled at 40 K min^{-1},	7·27	460	1.8	[18]
Astaloy Mo+1.0%G		7·14	555	1.1	
Astaloy CrL+0·5%G	Sample was compacted under 600 MPa and sintered in 1120 °C for 30 min. then sized at 200 °C (the better procedure)	6.97	380	2.4	[19]
NC100.24 +0.4%G	All samples compacted under 400 - 490 MPa and hot-forged after heating at 1100 °C for 10 min. and leaved to cool in air.	7.72	430	4.84	This study
NC100.24 +0.6%G		7.69	465	3.81	
NC100.24 +0.4%F		7.78	500	6.3	
NC100.24 +0.6%F		7.75	543	5.6	
NC100.24 +1.0%F		7.71	590	4.4	

4.3.Microstructural Properties:

The structures of the samples with the addition of fullerene and graphite have no significant differences. They are a combination of areas of ferrite, sorbite, troostite and

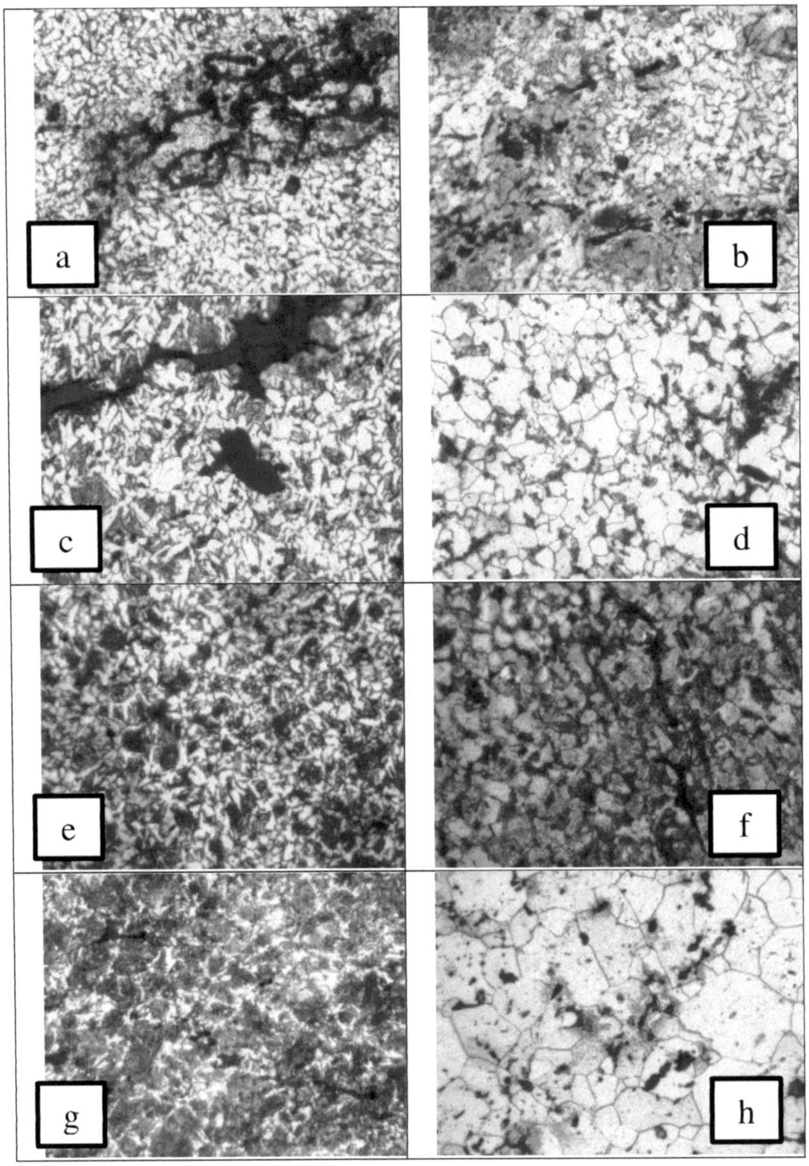

Figure 5. Structures of carbon as-forged powder steels, x400. a) 0.4% fullerene; b) 0.4 graphite, c) 0.6 % fullerene;d) 0.6% graphite;e) 0.8% fullerene; f) 0.8 % graphite; g) 1,0 % fullerene and h) 0.2 % graphite.

bainite (Fig.5). With an increase in the carbon content, the number of sites and the area occupied by eutectoid structures increases. In table 3, the results of measurements of microhardness of steel samples in the state after HF are presented.

Table 3. Microhardness of the structural components of as-forged steel samples with fullerene and graphite additives, individually.

Carbon Additives	**Microhardness, MPa**

	Ferrite	Perlite	Sorbite	Troostite	Bainite
Fullerene	1400-1600	3000-3400	4000-4200	5000-5500	6000-6500
Graphite	1100-1300	2800-3000	3000-3200	4000-4500	5000-5500

It was noted that the microhardness phases of fullerene steels samples (ferrite, perlite, sorbite, troostite and bainite) increased effectively in a range from 200 to 1000 MPa more than those with graphite . The samples of the surface rolling material with 0.8 wt. % carbon have the highest surface hardness (340 HV0.1)[20]. In work [21] were a hot-compacted modified chromium-molybdenum steels' samples by fullerenes and carbon nanotubes additions showed higher ultimate tensile strength and microhardness (HV) but with a lower plasticity than unmodified samples.

5.Conclusions:

1-Hot forging powder materials technology had effective role in improving mechanical properties of steels' samples.
2-The addition of fullerene (C60) to iron-based powder materials contributes to an increase in the strength characteristics of samples (higher density 0.81% and higher strength 25% than for graphite steels' samples with a good plasticity).
3-There are no significant differences between the structures of the samples with fullerene and graphite additions but the first one had higher microhardness (HV10) than the samples with graphite.
4-Higher strength of steel samples with fullerene additions came from martensitic and austenitic transformations.

References:

[1] Bal'shin M Y 1972 Scientific bases of powder metallurgy and fiber metallurgy *Moscow Metal.*

[2] Dorofeev Y G 1977 Dynamic Hot Pressing of Porous Powder Blanks *Mashinostroenie, Moscow*

[3] Dorofeev Y G, Gasanov B G and Dorofeev V Y 1990 Industrial Technology of Hot Pressing of Powder Articles *Metall. Moscow.*

[4] G. A. Libenson, V. Yu. Lopatin, and G. V. Komarnitsky2002 Powder Metallurgy Processes. In 2 volumes. Vol. 2. Molding and sintering .

[5] Kiparisov S S and Libenson G A 1980 Powder metallurgy: study guide *Moscow Metall*

[6] Rakov E G 2006 Nanotubes and fullerenes *Logos, Moscow* 141–145.

[7] Grigoriev S N, A.A. Gribkov, and S.V. Aleshin 2008 Nanoprocessing Technologies

[8] Turan M E 2019 Investigation of mechanical properties of carbonaceous (MWCNT, GNPs and C60) reinforced hot-extruded aluminum matrix composites *J. Alloys Compd.* **788** 352–360.

[9] O. P. Tchernogorova, O. A. Bannykh, V. M. Blinov, E. I. Drozdova, A. A. Dityat'ev, and N. N. Mel'nik, 'Superhard carbon particles forming from fullerites in a mixture with iron powder', *Mater. Sci. Eng. A*, vol. 299, no. 1–2, pp. 136–140, 2001.

[10] Job R C 1995 Method of producing diamond crystals from metallfullerite matrix and resulting product.

[11] Antsiferov V N, Latypov M G and Shatsov A A 1998 Isothermal decomposition of austenite in concentration-inhomogeneous nickel steels *Met. Sci. heat Treat.* **40** 197–201.

[12] LIBENSON G A 1982 The manufacture of sintered products(Russian book) *Moscow, Izd. Metall. 1982, 256.*

[13] Yu. G. Dorofeev 1977 Dynamic hot pressing of powder blanks Moscow Metallurgy

[14] Ruoff R S, Thornton T and Smith D 1991 Density of fullerene containing soot as determined by helium pycnometry Density of fullerene containing soot as determined by helium pycnometry Formation of endohedral carbon-cluster noble-gas compounds with high-energy bimole . cular reactions : C 60 He n + (n = 1 , 2)*Chem. Phys. Lett.* **186** 456–458

[15] Antsiferov V N, Grevnov L M, Torsunov M F and Boyarshinov V A 2014 Structure of fullerene-containing deformed powder steels *Russ. J. Non-Ferrous Met.* **55** 167–171

[16] Höganäs 2015 *Sintered Iron-Based Materials* 45–90.

[17] Glebov V A, Bakulina A S, Efremov I V, Shchetinin I V, Yagodkin Y D, Glezer A M, Rashkovskii A Y and Vainshtein D L 2010 Study of the structure of steel 12Kh12M1BFP modified with additions of fullerenes and carbon nanotubes *Met. Sci. Heat Treat.* **52** 321–324.

[18] Youseffi M, Wright C S and Jeyacheya F M 2000 Effect of carbon content, sintering temperature, density, and cooling rate upon properties of prealloyed Fe–1· 5Mo powder *Powder Metall.* **43** 270–274.

[19] Recknagel C, Marquardt A, Langer I, Müller S and Kieback B 2011 Higher densities of PM-steels by warm secondary compaction and sizing *Proc. EURO PM 2011.*

[20] Zhao Y, Chen D, Li D, Peng J and Yan B 2018 Effect of Carbon Content on the Properties of Iron-Based Powder Metallurgical Parts Produced by the Surface Rolling Process *Metals (Basel).* **8** 91.

[21] Kaputkina L M, Shchetinin I V, Yagodkin Y D, Savchenko A G, Gorshenkov M V and Glebov A V 2014 Structure and properties of chromium-molybdenum steel modified by the addition of fullerenes and carbon nanotubes *Inorg. Mater. Appl. Res.* **5** 334–339.

Acknowledgments
The study was carried out with the financial support of the Russian Foundation for Basic Research, grant No 19-08-00107 A.

Determination of Uranium Concentration in Blood Samples for Patients with Kidney Failure in Salah din Governorate

Asmaa A Aziz[1], Fareed M Majed [2] and Nada F Tawfiq [3]
[1]Department of Physics, College of Education for Science, Tikret University, Tikret, Iraq
[2]Department of Physics, College of Science, Tikret University, Tikret, Iraq
[3]Department of Physics, College of Science, Al-Nahrain University, Baghdad, Iraq

Email: asmaa.jamal@tu.edu.iq

Abstract. Analysis of human biological samples, such as blood, is generally used to verify human exposure to uranium. The uranium content in the blood of patients with kidney disease in Salah al-din governorate was determined using the Fission Track Analysis (FTA) of the detector CR39. Uranium concentrations of blood samples taken from kidney failure patients ranged between 1.636 ppb to7.477 ppb, with a mean value of 5.496 ppb. And the health group values ranged between 0.301 ppb to 2.332 ppb with a mean value 1.089 ppb.

1. Introduction

The blood's job is to deliver nutrients, hormones, and oxygen to tissues. The wastes of cellular metabolism are eliminated, and specialized cells are delivered to tissues external environmental protection agent prevents leakage by closing the holes in the blood vessels [1]. The total blood volumes in woman's are about (4-5) litter, while man are (5-6) litter, cell and cell fragmentation was 55%. Most blood cells grow from stem cells or precursors produced by the bone marrow [2]. The blood consists of cell-free plasma (50-60%) and erythrocytes (45%). And all blood cells, white cells (WBC) and platelets are 1% by volume [3].

Due to its radioactivity, uranium is considered a dangerous element. Therefore, uranium and its compounds pose a threat pollute the climate and endanger people's wellbeing [2]. Uranium is taken into account one of the most dangerous pollutants due to its radioactivity and hazardous substances that endanger public health and environmental balance. [4]. It absorbed after inhalation, oral ingestion, or skin contact, and the amount absorbed depends on the dissolution of the uranium compounds. These soluble compounds are absorbed for days while the less absorbed compounds settle in the lung and lymphocytes for long periods of time [5].

General, the kidneys are affected by the toxic effect of uranium, and some cells can be damaged while affected by the body, and cancer is the result of a physical mutation in the cell. The annual dose of natural radiation is about 1 mSv or more for the average citizen of the world [6]. Solid nuclear trace reagents (SSNTDs), technique for measuring the level of uranium in blood. Fleischer et al. proposed the fission pathway technique (FTA) [7]. This technique is suitable for uranium in blood quantification [8, 9]. Solid-state nuclear path detector (SSNTDs), and fission path technology were used in Salah al-din governorate, the CR-39 detector was used to determine the amount of uranium in patients with kidney disease.

The aim of the research to find levels uranium blood patients with kidney failure in this governorate, which witnessed a significant increase in the rates of kidney failure recorded after ISIS events (2014-2016).

2. Materials and Methods

2.1. Sample collection

This research was carried out on 80 male and female blood samples collected from two groups. First group included 60 samples of kidney failure patients from Dialysis Center at Tikrit Teaching Hospital in Salah al-din governorate. And the second group included 20 samples of healthy people living in these governorates. Figure1 and Figure 2 it depicts the cities of Salah din governorate as well as the number of kidney disease patients in this governorate.

.

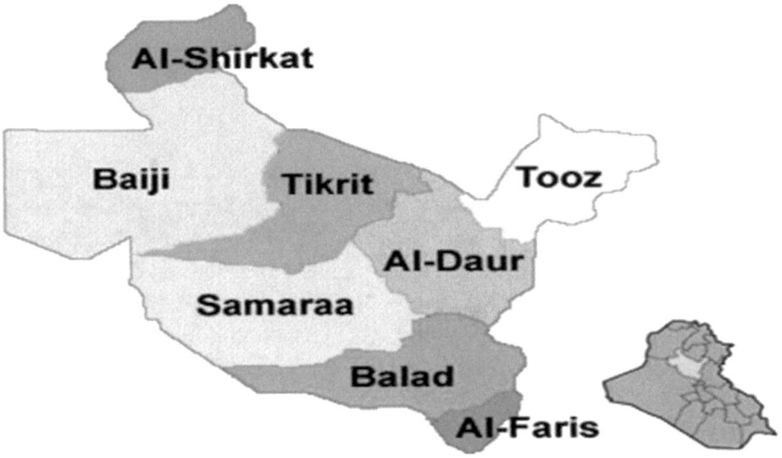

Figure 1. City locations in Salah al-Din governorate.

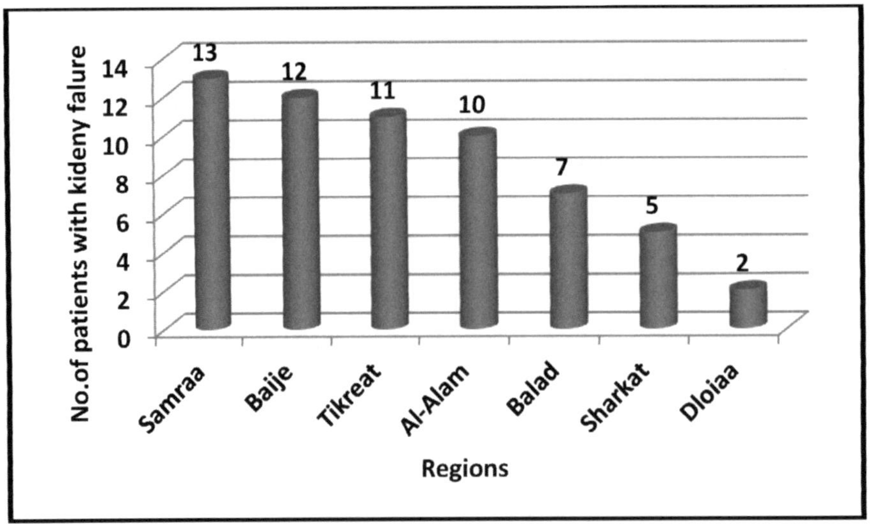

Figure 2. Number of patients in Salah Al-Din regions.

IOP Conf. Series: Earth and Environmental Science **877** (2021) 012010 doi:10.1088/1755-1315/877/1/012010

Table 1 shows characteristics of the patient group (number and gender), and the healthy group are shown.

Table 1. Descriptive statistics of samples of patients group and healthy group.

Classification	Kidney failure group	Healthy people
No. of samples	60	20
No. of man	31	7
No. of woman	29	13
Age range (years)	20-75	20-70

2.2. Experimental work

The presence of uranium in the blood of patients was discovered by using the fission path technique. Where a drop of blood (70 μl) is placed on the detector using a pipette and dried on an area about (1.5 x 1.5) cm^2 of the detector CR-39 [8,9]. Irradiation in (Ibn Al Haytham College, University of Baghdad, Iraq).Thermal neutrons from a neutron source is used to irradiate the blood (Americium - Barium) For a period one week in thermal neutrons flood of ($3.024x10^9 n/cm^2$) to produce fission tracks in the detector (CR-39).

After the reagents were exposed to a flood of neutrons, the reagents were etching from solutions of 6.25 N NaOH at 60 °c for about more four hours. Then clean etching detectors with water. The microscopic inspection was done using a light microscope. Calculate the intensity of traces (ρ using the equation:

$$\text{Fission path density } (\rho) = \frac{\text{Total track}}{\text{Area of view}} \quad (1)$$

The level of uranium in the blood was found by the following relationship:

$$C_x = \rho_x . (C_s/\rho_s) \quad (2)$$

C_x (ppb): Level of uranium in unknown samples, C_s: level of uranium in standard samples. ρ_x and ρ_s : Traces intensity of unknown and standard samples.

The slope between the trace intensity and the uranium concentration of the standard sample in Figure 3 which represents relationship trace density and uranium level (ppb) for a standard samples, the equation becomes:

$$C_x = \frac{\rho_x}{slope} \quad (3)$$

3. Analyze samples statistically

All results obtained from the uranium level (ppm) in blood of patients were analyzed by Statistical program.

4. Results and discussion

Lower, higher, and average level of uranium (ppb) using the density of fission fragments in the sample of patients in kidney failure and healthy group as shows in Table 2.

We note from the figure (4). The average concentration of the infected group has a higher concentration than the normal individuals on average. As level concentration the blood the patient group is 5.00 times more than the concentration it from healthy group.

•

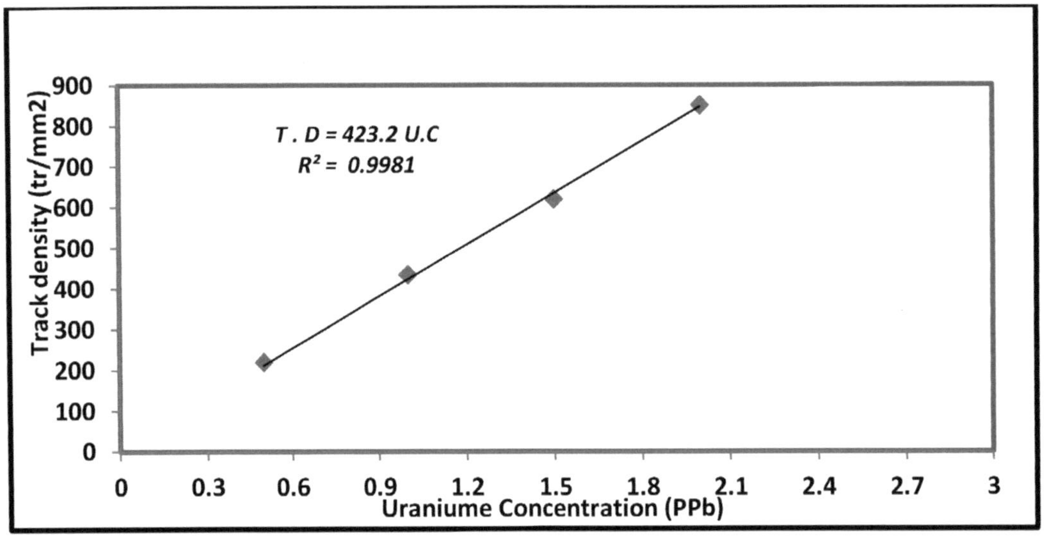

Figure 3. Path density (tr/mm^2) and uranium concentration (ppb) of standard blood samples

Table 2. Definition of data for concentration (ppb) in the blood of patients

Statistical values	Patients group	Healthy group
No. subjects	60	20
Minimum	1.636	0.301
Maximum	7.477	2.332
Av. ± Std. Errer	5.496 ± 0.850	1.089 ± 0.150
deviation	1.593	0.128
Value of P *	P = 0.004	P = 0.004

* One - way independent sample ANOVA

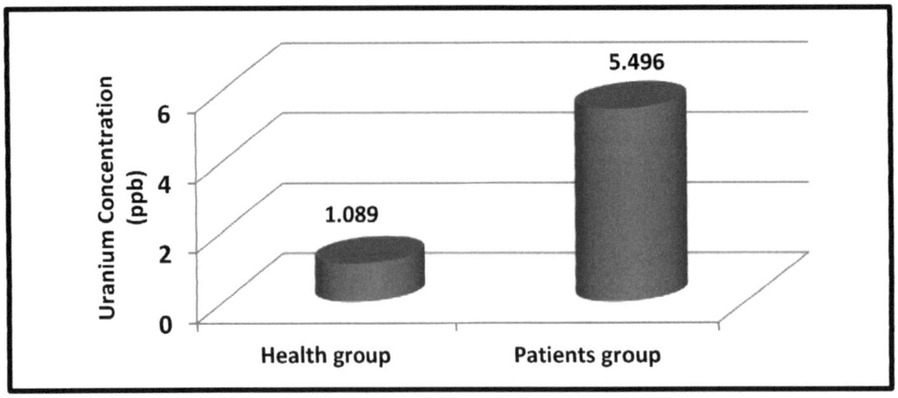

Figure 4. A mean value of uranium (ppb) of patients group and healthy group.

Table 3 Shows effect of a gender bias on uranium concentration in patients and healthy subjects is demonstrated. Figure 5 shows the average values it in man and woman patients and healthy subjects. And shows that average level for the two groups in man is higher than in woman. The reasons for that, the blood volume of woman's are (4.-5) liters, and in mans the blood volume are (5-6) liters [3].

Table 3. Uranium levels (ppb) as a function of gender patients and health.

Group type	Gender	Number of samples	Mean ± Std. Error	P-Value*
Kidney failure	Man	31	6.320±1.850	P = 0.004
	Woman	29	4.672±1.340	
Healthy people	Man	7	1.101±0.814	P = 0.004
	Woman	13	1.077±0.635	

* One - way independent sample ANOVA

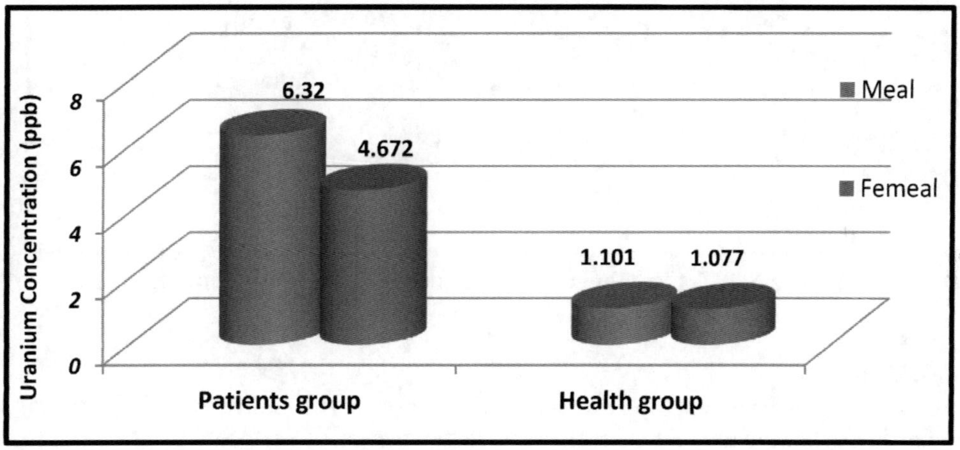

Figure 5. Uranium level (ppb) for kidney failure patients and a healthy people as a result of sex

Table 4 and figure 6 shows the average blood level of uranium in renal failure patients and healthy patients with age. Note that largest level of it in patients with kidney failure in the age group 20-40 years. In healthy group, we note that the level of uranium increases with age. Many researchers found these results [8-10].

Table 4. Uranium levels (ppb) as a function of age group for the two groups.

Group type	Age (years)	Number of samples			Mean ± Std. Error
		man	woman	total	
Kidney failure n=60	20 - 40	8	8	16	7.870 ± 2.24
	41 - 60	11	9	19	4.698 ± 1.109
	61 - 80	13	12	25	4.654 ± 1.465
Healthy people n=20	20 - 40	3	6	9	0.580 ± 0.041
	41 - 60	2	5	7	1.103 ± 0.1738
	61 - 80	2	2	4	2.193 ± 0.1716

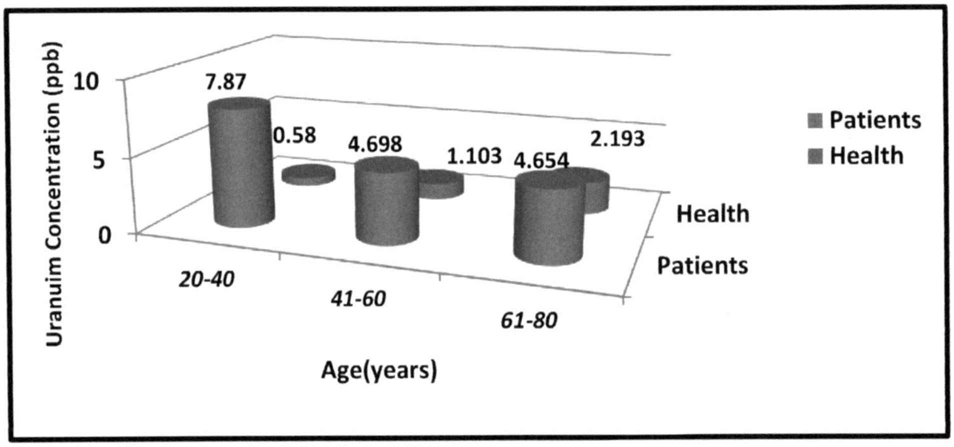

Figure 6. Level uranium (ppb) in patients and a healthy, with the age.

5. Conclusions

The results obtained showed that the average uranium level in the patient group's blood samples was 5.00 times higher than the uranium level in the stable group's blood samples. The male participants (both kidney disease patients and healthy people) had higher uranium levels than the females. In stable individuals, concentrations rose as the group became older.
The results showed that the uranium concentration in the blood of patients and a healthy group was less than to the permissible limit from the ICRP 0.115 ppm (115 ppb).

6. Reference

[1] Emmanuel CB 1992 *Hematology Introduction and Principles of Hematological Diagnosis* London, Sydney
[2] Berence M R and Veny M 1987 *Physiology* Sanders Company.
[3] Fox S I 2003 *Human physiology* McGraw-Hill Science NewYork
[4] Zou W, Zhao L, Li K and Han R 2011 Characterization and properties of zeolite as adsorbent for removal of uranium (VI) from solution in fixed bed column *Journal of Radioanalytical and Nuclear Chemistry* **288** 779–788
[5] Valentin J 2002 Basic anatomical and physiological data for use in radiological protection: reference values: ICRP Publication 89 *Annals of the ICRP* **32**(3-4) 1-277

[6] World Health Organization *WHO 2001 Depleted Uranium* Sources Exposure and Health Effects
[7] Fleischer R L and Price R M 1975 *Nuclear track in solid* University of California
[8] Tawfiq N F, Ali L T and Al-jobouri H A 2012 Uranium concentration measurements in human blood for some governorates in Iraq using CR-39 track detector *J. Radioanal Nuclur Chem.* **295** 671- 674
[9] Abed M M, Mahdi K H and Hamadany W S A 2020 Estimation of Uranium Concentration in Urine Samples of Kidneys Failure Patients in Al-Muthanna Governorate *Ibn AL-Haitham Journal For Pure and Applied Science* **33**(1) 11-16
[10] Segovia N, Olguin M E and Romero M 1986 Studies of U in the blood of two population samples. International Journal of Radiation Applications and Instrumentation *Part D. Nuclear Tracks and Radiation Measurements* **12**(1-6) 797-800

IOP Conf. Series: Earth and Environmental Science **877** (2021) 012011 doi:10.1088/1755-1315/877/1/012011

Study of haemoflagellates Trypanosoma sp. infection in some fish of Iraq marshes and relationship of leukocytes with inflammatory response

H M Jarallah [1*]

[1] Marine Science Center - Basrah University- Iraq
* Corresponding author: hind.jarallah@uobasrah.edu.iq

Abstract

Fresh water fishes infected with different species of genus Trypanosoma parasite that is similar to that same genus in blood of mammals. Trypanosomiasis or sleeping sickness disease that causes by protozoa flagellate parasites which live inside cell called Trypanosoma. Fish parasites play an important role in regulating the population of their hosts. This study was implemented to verify the infection with Trypanosoma sp. report the cellular intervening through inflammatory response in blood smears of infected fishes. A total of 190 fishes of three species Carasobarbus luteus, Aspius vorax and Liza abu were collected from Al- Hamare and Al-Chibayish marsh south of Iraq. The flagellate parasite Trypanosoma sp. observed in blood of Carasobarbus luteus 41.42% and Aspius vorax 12.28% fishes, while no infection was recorded with Trypanosoma sp. in Liza abu fishes. Prevalence and mean intensity of infection were calculated. The different counts and various types of inflammatory cells were monitored, there are increased in lymphocytes in the infected fishes. Statistically, there are significant differences ($p<0.05$) among species of infected fish with Trypanosoma sp.

Keywords

Fish, Parasites, Trypanosoma sp., Iraq marshes.

1.Introduction

Fresh and marine water fishes infected with different types of parasites genus Trypanosoma which is similar to that same genus in blood of mammals by a blood sucking insect tsetse fly (1). Trypanosomes are haemoflagellates having a single free flagellum at the anterior end of their body (2). Trypanosoma infection prevails against the host immune system resulting in pronounced diseases conditions and ultimately death of the host (3, 4). Fish health sector is considered as one of the main principal sectors that most of Arab regions adopt it, it is a necessary factor in the strategy of Arab food safety. lately, the Arab Organization for Agricultural Development (AOAD) Tackled, for the first time the problem of fish diseases aiming at improving and developing these important natural resources (5). Like all animals, fish can infect with diseases, the ecological factors and human interference would allow for entry of pathogens (6, 7). Olsen demonstrated that the worms belong to genus Piscicola and Hemiclepsis were transferring the infection among fresh water fish, while the worms belong to genus Pontobdella and Trachobdella were transferred the infection among marine water fish (8). There is

IOP Conf. Series: Earth and Environmental Science **877** (2021) 012011 doi:10.1088/1755-1315/877/1/012011

cellular intervention in the inflammatory response, neutrophils followed by macrophages (9). All non-mammals such as fishes have nucleated erythrocytes, while the leukocyte is similar to birds and mammals cells. Eosinophils number is thought to be associated with parasitism defense (10, 11, 12). There are few studies on fish protozoa parasites, therefore, the present study was carried out to investigate the infection with Trypanosoma sp. parasite in some fish in Al-Hammar marsh and to report the relationship between infection and inflammatory immune response through blood smears of infected fishes.

2. Materials and Methods

2.1.Fishes

One hundred ninety (190) of three species fishes Carasobarbus luteus, Liza abu and Aspius vorax were collected from Al- Hamare and Al-Chibayish marsh south of Iraq during October 2018 to November 2019 and brought to lab of Marine Science Center-Basrah University in plastic aquariums.

2.2.Blood sample collection and experiment

Blood samples (1ml blood by using 1 ml a syringe) were collected directly from heart of fishes and then stored in tubes with anticoagulant (EDTA) to determine heamatological values. Blood smears were fixed in 95% methanol for 5 min. and stained with Giemsa stain (13). Photos were taken under compound microscope. Fish blood examined by oil immersion, showed the organisms to be Trypanosoma. The relative number (percentage) of each white blood cell type was calculated based on the number of cells of that type counted divided by the total number of leukocytes counted (100 fields of view had been examined). Scientific photos were taken under compound microscope.

2.3.Statistical analysis

The parasites distributions were described using prevalence and mean intensity (14) of infection were calculated. Statistically, the data were given as Mean. The differences were recorded as significant whenever probability (p) was less than 0.05 SPSS, SPSS (15).

3.Results

Among 70 Carasobarbus luteus there are 29 were infected with the protozoa parasite Trypanosoma sp., the intensity of infection was 0.31, while the infected fish was 7 from 57 in Aspius vorax within intensity of infection 0. 28. No infection was recorded with Trypanosoma sp. in Liza abu fish Table 1. The flagellated parasite Trypanosoma sp. observed in blood of Carasobarbus luteus and Liza abu fishes. It is appears as monomorphic like S and C letters Figure 1.

Table 1. Prevalence and intensity infection with Trypanosoma sp. in some fish of Iraq marshes.

Fish	**Total No. of Fish**		**Total No. of parasites**	**%**	**Mean intensity of infection**
	Exam.	Infec.			
Carasobarbus luteus	70	29	77	41.42	0.31
Aspius vorax	57	7	11	12.28	0.28
Liza abu	63	0	0	0	0

(p>0.05) Methodology

IOP Conf. Series: Earth and Environmental Science **877** (2021) 012011 doi:10.1088/1755-1315/877/1/012011

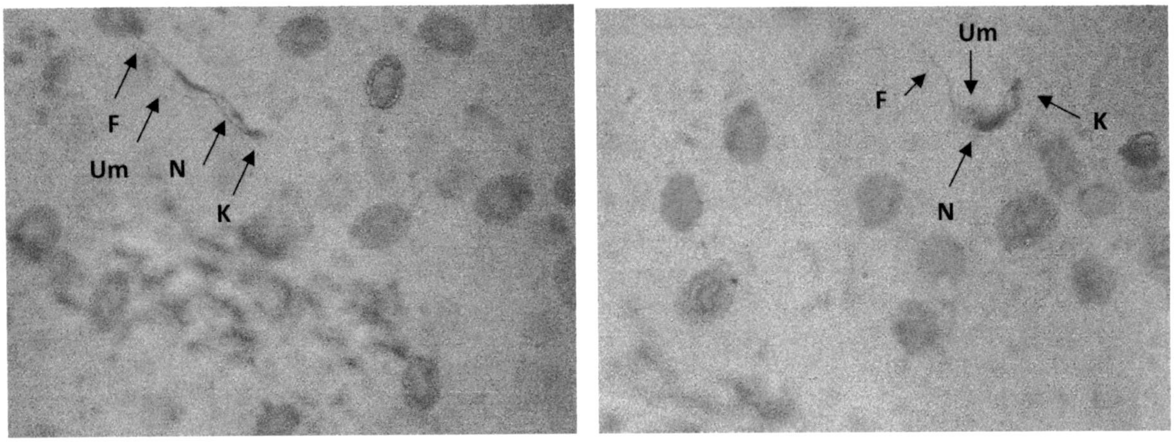

Figure 1. *Trypanosoma* parasites in blood smears, stained with Giemsa's stain. F: flagellum, N: nucleus, K: kinetoplast, UM: undulating membrane.

The inflammatory cells have been seen to increase in number in infected fishes, different count and various types of inflammatory cells were monitored, there are increased in Lymphocytes in the parasitized fishes. The mean values and the prevalence of the leukocytes for infected species of fishes are shown in Table 2. Lymphocytes, monocytes, neutrophils and eosinophils were found in Figure 2. Statistically, there are significant differences ($p<0.05$) among species of infected and different kind of leukocytes in blood smears in infected fish with *Trypanosoma* sp.

Table 2. The number and percentage of different type of leukocytes (WBC) in blood smear in some fish infected with *Trypanosoma* sp. (50 fields were scanned).

Leukocytes types	Fish type: Carasobarbus luteus		Aspius vorax		Control	
	No.	%	No.	%	No.	%
Monocytes	41	25.7	19	24.05	21	32.81
Lymphocytes	92	57.86	51	64.55	30	46.87
Basophils	3	1.88	1	1.26	1	105
eosinophils	12	7.54	6	7.59	7	10.93
Neutrophils	11	6.91	2	2.53	5	7.81

($p<0.05$)

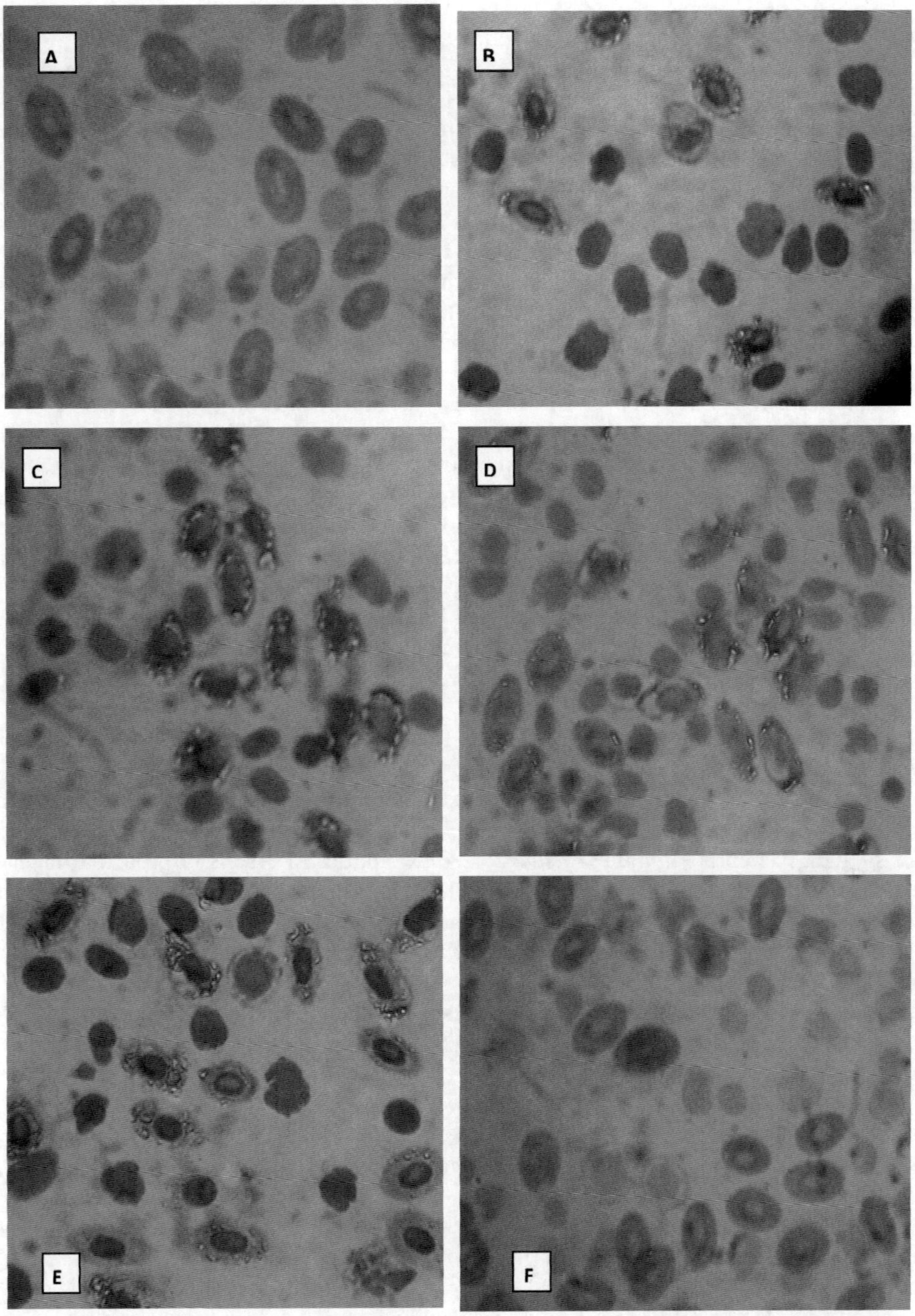

Figure 2. Smears of carp fish blood cells Carasobarbus luteus stained with Giemsa's stain (A: Erythrocyte, large lymphocyte and monocyte, B: Monocytes, C: Large lymphocyte and basophil, D:Neutrophil, monocyte and small lymphocyte, E: Eosinophil, large and small lymphocytes, F: Eosinophil).

4. Discussion

Trypanosomiasis or sleeping sickness disease that caused by protozoa flagellate which live among blood cells called *Trypanosoma* [16]. Fish parasites play an important role in regulating the populations of their hosts [6, 17]. Some of parasites consist of communicable diseases between mammals and fish which called zoonosis [7]. This study shows an increase in the prevalence was accompanied with increase mean intensity of infection and when decreased in prevalence of infection that perform to decreased in mean intensity of infection [18, 3, 12]. This study indicated Avery interesting features that accounts for the restlessness of infected fish and the different types of changes in haematological parameters in fish, theses results are agreement with Ramkrishanan which were quite comparable to those observed in mammals including man [19]. Fish suffer from diseases and parasites like human and other animals, but the automatism of fish immunity is similar to nethermost vertebrates for that if pathogens reach the defenses of fish; Parasites induce inflammation and modification of construct and functionality of fish [20, 21, 22].

The infection with parasites *Trypanosoma* sp. was slight in blood fishes of *Aspius vorax*, 7(12.28%), in this case the slight infection may be due to good healthy of hosts fishes. This result was indicated by Post [23] and Al-Daraji [24]. While the infection with these parasites *Trypanosoma* sp in *Carasobarbus luteus* blood was 29 (41.42%). Many investigations led to the demonstration of *Trypanosoma* sp. Infection in several species of fish. The first record of *Trypanosoma* sp. in fish blood of *Aspius vorax* in Iraq by Al-Daraji [24]. Leukocytes are the first line of defense for immune system; immune defense provide an important symbolizing of cells defense throughout the body [25, 26, 12]. The some more primary routes to evaluating the immune protocol is by surveying changes in the number of fish distribution leukocytes. This study has confirmed the previous findings that leukocytes are directly related with the increase in parasitemia during the *Trypanosoma* infection in fishes [4]. Also, theses results were reported by recent study in rats infected with *Plasmodium* parasites [27]. Immune functionality may be reflected by the lymphocytes and neutrophils counting because of important cell types within the fish immune system, lymphocytes and neutrophils cells have been seen to increase in numeral during infected fish and it constitute the primary immune system.

Fish can develop inflammatory responses that increase the flow of blood to infected areas and deliver erythrocytes that attempt to destroy the pathogens. The inflammation consists of mechanisms involving the immune, nervous and circulatory systems that contribute in the defense system against health and environmental factors [28, 21, 29]. Fish have relatively simple system for that they are used as model for studying vertebrate immune systems [7]. Descriptions of effects of protozoa flagellates on fish are rare and limited observations include the definition of the inflammatory cells associated with host reaction [30]. Eosinophilic granule cells are morphologically and functionally similar to their mammalian counterparts. Eosinophils number is thought to be associated with parasitism defense [6, 11]. Statistically, there is no significant differences $p>0.05$ were noticed in the percentage of prevalence and mean intensity of infection of all species of fish. More studies are needed in this scope of research for knowledge the associating of these granulocytic in immune defense against protozoa infection in fish.

References

[1] Kudo RR 1971 Protozology. 5th ed., Charles. C Thomas *Publ., Spring Field, Illinois,* **1174**

[2] Shahi1 N, Yousuf AR, Rather MI, Ahmad F and Yaseen T 2013 *Open V. J.* **3**(2): 89-95.

[3] Khan R A 1985 Pathogenesis of Trypanosoma murmanensis in marine fish of the northwestern Atlantic following experimental transmission *Can. J. Zool.* **63** 2141–2144

[4] Ahmed M S, Shafiq K, Ali H and Ollevier F 2011 Pathogenic effects associated with Trypanosoma danilewskyi strain FCC 1 infection in juvenile common carp, Cyprinus carpio L *J Anim Plant Sci* **21** 800–806

[5] Food Agriculture Organization (FAO) World fish center 2005 June **6**- 16.

[6] Noga E J 2000 Iowa State University Press/Ames.

[7] Jarallah HM 2018 *LAP /Lambert Academic Publishing*. Pp 117.

[8] Olsen OW 1974 *Univ. Park Press Baltimore,* 562

[9] Reite O B and Evensen Ø 2006 Inflammatory cells of teleostean fish: a review focusing on mast cells/eosinophilic granule cells and rodlet cells *Fish Shellfish Immunol.* **20** 192–208

[10] Burrows A S, Fletcher T C and Manning M J 2001 Haematology of the turbot, Psetta maxima (L.): ultrastructural, cytochemical and morphological properties of peripheral blood leucocytes *J. Appl. Ichthyol.* **17** 77–84

[11] Kiesecker J M 2002 Synergism between trematode infection and pesticide exposure: a link to amphibian limb deformities in nature? *Proc. Natl. Acad. Sci.* **99** 9900–9904

[12] Jarallah H M and Aabadi H I 2012 Use of the recombinant rk39 antigen detection for diagnosis of visceral leishmaniasis in Maysan children, Iraq: with blood parameters *J Env. Bio Sci* **26**

[13] Dacie J V and Lewis S M (1975) Practical haematology *Churchill Livingstone, Edinburgh, London*

[14] Ford SE 1988 *American Fish Society Special Publication*, **18**: 206-224

[15] SPSS A W S 1999 SPSS 10 für Windows

[16] Mhaisen FT 1982 *Basrah University* 227.

[17] Food Agriculture Organization (FAO) 2010 fisheries Department, Rome.

[18] Khamees NR 1983 *M.Sc. thesis, Coll. Agri. Univ. Basrah. Iraq.*

[19] Ramkrishanan SP 1950 *Mosby Wplfe, Iowa State University Press*. Pp: 95-100.

[20] Castro G A 1992 Intestinal physiology in the parasitized host: Integration, disintegration, and reconstruction of systems. *Ann. N. Y. Acad. Sci.* **664** 369–379.

[21] Stosik M, Deptula W and Travnicek M 2001 Studies on the number and ingesting ability of thrombocytes in sick carps (Cyprinus carpio L.) *Vet. Med.* **46** 12–16.

[22] Tavares-Dias M and Moraes F R de 2007 Leukocyte and thrombocyte reference values for channel catfish (Ictalurus punctatus Raf), with an assessment of morphologic, cytochemical, and ultrastructural features *Vet. Clin. Pathol.* **36** 49–54

[23] Post GW 1983 *T.F.H. Publ.* Pp 256.

[24] AL-Daraji SA 1986 *M.Sc. thesis, Coll. Agri. Univ. Basrah Iraq.*

[25] Affonso E G, da Costa Silva E, Tavares-Dias M, de Menezes G C, de Carvalho C S M, Nunes É da S S, Ituassú D R, Roubach R, Ono E A and Fim J D I 2007 Effect of high levels of dietary vitamin C on the blood responses of matrinxã (Brycon amazonicus) *Comp. Biochem. Physiol. Part A Mol. Integr. Physiol.* **147** 383–388

[26] Pavlidis M, Futter W C, Katharios P and Divanach P 2007 Blood cell profile of six Mediterranean mariculture fish species *J. Appl. Ichthyol.* **23** 70–73

[27] Wankhede HJ, Shaikh KM, Nirmale MS, ReddyYR and Dongare VK 2007 **Nat. J. Life. Sci.** 463-466.

[28] Sharkey K A 1992 Substance P and Calcitonin Gene-Related Peptide (CGRP) in Gastrointestinal Inflammation a *Ann. N. Y. Acad. Sci.* **664** 425–442

[29] Passantino L, Cianciotta A, Patruno R, Ribaud M R, Jirillo E and Passantino G F 2005 Do fish thrombocytes play an immunological role? Their cytoenzymatic profiles and function during an accidental piscine candidiasis in aquarium *Immunopharmacol. Immunotoxicol.* **27** 345–356

[30] Dezfuli B S, Lui A, Boldrini P, Pironi F and Giari L 2008 The inflammatory response of fish to helminth parasites *Parasite* **15** 426–433

Ultrafiltration Membrane Technology for Oily Wastewater Treatment

Zahraa N. Mahbouba[1], Dr. Abdulkhalik K. Mahmood [2], Prof Dr. Musa H. Alshammari [1]

[1] Civil Engineering Dept., University of Karbala, Iraq
[2]Civil Technical Dept., Kut Technical Institute, Middle Technical University, Iraq
E-mail: abdulkhalik.kamal@mtu.edu.iq , E-mail: eng.zahra92@gmail.com

Abstract. Oil and gas sectors generate large amounts of oily wastewater, which is called produced water. In which, it contains high concentrations of hazardous organic and inorganic pollutants. This paper attempts to evaluate the performance and quality of using a polyethersulfone ultrafiltration membrane (UFM) to treat the produced water of Al-Ahdab oil field (Wassit, Iraq). 8 rectangular flat sheets of polyethersulfone ultrafiltration membrane were used. The area of each is 60 cm2 and pore size about 15 nm used in the experimental work. Prepared UFM is characterized by determining the surface morphology by scanning electron microscopy (SEM). The result showed that the UFM indicated high removal efficiency in all parameters and especially oil and grease and total suspended solid but in general it still less than the requirement of water reuse. The results showed that, a combination of a conventional treatment method and UFM technology have higher efficiency than using UFM only.

1. Introduction

Oily wastewaters are produced in huge quantities by many different industries with the main contributor being the petrochemical sector. The oil/gas industry generates a large amount of wastewater during the operation of extract the hydrocarbons, processing and refining [1]. Therefore, the water called produced water (PW), which comprises both formation water and injected water. Formation water naturally occurs with the hydrocarbons in the reservoir, while injected water is being withdrawn from an external source and injected in the well to improved oil recovery [2]. As a result of the increasing volume of produced water all over the world, it becomes a big concern on the environment, that's due to a complex composition of toxic organic and inorganic compounds [3]. Organic matter of the produced water includes dispersed oil (hydrocarbon) and non-hydrocarbon organic material [4]. The inorganic compounds exist as dissolved anions and cations, cations such as Na+, K+, Ca2+, Mg+2, Sr+2, Ba+2, Fe+2, anions SO42−, Cl−, CO32−, HCO3 and other substances such as heavy metals (e.g., barium, uranium, cadmium, chromium and lead), and can contain any chemicals added during the drilling and production processes [5]. See table 1.

Table 1: Characteristic of produced water from Al-Ahdab oil field comparing with Typical oil field In the world.

Parameter	AL-Ahdeb oil field	Typical oil field [6]
Density (kg/m^3)	1,100	1,014-1,140
Oil content (mg/L)	10,000	2- 565
pH	5.8	4.3-10
EC (μs/cm)	227,000	4,200-58,600
T.D.S (mg/L)	243,199.47	100-400,000
T.S.S (mg/L)	2,500	1.2-1,000
So_4 (mg/L)	160	77-3,990
Fe (mg/L)	5.84	0.1-100
CL^- (mg/L)	132,937	80-200,000
Na (mg/L)	68,500	132-97,000
Mg (mg/L)	5,687	8-6,000

Table 1 shows the optimum range of physical and chemical analysis for raw produced water from Al-Ahdab oil field. Located between Numania and Kut about 180 km to the south-east of Baghdad/Iraq. Oil and grease are the most important constituent in produced water. It's not referring to a single chemical, but many families of organic chemicals. Another important constituent is the salt content of produced water which it effects on soils, water quality, and ecosystems [7].

Treatment of the produced water conventionally done by physical and chemical processes. Principles of these treatments including (de-oiling, desalination, sand and suspended particles removal, soluble organics removal, gases removal, and NORM removal) [8]. These principles were applied using several methods such as (gravity separation, adsorption, electrochemical process, filtration, and flotation) [9]. These methods may be costly and generate a burden on producing companies. Therefore, membrane technologies can be used for produced water treatment, which considers an efficient and widely used treatment method. Its advantages are decreasing sludge, require a small space, ease of operation, reasonable capital costs and possibility of total recycle water systems. All of these advantages make membrane treatment an economical alternative to traditional treatment methods or addition for it [10,11].

Membranes classified into four types according to pore size. It includes microfiltration (MF), ultrafiltration (UF), Nano filtration (NF) and reverse osmosis (RO). MF has the largest pore size and decreases in UF then NF and RO at the end. Therefore, hydrodynamic resistance for the liquid passage increases from MF to RO [12]. UF is one of these types, Its pore size about (2-100 nm). It discards virus and macromolecular colloids. Unfortunately, almost more dissolved ionic species may pass through the membrane [13]. Also, UF is considered an effective method for produced water and oily wastewater treatment. The efficiency of oil removal, no chemical additives needed, energy cost is low, operate at low transmembrane pressure about 1–30 psi and space requirements small are the main advantage of it use [14].

This study is a practical application for using polymeric ultrafiltration membrane in the treatment of produced water from Al-Ahdab oil field, Wassit, Iraq.

2. Material and Methods

2.1. Proposal System
Figure 1 shows a schematic diagram of the produced water treatment system used in experimental work. It consists of the following parts.

- Treatment system contains: (feed container, mixer, heater, pump, and thermometer).
- Membrane cell.
- Washing system contains: (container, mixer, heater, pump, and thermometer).
- Container (3) used for collecting the treated produced water.

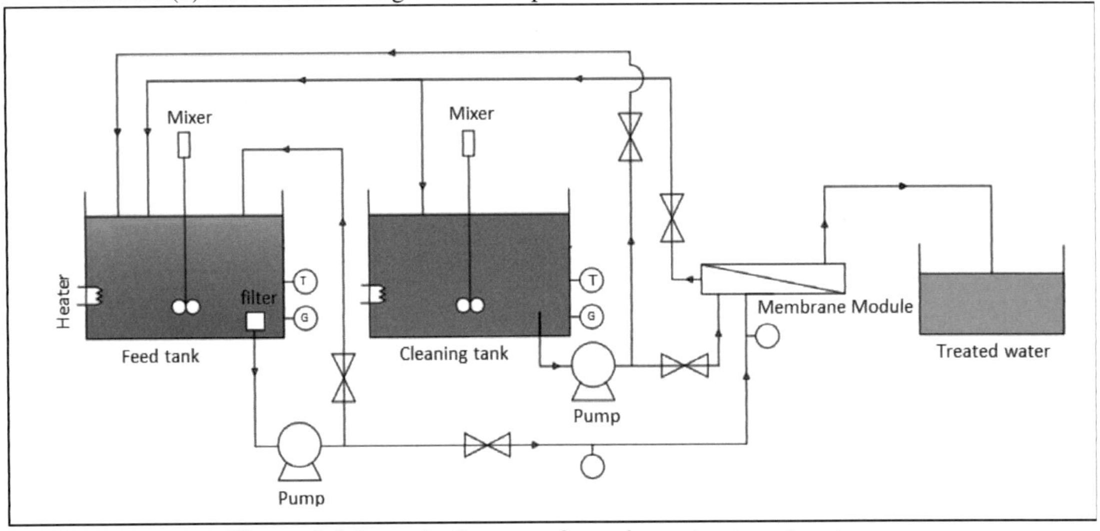

Figure 1 Schematic diagram of membrane treatment system.

2.2. Membrane Preparation
The membranes used in their experiments were (8) rectangular sheet polyethersulfone ultrafiltration with a pore size of 15 nm, and dimensions of 10*6 cm, prepared in the chemical engineering laboratory of the university of technology. The membrane is porous and asymmetric.

The phase inversion casting method used to prepare a polyethersulfone ultrafiltration flat sheet membrane. 15 gm. of polyethersulfone pellets (Sigma–Aldrich, St. Louis, MO, USA) heated for 5 hours at 150 C° to degas. Further, it melted with 85 gm. of DMSO (Dimethyl sulfoxide) (Sigma–Aldrich, St. Louis, MO, USA) for the 15.0 wt. % concentrated solution. 1.0 wt. % of polyvinyl pyrrolidone (PVP) (Sigma–Aldrich, St. Louis, MO, USA) added to the polyethersulfone solution as additive. Figure (2) shows the solution mixing glass flask and all the necessary laboratory tools required to prepare a polymer solution. The mixer was rotating at (500r/min) for 24 hours at a temperature of 50 C° to sure that polyethersulfone was dissolved in DMSO and formed a homogeneous solution. To reduce the number of bubbles in the solution it kept in the dark. Nonwoven polyester fabric fixed onto a glass plate and then the solution casting using a stainless-steel knife. Then lets for 1 minute to get a uniform surface of the support layer. The glass plate was put into a water bath for 1 hour at 23 C°. The membrane was stored in distilled water 18.2 MΩ. cm at 40 C°. Figure 3 shows the casting of polyethersulfone layer. The thickness was approximately 200 μm [15].

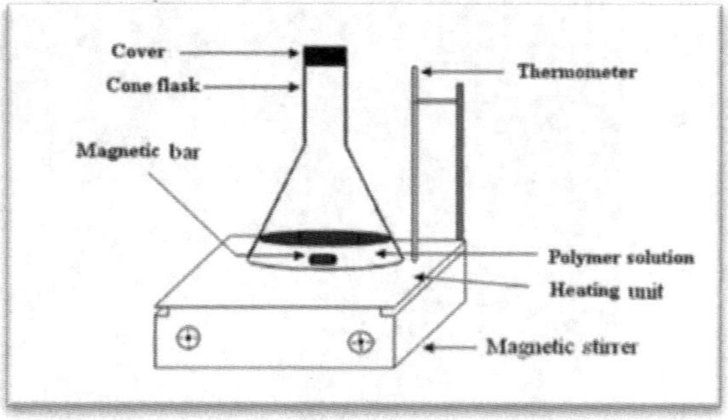

Figure 2. Tools used for preparing the solution of membrane.

Figure 3. Polyethersulfone membrane after casting.

2.3. Scanning Electron Microscopy (SEM)

A scanning Electron Microscopy (SEM) also known as SEM microscopy is a test using an electron beam to scans a sample of the membrane to get a high-resolution superficial image (1 nm) and provide information about pore size, pore size distribution, pore density, roughness. It is efficiently in microanalysis solid inorganic materials. Figure 4 a, b, c and d below show SEM analysis of membrane used in experimental work [16].

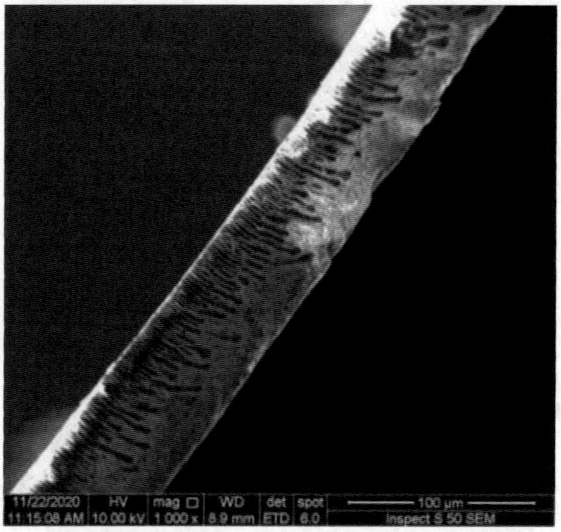

Figure 4 (a). Cross-sectional view of membrane with magnification×1000.

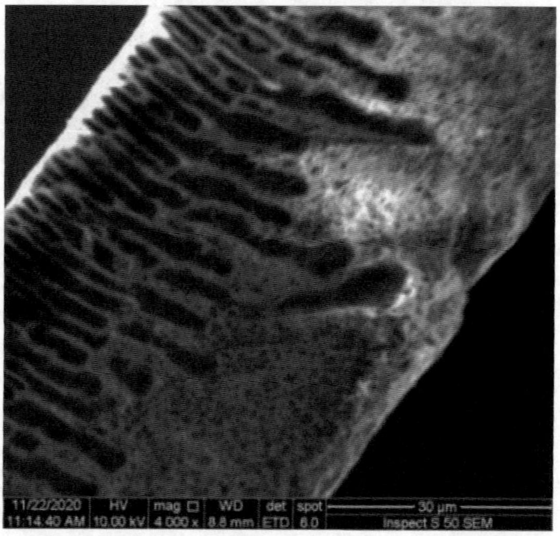

Figure 4 (b). Cross-sectional view of membrane with magnification×4000

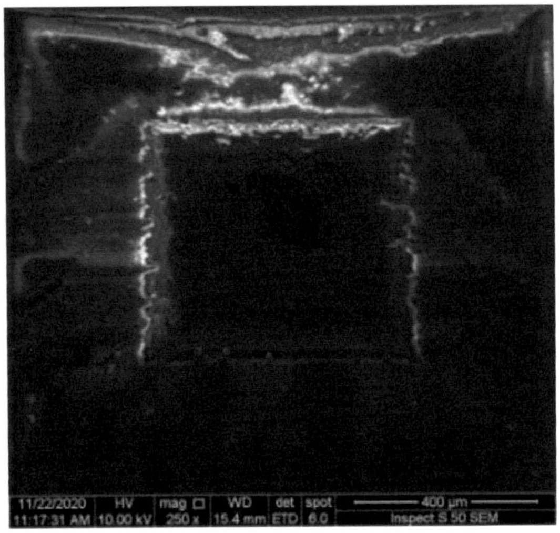

Figure 4 (c). Cross-sectional view of membrane with magnification×250

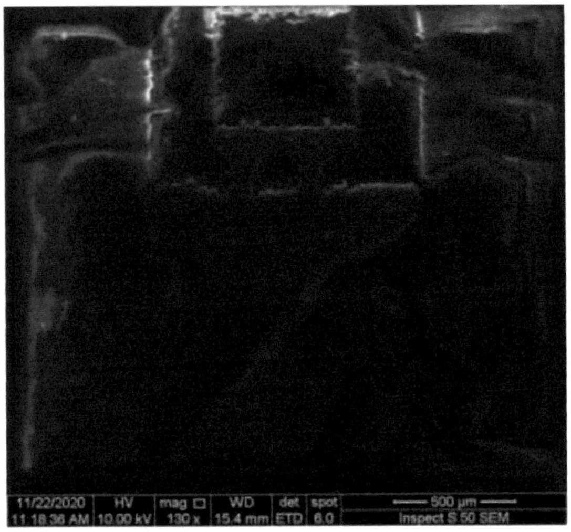

Figure 4 (d). Cross-sectional view of membrane with magnification×130

Figure 4 (a) and (b) showed the cross-sectional view and Figure 4 (c) and (d) showed the top surface of different membranes. From Figure 4 (a) and (b) the membranes have an asymmetric structure and seem to have finger-like pores

2.4. Feed water

Two Type of the samples are taken from Al-Ahdab oil field to use in the experiment:

- Raw produced water.
- Treated water by conventional treatment method in the field [17].

2.5. System operation

The experiments were carried out based on the cross-flow membrane system. The schematic flow diagram of the membrane treatment system shown in Figure 1. The system consists of the membrane treatment system and membrane wash system every one includes (feed container, mixer, heater, pump and thermometer), membrane cell and container (3) used for collecting the treated produced water. The feed water pumped from the container of produced water to the inlet of the membrane cell. The pump used in the system with a maximum flow rate = 480 L/h, temperature = 60 C° and transmembrane pressure = 100 bar. A heater in the feed produced water container used to stabilize the temperature of it about 40 C°. Operating pressure and flow rate controlled by the valves. The membrane cell was epoxy resin rectangular chamber. Flat sheet membrane with an area of 60 cm2 fixed in the cell with sub-layer to save the membrane from damaged under the applied pressure. Variation of the oil pressure gauge about (2 bar) before and after the membrane cell. The surplus feed is recycled back through the by-pass stream to the feed produced water container. The permeation flux from the membrane was measured for 2000 min and collected in the container (3). Five levels of pressure were used (1, 2, 3,4 and 5 bar).

3. Results and Discussion

3.1. Evaluating Of Ultra-Filtration Membrane flux

Effect of TMP On Permeation Flux

Under a constant period of filtration equal to 30 min and when increase TMP gradually from 1 bar to 5 bar amount of permeation flux were increase significantly due to the increase in driving forces across the membrane, as seen in figure 5. However, with increasing TMP, fouling can occur at a faster rate when oil droplets become more compact on the membrane surface and block the pores.

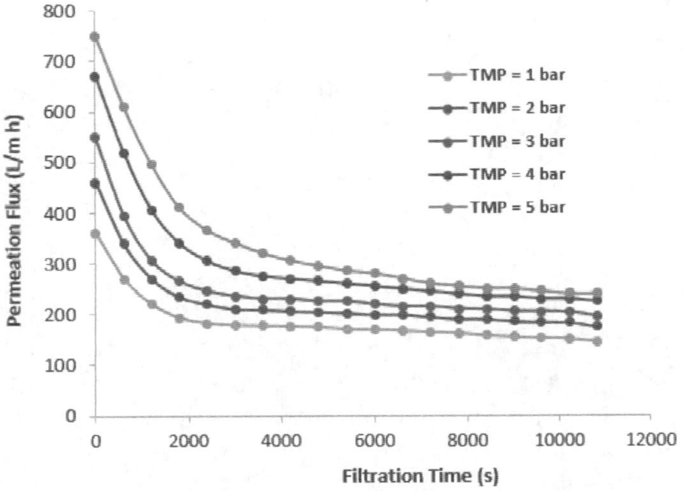

Figure 5. Relationship between TMP and permeation flux.

Effect Of Membrane Fouling On Permeation Flux

Under constant TMP (3 bar) and as time go on, permeation flux decreased gradually, it indicates accumulation of pollutants on the membrane. Due to oil droplet adsorption at the membrane surface and pores. See figure 6.

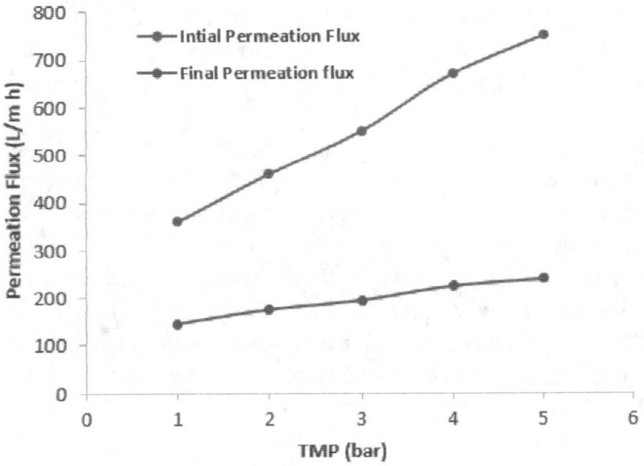

Figure 6, difference between initial and final permeation flux due to fouling of membrane.

3.2. Chemical and Physical Analysis of Treated Water

Table (2) shows the results of the treatment of produced water by ultrafiltration membrane technology and conventional methods.

Table 2: Comparisons of different Methods for Produced Water Treatment

Parameter	Feed water	UFM treatment	Conventional treatment	Conventional +UFM treatment
pH (value)	6.34	6.41	7.32	7.18
E.S (ms/cm)	132	62.66	18.89	5.477
T.D.S (mg/l)	111,400	37,681	11,328	3286
T.S.S (mg/l)	2,500	0.014	36.48	0.0006
Sp.Gr	1.062	1.021	1.02	1.016
Salinity (PPT)	140.5	38.77	10.41	2.39
Ca+2 (mg/l)	5,730	1,896	1,021	278
Mg+2 (mg/l)	1,142	385	208	76.4
Na+1 (mg/l)	11,560	3,648	1,165	266
K+1 (mg/l)	1,860	482	50.85	12.65
So-2 (mg/l)	12,860	4,310	2,260	642
Co3-2 (mg/l)	18,226	6,132	3,398	1,062
Cl-1 (mg/l)	13,432	4,576	1,993.32	498.3
Oil content	644	N	20.75	N
T.O.C (mg/l)	121	0.076	0.082	0.058
Cd (mg/l)	0.02	N	0.012	N
Cu (mg/l)	0.485	0.143	0.329	0.0042
Fe (mg/l)	0.3120	0.2641	0.0866	0.0652
Zn (mg/l)	1.28	1.232	0.1119	0.0668
Cr (mg/l)	0.0061	N	N	N
Ni (mg/l)	0.0542	N	N	N
Pb (mg/l)	0.031	N	0.078	N
COD (mg/l)	380	68.4	87.6	43.7

The results in the above table showed high removal efficiency for almost all parameter. In this issue about 100% removal for the three parameters oil content, total suspended solids and heavy metal (which consider a big concern of the environment and the human body). The results obtained for heavy metal are in the range of authorised limits. Unfortunately, it is not proper for irrigation or other beneficial used because of some dissolved materials.

Furthermore, when we compare a UF method with Current treatment methods in the field, it seems that it is an ineffective treatment. Lower efficiency and membrane fouling problem are the major drawbacks of UF. Therefore, a combination of the two previous methods was result water with high efficiency than any one of them only.

Anion and cation produced when the salts dissolve in water. These ions make up the basis of conductivity in water. The conductivity value is 5400 μS/cm may consider an acceptable range for poultry drinking water that limited to about 6000 μS/cm and 10000 μS/cm for the most livestock. The recommended TDS for beneficial uses such as stock ponds or irrigation about 1000–3000 mg/L. While the result of TDS equal to 3286 mg/l, which mean it is suitable only for crops that can tolerate high salinity such as wheat and barley.

SAR is a term of sodium adsorption ratio used to evaluate the propriety of water in irrigation. The high SAR effects on the soil by replacing the magnesium and calcium by sodium, which decrease its the ability to form stable aggregates. The permissible ratio for SAR is less than 3 while SAR for the resulting water is 1.3 that's mean there is no effect.

COD used as a measure of the oxidation of the organic and inorganic components in the produced water. COD for the raw produced water was decreased from 380 to 68.4 (mg/l) by using UFM only and, 43.7 mg/l for pre-treated produced water and this value within the permissible range.

4. Conclusion

This paper was a study of the treatment of produced water from Al-Ahdab oil field. (8) a rectangular flat sheet of polyethersulfone ultrafiltration membrane (UFM) used in the experimental work. Preparation of the membrane and Scanning electron microscopy (SEM) in the chemical engineering department in University of technology, Baghdad/Iraq.

The results show a combination of conventional treatment and UFM will improve the characteristic of the resulting water to meets the requirement of irrigation crop that tolerate high salinity and drinking for poultry and livestock. But it still not meet the quality of potable water. On the other hand, pre-treatment before membrane technology reduced fouling of the membrane. Which, considers the major setback of these technologies. Therefore, single technology cannot meet suitable effluent characteristics .

5. References

[1] Jiménez S, Mico M, Arnaldos M, Medina F and Contreras S 2018 State of the art of produced water treatment *ChemospHere* **192** 186-208.

[2] Ekins P, Vanner R, and Firebrace J 2007 Zero emissions of oil in water from offshore oil and gas installations economic and environmental implications *Journal of Cleaner Production* **15** 1302-15.

[3] Pichtel J. 2016 Oil and gas production wastewater: soil contamination and pollution prevention *Applied and Environmental Soil Science* **2016.**

[4] Hedar Y 2018 Pollution impact and alternative treatment for produced water *InE3S Web of Conferences* **31** 03004.

[5] Hansen B and Davies S 1994 Review of potential technologies for the removal of dissolved components from produced water *Chemical engineering research & design* **2** 176-88.

[6] Fakhru'l-Razi A, Pendashteh A, Abdullah LC, Biak DR, Madaeni SS and Abidin ZZ 2009 Review of technologies for oil and gas produced water treatment *Journal of hazardous materials* **170** 530-51.

[7] Echchelh A, Hess T and Sakrabani R 2018 Reusing oil and gas produced water for irrigation of food crops in drylands *Agricultural Water Management* **30** 124-34.

[8] Arthur JD, Langhus BG and Patel C 2005 Technical summary of oil & gas produced water treatment technologies *All Consulting, LLC, Tulsa, OK.*

[9] Nonato TC, Alves AA, Sens ML and Dalsasso RL 2018 Produced water from oil-A review of the main treatment technologies *J. Environ. Chem. Toxicol* **2** 23-7.

[10] Igunnu ET and Chen GZ 2014 Produced water treatment technologies *Int. Journal of Low-Carbon Technologies* **9** 157-77.

[11] Xu P, Drewes JE and Heil D 2008 Beneficial use of co-produced water through membrane treatment: technical-economic assessment *Desalination* **225** 139-55.

[12] Madaeni SS 1999 The application of membrane technology for water disinfection W*ater Research* **33** 301-8.

[13] Drewes J, Cath T, Debroux J and Veil J 2009 An Integrated framework for treatment and management of produced water: technical assessment of produced water treatment technologies *Colorado School of Mines* (Colorado) 8-128.

[14] He Y and Jiang ZW 2008 Technology review: Treating oilfield wastewater *Filtration & Separation* **45** 14-6.

[15] Abdulkhalik KM 2015 Fabrication of nanocomposite membrane containing nanomaterial for water purification. PhD Thesis *University of Technology* (Baghdad).

[16] Mohammed A and Abdullah A 2018 Scanning electron microscopy (SEM): A review *Proc. of the 2018 Int. Con. on Hydraulics and Pneumatics—HERVEX* (Băile Govora, Romania) 7-9.
[17] Mahbouba ZN, Abdulkhalik MK and Mussa JH 2021 Characteristics and management of produced water in Al-Ahdab oil field. *IOP Con. Series INTCSET: Materials Science and Engineering* **1094** 012090.

Chemical and biological compatibility of different injection waters with Zubair Formation Water of Upper Sand Member in Zubair Oil Field, South of Iraq

Inass Abdal Razaq Almallah [1*], Fahad Al Najm [1], Zainb Ali Husain [2]

[1] Geology Department, Basrah University, Iraq
[2] Basrah Oil Company, Basrah, Iraq

* Corresponding author: inass.abdal_qader@uobasrah.edu.iq

Abstract. Water injection by water flooding was used to enhance and increase oil production in Zubair oil field, southern Iraq. Physical-chemical and biological analysis of five water samples from different sources were collected to evaluate its compatibility with formation water using biological experiments and chemical compatibility simulation. The results show that injection water is classified weakly acidic-weakly alkaline and saline water, whereas surface water samples are considered weakly acid- weakly alkaline. The total dissolved solids results show brackish types accept for Formation water which classified weakly acid and Brine water. All the studied water samples contain bacteria colonies of Escherichia coli and Coliform expect for one sample, while Sulfate Reducing Bacteria was founded in all studied samples. Mathematical model of chemical compatibility between studied water samples and Zubair Formation water of the scale prediction model show that there are no needs for any inhibition treatments of all scales except for Geothite and Dolomite that should be treated before water injection. The biological compatibility experiments results show Formation damage about (61%) and (69%) in the studied core samples, while Bactria in water injection caused formation damage about (20%) and (51%).

Keywords: Zubair Formation, water flooding, hydrogeochemical modeling, biological compatibility, Iraq.

1.Introduction

Water flooding considered as a secondary production approach that help in push the hydrocarbons towards the produced wells [1] to increase the oil production. Injection surface water or sea water with low salinity or ions were used in order to change the water viscosity to become higher than oil viscosity and increase the prodictivityof oil [2]. Basrah Oil Company used water flooding approach in Zubair Formation after water treatment. The Company used the water of Nahran Ummar aquifer to injection purposes since 2011in order to improving oil production in the Mishrif and Zubair Formations in Zubair Oil Field [3]. The aims of the present study is to evaluate the physical, chemical and biological properties of the injection waters (Shatt Al-Arab River water and two injection station waters: Qarmat Ali river station and Nahran Ummar aquifer) that used in the Zubair Formation and assess them compatibility with water reservoir using chemical compatibility simulation, in addition to determine the effects of water injection on the reservoir properties and Formation damage by chemical and biological compatibility with Zubair Formation water.

2.Geological and stratigraphic setting

Lower Creataceous depositional cycle in Iraq consist from many Formations, Zubair Formation is one of them [4]. Its thickness about 389.33m according to its section in Zubair area well Zu-24 [5]. Figure (1) shows the five members of Zubair Formation where [6]. The petrophysical properties of the upper sand member in Zubair Formation was evaluated by [7] [8], where the results reveal improved in its properties in the northern parts of Zubair oil field, additionally, the best units as good reservoir units were AB, DJ, and LN in Zubair Formation.

3.The general description of the study area

Zubair oil field is located about twenty kilometers far from Basra city between (47° 32' – 47° 45') Latitude and (30° 05' – 30° 42') Longitude Figure (1). The Field has extends over an area of 1170 Km2 [7]. The upper sand member considered as the 3rd pay reservoir which consisting of complex sand and shale where 55meter of its thickness containing oil.

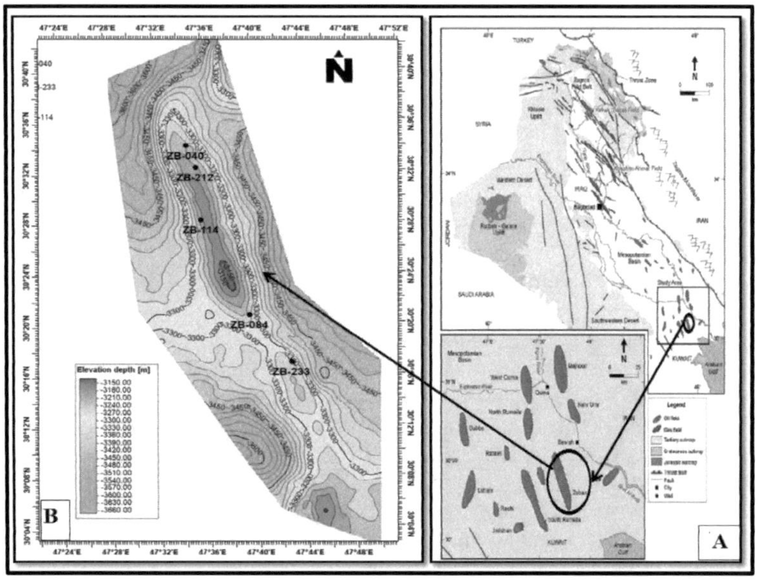

Figure 1. location and Structural maps of upper sandstone member[8].

4.Methodology

In order to achieve the aims of the present study, three surface water samples (S1, S2, and S3) and two injection water samples (I1 and I2) that described in table (1), were collected to analyzed the physical-chemical and biological properties according to the routine approaches in the Central Laboratories of Basrah Oil Company (Figure 2). Four core samples (plugs) (three inch height to one and half inch diameter) were taken from two well core samples that stored in Basrah Oil Company. The permeability in addition to the chemical and biological laboratory compatibilities were tested in the Basrah Oil Company.

Table 1. The core and water samples used in the present study.

Formation	Core well Samples	Shatt Al Arab water samples	Injection water samples

	(plugs)		
Upper sandstone member in Zubair oil field	ZB-40	S1	I1
	ZB-233	S2 S3	I2

Chemical compatibility simulation was done according to the calculation of the mineral saturation indices (SI) using Pitzer equations that represent the activity in high saline water [9]. The hydrogeochimical PHREEQC Version 3 computer program according to the equation below:

$$(SI) = \frac{IAP}{Ksp} \qquad (1)$$

Where SI= saturation index of minerals
IAP = ion activity product
Ksp = solubility product.
The calculation of (SI) could confirms the dissolution or precepitation processes resulting from mixing between surface testing injection waters and Zubair Formation water by assess the minerals scales that dissolved or precepitate[10].
The compatibility between water injection and reservoir rocks were done, firstly by testing the effect of suspended material in water injection on the permeability, and the second experiment evaluate the effect of Bactria in water injection and clay movemente on the permeability in the reservoir rocks when saturated with water injection, theis two experminants were done in Basrah Oil Company labortories. After calculating the porosity and air permeability of the samples(plugs), the experiments were done for two samples of well (ZB-40) at depth (2361.14), and(3334.63) using the processor injection water only, with many stages , in addition to the particle size of suspended materials is measurd in the injection water before and after pumping. Permeability of liquid calculated accordance to Darcy Law as followed :

$$KL = \frac{245V\mu L}{A\Delta Pt} \qquad (2)$$

Where:
KL = permeability of the liquid (Milli Darcy)
μ = the viscosity of the liquid (cp).
V= Sample size (Ml)
L = Sample length (cm).
A = Sample section space (cm^2).
ΔP= Pressure on the sample (PSI).
T= Fluid flow time (Min) .
Graphic forms are drawn to explain the change in relative permeability(Krw) with the amount of injected water

$$Krw = \frac{Kl}{Kwi}\ \% \qquad (3)$$

The following equation is used for computing the Formation damage (Dr%)

$$Dr = 100 - \frac{Klf}{Kwi} * 100 \qquad (4\text{-}5)$$

Klf:finaly permeability , Kwi=Initial water permeability.

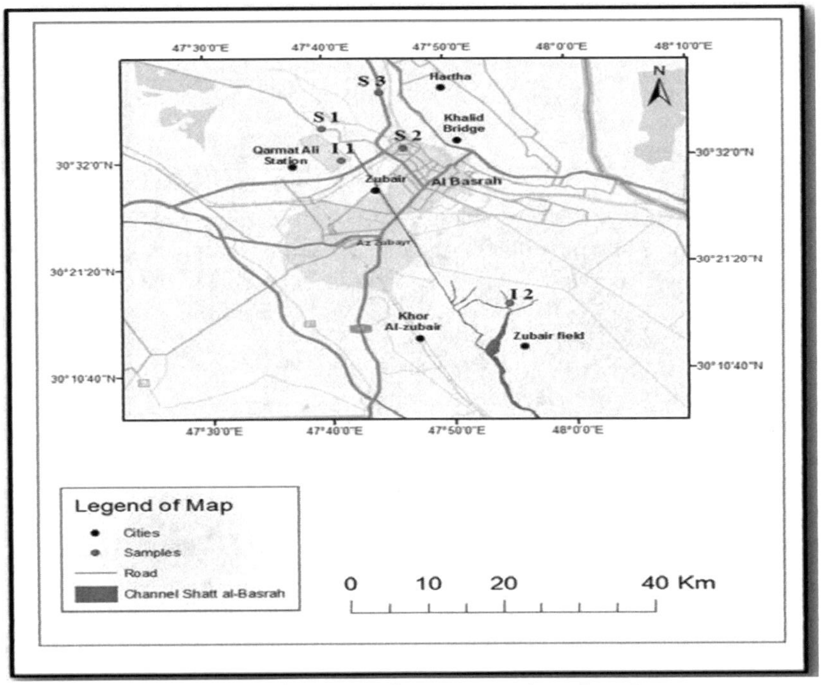

Figure 2. Location map of water samples of study area.

5.Results and Discussion

5.1 Physical properties of Water

The descriptive statistics of physical water samples are shown in (Table 2). PH results show weakly acidic – weakly alkaline according to [11]. Total dissolved solids TDS results show brackish water type according to [12] and [13] for surface water, while the I1 and I2 water samples considered as saline and brine water types.

Total Suspending Salts TSS may found in petroleum a associated waters as organic and inorganic matter [14]. Its range between (11-24mg/l), (53-104mg/l), and (153mg/l) in surface, injection and formation waters respectively. EC(μs/cm) results show the same like TDS. The values of specific gravity were ranged between (1.0202-1.0205), (1.13-1.14), and (1.176) in surface, injection, and formation waters respectively. Turbidity ranged between (3-18.7NTU), (31.9-268 NTU), and (153 NTU) in the same water samples order.

Table 2. The descriptive statistics of the water samples.

Type water	Sample no.	pH	TDS(mg/l)	TSS(mg/l)	EC(μs/c)	SP. Gravity (15°c)	Turbidity NTU
surface	S1	6.4	12590	11	18.87	1.0205	4
	S2	7.2	8450	7	13.44	1.0202	3
	S3	7.2	6420	24	10	1.0203	18.7
	MAX.	7.2	12590	24	18.87	1.0205	18.7
	MIN.	6.4	6420	7	10	1.0202	3
	AVR.	6.9	9153	14	14.1	1.0201	8.56
Injection Water	I1	7.4	15690	53	23.9	1.1351	31.9
	I2	6.2	234460	104	207	1.141	268
	AVR.	6.8	125075	78.5	115.45	1.138	149.95
Formation Water	FW	6.3	120120	85	109	1.1764	153

5.2 Chemical composition of water samples

The Calcium concentration is considered good indicators of erosion of pyroxene; amphibole and feldspar minerals [15]. Calcium Concentration in the Zubair oilfield is the highest between the cations concentrations ranged between 13600-15900 mg/l (Table 3). These high values confirm dissolving of carbonate in upper sandstone member.

Mg^{+2} concentrations results show ranged between (317-477.1 mg/l), (623.4-1805 mg/l), and (458 mg/l) in surface, injection and formation waters, respectively. The value of the Na^+ concentration are varied between (1760-4578 mg/L) in the surface water, and in I1, I2 are varied (172.2-1717.5 mg/L). The K^+ concentrations were ranged between (54.75-121.2 mg/l), (172.2-1717.5 mg/l) and (665 mg/l) in surface, injection and formation waters, respectively. The low solubility of minerals that contain K^- feldspar is the main reason for lowering potassium concentration in groundwater [16]. Chloride concentration in the present water samples is more than the other cations due to its easer of solubility in the water [17]. Sulfates in oil reservoirs have important relationship with Strontium and Barite in oil reservoir that make negative effects on reservoir at it will ploges the pores and reduce the fluid permeability through the reservoir rocks (Al- Atabi, 2009). The SO_4^{2-} concentrations were ranged between (947-1508 mg/l), (323-1669) and (616 mg/l) in surface, injection and formation waters, respectively. The bicarbonate concentration in the present water samples are varied between (232-238 mg/l), (222-232 mg/l), and (227 mg/l) in the same samples order (Table 3).

Table 3. The chemical conposition of studied water samples.

Type of water	Sample no.	Ca^{2+} (mg/l)	Mg^{2+} (mg/l)	Na^+ (mg/l)	K^+ (mg/l)	Cl- (mg/l)	HCO_3^- (mg/l)	CO_3^{2-} (mg/l)	SO_4^{2-} (mg/l)
Surface Water	S1	275.4	477.1	4578	121.2	6262	238	Nil	1508
	S2	213.9	317	2394	87.8	4088	238	Nil	1012
	S3	200	246.55	1760	54.75	2957	232	Nil	947
	MAX.	275.4	477.1	1760	121.2	6262	238	Nil	1502
	MIN.	200	246.55	2394	54.75	2957	232	Nil	947
	AVR.	229.76	346.88	2910.66	87.916	4435.6	236	Nil	1155.6
Injection Water	I1	327	623.4	4400	172.2	8002	222	Nil	1669
	I2	13450	1805	71250	1717.5	139160	232	Nil	323
	AVR.	6888.5	1214.2	37825	944.85	73581	22	Nil	996
Formation water	F	6750	458	38000	665	72853	227	Nil	616

5.3 Chemical Compatibility simulation

Chemical compatibility simulation is important in oil production as good technique to understand the suitability of two water types mixed and evaluate the types of mineral scales that results from mixing process. When testing the compatibility of these water samples with formation water sample, it will considered as compatible if they mixed without precipitation, whereas they considered incompatible waters if precipitate minerals [18]. Before starting the injection operation it is important to know the suitability of water that used in injection purposes with Formation water [19]. The scales known in the oil reservoirs like calcium carbonate, Calcium sulfates, Gypsum, Anhydrite, Hemihydrate, Barium and Strontium sulfates are the main scales[20].

6.The model of chemical compatibility results

6.1 Chemical compatibility results of water samples

Figures (3,4,5,6 and 7) show the results of simulation mixing ratios between Formation water and present water samples S1, S2, S3, I1, and I2. The chemical compatibility simulation model of Zubair Formation water with Shat AL-Arab and injection water samples that results show Anhydrite, Gypsum, Siderite, Strontianite, and Witherite scales in all mixing ratio were under saturation due to low saturation index and indication no need for them inhibition treatments. Geothite scale was needed permanent inhibition because of high saturation index and precipitation will be possible. All the scales results show no need for treatments due to low saturation index accept for dolomite which needs periodic inhibition [10].

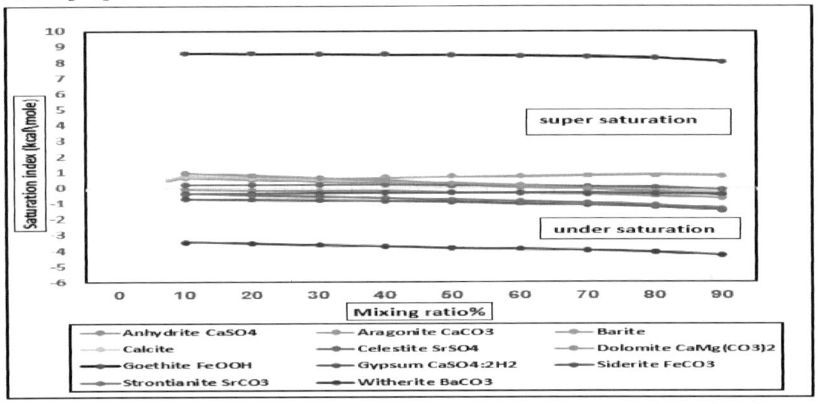

Figure 3. Mineral Scales results from mixing in (S1).

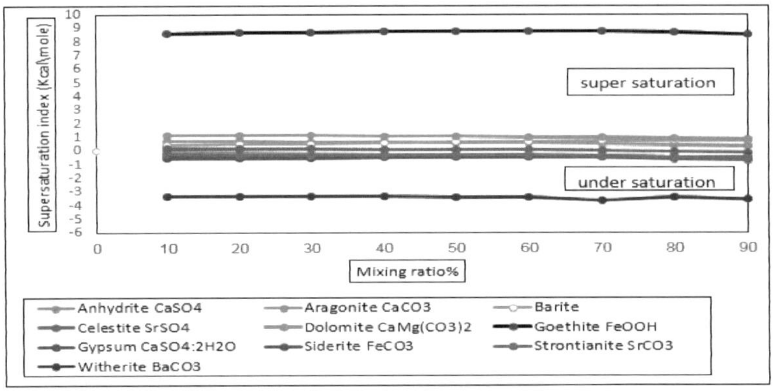

Figure 4. Mineral Scales results from mixing in (S2).

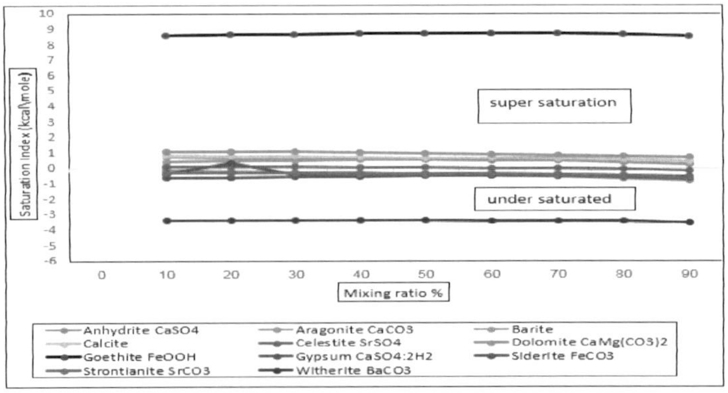

Figure 5. Mineral Scales results from mixing in (S3).

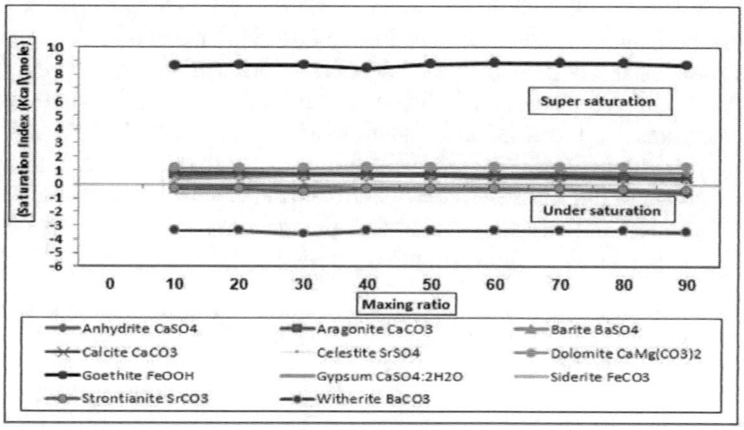

Figure 6. Mineral Scales results from mixing in (I1).

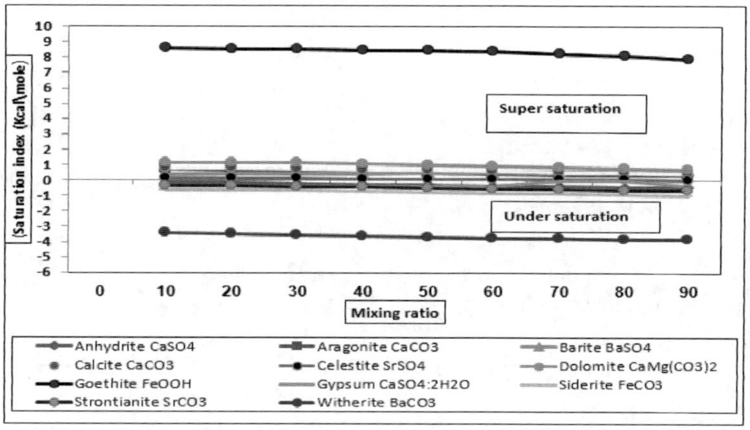

Figure 7. Mineral Scales results from mixing in (I2).

7.Compatibility between water injection and reservoir rock

It will be able from these experiments results to obtain the size of the suspended material of the injection water that allowed which have no effect on water flow rate or cause any Formation damage . In addition, we will be able to assess the permeability due to the effect of clay minerals swelling and movement with the increasing the saturation rates by increasing the injection water rates with time of testing sample.

7.1 Experiments the effect of suspended material in water injection on permeability

Table (4), Figures (8)and (9) show the results of the experiment and the graphic forms of permeability of the air (Ka)after and befor water injection, The suspended material in the injection water has a great impact on reducing the flow of water in the rocks by reducing the permeability. The percentage of reducing is depending on the shape, size and amount of the suspended material [21]. Figures (8) and (9) show the results of the suspended materials effect experiment on reservoir permeability, where firstly through pumping injection water of two samples without a filter, the relative permeability will reduced from (0.92) to (0.18) and from (0.88) to (0.20) in the first and second samples respectively .Through the second stage of the experiment when pumping with filtered injection water in the opposite direction. The relative permeability result became 0.66 and 0.63 in the first and second sample respectively because this enabled the samples to remove the suspended materials that could not influence inside it because of its large diameters [22].The relative permeability of water would be decreased when pumping the filtered injection water in the normal direction where it became 0.35 in the first sample and 0.27 in the second sample. Figure (10) and (11) shows the results of the diameter of the suspended material in the treated water injection before and after injection in the sample (ZB - 40) (S1) respectively, where the primary diameter was (3.3-16.7 µm), after pumping the diameter became (2.3µm) indicates that the suspended material larger than the (2.3µm) remained in the pores of reservoir rock reducing the permeability. At well sample (ZB-40) (S2) the diameter of the suspended material in the treated water injection was (2.3-5.2 µm). After injection, the diameter was (2.3 µm) due to remained the suspended particles larger than (2.3 µm) to remain in the pores of the rock reservoir causing reducing in permeability as shown in the (Figures 12 and 13).

Table 4 : The result of compatibility between samples of reservier rock and injection water .

Well No.	**ZB-40**	**ZB-40**
Samples No.	1	2
Depth (m)	3261.14	3334.63
K-air befor injection (md)	452.7	366.9
K- air after injection and drying (md)	426.3	348.6
Q-befor injection %	21.2	21.7
Q-after injection %	21.2	21.9
V.p befor injection (cc)	12.42	12.3
V.p after injection (cc)	12.46	12.36
Kw origin with filter (md)	356.94	298.35
KL(with out filter) in first handred	(328.26)	(261.16)
KL(with out filter) in last handred	(64.28)	(59.67)
Krw(with out filter) in first handred	0.92	0.88
Krw(with out filter) in last handred	0.18	0.20
KL(with filter inverse the sample) (md)	240.30 in 400cc 236.65 in 500 cc	162.42 in 400cc 189.34 in 500cc
Krw(with filter inverse the sample)	0.67 in 400cc 0.66 in 500 cc	0.54 in 400cc 0.63 in 500 cc
KL(with filter in normal direction) (md)	132.83 in 400cc 125 .99 in 500 cc	84.72 in 44cc 82.01 in 500 cc
Krw(with filter in normal direction)	o.37 in 400cc 0.35 in 500 cc	o.28 in 400cc 0.27 in 500 cc
Dr%	61	69
Partical dimeter –befor injection (µm)	3.3-16.7	2.3-12.9
Partical dimeter -after injection (µm)	2.3	2.3

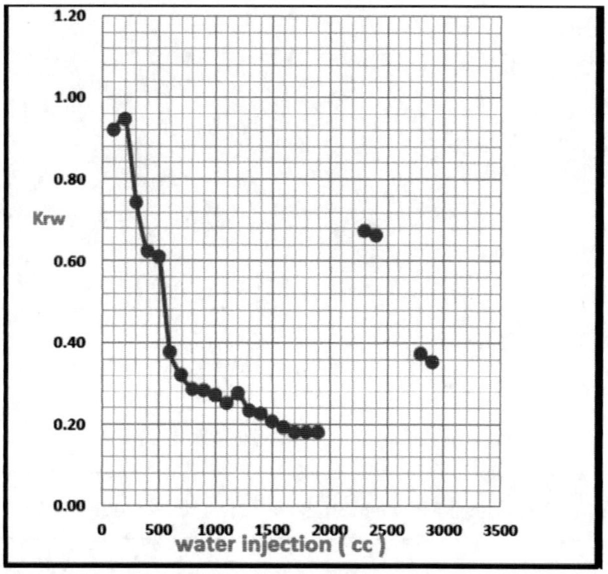

Figure 8. The change in relative permeability (Krw) with the amount of injected water (ZB-40,S1) depth (3261.14m).

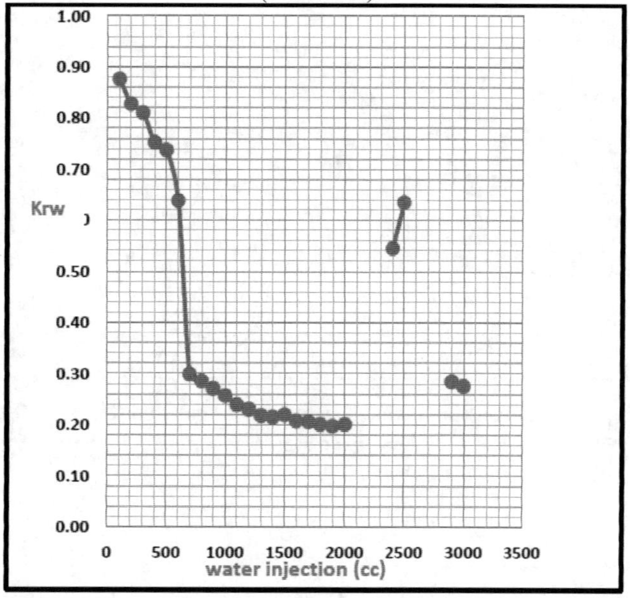

Figure 9. The change in relative permeability (Krw) with the amount of injected water (ZB-40 ,S2) depth(3334.63m).

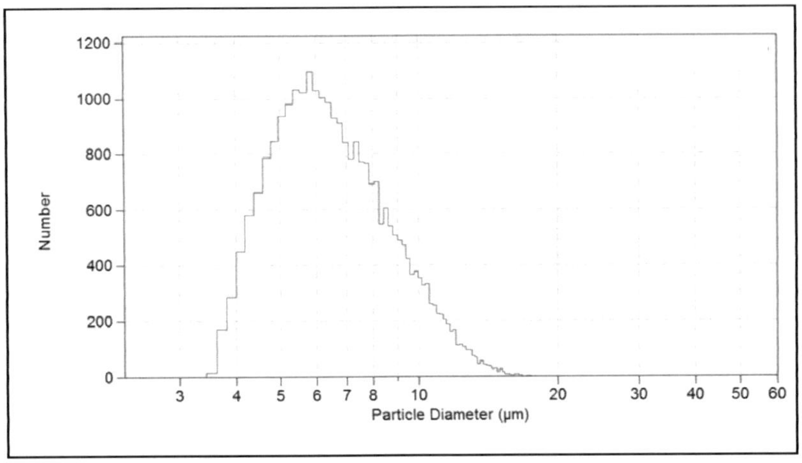

Figure 10. Particle of suspended material of water injected before injected it in (ZB-40, S-1) depth (3261.14m).

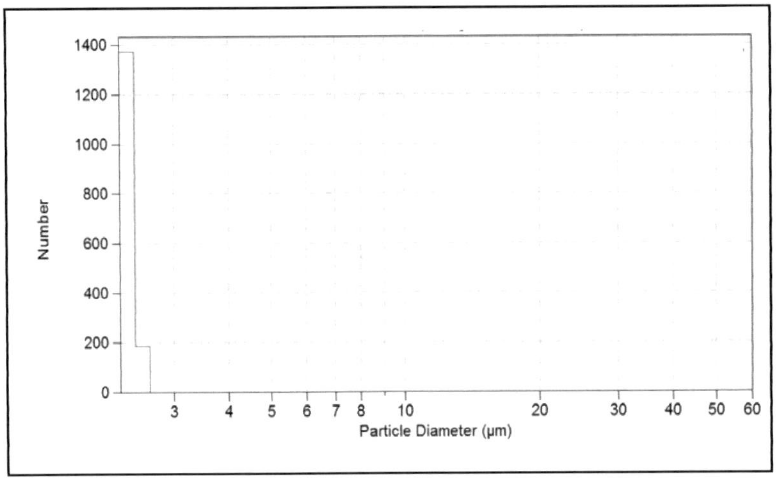

Figure 11. Particle of suspended material of water injected after injected it in(ZB-40, S-1) depth (3261.14m).

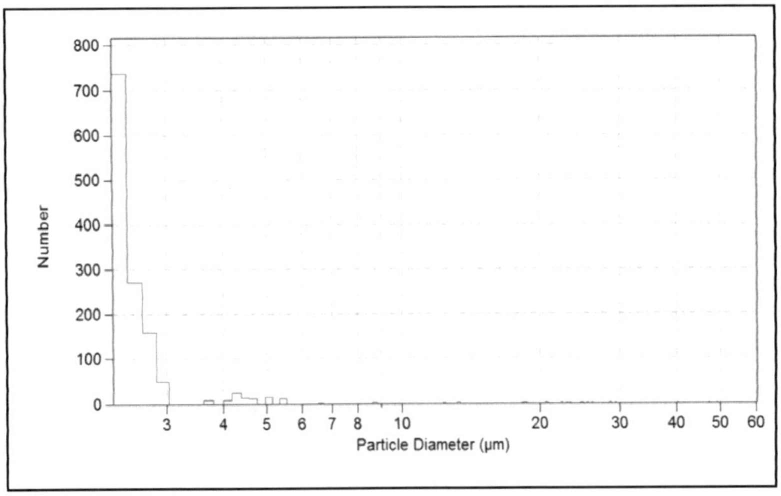

Figure 12. Particle of suspended material of water injected before injected it in(ZB-40 ,S2) depth (3334.63m).

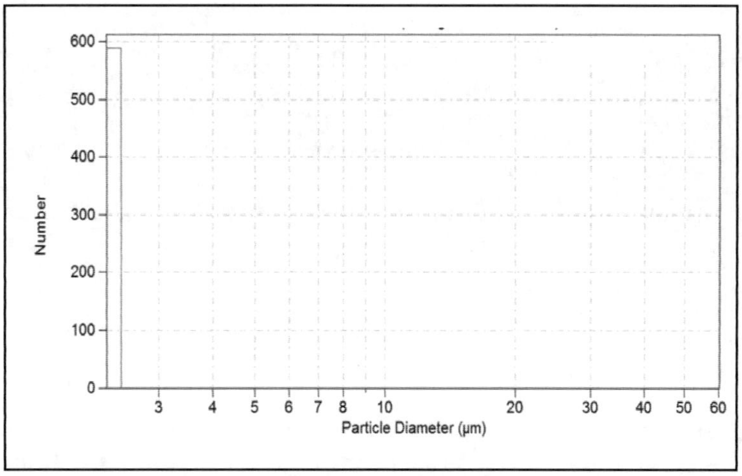

Figure 13. Particle of suspended material of water injected after injected it (ZB-40 ,S2) depth (3334.63m).

The air permeability of the two samples were damaged and did not return to the original value prior to injection, whereas the porosity remained without any affected. The above results revealed reducing in permeability and an increase in the Formation damage, where the damage results was (61%) and (69%) in the first and second sample respectively Table (4).

7.2 The effect of Bactria on rock reservoir Experiment

After determining the air permeability and porosity of the samples (plugs) two Experiments were done for two samples of well (233) in Zubair oil Formation (ZB-233) at depth (3323.45) and (3361.29) using the brine water and processor injection water with the many stages.The result effect of Bactria in water injection on the permeability after the injection of water in those rocks shows in table (5). Draw the results to explain the change in relative permeability (Krw) with time during the amounts of injected water.

$$Krw=\frac{Kl}{Kwo}\% \qquad (4\text{-}6)$$

Take samples of injected water and examine it a biologically to determine the growth of bacteria inside the rocks.Depending on the results shown in table (5), we found that the permeability have been reduced with continuing pumping water treatment from (453.57 md) to (332 md) and from (99.06 md) to (67.01 md) for the first and second samples respectivel . Whereas in pumping brine water the permeability was decreased from (392.15 md) to (356.17 md) and from (62.92md) to (47.61 md) for the first and second samples respectively. The relative permeability in pumping water treatment with filter was reduced from (0.72) to (0.53) in the first sample and from (2.20- 1.49) in the second sample (Figure 14 and 15).

Table 5. The result effect of Bactria in water injection on the permeability after the injection of water in those rocks.

Well No.	ZB-233	ZB-233
Samples No.	1	2
Depth(m)	3323.45	3361.29
K_{air} before saturation (md)	680.1	155.9
K_{air} after injection and drying (md)	477.1	52.0
Φ before injection%	20.9	16.3

IOP Conf. Series: Earth and Environmental Science **877** (2021) 012013 doi:10.1088/1755-1315/877/1/012013

Φ after injection %	11	9.4
$K_{W\ origin}$ (with filter) (md)	628.53	45.11
K_L water treatment)(with filter) (md)	453.57 → 332 Continuing test	99.06 → 67.01 Continuing test
K r w water treatment)(with filter)	0 .530 → 0.72 Continuing test	2.20 → 1.49 Continuing test
K_L brine (with filter) (md)	392.51 → 356.17 Continuing test	62.92 → 47.61 Continuing test
K_{rW} brine (with filter)	0.57 → 0.62 Continuing test	1.06 → 1.39 Continuing test

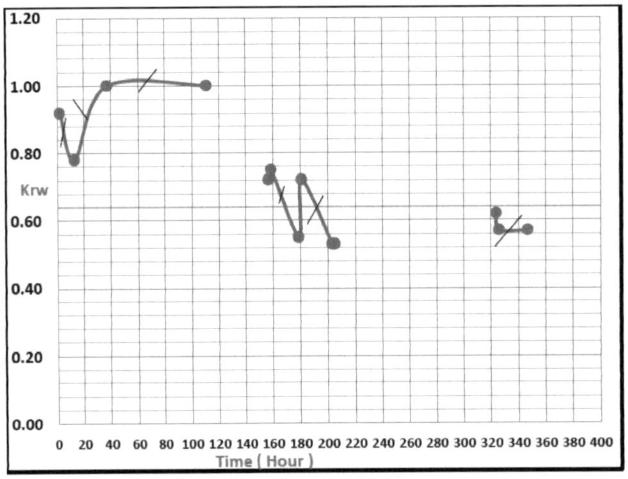

Figure 14. The change in relative permeability(Krw) with time during the amounts of injected water (ZB-233,S1)depth(3323.45m).

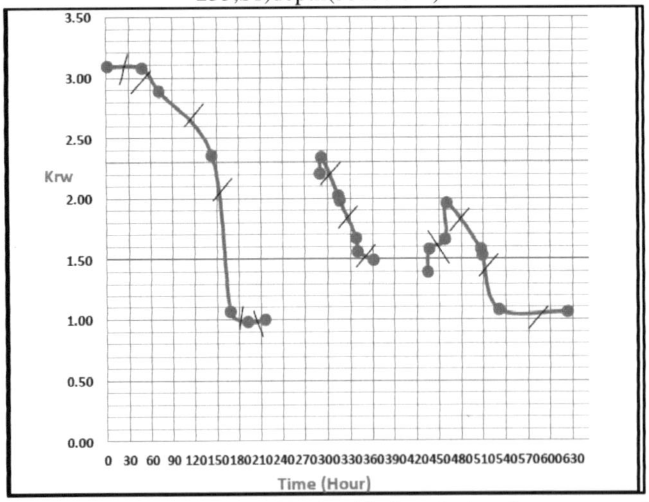

Figure 15. The change in relative permeability(Krw) with time during the amounts of injected water (ZB-233,S2) depth(3361.29m).

The results of porosity and air permeability of both samples reveal a decrease in its value without returning to their original value before injection due to the movement of clay minerals within the samples and growth in Bacteria [23]. The growth Bacteria is shown in the tables 6 and 7. The results show that there are damage in water relative permeability about 20 % and 51 % in the first and second tested samples respectively.

Table 6. Growth of Bacteria in injection water samples before and after injection it in reservoir rock (ZB-233,S1) depth (3323.45).

Type of Bacteria(col/ml)	Treated water Before injection	Treated water After injection Day -1	Treated water After injection Day -2	Treated water After injection Day -3
General aerobic Bacteria	2135	2500	2315	1600
An aerobic Bacteria	1915	Uncountable(very high)	1095	Uncountable(very high)
Sulfate Reducing Bacteria	460	336	280	Uncountable(very high)

Table 7. Growth of Bacteria in injection water samples before and after injection it in reservoir rock(ZB-233). depth (3361.29)m .

Type of Bacteria(col/ml)	Treated water Before injection	Treated water After injection Day -1	Treated water After injection Day -2	Treated water After injection Day -3
General aerobic Bacteria	2135	1305	1000	1020
An aerobic Bacteria	1915	Uncountable(very high)	Uncountable(very high)	Uncountable(very high)
Sulfate Reducing Bacteria	350	Uncountable(very high)	Uncountable(very high)	Uncountable(very high)

8 Conclusions

1- Water flooding by water injection was used for enhancement and increasing oil production in Zubair oil field.
2- According to pH values the reservoir Formation indicating weakly acid that have a positive effect on reservoir petro-physical characteristics (porosity and permeability), while the pH values of all surface and injection water samples reveal weakly acid - weakly alkaline .
3- The surface and injection water samples were considered as brackish and saline water type respectively according to TDS. Sodium and Chlorite are the predominant ions in all water samples .The hydrochemical formula show the cations in order: $Na^{+} > Mg^{2+} > Ca^{2+} > K^{+}$ and of anions in order $Cl^{-} > SO_4^{2-} > HCO_3^{-}$.
4- Chemical compatibility simulation model of Zubair Formation water with Shat AL-Arab and injection water samples results show no need for scales inhibition treatments except for Geothite scale which needed permanent inhibition because of high saturation index and precipitation will be possible. In addition to periodic treatment for Dolomite scale.
5- 61% and 69% Formation damage in the first and second core samples respectively were found through the experiment of affected of suspended material in water injection on permeability that reduced even with using filter during the experiment.
6- During the experiment of the effect of Bacteria in water injection and the effect of the clay mineral of core samples on Formation permeability, the results show that permeability will reduce from (453.57 md) to (332 md) in the first sample and (99.06 md) to (67.01 md) in the second sample. During pumping brine water the permeability did not increase because of movement of clay minerals and growth of Bacteria. Finally the Formation damage was approximately 20% and 51% at the first and second core samples respectively.

9.References

[1] Jreou G N 2012 Increasing of oil field productivity by implementation of re-entry horizontal injection well, case study *Iraqi J. Chem. Pet. Eng.* **13** 17–34

[2] Morrow N and Buckley J 2011 Improved oil recovery by low-salinity waterflooding *J. Pet. Technol.* **63** 106–12

[3] ZFOD, Zubair Field Operating Division 2017 Annual Report of Zubair Formation. (internal Report unpublished) .

[4] Al-Sayyab A 1989 Geology of petroleum 472

[5] Owen R M S and Nasr S N 1958 Stratigraphy of the Kuwait-Basra Area: Middle East

[6] Al-Naqib K M 1967 Geology of the Arabian peninsula *USGS Prof. Pap. No*

[7] Zainab M H A L, Almallah I A and Al-Najm F M 2019 Petrophysical properties evaluation using well logging of the upper sand member of Zubair Formation in Zubair oil Field, Southern Iraq. *Basrah J. Sci.* **37** 457–480

[8] Al-Ameri T K and Batten D J 1997 Palynomorph and palynofacies indications of age, depositional environments and source potential for hydrocarbons: Lower Cretaceous Zubair Formation, southern Iraq *Cretac. Res.* **18** 789–797.

[9] Pabalan R T and Pitzer K S 1991 Mineral solubilities in electrolyte solutions *Act. Coefficients Electrolyte Solut.* **2**.

[10] Ghalib H B and Almallah I A R 2017 Scaling simulation resulting from mixing predicted model between Mishrif formation water and different waters injection in Basrah oil field, southern Iraq *Model. Earth Syst. Environ.* **3** 1557–1569

[11] Komatina M 2004 *Medical geology: effects of geological environments on human health* (Elsevier).

[12] Freeze RA, Cherry JA 1982 Groundwater. Prentice-Hall, Englewood Cliffs.

[13] Todd D K and Mays L W 2004 *Groundwater hydrology* (John Wiley & Sons)

[14] Rainwater F H and Thatcher L L 1960 *Methods for collection and analysis of water samples* (US Government Printing Office)

[15] Pradhan B and Pirasteh S 2011 Hydro-chemical analysis of the ground water of the basaltic catchments: upper Bhatsai region, Maharastra *Open Hydrol. J.* **5** 51-57.

[16] Al-Yasiri A A 2000 A Study of geochemistry and hydrogeochemistry of the Upper Sandstone Member–Zubair Formation in South Rumaila Field/Sothern Iraq *Unpubl. Thesis, MS ((In Arab. Coll. Sci. Univ. Basra, p. 77*

[17] Mason, B. and Moore, C.B., 1982:Principle of Geochemistry, 4th ed., John Wiely and Sons. INC. New York. 344p.

[18] Henkel H O 1953 Surface and underground disposal of chemical wastes at Victoria, Texas *Sewage Ind. Waste.* 1044–1049.

[19] Collins, A.G., and Bryant, R.S., 1985: Potential MEOR environmental hazard. Unpublished Report. NIPER-25. 13p.

[20] Moghadasi J, Jamialahmadi M, Müller-Steinhagen H, Sharif A, Ghalambor A, Izadpanah M R and Motaie E 2003 Scale formation in Iranian oil reservoir and production equipment during water injection *International Symposium on Oilfield Scale* (Society of Petroleum Engineers)

[21] SOC, (South Oil Company), -Basra-1980: central laboratory department, Compatibility the injection water of Qarmat –Ali with the reservoir water ,and reservoir rock at Zubair oil field.(internal Report unpublished).

[22] SOC, (South Oil Company),-Basra-1983: central laboratory department, decrees the permeability during injection water ,causes and treatment . (internal Report unpublished).

[23] SOC, (South Oil Company),-Basra -2012: An updated study of reservoir water analysis. Internal report. Baghdad. Iraq (Unpublished)

IOP Conf. Series: Earth and Environmental Science **877** (2021) 012014 doi:10.1088/1755-1315/877/1/012014

Constructing the optimal power algorithm for AS/RS systems using multi-mobile robots

H M Hameed[1*], A T Rashid[2], M T Rashid[3], K A Al Amry[4]

[1] Teacher Assistant, Electrical Engineering Department, University of Basrah, Iraq
[2] Professor Assistant, Electrical Engineering Department, University of Basrah, Iraq
[3]Professor, Electrical Engineering Department, University of Basrah, Iraq
[4]Professor Assistant, Electrical Engineering Department, University of Basrah, Iraq

* Corresponding author: hananhameed@uobasrah.edu.iq

Abstract. For any mobile system there is especially advantageous from a business perspective to develop energy-saving techniques that also extend to existing production processes. Therefore, looking for ways to enhance the energy efficiency of robot operations to maximize energy consumption efficiency is of considerable significance, a route-planning issue that refers to finding the shortest path to meet the predetermined goal location in a certain complex environment. So, one of the energy saving methods for a multi-mobile robot environment is to find the optimal path for a mobile robot that can improve power consumption. In this paper, an optimal power algorithm for "automatic storage and retrieval system" using multi-mobile robots is introduced based on efficient motion planning among a group of multi-mobile robots that gives a significant improvement in the level of energy consumption. Energy mechanism can be achieved using electrical power quantities on real robots, models or analytical equations based on robots' physical model. The simulation results indicate that the algorithm enhanced power consumption efficiency.

1. Introduction

Automatic storage and retrieval system (AS/RSs) had been represented as controlled through a computer system for automatic storage and retrieval from specialist storage locations by using an automatic system to handle operations and eliminate human intervention. To do this, a program system is used to work with automated retrieval and recovery mechanisms [1]. The robot is driven by electric motors powered by a battery and the average speed throughout the mission remains constant. Minimum power formulation may be an interesting approach for mobile robots, particularly in applications where battery weight is a critical issue. Minimum power problems can be difficult to solve with classical methods because they involve both the nonlinear dynamics of the robot and a combination of

Restrictions. One way to reduce energy for robots focuses on motion planning to reduce the force of motion. Therefore, it is imperative to consider a specific solution when the battery is running out of power. One such solution is to replace a robot that has run out of battery with another robot with a full battery. In robot competition, energy and time are the main factors that determine the performance Hence the importance of multi-mobile robots within the same environment, noting that there is another benefit from using multiple mobile robots, and among those benefits such as increasing speed,

[5]To whom any correspondence should be addressed.
increasing stability to resist system failure and many features that increase the performance efficiency of a multi mobile robots versus single robotAS [2].

In general, a multi-Robot System may typically be defined as a group of robots running in the same environment. Usually robotic devices, though, can range from basic sensing, data collection and analysis to sophisticated human-like robots capable of communicating in very complex ways with the environment. [3]. From a classical point of view, mobile robots operate on a fixed energy supply, provided by some batteries. The constraint of the energy supply demanded human interface to recharge the batteries where mobile robots are devices distinguished by their ability to traverse diverse environments. There is a need for planning the motion of robots in order to get the optimal path that leads to optimal power consumption. Thus, mobile robots route preparation is a central feature of mobile robots science. The key aim of route planning is to find a path free from collisions, through an obstacle environment, from a specified starting place to a target destination while following certain criteria [4]. Using either static or dynamic environments, the object storage structures may be configured. In the static environment, only the robot can move while all other objects in this environment have no movement; while in the dynamic environment, several different entities (such as moving vehicles and moving paths) have the potential to move. Overall, the dynamic environment has more concerns than static route planning as the robot uses sensory input to sense the actual location and trajectory of moving objects in the environment [5]. In today's times, computational optimizations and simulations are the two most used methods to modelling and learning AS/RS. Researchers such as Lerher, Šraml, Borovišek, Potrč [6], Yang, Miao, Xue, and Qin [7] studied optimization of AS / RS operation using 'modern optimization methods' to improve warehouse system performance. De Koster, Le-Duc and Roodbergen [8], Gu, Goetschalckx, McGinnis [9], Baker, Canessa (2009) [10] present a general study of warehouse engineering and control methods suggested by De Koster, Le-Duc and Yugang [11] Kuo, Krishnamurthy, and Malmborg are Mathematical Concept Models for Efficient Autonomous Vehicle Action (AVS / RS) centered around the AS / RS Loader [12], and Manzini, Gamberi and Regattieri provide a multi-parameter hierarchical model for the product storage device to the picker with product customization On the basis of category [13]. Fukunari, Malmborg, Hur, Nam, Yin, Rau, Dooly, Lee et al. Presented comprehensive studies. Fukunari and Malmborg coined the term "interleaving" related to coupling storage and retrieval parameters to generate DC loop loops in the same cycle [14]. Hor and Nam proposed stochastic methods for estimating the AS / RS performance of a unit load [15], Yin and Rau analyzed the dynamic range of the AS / RS-based sequencing rules of the unit load [16].Twin-shuttle AS/RS [17], Dooly and Lee raised a shift-based sequencing problem. In an overall state-of-the-art analysis, authors Roodbergen and Vis [18] found that AS/RS research could best leverage the power of simulation by comparing various designs. While taking more aspects of architecture into account, especially in conjunction with numerous control policies. A small amount of energy is taken in by mobile robots. To increase the running time of the robot before the batteries are discharged, it is necessary to improve the energy efficiency of the robot. Modelling the energy consumption of a robot has many benefits. One of the major drawbacks of preparing a challenge is sufficient energy. If the estimated energy demand for executing the task exceeds the energy available, an energy-conscious robot will assign the task of recharging the batteries the highest priority.

In addition, it will be necessary for the robot to identify if the mission can be performed with the minimal energy available to the robot [19]. Robotic cell energy optimization also focuses on optimizing the individual robot paths (local optimization) with regard to the physical shortcomings of robots and the obstacles to be avoided [20]. Currently, there is a growing interest in new innovative methods and tools for optimizing robotic energy consumption. From a business viewpoint, it is Particularly beneficial to develop energy-saving strategies that also apply to current manufacturing processes and are responsible for minimal change possibilities. The remainder of the paper is organized as follows:

- The system description will be given in a way of how the mobile robot can choose the optimal path for power optimization,

- Finding the optimal path that leads to a power optimization of the mobile robot will be mentioned with the mathematical model via a formulated equation that calculates the shortest path finding problem, then the algorithm can be used to seek the features of the optimal path and then choose the shortest path of the mobile robot that provides the optimal power consumption,
- Analysis results with the *drawings* are mentioned,
- Some conclusions are mentioned.

2. System description

This study deals with the design of a system to store and retrieve a group of materials organized in rows and columns within an area. Its intersections reflect locations of storage. As shown in Figure 1, the title of any content is defined by the row and column number inside it, also a single delivery and receiving port. A method of storing and retrieving a batch of orders (materials) by using a number of robots are suggested here.

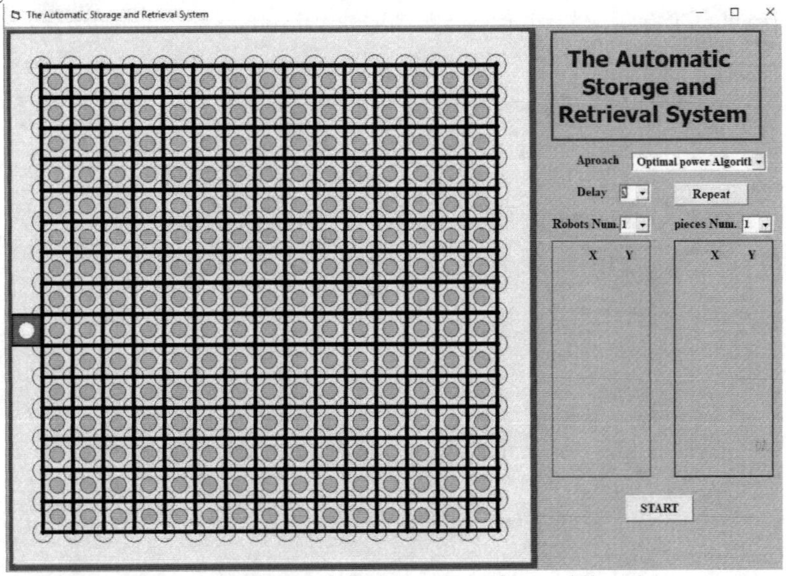

Figure 1. Environment for mobile robot.

3. Optimal power algorithm

First, the notion of this algorithm is to determine the distance of order individually from the point of delivery and, second, to determine the distance of each robot from the delivery point, and then calculate all delivery possibilities for each robot for all orders at coming and at going to the delivery point. All probabilities are calculated using a combination algorithm by calculating the sum of the paths for the delivery of all orders. In each probability distribution of orders to robots, the shortest path is taken as the best case for all probabilities, which represents the shortest distance traveled for all robots, this represents the least power consumed to implement the task (Optimal power) and complete it in the least time.

The steps for implementing the combination algorithm by following the following:

Point 1: Distance guess between the orders and the point received: the following equation is used

Disrcv(i)=Abs(Xor(i)-Xrec) + Abs(Yor(i)-Yrec)

Point 2: The distance between the orders and all the robots is estimated by:

Disrbt(I,j)=Abs(Xor(i)-Xrob(j) + Abs(Yor(i)-Yrob(j))

Point 3: The computation of the length of total paths, use one of the permutation algorithms.

The distance between each robot is combined with the duration of the routes back and forth between the order point and the receiving point.

Select the total distance for each chance.

IOP Conf. Series: Earth and Environmental Science **877** (2021) 012014 doi:10.1088/1755-1315/877/1/012014

For this work from the first robot to the last one in each probability

$$Total(I) = Total(I) + 2 * disrcv(i) + disrb(j)$$
$$Next\ i$$

Point 4: choose the minimum distance. Form all the last probabilities choose the minimum distance to be the optimal power spent

Point 5: move the robot as to the arranged sequence.

Point 6: Collision avoidance among robots when they move:

1. Arrange the robots in a decreasing manner according to their distances from the received place.
2. Through the movement of the robots, each robot checks the distances with the other robots.
3. If the distance is less than the length of one cell in the environment, then the robot with the higher order must stop its movement.
4. The last step is repeated until the distance with the other robot is greater than the size of one cell.

The last two steps are repeated until all robots complete their tasks.

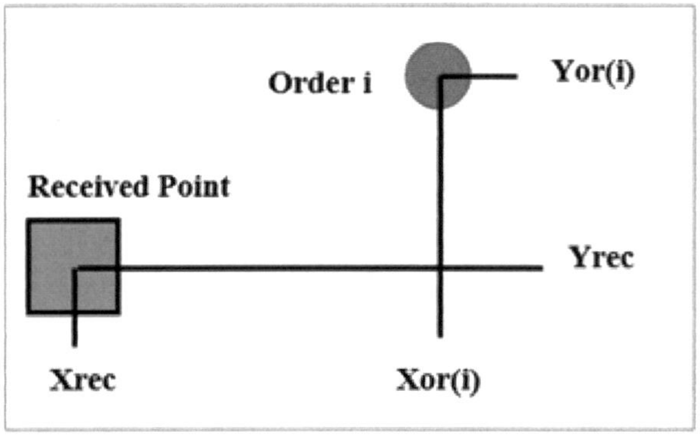

Figure 2. "The assortment axis of the request and the receiver point".

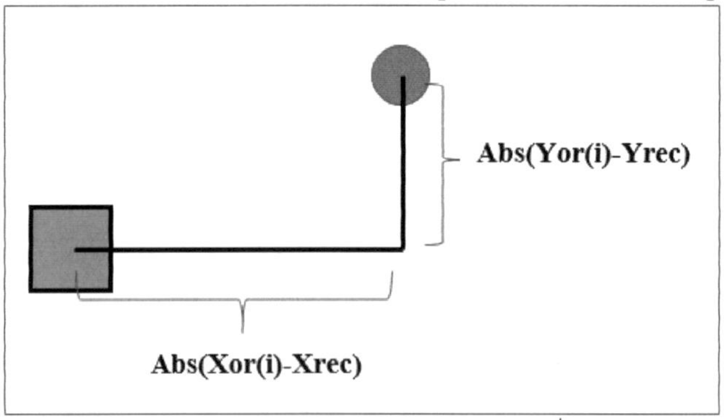

Figure 3. "The length between any request i and the receiver".

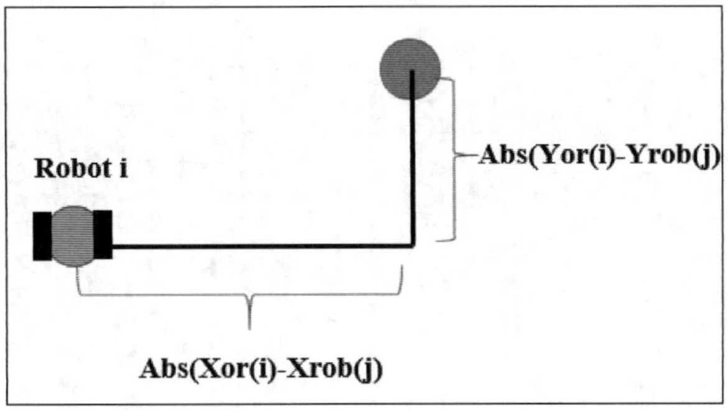

Figure 4. "The long between request I and the robot j".

4. Simulation results

The proposed algorithm (optimal power algorithm) is simulated to analyse the process of storage and recovery using Visual Basic Program Language and to evaluate using laptop contain an Intel Core i5 in the Windows environment. To simulate the performance of the proposed algorithm and compare it with a previously developed algorithm, two methods were implemented. The first is to apply the optimal power algorithm, assuming that robots will not collide when performing tasks. The second is to apply the optimal route of the previous algorithm to a collision between robots. In order to determine the length of the path to complete the mission, the second simulations are performed as well as for different order goals and for both algorithms. The first simulation is to determine the optimum time to complete the task, while the second simulation is to evaluate the average distance of the path during the job execution. In this section, two simulation experiments are added. The first is to use the optimal Path algorithm to measure the arrival time average to complete the task, while the second simulation is performed by measuring the arrival time average during the execution of the tasks using the optimal Power algorithm. The following success metrics have been used for a measurable study of these approaches:

1-Storage and retrieval completion path (L): The usage of this metric is for calculating the total path to the order number for the storage and retrieval process.

2- Storage and recovery completion time (t): The usage of this metric is for calculate the number of orders by the percentage of completion time.

The simulation screenshots at different time stages using the optimal path algorithm are represented in Figure 5 (a)-(e) (No collision among the robots). Figure 6 (a)-(e) shows the simulation screenshots using the optimum power algorithm at different time levels (With collision among the robots). The primary objective of this simulation is to demonstrate the relationship between the orders number and the store and retrieval period of achievement for both circumstances.

A comparison of the arrival time average for both the optimal power algorithm and the optimal path algorithm is shown in Figure 7. The average of the total paths for an optimal power algorithm is shown in figure 8 (a to d). a) 2 robots b) 3 robots c) 4 robots d) 5 robots. The average total path for the optimal path algorithm is shown in figure 9. A) 2 robots b) 3 robots c) 4 robots d) 5 robots. A comparison of the average total path for both the Optimal Path Algorithm and the Optimal Power Algorithm is shown in figure 10.

IOP Conf. Series: Earth and Environmental Science **877** (2021) 012014 doi:10.1088/1755-1315/877/1/012014

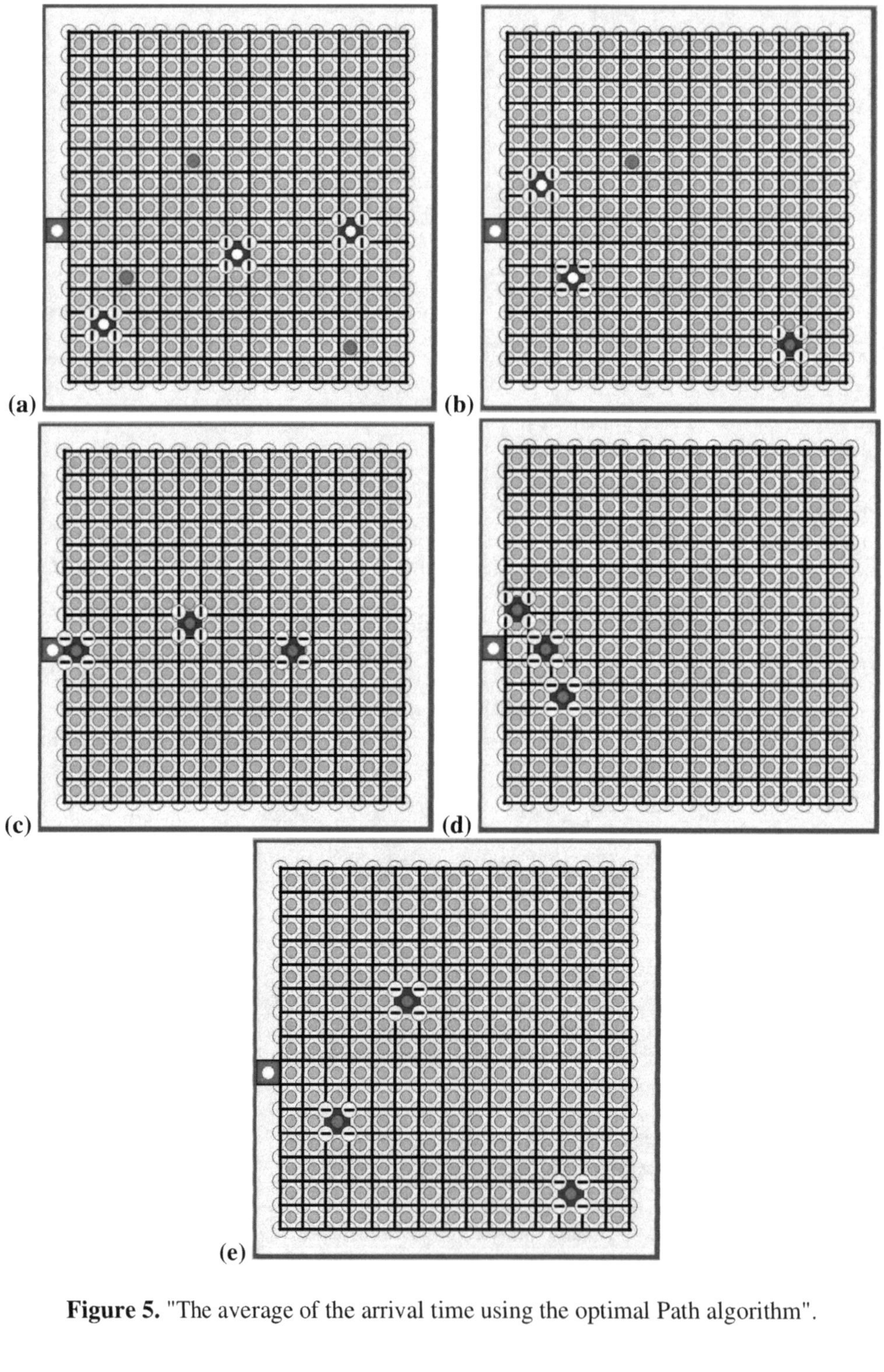

Figure 5. "The average of the arrival time using the optimal Path algorithm".

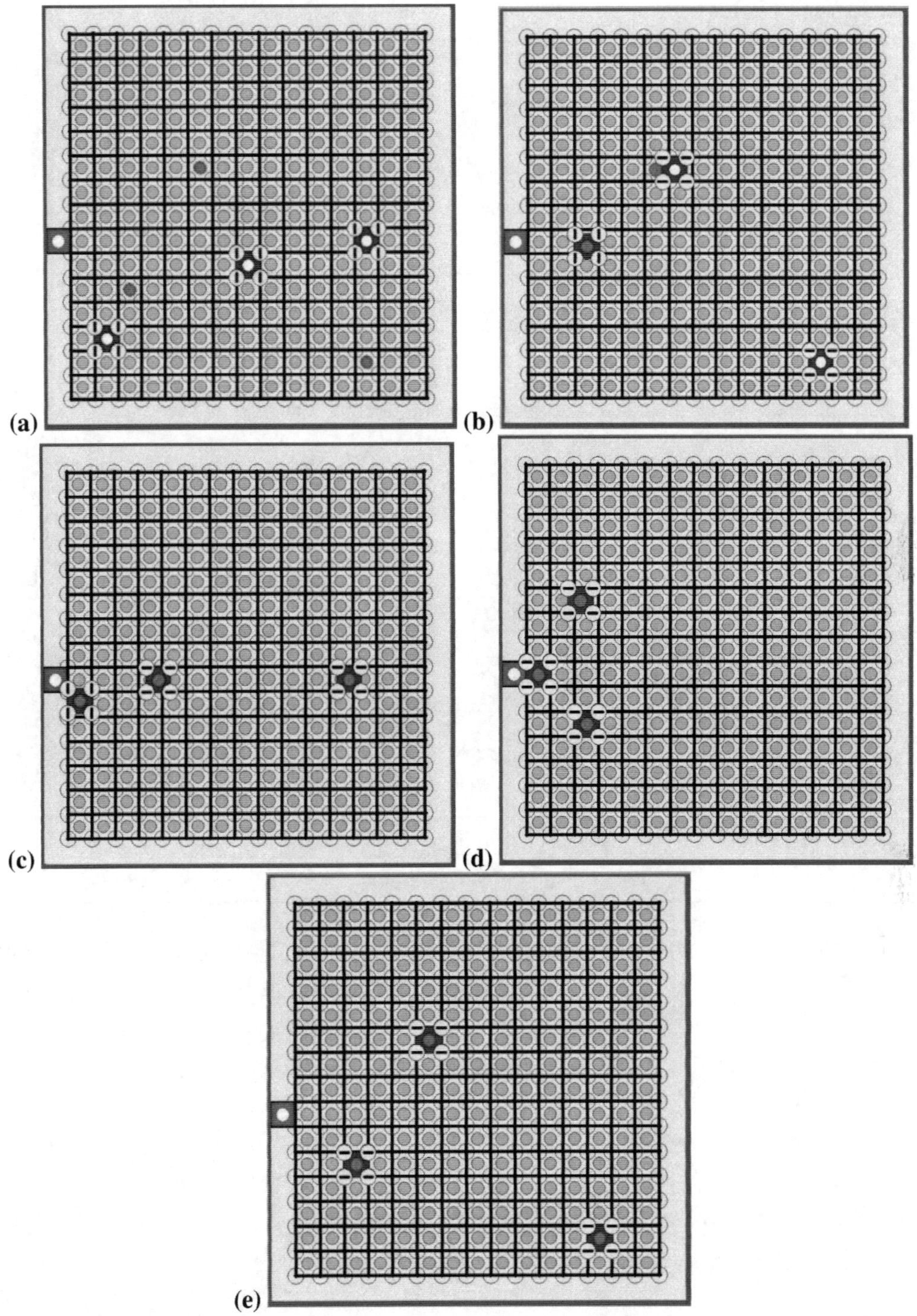

(a) (b) (c) (d) (e)

Figure 6. "The average of the arrival time using the optimal Power algorithm".

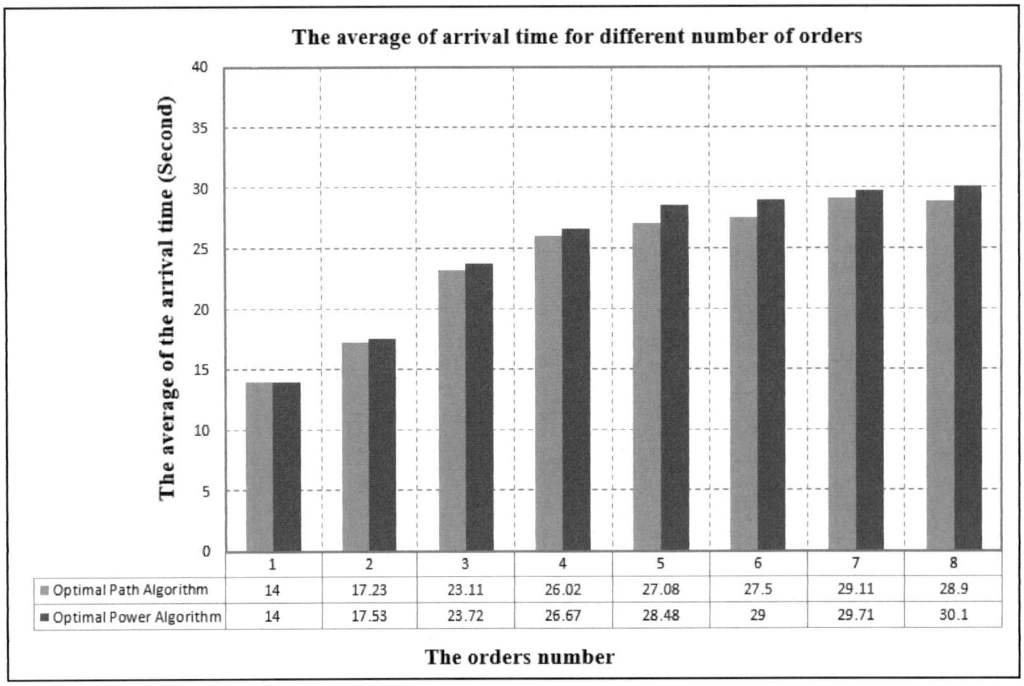

Figure 7. Comparison of the estimated time of arrival for both the optimal power algorithm and optimal path algorithm.

IOP Conf. Series: Earth and Environmental Science **877** (2021) 012014 doi:10.1088/1755-1315/877/1/012014

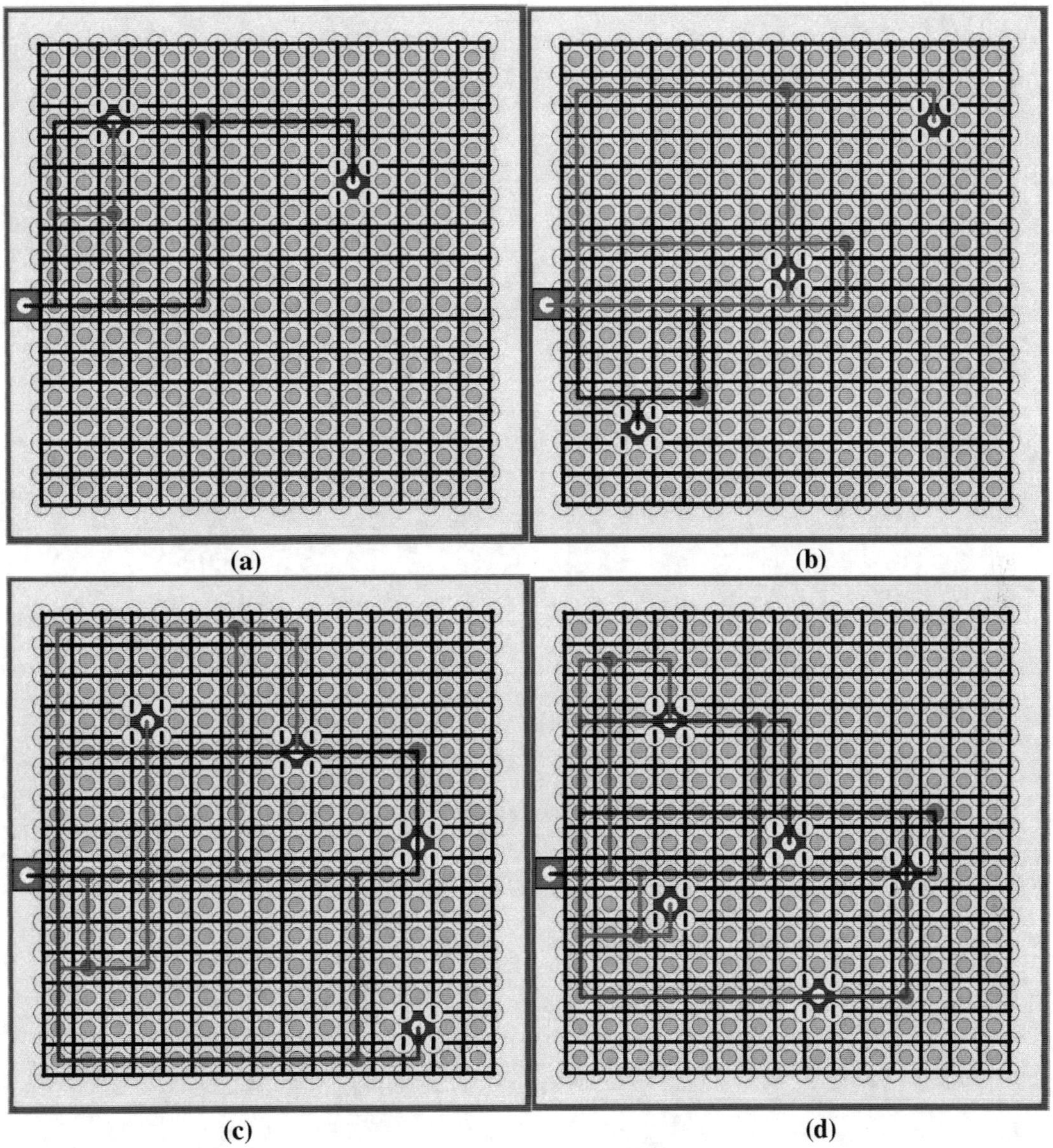

Figure 8. The mean of the total path for the optimal power algorithm. a) Two robots b) Three robots c) Four robots d)Five robots.

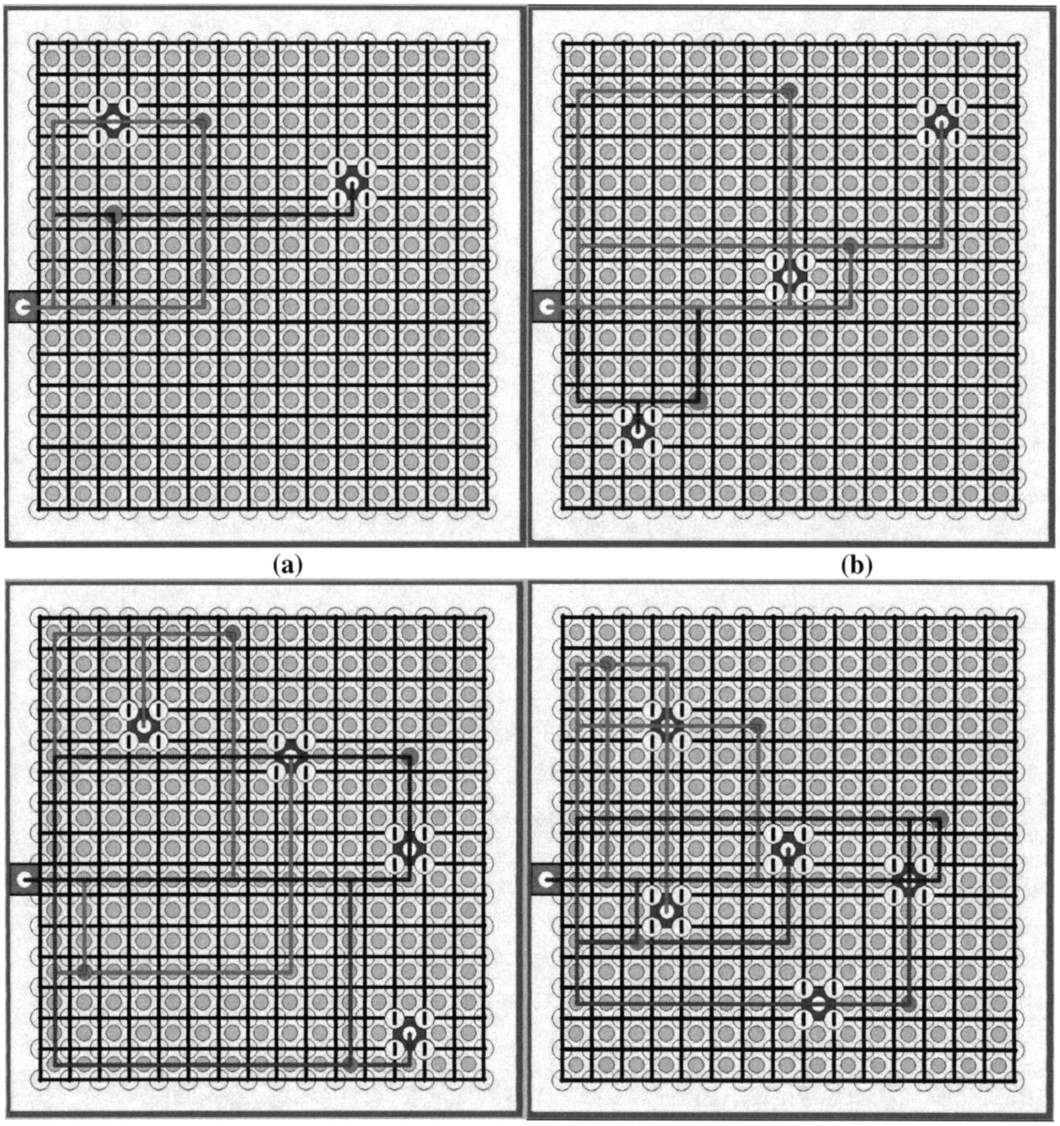

(a) (b) (c) (d)

Figure 9. Total average route for the optimal Path algorithm's total path. a) Two robots b) Three robots c) Four robots d) Five robots.

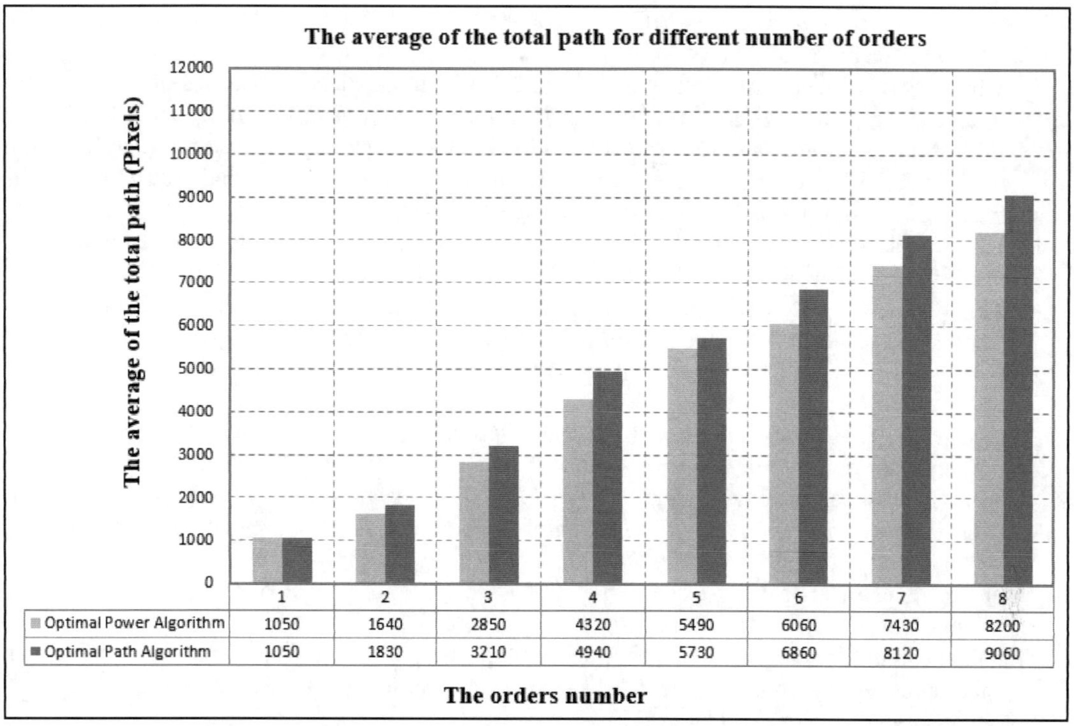

	1	2	3	4	5	6	7	8
Optimal Power Algorithm	1050	1640	2850	4320	5490	6060	7430	8200
Optimal Path Algorithm	1050	1830	3210	4940	5730	6860	8120	9060

Figure 10. Comparison of the overall route average for both the Optimal Path Algorithm and the Optimal Power algorithm.

5. Conclusions

In this paper, using several numbers of orders with many numbers of mobile robots, a storage and recovery method (optimal power algorithm) in the static setting is suggested. In an area with various numbers (1 to 8) of requests and robots, simulation findings are applied. The findings shows that as increasing the orders number, the time proficient among the robots for the two the optimal path and optimal control increases too. The optimal power algorithm offers the best output in an achieved time than the optimal line one. From the results, it is found that as the number of request raise the achievement path increases, also. In other words, the optimal power algorithm leads to a reduction in time arrival, which means faster, and a decrease in the overall direction, which means faster with low power consumption.

References

[1] Hameed H M, Rashid A T and Al Amry K 2019 Survey of automatic storage and retrieval system *International Journal of Computer Applications* **177** 0975 – 8887

[2] Hameed H M, , Rashid A T and Al Amry K 2020 Automatic storage and retrieval system using a single mobile robot *Proc. of the 2nd Int. Conf. on Electrical, Communication and Computer Engineering (ICECCE)* 14-15 April, Istanbul, Turkey

[3] Hameed H M, , Rashid A T and Al Amry K 2020 Automatic Storage and Retrieval System using the Optimal Path Algorithm *The 3rd Scientific Conf. of Electrical and Electronic Engineering Researches (SCEEER)* 15-16 June, Basrah, Iraq DOI: 10.37917/ijeee.sceeer.3rd.18

[4] Hussein A, Mostafa H, Badrel-din M, Sultan O and Khamis A 2012 Metaheuristic optimization approach to mobile robot path planning *2012 international conference on Engineering and Technology ICET* Cairo, Egypt 10-11Oct. 2012 October, DOI: 10.1109/ICEngTechnol.2012.6396150

[5] Rashid A T, Ali F R. and Rashid O T 2018 Software implementation of a static store system using the digital differential analyzer algorithm *international conference on Engineering Technology*

and their application IICETA 8-9 May 2018 Al-Najaf, Iraq DOI: 10.1109/IICETA.2018.8458087

[6] Lerher T, Šraml M, Borovinšek M and Potrč I 2012 Multi-objective optimization of automated storage and retrieval systems *11th Int. Conf. on Industrial Logistics*, ICIL, Zadar, Croatia

[7] Yang P, Miao L, Xue Z and Qin L 2014 Optimal storage rack design for a multi-deep compact AS/RS considering the acceleration/deceleration of the storage and retrieval machine *International Journal of Production Research* **53** 929-943

[8] De Koster R, Le-Duc T and Roodbergen K J 2007 Design and control of warehouse order picking: A literature review *European Journal of Operational Research* **182** 481–501

[9] Gu J, Goetschalckx M and McGinnis L F 2007 Research on warehouse operation: A comprehensive review *European Journal of Operational Research* **177** pp 1–21

[10] Baker P and Canessa M 2009 Warehouse design: A structured approach *European Journal of Operational Research* **193** 425–436

[11] de Koster R, Le-Duc T and Yugang Y 2008 Optimal storage rack design for a 3-dimensional compact AS/RS *International Journal of Production Research* **46** 1495–1514

[12] Kuo P H, Krishnamurthy A and Malmborg C J 2007 Design models for unit load storage and retrieval systems using autonomous vehicle technology and resource conserving storage and dwell point policies *Applied Mathematical Modelling* **31** 2332–2346

[13] Manzini R, Gamberi M and Regattieri A 2006 Design and control of an AS/RS *International Journal of Advanced Manufacturing Technology* **28** 766–774

[14] Fukunari M and Malmborg C J 2009 A network queuing approach for evaluation of performance measures in autonomous vehicle storage and retrieval systems *European Journal of Operational Research* **193** 152–167

[15] Hur S and Nam J 2006 Performance analysis of automatic storage/retrieval systems by stochastic modeling *International Journal of Production Research* **44** 1613–1626

[16] Yin Y L and Rau H 2006 Dynamic selection of sequencing rules for a class-based unit-load automated storage and retrieval system *International Journal of Advanced Manufacturing Technology* **29** 1259–1266

[17] Dooly D R and Lee H F 2008 A shift-based sequencing method for twin-shuttle automated storage and retrieval systems *IIE Transactions (Institute of Industrial Engineers)* **40** 586–594

[18] Roodbergen K J and Vis I F A 2009 A survey of literature on automated storage and retrieval systems *European Journal of Operational Research* **194** pp 343–362

[19] Pellicciari M, Berselli G, Leali F and Vergnano A 2011 A minimal touch approach for optimizing energy efficiency in pick-and-place manipulators *15th Int. Conf. on Advanced Robotics* Tallinn University of Technology Tallinn, Estonia, June 20-23, 978-1-4577-1159-6/11/$26.00 ©2011 IEEE

[20] Bukata L and Sucha Premysl 2017 Energy Optimization of Robotic Cells *IEEE Transactions on Industrial Informatics* **13** 92-102

IOP Conf. Series: Earth and Environmental Science **877** (2021) 012015 doi:10.1088/1755-1315/877/1/012015

Energy Harvesting of Ambient E.M Waves by Employing Different Types of R.F. Antenna

Ibtesam O. Radi[1*], Samira A. Mahdi[1], H. J. Motlak[2]

[1] Physics Department, Collage of Science, University of Babylon, Hilla, Iraq
[2] Department of Electrical Engineering, College of Engineering, University of Babylon, Hilla Iraq

* Corresponding author: sci.ibtsam.omran@uobabyion.edu.iq

Abstract. Used to operate small electronic devices, such as the cell phones. This work will deal with the electromagnetic signals as a renewable energy sources. An analog circuit will prepare using diodes and a charged capacitance and antenna to convert the energy of ambient E.M. waves (Radio Frequencies RF) to a D.C. voltage. The measurements will be demonstrated using different kinds of antenna. In addition, the ambient RF frequency according to Iraq/ Babylon/ Hilla city will be collected from the communications tower and measured using a receiver circuit.in this paper used three kinds of antenna, wire antenna, air antenna, dish antenna, Measuring the DC Power. the amount of the DC Power that obtained using a dish antenna is approximate the energy that collected using an air antenna. The amount of the DC. Power could be controlled by the charging time (RC) of the capacities.

1. Introduction

Finite electrical battery life enables businesses and researchers to come up with creative concepts and developments for an endless or increased period of time to operate electronic mobile devices. Popular resource restricted wireless devices that should be recharged when they run out of battery. We need the main supply and battery to charge empty cell phone batteries or aor any removable phone batteries for this kind of substance. It is not technically possible to bear chargers everywhere we go and even to expect power supply availability anywhere. Any kind of alternative should be given to prevent those drawbacks, and it can be wireless charging of cell phones. If the smartphone can collect RF power signals from the mobile towers, this can be achieved by the RF energy harvesting system or technology called [1]. The RF wave energy used by systems can be harvested and used to perform more reliably and effectively [2], Ambient energy harvesting has been an attractive area of research due to its prospective applications in several modern low-power systems [3]. As this emerging technology enables proactive energy replenishment of wireless devices, it is advantageous in supporting applications with quality of service (QoS) requirements [2], radio frequency (RF)energy harvesting energy from ambient radio frequency signals radiated by different communication sources [3], this is the power to transform the RF signals received into electricity. A potential approach for powering energy-constrained wireless networks is this technique. Conventionally, wireless energy-restricted networks, such as wireless sensor networks, have a brief lifetime that largely limits the efficiency of the network. [4]. In the other hand, the RF Energy Harvesting Network (RF-EHN) has a renewable energy source from the radio environment. The RF energy harvesting capability therefore enables wireless devices to harvest energy from RF signals for the processing and transmitting of their information. As a result, RF-EHNs have rapidly found their applications in diverse ways, such as wireless sensor networks [5]. In RF energy harvesting, radio signals with frequency range 300GHz to as low as 3kHz are used as a medium to carry energy in a form of electromagnetic radiation [6].

2. Implementation

Radio waves are available everywhere and they are used for Cable, radio, cell phone signal communications, etc. The key components used in communication systems to relay RF power in the KW range are Omni directional antennas. Energy harvesting from ambient electromagnetic (EM) waves is one of the most advantageous technologies for providing these standalone devices with continuous power. [5,6], In fact, As the receiver sensitivity of cell phone antennas is very high, very little milliwatts of RF power can be extracted from the mobile communication environment. The major cause for such a significant reduction in the transmitted power is the absorption of bodies (i.e. obstacles) present in the path of the RF waves and also the lack of power in the form of heat in materials where it is present [5]. For their operation in sleep and active modes, most wireless devices such as cell phones use just microwatts to mill watts in the power spectrum. So, using the scavenging circuit, we can conveniently access the RF power available in the external world and use it to run our cell phones. Now, to achieve such features, we can see our proposed circuit [6]. The block diagram of different ingredients for the construction of our suggested circuit is shown in figure (1).

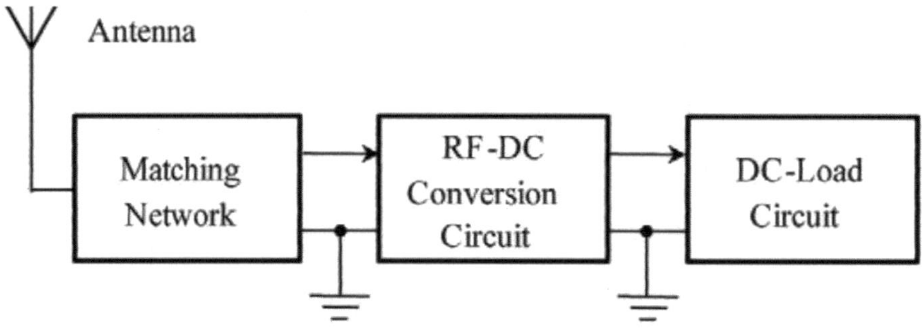

Figure 1. Schematic view of a RF energy harvesting system.

The block diagram displays the circuits that form the rectenna. The antenna signal received is balanced by the circuit of the voltage multiplier, then the circuit of the high output voltage multiplier is used [6]. It was proposed to supply electricity for wireless sensor nodes or other applications from commercial RF transmitting stations such as TV, WIFI, cellular tower or Radar[7].

In recent years, the concept of using A great deal of attention has been paid to radio frequency (RF) resources for low-power electronic devices. To substitute the battery and reduce maintenance costs, [8]. Multiple RF energy sources can be classified into three categories: intentional sources, expected sources for the atmosphere and uncertain sources for the environment. To start up battery-less devices directly and for battery activation, RF energy harvesting or RF energy scavenging may be used. It can also be used to recharge a remote batteryTo power wireless sensor networks (WSN) and to wake up sensors in sleep mode, like a mobile phone. [9]. Using a rectifying antenna called a rectenna, this can be done by translating the ambient EM waves to a workable DC voltage (i.e.,antenna rectifier) [10].

The energy supply for wireless sensor nodes or other devices has been suggested from commercial RF transmitting stations such as TV, WF, cellular tower or Radar. Figure (1) The DC power depends on the RF power and RF/DC conversion performance available. How to cope with the frequency-dependent impedance of the antenna and the rectifier is a big issue in the rectenna model. The impedance of an antenna must meet the impedance of the rectification circuit for optimal power transmission. As seen in Figure (1) A rectenna system can be divided into three parts, which are a transmitting antenna, a low-pass microwave filter and a rectifying circuit. The signal is received by an antenna, the operating frequency of which must be the same as that of the signal. The power of the RF incident signal is transformed by the rectifier circuit to a dc signal. An impedance balancing network of inductive and capacitive components is implemented between the antenna and the rectifier, That guarantees full power supply from the antenna to the rectifier[9]. The RF energy harvesting circuit is generally known as the impedance matching network and rectifier circuit. As seen in the diagram, a typical rectifying antenna device typically consists of five separate sections. [10,11].

1- A transmitting antenna that is designed to receive RF signals from random signals in the ambient environment (WEH). The input impedance of the antenna is normally calibrated to norm 50 Ω. The antenna is an integral part of the rectenna and should have some criteria for harnessing and translating the ubiquitous RF energy into a functional DC output voltage, so that the antenna can receive all the ambient RF energy available in the surrounding multiple frequency sources.

2- To reject the higher order harmonic signals produced by the rectifier by a band pass filter, Since the signals could be radiated by the antenna, which could decrease the overall efficiency of conversion and cause interference. The filter may either be combined with the antenna to create a filtering antenna structure [12].

3- An impedance matching network that is designed to adjust the rectifier's complex impedance to a resistive port (e.g. 50 Ω). The power of the received signals will then be entirely transferred to the rectifier. mpedance-matching network is an integral requirement in a rectenna design , The Rectenna's RF-DC conversion efficiency relies strongly on its impedance-matching network. The layout of the matching network is difficult and requires many factors. This is due to the fact that both the rectifier circuit and antenna input impedance values change with frequency .

4- A rectifier that is tailored to change power from RF to DC. The rectifier's input impedance ranges in a wide variety of values and the impedance is very responsive to frequency, input power and load impedance variations, a rectifier should have good performance of RF to DC conversion. The choosing of diode is usually carried out by one or more diodes, since it can be significant reason for failure and its efficiency dictates the overall efficacy of the device[13].

5- A load that may normally be a resistor or a super capacitor for energy storage

A figure displays the high-level model flow for a simple RF energy harvesting circuit (2). This particular configuration flow is for the purpose of charging a Lithium Ion battery, but these measures are also required to feed a low-power system using RF energy harvesting. The architecture in this research focuses on the circuitry of processing, which includes filtering (C_3, C_4 and D_1, D_2), switch mode power conversion (C_1, C_2, and D_3, D_4). C_1,C_2=100 µF, C_3,C_5=20 nf ,D_1,D_2,D_3,D_4= Single diode, C_1,C_2= electrolytic capacitor and C_3,C_4=ceramic capacitor

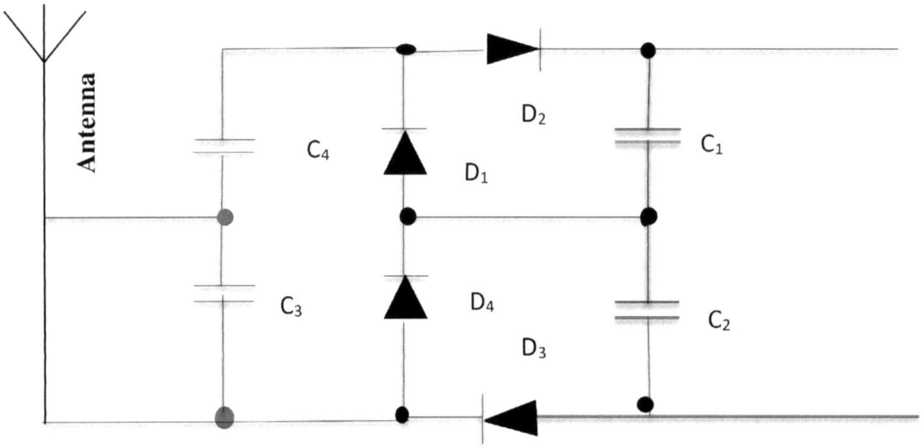

Figure 2. design RF energy harvesting circuit.

The circuit itself exhibits nonlinearity, because the energy harvesting circuit consists of diodes, which are nonlinear devices [13]. This ensures that the energy harvesting circuit's impedance changes with the amount of antenna power produced. Because when the circuit is combined with the antenna, the full power transfer occurs, the impedance matching is typically achieved at the same input power. The impedance matching network converts impedance to ensure optimal power distribution [14].
In the ideal case, the highest output voltage is achieved at the resonant frequency of antenna having inductance L [15].

$$f = \frac{1}{2\pi\sqrt{LC}}$$

3. Experimental Results

3.1. Without Antenna

Figure (3-a) shows the circuit connection without using an antenna, Figure (3-b) displays the amount of energy (DC. voltage) on the screen of the oscilloscope at T= 0 minute and Figure (3-c) shows the amount of energy T= 3 minutes. This is representative of the exponential recovery function, which is the predicted charging characteristic of the capacitor. These findings show the RF energy harvester's proof-of-concept, as well as validating the processing circuitry's ability to rectify the RF signal received. These findings, however, indicate that the boost converter can not work as planned. The booster converter was engineered to output a voltage of approximately 5V, which is far higher than the real voltage calculated through the capacitor. The potential explanation why the booster converter does not produce the desired output is that the cool-start voltage of the converter is 215 mV and that the RF energy harvester is unable to reach this threshold due to the combination of the converter: insufficient power received poor power transfer.

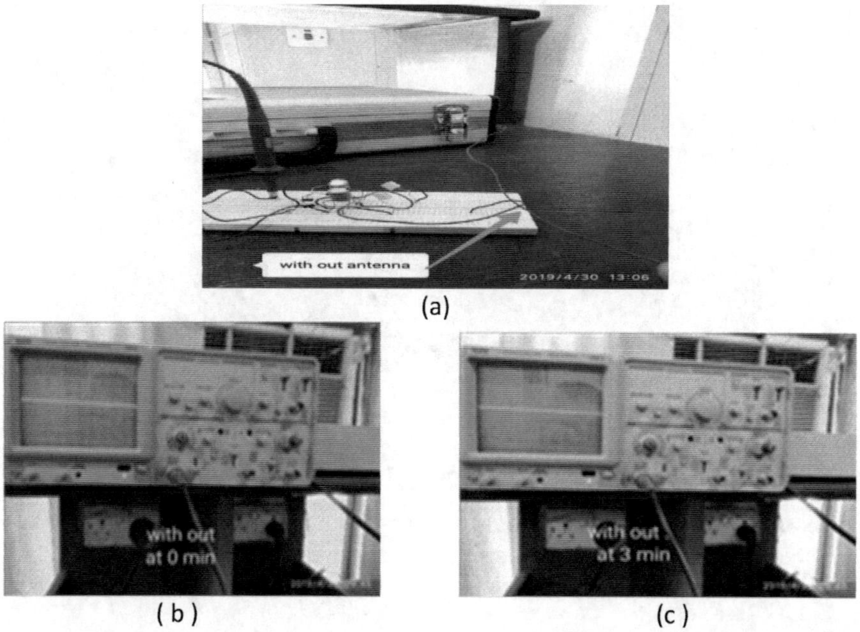

Figure 3. a) circuit design, b) The amount of voltage on the oscilloscope screen at T=0 min., and c) The amount of voltage on the oscilloscope screen at T=3 min.

Table 1. The amount of DC. Voltage.

Antenna	V(mv)	T(min)
Without antenna	20	0
	22	3

3.2. With Wire Antenna:

The wire antenna has been suggested to get more energy from RF signal. Figure (4-a) show circuit design, and the amount of the DC voltage on the oscilloscope screen at time (0,3) minute shows in Figure (4-b, c) respectively. Although the input power obtained from the antenna varies over time,The RF energy harvester will have to sustain a voltage level of at least 625mV at the input of the boost converter even though the threshold is exceeded. Using an electrolytic capacitor for research is an added challenge that may restrict the maximum output voltage reached by the circuit. In a low power application like the one discussed here, the leakage current of the capacitor may have a substantial negative effect on the output voltage. If the output current decreases to a degree equivalent to the capacitor's leakage current, Then the capacitor no longer receives a charge. Many design changes are required to make the processing circuitry a feasible solution for the processing of RF signals obtained from an antenna for the purpose of charging a lithium-ion battery with all of the above problems.

IOP Conf. Series: Earth and Environmental Science **877** (2021) 012015 doi:10.1088/1755-1315/877/1/012015

Figure 4. The circuit and the amount of the DC voltage on the oscilloscope screen (a) circuit design, (b) at time (0)minute and (c) after (3)minute.

Table 2 . The amount of voltage.

Antenna	V(mv)	T(min)
wire antenna	50	0
	75	3

3.3. With air antenna:

The results of air antenna recorded at t= (0, 3) min. as shown in Figure (5) and Table (3). The amount of voltage has been increased up to 400mv at t=0, and 700 mv at t=3min. That's because the air antenna has an ability to capture the ambient E.M. waves more than the wire antenna.

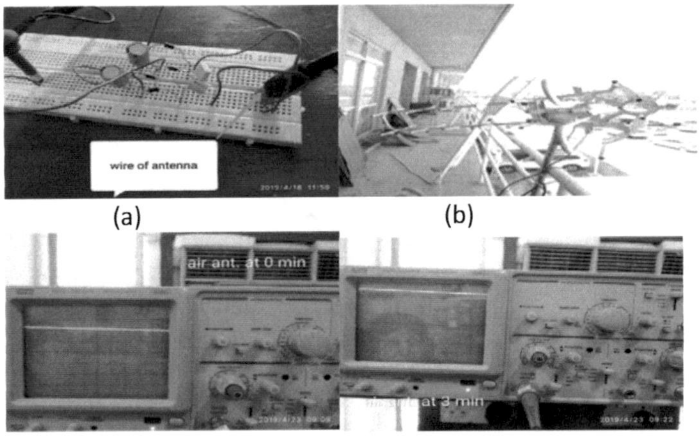

Figure 5. The circuit and the amount of the DC voltage on the oscilloscope screen (a) air antenna, (b) circuit design,(c) at time (0)minute and (d) after (3)minute.

Table 3. The amount of voltage.

Antenna	V(mv)	T(min)
air antenna	400	0
	700	3

3.4. With Dish Antenna:

The results of dish antenna have been recorded at t= (0, 3) min. as shown in Figure (6) and Table (4). The amount of voltage has been increased up to 280mv at t=0, and 800 mv at t=3min. obviously, the value of the energy gain in this case is approximately equal the value of the energy gain in case of air antenna. It may be referred to the size of the dish antenna as compare as with the size of air antenna.

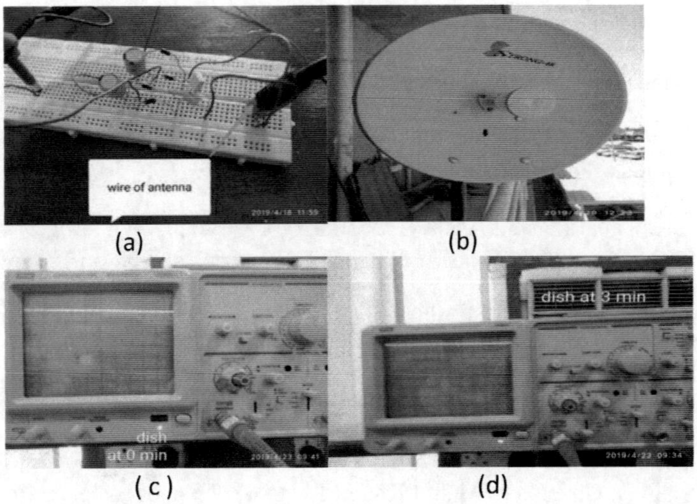

Figure 6. The circuit and the amount of the DC voltage on the oscilloscope screen (a) circuit design, (b) air antenna, (c) at time 0 minute and (d) after 3minute

Table 4. The amount of voltage.

Antenna	V(mv)	T(min)
Dish antenna	280	0
	800	3

4. Conclusion

1. The ambient RF frequencies are a considerable, powerful, and promised source of energy.
2. The amount of the DC. Voltage that collected from the RF frequencies without antenna can be consider and employed after using an amplifier circuit.

3. The wire antenna gives a significant improvement in the value of DC. Voltage as compare as with of the kids of antenna.
4. Although the air antenna raised the DC. Power to a high level, the shape and the large size of the antenna could be limit to its use.
5. The amount of the DC. Power that obtained using a dish antenna is approximate the energy that collected using an air antenna.
6. The amount of the DC. Power could be controlled by the charging time (RC) of the capacities.

References:

[1] Sivaramakrishnan A and Jegadishkumar K J 2011 A highly efficient power management system for charging mobile phones using RF energy harvesting *Int. J. Inf. Technol. Converg. Serv.* **1** 21

[2] Ghazali R, Hussain A J, Al-Jumeily D and Merabti M 2007 Dynamic ridge polynomial neural networks in exchange rates time series forecasting *International Conference on Adaptive and Natural Computing Algorithms* (Springer) pp 123–32

[3] Khalaf M, Hussain A J, Keight R, Al-Jumeily D, Fergus P, Keenan R and Tso P 2017 Machine learning approaches to the application of disease modifying therapy for sickle cell using classification models *Neurocomputing* **228** 154–64

[4] Visser H J and Vullers R J M 2013 RF energy harvesting and transport for wireless sensor network applications: Principles and requirements *Proc. IEEE* **101** 1410–23

[5] Nishimoto H, Kawahara Y and Asami T 2010 Prototype implementation of ambient RF energy harvesting wireless sensor networks *SENSORS, 2010 IEEE* (IEEE) pp 1282–7

[6] Ramachandran M, Ramanadhan I, Elavazhagan S R and Shanmugham B 2017 Rectenna design of GSM band signal for energy harvesting *Int. J. Sci. Technol. Eng.* **3** 221–6

[7] Shrestha S, Noh S-K and Choi D-Y 2013 Comparative study of antenna designs for RF energy harvesting *Int. J. Antennas Propag.* **2013**

[8] Saeed W, Shoaib N, Cheema H M and Khan M U 2018 RF energy harvesting for ubiquitous, zero power wireless sensors *Int. J. Antennas Propag.* **2018**

[9] Akter N, Hossain B, Kabir H, Bhuiyan A H, Yeasmin M and Sultana S 2014 Design and performance analysis of 10-stage voltage doublers RF energy harvesting circuit for wireless sensor network *J. Commun. Eng. Networks* **2** 84–91

[10] Alloghani M, Hussain A, Al-Jumeily D and Abuelma'atti O 2015 Technology Acceptance Model for the Use of M-Health Services among health related users in UAE *2015 International Conference on Developments of E-Systems Engineering (DeSE)* (IEEE) pp 213–7

[11] Zhang J-W, Huang Y and Cao P 2014 An investigation of wideband rectennas for wireless energy harvesting *Wirel. Eng. Technol.* **5** 107

[12] Song C, Huang Y, Zhou J, Carter P, Yuan S, Xu Q and Fei Z 2016 Matching network elimination in broadband rectennas for high-efficiency wireless power transfer and energy harvesting *IEEE Trans. Ind. Electron.* **64** 3950–61

[13] Ma Z and Vandenbosch G A E 2013 Wideband harmonic rejection filtenna for wireless power transfer *IEEE Trans. Antennas Propag.* **62** 371–7

[14] Ladan S, Ghassemi N, Ghiotto A and Wu K 2013 Highly efficient compact rectenna for wireless energy harvesting application *IEEE Microw. Mag.* **14** 117–22

[15] Al-Kafri A S, Sudirman S, Hussain A, Al-Jumeily D, Natalia F, Meidia H, Afriliana N, Al-Rashdan W, Bashtawi M and Al-Jumaily M 2019 Boundary delineation of MRI images for lumbar spinal stenosis detection through semantic segmentation using deep neural networks *IEEE Access* **7** 43487–501

IOP Conf. Series: Earth and Environmental Science **877** (2021) 012016 doi:10.1088/1755-1315/877/1/012016

Review: The flow and heat transfer investigation inside tapered and straight two pass channels with rib turbulators

Barakat Hassan and Riyadh AL-Turaihi
Mechanical Engineering Department, University of Babylon, Iraq

Email: barakat.alaiwi@student.uobabylon.edu.iq, drriyadhalturaihi@yahoo.com

Abstract. This paper is a review of the number of experimental and CFD experiments performed with rib turbulators about the heat transfer and the at two. pass channel . In respect to achieve higher thermal. efficiency of gas. turbines, efforts are made to raise the inlet temperature. The secondary fluxes resulting from the rib and U-shaped curvature play an important role enhancing heat transfer in the two pass ribbed channels. The ribs on the internal surface of cooling passage will increase the strength of heat transfer. This paper deals with the effect of tapered and straight two-pass channels with the influence of the variable rib cross-section on flow and heat transfer enhancement

1. Introduction

Modern gas turbines operate at a very high entry temperature that may reach 1370 degrees Celsius, which is crossing the highest permissible temperature, and the gas turbine blade may reach. In circulating the air at the inner passages, cooling is achieved. These internal passageways are detached by a 180° bend . The switch triggered secondary flow is an more important to inner turbine blade cooling. Various turn design have several the flow fields in turn zone of the mu ti pass duct. The Properties of the flow varies about the turn due to secondary flow, isolation flow, flow Collision flow and turn flow therefore play a significant role at the improvement of the pressure and the heat transfer for inner cooling of the turbine blade. The modern design of gas turbine requires lower pressure drops across its total duration and improved heat transfer improvements.

Most of the experiments focused in the inland ducts of the gas turbine blade on fixed cross-sectional areas from entry to rotation. To reduce heat packing, gas turbine blades are usually tapered from hub to tip. These channels occur inside the blades of the high performance turbine to supply the outer surface of the rotor, which is subjected to a high temperature gas flow, with effective cooling.

Many variables affect the degree of heat distribution improvement in the ribbed chanals:

- Orientation of rib
- Spacing. of rib.
- Height of rib
- Cross-section of rib (square rib, sharp corners rib, rounded corners rib , etc.).

Together with a large range of the channel sizes, these factors will contribute to very different degrees of enhanced heat transfer through the cooling Channels (aspect ratios).

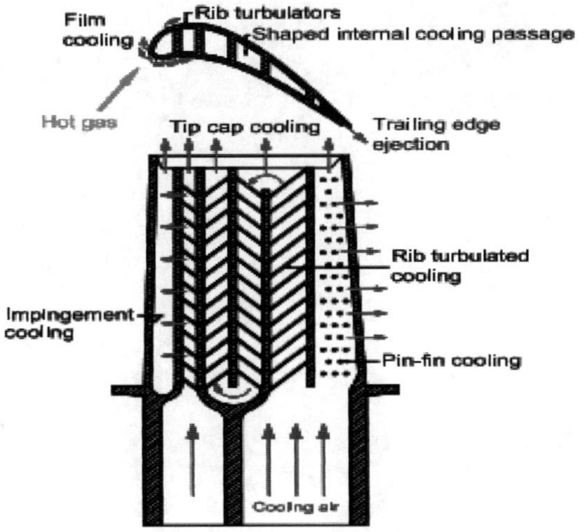

Figure1. Internal cooling of gas-turbine blades [6].

2. Straight channel

2.1. Effects of rib orientation on heat transfer enhancement

The combined effects at the flow and heat transfer through two pass ducts with ribs for rib directions (N type, p type) and rib angles (30° -75°). The transfer of additional secondary flows through the downstream passage from the weakening of the energy loss and local flow secondary of the upstream flow secondary is primary factor for improving local heat transfer [1]. Gao et al. study the pairing of nine distinct rib orientations and the 180° bend in the U shaped passage of the ANSYS CFX industrial CFD at Re = 30000 on total friction loss and forced convection Qualitative findings demonstrate that the Nusselt ratio and the average loss of pressure in the downstream passage are heavily influenced by the upstream geometry [2]. Mochizuki and Murata numerically investigated studied the influence of channel rotation and rib orientation on heat transfer in a squre two pass channel at180° sharp turn . The heat transfer in and after bend was improved in the stationary state, and the sharp-bend induced flow field dominated it. In both the arrangement of the rib and the rotation of the tube, the friction factor was more sensitive than trans-heat [3]. In the revolving two pass square channel, the influence of different 45° rib angle turbulator arrangement on the Nu number ratio was studied with rotated number up to 0.11 , and with two channel orientations with respect to axis of rotated (β = 90 , 135deg). At an angle (+45° or -45°) to the main stream, five separate structures of rib turbulators are mounted on trailing and leading surfaces [4]. Tao and Zhao studied the influence of five angles of the rib (45°, 60°, 90°, 120° and 135°) of heat transfer properties in the U-shaped stationary ducts has been experimentally tested. The findings revealed that in a 60° ribbed channel, followed by 45°, 90° and 135°, the optimum Sherwood number ratio was identified ; the rib angle and inclination of having major flow on local overall heat transfer [5].

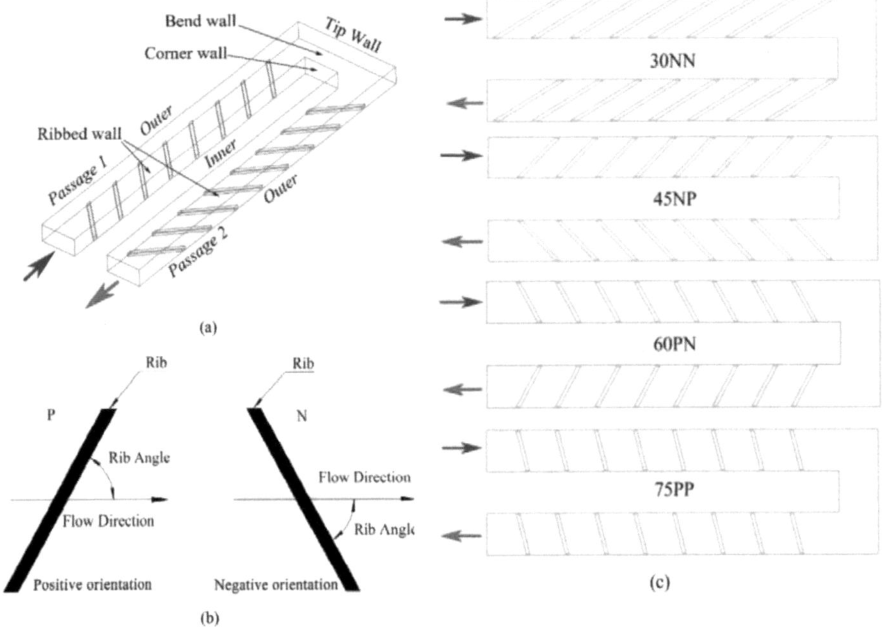

Figure 2. (a) Geometric model; (b) Rib orientation definition; (c) Four rib channel with rib orientation [1].

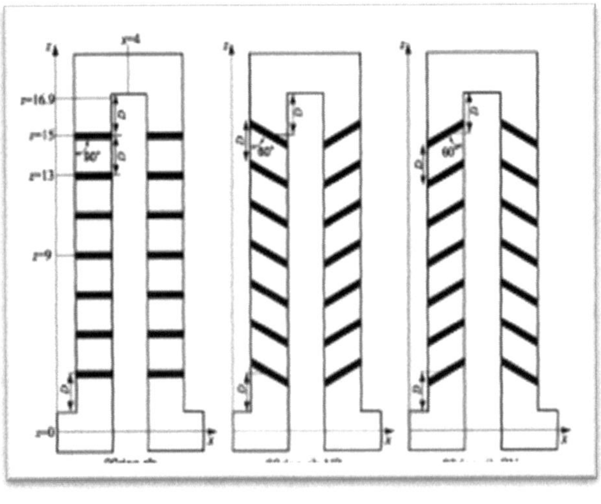

Figure 3. arrangement of ribs [3].

2.2. Effect of the rib height and the rib Spacing on Heat Transfer Enhancement

Experimentally examines the effect of rib spacing in both non-rotating and rotating rectangular cooling ducts on pressure penalty heat transfer improvement and the overall thermal efficiency. The leading and trailing surfaces, 45° angled ribs are mounted in the 1:2 rectangular channels [6] .The rib pitch varies, 10, 7.5, 5, and3 rib pitch to-height (P/e) ratios are considered. The influence of rib spacing in pressure penalty and heat transfer in stationary and the rotary U shaped channels has been experimentally studied. The findings showed the rib spacing of p/e = 3 achieved greatest heat transfer improvement, while .p/e = 5 performed the maximum pressure loss [7]. The influence of the spacing of rib and height of rib on heat flow in the 1:4 tube is studied. The three variants in the rib spacing 1(p/e=2.5,5,10) with the ratio 0.078 e/D .Rib structure with e/D ratio=0.156 and a P/e ratio =10 was considered to investigate the effect of the rib height [8]. Experimental analysis was studied the influence rib spacing on heat transfer and friction in a revolving two1pass square tube. On the pressure side, the rib pitch-to-height (p/e) ratio ranges from 3.8 to 14.4 in the positive rotational direction, while the suction side holds a constant value of 10.The findings show that the rib spacing influence is more pronounced than the second passage in the first radially outward flow passage [9]. In a revolving two passage channe1 (aspect ratio 2:1), the spacing of rib influence on heat transfer was tested at an orientation angle of 135°. At a 45° flow angle, parallel ribs were added to the trailing and the leading of the revolving channe1 .Rib pitch to rib1height ratios tested (5, 7, 5 and 10) [10].

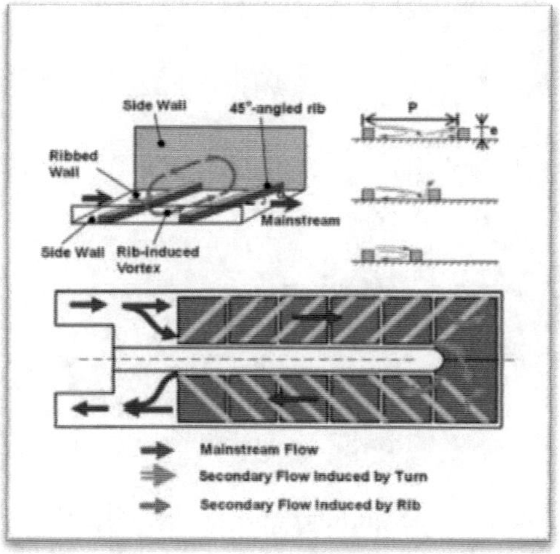

Figure 4. Flow patterns around turn and ribs [10].

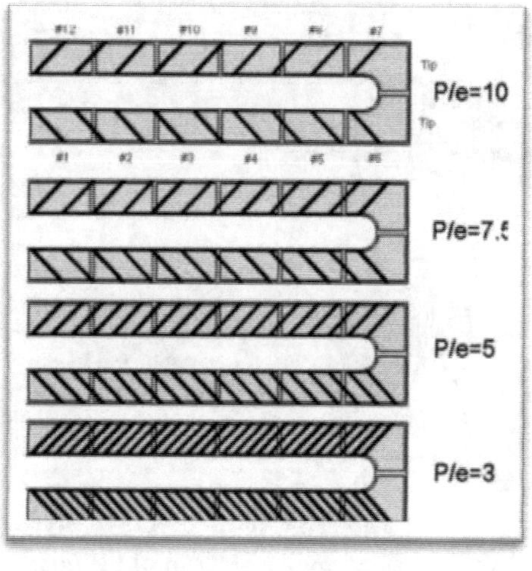

Figure 5. Rib -Spacing Considered in the Present Study [6].

2.3. Effect of the rib cross-section on Heat Transfer Enhancement

2.3.1 Solid rib. In rectangular channel with various forms of ribs truncated prismatic on bottom surface, precise flow field and heat transfer properties were analyzed. The truncated prismatic ribs give a lower penalty for pressure (54.65 percent) than square ribs. The truncated prismatic ribs give improved thermohydraulic efficiency at lowest and 25-53 percent higher than the square ribs [11]. Fifi NM Elwekeel et al. study performed in horizontal air/mist and the air of rectangular duct, with the separate shaped ribs, to explore the flow friction, forced convection , and efficiency factor. Calculations are made for triangular , square ribs and cross sections of trapezoidal ribs. The average mist cooling improvement is found to be around 1.8 times. In air/mist offers the greater enhancement of heat flow over the air with trapezoidal ribs (38) [12]. Ali et al. studied the effect of the change in the trapezoidal with theangle (5°, 10°, 15° and 20°) trapezoidal lrib with decrease in the flow path height on the flow, and its consequent influence on the rise in heat transfer at the Re numbers(9400, 27,120, 44,600 and 61,480)used LCT and PIV techniques .The area of the secondary recirculation bubble has been found decrease at higher Re and higher trapezoidal angles, contributing the elimination of the hot spot directly downstream of the rib [13]. Numerical analysis records fluid flow and heat transfer properties in a cooling canal with separate crescent ribs mounted at one wall. In order to optimize thermal efficiency of cooling tube, three types of ribs are considered, i.e. a straight rib, a crescent rib concave to a stream-wise direction, a crescent rib convex to a stream-wise direction to a stream-wise direction [14]. The influence of differing pentagonal (5 ° to 20 °) angle and the rib p.The result shows that the best thermal efficiency is the case with crescent ribs concave pitch to height (6-12) ratio on the friction factor characteristics and local heat transfer were analyzed inside the rectangular channels roughened by the one main wall with pentagonal ribs. LCT was used to calculate the distribution of surface temperature and finally to show the local HTC at distinct Re (9400-58,850) over the ribbed surface .A major increase in the heat transfer augmentation directly behind the pentagonal ribs has been discover at higher Reynolds number and pentagonal angle, contributing to prevention of the hot spots [15].

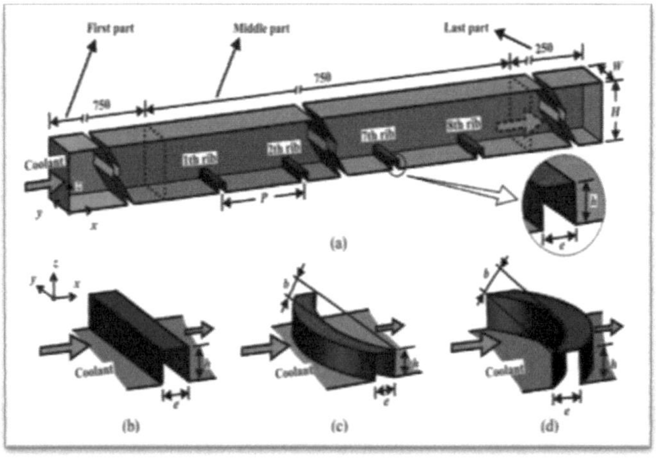

Figure 6. (a) overview of cooling-channel (b) straight-rib (c) crescent Rib concave to the stream wise .direction (d) crescent rib-convex to- the stream wise direction [14].

Sixteen rib structures with varying cross-sectional shape were numerically analyzed and the boot-shaped rib turbulators were found to have the best of the heat transfer output with the pressure drop equal to that square rib .The researchers have stated that the front surface slope of the rib is a crucial factor in deciding the achievement of the heat transfer because it influenced directly the size of the recirculate region [16]. Lei Wang and Bengt Sunde´n study the ribs of trapezoidal shape with reducing height there flow direction have been shown to provide the largest improvement in friction factor and heat transfer between the triangle, square and trapezoidal ribs with rising height in flow direction and decreasing height ,and perhaps helpful in preventing creation of hot-spots [17].

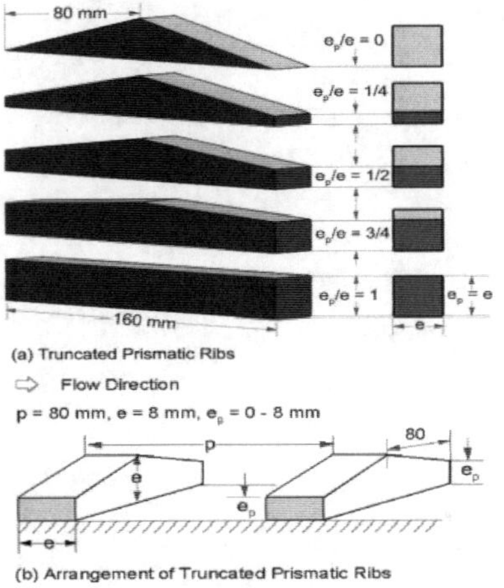

Figure 7. Different rib configurations [11].

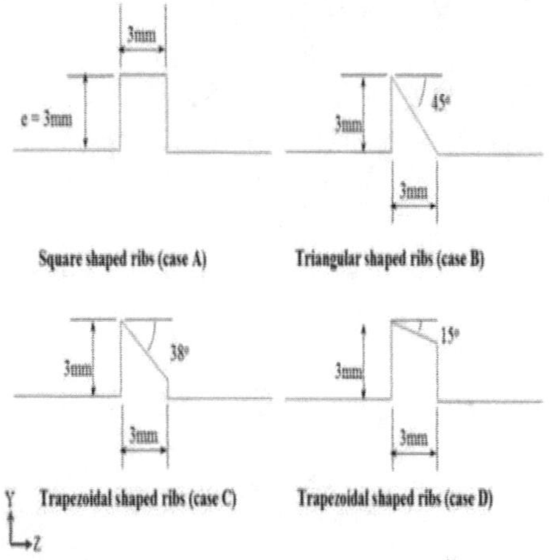

Figure 8. Profile of the shaped ribs [12].

2.3.2. Slit rib. Convective fluid flow and heat1 transfer in ribbed ducts using PIV and LCT techniques, respectively, The results indicate that the cribriform the rib of square shape Effective way obviate the forming hot spots and also decrease the reattachment length 45% [18]. In with detail turbulent kinetic energy (TKE) and turbulent statistics budget for same the permeable Shorter reattachment duration for the maximum stresses of Reynolds, TKE output and pressure transport was observed among all the tested configurations for incline split-slit rib [19].The ability of PIV and LCT method was used to investigate influence of differ the trapezoidal angle of a continuous slit trapezoidal-rib (0°, 5°, 10°, 15°, and 20°) on the flow pattern and its corresponding effect onthe heat transfer improvement at the set of Reynolds 61,480.The trapezoidal angle variance has been shown to efficiently regulate the small scale corner vortices behind the firm rib, thereby helping to prevent hot spots [20]. The thermal efficiency of the permeable shape of ribs placed onthe bottom surface of the square two-pass duct was analyzed and show the split -slit design demonstrated the significant improvement in the heat transfer relative to solid and slit rib without any proportionate rise in the penalty pressure 2018 [21].Five different geometric models, include rectangular slit ribs, (V-shaped) slits , (anti -shaped) slits , (broken V-shaped) slits , and (broken anti V shaped) slits ribs, have been examined. Results obtained show that due to the higher turbulence strength, the ducts with the broken-slits provide greater heat transfer efficiency [22]. In the rectangular channel with the solid, converging slit and alternating solid slit ribs placed transversely at a bottom wall, the heat transfer and frictional characteristics were analysed . As predicted, the converging slit rib greatly improved the rate of heat transfer in downstream locality and prevented creation hot spot local [23] . Liu et al .study with the inclined angle of the holes varying from 0° to 45°, two pairs of perforated ribs are used and the cross-sections are, respectively. The total average number of Nusselt (Nu) for all inclined cases is significantly higher than for straight cases .The change ratio is around 1.85 percent - 4.94 percent . For the inclined hole situations, the average Nu in the half section toward the inclined path is expanded [24] .Experimental analysis the detailed of aerothermal properties in the duct with the set of the solid and the permeable pentagonal-ribs placed on bottom wall with the parallel and the inclined slit. Open area ratio set through studies is (0.125%, 12%, and 25 %), respectively .The result shows that the flow from the inclined-slit pentagonal rib has a major effect on magnitude of the stream velocity ,statistics of fluctuation and the vorticity [25].

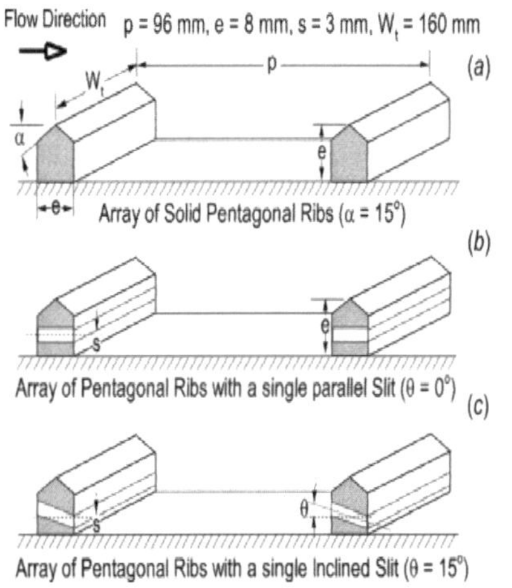

Figure 9. Rib configurations [25].

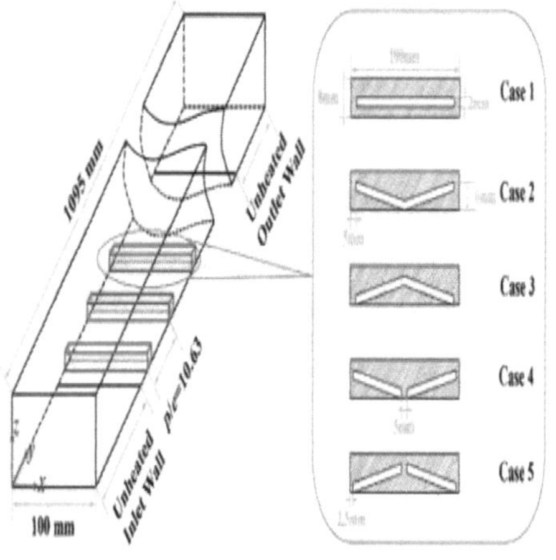

Figure 10. Schematic diagram of geometry [22].

2.4. Effect of channel aspect ratio on heat transfer enhancement

The heat transfer and flow in the rotational U-shaped smooth ducts with aspect ratios (AR=0.25, 0.5 and 1) were numerically analyzed. They findings revealed that the rotating influence on the heat transfer (AR) aspect - ratio duct in narrow was more important [26].Wen-Lung Fu L et al. studied experimentally analyzes influence of buoyancy force and duct aspect-ratio W:H on the heat transfer with smooth walls and 45° ribbed walls in rectangular two-pass revolving channels. 4:1, 2:1, 1:1, 1:2, and 1:4 are the duct aspect ratios. The findings showed that, because it has the smallest pressure penalty, the 1:4 channel has stronger heat transfer [27]. The influences of ribs on heat transfer efficiency and cooling air flow properties are numerically tested under various working conditions in different aspect-ratios (AR) in U-shaped ducts. The wide of the duct (AR = 2:1) has a stronger increase factor than the narrow channel (AR = 1:2) and the heat transfer weight of ribs increases with an increase in the Reynolds number at inlet [28]. A experimental research based on the influence of the aspect-ratio =1:2, 1:1, and 1:6 on heat transfer of smooth and rib-turbulated rectangular ducts .They stated that of 1:2 aspect-ratio channels, there is a linear variance between the Re number and the normalized Nusselt number. Their analysis found that for the smooth and the ribbed channel with an aspect-ratio = 1:6, the maximum heat transfer was obtained [29]. Inside the smooth duct with different aspect - ratios at vary divider to tip wall lengths, flow and transition are numerically analyzed. The findings revealed that both pressure drop and heat transfer at the bend and outlet the passage were affected by divider-to-tip wall size [30].

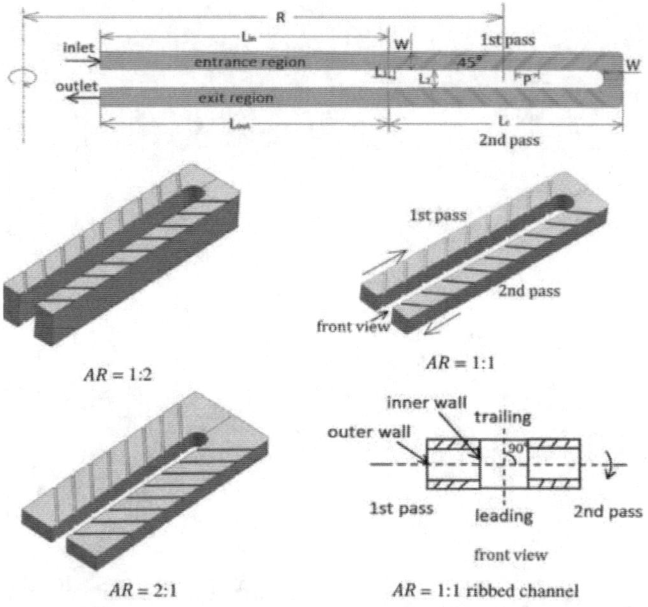

Figure 11. Schematic of U-shaped channel with ribs [28].

2.5. Effect of turn geometry on (turbulent fluid flow and heat transfer)

In smooth U-shaped tube, the effect of bend geometry on the flow characteristics and heat transfer has been studied. This research found that when the external wall of turn was closely linked without a rotund rim and the inner wall of turnwas semi-circular form, the best heat transfer efficiency was obtained [31]. The effect of turned vanes on pressure loss and heat transfer in two pass rectangle channel was numerically and experimentally studied. They found a decrease in pressure drop of 25 percent and could sustain the same degree of heat transfer with the use of inner- turning vanes in bend area [32]. The effect of the shape of the bend on overall heat transfer efficiency was investigated and proposed that symmetrical bulb structure increased a thermal effectiveness agent by 41% [33]. In t he time-resolved particle picture velocimetry system is used in square cross-section U-shaped channels and different bend section configurations to study the properties of the internal flow area as show in fig (12). A variety of essential conclusions are derived from the data. The structure of the section of the turn has an evident effect on the flow field properties for the main flow [34]. In two pass, square duct with 180°sharp turn as show in fig (13), empirical results for secondary flow patterns and heat transfer propagation are provided to explore the effects of the various turn design: straight ,rounded, and circular - turn. The measured results show that the straight case has the strongest turn induced shift in heat transfer at the turn [35].

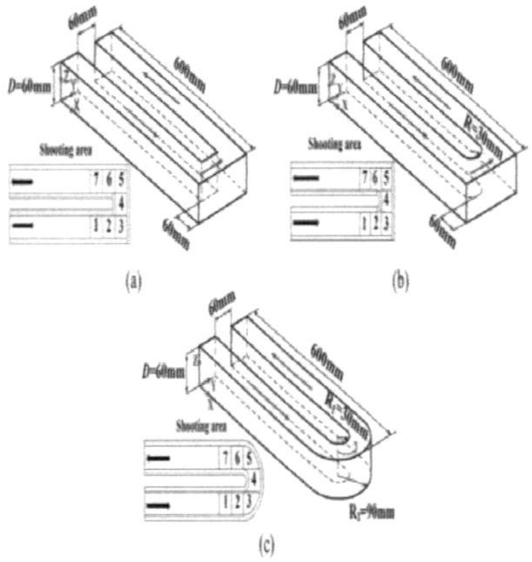

Figure12.U-shaped channels (a) Square square (b)Circle square (c) Circle circle[34].

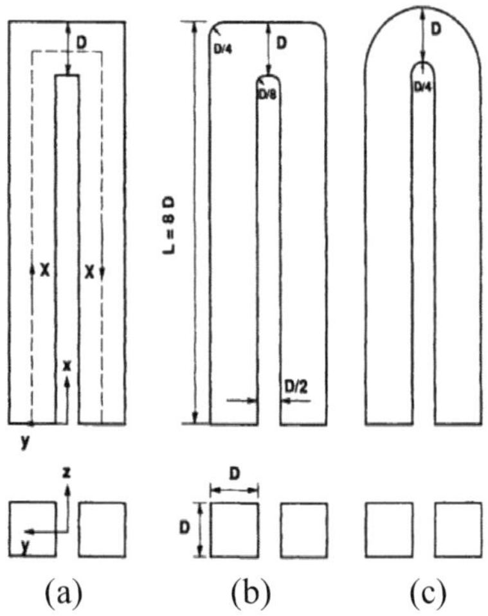

Figure 13. (a) Straight corner, (b) Rounded corner, and (c) Circular corner [35].

3. Tapered channels

The heat transfer influence of a tapered channel has been studied. The area of tapered ducts cross-section various frequently. The findings revealed that the tapered channelwith ribs gave greater gains in heat transfer than straight ducts. However, no research on exit mass flow rate inside the variable cross section channel with ribs has been published on the Nusselt number distributions [36]. Liu and Feng investigated the heat effect within inclined ducts or ducts with bleeding holes, researcher noted that the inclined channel or channel with bleeding holes would provide a greater effect of heat transfer in a way [37]. The pressure drop and heat transfer data were obtained by maintaining constant rib height , side height to the mean diameter ratio equal to (0.081), and the side slope height ratio equal to (6 17.8) for standard flow ribs. Heat transfer optimization was introduced on the basis of modified constant pumping capacity, constant pressure, and constant mass flow, drop, and performance of ribbed channels ranged from (0.8 to 0.6) times that of static air channel [38].

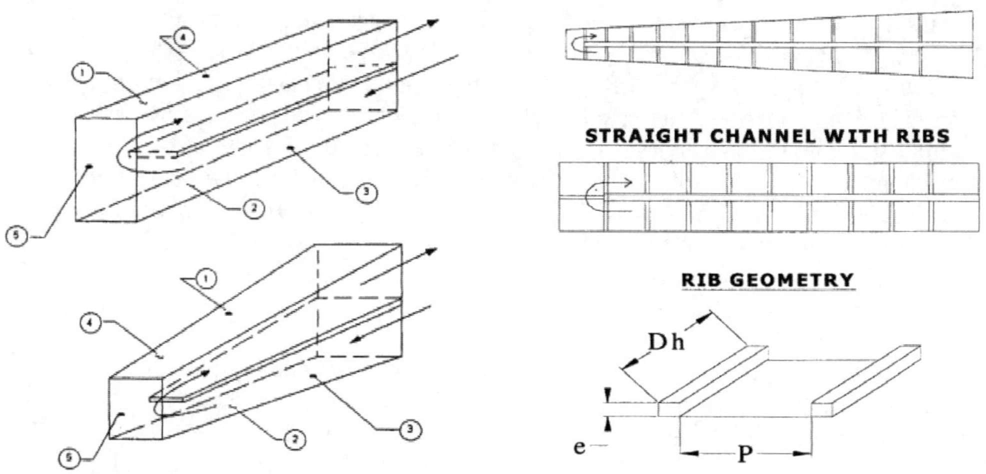

Figure 14. Graphic of the straight and tapered two-pass channels and the distribution of ribs direction [36].

4. Conclusion

It is possible to derive some concluding conclusions from these experimental and numerical studies:

- Rib mediated secondary flow has a significant effect on the transfer of heat to the channel. It is also important to refine the configuration of the rib to take advantage of the secondary flow caused by the rib. For eg, the rotation of the turbulator, such that it is not orthogonal to the mainstream flow, The heat transfer coefficients are greatly increased due to the angle of rib caused by secondary flow . The ideal space for ribs has been found to be about P/e =10.
- The slit ribs have the feature of preventing hot spot and the lower pressure than the solid rib. Reynolds number corresponding to onset of the flow permeability decreases with the increasing in open field ratio. The increasing in heat transfer with increasing the angle of perforation, the ratio of the open area, with the decrease in rib degree to the ratio of hydraulic diameter(Dh) and Re number
- Comparison of the average number of Nusselt and the friction factors of the different turn design shows that the computational simulations adopt the same behaviour as the experimentally revealed but the chosen turbulence model predicts an increase in heat transfer in the bend area. The overall thermal efficiency (sharp 180° bend with circular divider) is higher than in other situations.
- The average level of the heat transfer change for all ribbed channels is comparable. However, there are major variations in the pressure losses suffered inlthe duct. The 1:4 duct obtained the lowest pressure penalty thus, the thermal efficiency of the 1:4 duct is higher than the 1:2, 1:1 and 2:1 ducts respectively.
- The researchers revealed that tapered duct with the ribs has more heat transfer benefits than straight duct.

5. Reference

[1] Gao T, Zeng J, Zhu J, Li J and Gong J 2018 Effects of rib angle and rib orientation on flow and heat transfer in two-pass ribbed channels *Journal of Mechanical Science and Technology* **32** 513-526

[2] Gao T, Zhu J, Li J and Xia Q 2018 Numerical study of the influence of rib orientation on heat transfer enhancement in two-pass ribbed rectangular channel *ENGINEERING APPLICATIONS OF COMPUTATIONAL FLUID MECHANICS* **12**117–136

[3] Murata A and Mochizuki S 2004 Effect of rib orientation and channel rotation on turbulent heat transfer in a two-pass square channe l with sharp 180 turns investigated by using l arge eddy simulation International Journa l of Heat and Mass Transfer 47 2599–2618

[4] Al-Hadhrami L and Han J 2003 Effect of rotation on heat transfer in two-pass square channels with five different orientations of 45°angled rib turbulators *International Journal of Heat and Mass Transfer* **46** 2003 653–669

[5] Zhao C Y, Tao W Q 1997 Effect of rib angle orientation on local mass transfer distribution around sharp 180 deg turn with rib-turbulators mounted in entire two-pass channels Heat and Mass Transfer 32 325–332

[6] Yao-Hsien L, Lesley M W, Wen-Lung F, and Je-Chin H 2006Rib Spacing Effect on Heat Transfer and Pressure Loss in a Rotating Two-Pass Rectangular Channel (AR=1:2) with 45-Degree Angled Ribs *Proceedings of GT2006 ASME Turbo Expo*

[7] Yao-Hsien L, Lesley M. W,Wen-Lung F and Je-Chin H,2007 Rib Spacing Effect on Heat Transfer in Rotating Two-Pass Ribbed Channel (AR= 1:2) JOURNALOF THERMOPHYSICS AND HEAT TRANSFER 21

[8] A Dissertation ,2009"HEAT TRANSFER IN SMOOTH AND RIBBED RECTANGULAR TWO-PASS CHANNELS WITH A DEVELOPING FLOW ENTRANCE AT HIGH ROTATION NUMBERS"

[9] Xu G, Li Y and Deng H 2015 Effect of rib spacing on heat transfer and friction in a rotating two-pass square channel with asymmetrical 90-deg rib turbulators *Applied Thermal Engineering* **80** 386-395

[10] Jiang L, Je-Chin H and Michael H 2012 Effect of Rib Spacing on Heat Transfer in a Two Pass Rectangular Channel (AR 5 2:1) at High Rotation Numbers *Journal of Heat Transfer* **134**

[11] Sharma N, Tariq A and Mishra M 2018 Detailed heat transfer and fluid flow investigation in a rectangular duct with truncated prismatic ribs *Experimental Thermal and Fluid Science* **96** 383-396

[12] Fifi NM Elwekeel, Qun Zheng and Antar MM Abdala ,2013,"Air/mist cooling in a rectangular duct with varying shapes of ribs *J Mechanical Engineering Science* **228** 1925–1935

[13] Ali M S, Andallib T and Gandhi B K 2013Flow and heat transfer investigation behind trapezoidal rib using PIV and LCT measurements *Exp. Fluids* **54**

[14] Xie G, Liu X, Yan H and Qin J 2017 Turbulent flow characteristics and heat transfer enhancement in a square channel with various crescent ribs on one wall *International Journal of Heat and Mass Transfer* **115** 283-295

[15] Sharma N, Tariq A and Mishra M 2019 Experimental investigation of heat transfer enhancement in rectangular duct with pentagonal ribs *Heat Transfer Engineering* **40** 147-165

[16] Mi-Ae Moon, Min-Jung Park,and Kwang-Yong Kim,2014,"Evaluation of heat transfer performances of various rib shapes" http://dx.doi.org/10.1016/j.ijheatmasstransfer.2013.12.026.

[17] Wang L and Sundén B 2007 Experimental investigation of local heat transfer in a square duct with various-shaped ribs *Heat and Mass Transfer* **43** 759-766

[18] Sundén B 2011 Convective heat transfer and fluid flow physics in some ribbed ducts using liquid crysta l thermography and P1V measuring techniques Heat and mass transfer 47 899-910

[19] Panigrahi P K, Schroeder A and Kompenhans J 2008 Turbulent structures and budgets behind permeable ribs Experimental Thermal and Fluid Science 32 1011–1033

[20] Ali M S, Tariq A and Gandhi B K 2012 LCT and PIV investigations behind trapezoidal-rib with a slit mounted on bottom wall of a rectangular duct *In Gas Turbine India Conference* **45165** 313-325

[21] Qayoum A Panigrahi P 2019 Experimental investigation of heat transfer enhancement in a two-pass square duct by permeable ribs *Heat Transfer Engineering* **40** 640-651

[22] Zheng D, Wang X and Yuan Q 2019 Numerical investigation on the flow and heat transfer characteristics in a rectangular channel with V-shaped slit ribs *Infrared Physics and Technology* **101** 56-67

[23] Sharma N, Ali M S, Tariq A and Mishra M 2018 Detailed heat transfer and friction factor characteristics in a rectangular duct with alternate solid and converging-slit ribs *Experimental Heat Transfer* **31** 552-570

[24] Liu J, Hussain S, Wang W, Wang L, Xie G and Sundén B 2019 Heat transfer enhancement and turbulent flow in a rectangular channel using perforated ribs with inclined holes *Journal of Heat Transfer* **141**

[25] Tariq A, Sharma N and Mishra M 2019 Aerothermal Characteristics of Solid and Slitted Pentagonal Rib Turbulators *Journal of Heat Transfer* **140**

[26] Su G, Chen H C, Han J C and Heidmann J D 2004 Computation of Flow and Heat Transfer in Two-Pass Rotating Rectangular Channels (AR= 1: 1, AR= 1: 2, AR= 1: 4) With 45-deg. Angled Ribs by Reynolds Stress Turbulence Model *In Turbo Expo: Power for Land, Sea, and Air* **41685** 603-612

[27] Fu W L, Wright L M and Han J C 2006 Rotational buoyancy effects on heat transfer in five different aspect-ratio rectangular channels with smooth walls and 45 degree ribbed walls Turbine Heat Transfer Laboratory, Department of Mechanical Engineering, Texas A&M University

[28] Wang L, Wang S, Zhou X, Wen Fe and Wang Z 2017 Numerical Prediction of 45° Angled Ribs Effects on U-shaped Channels Heat Transfer and Flow under Multi Conditions" Int. J. Turbo. Jet. Eng. **37** 41-59

[29] Smith M A, Mathison R M and Dunn M G 2014 Heat Transfer for High Aspect Ratio Rectangular Channels in a Stationary Serpentine Passage With Turbulated and Smooth Surfaces *Journal of Turbomachinery* **136**

[30] Siddique W, Shevchuk I V, El-Gabry L, Hushmandi N B and Fransson T H2013 On flow structure, heat transfer and pressure drop in varying aspect ratio two-pass rectangular channel with ribs at 45 Heat Mass Transfer 49 679–694

[31] Erelli R, Saha A K and Panigrahi P K 2015 Influence of turn geometry on turbulent fluid flow and heat transfer in a stationary two-pass square duct *International Journal of Heat and Mass Transfer* **89** 667-684

[32] Schüler M, Zehnderm F, Weigand B, Wolfersdorf J and Neumann S O 2011 The Effect of Turning Vanes on Pressure Loss and Heat Transfer of a Ribbed Rectangular Two-Pass Internal Cooling Channel *Journal of Turbomachinery* **133**

[33] Saha K and Acharya S 2013 Effect of Bend Geometry on Heat Transfer and Pressure Drop in a Two-Pass Coolant Square Channel for a Turbine" *Journal of turbomachinery* **135**

[34] Liu R, Li H, You R and Tao Z 2020 Experimental investigation of turbulent flow in a two-pass channel with different U-shaped bends *AIP Advances* **10**

[35] Wang T S and Chyu M K 1994 Heat convection in a 180-deg turning duct with different turn configurations *Journal of Thermophysics and Heat Transfer* **8** 595-601

[36] Ekkad S V, Pamula G, Shantiniketanam M 2000 Detailed heat transfer measurements inside straight and tapered two-pass channels with rib turbulators *Experimental Thermal and Fluid Science* **22** 155-163.

[37] Liu X and Feng J 2009 Heat transfer performances in U-shaped variable cross-section channel with bleeding holes *In 2009 International Conference on Energy and Environment Technology* **1** 691-694

[38] Liang-Bi W, Wen-Quan T, Qiu-Wang W and Tsun T W 2001Experimental study of developing turbulent flow and heat transfer in ribbed convergent/divergent square ducts *International Journal of Heat and Fluid Flow* **22** 603-413

Deterioration characterization of land cover and land use in Muqdadiya district center for the period 1995 to 2019

Alaa Mahdi Saleh Al-Zuhairi[1] and Tanzeeh Majeed Hameed[2]

[1] The General Directorate of Education in Diyala, Iraq
[2] University of Diyala, College of Education for Humanities, Iraq

Email: n20162017386@uodiyala.edu.iq

Abstract. The research dealt with the analysis of environmental changes in the types of land cover and land uses within the Muqdadiyah district center in Diyala governorate and revealed between 1995 and 2019 in order to determine the environmental deterioration of the land cover and land uses and the resulting environmental changes. For the American satellite (Landsat). It turned out that there were changes during the research period, as the amount of change in the area of agricultural land amounted to (-91,16) km2, and the area of pastures decreased as the amount of change reached (-39.89) km2, while the amount of change in the area of urbanization was (44, 38) km2, while the change amount was recorded (100,19) km2 for barren lands, which means a decrease in the area of agricultural and pasture lands in favor of built-up lands and barren lands, which negatively affected the biodiversity within the study area.

Keywords. Land cover characterization.

1. Introduction

The distinction and identification of the types of land cover and land uses in the Muqdadiyah district, and by relying on satellite images, relied on a number of foundations of interpretation, namely: (shape, size, shadow, pattern, color intensity, degree of darkness, texture, and geographical location) [1] Therefore, The process of monitoring the development in the land cover and the uses of the earth by means of the data provided by the remote sensing data has become an important application of space data through which the changes that occurred in the land cover and the land uses of a region are determined by means of multi-date satellite visuals used to detect the change that occurred during a period Given time. As the detection of changes in land cover and land uses provides the basis for understanding the change that has occurred in the environment, which has negatively affected biological diversity and the local climate on it. The researchers used these data to study the changes that occurred in the land cover and land uses within the Muqdadiya district center in Diyala governorate and to reveal the change that occurred on Agricultural lands resulting from urban encroachment on it for the period using space data.

IOP Conf. Series: Earth and Environmental Science **877** (2021) 012017 doi:10.1088/1755-1315/877/1/012017

2. Research problem

* Is there a change in the land cover and land use within the Muqdadiyah district for the period from 1995 to 2019 AD?

3. Research hypothesis

* There are changes in the land cover and land use within the Muqdadiya district center for the period between 1995 - 2019 AD.

4. Research objectives

The aim of this study is to determine the changes that occurred to the land cover and land use within the Muqdadiyah district center between 1995 - 2019 AD, using geospatial tools and techniques (RS) Remote Sensing and (GIS)Geographic information systems to monitor these changes and know their size by producing digital maps that illustrate these changes.

5.The area of study

The spatial boundaries represent the boundaries of the Muqdadiya district center, which is located between the latitude (33.49 and 34.43) north and longitude (44.48 and 45.15) east, and the Muqdadiya district borders from the north and north-east the Khanaqin district and from the north and northwest the district.Al-Khalis is bordered from the south by Baladruz district, while Al-Wajihiya district borders the Muqdadiya district center from the south and southwest, while Abu Saida borders the study area from the west, Map 1.

6. Temporal limits of research

The temporal limits of the research are represented by studying the changes in land cover and land use in the Muqdadiya district center between 1995 and 2019.

7. Research methodology

The two researchers relied on the descriptive approach and quantitative analysis to determine the types of land cover and land uses in the Muqdadiya district center and the changes that occurred in it between 1995 and 2019.

8. Methods and techniques for change detection

The study began to detect change in the use of multi-band remote sensing visuals, as they provide sufficient information to determine the magnitude and direction of change in land cover and land use.The methods of detecting change for multi-band remote sensing images can be divided into the method of detection by visual analysis, the method of detection by algebra, the method of subtraction of the image, the method of image ratio, and finally the method of detecting the change after classification [2], which is the method used by the researchers due to its advantages. In addition, it is the most common.

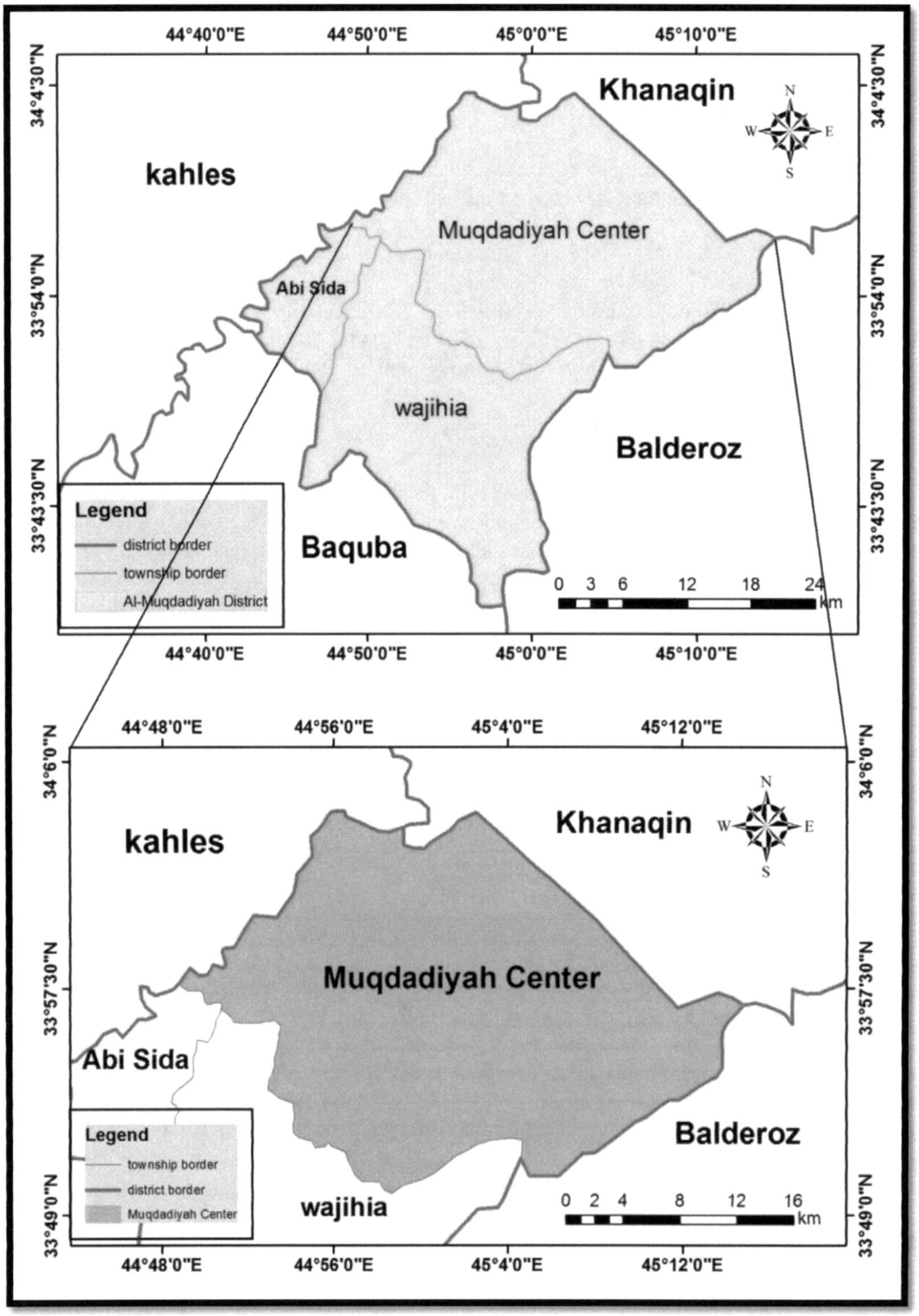

Map 1. Of the location of the study area in relation to the district of Muqdadiyah.

9.Results

Satellite visuals are a major source of geographic information and are a product of satellite imagery, as electromagnetic radiation reflected from the Earth's surface is measured and recorded in multiple spectral ranges [3].Also, US satellite images (Landsat) are one of the research methods in the field of studying environmental resources and land uses through monitoring, following up and evaluating them. The researchers used two satellite images in the US satellite series (Landsat), as in Table 1. and Figure 1.

Table 1. the satellite visuals used in the study

N	the moon Industrial	Allergy type	History Visual capture	Row	Path	The number of spectra
1	Landsat 5	TM	10\8\1995	(36) (37)	168	7
2	Landsat 8	OLI-TIRS	27\7\2019	(37)	168	11

10. How to detect change after classification

The post-classification change detection method is the simplest of the classification-based change-detection analysis technique. After the classification process, the comparison method can be used with two or more images, as each of the images with multiple time periods are classified separately, and then the images resulting from the classification process are compared. One of the disadvantages of this method is the high requirements and the many details for classifying the required categories for space visualization [4]. While it is characterized by the ability to calculate quantitative data (areas) for the total change in the land cover and land uses within the study area, as well as the ability to calculate the areas of each cover or use separately. As in Map 2. and 3. The directed classification was also used, which is the most used by researchers, as it is based on prior information about the spectral reflectivity characteristics of the land cover features to visualize the study area. And that is through geographical knowledge, field visits, or aerial photograph of the region [5]

11. Land cover change and land use for the period (1995-2019)

The changes that occurred during the period from 1995 to 2019 were monitored by comparing the results of the classification directed to the satellite visual that covered the study area for the year 1995 and that resulted from the mosaic process with the results of the same classification of the satellite visual that covered the study area for the year 2019 and it was found that the amount of the total change In the land cover and land use between these two years within the Muqdadiya district center, it has reached (289.14) km2 with a change rate of (51.54%). In addition, the percentage of change that occurred in some types of land cover and land uses is very high and will take place in the next Speak a review of these percentages and the amounts of change as shown in Table 2. and Figure 2. as follows:

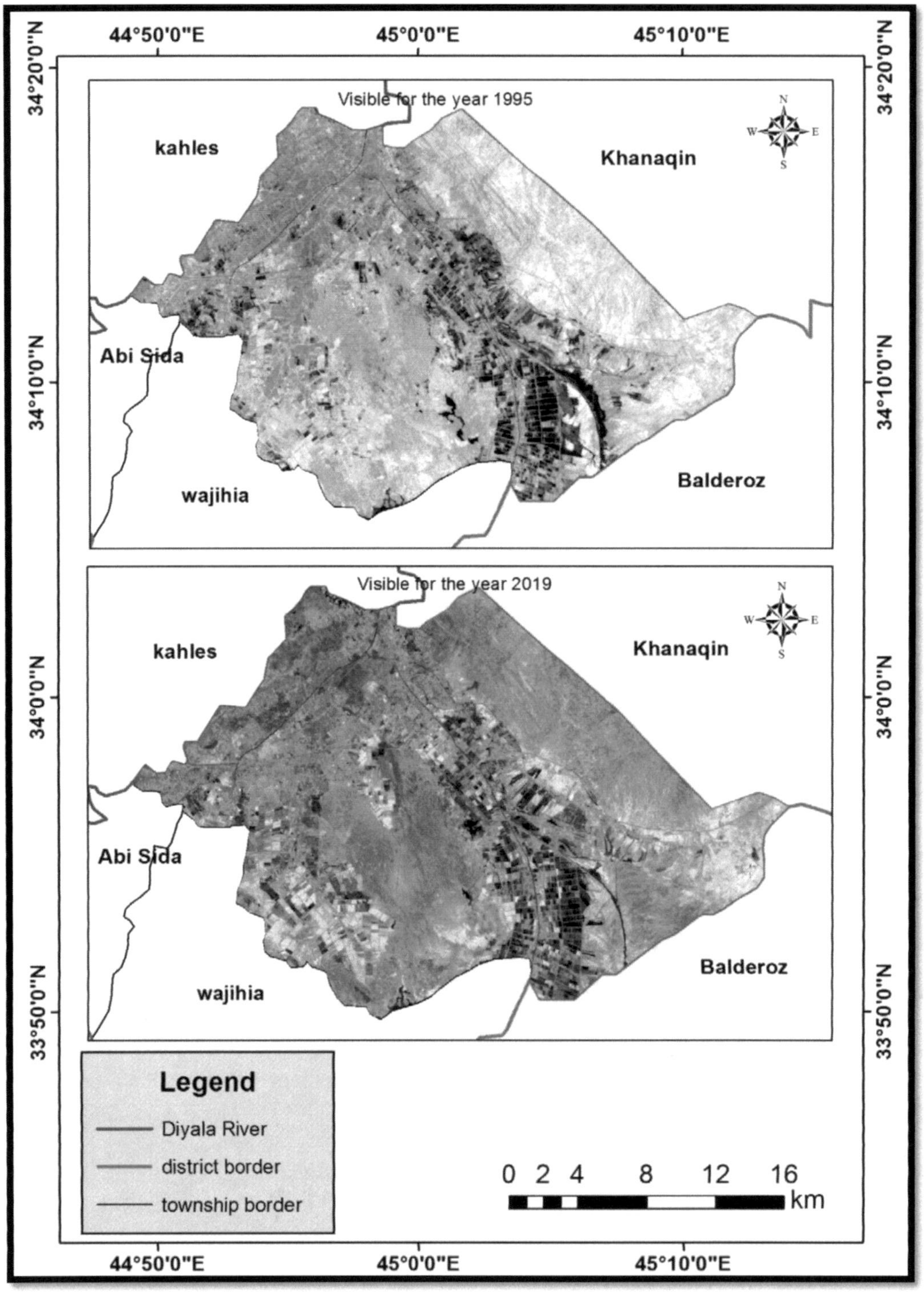

Figure 1. Visuals used in the study

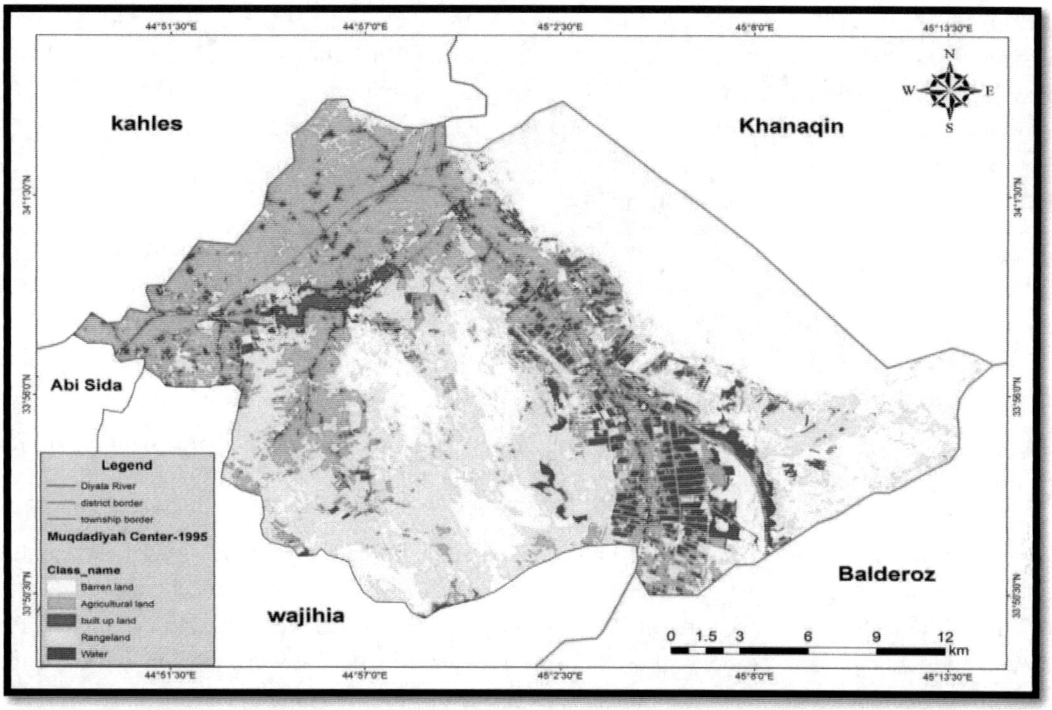

Map 2 . classification of land cover and land use in Muqdadiyah district center for the year 1995

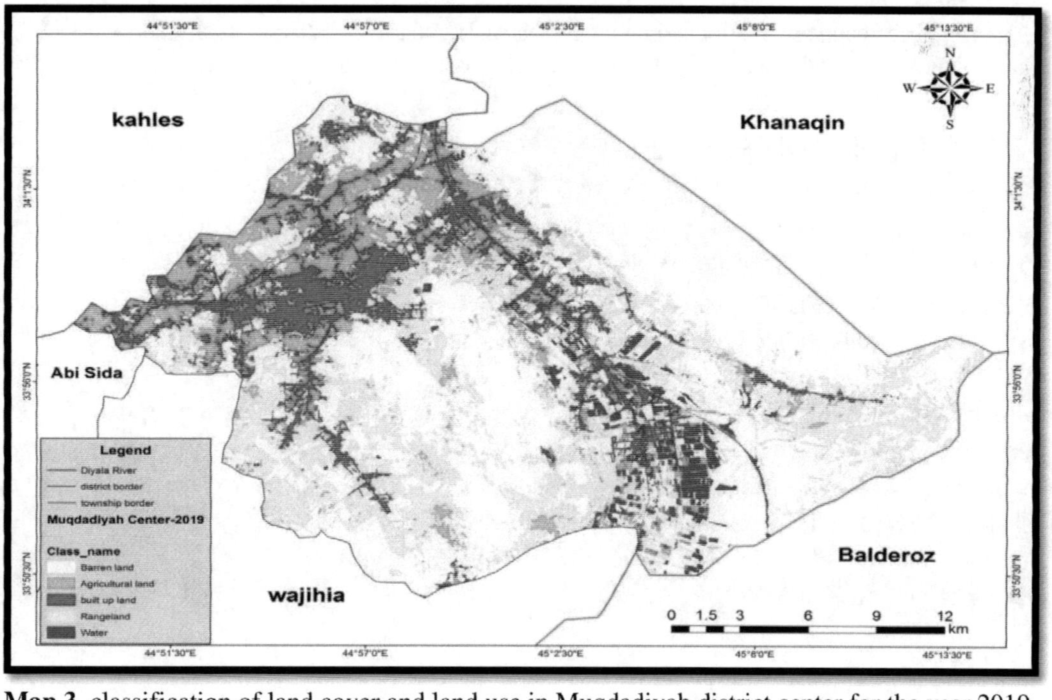

Map 3. classification of land cover and land use in Muqdadiyah district center for the year 2019

11.1 Built-up Land: The change detection process indicates an increase in built-up lands (urbanization), as it is evident from Table 2. and Figure 2. that the amount of change in built lands between 1995 and 2019 amounted to (44.38) km2 with a positive change rate of(204,42).After the built-up land covered an area of (21.71) km2 in 1995, it increased significantly to become in 2019 (66.09) km2.The reason for this urban expansion and the major change in the built up area (urbanization) is due to several reasons, including:

A- The increase in population numbers within an area for study, whether by natural increase or migration, as its population reached (174791) people in 2019 after its population did not exceed (109,106) people in 1997 [6], and this increase in turn requiredAn increase in the number of housing units in order to accommodate the increase in the number of families, as well as the need for additional service, commercial and industrial institutions.

B - The improvement in the economic level of the population and the accompanying need for commercial and industrial establishments, in addition to other land uses.

11.2 Agricultural Land Table 2. and Fig 2. indicate a decrease in the agricultural land area within the district of Muqdadiyah during the period between 1995 - 2019, as the amount of change reached (-91.16) km2 with a negative change rate of (-56.58), after the agricultural lands used to occupy an area that amounted to 161.12 km2 of the total area of the study area in 1995 (69.96) km2 in 2019 and the reason for this apparent decrease in agricultural lands is due to a number of factors, including:

*- Agricultural lands have been subjected to urban sprawl and expansion at its expense, as a result of the reluctance of many peasants to cultivate and abandon their lands for several reasons, including:

1. The deteriorating security situation in the center of Al-Muqdadiya district, which had its effects on agricultural lands and fruit orchards, as a result of the migration of farmers and leaving their agricultural lands.
2. The high costs of producing agricultural crops offset by their low economic return.
3. Entering all kinds of imported agricultural crops at low prices that compete with the local crop.
4. Most of the population engaging in productive trades such as trade and industry and other trades and businesses that provide a better financial return than what agriculture provides.
5. The enrollment of a large number of young people, especially the countryside, who prepare the labor force in agriculture in the ranks of the army and the police, in order to secure stable income for them and their families.

11.3 Almarai (Rangeland) Table 2. and Figure 2. show that the pastures witnessed a decrease in their area during the period between 1995 and 2019 within the Muqdadiyah district center.As it recorded an amount of change of (- 39.89) km2, and a negative rate of change of (- 27.05%).As pastures formed an area of (147.49) km2 in 1995, this area decreased to reach (107.6) km2 in 2019.This is due to the prevalence of drought in general and within the date of visual capture in particular, as well as to overgrazing as well as urban sprawl

Table 2. Land cover change and land use between 1995 – 2019

N	Cover type and land uses	Area, km2 for the year 1995	Area, km2, for the year 2019	The amount of change (*) km2 between 1995-2019	Percentage change (**)%	Type of change
1	Urbanism	21.71	66.09	44.38	204.42	a plus
2	Farmland	161.12	69.96	-91.16	-56.58	Shortage
3	Pastures	147.49	107.6	-39.89	-27.05	Shortage
4	Water	32.93	19.41	-13.52	-41.06	Shortage
5	Barren lands	197.75	297.94	100.19	50.66	a plus
6	Total	561	561	289.14	51.54	-

(*) - The amount of change: it is the difference in area between two different years and the researcher was able to obtain it through the following formula:

(Amount of change = area in the newest year - area in the oldest year).

(**) - percentage change: the percentage change was extracted through the following formula:

$$x = \frac{A2 - A1}{A1} * 100$$

whereas :

X = percentage change

A1 = Land cover area in the oldest.

A2 = Area of land cover in the most recent year.

See: DalalZureikat, The change of land cover and land use in Azraq district, Zarqa Governorate between 1978-2015 using geographic information systems and remote sensing, Jordan Journal of Social Sciences, Volume 9, Issue 3, 2016, p. 290.

11.4 Water Show the change in the area of water to decrease, as the amount of change reached (-13.52) km2 with a negative change rate of (- 41.06) due to the change in climatic conditions that characterized the fluctuation of the amount of rain falling in the study area On the one hand [7]. The area of agricultural lands, especially the rice farms that appear in the form of water basins, is declining due to the coincidence of its cultivation with the date of visual capture on the other hand. Which contributed to increasing the water area in the study area in the year 1995. After the water used to occupy an area that amounted to (32.93) km2 of the total area of the study area in 1995, it decreased to reach in 2019 (19.41) km2 of the total area of the study area as shown in Table 2. and Figure 2.

11.5 Barren Land: Barren lands continued to increase and its area expanded within the district of Al-Muqdadiya between 1995 and 2019, as it recorded an amount of change of (100.19) km2 and a positive change rate of (50.66%).After it covered an area of (197.75) km2 in 1995, its area expanded in 2019 to constitute (297.94) km2, as shown in Table 2. and Figure 2.As a result of converting agricultural lands and fruit orchards to residential land, as well as changing climatic conditions and the prevalence of drought.

12. Conclusions:

12.1 The research revealed that there are five types of land cover and land uses within the Muqdadiya district center.

12.2 The decline of agricultural lands, pastures and water, which was matched by the expansion of barren lands and urbanization, which created environmental changes that had their effects on the environment and local climate of the study area.

13. Recommendations:

13.1 Limiting the phenomenon of urban expansion on agricultural lands and green areas and working on the sustainability of agricultural lands within the study area.

13.2 The trend towards vertical housing (vertical expansion) for urbanization in order to preserve agricultural lands and limit the phenomenon of urban expansion in its direction.

13.3 Work to enact strict laws against transgressors of agricultural lands.

14.References

[1] SabaaA M and Najm A H 2018 Observing changes between land cover and landuse in the Amiriyat al-Fallujah district*Tikrit University for Human Sciences Journal***25** 189
[2] CampbellJ B 2007 *Introduction of Remote sensing* New York: Guilford Press 325
[3] Campbell D B 1996 *Introduction of Remote sensing* New York: Guilford Press 149
[4] StuartM R 2007 *A remote sensing change detection study in the arid Richtersveldregion of South Africa* Cape: University of Western 20.
[5] ShaoqingZ2008The comparative study of three methods of remote sensing imagechange detection*magazine the international archives of the photogrammetry remotesensing and spatial information sciences*1596 – 1597
[6] Ministry of Planning. Central Bureau of Statistics *Diyala Statistics Directorate*unpublished data
[7] Al-Dzai S A 2013 *Iraq's Old and Contemporary Climate*Baghdad: House of GeneralCultural Affairs 317

Mitigating the impact of climate change on UK buildings through zero energy strategies

Egglestone Kayleigh[1], Abdellatif, Mawada[2], Amoako-Attah Joseph[2], Saif S. AlQuzweeni[3], Khalid Hashim[2,4]

[1] Student Civil Engineering Department, Liverpool John Moores University, Liverpool, UK.
[2] Built Environment and Sustainable technologies (BEST) Research Institute, Liverpool John Moores University, Liverpool, UK.
[3] Civil Engineering Department, University of Babylon, Iraq.
[4] Department of Environmental Engineering, University of Babylon, Babylon, Iraq

Email: K.Egglestone@2018.ljmu.ac.uk

Abstract. It is currently established that one of the paramount concerns in the built environment is the energy efficiency of new and existing UK dwellings, respective to the unfavourable impacts posed to climate change. The Department for Business, Energy and Industrial Strategy in the United Kingdom have reported that the UK's highest recording temperatures have transpired in the years since 2002. With over 90% of England homes currently in use of high carbon systems for space heating and domestic hot water. Contributing to increased atmospheric carbon emissions in the dependency on fossil fuel burning; alluding to human-produced atmospheric temperature increase. To help tackle these issues in the residential sector, the capacity of zero-energy technologies has been introduced. Zero-energy implementation has potential to revolutionise the power system, with on-site power generation at the forefront of this. This paper will explore the influence of zero-energy implementation on two UK residential dwellings of disparate locations, using Integrated Environmental Solutions Virtual Environment (IE-SVE) by focusing on renewable on-site micro-generation systems. The ASHRAE climate zones of Edinburgh and London Gatwick has been selected to examine the performance of the building over varied regional climates of disparate locations. The selected design variables were finally implemented in combination for building simulation in IESVE and compared with a basic model dwelling. The processed simulation results showed a reduction in the buildings energy consumption of 43.4538MWh (71%) for Edinburgh and 33.9929MWh (64%) for London respective to the baseline model. The greatest savings in mitigation of UK climate change can be evaluated in relation to reduction of carbon emissions, which were 7880kgCO_2 (46%) and 5423kgCO_2 (36%) respectively.

Keywords: Energy efficiency; zero-energy; IESVE; energy consumption; carbon emissions.

1. Background

Evidence of global warming presents the years since 2002 as the UK's highest recording annual temperatures according to the reports of the Department for Business, Energy and Industrial Strategy, UK. With recorded sunlight intensity in 2017, presenting 100% greater than the 1981-2010 average [1]. Addressing anthropogenic atmospheric temperatures from UK domestic energy consumption is vital in mitigation of this to rectifying the potentiality of this magnifying. Governmental decarbonisation objectives hold the authority to the correction of UK climate change, moving forward with legally binding national targets for reduction in carbon emissions. It must be highlighted the climate change is a global problem [2, 3] that causing an unexpected increase in weather temperature [4, 5], uneven distribution of rain [6, 7] and a severe freshwater shortage [8, 9]. This problem (climate change) is a result of several reasons; the first and most imp[ortant reason is the industrial emissions [10, 11], such as the emissions of cement factories [12-14].

These legally binding targets for greenhouse gas (GHG) reduction were approved by the 2008 Climate Change Act, operating through the introduction of carbon budgets. Specifically, with the point of lowering the UK's GHG emissions by 80% by 2050, relative to the 1990 baseline [15]. The Energy Company Obligation (ECO) outlines the importance of household energy efficiency in the achievement of this. It states in its most recent addition, that between 2018-2022, the housing stock is to account for "31% of the UK's total primary energy consumption". This high figure is in accordance with the basic dwelling energy demand requirement in the assurance of substantial thermal comfort for dwelling occupants. According to the Department for Business, Energy and Industrial Strategy, UK; almost 90% of homes in England currently use fossil fuels for heating and hot water, with fossil fuels being a huge contributor to the output of carbon emissions into the atmosphere. This accumulation of carbon emissions from fossil fuels is a key GHG source, driving the heating of the lower layers of the atmosphere and penultimately causing increased temperatures. Therefore, presenting the desirability for alternative energy production, in terms of renewable micro-generation systems, to reduce dependency on natural resources. Thus, mitigating the irreversible consequences of climate change through safe and secure energy resources.

Considering the current attainability of this, the energy efficiency of UK homes are not in the facilitation of this. Evidenced figures present that 66% of total UK homes, around 16 million, have an Energy Performance Certificate (EPC) of D or worse. Department for Business, Energy and Industrial Strategy, UK stated, in 2018, with dwelling band A certificates depicting that of being the most energy-efficient. Therefore, exclaiming a national energy performance inadequacy, working in opposition to the governmental objectives by 2050. Calling for urgency in the retrofit of new and existing homes, with zero-energy strategies as a means of upgrading the EPC and subsequent energy consumption performance of UK homes. Achieving this will ensure the current housing stock is performing in line with national emission targets set by the fourth and fifth carbon budgets. This requires conformity to new and existing rigorous building regulations, along with passive zero-energy considerations in the design stages. The promotion of on-site renewable energy deploys low-carbon techniques to replace fossil fuels, operating towards carbon-neutral dwelling energy systems [16]. Stabilising the reliance upon non-renewable natural resources for dwelling energy production; providing maximised dwelling thermal properties, whilst ensuring occupant comfort. The importance of this is presented in the statistics of UK housing. Hereby, housing demand is increasing consistently per annum, in line with natural population growth. 2019 evidenced the completion of 173,660 newly built dwellings, which is up 20,000 from 2015 figures [17]. This was exclaiming the urgency of sustainable construction to entitle the progression of future housing developments in line with increased population demand; while committed to national carbon emission reductions by 2050

In the context of implementing these zero-energy strategies for low energy homes in the UK, this paper presents the methodology and findings from a dynamic thermal simulation. This simulation process is performed by Integrated Environmental Solutions Virtual Environment (IESVE), considering energy generation and conservation of the residential dwelling before and after micro-generation implementation. Zero-energy modification examples refer to the energy-efficient retrofits which focus on lowering building emittance and energy consumption, specifically solar PV, wind power and heat pumps. Holding analysis to the sensitivity of building performance respective to the regional climates of the disparate UK locations, in terms of dwelling total energy, carbon emissions and total electricity. Providing conclusion to the feasibility of the specified micro-generation measures, in corroborating with national targets for the future successfulness of green building in the mitigation of UK climate change.

2. Methodology

2.1 IESVE modelling

The data used is the ModelIT two-storey residential-detached building. The building drawings consist of a ground floor plan, first-floor plan and roof arrangement plan. Figs. 1(a)-(c) below indicate the plan of the house used in this work. In the current study, table 1 presents standard construction template U-Values utilised consistently throughout the modelling, in line with the 2013 Building Regulations.

Table 1 Standard construction specifications in IESVE.

	Basic dwelling		Improved dwelling	
	Material	U-Value (W/m^2K)	Material	U-Value (W/m^2K)
Roof	Uninsulated Roof	2.1442	2013 Roof	0.1800
Door	Wooden Door	2.1944	2013 Door	2.1997
External Window	Small Double Glazed	3.1704	2013 External Window	1.6000
Ground/Exposed floor	Uninsulated Concrete	0.8871	2013 Ground/Exposed floor	0.2200
Internal Ceiling	Timber-Joist	1.2585	2013 Internal Ceiling	1.0866
External Wall	Brick Cavity with Dense Plaster	1.4917	2013 External Wall	0.2599
Internal Partition	2013 Internal Partition	1.7888	2013 Internal Partition	1.6598

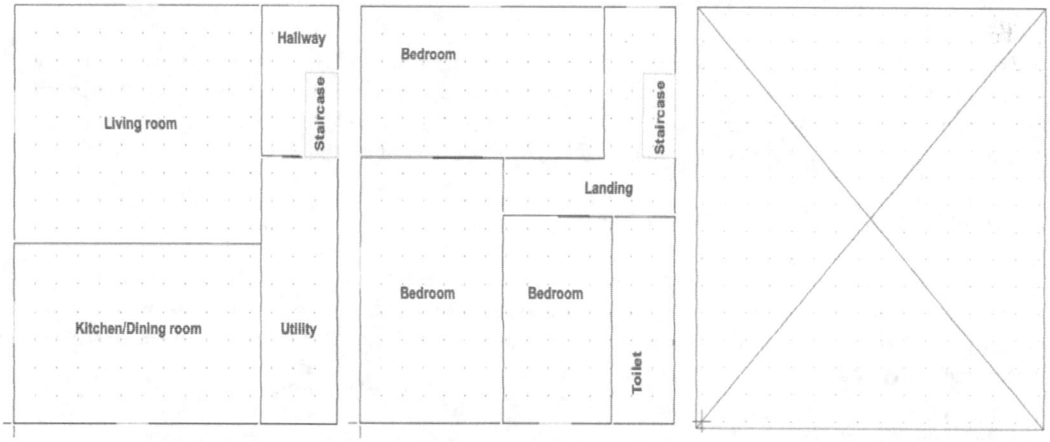

(a) Ground floor plan (b) First floor plan (c) Roof arrangement

Figure 1. Architectural plan

2.2 Modelling process

2.11 Baseline model

The baseline model presents a basic dwelling construction template in line with the requirements set by the 2013 Building Regulations. Hereby, there is no implementation of on-site micro-generation systems. This means that the dwelling is solely governed by the building fabric U-Values outlined in table 1, thermal mass, as well as internal and external heat gain. The reliance on a conventional heating system is presented, adopting a gas boiler for central heating using water to heat the dwelling.

2.21 Solar PV

Roof installation of three 'Monocrystalline silicon solar cells' measuring 10m^2 was presented for both dwellings in London Gatwick and Edinburgh. This PV system is selected in paramount, over a 'Polycrystalline silicon panel' in the achievement of the highest photovoltaic efficiency, with electricity conversion rates said to be higher. Concisely working in the conversion of the sun's energy into available dwelling electricity as an inexhaustible renewable energy source.

2.22 Wind power

The implementation of a small wind turbine was consistently applied over both London Gatwick and Edinburgh. These turbines, which were used in simulations, hold a power rating of 10kW. In the current study, wind power was used to harness kinetic energy to convert into a free and inexhaustible source of dwelling electricity.

2.23 Air-Source heat pump (ASHP)

An electric ASHP was applied to the baseline model for both locations in this processed simulation, reversing the natural flow of heat. This transition from gas to electric heating holds promise to an improved dwelling carbon footprint; attainable due to the refined energy efficiency conversion rate of the ASHP. Specifically selected over the ground-source heat pump due to improved economic viability through reduced initial capital and operation costing.

2.24 Combination of all three design considerations

Hereby all the three respective dwelling zero-energy measures were combined into one concise model and combined with the respective baseline model. This with the point of exploiting solar pv, heat pump and wind energy to improve the building performance for enhanced dwelling energy consumption. Presenting the efficacy of measures in the attainability of decelerated progression of UK climate change.

2.3 Simulation process

IESVE as a dynamic simulation modeller, analyses energy and different building performance workflows. The building performance simulation requires a relevant selection of modelling parameters and assumptions; these have been fulfilled following 2013 Building Regulations and outlined in table 1. The other simulation parameters of occupancy profile and DHW consumption were populated to simulate and reflect true post-construction behaviours. These modelling assumptions can be taken as 5.8l/h (max) for DHW, operating to a consistent occupancy profile shown in figure 2. This is in accordance with the understanding that the dwelling occupants are to be periodically vacated from the dwelling during

IOP Conf. Series: Earth and Environmental Science **877** (2021) 012018 doi:10.1088/1755-1315/877/1/012018

standard working hours. Most existing buildings in the UK are naturally ventilated, and thus, this was the ventilation strategy implemented in the case of the modelled residential dwelling.

The modelling and simulation parameters of occupancy profile, DHW consumption and internal conditions were kept consistent throughout the process, with the only variant being the weather data used. This is due to the differential weather associated with the Northern location of Edinburgh and the Southern location of London Gatwick. The thermal simulation process is shown in Fig. 3, with its associated modelling parameters outlined in table 1.

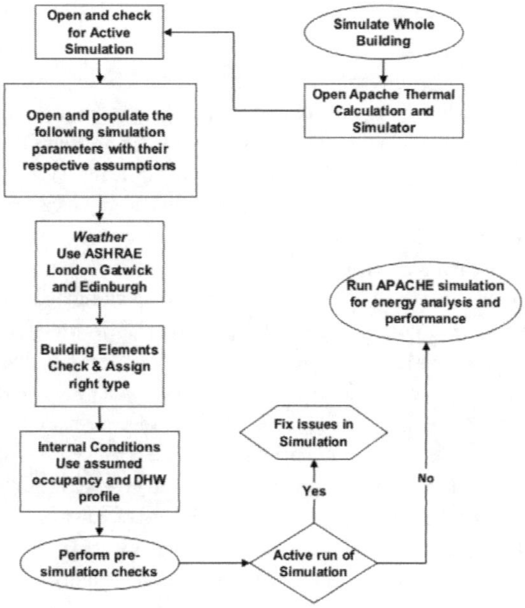

Figure 2. Consistent occupancy profile for the residential dwelling.

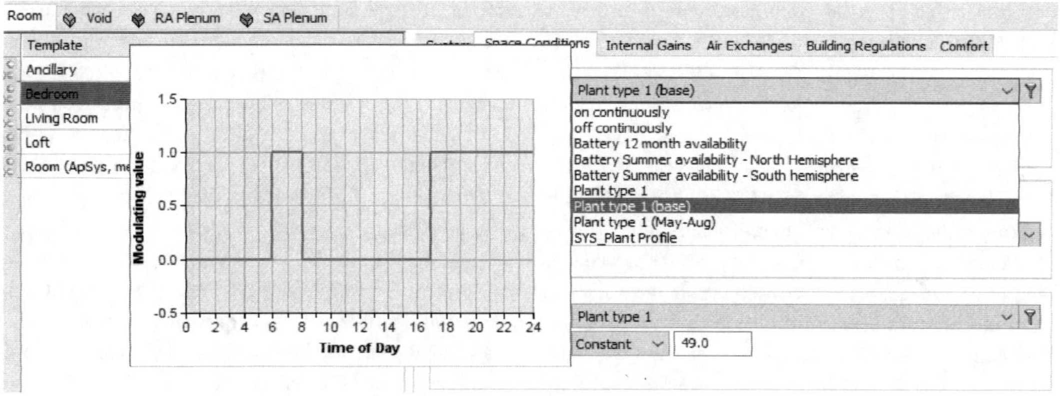

Figure 3. Thermal simulation process.

3. Results and Discussion

3.1 Combination of all three design considerations

The simulation process of the three combined design considerations was carried out by IESVE. The simulation was evaluated in respect to the parameters 'Total Energy', 'Total Carbon Emissions' and 'Total Electricity'. The statistical analysis carried out in figure 4(a-c) suggests the present relation between zero-energy design modifications and improved building performance is established. Figures 4(a)-(c) show comparative statistical analysis to the deviation from the baseline model, after zero-energy implementation at both disparate locations. The general trends consistently evidence disparity in dwelling energy consumption over the three tested parameters, with London persistently of a lower value. This change in consumption yielded is in response to the variation of regional climates at the separate northern and southern sites, as the basic construction and thermal template remain consistent for both. The impressionability of location to dwelling energy performance is presented through the colder northern climate of Edinburgh, to London Gatwick, in closer proximity to the equator. Hereby both dwellings possess variability in energy requirements in the maintenance of comfortable occupant thermal comfort. With lower external temperatures permitting increased energy demand for dwelling space and water heating, possessing adequacy throughout the annum. The annual energy consumption results indicated in figure 4(a), evidence a significant decrease in energy for both locations after zero-energy modification. A declining trend is in respect to the improved dwelling energy efficiency throughout the year, in the presence of on-site energy generation systems. This is attributed to a reduction of 43.4538MWh for Edinburgh, performing 71% more efficiently than the baseline model. Whilst London presents a 64% lower efficiency than Edinburgh, with a total energy reduction of 33.9929MWh throughout the annum comparatively. This improved energy points towards the prosperity of enhanced energy efficiencies and subsequent EPC ratings upon the achievement of net-zero ambitions. Hereby, the renewable house stands paramount when compared to the non-renewable basic dwelling, satisfying lower operational efficiency requirements cohering to government targets. This encourages long-term sustainability in the sector over the estimated service lifespan of the building; whilst amending the depletion of natural resources from dwelling HVAC conditions. Although heat pumps incur a higher capital cost than conventional gas boilers, they can provide increased emission reductions from dwelling heat demand. The dwelling total carbon emissions results indicated in figure 4(b) show an observable decrease in values over both locations, after zero-energy implementation. Unlike the evidenced trend in figure 4(a), the margin of emission deviation from the baseline model is less significant, with extreme value difference less prominent throughout the annum. With both dwellings presenting the lowest emissions in June, exhibited as almost fully carbon-neutral over the months of June-September. This is indicated by an emission decrease from January to June in Edinburgh and London, by 91% and 94% respectively. This can be attributed to the COP of an ASHP being greater in the summer relative to the winter months [18]. The overall annual savings in carbon emissions comparative to the baseline model, can be said to be the greatest in contribution to mitigation of UK climate change. This is evaluated by a reduction of total carbon emissions of 7880kgCO_2 (46%) for Edinburgh and 5423kgCO_2 (36%) for the Southern counterpart. This fall in dwelling emissions gives indication to the potential alleviation to environmental pollution in the replacement of conventional high-carbon and non-renewable dwelling generation systems. Holding authority to this scenario, in providing the greatest offering to government decarbonisation targets, adhering to an 80% reduction relative to the 1990 baseline. Analysis of total electricity in figure 4(c), presents the baseline consumption as almost completely consistent throughout the annum. After full zero-energy implementation, the variability in observed values across both locations is sustained. Hereby, London is evidenced as consuming 1.3677MWh more electricity throughout the annum than the Northern equivalent. This increase can be accredited to characteristic location factors, in the reduced available wind energy and restricted potential return from turbine installation. Edinburgh is located within a mountainous topography, whereby mountains serve as natural towers increasing the distance above sea level for magnified exposure to higher wind speeds. This ensures maximised energy generation for the Edinburgh dwelling, compared to London, which is situated within a more sheltered and low-lying terrain. It can be said that the fall in electricity is consistent with previous trends of seasonal dwelling output, emphasising the efficiency of the systems in the summer months to yield greater electricity cuts. However, this fails to remain consistent for a sufficient duration, holding an incapacity in advocating a year-long enhanced electricity efficiency. Evidenced is a higher overall electricity consumption of 4.9702MWh for Edinburgh and 6.3379MWh for London compared to the baseline model. A potential reason for this could point to the implementation of an electric ASHP, whereby operation relies on

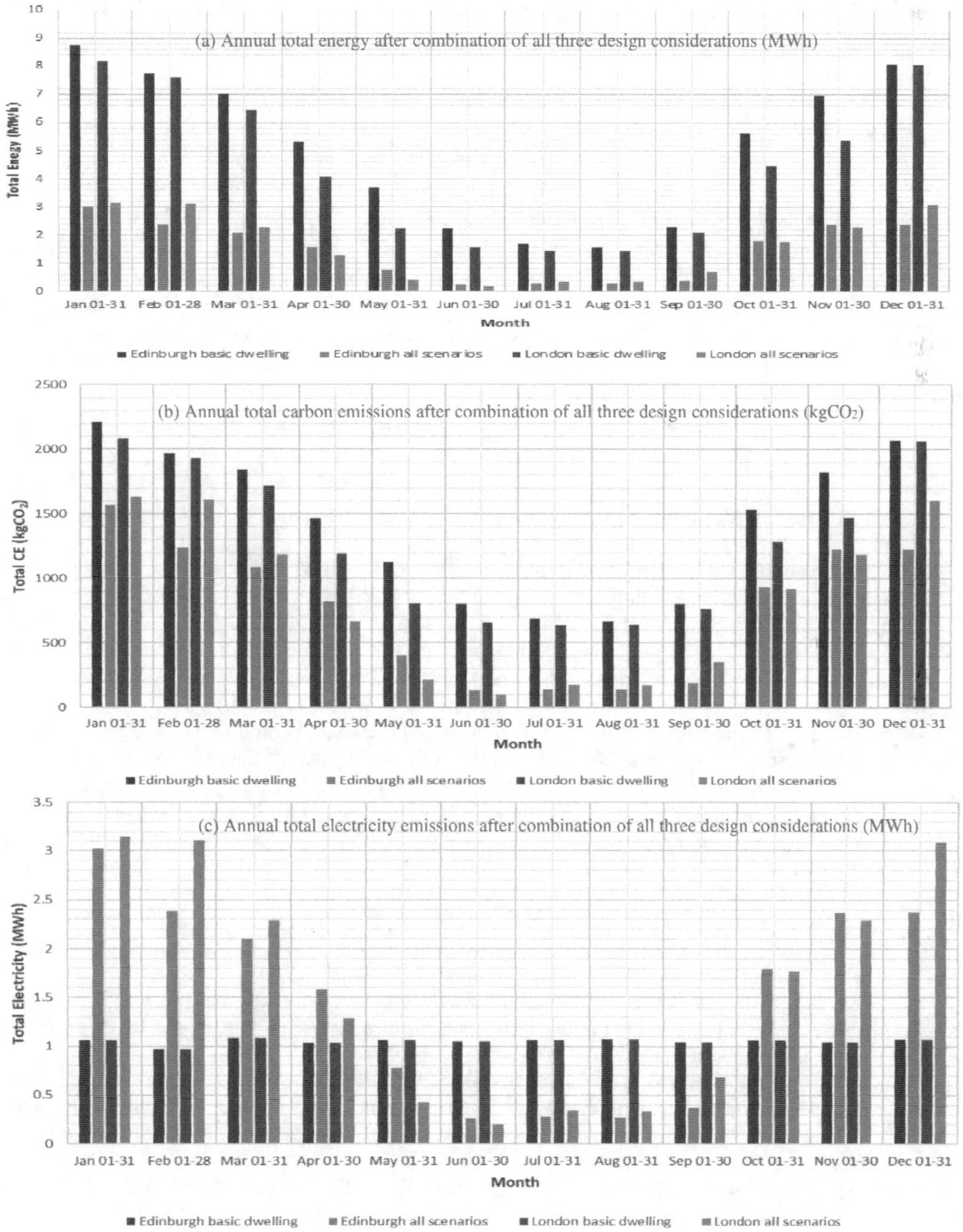

Figure 4. Annual analysis after full zero-energy implementation of both locations.

electricity to move heat from a cool to a warm space. This additional electricity demand is a disadvantage in the installation of ASHPs. Despite the installation of wind and solar electricity generation systems, this still evidences insufficient in offsetting the increased demand. Hereby consideration of the equality of utility costing is emphasised, whereby the well-developed gas networks and cheap natural gas provide a challenge to the acceptance of electric ASHPs.

4. Conclusions

This study empirically quantifies domestic energy, carbon emissions and electricity consumption through simulation of implemented zero-energy strategies respective to a baseline model. Results show that domestic energy consumption differs in different locations and changes appropriate to seasonal variance. This highlighted the influence of location and regional climate on dwelling energy patterns. Generally, it was found that zero-energy efficacy explicitly offered the greatest savings in regards to total energy and total carbon emissions across both Edinburgh and London. As expected, there was a slight increase in dwelling electricity consumption, as the combined simulation held installation to an electric ASHP. However, it should be noted that in line with prospective mass electrification, the inequalities associated with this increased figure are to be minimised. The results of this paper should encourage the acceptance of zero-energy strategies, particularly due to their positive alleviation of environmental pollution through the depreciation of carbon emissions. Further studies are desirable to better understand the role of zero-energy strategies in developing their markets and policy frameworks to increase public acceptance and education into more extensive national deployment.

References

[1] M. Kendon, M. McCarthy, S. Jevrejeva, A. Matthews, and T. Legg, State of the UK climate 2017, International Journal of Climatology. 38 (2018) 1-35.

[2] S.L. Zubaidi, P. Kot, K. Hashim, R. Alkhaddar, M. Abdellatif, and Y.R. Muhsin, Using LARS–WG model for prediction of temperature in Columbia City, USA, IOP Conference Series: Materials Science and Engineering.584, (2019) 012026.

[3] L. Zubaidi Salah, H. Al-Bugharbee, S. Ortega Martorell, S. Gharghan, I. Olier, K. Hashim, N. Al-Bdairi, and P. Kot, A Novel Methodology for Prediction Urban Water Demand by Wavelet Denoising and Adaptive Neuro-Fuzzy Inference System Approach, Water. 12 (2020) 1-17.

[4] S. Zubaidi, S. Ortega-Martorell, H. Al-Bugharbee, I. Olier, K.S. Hashim, S.K. Gharghan, P. Kot, and R. Al-Khaddar, Urban Water Demand Prediction for a City that Suffers from Climate Change and Population Growth: Gauteng Province case study, Water. 12 (2020) 1-18.

[5] Z. Salah, I.H. Abdulkareem, K.S. Hashim, H. Al-Bugharbee, H.M. Ridha, S.K. Gharghan, F.F. Al-Qaim, M. Muradov, P. Kot, and R. Alkhaddar, Hybridised Artificial Neural Network model with Slime Mould Algorithm: A novel methodology for prediction urban stochastic water demand, Water. 12 (2020) 1-18.

[6] Z. Salah, S. Ortega-Martorell, P. Kot, R.M. Alkhaddar, M. Abdellatif, S.K. Gharghan, M.S. Ahmed, and K. Hashim, A Method for Predicting Long-Term Municipal Water Demands Under Climate Change, Water Resources Management. 34 (2020) 1265-1279.

[7] S. Zubaidi, H. Al-Bugharbee, Y.R. Muhsin, K. Hashim, and R. Alkhaddar, Forecasting of monthly stochastic signal of urban water demand: Baghdad as a case study, IOP Conference Series: Materials Science and Engineering.888, (2020) 012018.

[8] S.L. Zubaidi, H. Al-Bugharbee, Y.R. Muhsen, K. Hashim, R.M. Alkhaddar, D. Al-Jumeily, and A.J. Aljaaf, The Prediction of Municipal Water Demand in Iraq: A Case Study of Baghdad Governorate, 12th International Conference on Developments in eSystems Engineering (DeSE), Kazan, Russia (2019) 274-277.

[9] Z. Salah, K. Hashim, S. Ethaib, N.S.S. Al-Bdairi, H. Al-Bugharbee, and S.K. Gharghan, A novel methodology to predict monthly municipal water demand based on weather variables scenario, Journal of King Saud University-Engineering Sciences. 32 (2020) 1-18.

[10] R.A. Grmasha, O.J. Al-sareji, J.M. Salman, K.S. Hashim, and I.A. Jasim, Polycyclic Aromatic Hydrocarbons (PAHs) in Urban Street Dust Within Three Land-Uses of Babylon Governorate, Iraq: Distribution, Sources, and Health Risk Assessment, Journal of King Saud University - Engineering Sciences. 33 (2020) 1-18.
[11] O.J. Al-Sareji, R.A. Grmasha, J.M. Salman, I. Idowu, and K.S. Hashim, Street dust contamination by heavy metals in Babylon governorate, Iraq, Journal of Engineering Science and Technology. 16 (2021) 3528 - 3546.
[12] A.A. Shubbar, M. Sadique, M.S. Nasr, Z.S. Al-Khafaji, and K.S. Hashim, The impact of grinding time on properties of cement mortar incorporated high volume waste paper sludge ash, Karbala International Journal of Modern Science. 6 (2020) 1-23.
[13] A.A. Shubbar, M. Sadique, H.K. Shanbara, and K. Hashim, The Development of a New Low Carbon Binder for Construction as an Alternative to Cement. In Advances in Sustainable Construction Materials and Geotechnical Engineering, 1st ed. Berlin: Springer, 2020, pp. 205-213.
[14] A. Kadhim, M. Sadique, R. Al-Mufti, and K. Hashim, Developing One-Part Alkali-Activated metakaolin/natural pozzolan Binders using Lime Waste as activation Agent, Advances in Cement Research. 33 (2021) 342-356.
[15] H. Government, Implementing The Climate Change Act 2008, (2011) 6.
[16] B. Chanchpara, Zero energy building, Master in Energy Management in the School of Engineering. New York Institute of Technology (2019)
[17] Ministry of Housing, Communities & Local Government, 2019. House Building; New Build Dwellings, England: June Quarter 2019. pp.5-8.,
[18] S. Tangwea and K. Kusakana, Qualitative Methodology for comparison of Performance of Air Source Heat Pump Water Heaters, 2020 International Conference on Smart Grid and Clean Energy Technologies (ICSGCE) (2020) 96-102.

The Success of the Construction Industry's Adoption of the Carbon Assessment Strategy PAS2050

Holly Smith[1], Amoako-Attah Joseph[2], Raad K. Hashim[3], Khalid S. Hashim[2,3]

[1] Student at Civil Engineering Department, Liverpool John Moores University, Liverpool L3 3AF, UK.
[2] Built Environment and Sustainable Technologies (BEST) Research Institute, Liverpool John Moores University, Liverpool L3 3AF, UK.
[3] Engineering Faculty, University of Babylon, Iraq.

Email: H.M.Smith@2018.ljmu.ac.uk

Abstract. PAS2050 was introduced in order to help combat the extremely high levels of emissions that the construction industry emits, as well as the large volume of natural resources they use, through the life cycle of a project. It was developed to create a consistent method of assessing the life cycle of Greenhouse gases as well as a method to increase understanding of where they arise from within a supply chain. This is essential in today's world as without change, the climate will continue to rise in temperature and contribute negatively to the already dramatic effects of climate change that have already been seen. By implementing PAS2050 the industry aims to allow stakeholders to realise the impacts and encourage movement towards a more sustainable future. This research aims to determine how well the construction industry has adopted the carbon assessment strategy covered by PAS2050 through informed opinions, data and case studies. Using journals and research, alongside a questionnaire which will be distributed to individuals within the industry, an informed decision will be made to determine how well the construction industry has adopted PAS2050. The results found that the industry is having difficulty in adapting to the change at all levels and scopes. The findings of this research show that the industry has only touched the surface and although the larger organisations are leading the charge, enough is not being done by smaller organisations and the supply chain.

Keywords: Carbon; PAS2050; AHP; Climate change.

IOP Conf. Series: Earth and Environmental Science **877** (2021) 012019 doi:10.1088/1755-1315/877/1/012019

1. Background

Climate Change influences all aspects of our lives [1, 2], it could be considered the greatest threat to the natural world due to several reasons, including the increase in the atmospheric temperature [3, 4], high consumption of potable water [5, 6], and high intensity precipitations in some area that could cause floods [4, 7], and at the same time it causes drought in many areas [8]. The main reported reason for the climate change is the CO_2 emission from man activities, such the manufacturing of cement [9-11] and other construction related activities [12-14]. PAS2050 was introduced and was expected "to boost international efforts to carbon footprint products" [15]. Benefits were expected to be 'invaluable' as it's a positive move towards 'greater appreciation of our environmental impact' which would be essential for long-term growth" [15]. There is an essential need to reduce our carbon footprint, especially in the construction industry that contributes to around '47% of the UK's total carbon dioxide emissions' [16]. Carbon footprints can have an effect on the environment, wildlife, human health and the economy; affects to any would be detrimental for a construction organisation and the stakeholders involved. The demand for change is more important than ever as the EU have set reduction targets for 2050 that require carbon trading levels to be 80-95% below the 1990 levels by 2050 [17]. The international construction industry has a duty to everyone, specifically its stakeholders, to do what they can to change. Arup stated that without immediate action there would not only be 'long term rises in materials, energy and waste disposal' but also a 'risk to reputation, potential loss of market share', [18], which would be difficult to reverse. As knowledge grows on the subject new ways to develop are coming into play; with carbon management, low carbon construction and life cycle analysis being just a few efforts made. The implementations of life cycle analysis and carbon analysis through PAS2050 allows for evaluating the environmental impact of all the processes and products. This means all elements that are used in construction, from extracting materials to disposal of the built structure, are analysed so that it can be indicated where and what can be changed or altered. Through life cycle analysis and following PAS2050 principles there will be a concise translation from inventory flows into a profile with 'a number of environmental impact scores', [19], which allows for better management and control and transparency; PAS2050's guideline to share details with appropriate third parties. PAS2050 works alongside PAS2080, which looks at the whole life cycle of the carbon used within projects [20]. Within this paper information will be gathered in order to determine how well these 'invaluable' policies have been implemented through the industry. It will be explored to critically evaluate how well the construction industry has adopted PAS2050 and its criteria and understand the industries opinion on the adoption process. Ultimately understanding what stage of adoption the industry is at; the progress made so far and how the industry has taken it on board. All of this being essential as if we do nothing and to not reduce emissions soon, 'the Earth's temperature will rise, the ozone layer will disappear, the Antarctic ice sheet will melt - causing sea levels to rise and flood most of the world' [21].

2. Methodology

2.1 Strategy and Methods

A questionnaire was introduced to understand the significance of the changes and how they have so far impacted the industry by exploring stakeholder and industry opinions. The questionnaire was developed in order to collect data as it provides a direct response from stakeholders and professionals within the industry. It allows for collection of qualitative and quantitative data from both open and closed questions. Each question had a specific purpose. The ideal questionnaire sample size was set at 50-100 participants. As the research requires an industry overview in order to determine an answer, the whole construction industry population was targeted through LinkedIn with simple random sampling. There was no specific requirement, no knowledge required, just being within the industry.

A case study was also created for in-depth understanding of one organization's adoption process. It will allow for identification of how one organization has changed and implemented processes or methods in order to meet requirements of PAS2050. This understanding will be found by determining the new rules and guidelines that were introduced and the program made. In order to understand the adoption of PAS2050 further, an in-depth a short case study was produced on the actions taken by a given organization within the construction industry through questionnaire answers and discussions with a carbon manager within the organization itself. The case study aimed to answer specific questions surrounding how the organization has implemented plans to tackle climate change and implement PAS2050 but also to discover how they see and compare with the industries adaptations.

2.2 Data Analysis

The case study is being created in order to determine how a singular entity within the industry has taken steps to adapt. The analysis within this will be used to determine how well adjusted they are and if their steps will help make a change to their carbon footprint, at all levels, as well as their impact on the environment. This will be done through the primary researcher gathering information and processing it in a similar way to that of the questionnaire in order to produce a minimally subjective opinion on how well the industry has adopted PAS2050.

Analytical Hierarchy Process (AHP) will be used for analysis of the questionnaire answers gathered in this study. AHP is a structured multi-criteria decision-making tool for assessing and analyzing complex decisions [22]. It uses analysis of elements in a hierarchy for creating the best decision as seen in Figure 1.

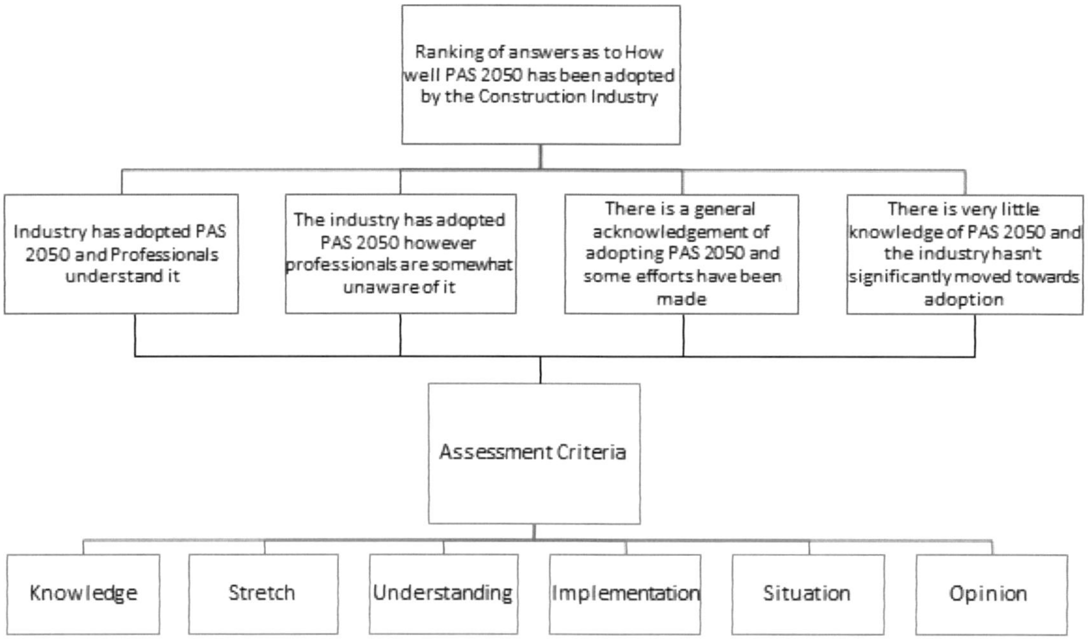

Figure 1: Hierarchical Structure of Ranking and Selection of answers

This study answers the key research questions "Has the industry adopted PAS2050? To what extent? What was the industries opinion?" The two main steps of this study are to determine (a) weights of criteria for the assessment of how well the industry has adopted PAS2050, and (b) the scores of how everyone has been impacted by it. We do this based on certain criteria as seen in Table one. The process begins by ranking different conclusion alternatives according to multiple criteria and determines the appropriate answer for 'how has the industry adopted PAS2050'. The criteria used and displayed in Table 1 applies to the data supplied through an administered questionnaire.

Table 1: Assessment Criteria Description

Criteria	Description
Knowledge	Is there knowledge of it at all levels within the industry; not just within decision-makers, includes lower job levels too.
Stretch	Does the knowledge stretch across various employers, enough to indicate for the whole industry?
Understanding	Significance of understanding of PAS2050 within the industry.
Implementation	Number of implemented regulations or schemes (on average).
Situation	Stage or situation at which the schemes are at.
Opinion	Positive or Negative Opinions

Through AHP there is the allowance for data to be compared in a rational and consistent format for processing an informed decision about the progress of implementation. We can also determine the successfulness so far. Using the 6 criteria for the assessment, the next step is to determine weights for each criterion using the AHP. The questionnaire combines both quantitative and qualitative data. The opinions from stakeholders within the industry will be determined to be negative or positive in order to rank the ratio and understanding of PAS2050 will be ranked on the primary researcher's authority. The performance scores displayed in Table 2 are collected, normalized, and analyzed in this study to give an overall decision.

Table 2: Performance Scores of Criteria

Criteria	Scores	
	0	**10**
Knowledge	No knowledge of PAS2050	Knowledge of PAS2050 at all levels
Stretch	1 organization surveyed	10+ companies surveyed
Understanding	No understanding of PAS2050	Complete and intense understanding of PAS2050
Implementation	Zero schemes known to be introduced (on average)	10+ schemes known to be introduced (on average)
Situation	No schemes completed or in operation	All schemes fully operational
Opinion	All negative responses	All positive Responses

The AHP obtains experts' input through comparisons of the criteria and examines the consistency of answers. Inputs from multiple experts are compiled and used to calculate the weights of criteria. Numerical priorities are derived to determine which best suits the decision goal. These criteria are then used to assess PAS2050s implementation in the industry. The analysis of the questionnaire will determine one of the following answers: Industry as adopted PAS2050 and professionals understand it/ The industry has adopted PAS2050 however professionals are somewhat unaware of it/ There is general acknowledgement of adoption and some efforts made/ There is very little Knowledge and movement towards adoption. This will be determined through assessments of decisions in Table 3 which categories the requirements for each decision.

Table 3: Assessment of Decision

Decision	Criteria					
	Knowledge	Stretch	Understanding	Implementation	Situation	Opinion
Industry as adopted PAS2050 and professionals understand it	8-10	0-10	8-10	5-10	5-10	5-10
The industry has adopted PAS2050 however professionals are somewhat unaware of it	4-9	0-10	4-9	3-7	3-7	3-7
There is general acknowledgement of adoption and some efforts made	2-5	0-10	2-5	2-5	2-5	2-5
There is very little knowledge and movement towards adoption.	0-5	0-10	0-2	0-2	0-2	0-2

3. Results and discussion

3.1 Questionnaire Findings and Analysis

The questionnaire responses were fully analysed by the primary investigator. In total 5 organizations were surveyed; the organizations are labelled A-D for easy analysis. Job level ratings were determined by the primary researcher based on the levels, ranging 1-6. In total, sixteen participants responded to the questionnaire, their results can be seen below.

Table 4: Questionnaire response analysis

Response ID	Criteria					
	Job Level	Organization	Understanding	Implementation	Situation	Opinion
72656642	6	D	7	2	3	4
72767286	5	C	9	10	10	7
72774404	5	C	8	5	9	9
72786629	5	C	3	0	3	6
72786486	3	C	8	10	10	9
72787863	5	C	2	2	3	5
72788426	5	C	6	10	10	9
72790265	5	C	3	2	6	8
72790926	4	C	5	2	6	8
72791879	3	C	1	0	0	4
72800875	5	A	8	8	6	8
72818786	4	C	7	10	3	10
72832110	2	B	0	0	0	0
72873977	4	B	7	2	10	9
72934509	4	C	1	2	3	5
73699789	5	E	3	0	0	6

In terms of the stretch of research, the questionnaire reached a total of 5 companies through its distribution. This would quantify stretch to value 5. In terms of knowledge, it is hard to determine the extent in the industry as there was few responses from the lower ranking job levels. However, using the difference in responses that clearly comes within organization 'C', knowledge is varied in all levels. As there was evidently contrasting levels of understanding at the same levels and inability to judge that of the lower ranking levels the knowledge will also be valued at 5. With all data collected from questionnaire presented in Table 4, it was averaged out using basic mean average calculations. The results for all calculations and criteria can be seen in Table 5.

Table 5: Results for final criteria values

Criteria					
Knowledge	Stretch	Understanding	Implementation	Situation	Opinion
5	**5**	4.875= **5**	4.063= **4**	5.125= **5**	6.688= **7**

Using this data and implementing it into Table 3: Assessment of Decision, determines the outcome of the information received from the questionnaire. By using the results from the questionnaire and then the finalised criteria value above they can be applied to the decision table. The values that can apply to a given criteria will be highlighted red. This display will then determine and ultimate decision: the decision that fully fits the criteria

Decision	**Criteria**					
	Knowledge	Stretch	Understanding	Implementation	Situation	Opinion
Industry as adopted PAS 2050 and professionals understand it	8-10	0-10	8-10	5-10	5-10	5-10
The industry has adopted PAS 2050 however professionals are somewhat unaware of it	4-9	0-10	4-9	3-7	3-7	3-7
There is general acknowledgement of adoption and some efforts made	2-5	0-10	2-5	2-5	2-5	2-5
There is very little Knowledge and movement towards adoption.	0-5	0-10	0-2	0-2	0-2	0-2

model will be the ultimate result.

Figure 2: Determination of the decision gathered from AHP

As can be seen by the red highlighted section in Figure 2 only one decision criteria is met 6 out of 6 suggesting that the finalised decision is 'that the industry has adopted PAS2050 however professionals are somewhat unaware of it'. Although due to the answers and a 5 five out of the 6 criteria match, it could also be considered to meet or fall closer to the other decision which states 'there being a general acknowledgement of adoption and some efforts made' but PAS2050 is not fully adopted. The opinions provided by the questionnaire offer a further insight into the adoption of PAS2050 into the construction industry. Although varied, responses provide an insider's perception. One question asked was 'what do you think the industry's reaction is to PAS2050 and new carbon assessment strategies?'. In response to this participant 72656642 stated that the industry's reaction has been poor so far and the commercial impact has not been embraced to the fullest extent. This suggests that although there has been an uptake in some procedure, not all parts of the business have taken a move towards adopting PAS2050. Participant 73699789 described the reaction to be good in part and 'those with specialist knowledge are well tuned in and trying to promote PAS2050 so that it makes a change, but it is not a widespread movement across the industry'. This participant stated that with the pressures of time/ knowledge/ cost will make it hard to implement.

This shows that implementation is possible however the industry is not offering enough time or money to educate its employees. Participant 72786486 stated that so far it appears that "the larger companies are taking their responsibilities seriously and have employed the required qualified staff to assist in preparing PAS2050 assessments and reduction strategies". This again shows key players in the industry are looking for change however, without full support of the smaller organisations and supply chain, the change cannot escalate to the required scale. A number of participants stated the process needs to be started properly, but we are only just starting to touch the surface and there is not much difference yet. It was also said that most companies in the industry have made tremendous efforts to make steps within scope 1 and scope 2 reduction, but scope 3 had been mostly ignored. These opinions support each other; without everyone changing, the change will not be enough.

Understanding impact and significance is also essential to understanding how well PAS2050 has been adopted. In asking for opinions on this matter there was again a varied response. Participant 72767286 stated it is "significant impact because clients require demonstration of reductions in the carbon impact of projects now". This shows that clients are requesting it and could indicate to the smaller organisations and supply chains causing the issues. However, one participant, 73699789, said in response to significance, "I would say not significant, simply as it is not commonly talked about by clients, in bids, but on projects there is specific requirements/ target but it's not specifically PAS2050". This states that although there is requirements of the same trajectory, there is not any specific to PAS2050. One participant (72786486) responded to the questionnaire stating, "initiatives put in place by the larger companies are having a significant effect on their GHG emissions but the whole industry needs to come together, and all adopt similar initiatives". They also provided a recommendation in saying 'fully qualified GHG reduction specialists either need to be in place in all companies or available as consultants to help drive down the industries GHG emissions.

3.2 Case Study and Analysis

3.2.1 Case Study

The organisation set out to map out their total carbon outlay and determined which areas have the largest number of problems, using this they developed an action plan to determine how to get to net zero emissions through tweaking elements in the whole life cycle of their products. This has created a 15-year action plan which aims to get them to a position of net zero emissions by 2035. The plan, if implemented correctly, is expected to eliminate over 1,250,000 tonnes of carbon dioxide equivalent (tCO2eq) from their carbon footprint and expected to help save £6million.

For maximum progress, the organisation states there is essential movements that need to be taken in order to engage not just staff but clients and partners. This means they have taken steps into educating those stakeholders thoroughly on the topic for increased awareness and delivering a concise structure on the strategy for improvement of their processes. This step aims to clean up not just scope one and two, but scope three as well making it a whole life cycle change. Educating is essential for these organisations plan as they believe for many people carbon management is seen as a significant financial burden when in reality it can save money, cut carbon and reduce things such as deliveries which can help reduce air pollution, noise and disruption to local communities.

Within the 15-year plan there is many key steps mostly surrounding PAS2080 but the plan ultimately involves reducing their footprint to net zero and offering low carbon solutions to clients. They way in which they are monitoring their life cycle means, indirectly, they are also adopting PAS2050. The reasoning for this implementation is to meet the UK governments target to be net zero by 2050 as well as to address and tackle the greenhouse gas emissions associated with the built environment. Hence the need for reduction of emissions at every life cycle stage: design; delivery; usage of assets and end of like of the assets. The organisation has already closed on year one of the plan and stated 6 key progressive movements. These involved training senior members of staff; reducing vehicle emissions and creating a plan to transition the fleet of vehicles to electric or hydrogen powered alternatives by 2030; optimising designs aligned with PAS2080; accelerating clean growth strategies; integrating new systems to help improve life cycle management.

The organization's carbon manager stated that climate change needs to be tackled in steps. 'It is not possible for an overnight transformation of any industry', not only because it is a slow process but primarily because the population isn't educated thoroughly enough on how to do it efficiently. As can be seen from the organizations policies PAS2080 was closely aligned, this was stated to be because of its maturity and understanding however it was suggested that 'PAS2050 will follow a similar trajectory'. Maturity is to be built with PAS2050 and as it becomes legal, maturity in all areas will jump as there is a necessity to it rather than a choice. This jump should come with the 2022 carbon tax changes, which ultimately states that there will be a large increase in costs of material with high carbon costs. The industry will subsequently start working towards using lower impacting materials to lower taxes and increase potential profits. Furthermore, with the industry legally liable for making it publicly available what they are doing to eliminate their impact, which includes publishing the numbers to match, those organisations that had previously held back from making changes will make change for fear of carbon callouts and subsequent loss of business. This also means they will now be complying with mandatory GHG emissions in scope 1 & 2.

In the questionnaire it was said by one employee that many companies have plans in place for committing to PAS2050, however they are struggling to but those plans in action, unlike the organization discussed above that was described as 'one of the few that have' made steps into the role out process. It was also said by one participant that the organisation has supported the PAS2080 guidance to the industry and one project was cited as a great example of how early engagement and collaboration can reduce embodied carbon and drive sustainability.

3.3 Discussion

The finalised decision using AHP was 'that the industry has adopted PAS2050, however professionals are somewhat unaware of it' however as previously discussed it narrowly aligned with 'there being a general acknowledgement of adoption and some efforts made' within the industry meaning that PAS2050 is not fully adopted. By using this information and comparing it with the questionnaire's qualitative responses and case study data, it can be determined an exact decision for how well PAS2050 has been adopted. The opinions gathered through the questionnaire suggested most commonly that although there has been an uptake in some procedure, not all parts of the industry have taken a move towards adopting PAS2050. It was also determined that without full co-operation impact to climate change would be minimal. As discovered through the case study it is essential that all areas of the industry engage in the changes and in order to make this possible, education is the only key. Without it at all levels and sectors of the industry implementation is not fully viable. When comparing this with the case study it was clear education is not in the lower ranks and until effort is put into allowing time do educate and improve knowledge about the impact on everything including costs and the processes required, uptake will not increase. As many stated, the supply chains were slow to react, which will only hinder progress. It was made clear through all responses and information that its more than PAS2050 making a difference, a lot of the organisational improvements discussed in the case studies are centred around PAS2080 which is said to run alongside PAS2050 and other similar standards. This comes as maturity is key to implementation hence why PAS2080 is much more understood in the industry. As described, PAS2050 is set to follow a similar trajectory but this will only begin once there is necessity, a necessity that comes with the 2022 tax initiatives. It is clear that we are only just starting to touch the surface of the problem, and although organisations have made significant changes and have committed to net zero targets, many still do not have a viable plan in how to achieve it. This means there is not much in the way of adoption or change yet. When understanding the organisation within the case study its clear the process towards a carbon secure future is slow and involves lots of steps, this is likely why the industry is still processing it.

Despite how the one organization investigated has helped make huge step to lead the industry, one organization is not enough. When comparing these finding with those decisions gathered through AHP the alignment in all opinions is predominantly with there being a general acknowledgement of the adoption and PAS2050, however there is only a few efforts made and PAS2050 is not fully adopted. This decision was made as the information clearly points towards too many organisations being neglectful of the adoption therefore giving the decision that PAS2050 was adopted would not be representative of all sectors including supply chain.

4. Conclusions

The present project is about the adoption of PAS2050 into the construction industry. Results showed that generally there is movement being made into a sustainable future however not enough. It was found that although many big organisations are making strides and beginning the teaching process, many within the supply chain have not due to the lack of legal necessity. This is a damaging discovery for this moment in time as it was determined that without significant teaching, progress will not reach its full potential. The results of this project showed that the industry still has tremendous strides to take in teaching and implementing for PAS2050 to be implemented and make a difference, and although steps are being made with PAS2080, without all efforts, it will not be enough.

References

[1] S.L. Zubaidi, P. Kot, K. Hashim, R. Alkhaddar, M. Abdellatif, and Y.R. Muhsin, Using LARS–WG model for prediction of temperature in Columbia City, USA, IOP Conference Series: Materials Science and Engineering.584, (2019) 012026.

[2] Z. Salah, S. Ortega-Martorell, P. Kot, R.M. Alkhaddar, M. Abdellatif, S.K. Gharghan, M.S. Ahmed, and K. Hashim, A Method for Predicting Long-Term Municipal Water Demands Under Climate Change, Water Resources Management. 34 (2020) 1265-1279.

[3] S. Zubaidi, S. Ortega-Martorell, H. Al-Bugharbee, I. Olier, K.S. Hashim, S.K. Gharghan, P. Kot, and R. Al-Khaddar, Urban Water Demand Prediction for a City that Suffers from Climate Change and Population Growth: Gauteng Province case study, Water. 12 (2020) 1-18.

[4] L. Zubaidi Salah, H. Al-Bugharbee, S. Ortega Martorell, S. Gharghan, I. Olier, K. Hashim, N. Al-Bdairi, and P. Kot, A Novel Methodology for Prediction Urban Water Demand by Wavelet Denoising and Adaptive Neuro-Fuzzy Inference System Approach, Water. 12 (2020) 1-17.

[5] Z. Salah, I.H. Abdulkareem, K.S. Hashim, H. Al-Bugharbee, H.M. Ridha, S.K. Gharghan, F.F. Al-Qaim, M. Muradov, P. Kot, and R. Alkhaddar, Hybridised Artificial Neural Network model with Slime Mould Algorithm: A novel methodology for prediction urban stochastic water demand, Water. 12 (2020) 1-18.

[6] Z. Salah, K. Hashim, S. Ethaib, N.S.S. Al-Bdairi, H. Al-Bugharbee, and S.K. Gharghan, A novel methodology to predict monthly municipal water demand based on weather variables scenario, Journal of King Saud University-Engineering Sciences. 32 (2020) 1-18.

[7] S. Zubaidi, H. Al-Bugharbee, Y.R. Muhsin, K. Hashim, and R. Alkhaddar, Forecasting of monthly stochastic signal of urban water demand: Baghdad as a case study, IOP Conference Series: Materials Science and Engineering.888, (2020) 012018.

[8] S.L. Zubaidi, H. Al-Bugharbee, Y.R. Muhsen, K. Hashim, R.M. Alkhaddar, D. Al-Jumeily, and A.J. Aljaaf, The Prediction of Municipal Water Demand in Iraq: A Case Study of Baghdad Governorate, 12th International Conference on Developments in eSystems Engineering (DeSE), Kazan, Russia (2019) 274-277.

[9] A. Kadhim, M. Sadique, R. Al-Mufti, and K. Hashim, Long-term performance of novel high-calcium one-part alkali-activated cement developed from thermally activated lime kiln dust, Journal of Building Engineering. 32 (2020) 1-17.

[10] M.K. Obaid, M.S. Nasr, I.M. Ali, A.A. Shubbar, and K.S. Hashim, Performance of green mortar made from locally available waste tiles and silica fume, Journal of Engineering Science and Technology. 16 (2021) 136-151.

[11] H.S. Majdi, A. Shubbar, M.S. Nasr, Z.S. Al-Khafaji, H. Jafer, M. Abdulredha, Z.A. Masoodi, M. Sadique, and K. Hashim, Experimental data on compressive strength and ultrasonic pulse velocity properties of sustainable mortar made with high content of GGBFS and CKD combinations, Data in Brief. 31 (2020) 105961-105972.

[12] A.A. Shubbar, M. Sadique, H.K. Shanbara, and K. Hashim, The Development of a New Low Carbon Binder for Construction as an Alternative to Cement. In Advances in Sustainable Construction Materials and Geotechnical Engineering, 1st ed. Berlin: Springer, 2020, pp. 205-213.

[13] D. Al-Jumeily, K. Hashim, R. Alkaddar, M. Al-Tufaily, and J. Lunn, Sustainable and Environmental Friendly Ancient Reed Houses (Inspired by the Past to Motivate the Future), 11th International Conference on Developments in eSystems Engineering (DeSE), Cambridge, UK (2019) 214-219.

[14] A.A. Shubbar, H. Jafer, A. Dulaimi, K. Hashim, W. Atherton, and M. Sadique, The development of a low carbon binder produced from the ternary blending of cement, ground granulated blast furnace slag and high calcium fly ash: An experimental and statistical approach, Construction and Building Materials. 187 (2018) 1051-1060.

[15] BSI, Newly revised PAS 2050 poised to boost international efforts to carbon footprint products, (2011) https://www.bsigroup.com/en-GB/about-bsi/media-centre/press-releases/2011/9/NEWLY-REVISED-PAS-2050-POISED-TO-BOOST-INTERNATIONAL-EFFORTS-TO-CARBON-FOOTPRINT-PRODUCTS/.

[16] D. Marshall, D. Worthing, N. Dann, and R. Heath, The Construction of Houses. Oxon; New York: Taylor & Francis, 2013.

[17] S.W. Zhou, Carbon management for a sustainable environment. Springer Nature, 2020.

[18] ARUP, "Carbon Management PAS 2050: A Business Opportunity," [Online]. Available: https://www.arup.com/perspectives/publications/research/section/carbon-management-pas-2050

[19] M.Z. Hauschild and M.A.J. Huijbregts, LCA Compendium – The Complete World of Life Cycle Assessment. Springer, 2015.

[20] BSI, PAS 2080: Carbon Management in Infrastructure Verification, https://www.bsigroup.com/en-GB/our-services/product-certification/product-certification-schemes/pas-2080-carbon-management-in-infrastructure-verification/ (2013)

[21] Mashable, Planet Dearth, https://mashable.com/ad/feature/stop-climate-change, Mashable. (2016)

[22] G. Budak, X. Chen, S. Celik, and B. Ozturk, A systematic approach for assessment of renewable energy using analytic hierarchy process, Energy, Sustainability and Society. 9 (2019)

Parametric Theoretical Study of Solar Assisted Cooling System Using Lithium Bromide- Water Pair

Ahmed A Shahhath*, Haroun A K Shahad and Alaa A Mahdi
Department of Mechanical engineering, College of Engineering, University of Babylon, Babylon, Iraq.

*Corresponding E-mail: Eng.haroun.abdul@uobabylon.edu.iq

Abstract: In this paper, the effect of parameters of solar absorption system such as evaporator, absorber, condenser, generator temperatures and the mass of the solution on the evaporator cooling load and the coefficient of performance has been explained theoretically. The results show that, increasing of evaporator and condenser temperatures increase the evaporator cooling load, performance coefficient and the Ratio of Circulation while increasing the temperature of condenser and absorber decreases the evaporator cooling load, performance coefficient and the Circulation Ratio. In addition, increasing the solution mass increases the refrigeration power while the performance coefficient and the Circulation Ratio was constant at increasing the solution mass. The reached maximum cooling load was (1.932 kW) at 15 kg solution mass and 100 °C generator temperature, the maximum COP was 0.774 at (10 °C) temperature of evaporator and the peak Circulation Ratio was 0.3066 at (30 °C) temperature of absorber and (100 °C) temperature of generator.

Keywords: Water; Lithium Bromide; Generator Temperature; Evaporator Cooling Load; Coefficient of Performance

1. Introduction

Solar power is one of the most major renewable energy resources. In Comparison to other sources, solar power can be described by the reality that it is obtainable, almost, anywhere. Moreover, it is neat and has no poor impacts on the atmosphere. It could be used in one of the next two methods: by converting irradiance directly to electric power, by utilizing photovoltaic cells, or by converting irradiance into frontal thermal power utilizing collectors. Climatic conditions decrease the level of solar irradiance that hits the Earth's surface before dissipating about 20% to 30% of the incident light and adjusting its spectrum. After crossing meantime, the atmosphere of Earth, about one half of the solar radiation is in the visible electromagnetic spectrum with the other half is mostly in the infrared and ultraviolet spectrum [1].

Gonzalez-Gil et al. [2] assessed the efficiency of an absorption cooler cooled directly by air in summer 2010, centered on test outcomes of many days in Madrid. The system was working effectively, with COP of about 0.6. Refrigeration capacity ranged from 2 kW to 3.8 kW, that accounted for around 85% of the total output of the prototype. The temperatures of cooled water varied mainly between 14°C and 16°C with 12.8 °C minimum calculated value. The system was capable to fulfill an average of 65% of the refrigeration request corresponds to 40 m^2 space. During around one hundred hours of process no signs of crystallization were found.

Shirazi [3] concentrated on the combination of LiBr-H_2O single, dual and triple-acting absorption coolers with collectors. He observed that the SHC triple-acting cooler has the more power-sufficient and eco-friendly environment quality which led to decrease energy consumption between 47% and 62%. It was revealed that systems could provide satisfying payback durations within 2 to 6 years. The parabolic pipe collectors and evacuated flat plate collectors have been proposed for such implementations and the chiller worked with a performance coefficient of about 0.9-1.1 at a temperature of steady supply of ~168 °C, generating 6-7 kW cooling.

Hussein [4] developed and simulated the unit of absorption cooling using a solar receiver of flat-plate type with a mixture of (LiBr-H_2O) as an operating fluid, and to choose the best possible form of absorption refrigeration system that is suitable for use in Iraq. The weather conditions for the city of Erbil has been chosen. Performance Coefficient(COP) for single, double (series and parallel), triple and half absorption cooling impacts were computed and configured. Programs of TRNSYS and EES were used to simulate and represent COP Estimation, hot water temperature and rate of heat transfer. The receiver effective field of solar flat-plate has also been calculated and optimized for five proposed areas (140, 150, 160, 170 and 180 m^2). The indicated results that the preferred form of the four absorption-chilling systems had a single effect with a COP of 0.80 at a temperature of generator 81 °C. The outcomes also demonstrated that the optimal field of the receiver of solar flat-plate was 160 m^2 since it had accomplished the peak cooling and consuming capability of the generator (50 and 62.5 kW respectively).

Barrera et. al [5] investigated the analysis of the experimental examinations done with the advanced Solar-GAX absorption cooling device, configured for chilling capacity of 10.5 kW (3 ton) with the mixture of aqua-ammonia. The device developed to work at about 200 °C heat source temperatures, comprised of a generator and absorber of a falling layer kind and an absorber and condenser cooled by air of the finned tube type, which is a choice to areas with water lack. Temperatures of energy source of 160 °C have been developed to simulate the circumstances in which solar thermal concentration technology was used to provide heat to the system. A 3.17 kW partial load with a 0.15 COP have achieved when the Solar-GAX device was controlled at a 120 °C temperature of generator at a 200 °C generator operating temperature with a10.5 kW design power. The absorption device that was air-cooled has been tested at temperatures of ambient higher than 30°C. In addition, at temperatures of ambient 28 °C, a cooling power of 7 kW was reached, with such a 0.35 kg/min flow rate of ammonia mass and a rate of water flow 15 kg/min, for diluted solution temperatures at 140 °C, a 0.2 to 0.3 COP on the side of water and 0.25 to 0.45 on the side of ammonia was obtained under the conditions.

Mezher et al [6] built, installed and operated a prototype of solar refrigeration device using the flat plate collector. The studied device displayed a chilling power of 0.15 ton with a performance coefficient of 0.48, and the obtained minimum temperature of evaporator was 14.2 °C. The outcomes provided a better explanation for the usage of methanol as a refrigerant with the solar absorption device, and the device could operate in a continuous cycle of operation. That analysis provided a fundamental explanation for the solar refrigeration device design. The study outcomes were employed to model a one-ton air cooling device utilizing solar power and the refrigerant used was the methanol.

Prakash et al [7] designed, constructed and analyzed the output of a one stage Vapour absorption system with 1kW refrigeration power. The installed heat exchanger in the unit was of spiral tube and shell kind. The employed condenser was of forced horizontal tube heat exchanger kind in the unit configuration. The unit generator was a pool boiling kind. The theoretical outcomes that include the coefficient of overall heat transfer were compared with the examination outcomes for the unit with 1kW output. The unit real COP was enhanced by 15% with the solution concentration reduction at absorber inlet.

Mussati et al [8] conducted a results for optimization which gained for a dual-acting H_2O-LiBr absorption cooling unit with regard to the overall cost as a criterion for minimization, for a wider range of refrigeration power values. It has noted that the overall cost has tremendously decreased when the heat exchanger working at lower temperature is cancelled in comparison to the construction which included it. As well as, it has discovered that the acting of deleting the heat exchanger was comparatively most important at raising refrigeration power levels. A decreasing of 9.8% in the overall cost was gained for a refrigeration power of 16 kW (11,537.2 $·year^{-1} vs. 12,794.5 $·year^{-1}), while a decreasing of 12% has gained for a 100 kW refrigeration power (31,338.1 $·year^{-1} vs. 35,613.9 $·year^{-1}).

Moreover, Fischer et al [9] developed a thermodynamic modeling of a single acting absorption cooler focused on mass, energy and species equilibriums in steady state condition for analyzing the acting of the external rates of flow in the cooler. The enhancement of the flow rate of hot water or cooling water produced a good output of the cooler. They found that, if the flow rate of chilled water reduced, the output improved, the best output construction was with 120% of flow rate of hot water and chilling water, 80 % of flow rate of chilled water and the higher temperatures of inlet hot water resulted in a high coefficient of performance (COP) and heat removal as well as low temperature of inlet chilling water.

Uçkan et. al. [10] designed a single-acting solar-powered unit and used lithium bromide as an absorber and water as a refrigerant at 35.17 kW cooling power, consisting of an evacuated pipe collector using 95 m^2 space of aperture, and a collector's inclination of 30° with the horizon. The hot water storage capacity was 3 m^3. The system cooled a ground area of 270 m^2. The outcome was that the COP of the chiller oscillated between 0.69 and 0.67 between 8:00 a.m. and 14:00 pm, afterwards, the COP of chiller gradually decreased from 14:00 to 18:00 between 0.69 and 0.57. The cooler absorption average output was noticed to be 0.63. The unit optimization results revealed that the unit provided a building air cooling of 270 m^2. The mean absorption COP was about 0.62.

Zeiny et al. [11] investigated the proposal of mixing the solution of lithium bromide (LiBr) with nanoparticles. The experiments were accomplished for appraising the nanoparticles acting addition, i.e. carbon black (CB) and multiwall carbon nanotubes (MWCNTs), on the optical properties, viscosity, thermal conductivity, steam absorption and photo-thermal conversion of LiBr solutions. On the Contrary to oncer reported, the absorption rate of steam was not improved by addition of nanoparticles to the aqueous LiBr solution, which disclosed that grazing and Brownian motion have neglected acting on the mass transfer. A neglected nanoparticle acting on the effective thermal conductivity was also remarked. However, a lower nanoparticle concentration of 0.005 wt.% reduced the aqueous LiBr solution transparency to 0%, that importantly enhanced the photo-thermal conversion of the samples. On the contrary to CB, MWCNTs enhance the viscosity importantly. Therefore, seeding cheap commercial CB in aqueous LiBr solution has the possibility for contribution into the evolution of solar-led cooling technique.

Aramesh et al [12] analyzed the output of a dual-acting LiBr-H_2O absorption cooling loop and the unit was enhanced by applying solar energy and using nanofluids. A trough collector has utilized for preheating the operating fluid before entering the cycle generator. In addition, four various nanofluids were considered as the heat transfer fluid of the collector: Al_2O_3, Ag, Cu, and CuO. Various concentrations of the nanoparticles from 0 to 2.5% were used for the nanofluids. The outcomes showed that, in all the concentrations, Ag nanoparticles would have a good output in comparison to other kinds. Also, it was deduced that the high nanoparticles concentrations and along with it the higher inlet generator temperature would reduce the generator heat production rate up to 4%. As well as, considering the cycle constant refrigeration power, utilizing of the Ag nanoparticles in the concentration of 2.5% enhanced the value of COP up to 3.9%, with respect to the pure water.

Abdul Ameer et. al [13] carried out a theoretical and experimental study for the performance of a solar-powered continuous flow absorption refrigeration unit using a compound parabolic concentrator (CPC) solar collector and two types of working fluids, LiBr-H_2O and diethyl ether-ethanol in Iraq. The experimental outcomes revealed that the peak thermal efficiency of the CPC was obtained at approximately 12:00 p.m. at 21st day of each month from January to June (0.48, 0.512, 0.57, 0.606, 0.618 and 0.67) respectively when the water flow rate was about 0.0277 kg/s and the efficiency was boosted at higher water flow rates. Hot water temperature as high as 98 °C was obtained. In the mathematical model, the thermal efficiency at 12:00 p.m. for the 21st day of each month ranged from (0.67-0.73) at water flow rate of (0.0277 kg/s) at the same climate conditions. The theoretical and experimental coefficient of performance of the absorption system for two pairs and at different solution concentrations are ((0.31-0.72) for lithium bromide/water while (0.6-.82) for Diethyl Ether/Ethanol).

Hamzah et al [14] manufactured an interrupted absorption cooling unit under Hillah weather in Iraq. The absorption system comprised of parabolic trough solar concentrator (PTSC) was utilized as a reflector of solar mirror with 2 m^2 aperture area, carbon steel tube, a vacuum glass envelope with a 1.5 in diameter as a receiver, a tank for water storing, condenser and an evaporator. The solution of ammonia-water (NH_4OH) was utilized as an operating fluid with various concentrations (25%, 30%, 35%, 40%). The quality and vision of the system were assessed during a year from May 2014 to July 2015 by measuring pressure and temperature at various components of the unit. The peak temperature and pressure at generator were approximately 120 °C and 12 bar respectively. The performance coefficient was about from 0.01 to 0.09.The performance coefficient was about from 0.01 to 0.09. As well as, Bellos et. al [15] tested numerically the efficiency of Linear Fresnel Reflector thermal power in existence of the Nanofluid Syltherm 800/CuO for the 2, 4 and 6% concentration by using CFD technique. Collector reflectors are flat mirrors of a 27 m^2 total pure area of aperture, while the supplementary reflector has a parabolic form. The temperature of inlet was between 350 K and 650 K for the study. He noticed that the thermal efficiency improvement was approximately 0.28 % when nanofluids were used as a working fluids.

Ammonia-water absorption refrigeration device with parabolic trough collector, which concentrates solar power in a receiver tube for heating water was tried by Stanciu et al [16]. Time-dependence of refrigeration load was regarded for a domestic two-story home cooling. The outcomes stated that there was a particular size for storage tank combined with a particular collector dimension that guarantee the longest continual refrigeration device working at considered permanent rates of mass flow within the device.

The low residential silica gel-water adsorption refrigeration efficiency of a device in existence of a reflective concentrator parabolic collector domain powered adsorption (silica gel-water) refrigeration device at hot and arid climate experimentally investigated by Reda et al [17]. The system consisted of a 36 m^2 entire space, an 8 kW refrigeration power adsorption cooler with silica gel-water. The findings of the domain test of the summer period showed that at daily solar irradiance ranging from 21 to 27 MJ/m^2, the utilized collectors had a higher and often constant efficiency of thermal power. The daily output of solar collectors through the device operation varied from approximately 50-78%.

In Iraq, due to the high temperatures encountered, an important share of generated electric power is employed for air conditioning sector. Due to Iraq geographical location the availability of maximum solar radiation and the maximum power demand approach simultaneously peak rates. Iraq obtains a large quantity of energy from the sun more than (6.5-7) kWh/m^2, in comparison to the United States that is 3.6 kWh/m^2 and Europe's 2.5 kWh/m^2 [18]. The refrigeration systems that depend on solar radiation have a great potential since the solar radiation is obtainable all year round. The advantage of utilizing solar radiation to operate the units of air conditioning is that the solar radiation availability and the demand for cooling have reached peak rates simultaneously and relatively.

It is noticed from the above sighted literature that most papers concentrated on a single operating parameter and its effect on performance. This work shall involve parameters that affect the absorption cooling system as a whole and each single components of system such as evaporator, absorber, condenser and generator. The EES software will be utilized for analyzing the solar refrigeration loop.

2. Modeling of Used System

The used lithium bromide-water absorption cooling loop is explained in Figure 1. In this loop, high pressure Liquid water passes the expansion valve where its pressure decreases and its condition become a mixture of liquid and gas. This mixture enters the evaporator and absorbs heat from the volume to be cooled and it becomes water vapor at low pressure. The low pressure steam passes to absorber and lithium bromide absorbs it. The mixture crosses the pump with increase in its pressure. The weak lithium bromide mixture with high pressure gains heat in the heat exchanger and leaves to the generator. This solution gets heat from the solar power and higher pressure steam and strong lithium bromide mixture leaves the generator. This strong lithium bromide mixture crosses to the heat exchanger and the higher pressure steam passes to the condenser. At the condenser, a heat transfers from the higher pressure steam to the circumstances and the higher pressure steam becomes a liquid.

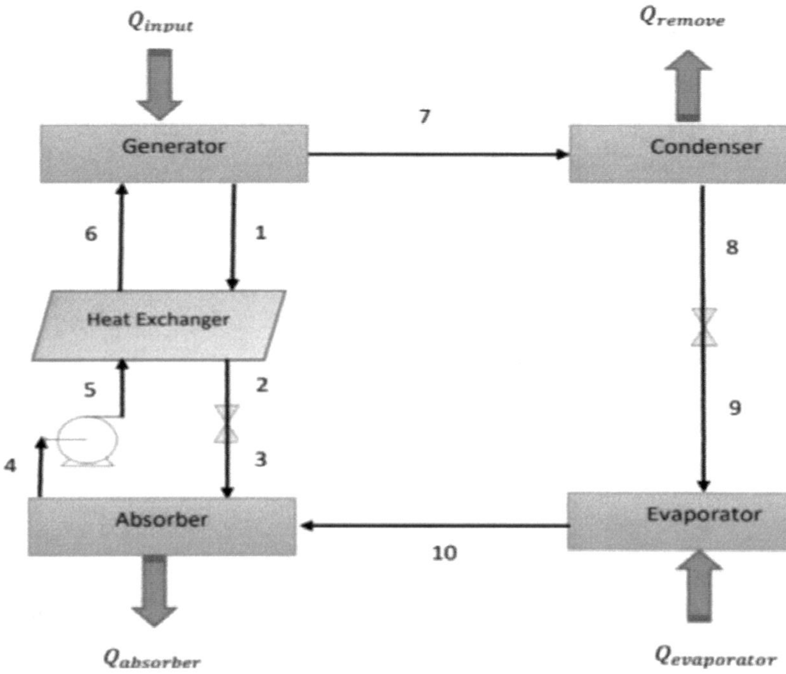

Figure 1. A scheme of an absorption refrigeration cycle [13]

3. Thermodynamic Analysis

The system is analyzed by considering the main elements i.e. the generator, condenser, evaporator, absorber and heat exchanger as a single control volume. The adding or removing of heat at each element is gained by considering an energy equilibrium at each part.

The design calculation is based on the following assumptions:

1. Neither the generator releases any heat to the surroundings nor the evaporator receive heat from the surroundings.
2. The refrigerant is considered to be in the saturated vapor state at the evaporator exit and generator exit.
3. The refrigerant is considered to be in the saturated liquid state at the condenser exit.
4. The pump work is negligible.
5. There is no pressure drop in the pipes.
6. Generator and condenser are assumed as same pressure and equal to high pressure of the system.
7. Evaporator and absorber are assumed as same pressure and equal to low pressure of the system.
8. Lithium bromide – water solution is used as working pair in design system.
9. The fluid temperature is constant inside the components.

The heat transfers rate from condenser to surrounding is given by,

$$\dot{Q}_{cond} = \dot{m}_7 \times (h_7\text{-}h_8) \quad (1)$$

where " $\dot{m}$" is the rate of mass flow of water, and "h" is the enthalpy. At the expansion valve, the expansion occurs at constant enthalpy,

$$h_8=h_9 \quad (2)$$

The evaporator heat removing rate is obtained as

$$\dot{Q}_{evap} = \dot{m}_9 \times (h_{10}-h_9) \quad (3)$$

The absorber heat transfer rate to surrounding is gotten from

$$\dot{Q}_{abs}= \dot{m}_{10}\, h_{10} + \dot{m}_3\, h_3 - \dot{m}_4\, h_4 \quad (4)$$

The mass balance in the absorber is,

$$\dot{m}_4 = \dot{m}_{10} + \dot{m}_3 \quad (5)$$

The input power for pump is obtained by

$$\dot{W}_p= \dot{m}_4(h_5-h_4) = \dot{m}_4 \times v_4\, (P_5-P_4) \quad (6)$$

where "v" is the specific volume, and "p" is the pressure. For heat exchanger, the energy balance can be obtained as

$$\dot{m}_5(h_6-h_5) = \dot{m}_1(h_2-h_1) \quad (7)$$

The heat exchanger effectiveness can be put in the form

$$HX=\frac{T_G-T_2}{T_G-T_5} \quad (8)$$

The rate of heat transfer in generator is found by

$$\dot{Q}_{gen}= \dot{m}_7\, h_7 + \dot{m}_1\, h_1 - \dot{m}_6\, h_6 \quad (9)$$

The mass balance equation in the generator is,

$$\dot{m}_6 = \dot{m}_1 + \dot{m}_7 \quad (10)$$

The governing equations used to evaluate the solution mass in each component is,

$$m_{ws} = m_{ss} + m_w \quad (11)$$

This equation is the same as the mass balance equation in the generator (10).The coefficient of performance is expressed as:

$$COP = \frac{Q_{Evaporator}}{Q_{Generator} + \dot{W}_P} \quad (12)$$

4. Results and Discussions

4.1 Evaporator Cooling Load

Figure 2 shows the relation between the coaling load and the temperature of generator at various refrigerant temperatures at exit of evaporator. It can be noticed that raising the temperature of generator enhances the cooling load at all refrigerant exit temperatures and the increase of refrigerant exit temperatures at evaporator also raises the refrigeration capacity because of raising the enthalpy difference at the evaporator, so the maximum value of the cooling load at temperature of generator of about (100 °C) and refrigerant exit temperatures of (6 °C) that is about (1.685 Kw). Also, Figure 3 explains the relation between the coaling load and the temperature of generator at various temperatures of condenser. It was noted that increasing the temperature of condenser decreases the refrigeration load because of decreasing the enthalpy difference at the evaporator, so the maximum value of the cooling load at temperature of generator of about (100 °C) and temperature of condenser of (30 °C) of about (1.689 kW). Moreover, Figure 4 shows the relation between the coaling load and the temperature of generator at various temperatures of absorber. It was remarked that the increase of absorber temperature decreases the cooling load, so the maximum value of the cooling load at temperature of generator of about (100 °C) and temperature of absorber of (30 °C) of about (1.853 kW). As well as, in Figure 5, the relationship between the coaling load and generator temperature at different solution masses at constant other parameters has been explained. It can be remarked that, increasing the mass solution increases the cooling load because of increasing the vaporized water at evaporator, so the maximum value of the cooling load at temperature of generator of (100 °C) and mass solution of (15 Kg) of about (1.932 Kw).

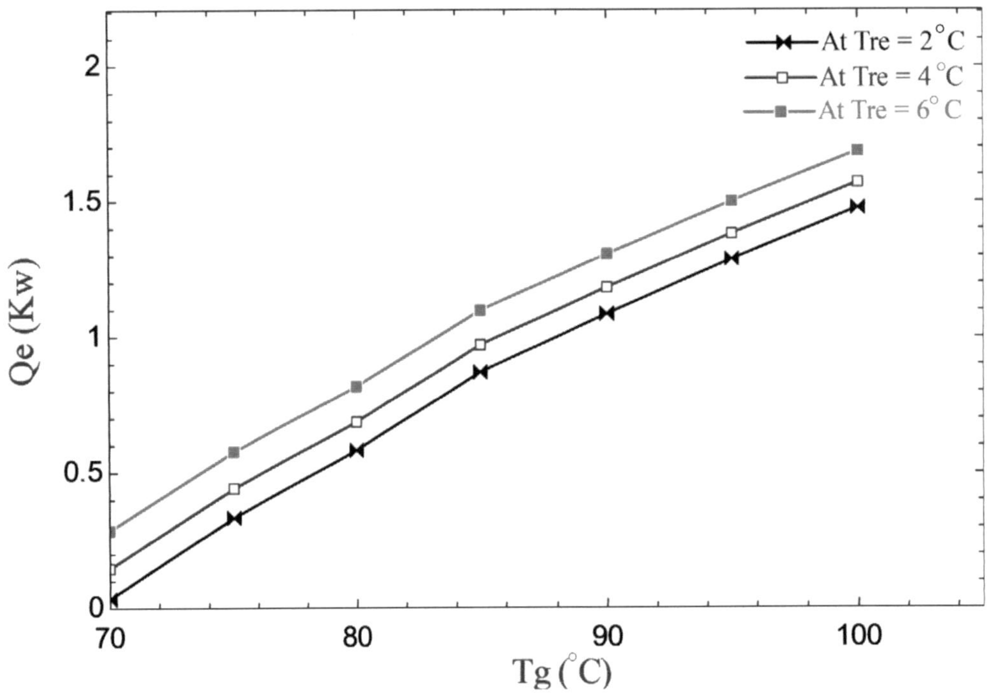

Figure 2. The relation between the coaling load and temperature of generator at various refrigerant temperatures at exit of evaporator.

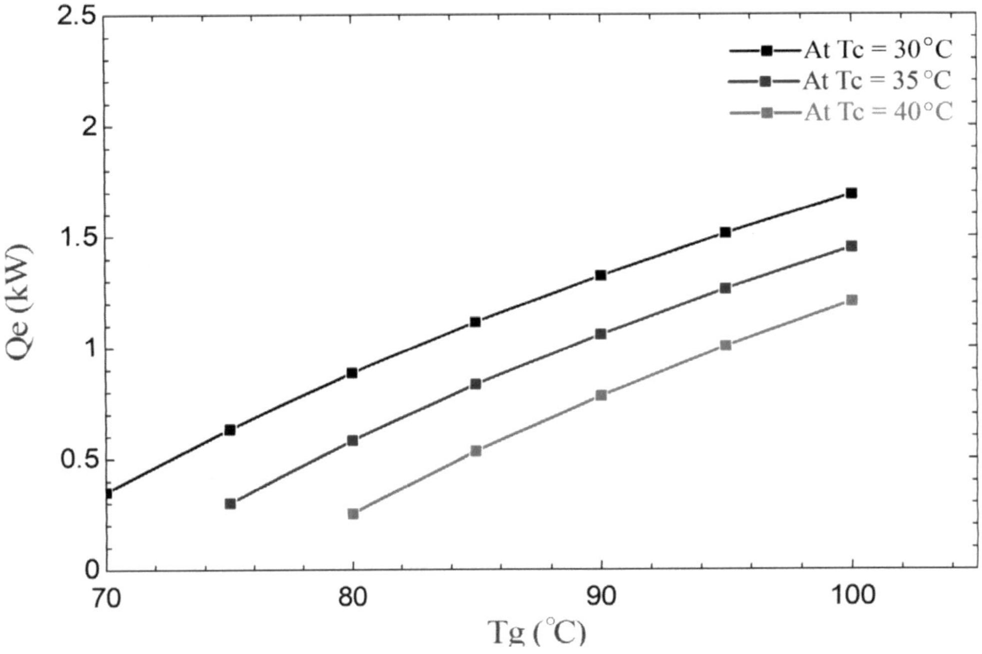

Figure 3. The relation between the coaling and temperature of generator at various temperatures of condenser.

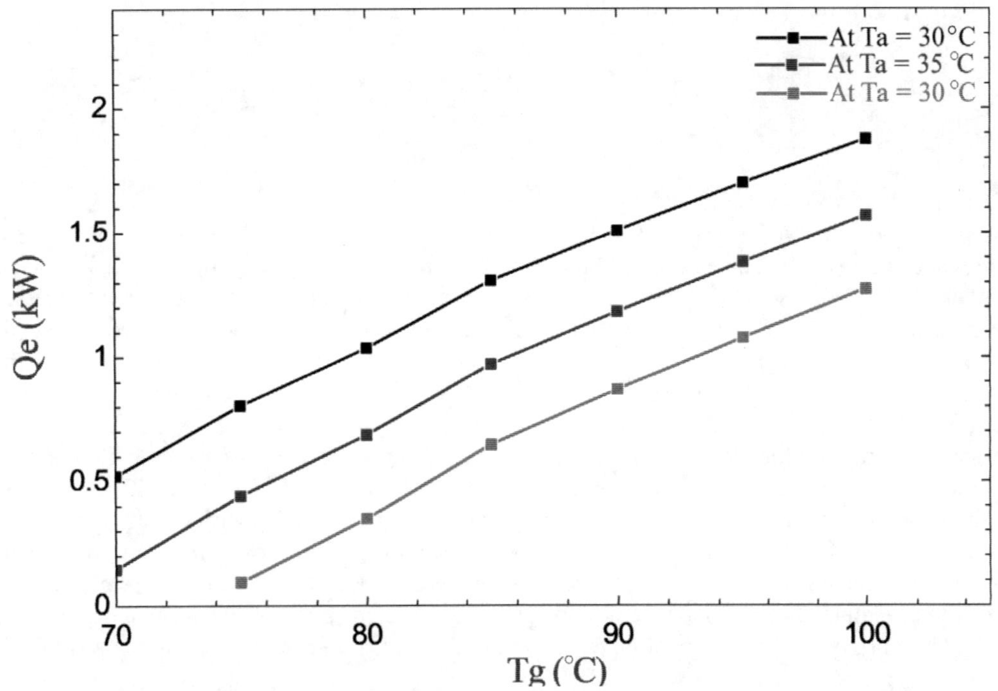

Figure 4. The relation between the coaling load and temperature of generator at various temperatures of absorber.

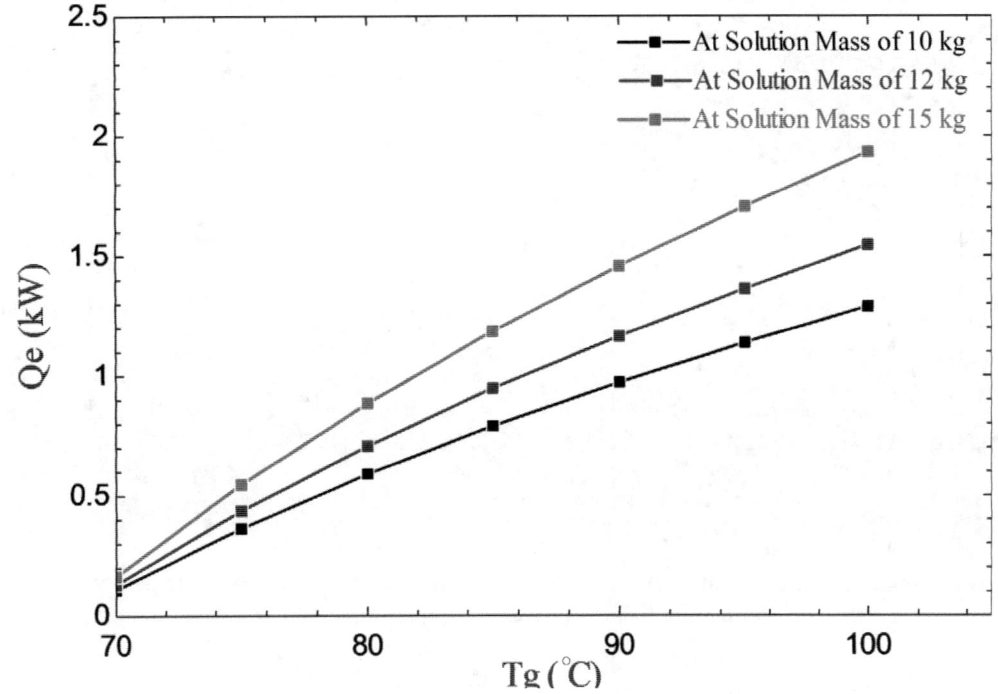

Figure 5. The relation between the coaling load and temperature of generator at various mass solutions.

4.2 Coefficient of Performance

Figure 6 shows the relation between COP and generator temperature at different refrigerant exit temperatures at evaporator. It can be noticed that raising the temperature of refrigerant at evaporator exit raises COP due to the increase of water enthalpy difference at evaporator, and the increase of generator temperature also increases COP because of increasing the vaporized water at the generator, so the maximum value of COP at (100 °C) temperature of generator and (6 °C) temperature of evaporator that is about (0.762). Also Figure 7 explains the relation between COP and temperature of generator at various temperatures of condenser. It can be noted that increasing the condenser temperature reduces COP because of decreasing the enthalpy difference at the evaporator, so the peak value of COP at (100 °C) temperature of generator and (30 °C) temperature of condenser that is about (0.762). Moreover, Figure 8 shows the relation between COP and temperature of generator at various temperatures of absorber. It was remarked that the increase of absorber temperature decreases COP due to decreasing the vaporized water at the generator, so the maximum value of COP at (100 °C) temperature of generator and (30 °C) temperature of absorber that is about (0.765). The increase of solution mass does not increase COP because of the proportional increase of cooling load and generator heat at constant COP.

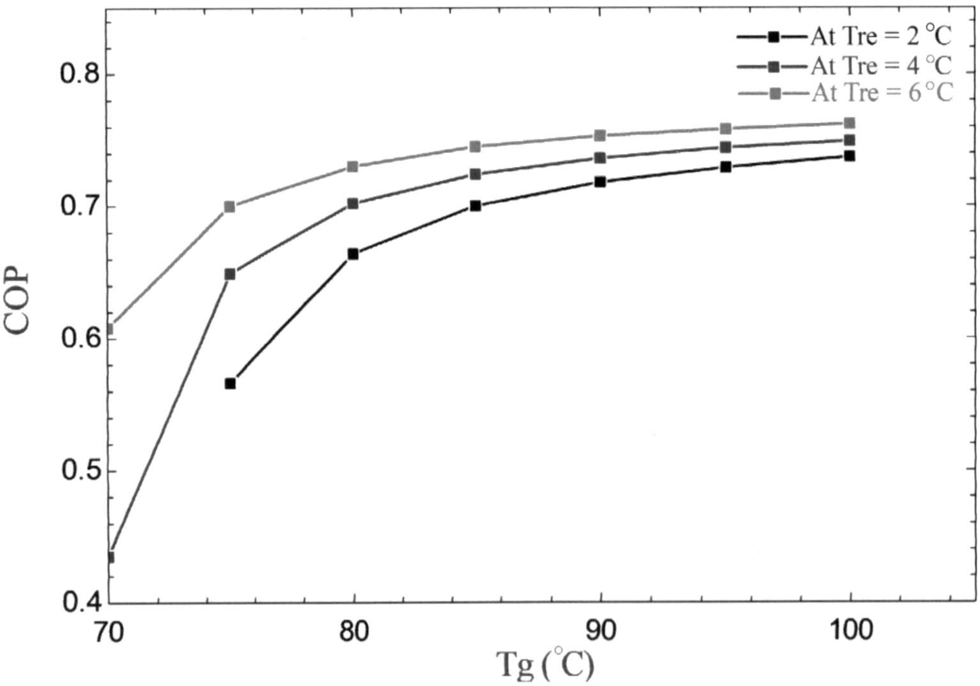

Figure 6. The relationship between COP and temperature of generator at various refrigerant temperatures at exit of evaporator.

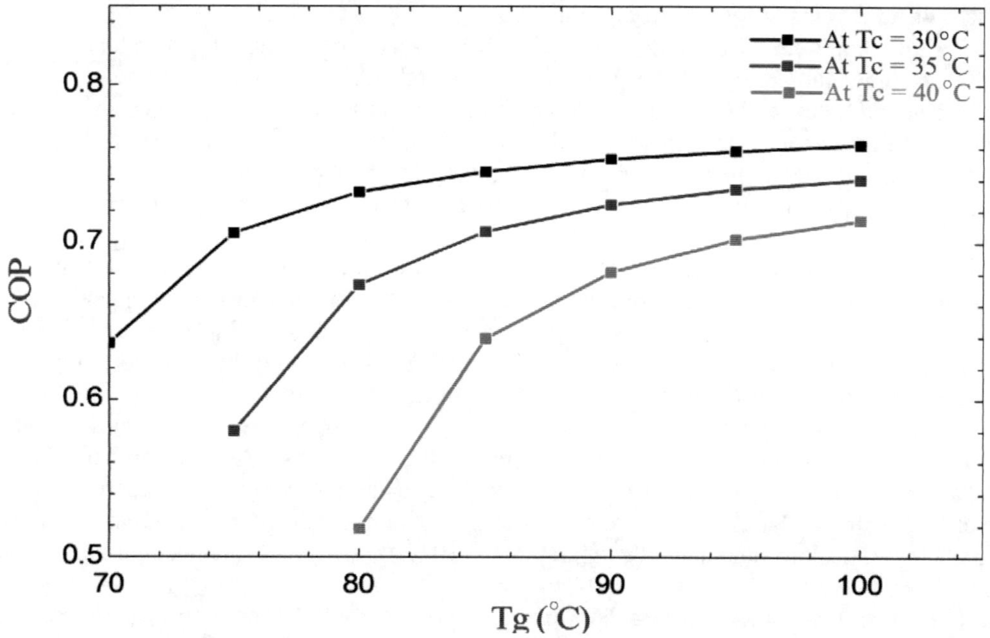

Figure 7. The relation between COP and temperature of generator at various temperatures of condenser.

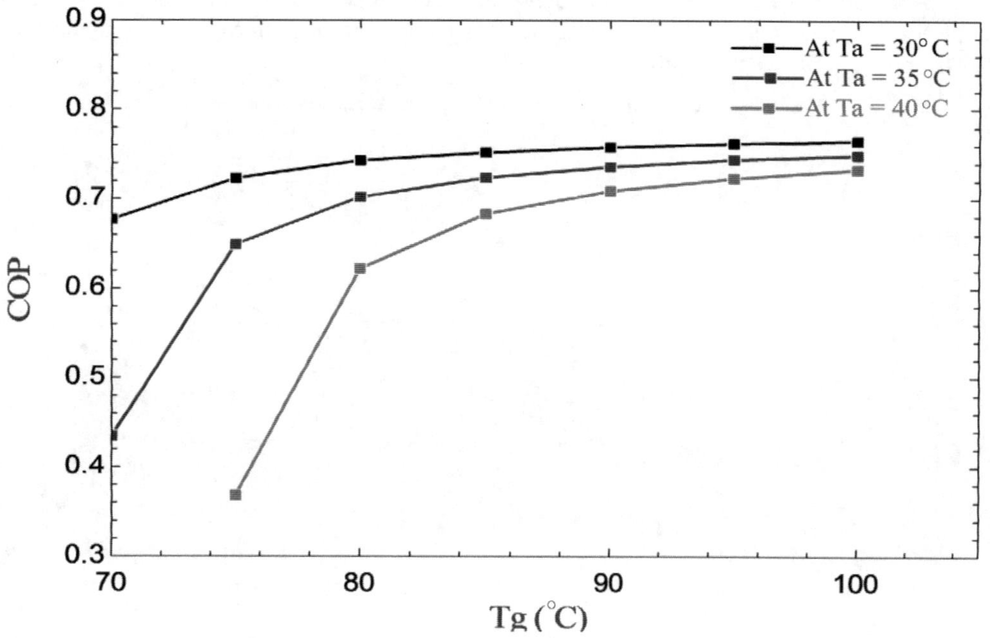

Figure 8. The relation between COP and temperature of generator at various temperatures of absorber.

4.3 Circulation Ratio

The ratio of circulation can be expressed as (the proportion between the rate of mass flow of refrigerant to the rate of mass flow of strong solution).

Figure 9 shows the relation between Circulation Ratio and temperature of generator at various refrigerant exit temperatures at evaporator. It can be remarked that, raising the temperature of generator raises the Circulation Ratio at all refrigerant exit temperatures because of increasing the vaporized water, and the increase of refrigerant exit temperatures also raises the Circulation Ratio, so the maximum value of Circulation Ratio at (100 °C) temperature of generator and (6 °C) temperature of evaporator that is about (0.2706). Also, Figure 10 explains the relationship between Circulation Ratio and generator temperature at various temperatures of condenser. It was noted that the increase of condenser temperature decreases Circulation Ratio due to decreasing the vaporized water, so the maximum value of Circulation Ratio at (100 °C) temperature of generator and (30 °C) temperature of condenser that is about (0.2701). Moreover, Figure 11 shows the relation between Circulation Ratio and temperature of generator at various temperatures of absorber. It was remarked that raising the absorber temperature reduces the Circulation Ratio due to decreasing the vaporized water, so the maximum value of the Circulation Ratio at (100 °C) temperature of generator and (30 °C) temperature of absorber that is about (0.3066). As well as, in Figure 12, the relation between the Circulation Ratio and refrigerant exit temperatures at evaporator has been shown. It was noted that, raising refrigerant exit temperatures raises the Ratio of Circulation because increasing the vaporized water at the generator, so the maximum value of the Circulation Ratio at refrigerant exit temperature of (10 °C) and generator temperature of (100 °C) of about (0.1761).

Figure 13 explains the relation between the Circulation Ratio and condenser temperature. It was seen that, increasing the condenser temperature decreases the Circulation Ratio due to decreasing the vaporized water at generator, so the maximum value of the Circulation Ratio at condenser temperature of (25 °C) of about (0.1717). Also, Figure 14 shows the relationship between the Circulation Ratio and temperature of absorber. It was noted that, increasing the temperature of absorber decreases the Circulation Ratio due to decreasing the vaporized water, so the maximum value of the Circulation Ratio at absorber temperature of (25 °C) of about (0.2219).

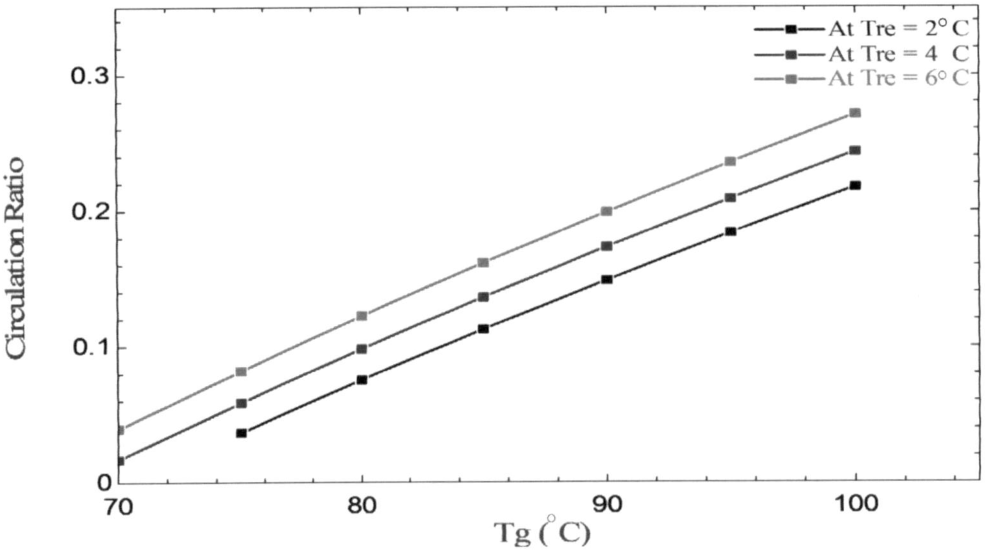

Figure 9. The relationship between the Circulation Ratio and temperature of generator at various refrigerant temperatures at exit of evaporator.

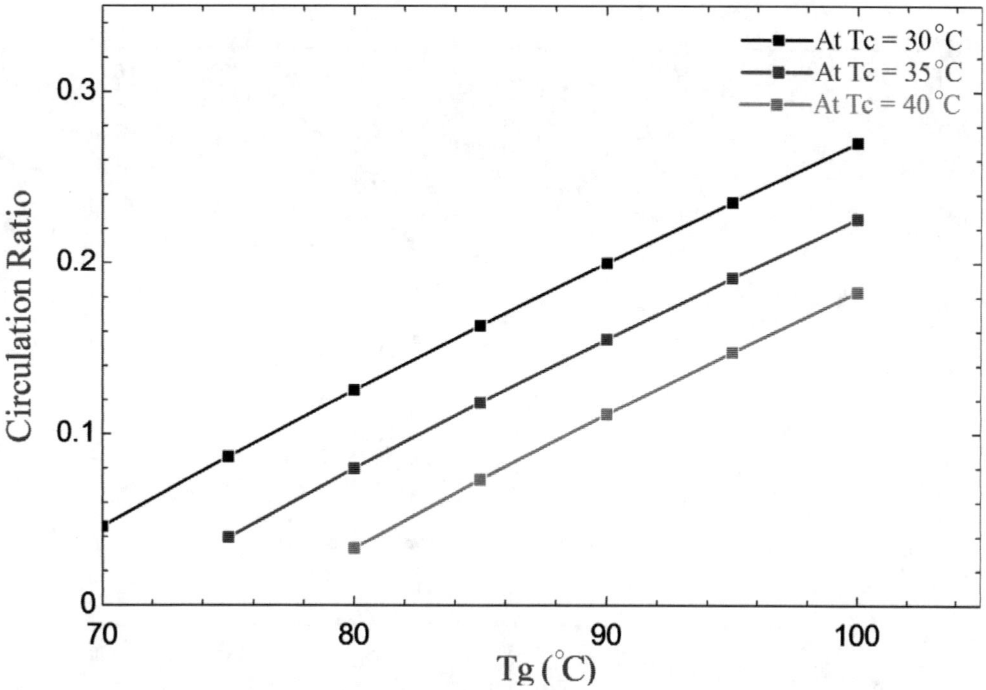

Figure 10. The relationship between the Circulation Ratio and temperature of generator at various temperatures of condenser.

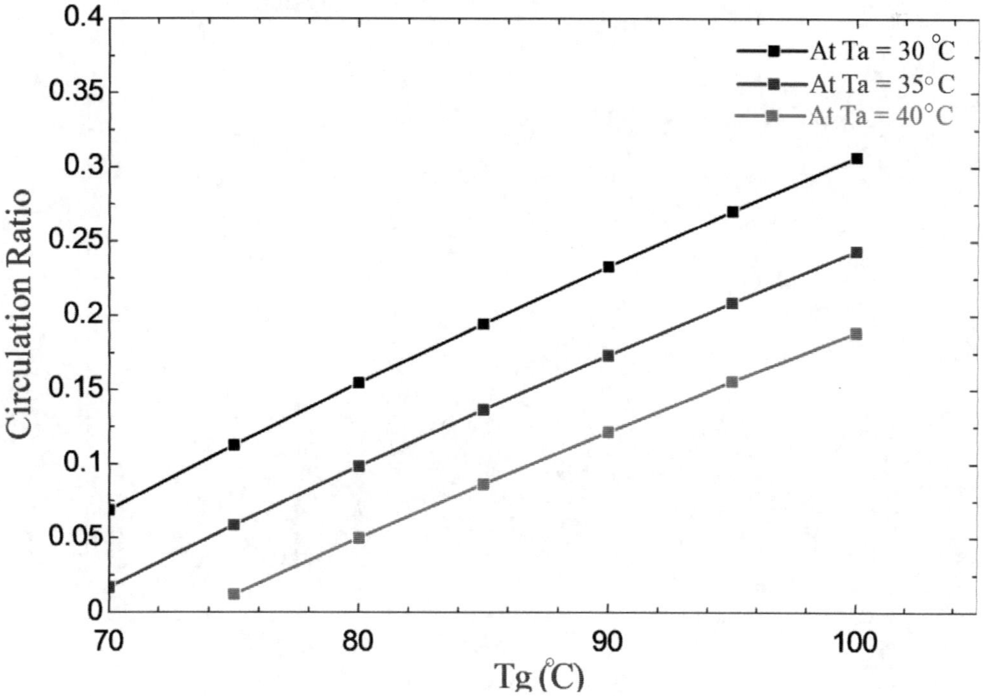

Figure 11. The relationship between the Circulation Ratio and temperature of generator at various absorber temperatures.

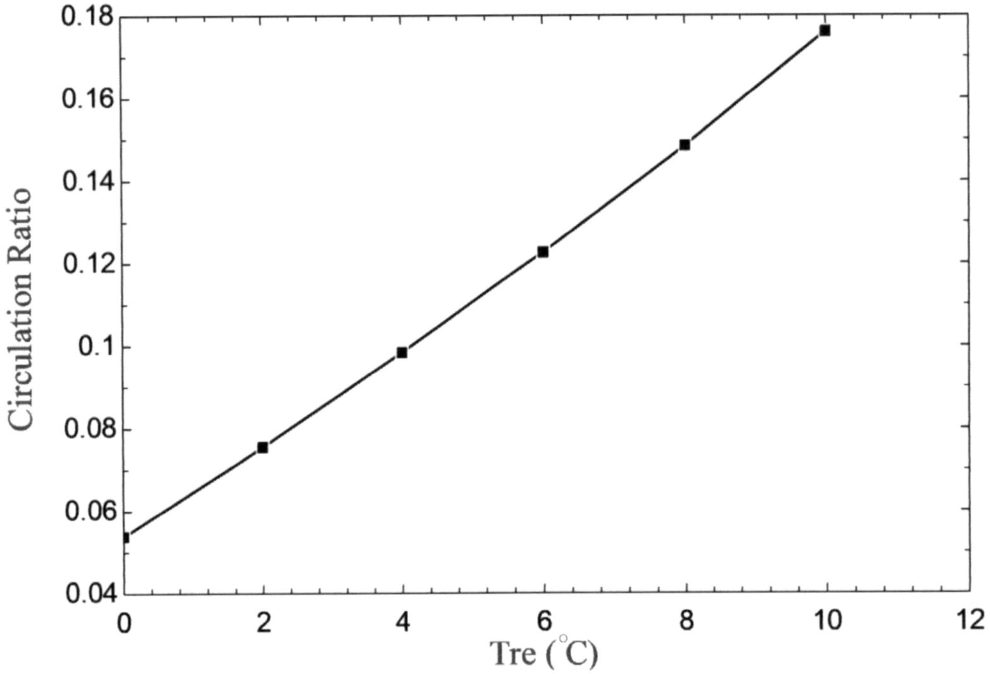

Figure 12. The relationship between the Circulation Ratio and the refrigerant exit temperatures from evaporator.

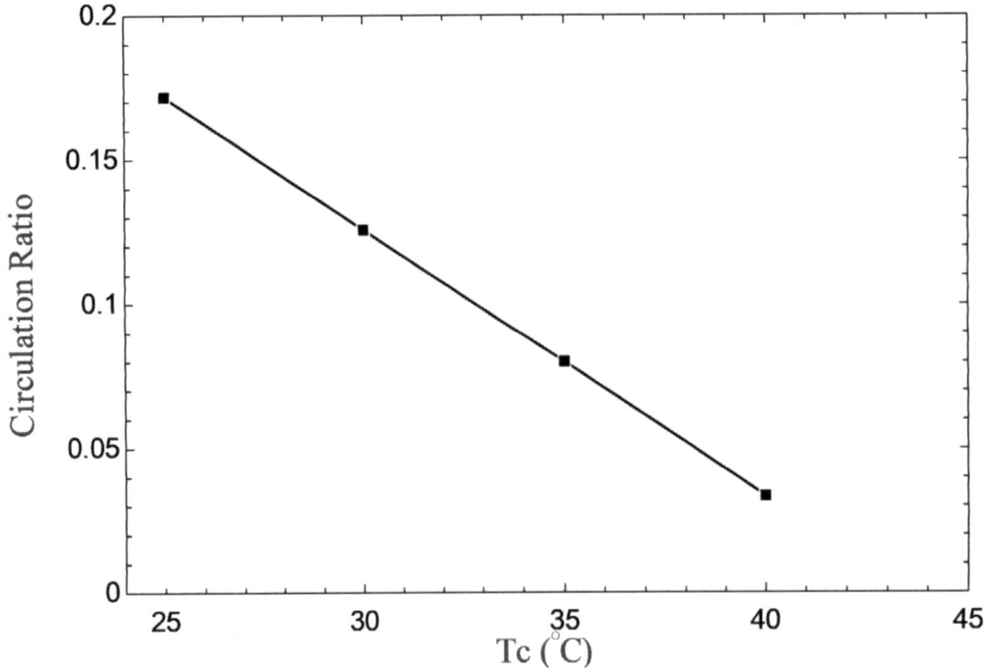

Figure 13. The relation between the Circulation Ratio and the temperature of condenser.

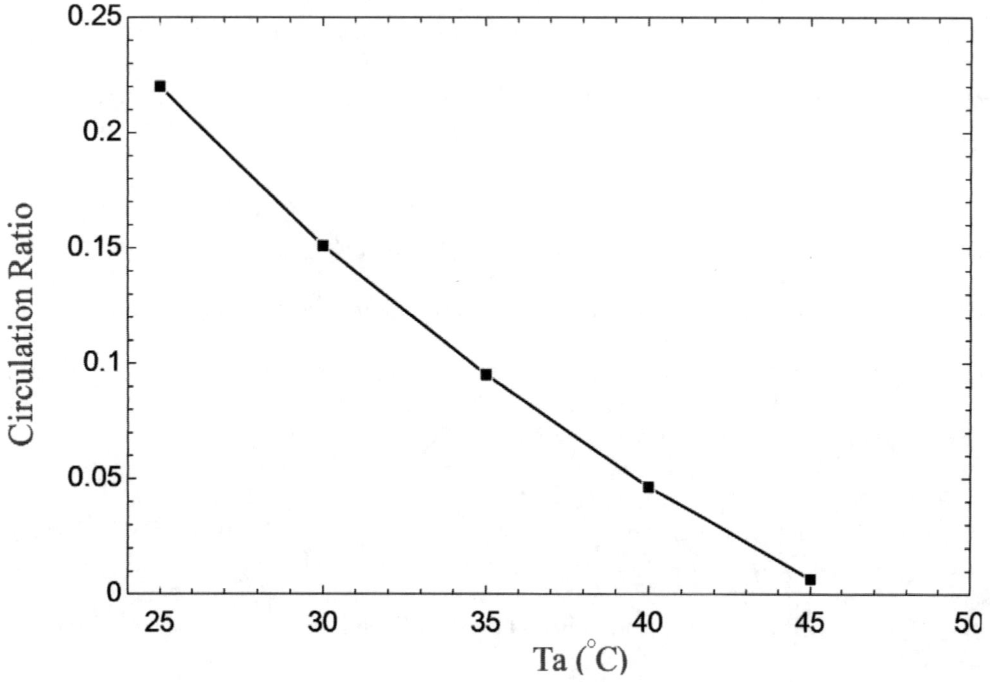

Figure 14. The relation between the Circulation Ratio and the temperature of absorber.

5. Conclusions

The absorption cooling system is studied theoretically using lithium bromide- water mixture using EES software under Iraq weather conditions. Different parameters that affect the absorption cooling system are considered such as the variation of temperature of generator, condenser, evaporator and absorber. Moreover, the effect of rising the size of the system is considered. The maximum obtained cooling load is (1.932 kW) at 15 kg solution mass and 100 °C generator temperature, the maximum COP was 0.774 at refrigerant exit temperature from evaporator of (10 °C) and the peak Circulation Ratio was 0.3066 at 30 °C absorber temperature and 100 °C temperature of generator. From the results, it was concluded that:

1- Increasing of generator temperature improves the cooling load, system COP and circulation ratio due to increasing the vaporized water at the generator.
2- Increasing of refrigerant exit temperature at evaporator improves the refrigeration capacity, system COP and circulation ratio due to increasing the enthalpy difference at the evaporator.
3- Increasing condenser temperature reduces cooling load, system COP and circulation ratio due to decreasing the enthalpy difference at the evaporator.
4- It is also concluded that increasing absorber temperature reduces both system COP and circulation ratio due to decreasing the vaporized water at the generator.

5- In addition, raising the size of the system improves the refrigeration capacity while the performance coefficient and the Circulation Ratio remain constant. **6.**

References

[1] Wald L 2007 Solar radiation energy (fundamentals) *Solar Energy Conv. and Photoenergy Systems* **I**.

[2] Gonzalez-Gil A Izquierdo M Marcos J D and Palacios E 2011 Experimental evaluation of a direct air-cooled lithium bromide-water absorption prototype for solar air conditioning *J. Appl. Thermal Eng.* **31** 16 pp 3358–3368.

[3] Shirazi A 2017 *Feasibility of Solar-Assisted Absorption Chillers for Air-Conditioning Applications* PhD dissertation School of Mechanical and Manufacturing Engineering University of New South Wales.

[4] Hussein H A 2017 Modeling and Simulation of an Absorption Solar Cooling System Under Iraqi Weather Conditions *Eng. Tech. J.* **35** 4 pp 356-364.

[5] Barrera M A Best R Gómez V H García-Valladares O Velázquez N and Chana J 2012 Analysis of the performance of a GAX hybrid (Solar - LPG) absorption refrigeration system operating with temperatures from solar heating sources *Energy Procedia* **30** pp 884 – 892.

[6] Mezher W M Awadh A A and Mohammed A A 2014 Solar Powered Air-Conditioning Using Absorption Refrigeration Technique *J. Eng.* **20** 10 PP 47-59.

[7] Prakash CH J Kumar K D Rao T S 2019 Fabrication and Experimental Analysis of Absorber Based LiBr – Water Absorption Refrigeration System *Int. J. Eng. Adv. Techn.* **8** 6 pp. 4094-4103.

[8] Mussati S F Mansouri S S Gernaey K V Morosuk T and Mussati M C 2019 Model-Based Cost Optimization of Double-Effect Water-Lithium Bromide Absorption Refrigeration Systems *Processes* **7** 1 pp. 1-16.

[9] Fischer Y R Silva I S Dutra J C C 2019 the influence of the external flow rates on an absorption chiller *Aust. J Basic Appl. Sci.* **13** 9 pp. 92-98.

[10] Uçkan I and Yousif A A 2019 Simulation of a solar absorption cooling system in Dohuk city of the Northern Iraq, Energy Sources, Part A: Recovery, Utilization, and Environmental Effects pp1-17.

[11] Zeiny A Haruna M A and Wen D 2019 Aqueous lithium bromide nanosolution for solar absorption refrigeration systems *AIP Conf. Proceed.* 2123, 020088.

[12] Aramesh M Pourfayaz F Haghir M Kasaeian A and Ahmadi M H 2019 Investigating the effect of using nanofluids on the performance of a double-effect absorption refrigeration cycle combined with a solar collector *j. power energy* pp. 1-13.

[13] Abdul Ameer S A 2018 *Experimental and Theoretical Study of Absorption Cooling System Integrated with Solar Concentrated Collector* PhD dissertation Department of Mechanical Engineering College of Engineering University of Babylon.

[14] Hamzah D A and Shahad H A K 2015 Investigation on an Intermittent Absorption Refrigeration prototype powered by Solar Irradiation Al-Qadisiyah *J. for Eng. Sci.* **8** 4 PP. 513-525.

[15] Bellos E Tzivanidis C and Papadopoulos A Enhancing the performance of a linear Fresnel reflector using nanofluids and internal finned absorber *J. Thermal Analysis and Calorimetry* PP.1–19.

[16] Stanciu C Stanciu D and Gheorghian A T 2017 Thermal Analysis of a Solar Powered Absorption Cooling System with Fully Mixed Thermal Storage at Startup *Energies* **10** pp.1-19.

[17] Reda A M Ali A H H 2017 Performance of a small-scale solar-powered adsorption cooling system, *Int. J. Green Energy* **14** 1 pp.75–85.

[18] Abass K I 2007 Is Iraq Ready to Use Solar Energy Applications: A Review, *Int. J. Eng. Sci. Inv.* **6** 10 pp. 27-42.

IOP Conf. Series: Earth and Environmental Science **877** (2021) 012021 doi:10.1088/1755-1315/877/1/012021

The Use of Electrocoagulation to Remove Fluoride, Nitrates and Phosphorous from Water

Rusul K Abdulkhadher[1*], **Ali j Jaeel**[1]

[1] Department of Civil Engineering, College of Engineering, Wasit University, Iraq

* Corresponding author: rkareem@uowasit.edu.iq

Abstract. In recent times, the concentrations of fluoride, nitrates and phosphates in the water have increased as a result of a decrease in the quantities of water and an increase in industrial waste dumped into the rivers. In the current study, a method of electrocoagulation was depended on to treat water contaminated with fluoride, nitrates, and phosphates, using aluminum electrodes as anode and cathode electrodes. According to the obtained results, it can be noticed that the best fluoride, nitrates, and phosphates removal were 99%, 99%, 100%, respectively. It was obtained with a Hydraulic retention time HRT = 3 hours, an applied voltage = 40 volts, a gap between the used electrodes=2 cm, the anode's active area= 792 cm^2, and the flow angle= zero. By using an initial concentration (5 mg F/l, with pH= 4) for fluoride, (75 mg NO_3/ l, with pH= 6) for nitrates, and (0.5 mg PO_4 / l, with pH=9) for phosphates.

1. Introduction

Although water is an essential element for life on planet Earth. However, only 1% or less is drinkable [1-3]. Due to population growth, one of the main challenges currently facing human civilization is the provision of sufficient quantities of potable water [4,5]. Therefore, the ability of mankind to obtain sufficient sources of the required water should be considered as an essential global problem. Achieving sustainable solutions for water management is the possible solution for this challenge. The population of the world is over 7 billion and is estimated to reach 10 billion by 2040 [6]. Climate change also negatively affects fresh water resources in the middle of cities, which in turn affects the sustainable development of water availability and thus affects economic and social activities [7]. Additionally, several studies have shown that freshwater resources are commonly influenced by contamination [8,9]. Water scarcity situations are faced in different regions around the world, which means that the gap between water supply and demand is likely to grow in the future. In 2010, the European Environment Agency reported that municipal water consumption is driven by complex interactions between human and natural system factors on multiple spatial and time scales [10]. There is an inevitable need to maximise wastewater treatment. In order to avoid potential global crises. Due to depleting available resources of clean and potable water, making use of unlimited available saltwater in the oceans, and reducing discharges from water treatment plants. Thousands of years ago, people attempted and succeeded in improving the water quality through

processes of treatment such as the chemical alum addition, boiling, straining, and filtering through charcoal to separate suspended solids from water [11]. In recent decades the wastewater treatment industry has identified discharges of inorganic anions such as nitrates, phosphates, and fluorides in waterways pose a great danger to the natural environment, which has also been shown to be dangerous to humans. To remove inorganic pollutants from water supplies, many common treatment techniques are nowadays used. For instance, coagulation and precipitation, this treatment system has some disadvantages, such as dealing with large amounts of chemicals and the formation of great quantities of sludge that may be harmful to the environment and difficult to get rid of [12]. While in adsorption, ion exchange, and membrane separations, there are some problems such as the additional costs connected with elimination of utilised regenerants materials and regeneration [13,14]. Nutrients (nitrates and phosphates) and fluoride are inorganic anions, which are present in different concentrations in water where the excess of concentration leads to contamination of water and wastewater.
The presence of fluoride in drinking water in a small amount is essential due to it strengthens the teeth and bones [15]. Naturally present amounts of phosphates in ground water and surface water are not dangerous to the environment, animals, or human health [16]. For all life forms, phosphate (PO_4^-) is an essential nutrient element. For the living cells, it plays a fundamental role in the generation of deoxyribonucleic acid (DNA), ribonucleic acid (RNA), adenosine di- and triphosphate (ADP and ATP) [17]. The presence of phosphates and nitrates in a quantity that is greater than the permissible quantities leads to the occurrence of the eutrophication phenomenon that causes an imbalance in the environment [18]. Excessive doses of nitrates in drinking water also cause blue baby syndrome and stomach cancer. High phosphate concentrations also lead to digestive problems [19,20]. Among the diseases caused by excessive amounts of fluoride are brain damage, arthritis, and osteoporosis [21]. Due to the above problems the three anions need to be removed from water using the appropriate methods. The main purpose of the present research is the implementation of the electrocoagulation technique to remove fluoride, phosphorus, and nitrate from drinking water. Using the EC system to provide environmentally acceptable water. Test operating parameters through a continuous system running condition to enhance the removal efficiency performance of the EC.

2. Experimental work

2.1. Material and methods

Synthetic water samples, containing fluoride, nitrates and phosphates, were treated using electrocoagulation cell at different the pollutant's initial concentration (C_i), the acidic function (pH), the applied voltage(V), the gap between the used electrodes (S), the anode's active area(A_{eff}) and the flow angle(θ) in addition to the hydraulic retention time. The desire pH value (4 to 9) was provided using HCl and NaCl. Voltage value was provided using a DC power supply named LW Longwel (LW-K3010D), (0-10 A, 0–40 V). While the electrode distance and the anode's active area were manually adjusted to the required value (2 to 8 cm), (198 to 792 cm^2), respectively. In the EC reactor, at a flow rate of = 25 ml/min, the electrolysis process was carried out by continuously pumping the synthetically prepared water samples. The HRT time changed from (1-3) hours. 10 mL of water samples were taken from the EC outlet reactor on a regular basis to track the removal process. Pollutant concentrations are measured spectrophotometrically by using single beam UV spectrophotometer (SpectroDirect). In each step, one variable was changed and then their optimum value was proposed for the next part of the experiment. All experiments were carried out at room temperature (25±5C). The removal efficiency of Fluoride, Nitrates and Phosphor in the artificial polluted water treated by electrocoagulation is calculated as follows:

$$\eta\% = \frac{(C_o - C_e)}{C_o} \times 100 \qquad (1)$$

Where η% is removal efficiency of fluoride, nitrates and phosphor, C_i is the initial concentration of pollutants (mg/l), and C_e is final pollutant concentration (mg/l).

2.2. Electrocoagulation cell

The removal of fluoride, nitrate and phosphorous was carried out using an EC reactor. The electrocoagulation cell is rectangular in dimensions 20 cm ×16cm×12cm with 6mm thickness made of transparent plastic served as the EC reactor. Eight aluminum electrodes size 20 cm ×6cm×0.09cm. As shown in Figure 1.

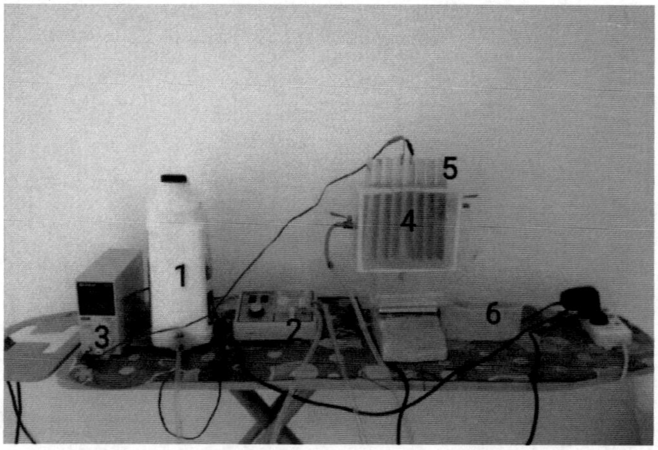

Figure 1. A photograph of the EC cell 1. Influent tank (polluted water). 2. Pump. 3. Power supply. 4. EC cell. 5. Anode and cathode. 6. Final storage.

3. Results and discussion

3.1. Influence of HRT on F, NO_3, PO_4 elimination relative to the initial concentration of F, NO_3, PO_4 (C_i)

The influence of HRT on removal of F, NO_3, PO_4 which has been examined for solutions with initial concentrations (C_i) of 5, 10, 15, and 20 mg/L for F, 75, 100, 150, and 200 mg/L for NO_3, 0.5. 1, 1.5, and 2 mg/L for PO_4. While HRT was changed to 1, 1.5, 2, 2.5 and 3 hours. When pH environment, the applied voltage (V), the distance between electrodes (S), the anode effective area (A), and the angle of flow (θ) were kept constant at 7, 20 volts, 4 cm, 198 cm^2, and 90°, respectively. Figure 6 showed that higher initial concentration extended the treatment duration. The explanation for this decrease in removal with an increase in the initial concentration is the fact that this may be due to the formation of an insufficient amount of coagulation complexes [22,23].

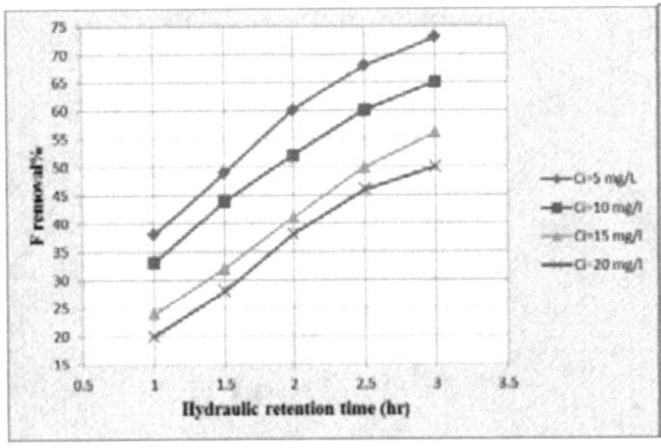

Figure 2. Influences of the C_i on the removal of F.

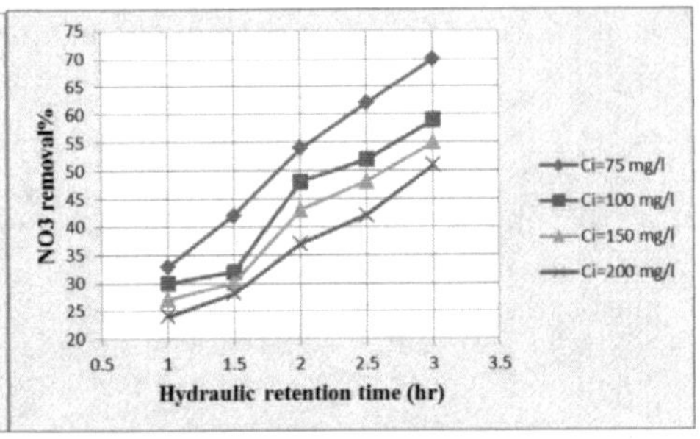

Figure 3. Influences of the C_i on the removal of NO_3.

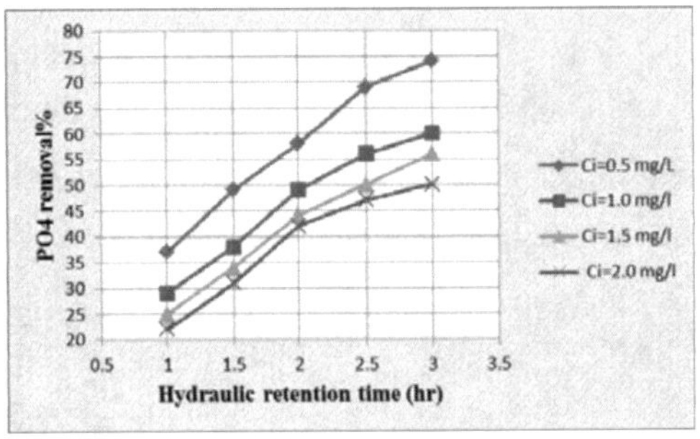

Figure 4. Influences of the Ci on the Removal of PO_4.

3.2. *The influence of HRT on F, NO_3, PO_4 removal relative to pH*

To examine the influence of (HRT), the pH was fixed at a certain value (4, 5, 6, 7, 8 and 9) while HRT was changed to 1, 1.5, 2, 2.5 and 3 hours. When the C_i, V, S, A_{eff}, and the angle of flow (θ) were maintained at a steady level of (5 F mg/l, 75 NO_3 mg/l, 0.5 PO_4 mg/l) 20V, 4cm, 198 cm^2 and 90° respectively. A decrease of fluoride removal can be observed when the primary pH value raised from 4 to 8, and it reached the smallest value at primary pH 8. While by increasing the pH from 4 to 6, the removal efficiency of nitrates would increase. After pH 6, the removal efficiency gradually decreased. While, the highest removal efficiency of phosphate can be achieved near pH 9, this agrees with [24,25].

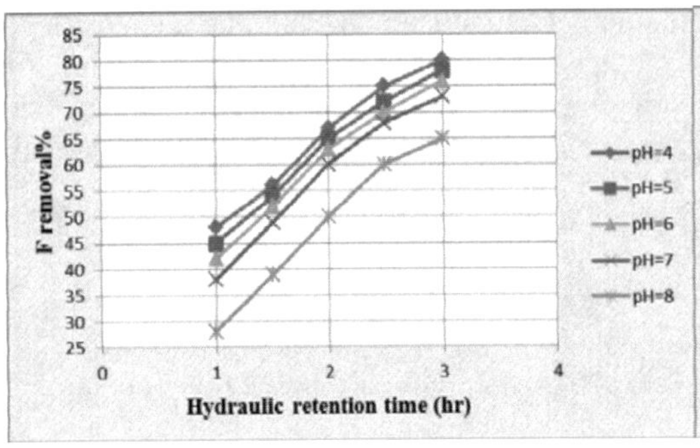

Figure 5. Influences of the pH on the removal of F.

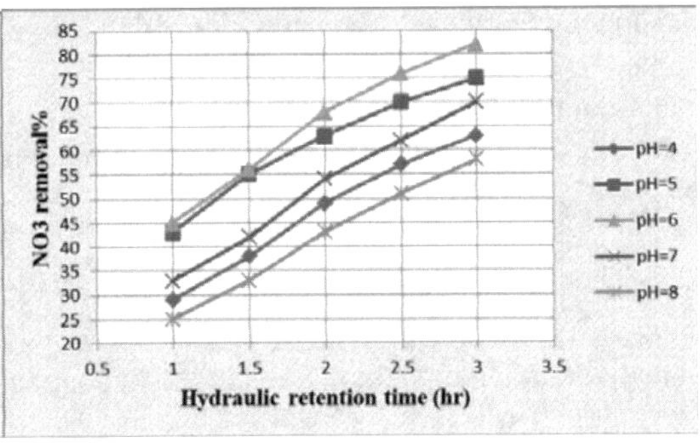

Figure 6. Influences of the pH on the removal of NO_3.

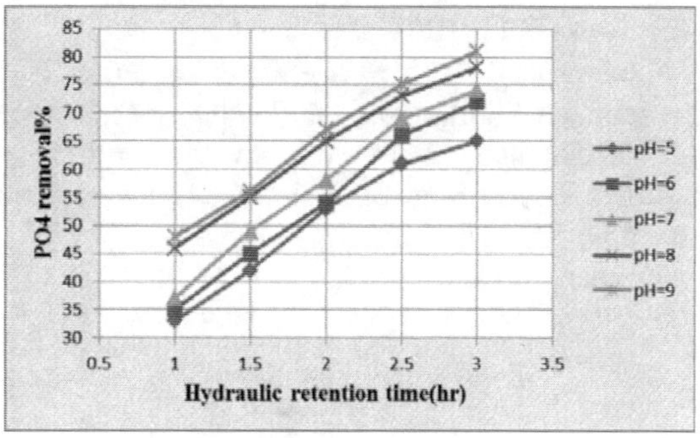

Figure 7. Influences of the Ci on the Removal of PO_4.

3.3. *Influence of HRT on F, NO_3, PO_4 elimination relative to the voltage applied (V)*

The HRT was changed to 1, 1.5, 2, 2.5, and 3 hours while the voltage applied (V) was set at a specific value (10, 20, 30, and 40 V). When the C_i, pH, S, A_{eff}, and the angle of flow (θ) were kept constant at (5 F mg/l, pH 4), (75 NO_3 mg/l, pH 6), (0.5 PO_4 mg/l, pH 9), 4, 4 cm, 198 cm^2, and 90°, respectively. Removal efficiency increases with increasing voltage. Actually, the supplied voltage governs both the rates of bubble generation and dosage of coagulant, it also affects the mixing of solution and transfer of mass at the electrodes [26].

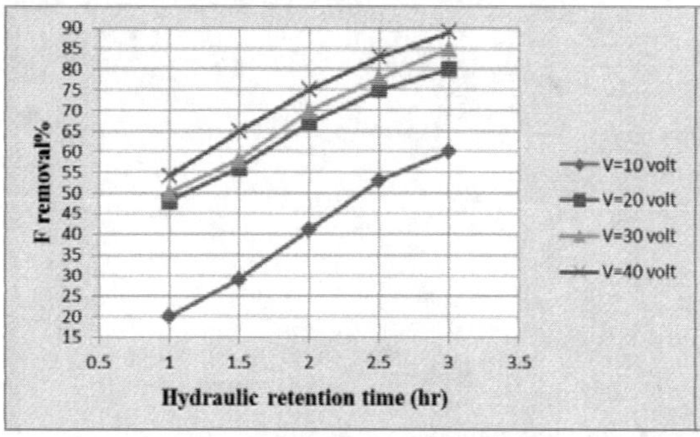

Figure 8. Influences of the pH on the removal of F.

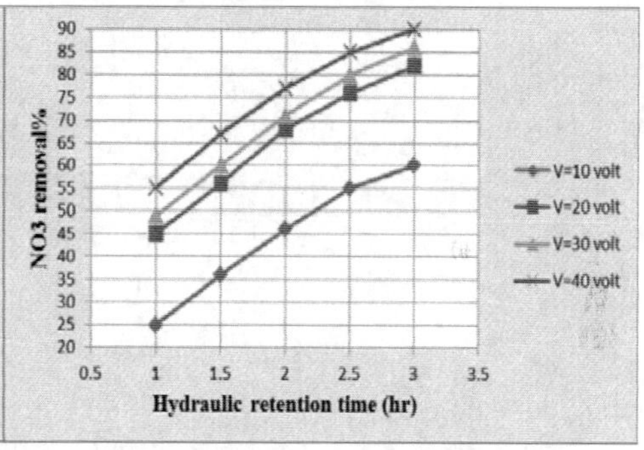

Figure 9. Influences of the pH on the removal of NO_3.

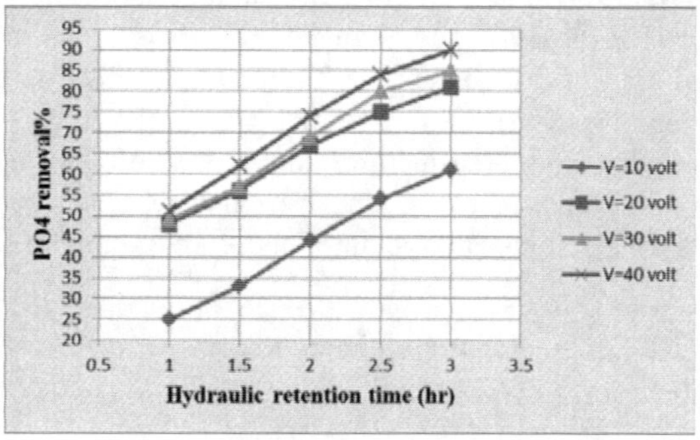

Figure 10. Influences of the Ci on the Removal of PO_4.

3.4. *Influence of HRT on F, NO_3, PO_4 elimination relative to electrode distance (S)*

To examine the influence of (HRT), HRT was changed to 1, 1.5, 2, 2.5 and 3 hours while the electrode distance (S) was fixed at a specific value (2, 4, 6, and 8 cm). When the C_i, pH, V, A_{eff}, and the angle of flow (θ) were kept constant at (5 F mg/l, pH 4), (75 NO_3 mg/l, pH 6), (0.5 PO_4 mg/l, pH 9), 40 V, 198 cm^2, and 90° respectively. The distance = 2 cm represents the optimal distance in this study. This can be attributed to the face that as the value of S raises, resistance of the cell and the development of anodic passive film increase significantly. Thus, the current reduces, and the quantity of the generated floc reduces as well, so the efficiency changes [27].

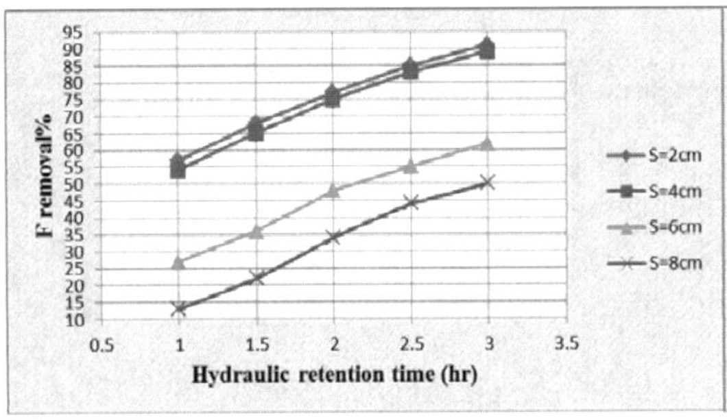

Figure 11. Influences of the S on the removal of F.

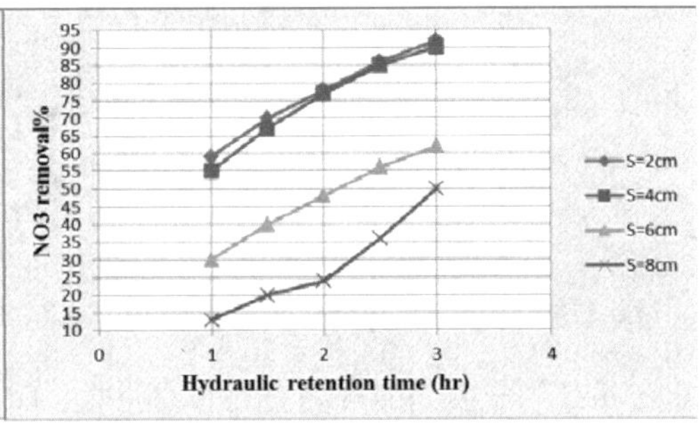

Figure 12. Influences of the S on the removal of NO_3.

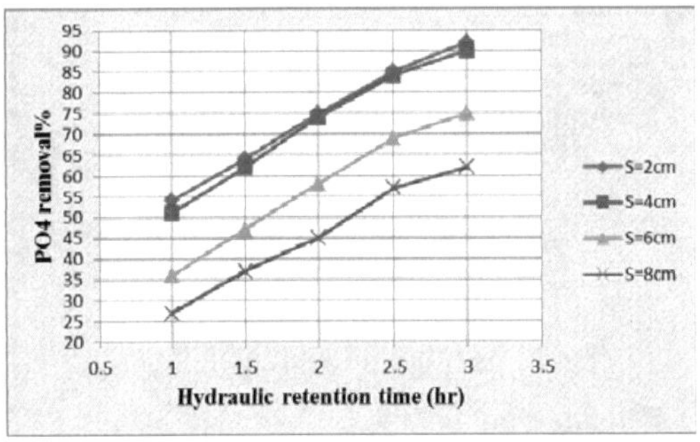

Figure 13. Influences of the S on the Removal of PO_4.

3.5. HRT's impact on F, NO_3, PO_4 elimination in relative to an anode's effective area (A_{eff})

To examine the influence of (HRT), HRT was changed to 1, 1.5, 2, 2.5 and 3 hours while an anode's effective area (A_{eff}) was fixed at a specific value (198, 396, 594 and 792 cm^2). When the C_i, pH, V, S, and the angle of flow (θ) were kept constant at (5 F mg/l, pH 4), (75 NO_3 mg/l, pH 6), (0.5 PO_4 mg/l, pH 9), 40 volt, 2 cm, and 90° respectively. Increasing the active area of the anode from (198 to 792 cm^2) enhanced removal efficiency and reduced processing time that was required. This can be explained by that the use of a larger active area generates large amounts of aluminum which traps more pollutants and improves the removal efficiency [28,29].

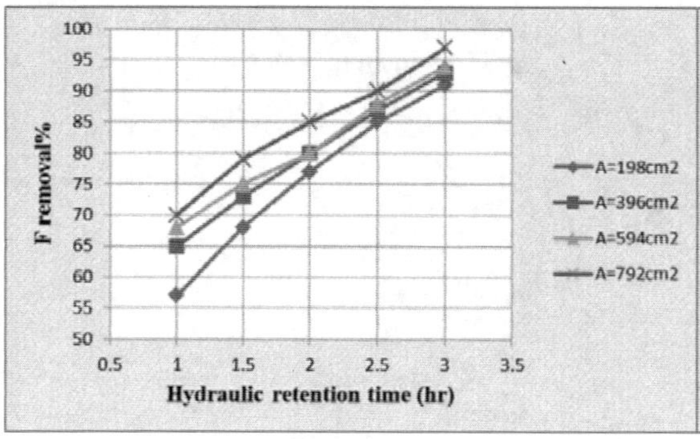

Figure 14. Influences of the A_{eff} on the removal of F.

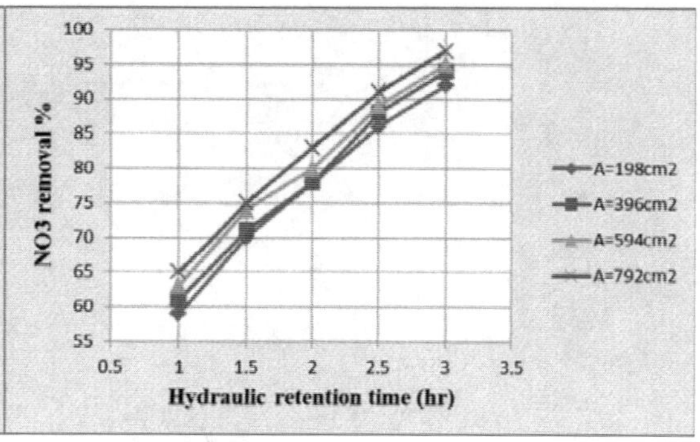

Figure 15. Influences of the A_{eff} on the removal of NO_3.

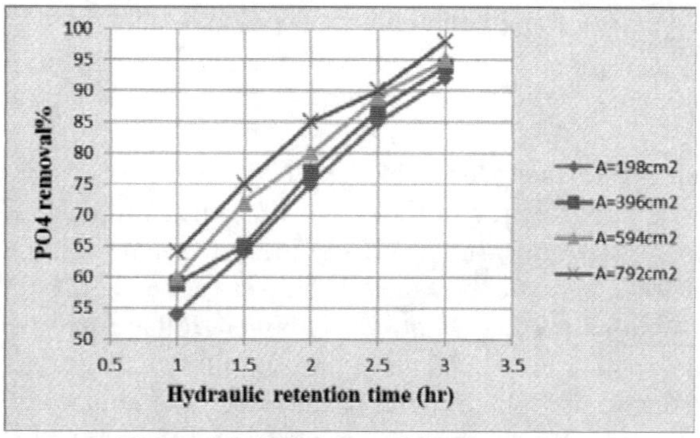

Figure 16. Influences of the A_{eff} on the Removal of PO_4.

3.6. *Influence of HRT on fluoride elimination relative to flow direction*

To examine the influence of (HRT), the angle of flow was set to a specific value (the original direction where the flow is perpendicular to the surface area of the anode electrode, i.e., θ=90° and the other direction where the flow is parallel to the surface area of the anode electrode, i.e. θ=0°). While HRT was changed to 1, 1.5, 2, 2.5 and 3 hours. When the C_i, pH, V, S, and A_{eff} have been kept constant at (5 F mg/l, pH 4), (75 NO_3 mg/l, pH 6), (0.5 PO_4 mg/l, pH 9), 40V, 2cm, 792 cm^2, respectively. The removal efficiency increased when the direction of flow was changed from perpendicular (θ=90°) to parallel (θ=0°) relative to anode electrode surface area plane. This can be attributed to the fact that the horizontal flow, i.e. with angle zero, enhances the homogeneity of the bubbles over the cross-section and leads to an increase in the diffusion rate of Al^{3+} dissolved from the anode to the emulsion, this agrees with [30,31].

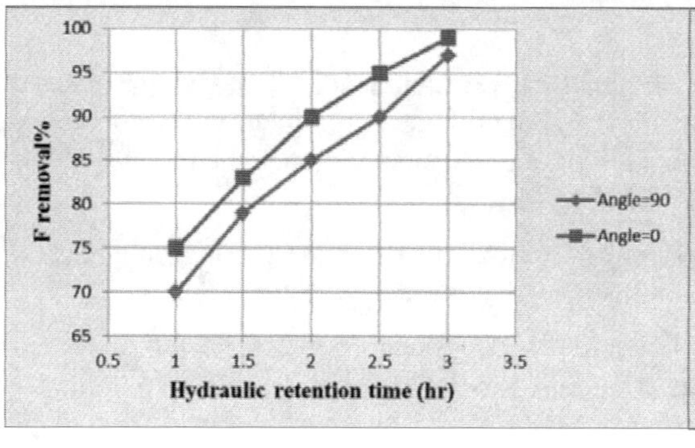

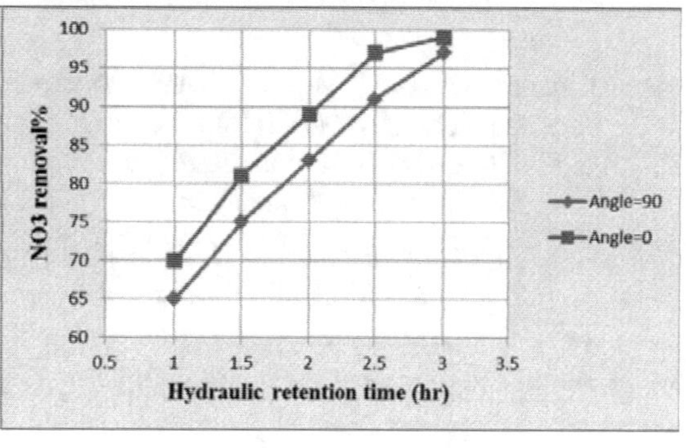

Figure 17. Influences of the θ on the removal of F.

Figure 18. Influences of the θ on the removal of NO_3.

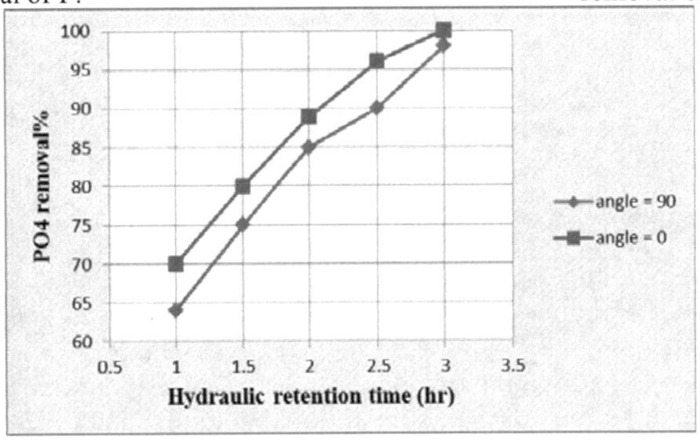

Figure 19. Influences of the θ on the Removal of PO_4.

4. Conclusions

In this study, aluminum electrodes were used in the EC process, based on previous studies. As these electrodes proved their cost and efficiency in removing fluoride, nitrates, and phosphorus from water to obtain environmentally acceptable water. It was clearly indicated that the fluoride, nitrate, and phosphate removal efficiency of 99%, 99%, and 100%, respectively, were achieved from the generation of aluminum hydroxide in the EC cell. Within a HRT = 3 hours, an applied voltage =40 volts, a gap between the used electrodes=2 cm, the anode's active area= 792 cm^2, and the flow angle= zero. The aluminum hydroxide removes the fluoride, nitrate, and phosphate found in the water and reduces the concentration of fluoride, nitrate, and phosphate to less than 1.5, 50, and 0.1 mg/l, respectively and makes it within the acceptable range for drinking water as mentioned by WHO guidelines.

5. References

[1] Al-Saati N H, Hussein T K, Abbas M H, Hashim K, Al-Saati Z N, Kot P, Sadique M, Aljefery M H and Carnacina I 2019 Statistical modelling of turbidity removal applied to non-toxic natural coagulants in water treatment: a case study *Desalination and Water Treatment* **150** 406-12.

[2] Omran I I, Al-Saati N H, Hashim K S, Al-Saati Z N, Patry K, Khaddar R A, Al-Jumeily D, Shaw A, Ruddock F and Aljefery M 2019 Assessment of heavy metal pollution in the Great Al-Mussaib irrigation channel *Desalination and Water Treatment* **168** 165-74.

[3] Zubaidi S L, Kot P, Hashim K, Alkhaddar R, Abdellatif M and Muhsin Y R 2019 Using LARS–WG model for prediction of temperature in Columbia City, USA *IOP Conference Series: Materials Science and Engineering* 584.

[4] Emamjomeh M M, Mousazadeh M, Mokhtari N, Jamali H A, Makkiabadi M, Naghdali Z, Hashim K S and Ghanbari R 2020 Simultaneous removal of phenol and linear alkylbenzene sulfonate from automotive service station wastewater: Optimization of coupled electrochemical and physical processes *Separation Science and Technology* **55** 3184-94.

[5] Zubaidi S L, Al-Bugharbee H, Muhsin Y R, Hashim K and Alkhaddar R 2020 Forecasting of monthly stochastic signal of urban water demand: Baghdad as a case study *IOP Conference Series:Materials Science and Engineering* 888.

[6] Coleman D 2004. V. world population in 2300: A century too far?. *World Population to 2300.*

[7] Zubaidi S L, Al-Bugharbee H, Muhsen Y R, Hashim K, Alkhaddar R M, Al-Jumeily D and Aljaaf A J 2019 The Prediction of Municipal Water Demand in Iraq: A Case Study of Baghdad Governorate *12th International Conference on Developments in eSystems Engineering (DeSE)*

[8] Hashim K S, Al-Saati N H, Alquzweeni S S, Zubaidi S L, Kot P, Kraidi L, Hussein A H, Alkhaddar R, Shaw A and Alwash R 2019 Decolourization of dye solutions by electrocoagulation: aninvestigation of the effect of operational parameters *First International Conference on Civil and Environmental Engineering Technologies (ICCEET)* 584.

[9] Hashim K S, Hussein A H, Zubaidi S L, Kot P, Kraidi L, Alkhaddar R, Shaw A and Alwash R 2019 Effect of initial pH value on the removal of reactive black dye from water by electrocoagulation (EC) method *2nd International Scientific Conference.*

[10] Hashim K, Kot P, Zubaid S, Alwash R, Al Khaddar R, Shaw A, Al-Jumeily D and Aljefery M 2020 Energy efficient electrocoagulation using baffle-plates electrodes for efficient Escherichia Coli removal from Wastewater *Journal of Water Process Engineering* **33** 101079-86.

[11] Nguyen D D, Yoon Y S, Bui X T, Kim S S, Chang S W, Guo W, and Ngo H H 2017 Influences of operational parameters on phosphorus removal in batch and continuous electrocoagulation process performance *Environmental Science and Pollution Research* 24(32), 25441-25451

[12] Duan, J., & Gregory, J. 2003 Coagulation by hydrolysing metal salts. *Advances in colloid and interface science*, **100** 475-502.

[13] Li Y, Liu C, Luana Z, Peng X, Zhu C, Chena Z, Zhang Z, Fan J, Jia Z 2006 Phosphate removal from aqueous solutions using raw and activated red mud and fly ash, *Journal of hazardous materials* **137** 374-383.

[14] Lin L, Wan C, Lee D J, Lei Z, and Liu X 2014 Ammonium assists orthophosphate removal from high-strength wastewaters by natural zeolite *Separation and Purification Technology* **133** 351-356.

[15] Li Y, Jiang Y, Wang T J, Zhang C, and Wang H 2017 Performance of fluoride electrosorption using micropore-dominant activated carbon as an electrode *Separation and Purification Technology* **172** 415-421.

[16] Jiang C, Jia L, He Y, Zhang B, Kirumba G, and Xie J 2013 Adsorptive removal of phosphorus from aqueous solution using sponge iron and zeolite *Journal of colloid and interface science* **402** 246-252.

[17] Ohtake H, Kuroda A, Chandrasekaran M, Wu H, Tanaka S, Morohoshi T, and Takiguchi N 2001 Molecular Genetics of Bacterial Polyphosphate Accumulation to Better Understand the Mechanism Underlying Biological Phosphorus Removal *Environmental Monitoring and Biodiagnostics of Hazardous Contaminants* Springer 181–196.

[18] Jorgensen S E, & Williams W D 2001 Water quality: The impact of eutrophication. *UN Environment Programme.*

[19] Chiu H F, Tsai S S, and Yang C Y 2007 Nitrate in drinking water and risk of death from bladder cancer: an ecological case-control study in Taiwan, *Journal of toxicology and environmental Health, Part A*, *70*(12), 1000-1004.

[20] Kumar E, Bhatnagar A, Ji M, Jung W, Lee S H, Kim S J, Lee G, Song H, Choi J Y, Yang J S, Jeon B H 2009 Defluoridation from aqueous solutions by granular ferric hydroxide (GFH), *Water Research*, *43*(2), 490-498.

[21] Ün Ü T, Koparal A S, Öğütveren Ü B, and Durucan A 2010 Electrochemical process for the treatment of drinking water *Fresenius Environmental Bulletin, 19(9), 1906-1910.*

[22] Vasudevan S, Lakshmi J, and Sozhan G 2009 Studies on a Mg-Al-Zn Alloy as an Anode for the Removal of Fluoride from Drinking Water in an Electrocoagulation Process *Clean–Soil, Air, Water,* 37(4-5), 372-378.

[23] Hendaoui K, Ayadi M T, and Ayari F 2021 Optimization and mechanisms analysis of indigo dye removal using continuous electrocoagulation, Chinese Journal of Chemical Engineering **29** 242-252.

[24] İrdemez Ş, Demircioğlu N, and Yildiz Y Ş 2006 The effects of pH on phosphate removal from wastewater by electrocoagulation with iron plate electrodes *Journal of hazardous materials* 137(2), 1231-1235.

[25] Shalaby A, Nassef E, Mubark A, and Hussein M 2014 Phosphate removal from wastewater by electrocoagulation using aluminium electrodes *American Journal of Environmental Engineering and Science* 1(5) 90-98.

[26] Maitlo H A, Lee J, Park J Y, Kim J C, Kim K H, and Kim J H 2019 An energy-efficient air-breathing cathode electrocoagulation approach for the treatment of arsenite in aquatic systems *Journal of Industrial and Engineering Chemistry* **73** 205-213.

[27] Hashim K S, Shaw A, Al Khaddar R, Pedrola M O, and Phipps D 2017 Defluoridation of drinking water using a new flow column-electrocoagulation reactor (FCER)-Experimental, statistical, and economic approach *Journal of environmental management* **197** 80-88.

[28] Sharma A K and Chopra A K 2017 Removal of nitrate and sulphate from biologically treated municipal wastewater by electrocoagulation *Applied Water Science* 7(3) 1239-1246.

[29] Amarine M, Lekhlif B, Sinan M, El Rharras A, and Echaabi J 2020 Treatment of nitrate-rich groundwater using electrocoagulation with aluminum anodes *Groundwater for Sustainable Development* 11 100371.

[30] Fouad Y O A, Konsowa A H, Farag H A, and Sedahmed G H 2009 Performance of an electrocoagulation cell with horizontally oriented electrodes in oil separation compared to a cell with vertical electrodes *Chemical Engineering Journal*, 145(3) 436-440.

[31] Liu Y, Jiang W, Yang J, Li Y, Chen M, and Li J 2017 Experimental study on evaluation and optimization of tilt angle of parallel-plate electrodes using electrocoagulation device for oily water demulsification *Chemosphere* **181** 142-149.

IOP Conf. Series: Earth and Environmental Science **877** (2021) 012022 doi:10.1088/1755-1315/877/1/012022

Greening ingredients for existing and new diplomatic buildings

Salih Khalaf Mohammed[1], Bahjat Rashad Shahin[1]

[1]Department of Architectural Engineering, College of Engineering, University of Baghdad, Baghdad, Iraq.

E-mail: s.muhemmed1304@coeng.uobaghdad.edu.iq; bahjatsha@yahoo.com

Abstract. Diplomatic buildings are considered a mirror that reflects the civilization and values of one country in the space of other countries, and therefore one of the most basic elements in planning these buildings in our time is related to achieving the environmental suitability of these buildings. As embassy buildings in different countries of the world should be a guest that respects the local context and environment in which he is present, and with the increasing number of diplomatic buildings in different countries , the need to adopt more attention has emerged to study their relationship with the environmental aspect,so this will be given The research pays great attention to environmental standards in diplomatic buildings To become the research problem: the lack of knowledge in previous literature about the environmental planning and design indicators of diplomatic buildings in general, and how they deal with the host country's resources , The aim of the research is: to extract the most important vocabulary and indicators of greening decisions in diplomatic buildings,Where the research results confirm the role of treatments for greening diplomatic buildings in preserving the host country's natural resources from depletion, which in turn contributes to strengthening diplomacy between the two countries, in addition to achieving significant savings in the costs of paying bills for energy and resource use.

Keywords: diplomatic buildings, local natural resources, host and guest country, environmental suitability, greening ingredients.

1. Introduction

The global community, in our present time, has become aware of the damage to the natural environment and its dire economic and ecological consequences, which led those interested in the field of environment and development to think about the reasons that contributed to stressing the environment. Which showed that buildings represent the major consumers of depleted energy in the world. As serious studies began to revolve around developing appropriate solutions to reduce environmental pollution and depleting its depleted resources. In 1983 AD the United Nations established the Committee on Environment and Development to resolve disputes arising around the aspirations of the developed and developing countries on sustainable development, and in 1989 the report of the United Nations Committee was published. (Brundtland) on the definition of sustainability, which is the most famous interpretation of sustainability globally, namely: "Development that meets the needs of the present without compromising the ability of future generations to meet their own needs." The components of green architecture and the idea of greening buildings in general, and embassy buildings in particular, will be addressed, with the aim of diagnosing The most important design standards and guidelines in this field. This research relied on a group of United Nations (UN) reports and reports of the Intergovernmental Authority on Climate

Change (IPCC), as well as the directives of the United States Environmental Protection Agency (USEPA), where we will refer to the most important indicators and goals of the initiative to green diplomatic sites in general, And embassy buildings in particular, with an attempt to put forward the most important internationally accredited environmental diplomacy programs as a basis for local accreditation.

2. The research problem

Lack of knowledge in local studies about the environmental planning and design indicators of diplomatic buildings in general, and how they deal with the host country's resources, in addition to the degree of their environmental compatibility with the local urban context.

3. Research methodology

The research relied on the descriptive and analytical approach to crystallize the knowledge framework and build an information base on the design and planning concepts of diplomatic buildings related to environmental determinants, by extracting the theoretical framework for vocabulary and indicators of design principles and facts from global and Arab experiences for the purposes of adopting them in assessing the level of sustainability and harmonization of the context in diplomatic buildings Local, and draw conclusions and recommendations, by means of comparisons and diagnostic observation.

4. Basic concepts

Before going into the details of the green embassy buildings, it is necessary to introduce green architecture in general, and its characteristics, in order to understand and understand green embassies and their environmental relationship, and accordingly a set of descriptions pertaining to them were taken from the viewpoints of their commentators, as follows:

Green architecture: It is defined as "architecture that is compatible with the surrounding environment, which integrates with all its surrounding environmental determinants and fills its deficiency or fixes its defect, and benefits from the phenomena and constituents of this ocean, and it is architecture that does not harm the environment with its waste, as it grows up to the stage of stability." Green architecture is also seen as (a highly efficient system compatible with its surroundings with minimal damage, as it resembles a plant in terms of shape and composition and coexists and grows to restrict everything around it. It is a living organism that has a life with a beginning and an end, with an emphasis that its properties and qualities are present across generations, and evolve according to The influences to appear clearly in certain practices and different periods of time [4].

Green architecture seeks to reduce the negative environmental impacts in buildings in general and diplomatic buildings in particular, by enhancing the efficiency of the use of materials and renewable energy. It is a building system that fulfills all the principles of sustainability, and each seeks to preserve the environment and ensure the right of future generations to energy and resources [5], Among the concepts and definitions developed in this field by senior architects interested in this field, on the level of ideas and application, we address the most important pioneers of architecture about their views on the concept of green architecture.

Architect Ken Yeang, a Malaysian architect, proposed the concept of green architecture from an environmental point of view. He criticizes the effect of buildings on natural systems, and asserts that green architecture, or sustainable architecture, must meet the needs of the present without neglecting the right of future generations, design decisions. Its effects are not limited to the contemporary environment only, but extend to future generations as well.

The architect William Reed, is an architect and thought leader who showed that green buildings are nothing but buildings that are designed, implemented and managed in a comparative manner that places the environment within its basic characteristics, and he believes that one of the concerns of green buildings appears to reduce the impact of the building on the surrounding environment besides Reducing the costs of its establishment and operation.

Greening: It means converting or renewing a building into a green building by planning, designing and using environmentally friendly materials, as this is a sustainable strategy in itself. Therefore, it is possible to recondition an existing structure and convert it into a new green building, while ensuring the use of less energy and materials, leading to a sustainable environmental practice. [10] Thus, re-greening of any building becomes a specific building approach, balanced and multifaceted in environmental contexts. It is a very new concept to reduce the impact of buildings on the ecosystem by adjusting energy outputs, technologies that in turn improve and enhance the overall energy efficiency of buildings, thus reducing carbon emissions.

5. Green architecture standards

After the spread of greening movements in contemporary architecture in abundance in the processes of construction and renovation, especially during the past decades, and its gaining more momentum around the world. It has become necessary to develop basic standards and approved standards for greening patterns and indicators, the most prominent of which is the Leadership for Energy Environment and Design program, or what is called for short (LEED). The LEED program has developed a strong set of standards for accrediting green architecture processes from the design stage to the last stage of the operations. [9] and these processes seek to achieve the goals of clean energy, and they can be set in mind for the purpose of ensuring the following standards in building green architecture in general, as green embassy buildings are part of that architecture and which are identical to some extent with the sustainability indicators, which we will refer to as the components of greening:

6. Green embassies and environmental diplomacy

The concept of green embassies: According to the regulations of the French project to interpret the green embassy, it represents a gradual and comprehensive initiative to reduce the environmental impact on diplomatic missions abroad. It aims to enhance adherence to international norms regarding sustainable development, in addition to making the diplomatic network a review of common experiences, improving the management and performance of embassies and achieving savings, and achieving the principle of carbon neutrality [12].

In short, it is a distinguished embassy that is a product of green design, which focuses on increasing the efficiency of the use of natural resources from energy, water and local materials to reduce the effects of buildings on human health and the environment in the host country during the life cycle of the diplomatic building, by improving design, construction, operation and maintenance ". Environmental diplomacy is defined as "the practice of managing international relations by facilitating and strengthening a common commitment to preserving depleted natural resources and promoting the use of renewable natural energies through sustainable operations and responsible environmental supervision." It is one of the art of diplomatic negotiation and the practice of international relations for the purposes of preventing environmental degradation and protecting Depleted natural resources, sharing of common water and energy resources, and policy and practices reinforce the components of environmental diplomacy and standards of shared values among nations committed to reducing their greenhouse gas emissions, preserving depleted natural resources, and increasing biodiversity. Contemporary environmental diplomacy is increasingly used to support efforts to prevent environmental pollution, build international trust, environmental governance, and maximize technological capacity in identifying environmental crimes and threats [15].

7. Guiding greening ingredients

The elements of greening contribute to ensuring the availability of safe buildings and a diplomatic post that represents governments in host countries, and supports employees during their work, to achieve foreign policy goals. It should also represent the Ministry of Foreign Affairs facilities represented by the diplomatic buildings the best in excellence in architecture, design, engineering, technology, sustainability, art and culture. In order to improve the environmental footprint of the administration and increase efficiency, it also requires harnessing experiences in politics and public diplomacy, in order to develop sustainability efforts and institutionalize them, and to include the following indicators: According to what is indicated on the official website of the World Green Building Council: [13]

- Implement best environmental practices in the design, construction and operation of diplomatic facilities.
- Utilizing the development of information technology to support high performance and efficiency.
- Reducing depleted energy consumption and increasing the use of renewable energy.
- Using local and environmentally friendly products, reducing waste flows and ensuring recycling.
- Protecting and preserving water resources and ensuring their recycling.

8. Climate change problems in the world

According to United Nations (UN) reports, the global average global temperature increased by 0.85 degrees Celsius during the time period from 1880 to 2012, and global sea level rise is expected to be 10 cm higher, with the possibility of a global temperature rise of 1.5 degrees Celsius by The year 2100. [2], and the 2013 IPCC report states that the construction sector has the broad potential to deliver rapid, deep and cost-effective efforts to mitigate greenhouse gas (GHG) greenhouse gases compared to any sector. else. The depleted energy consumption in both new and existing buildings is expected to be reduced by between 30% and 50% by adopting readily available technological technologies, sustainable environmental design, appropriate equipment, smart management systems and ensuring alternative generation solutions. And in view of the vastness of the joints of the diplomatic bag, and the wide spread of its buildings within the host countries, so a great role falls on the shoulders of those buildings to preserve the environment in the host countries. For example, the US government alone owns approximately 445,000 buildings with a total floor area of more than 300 million square meters, and rents about 57,000 buildings with a total floor area of more than 34.7 million square meters, and these buildings constitute 37% of the host government's total energy use. . Below we will include a set of executive orders of the United States announced in 2013, which concern the greening of US embassies abroad, the most important of which are: [7]

- A comprehensive annual inventory of greenhouse gas emissions should be reported, which includes purchased energy stocks, employee travel and commuting miles, solid waste disposal, and wastewater treatment.
- Ensure that the use of depleted energy is reduced by 30% by the year 2015, relative to the 2007 baseline.
- Ensuring that drinking water consumption will be reduced by 26% by the year 2020 relative to the 2007 baseline, at a rate of (2% per year).
- Reducing irrigation water use by 20% by 2020 using the 2010 baseline (2%) annually.
- Recycling 50% of non-hazardous solid waste by 2015.
- Reducing fossil fuel consumption from fleet equipment by 30% by 2020, relative to the 2005 baseline.

The new US Embassy of Sofia was the first American diplomatic mission to obtain a Leadership in Energy and Diplomatic Environmental Design (LEED®) certification in 2007. The current research confirms the necessity of commitment to issuing executive orders similar to the reality of Iraqi diplomacy.

9. The most important indicators of greening the diplomatic buildings

In general, all the aforementioned calendar areas can provide multiple benefits for greening embassies. For example, it is possible to improve air quality, expand the use of natural light or the idea of merging with nature, or guarantee internal comforts, or the work environment or the larger environment of the host country. The most important indicators of environmental diplomacy to form a good embassy lie in asking the following questions: [11]

- Transportation: How can diplomatic functions manage fleets, air travel, and transportation options for their workers to reduce greenhouse gas emissions?

- Site: What options are available for the ecosystem to manage, improve and display sustainable ocean landscapes, and irrigation practices to enhance biodiversity?
- Water: How can programming diplomatic posts reduce annual water consumption and cost, while managing rainwater to protect water resources?
- Energy: How can diplomatic jobs reduce annual spent energy consumption and costs? Also, how can we reduce dependence on fossil fuels and increase the use of renewable sources of energy?
- Materials: What options are available for the diplomatic functions to reduce the effects of over-procurement, reduce waste, and support local and regional business?
- Internal environment: How can diplomatic functions enhance internal environment indicators, maintain employee health, and improve productive work environments for their occupants?

The best example of the success of the principles of diplomatic greening occurred when the US embassy in Mongolia (Ulaan baatar) won in 2010 the GDI award, an award that the Greening Council of the US State Department prepares among the US embassies in the world. The first ever after obtaining the following achievements : [7]

- Calculating the carbon footprint, as it is the first US embassy to secure a carbon footprint.
- Conserving water, by drilling wells, resulting in a 16.94% availability in water bills.
- Recycling, resulting in a 22% savings in litter expenditures.
- Electrical improvements, resulting in 33.74% energy availability.

10. The goals of the Greening Diplomacy Initiative

The Greening Diplomacy Initiative (GDI), launched in 2009, aims to improve the environmental sustainability of global diplomatic management processes. The GDI challenges the diplomatic administration to develop and implement policies and procedures that reduce its overall environmental carbon footprint, reduce costs, and ensure sustainability at the forefront of foreign diplomatic policy in host countries. [16]The GDI initiative relies on specialized organizations such as the International Association for the Management of Diplomatic Facilities (IEMA) Institute of Environmental Management and Assessment by practicing training, as well as the League of Green Embassies network group, which consists mainly of US embassies. And the Green Embassies University (the university), a global network of more than 100 US and foreign diplomatic missions working to formulate a joint agenda for budgeting work for depleted energy efficiency, renewable energy and water conservation. Indeed, access to all of these diplomatic missions demonstrates concretely the strength of the link between environmental management and political will to strengthen the environmental diplomacy program. Member embassies undertake to provide innovative solutions, influence individuals and institutions in host countries, and play an important role in mobilizing public action, through the following principles: [19]

- The availability of environmental leadership in new and existing embassies.
- Work with appropriate government officials in seeking funding for embassy greening projects.
- Collaborate with counterparts and exchange ideas on energy conservation, and other strategies to reduce or reduce greenhouse gas emissions as much as possible.
- Encouraging dependence on increasing the use of renewable energy.
- Establishing recycling programs across embassies.

11. Strategic Reality of Environmental Diplomacy

11.1. Transport Strategies Program

Current transportation patterns globally generate significant economic, social and environmental damage, as transport consumes more than half of the liquid fossil fuels used, emits nearly a quarter of the energy-related carbon dioxide, and generates more than 80% of air pollution in urban areas in general and in Especially developing countries. This leads to more than 1.2 million fatal accidents annually, in addition to the suffocating traffic problems and harmful to humans and the environment

[1]. The ecological diplomatic functions of diplomatic buildings can be demonstrated by developing policies and implementing practices that support the goals of reducing transport emissions in their host countries. The most effective sequence to achieve the indicators of the environmental impact reduction strategy is as follows: [8]

1. **Pollution load reduction index:** It includes alternatives to using vehicles, using the transportation lines of large embassy employees instead of using small vehicles for each employee separately, or cycling, or even encouraging walking. Virtual meetings over the Internet can also reduce air travel by embassy personnel.
2. **Fuel Efficiency Index:** For the purpose of helping to achieve a clean embassy environment, it is assumed not to use old vehicles, which are not efficient because they consume more fuel than modern vehicles.
3. **Renewable fuel sources approval index:** The program includes the use of embassy building vehicles that use electricity, ethanol, biodiesel, natural gas, propane and hydrogen instead of the current fossil fuel vehicles available in the host country, which are cost-effective solutions.

11.2. A strategy program for reducing water consumption and recycling in the buildings of embassies and consulates

The locations of embassies and consulates provide opportunities to ensure that the citizens of the host country have a positive first impression of the guest country, to ensure the application of foundations based on respect for natural ecosystems, to promote biodiversity and to integrate technological innovations that demonstrate leadership in the wise use of resources in general and provide concrete examples of the principles of Environmental diplomacy. This is especially important in developing countries, which struggle to balance natural resource management and economic growth. The idea of preserving water represents one of the major diplomatic and development opportunities in the embassy building decisions. Therefore, the most effective sequence for monitoring water use efficiency indicators in buildings in general and in embassies in particular is: [7]

1. **Pollution load reduction indicator:** which could provide great benefits as a result of operational improvements to the existing water networks in embassy buildings. The base load of the systems can be greatly affected by consumer behavior, leakage problems, installation efficiency, and irrigation scheduling system, which do not take into account local climatic conditions.
2. **System efficiency index:** such as the use of high-efficiency, or double-flush toilets in embassy bathrooms, low-flow shower heads, and drip irrigation in embassy gardens, which can lead to significant water availability.
3. **Recycling Index:** Many final uses of water can be recycled by exploiting non-potable water sources, including rainwater or wastewater.

11.3. The program for the strategy of protecting the diplomatic site

Embassy sites usually consist of many elements, including buildings, green spaces, rainwater management infrastructure, parking lots, roads, patios, and a pedestrian sidewalk. The composition of the site varies greatly from country to country, and these differences in the proportions of the site elements, the total land area, the permeability of the surfaces, and the types of materials may result in a significant difference in the maintenance and operation needs. Therefore, the components of the embassy is an important factor in solving the problem of managing the site in general.

For example, small diplomatic sites with a high building density of a diplomatic building, as in the example of the US Embassy in Berlin, can be best served by structural solutions, such as green roofs or paved roads as in (Figure 1). The figure shows a comparison of the United States Embassy in Berlin with an area (13,721 square meters) and in Nairobi (37,350 square meters), which shows the division of construction areas and green spaces. In terms of the extensive on-site capabilities of the Nairobi Embassy, in promoting the Sustainable Landscape Index, and maximizing the foundations for environmental diplomacy [7].So Large diplomatic sites with extensive landscapes, such as the United

States Embassy in Nairobi, are a prime candidate for low environmental impact development strategies that use the principle of natural environmental alternatives, such as green lands, wetlands and water bodies, to reduce the environmental impact on the surrounding area, and reduce demand On the added resources are the most effective indicators for monitoring the improvement of the site's performance, as follows: [11]

1. **Infrastructure Load Reduction Indicator:** Non-native plants, weed areas, and large areas of solid areas can unnecessarily increase the demand for water loads at the site, which requires ensuring the availability of storage water more. Therefore, strategies must be devised to enhance landscape planning, and to reduce hard areas to reduce water demand.
2. **The indicator of choosing the least effective alternatives:** which includes alternatives to IPM methods, (Integrated pest management), external maintenance and cleaning, and sustainable landscaping works with low-impact technologies that require less water and fuel, and the speed of implementation compared to the non-sustainable traditional practices.
3. **Self-treatment on site indicator:** Self-treatment inside the diplomatic site can reduce the burden on the municipality's water treatment systems for the host country, for example by collecting rainwater from roofs and keeping it in tanks, as well as securing green roofs, and treating recycling of water Used, and discharged into ground water tanks. These technologies can reduce to a minimum or even eliminate the need to connect to the municipal water system, or even dependence on well water systems.

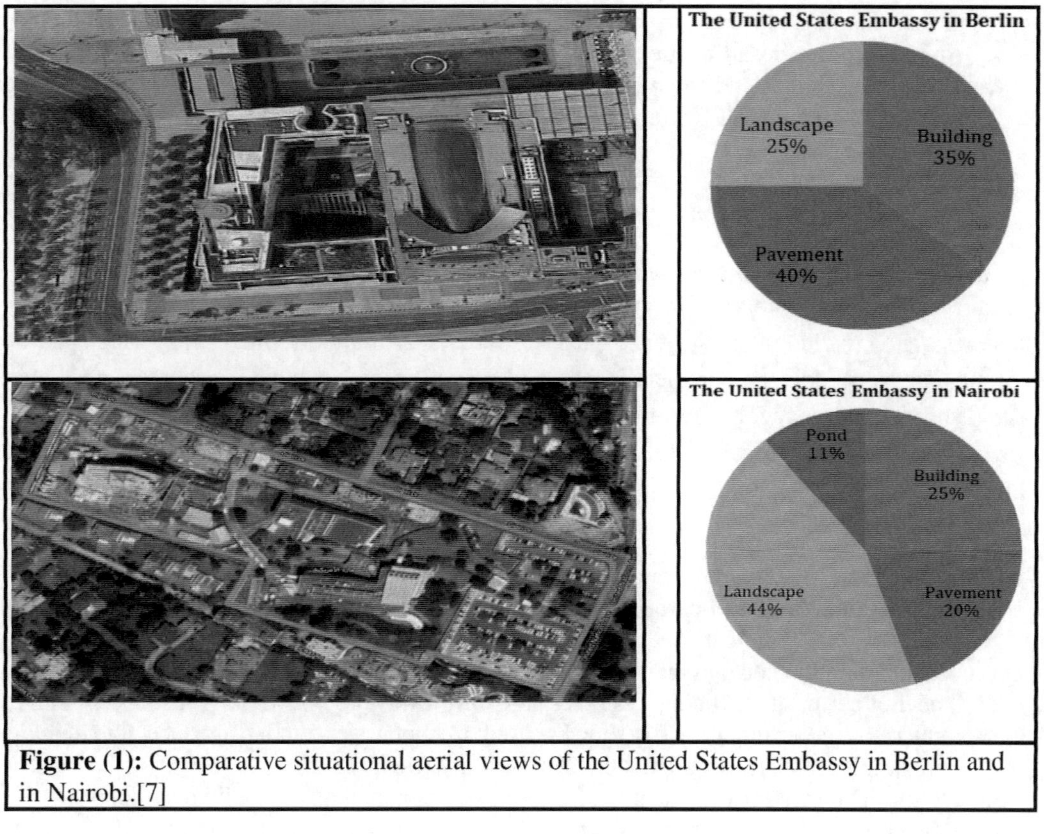

Figure (1): Comparative situational aerial views of the United States Embassy in Berlin and in Nairobi.[7]

11.4. A strategy program to reduce the burden of external thermal loads on the diplomatic building

The benefit of the program is to reduce the burden of external thermal loads on the Heating, Ventilating and Air Conditioning (HVAC) systems, thus reducing associated fossil energy use and payment bills, and regulating the environmental footprint. The phenomenon of the influence of external thermal loads is one of the phenomena that make crowded urban areas hotter than surrounding areas in rural areas, due to the amount of areas paved with materials with high thermal conductivity and the density of buildings that have replaced vegetative and permeable surfaces. Generally, solid and unshaded urban roofs absorb high solar energy loads and then radiate heat back into the atmosphere causing heat island phenomena, which raises air temperatures in general.

Thus, one of the most important design criteria for new and existing embassies projects are indicators of reducing the thermal loads of their buildings, where shading must be adopted, as well as the use of materials with a high white color (reflective or "cold") on the roofs of buildings and roofs. Various strategies can also be implemented to reduce heat in venues around existing buildings, and are particularly useful in locations with prolonged cold seasons. Therefore, the following indicators must be adhered to: [17]

1. The indicator of the use of types of local plants that are drought tolerant, while ensuring the availability of shade and organizing distinguished local agricultural projects on the territories of the embassies.
2. The indicator to ensure the installation of solar energy generation systems, such as photovoltaic (PV) or solar thermal cells, in the facades and rooftops of the building, as well as the surrounding exposed sites.
3. An indicator for installing green roofs and facades for the embassy buildings.
4. The use of natural and industrial lighting systems index: This means switching the lighting fixtures to the compact fluorescent lamp type (CFLs) or the LED type and returning the wires to smaller networks.
5. Daylight utilization index: This means architecturally designing the buildings of the embassy complex in order to increase the effectiveness of natural light in the workplace.

11.5. Strategies for Recycling Consumables Program

It is a very important program in support of green architecture and the concept of greening in embassies, as it includes all stages of the resource consumption life cycle, which starts with environmentally friendly procurement, reuse and recycling of consumed resources, and ends with waste disposal or recycling. For example, during the construction of the US Consulate in Dubai, 75% of the waste generated from landfills and incinerators was recycled. This abundant recycling was achieved by separating scraps of paper, plastic, wood and metals from each other, to be recycled separately. [14]

We refer to the most important indicators of embassy waste recycling [11]

1. **Review and audit indicator:** It is intended to ensure the availability of basic information about the procurement of materials needed for building embassies of all kinds, their consumption method and their waste specifications.
2. **The indicator of reducing resource consumption and ensuring reuse:** it is intended to reduce the consumption of resources used in embassies, and to reduce the quantities of purchasing products that are not environmentally friendly.
3. **Green Purchase Index:** It is intended to buy environmentally friendly materials.
4. **Recycling and fertilization index:** as it requires isolating organic materials from waste, and reusing them for fertilization and agricultural use.
5. **Hazardous waste disposal indicator:** it is intended to remove hazardous materials from landfills and reduce the impact of landfill waste on the environment.

11.6. The strategy program for reorganizing the internal environment of the embassy buildings

As it is known, the embassy employee spends an average of 90% of his time indoors during work, that is, in closed places. Consequently, IEQ Indoor Environmental Quality has a major impact on the health of employees, and thus affects their productivity and well-being. The indoor environment is formed as a result of the dynamic interactions between climate and site conditions, building regulations and activities conducted on site, indoor environment quality includes Indoor Air Quality (IAQ), thermal comfort, acoustic comfort, ergonomic ergonomics, visual quality, and connectivity. Naturally, through the principles of biophilia. [6]

For the purpose of implementing the program of reorganizing the internal environment of the embassy buildings and making the internal environment efficiency ((IEQ) high, the following practical steps must be followed: [11]

1. Reducing pollutants: that is, working to reduce the factors of internal pollution of embassy buildings .
2. Comfort improvements: that is, work to improve the operating efficiency of the HVAC system, and ensure the right amount of outside air in improving the most satisfying indoor climate conditions.
3. The use of low-emission materials: that is, setting up a policy to ensure the purchase of non-toxic and environmentally friendly building materials.
4. Ensure accuracy in designing interior decoration, so that rooms are of high acoustic quality, because background noise sources are a major contributor to affecting employee productivity and well-being in the workplace.
5. Adding interior decorations from nature that increases the association of embassy employees with nature (or the so-called biophilia) and this would positively affect the performance of embassy employees.

12. Examples of greening embassy buildings

12.1. An example of greening the existing Dutch embassy in Amman

Architect: Rudi Uittenhack

Location: Amman, Jordan, with an area of (1253 square meters)

Client: The Dutch Ministry of Foreign Affairs

Design team: Fumi Hoshino Frank Langhorst, Felix Reiter, Jaap Hikke, as well as the participation of a local consultant with the staff.

Project implementation year: 2010

The Dutch embassy in the Jordanian capital Amman is a unique example of the sustainability and greening of the existing embassies after they were modified and opened in 2010, as this embassy is located on a total area of (1253 square meters) only with all its buildings and annexes, including a villa for the ambassador's residence. It was the first building in Jordan to receive the prestigious international LEED-certification for green buildings with a silver rating. This environmentally-friendl building was constructed by transforming and renovating an old villa in Amman.

The design based on transforming and extending the one-story villa in inspiration from Jordanian architecture The Dutch Rudy Uytenhaak, undertook structural changes before starting the greening work, including removing the walls and ceiling on the first floor of the villa, the central staircase and a small piece of the first floor, while strengthening the existing concrete structure. For the building by adding a row of tall modern columns in front of the building, then he carried out some works aimed at greening the embassy according to (Figure 2), including: According to [3]

1. Adding protection to the "umbrella" system at the facade of the new building to control the balance of thermal radiation loads in summer and winter.
2. Adding a number of natural ventilation holes in the ceiling of the new first floor, which also includes ceiling lighting to increase the effectiveness of daylight in the spaces of the building.
3. Laying floating stone panels to ensure that space is left between the old construction ceiling and the floor of the new floor, as each panel is tilted in different directions so that there are sufficient spaces between them for natural ventilation through mesh openings in the facade, allowing the moderate night breeze to pass through to bring out the thermal storage gathered during the day on the roof of the building.
4. Install solar thermal energy collectors above the car park to ensure the availability of hot water that is transferred directly to the building or stored in the thermal tank (the swimming pool previously after roofing) and confirm the structural umbrellas system.
5. Supplying a movable shading system "umbrella", referring to Bedouin tents in the Levant, and it has been adopted at the southern façade to control solar heat gain for the process of closing and opening through the different seasons.
6. Putting a canvas covering parts of the roof to reduce the impact of direct thermal loads in the building.
7. Converting the current use of the swimming pool after it is covered into an underground thermal tank to store energy for heating, ventilation and air conditioning systems.

The skylights is designed to increase daylight inside the building. The garden is mainly rendered with hard landscaping features, while the applied materials are of high solar reduction impact value. Xeriscape garden that includes water efficient landscaping is used to reduce the water consumption by 50%. A plan for storm water management was implemented by harvesting the rain water from the roof and hardscape in winter, and from the condensate water from air conditioning units in summer, and stored in special tanks. The project captures and treats 90% of the annual rain fall and the collected rain water is treated by a suitable filtration system then reused in toilets and flushing.Therefore, the estimated savings for the running cost equals to JDs 1, 149 per year. Estimated annual energy produced from solar collectors equals 5,415 kWh/year, while the running cost for heating and cooling is about 7, 086 JDs/year, compared with the annual cost without green features that is about 10,935 JDs/year. [18], The Dutch embassy in Amman was distinguished by some green indicators and strategies and was classified within the main sustainability departments, which are as follows:

1. Keeping small windows on the ground floor to reduce heat gain, in addition to reflecting the ancient traditions of Jordanian heritage.
2. Automatically controlled movable fabric shade features for first floor windows.
3. The surrounding landscape features materials of high SRI value.
4. Ensuring rooftop solar thermal energy (PV) units for securing water heaters.
5. Executing a high specification double smart glass on both floors.
6. The use of stone and concrete for the general exterior of the building envelope to increase the effectiveness of the building's insulating thermal mass.
7. Providing a mechanical system for managing indoor air (heating and cooling) with individual self-control tools for thermal comfort (thermostats).
8. Using the original pool site after being covered as a thermal reservoir to support HVAC systems.
9. Providing natural and artificial lighting and electrical equipment needed for energy.
10. Insuring a reduction in water use.
11. Ensure low flow piping fittings.
12. Using natural plants that depend on collected rainwater in the rainy seasons, in order to reduce irrigation works from pure water.
13. Use plants with low water dependence.
14. A decrease in the embassy's energy consumption rate, ranging between (20-30%), compared to any similar building.

The indicators referred to above have reduced greenhouse gas emissions (greenhouses), which cause the phenomenon of global warming, and are associated with the escalation of the consumption of fossil energies. The project was certified by the American Green Building Council (LEED), while preserving the old shape of the building, and it is the second diplomatic building in the world to be certified. Digital visas refer to the following details:

1. The old structure that was reused within the new design program.
2. A cloth covering parts of the new roof to reduce the effect of direct thermal loads on the building.
3. Ceiling photovoltaic panels availability approx. 3% of building lighting requirements.
4. A movable shading "umbrella" on the southern façade to control solar heat gain or block it.
5. The canopy structure and its interlock with the new concrete columns on the southern façade.
6. Solar thermal energy complexes (P.V.) for the availability of hot water that is transferred directly to the building or stored in a special underground storage (pool site) that is currently roofed.
7. Reuse the existing swimming pool after roofing as a thermal tank for heating, ventilation and air conditioning systems.

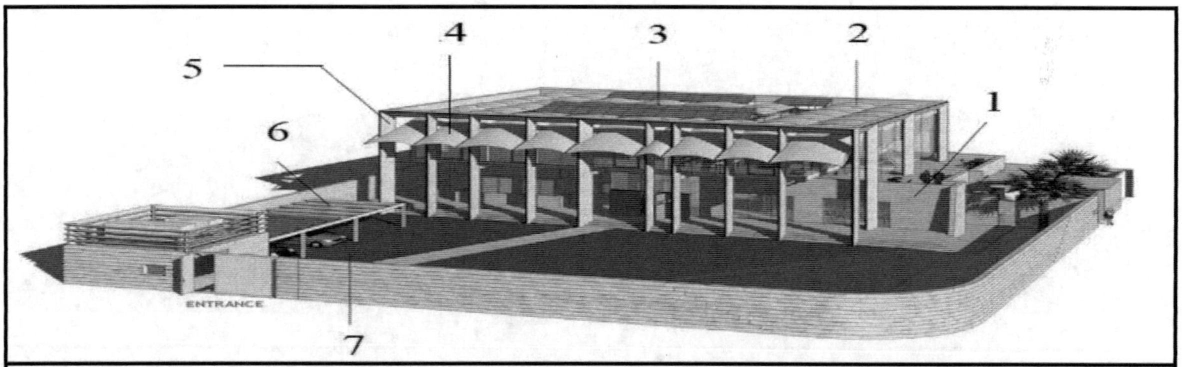

Figure 2. the general shape of the embassy with details of the principles of greening, which includes shading and the use of solar cells to provide hot water. While ensuring a beautiful front façade [3].

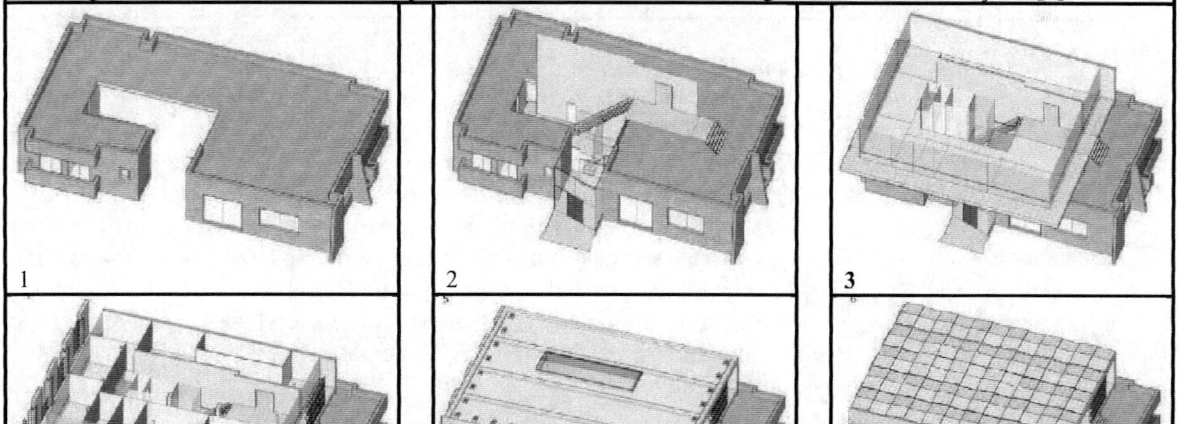

Figure 3. Steps of greening treatments used for the building of the Dutch embassy in Amman, starting from the work of adding the floor and reinforcing its old structure with concrete columns, and ending with the process of installing the canopies .[3]

Figure 4. Greening treatments in the building of the Dutch embassy in Amman, which has been certified by the American Green Building Council (LEED), which is the second diplomatic building in the world to receive this certification. [18]

12.2. The environmentally friendly Embassy of Finland in Washington turns green - and platinum

The decision of the Finnish Embassy in Washington to turn to the ideal environmentally friendly environment arose when, in the year 2000, the Director of Maintenance sought to address the reduction of the embassy's high energy costs. "We already had a very good building, but we have very high bills for energy and water consumption," said Finnish Ambassador Ritva Koukku-Ronde at the time. "As our manager started replacing old bulbs with more energy efficient bulbs, he started looking at our cooling system and especially our heating systems, as well as water taps, etc., to see how we can be more energy efficient." Also, wide glass walls allow natural daylight to enter the embassy, symbolizing the country's commitment to diplomatic transparency. Ultimately, the elegant diplomatic milestone has not only become more energy efficient but has also emerged as a model for sustainable green design among embassies in Washington.

In 2008, Finland became the first embassy in the United States to receive an Energy Star from the Environmental Protection Agency for high energy efficiency. In 2010, the building became the first embassy in the United States to achieve the prestigious Leadership Certificate from the US Green Building Council in Energy and Environmental Design (LEED), and then at the gold level. In December 2014, it became the first embassy in the United States to obtain Platinum status, and only second in the world. [20]

This aesthetic translated into pioneering environmentally friendly design. Ambassador Koukku-Ronde said that the embassy's greening - that is, making it more environmentally responsible - had been facilitated by the fact that it was well built in the beginning. Although the green building standards were not as widespread as they are now, some Finnish standards still apply.

When the embassy was built two decades ago, environmental aspects were not looked at as broadly as they are today, but the architects designed a building that not only met all the requirements in the capital and the United States, but was also using the normal sustainable Finnish method of building, " she explains. Ambassador Coco Rond. "They were using local materials and designed the building so that we could use natural daylight." This commitment to preserving the environment affected all the daily functions of the embassy. Toilets and faucets were equipped with water-saving devices, for example, and recycling became a top priority. [21]

To obtain the platinum certification, the embassy invested nearly $ 150,000, money the embassy recovered in the first year. A comparison of average annual energy spending between 2002 and 2004 showed an extraordinary decrease of $ 150,000 annually, electricity use in the building decreased by 50 percent and gas by 65 percent annually, and the embassy now blocks greenhouse gas emissions equivalent to consumption of 90 America's home of electricity ... and 30 percent less water compared to the mid-2000s. Renewable energy offsets 100 percent of the building's total energy consumption.

Roger Platt, President of the United States Green Building Council, praised Finns' dedication to achieving sustainability when he said, "This is the first embassy to receive an Energy Star from the US Environmental Protection Agency, the first in the United States to obtain LEED certification, and the first You will receive LEED Gold for an existing building. "And because the Finns believe that a green building should be alive and still green, they are the first to design an embassy in the United States to achieve LEED platinum." (Figure 5)

Figure 5. The Finnish Embassy in Washington, is the first in the United States to obtain an ENERGY STAR from the Environmental Protection Agency for superior energy efficiency. [20]

13. Conclusions

1. It requires maximizing the use of the components of the local natural energies to ensure the improvement of the internal climate efficiency and the reduction of the consumption of depleted energies.
2. Achieving sustainability, not only as a way to reduce the embassy's influence on the foreign country, and create a comfortable living and work environment, but also to declare the guest country's environmental protection standard.
3. Rationalizing consumption, relying on alternative energy, storing rainwater and using it later, reduces the costs of the bills paid, which are burdensome for the home country or the guest.
4. Diplomatic buildings are considered a guest in the host countries, which requires that this guest be a non-heavy burden on the host's resources, as a result of the reliance of diplomatic buildings on alternative energy, waste recycling, and rationalization of natural resources.
5. As a result of recent agreements related to climate and greenhouse gas emissions, the environmental aspect in diplomatic buildings represents an important trend, which has become in competition with the security and symbolic side, as developed countries took the initiative to pay attention to this aspect to produce green diplomatic buildings from the depth of nature and embrace it.
6. Environmental diplomacy is an exercise in managing international diplomatic relations, shared by a group of countries committed to reducing greenhouse gas emissions, preserving natural resources, and increasing biological diversity.
7. The materials and techniques used in greening the sovereign diplomatic buildings differed according to their suitability for their intended purpose, which all led to the production of buildings contributing to the preservation of natural resources, the use of clean energy and the preservation of the environment.
8. The greening of diplomatic buildings requires decision-making by the authority of the guest country or those who are the decision-makers in the building, such as ambassadors, so the greening process is directly affected by the decisions of the higher authority.

14. Recommendations

1. The engineers specialized in garden design should be involved during the design and implementation phase of diplomatic buildings to design the site perimeter in a way that suits the position of the guest and host country.
2. We must avoid choosing small sites that do not comply with the criteria for greening diplomatic buildings due to their small size.
3. You must take into account the selection of materials used in the implementation of the diplomatic building in order to achieve the largest thermal insulation to reduce energy consumption.
4. In the event that a country submits a request to the host country to open an embassy, this requires the adoption of a condition that cannot be waived, which is the submission of papers formulating its policy on the way to deal with the environment and climate of the host country for improvement, and a special focus on carbon emissions, and this applies to any ambassador when submitting His credentials include that ambassador's policy towards the environment of the host country.

15. References

[1] United Nations Environment Program, December (2011), Intergovernmental Panel on Climate Change, IPCC.

[2] United Nations Environment Program, (2012), the Intergovernmental Panel on Climate Change, IPCC, (Fifth Assessment Report of the United Nations Organization).

[3] R. Al Nasa'a, (2009), "What Works and What Needs to Be Done: The Embassy of the Kingdom of the Netherlands, Amman – Jordan", Business Council, Green Building Unit in Jordan.

[4] Mahdi, Rana Mazen, (2012), and others, "Green Architecture is an Ongoing Phenomenon," research in the Iraqi Journal of Architecture, Issue 25-26, October.

[5] Gad, Mona Rizk, (2015), "The concept of respiratory buildings to face climate change", Green Architecture Strategies for Reaching Zero Energy Buildings, Master's Thesis, Department of Architecture, Faculty of Engineering, Cairo University.

[6] Aya E Messelmani,(2018), "Biophilia, Humans & the Connections", A Thesis Submitted To The Department Of Architecture For The Masters Of Arts In Architecture Degree, Department Of Architecture The University Of Nicosia, NICOSIA, CYPRUS , JANUARY

[7] Bureau Of Overseas Buildings Operations,(2013)," Guide To Green Embassıes Eco-Diplomacy In Operatıon" , USA, Ministry of Foreign Affairs **.**

[8] Dan Chiras ,(2010), "Green Transportation Basics", New Society Publishers, CANADA**.**

[9] Jager, F.P (2010),Old and New Design Manual for Revitalizing Existing Buildings, Birkhauser .

[10] McDonough, W. & Braungart, M. (2002). Cradle to cradle: Remaking the way we make things. New York: North Point Press.

[11] U.S. Green Building Council,(2013), "Roadmap to Green Government Buildings",2101 L Street NW,Suite 500 ,Washington, Copyright 2011 by the U.S. Green Building Council, Inc. usgbc.org

[12] https://www.diplomatie.gouv.fr/en/the-ministry-and-its-network/missions-and structure/exemplary-administration-plan-pae/article/the-green-embassy-project ,Accessed**.** **25**/6/2020

[13] https://www.worldgbc.org/news-media/green-building-improving-lives-billions-helping-achieve-un-sustainable-development-goals. Accessed - 22/6/2020

[14] https://www.constructionweekonline.com/article-17844-us-consulate-in-dubai-achieves-leed-gold 7/2020

[15]www.ajlounnews.net/index.php?module=articles&id=11780. Accessed - 15/01/2014

[16] http://www.state.gov/r/pa/pl/156053.htm - 6/2020

[17] http://www.epa.gov/hiri/mitigation/coolroofs.htm 5/2020

[18] www.arch2o.com/dutch-embassy-in-amman 8/2020

[19] http://www.leagueofgreenembassies.org - 6/2020

[20] https://washdiplomat.com/eco-friendly-finnish-embassy-goes-green-and-platinum/

[21] https://finlandabroad.fi/web/usa/current-affairs/-/asset_publisher/h5w4iTUJhNne/content/embassy-of-finland-the-first-leed-certified-embassy-in-the-u-s-/384951

Concentration of some Polycyclic Aromatic Hydrocarbons in soil samples of Kirkuk province, Iraq

Yasin H Aoeed[1], Awaz B Mohammed[2*], and Attalah M. Hameed[1]

[1]Department of Biology, College of Education for Women, Tikrit University, Iraq
[2]Department of Biology, College of Science, Kirkuk University, Iraq

* Corresponding author: dr.awaz.eco@uokirkuk.edu.iq

Abstract

in order to evaluate the quantity of twelve PAHs (Naphthalene, Tetraphan and Acenaphthylene Fluorene, Phenanthrene, Anthracene, Pyrene, Benzo [a] Anthracene, Ovalene, Chrysene, Benzo [a] fluoranthene and DiBenzo [ah] Anthracene) in the soil samples from Kirkuk province, seven sites (Ras El-Gesr, Benja Ali, Wahed Hozeran, North Oil Company (NOC), Lillan, Kubri and Chimen) were selected using Gas Chromatography (GC) during the autumn 2017 and winter 2018. The results showed that the highest concentration of individual hydrocarbons during the autumn season was for the Acenaphthylene compound in the NOC site 131.19 µg/kg, and for Naphthalene compound, was 100.543µg/kg. The NOC recorded the highest concentration of total hydrocarbons 891.65 µg/kg. For the winter season, the highest concentration of polycyclic aromatic hydrocarbons was recorded for the Fluorene compound at the NOC site 79.19 µg/kg. Fluorene and Naphthalene compounds achieved the highest averages for the season (43.24 and 42.984 µg/kg) respectively, and recorded the location of the NOC, the highest summation of total hydrocarbons amounted to 609.77 µg/kg.

1. Introduction

PAHs in the atmosphere are constantly deposited into the ground by dry or wet deposition processes. Some of these compounds are from near sources, such as car exhausts from nearby roads. Other PAHs are from distant sources and transported over different distances through the air. At the same time, PAHs can be added to the soil if soil materials contain them. When PAHs and other compounds are deposited on the surface of the earth, they can become attached to the soil molecules [1, 2]. The most important factors affecting the movement of PAHs in the soil are the size of the attached particles and the size of the soil pores (this throat or pores can be defined as small spaces between individual minutes of soil) [3]. If the particles in which the PAHs cannot move through the soil, the movement of these compounds will be limited because they tend to be absorbed (impregnated) into these particles.
The soil contains measurable quantities of PAHs, primarily from airborne outputs. Reported levels of PAHs in soil near oil refineries are reported to 200,000 micrograms / kg of dehydrated soils. Levels of soil samples obtained near cities and areas with heavy traffic are usually less than 2,000 micrograms/ kg [4]. Thirteen of PAHs were measured in 46 soil samples from San Nicolás, an area that includes a brick manufacturing company in central Mexico. Total concentrations of PAHs ranged from 7 ng/ g to 1384 ng/ g with an average value of 220 ng/ g. Naphthalene (Nap) was more abundant followed by

Fluorene (Flu), chrysene (Chry), benzo [a] anthracene (BaA) and DiBenzo [a, h] anthracene (DahA). The average samples of naphthalene were 374 ng / g. It was also found that about 52% to 76% of the soil of San Nicolas had higher PAHs than those considered normal or non-serious for residential use. In addition, the total concentration of benzo [a] Pyrene in the soil of San Nicolas was higher than the upper limit. Of particular concern is the high volume of DahA and BaP, which are high-potential PAHs to cause cancer. All these factors mean a potential cancer risk for the exposed populations in San Nicolas [5].

In the study of [6] for the purpose of evaluating the levels of PAHs in the air of the thermal power plant in Baghdad, they indicated that the highest concentration of these compounds was for Naphthalene, with an average of 72-1573.5 ng / m3 during the summer season, 14.387 to 36.65 ng / m3 during the winter season.

The degradation of PAHs in the environment includes: biodegradation [7, 8], photolysis, chemical oxidation, adsorption of soil particles, leaching and bioaccumulation [9]. Each of these processes affects different PAHs in different ways. This is mainly due to the fact that each compound of the PAHs compounds has a unique structure and a range of physical, chemical and biological properties. On the other hand, the removal of these compounds from the atmosphere was studied by [10-12]. The gas phase of hydrocarbons is more efficiently removed from the atmosphere in colder conditions than in warm conditions [13, 14].

In the study [15] at north of Nijif city in Iraq, the total concentration of sixteen polycyclic aromatic hydrocarbons was 49.9 to 986.4 µg/ kg, by average 587.2 µg/ kg. The most abundant polycyclic aromatic hydrocarbon compound in the surface soil of the study area was Fluorene, followed by Acenaphthylene, Naphthalene, Chrysene, Phenanthrene, Benzo (b)fluoranthene, Fluoranthene, Benzo (a) anthracene, Benzo (a) Pyrene, Benzo (g, h, i) perylene and Dibenzo (a, h) anthracene, Indeno (1,2,3-cd) Pyrene, Pyrene, Acenaphthene, Anthracene and Benzo (k) fluoranthene. Syed et al. [16] noted that the total concentration of 15 hydrocarbon compounds studied in soil A and O for a number of Chinese forest soils ranged from 168-220 ng / g. The current study aims at evaluating the concentrations of PAHs in soil samples of seven different sites within Kirkuk provionvce during two different periods.

2. Materials and methods of study

2.1. Selection of study areas

The current study included a field survey of seven different sites of Kirkuk province showing intersections in GPS Table (1) during the autumn 2017 (average of September, October and November) and winter 2018 seasons (average of December, January and February). Soil samples analyzed for detecting twelve compounds of polycyclic aromatic hydrocarbons.. These sites and their characteristics are: first site – Ras El-Gesr (center of city): - It represents the crowded site with commercial activity. Second site - Benja Ali (center of city): -It is residential and contains a well-oil burning in the eastern side of the city. Third site - Wahed Hozeran (center of city): - represents the southwestern part of the city. Fourth - North Oil Company: - Located northwest of Kirkuk and includes a number of oil refining and refining facilities, in addition to a number of oil fields. The fifth site - Lillan: - includes Kirkuk cement factory and is located southeast of Kirkuk. The sixth site - Kubri: - is located north of kirkuk with high anthropogenic activity. Site seventh - Chimen area (as control): - A rural area free of industrial facilities, located northeast of Kirkuk.

Table 1. Geographical data of the studied sites using the GPS system.

No.	Site	North	South
1	Ras El-Gesr (center of city)	35.46°	44.39°
2	Benja Ali (center of city)	35.41°	44.43°
3	Wahed Hozeran (center of city)	35.40°	44.34°
4	North Oil Company	35.55°	44.36°

5	Lillan	35.34°	44.47°
6	Kubri	35.75°	44.14°
7	Chimen	35.31°	44.19°

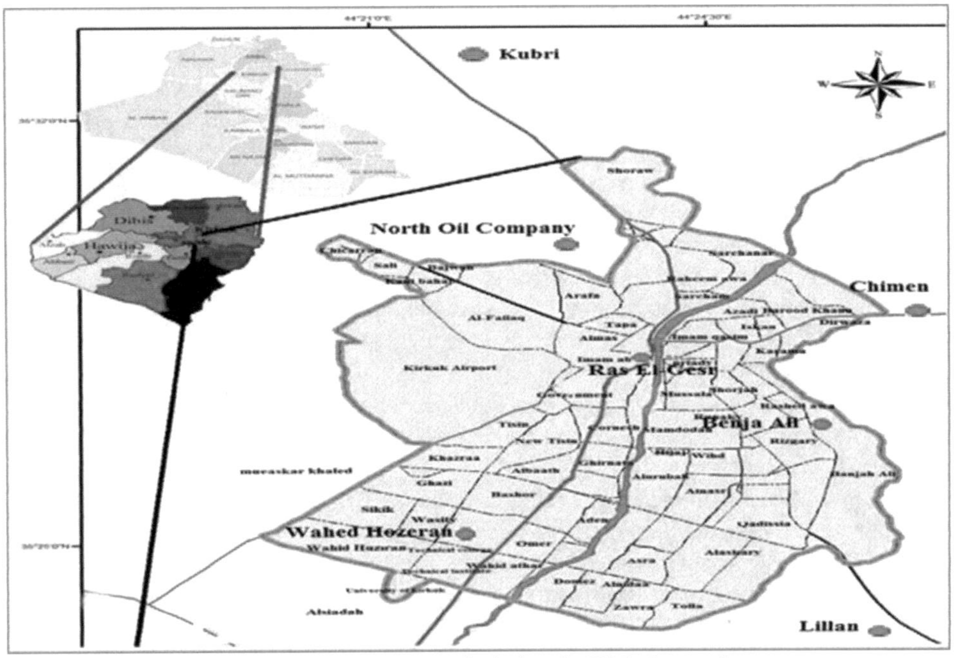

Figure 1. Map shows studied sites in Kirkuk province.

2.2. Soil sampling collection

Soil samples were taken from each of the seven selected sites from the upper soil layer of depth of 0-10 cm using a stainless steel drill. The soil samples were placed in foil aluminum paper [17]. Soil samples were cleaned, dried on paper, labeled and breezed at -20 C° till analysis, The samples were then sieved through 2mm sieve (EPA method 3052) to remove coarse particles (USEPA, 1995) [18] before extraction. Laboratory work extraction of samples was done as described by WMO / IOC, 1989 [19]. 10 g of soil drying was taken and homogeneous in a stainless steel bowl, then about 25 mL of acetone was added in the flask and mix with hand for five minutes. The sample is then placed in a dark, cool place overnight and then mixed with a shaker for an hour. The solution is separated and placed in a darkened glass container. This process is repeated three times. The solutions obtained from the previous step were then transferred and placed in centrifuge at 2500 rpm for 5 minutes. 50 ml hexane and 100 ml of free ion water were added to solution, two layers will be formed, the lower layer will neglect while the upper layer will be collected and add 50 ml of KOH will added. The volume is reduced to 10 mL by the rotary evaporator device , transferred to the silica gel column, dried and dissolved in 1 ml of Acetonitrile: Methanol at 90:10 and kept until measurement.

2.3. Aromatic hydrocarbons included in the study

The device has been calibrated using standard of twelve compounds of polycyclic aromatic hydrocarbons solution, as shown in Table 2.

2.4. Calculations and statistical analysis

PAHs were analyzed using a Chromatography Gas Model 2010-Shimadzo and FID ionized flame detector and used a column-type capillary30 (30-SE) 07 meters long and 7200 mm in diameter to separate compounds according to the following conditions [20]. The temperature of the injection area was 280 ° C. Reagent temperature was 330 ° C. The temperature of the separation column starts from 100 ° C and stays for 3 minutes and then increases 10 degrees every minute until reach 310 ° C. After the extraction process is completed, 1μl of the extract is injected with a device-type syringe (10 asn 1701 Hamilton), with a capacity of 10μl, and the injection is performed. Polycyclic aromatic hydrocarbons separated in the chromatogram of each sample with a standard chromatogram retention time, the concentrations of PAHs were calculated according to the following equation [21].

$$C = \frac{\mathrm{Cst} \times \mathrm{Asam}}{\mathrm{Ast}} \times \frac{\mathrm{D.F}}{\mathrm{Wt}}$$

Whereas C = concentration to be calculated for hydrocarbons per million. Cst = the standard concentration of the compound and equal 10. Asam = calculated area of peak. Ast = The standard area of peak is fixed for all compounds and is (15436). D.F = dilution factor equals 10. Wt = weight of sample.

Table 2. Twelve compounds of PolyCyclic Aromatic Hydrocarbon compounds (PAHs) standard solution.

No.	Compounds	simple	No. of Ring	R. Time
1	Naphthalene	Nap	2	7.962
2	Tetraphan	Tet	4	9.99
3	Acenaphthylene	Acy	3	11.11
4	Fluorene	Flu	4	11.836
5	Phenanthrene	Phe	3	13.801
6	Anthracene	Ant	3	15.834
7	Pyrene	Pyr	4	16.739
8	Benzo[a]Anthracene	B[a]Ant	4	16.861
9	Ovalene	Ova	10	17.545
10	Chrysene	Chry	4	18.171
11	Benzo[a]fluoranthene	BaF	5	18.723
12	DiBenzo[ah]Anthracene	DAH	5	19828

2.5.Statistical analysis

The results were statistically analyzed according to the statistical program (SPSS) .All data were treated with the analysis for significance differences among sites and study period using Duncan Multi-Range Test. The possible sources (Pyrogenic or Petrogenic origin) in soil samples may be assessed by the ratio of individual PAHs compounds [22], who stated that these ratios were based on the use of PAHs from the molecular mass 178 and 202. These ratios are used in the present study: (a) Ant / Ant + Phe: - According to this ratio it was assumed that the resulting values that are less than 0.1 indicate the petrogenic source of PAHs, while more than 0.1 to PAHs out of combustion. (B) Fluo / Fluo + Pyr: - It is assumed that the ratios ranging from 0.4 to 0.5 refer to pyrogenicsource, and the ratios of less than 0.4 refer to petrogenic sources. The ratio greater than 0.5 indicate the burning of wood and coal. Inengite et al., [23] also reported that the ratio of Australian crude oil is less than 0.4, and that some oils have very high proportions of fluoranthene. (C) BaA / BaA + Chry: - Assume that the ratio less than 0.2 referred to petrogenic origin, and the ratios of 0.2 to 0.35 with mixed sources. The ratio greater than 0.35 is due to combustion.

3. Results and discussion

Figures (2, 3, 4) and table (3) shows the concentrations of the twelve individual hydrocarbons during the autumn of the seven sites under study. The highest concentration of three-ring Acenaphthylene at the North Oil Company's site was131.9 µg / kg, followed by a four-ring Pyrene with a concentration of 130 µg / kg. Naphthalene was, with high concentrations at all sites except for the control site (Chemin), which was near to zero, with a maximum concentration of 128.6 µg / kg.. The results of seven site, North Oil Company were significantly higher for all compounds, followed by Kobri and Lillan sites with signifigance industrial activity. Barrán-Berdón at al., [5] were obtained total concentrations of PAHs in the soil samples from 7 µg / kg to 1384 µg / kg with an average of 220 µg / Kg. They noted that Naphthalene is more abundant followed by Fluorene, Chrysene, Benzo [a] Anthracene and Dibenzo [a, h] Anthracene. The average samples of Naphthalene were 374 µg / kg..The highest mean Naphthalene compound was 100.543 µg / kg and significantly higher than the average of the other compounds according to the results of the Duncan Multimodal Mean Test followed by Acryaphthylene triaclate (67.414 µg / kg), significant for some compounds and not significant for other compounds according to the Duncan Multi-Range Test. The site of the North Oil Company was recorded the highest concentration of total hydrocarbons with a value of 891.65 µg / kg, accounting for 25% of the total hydrocarbons. The availability of low molecular PAHs as Naphthalene compound may be due to their contribution to the formation of SOA molecules resulting from ozone degradation, which slowly characterizes evaporation and elevation of their viscosity and ensures longer atmospheric life. Increased viscosity provides a shield that protects PAHs from chemical degradation and evaporation, allowing the transport of toxic pollutants to distant ranges [24].

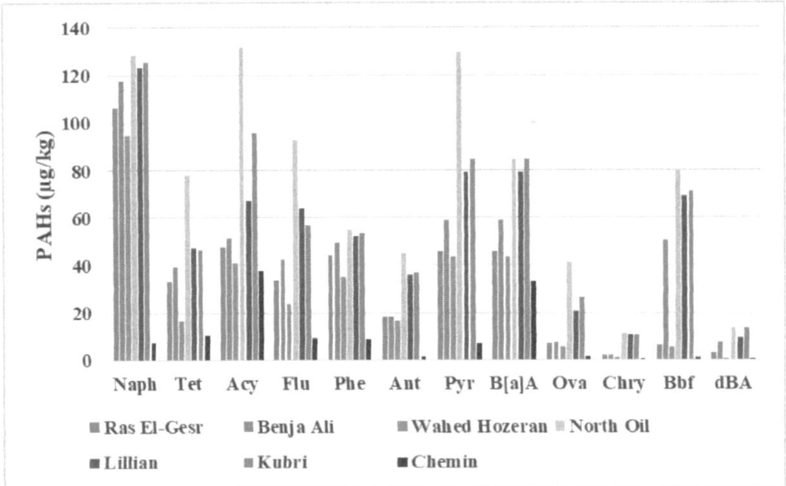

Figure 2. Individual hydrocarbon concentrations in the soil for the seven sites during the autumn season.

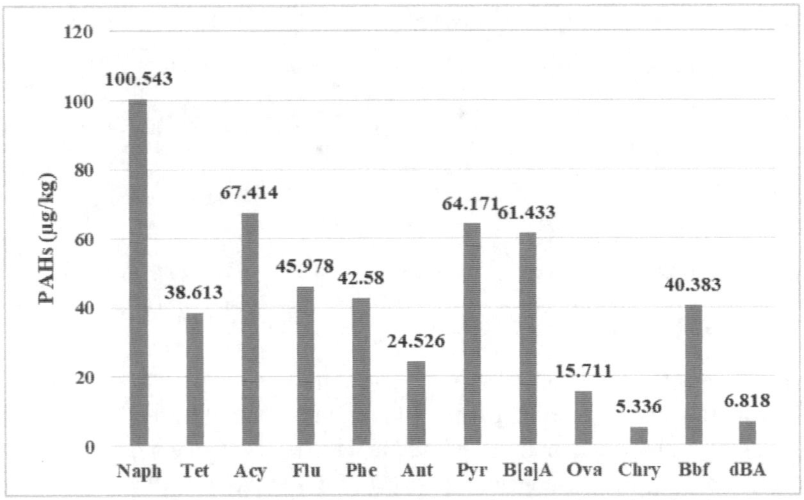

Figure 3. Means of PAHs in soil as average of seven sites during the autumn

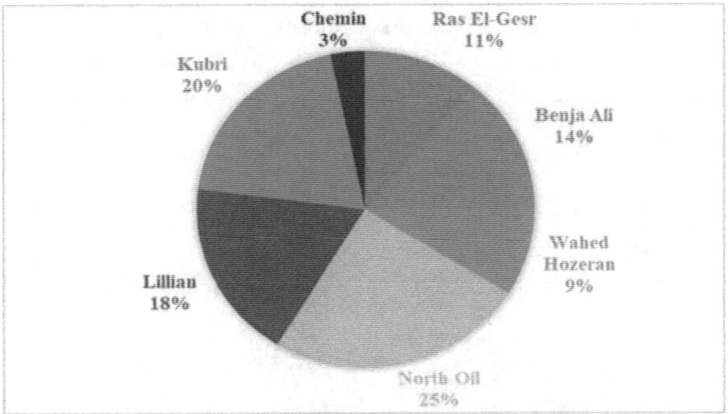

Figure 4. Total hydrocarbon concentrations in soil of seven sites during autumn.

Table 3. Concentrations of the compounds, their mean and their aggregates in the soil samples of the study sites during the autumn season.

Sites / Compounds	Ras El-Gesr	Benja Ali	Wahed Hozeran	North Oil Company	Lillan	Kubri	Chemin	Means of compounds
Naph	e106.6	d117.6	f94.7	a128.6	c123.3	b125.7	g7.3	a100.519
Tet	e33.01	d39.27	f16.28	a77.92	b47.23	c46.26	g10.32	de38.613
Acy	e47.6	d51.2	f40.7	a131.9	c67.3	b95.5	g37.7	b67.426
Flu	e33.76	d41.99	f23.47	a92.91	b63.81	c56.8	g9.11	cd45.979
Phe	e43.92	d49.35	f35.17	a54.89	c52.28	b53.56	g8.89	d42.58
Ant	e18.01	d18.11	f16.28	a44.94	c35.84	b36.89	g1.61	ef24.526
Pyr	e45.9	d58.9	f43.6	a130	c79.5	b84.6	g6.7	b64.169

B[a]A	d45.87	c58.89	e43.64	a84.56	b79.53	a84.56	f32.98	bc61.433
Ova	e6.95	d7.55	f5.5	a41.44	c20.66	b26.41	g1.47	fg15.711
Chry	e1.88	d2.03	f1.16	a11.08	b10.5	c10.48	g0.22	g5.336
Bbf	e6.06	d50.37	f5.2	a79.87	c68.93	b71.27	g0.98	de40.383
dBA	e2.89	d7.55	f0.59	b13.54	c9.31	a13.7	g0.15	g6.819
Total PAHs	392.45	502.81	326.29	891.65	658.19	705.73	117.43	

- Values followed by the same letter are not different from each other morally.

For the winter season, Figure (5, 6, 7) and Table (4) show the results of PAHs during winter. The highest concentration of PAHs was recorded for the Fluorine compound in the North Oil Company's site of 79.19 µg / kg, followed by Acenaphthylene and Naphthalene with a concentration of 69.94 and 68.23 µg / kg respectively. The lowest concentration was for Pyrene and Benzo [a] Anthracene, at some sites. It also noted that the high concentrations were recorded for all PAHs at North Oil Company site except Benzo [a] Anthracene, which exceeded significantly for other sites. For average PAHs, Naphthalene and Fluorene compounds recorded the highest mean at soil samples during the winter with concentratins of 43.24 and 42.984 µg / kg respectively according to the results of the Duncan Multimode Intermediate Test.

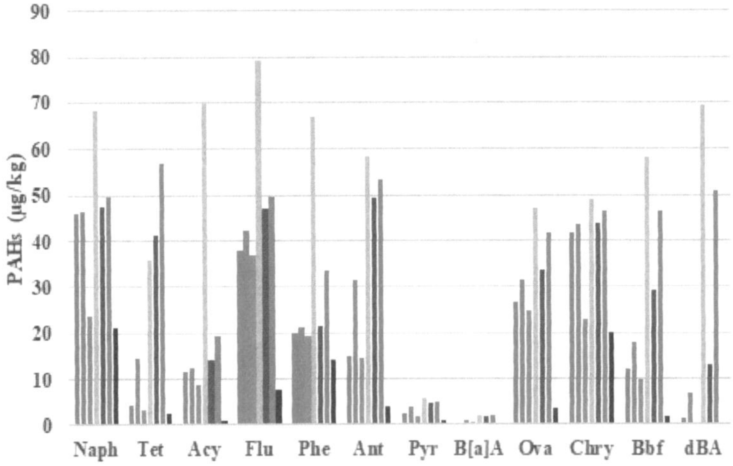

Figure 5. Individual hydrocarbon concentrations in soil for the seven sites during the winter season.

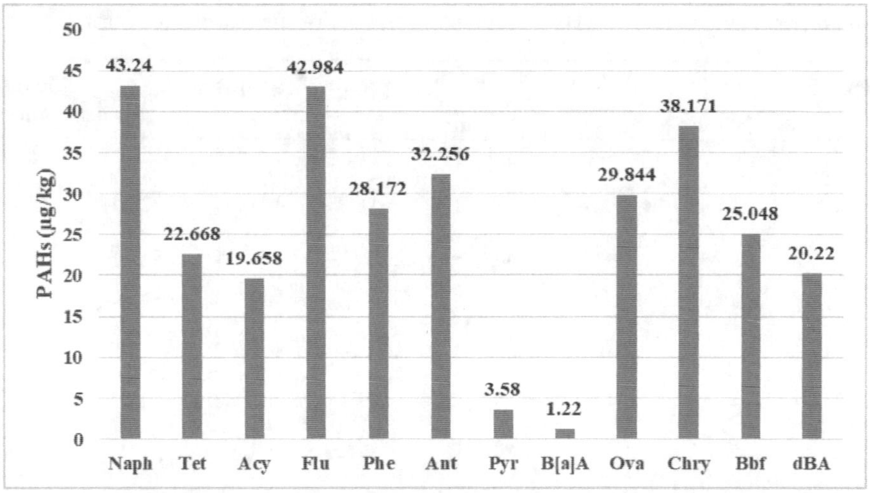

Figure 6. Mean of PAHs in soil at seven sites during the winter

The site of the North Oil Company was with highest concentration of 609.77 µg / kg, accounting for 28% Total PAHs of soil samples during the winter, while the differences were evident between the other sites. The control site (Chemin) recorded the lowest total hydrocarbons of 77.42 µg / kg, accounting for 4% of total PAHs in the soil samples during the winter season. Previous studies have found by Abed et al., [15] at north Biji City in Iraq showed that the total concentration of 16 polycyclic aromatic hydrocarbons ranged from 49.9 to 986.4 µg / kg at 587.2 µg / kg. The most abundant polycyclic aromatic hydrocarbon compound in the surface soil of the study area was Fluorene followed by Acenaphthylene and Naphthalene.

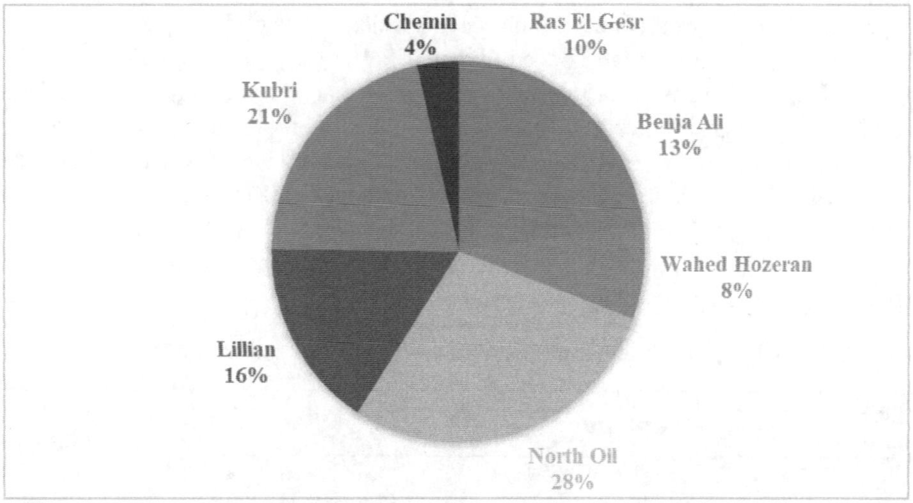

Figure 7. Percentage of PAH compounds in soil for the seven sites during winter.

Table 4. Mean concentrations of PAHs s in the soil samples of the study sites during winter.

Sites / compounds	Ras El-Gesr	Benja Ali	Wahed Hozeran	North Oil Company	Lillan	Kubri	Chemin	Means of compounds
Naph	e45.9	d46.56	f23.63	a68.23	c47.57	b49.54	g21.25	a43.24
Tet	e4.45	d14.45	f3.27	a35.71	c41.35	b56.94	g2.51	c22.383
Acy	e11.64	d12.49	f8.73	a69.94	c14.11	b19.51	g1.19	c19.659
Flu	e38.11	d42.42	f36.82	a79.19	c47.03	b49.73	g7.59	a42.983
Phe	e20.19	d21.24	f19.5	a66.71	c21.674	b33.62	g14.27	bc28.172
Ant	e14.89	d31.34	f14.56	a58.38	c49.13	b53.38	g4.11	abc32.256
Pyr	e2.58	d3.99	f1.79	a5.78	c4.89	b4.95	g1.08	d3.58
B[a]A	f0.09	d0.99	e0.8	b2.32	c1.96	a2.34	g0.05	d1.221
Ova	e26.71	d31.47	f24.74	a47.19	c33.57	b41.69	g3.54	bc29.844
Chry	e41.69	d43.35	f22.97	a48.89	c43.86	b46.38	g20.06	ab38.171
Bbf	e12.06	d17.95	f10.02	a58.11	c29.04	b46.42	g1.74	bc25.049
dBA	e1.48	d6.77	f0.22	a69.32	c13.04	b50.68	g0.03	c20.22
Total PAHs	219.79	273.02	167.05	609.77	347.224	455.18	77.42	

- Values followed by the same letter are not different from each other morally.

The results of the PAH concentrations in the soil samples during the two seasons of the study were summarized in Table 5. The highest concentration was recorded for Naphthalene during the autumn, which was 100.519 μg / kg with amean concentration of 71.879 μg / kg and shoes significantly different with all other compounds.. Seasonal averages show that the autumn season achieved the highest rate of 42.791 μg / kg, significantly different from the winter, according to the results of the multimodal Duncan method. The results of standard deviation showed a high dispersion between the sites for the concentration of each compound, where it is noted that all the mean of the compounds were significantly less than zero compared to their standard deviation values during the two seasons of study, except for the average of the two compounds Acenaphthylene during the autumn season and DiBenzo [ah] Anthracene during the winter season.

Table 5. Mean ± SD of 12 PAHs during autumn and winter for studied soil samples.

seasons / compounds	Autumn	Winter	Means of compounds
Naph	a100.519±39.630	cd43.24±14.972	a71.879
Tet	cde38.613±20.732	d-g28.383±20.269	bcd33.498
Acy	b67.426±32.070	f-i19.659±21.163	bc43.542
Flu	c45.979±25.826	cd42.983±19.594	b44.481
Phe	cd42.58±15.136	d-g28.172±16.636	bcd35.376
Ant	efg24.526±14.012	c-g32.256±20.076	de28.391
Pyr	b64.169±36.003	ij3.58±1.650	bcd33.874
B[a]A	b61.433±19.900	j1.221±0.917	de31.327
Ova	g-j15.711±13.370	c-f33.844±13.014	de24.778
Chry	hij5.336±4.669	cde38.171±10.777	ef21.754

Bbf	cde40.383±32.506	efg25.049±19.101	cde32.716
dBA	hij6.819±5.318	fgh20.22±25.994	f13.519
Means of seasons	a42.791	b26.398	

- Values followed by the same letter are not different from each other morally.

For the sources of PAHs, it is observed that, by application of weight equations according to molecular weight and the phenomenon in Table (6), Ant/(Ant+Phe) showed that PAHs were the source of combustion at all sites and studied periods. The results for Flu/(Flu+Pyr) showed that the source of the compounds was the pyrogenic source at all sites during the autumn, where the values ranged from 0.4 to 0.5, except for Wahid Hozeran site in the city center was petrogenic source, the value of less than 0.4, during the winter season, the ratios showed that the compounds were sourced from the burning of natural materials (weeds, trees, coal) and all the sites, worth more than 0.5. for B[a]A/(B[a]A+Chry) In the autumn season, the values of the PAHs were more than 0.35, wherefore may be due to combustion. During the winter season, the values of these compounds were petroleum All sites, with values of less than 0.2. Bayowa [25] states that the PAHs in the soil samples were oil-based as well as the combustion of oil, timber and coal during the dry and rainy seasons. Abed et al. [15] noted that burning fossil fuels and burning plants are the main sources of PAHs in the soil samples they studied.

Table 6. Ratios of some individual PAHs for studied soil samples.

Ratio / season sites	Ant/(Ant+Phe)		Flu/(Flu+Pyr)		B[a]A/(B[a]A+Chry)	
	Autumn	**Winter**	**Autumn**	**Winter**	**Autumn**	**Winter**
Ras El-Gesr	0.29	0.424	0.423	0.936	0.960	0.002
Benja Ali	0.268	0.596	0.416	0.914	0.966	0.022
Wahed Hozeran	0.316	0.427	0.349	0.953	0.974	0.033
North Oil Company	0.450	0.466	0.416	0.932	0.884	0.045
Lillan	0.406	0.693	0.445	0.905	0.883	0.043
Kubri	0.407	0.613	0.402	0.909	0.889	0.048
Chemin	0.153	0.223	0.576	0.875	0.993	0.002

The results of the study indicate that some soil samples contained relatively high concentrations of polycyclic aromatic hydrocarbons, especially at sites near to the sources of pollution, as North Oil Company site and the Kobri, which includes oil refining factories, And the site of Lillan, which includes the Kirkuk cement factory. Bayowa [25] states that 90% of the polycyclic aromatic hydrocarbons in the soil have a half-life longer than in the air and the plant. In the study of Ana et al [26] in Nigeria's coastal areas, concentrations of PAHs were obtained in the soils of the sites. The industrial giant is up to three times what it is in smaller industrial sites. As for the seasons of the study, the concentration of these compounds during the autumn was higher than in the winter. This is related to the physical and chemical properties of these compounds and their impact on environmental factors. Most of the researchers proved that the concentrations of these compounds in warm climates may due to the availability of the internal causes of movement by stirring electrons inside molecules, and the external movement associated with wind, and then the accumulation of these compounds and sedimentation to the soil by different sedimentation ways Bayowa [25], which found in high levels during the dry season than in the rainy season. During the winter, due to increased humidity, the degradation of these compounds is increased by biological, physical and chemical ways [27], Dandie et al [28], Fredslund et al [29] Who stated that bacteria can decompose any PAHs. These results are consistent with [9, 13, 14]. Musa Bandowe et al. [4] and Barrán-Berdón et al., [5], Afaj et al., [6] Abed et al., [15] and Syed et al. [16].

4. Conclusion

We concluded that in winter, Fluorene was the most abundant compound followed by Naphthalene, and according to the different individual PAHs ratios, the orifin of PAHs at soil samples was from pyrogenic and petrogenic sources. We recommend for further future studies to determine the non-point sources of PAHs in soil. Toxicological studies should be done to determine the impact of these compounds on human health.

References

1. Al-sareji OJ, Grmasha RA, Salman JM and Hashim KS 2021 *Journal of Engineering Science and Technology* **16**, 455 – 469.
2. Grmasha RA, Al-sareji OJ, Salman JM and Hashim KS. 2020 *Journal of King Saud* University-Engineering Sciences. Nov 14.
3. Riccardi C, Di Filippo P, Pomata D, Di Basilio M and Spicaglia S 2013 *Science of the Total Environment* **450** 13
4. Musa B, Sobocka J and Wilcke W 2011 *Environ Pollut* **159** 539
5. Barrán-Berdón A, González V, Aboytes G, Rodea-Palomares I, Arrillo-Chávez A, Gómez-Ruiz H and Cuéllar B 2012 *Rev. Int. Contam Ambie* **28** 277
6. Afaj A, Shanshal M, Almousawi I and Adnan S 2014 *Journal of Chemical and Petroleum Engineering* **15** 19
7. Abdel-Shafy H, Al-Sulaiman A and Mansour M 2014 *J. Water Process Eng.* **1** 101
8. Perera F, Wang S, Vishnevetsky J, Zhang B, Cole K and Tang D 2011 *Environ Health Perspect* **119** 1176
9. Nadarajah N, Van H, Pannu J, Singh A and Ward O 2002 *Appl. Microbial. Biotechnol.* **59** 540
10. Delgado-Saborit M, Aquilina N, Baker S, Harrad S and Harrison R 2010 *Anal. Methods* **2** 231
11. Zhong Y and Zhu L 2013 *Sci. Total Environ.* **444** 177
12. Zhu L, Lu H, Chen S, Amagai T and Amagai T 2009 *J. Hazard Mater.* **162** 1165
13. Tsal W, Cohen H and Sakugawa I 1991 *Environ Sci. Technol.* **25** 2012
14. Vardar N, Esen F and Tasdemir Y 2008 *Environ. Pollut.* **155** 298
15. Abed M, Ali S and Altawash B 2015 *Iraqi Journal of Science* **56** 2927
16. Syed J, Iqbal M, Zhong G, Katsoyiannis A, Yadav I, Li J and Zhang G 2017 *Scientific Reports* **7** 2692
17. Awofolu O 2005 *Environmental monitoring and Assessment* **105** 431
18. USEPA 1995 Vol. IA: Laboratory manual physical/chemical methods, SW 846, 3rd ed. US Gov. Print, Office, Washington D.C.
19. IOC/WMO Intergovernmental Oceanographic Commission / World Meteorological Office. (1982). Determination of petroleum hydrocarbons in sediments. Manuals and Guides, No.11. UNESCO Paris.
20. Martinez E, Gros M, Lacorte S and Barceló D 2004 *Journal of Chromatography A* **1047** 181
21. Mittendorf K, Hollosi L, Ates E, Bousova K, Phillips E, Hans-Joachim H and Chang J 2010 *Thermo Sci. J. Method* **51991**-E08/10M.
22. Brack W and Olajire A 2005 *International Journal of Environmental Science and Technology* **2** 343
23. Inengite A, Oforka N and Osuji L 2010 *International Journal of Applied Environmental Sciences*

24. Zelenyuk A, Dan G, Imre J, David M, Kaitlyn J, Manish Sh, Josef B, Alexander M, Kramerc A and Simonich S 2017 *J. Farady Discussions* 200
25. Bayowa A 2014 Thesis of Master, Univ. of South Africa. Nigeria
26. Ana G, Sridhar M and Emerole G 2011 *Journal of Environmental Health Research* **11** 77
27. Cerniglia C 2003 Recent advances in the biodegradation of polycyclic aromatic hydrocarbons by Mycobacterium species, in: V. Sasek (Ed.), *The utilization of bioremediation to reduce soil contamination: problems and solutions,* Kluwer Academic Publishers, The Netherlands, 2003, p. 51–73.
28. Dandie C, Thomas S, Bentham R and McClure N 2004 *J. Appl Microbiol.* **97** 246
29. Fredslund L, Sniegowski K, Wick L, Jacobsen C, De Mot R 559 Springael, D. 2008 *Res. Microbiol.* **159** 255

IOP Conf. Series: Earth and Environmental Science **877** (2021) 012024 doi:10.1088/1755-1315/877/1/012024

Environmental diagnosing of the new algal pollution of Tigris River in Iraq

Warqaa Y. Salih [1], Fikrat M. Hassan[2]

[1]Institute of Genetic Engineering and Biotechnology for Postgraduate Studies, University of Baghdad, Baghdad, Iraq
[2]Department of Biology, College of Science for Women University of Baghdad, Baghdad, Iraq

E-mail: fikrat@csw.uobaghdad.edu.iq

Abstract. The purpose of this study is to use eDNA in the biodiversity of the Tigris river's sediment. Algal samples were collected and examined under light microscopy. The collected algae were cultured, and after their growth, the DNA extractions were made from culture and amplified 16S ribosomal RNA gene partial sequences data by Polymerase Chain Reaction (PCR). Phylogenetic identification of species was conducted by the evaluation of obtained sequence analysis data by using computer software. *Leptolyngbya benthonica* (MN 714226.1) and *Nostoc paludosum* (MN 714225.1) were identified by molecular analysis and registered at NCBI and considered as a new record to the algal flora of Iraq. Implementing molecular data in the taxonomy of species will be essential to solve the taxonomic problems associated with microscopic methods.
Keywords: Environmental DNA, PCR technique, Cyanophyceae, Algae, Tigris River

1.Introduction

Blue greens algae are found in different habitat such as soil, aquatic ecosystem and another extern habitat. Soil ecosystem is more complex than freshwater and marine ecosystems [1, 2]. The majority of algal classes in terrestrial environments, including the blue green ones, are revolutionary micro-organisms and have a high capacity to adapt to different environmental conditions in terms of their morphology and physiology [3, 4].
DNA of blue greens algae is dispersed through the cytoplasm and not organized in chromosomes. They also possess thylakoids which are specialized inner membrane folds where the process of photosynthesis is carried out [5]. As a basis for classifying cyanobacteria, morphological features are considered. However, due to changes in environmental and developmental factors, the morphological features may be changed, whereas some Cyanobacteria are problematic because of the unclear features such *Nostoc* modified by Cyanobacteria [6]. The difficulties to identify these groups of algae by morphological concept led to use the molecular approach to resolve the resemble between the genera and species of blue-greens algae [7]. The *Nostoc* and *Leptolyngbya* are one of the most genera found within the community of the bank of river sediment and in both freshwater and saline water [8-10].
Nostoc is a filamentous, heterocysts blue-green alga. It found worldwide in environment and in the terrestrial ecosystems [11,12]. This genus considered an important source of for protein, vitamin and unsaturated fatty acids for another organism [13-15]. *Leptolyngbya* is one of the most widespread and widely found genus of cyanobacteria. Morphologically, species of Leptolyngbya are recognized by their thin trichomes (less than 3.5 μm wide) surrounded by individual sheaths and presence of parietal thylakoids [16,17]. Debnath et al. [18] explained that the morphological traits of this genera are

overlapping with other species (*Geitlerinema* and *Coleofasciculus*) and preferred to classified molecularly [19,20]. The filaments are never solitary, floating or attached, generally they are not attenuated at the ends, are not capitate, are potentially elastic and are occasionally pseudo-branched [21]. In addition, the absence of a conspicuous morphological characteristic in Leptolyngbya makes it difficult to differentiate it from other related morphotypes, suggesting that morphological work alone is not informative and accurate for this cyanobacterial genus [22].

Many authors emphasize using the DNA, which taken from the habits of the natural communities, and recently this approach was used in biodiversity studies [23,24]. Metabarcoding techniques provide good results of identification from isolated DNA [25,26]. Altermatt et al. [27] explain essential to use the eDNA in the monitoring tool for large scales and additional challenges of spatial patterns of biodiversity. Previous studies showed inconsistent between the traditional studies of biodiversity and the results of eDNA and they interpreted it due to abiotic and biotic factors [28,29]. Garcia-Martinez et al. [30] mentioned that the molecular approach is best solution for the taxonomic deconfliction of cyanobacterial taxa. This approach used the 16S and 23S rRNA gene which is differ from species to another and get a good result these algal group diversity. Thus, the gene sequence between the 16S and the 23S rRNA regions in the ribosomal operon is a tool for studying molecular systematics and population genetics at different taxonomic levels in cyanobacteria. The most common genes are 16S rRNA in this technique with PCR and used to identify different microorganisms in terms of evolutionary conserved sequences [31,32]. The molecular approach particularly environmental DNA in biodiversity studies has applied for different types of organisms because their accuracy and avoid the misidentification of most of algal taxa and other organisms by traditional classification [33,34].

In Iraq, a total of 508 Cyanophyceae taxa were recorded in different Iraqi ecosystems according to Maulood et al. [35], while only 13 taxa of Nostoc based on morphological features only identified. There is no record for Leptolyngbya while there are 43 Lyngbya taxa were identified in Iraq according to last checklist [35]. There are limited numbers of study on the identification of cyanobacteria and green algae [33,36] with the help of molecular techniques in Iraq. This study aimed to identify algae (non-diatomic algae) for samples taken from Tigris River by morphological and molecular characterizations.

2.Materials and methods

2.1.Study area: Five sites were selected over Tigris river within Baghdad city during March to May 2019 (Figure 1, Table 1).

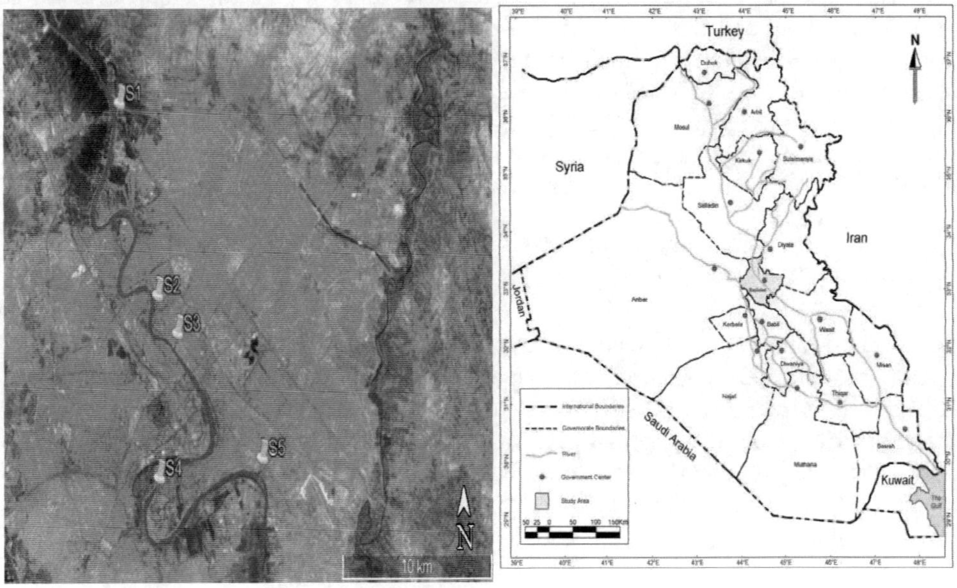

Figure 1. Map of the study areas.S1,S2,S3,S4,S5 represent the sites of study.

Table 1. The longitudinal and latitudinal of the study sites (GPS)

Sites	Longitudes (East)	Latitudes (North)
Al-Muthanna Bridge	44°34'55.50" E	33°42'83.22"N
Al-Sarafiya Bridge	43°99'43.6"E	36°91'33.4"N
Al-Shuhada Bridge	44°16'44.5"E	36°90'74.5"N
Al-Jadriah Bridge	44°22' 30.42" E	33°21'12.62" N

2.2.Sampling

The sediment sample was collected by spatula at a depth of 2-3 cm with a surface area of 50 m^2 below the sediment surface and placed in a nylon bag with several river water [37]. The collected samples were placed in petri dishes and algae were trapped by lens tissues as described in [38]. The examination of algal sample (non-diatoms) done by Genex compound microscope model GX-140105 at Algal and Environmental lab in Department of Biology, College of Science for Women- University of Baghdad. In the final step, the supernatant was discarded and the resulted pellets were stored at -20 °C, this step is to freeze the pellet in order to block the action of the enzymes like RNAase and protease. The pellets were kept for further use as recommended by Visco et al. [39].

2.3.Culturing of cyanobacteria and microscopic examination

The mixed cultures taken from natural habitats (sediment) were placed into test tubes and diluted with sterile water prepared by mixing 1 part of sediment with 2 parts of distilled water. Serial dilution method was used and followed the method described by Stein [40]. In this study, I ml of sample inoculated in BG11 (9ml) and this process were repeated to obtain unialgal species. The activity of colonial filaments was reactivated by maintaining them in BG11medium in 250 ml flasks [41]. The cultures were incubated in a controlled-environment cabinet at 25 ± 1°C with cool white fluorescent lights for 20 days (175 $\mu E/m^2/s$ and 26± 2, 12 h light/12 h dark). Microscopic examination was done by Genex compound microscope model GX- 140105 in the Advance Algal Laboratory of the Department of Biology, College of Science for Women at the University of Baghdad.

2.4. *Identification of samples using molecular method:*

2..4.1. *Genomic DNA extraction and Primer selection:*

The genomic DNA of algae was extracted by using a fast DNA Intron kit (Maxime PCR PreMix kit (i-Taq) 20µlrxn (Cat. No. 25025)) According to manufacturer's instructions. Two important factor taken in consideration (temperature and DNA concentration) for supplemental annealing. The stock solution concentration was 100 pmol/µl. The primer were lyophilized and dissolved into free ddH2O and stock was held at -20 to get 10 pmol/µl and followed the instruction of Integrated DNA Technology company, Canada. The isolated microalgae was identified by the amplification of specific primer 16s RNA of gene (Table 2).

Table 2. The identified algal by amplification of specific primer 16s RNA of gene

Primer	Sequence	Tm (°C)	GC (%)	Product size
Forward	5'- AGAGTTTGATCCTGGCTCAG- 3'	54.3	50.0	1250
Reverse	5'- GGTTACCTTGTTACGACTT- 3'	49.4	42.1	base pair

2.4.2.. DNA extraction and Polymerase chain reaction (PCR)
2.4.2.1. PCR:
The PCR amplification reaction was performed in a total volume of 25µl containing 2ng/µl DNA, (5µl) Taq PCR PreMix (Intron, Korea), and (1 µl) of each primer, DNA (1.5µl) and then distilled water (16.5 µl) was added into tubes. The thermal cycling conditions were performed as follows: Denaturation at 94 °C for 3 min, followed by 35 cycles of 94°C for 45s, 52°C for 1 min and 72°C for 1min with final incubation at 72°C for 7 min using a thermal Cycler (Gene Amp, PCR system 9700; Applied Biosystem).
2..4.2.2. Agarose gel electrophoresis of DNA
DNA segments were obtained via electrophoresis with standard DNA presence according Sambrook et al [42].

2..4.3. Preparation of sample
A 3 µl of the processor loading buffer (Intron/Korea) was mixed with 5 µl of the DNA to be electrophoresis (loading dye) and, after the mixing process, the loading process is now to holes of the gel. The electrical current of 5 vol/cm2 was exposed for 1:30 hours until the tincture reached the other side of the gel. The PCR products were separated by 2% agroseal gel electrophoresis and tested by ultraviolet light source (UV) with 336 nm after putting the gel in the pool containing 3µl of red safe nucleic acid staining solution and 500 ml of distilled water.

3.Results and discussion

3.1.The light microscope(LM) examination
The LM examination showed different groups of algae (non-diatomic). Two distinguish groups exists with diatoms, mainly Cyanophyceae and Chlorophyceae. The identification algae lists are listed in table 3. While the identified algae by molecular methods are not identified by LM.

Table 3. List of some identified algae (from natural community) by LM

Algal group	Font
Cyanophyceae	*Nostoc*
	Anabaena
	Oscillatoria
	Phormidium
	Spriulina
	Nodularia
	Anabaenopsis
Chlorophyceae	*Scendesmus*
	Pedistrum
	Tetradesmus

3.2. Sequencing and data analysis

Sequencing of 16S rRNA gene was performed by Macrogen (Korea), the national instrumentation center for environmental management (nicem) online at: http://nicem.snu.ac.kr/main/?en_skin=index.html, by Applied Biosystem and BioEdit program. Homology search was conducted using Basic Local Alignment Search Tool (BLAST) program which is available at the National Center Biotechnology Information (NCBI) online at (http://www.ncbi.nlm.nih.gov). The estimation of the number of times that predicted to be as similar as coincidental and lower the value of Expected (E), is called an expected value. This revealed that the degree of similarity between sequences was high; where the value close to zero, indicates that these sequences are identical. Bit Score is a statistical measurement for the similarity between sequences, and the high degree of similarity is represented by a higher value. The phylogenetic tree of aligned sequences was conducted using MEGA 6 program.

3.3. Identification cyanobacterial by Sequencing

Identification of cyanobacterial isolates was confirmed by sequence-based phylogenetic tree (aligned sequences was conducted using MEGA 6 program) structuring analysis using 16S ribosomal RNA (16S rRNA) gene sequencing (table 4); the resulted PCR products were subsequently sequenced to obtain DNA sequences, and a1250 base pair (bp) product were obtained. For Leptolyngbya benthonica , Nostoc paludosum , respectively in agarose gel electrophoresied for PCR product (Figure 2).

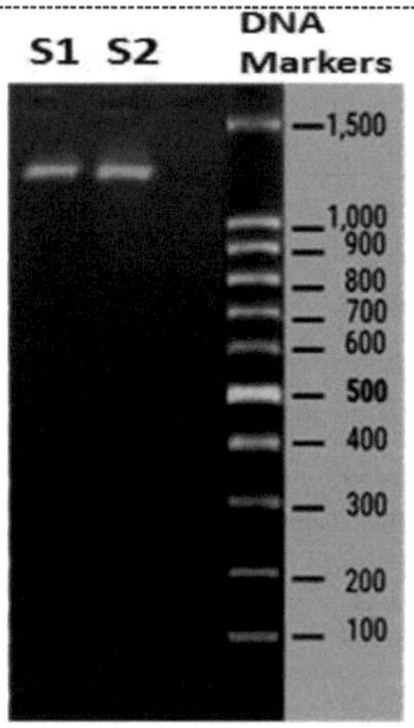

Figure 2. Amplified PCR product the band size 1250 bp. The product was electrophoresis on 2% agarose at 5 volt/cm^2. 1x TBE buffer for 1:30 hour.S1= *Leptolyngbya benthonica*, S2= *Nostoc paludosum*.

The amplicon was aligned using BLAST at NCBI, the 16S rRNA sequence of isolated alga of Leptolyngbya benthonica showed 99% homology with the existing NCBI database sequence of l. benthonica with accession number KM384743.1 in India (Diagram 1). Isolated alga of *Nostoc paludosum* showed 99% homology with the existing NCBI database sequence of *N. paludosum* with accession number KX423684.1 in Brazil (Diagram 2).

Table 4. Types of polymorphism of 16S rRNA gene from isolated algae

No. sample	Type of substitution	Location	Nucleotide	Sequence ID	Score	Expect	Identities	Source
1	Transvertion	349	A>C	ID: KM384743.1	789	0.0	99%	*Leptolyngbya benthonica*
	Transition	449	C>T					
	Transvertion	477	T>A					
	Transvertion	478	C>G					
2	Transition	153	G>A	ID: KX423684.1	1360	0.0	99%	*Nostoc paludosum*
	Transition	157	G>A					
	Transvertion	198	T>G					
	Transition	566	A>G					
	Transition	585	T>C					

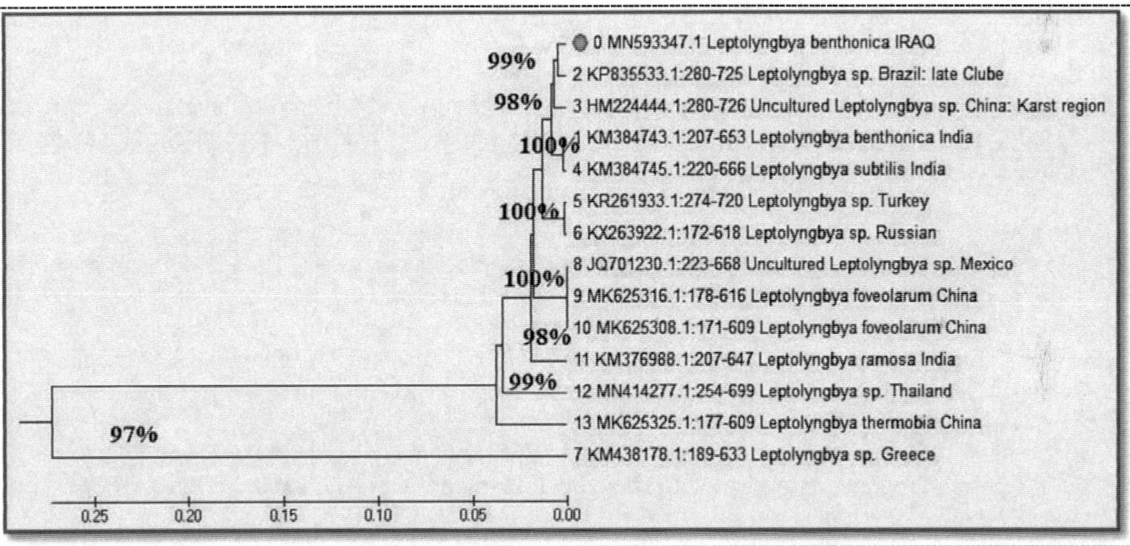

Digram 1. Phylogenetic tree of *Leptolyngbya benthonica* based on *16S rRNA* gene sequences conferred by GeneBank data base, aligned together with algae available in the NCBI were analyzed and aligned through BLAST from NCBI using the Neighbor-Joining Analyses of 1250 bp of corresponding position of *16S rRNA* gene sequence. MEGA 6 program was used for phylogenetic tree.

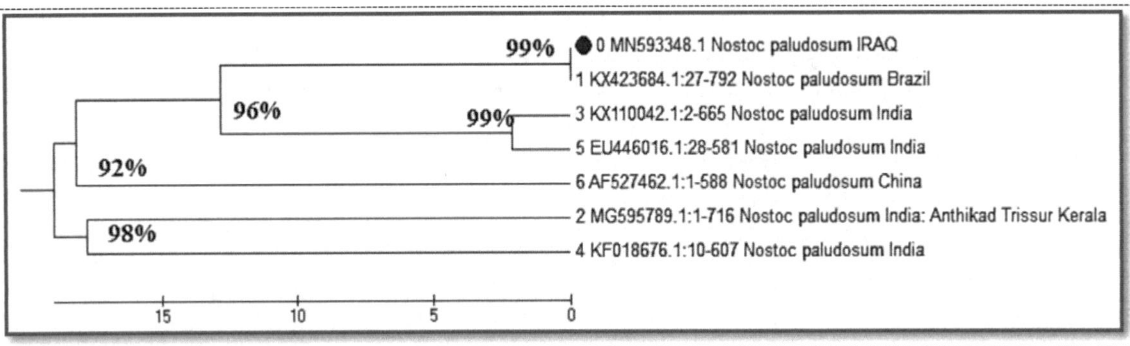

Digram 2. Phylogenetic tree of *Nostoc paludosum* based on *16 S rRNA* gene sequences conferred by GeneBank data base, aligned together with algae available in the NCBI were analyzed and aligned through BLAST from NCBI using the Neighbor-Joining Analyses of **1250** bp of corresponding position of *16S rRNA* gene sequence. MEGA 6 program was used for phylogenetic tree.

A few studies In Iraq use molecular data in the taxonomy of species (43,33, 34,44) and that will be essential to solving the taxonomic problems associated with microscopic methods due to the difficulties of identification by LM.

4. Conclusion

The light microscope (LM) examination did not identify the two new record of blue greens algae in this study. This study confirms the importance to use molecular analysis in algal classification to support the diversity studies in Iraqi aquatic ecosystems.

References

[1] Zenova GM, Shtina EA, Dedysh SN, Glagoleva OB and Likhacheva AA 1995 Ecological relations of algae in biocenoses *Microbiology (New York, NY) 64* **2** 121-133

[2] Weinbauer MG and Rassoulzadegan F. 1995 Are viruses driving microbial diversification and diversity?. Environmental microbiology. 2004 Jan;6(1):1-1.Zenova GM, Shtina EA, Dedysh SN, Glagoleva OB, Likhacheva AA, Gracheva TA Ecological Relations Of Algae In Biocenoses *Mikrobiologiya* **64** 121-133

[3] Hoffmann L 1989 Algae Of Terrestrial Habitats Bot Rev 55:77–105 Hu CX, Zhang DL, Liu YD (2004) Research Progress On Algae Of The Microbial Crusts In Arid And Semiarid Regions *Prog Nat Sci* **14** 289-295

[4] Lewin RA. 2006 Black Algae. *J. Of Applied Phycology* **18** 699-702 Https://Doi.Org/10.1007/S10811-005-9018-2

[5] Richmond A 2004 editor Hand book of microalgal culture: biotechnology and applied phycology. Oxford Blackwell Science

[6] Řeháková K, Johansen JR, Casamatta DA, Xuesong L and Vincent J. 2007 *Sep* Morphological and molecular characterization of selected desert soil cyanobacteria: three species new to science including Mojavia pulchra gen. et sp nov *Phycologia* 1 46 **5** 481-502

[7] Lyra C, Suomalainen S, Gugger M, Vezie C, Sundman P, Paulin Land Sivonen K. 2001 Mar Molecular characterization of planktic cyanobacteria of Anabaena, Aphanizomenon, Microcystis and Planktothrix genera. International Journal of Systematic and Evolutionary Microbiology. 151 **2** 513-26

[8] Saha SK, Das R, Bora KN and Uma L2007 *Sep* Biodiversity of epilithic cyanobacteria from freshwater streams of Kakoijana reserve forest, Assam, India. Indian Journal of Microbiology 47 **3** 219-32

[9] Gaysina LA, Saraf A and Singh P 2019 Cyanobacteria in diverse habitats. In Cyanobacteria Jan **1** 1-28 Academic Press.

[10] Komárek J, Johansen JR 2015 Filamentous cyanobacteria. In Freshwater Algae of North America Jan 1 (pp. 135-235). Academic Press.

[11] Dodds WK, Gudder DA and Mollenhauer D1995 The ecology of Nostoc ***Journal of Phycology*** 31 **1** 2-18
[12] Potts M, Olie JJ, Nickels JS, Parsons J and White DC1987 Variation in phospholipid ester-linked fatty acids and carotenoids of desiccated Nostoc commune (cyanobacteria) from different geographic locations. *Applied and Environmental Microbiology* 1 53**1** 4-9
[13] Gao K1998 Chinese studies on the edible blue-green alga, Nostoc flagelliforme: a review *Journal of Applied Phycology* 10 **1** 37-49
[14] Takenaka H, Yamaguchi Y, Sakaki S, Watarai K, Tanaka N, Hori M, Seki H, Tsuchida M, Yamada A, Nishimori T and Morinaga T. 1998 *Dec* Safety evaluation of Nostoc flagelliforme (Nostocales, Cyanophyceae) as a potential food *Food and chemical toxicology* 1 36 **12** 1073-7
[15] Liu XJ, Jiang Y and Chen F 2005 Jan Fatty acid profile of the edible filamentous cyanobacterium Nostoc flagelliforme at different temperatures and developmental stages in liquid suspension culture *Process biochemistry* 1 40 **1** 371-7
[16] Sciuto K and Moro I 2016 *Dec* Detection of the new cosmopolitan genus Thermoleptolyngbya (Cyanobacteria, Leptolyngbyaceae) using the 16S rRNA gene and 16S–23S ITS region. *Molecular phylogenetics and evolution* 1 **105** 15-35
[17] Kom_Arek J, Anagnostidis K 2005 Cyanoprokaryota 2.Teil: Oscillatoriales. In: Büdel B, Krienitz L, G€Artner G And Schagerl M (Editors). Süsswasserflora Von Mitteleuropa Munique *Elsevier* 1-759.
[18] Debnath M, Singh T and Bhadury P 2017 Aug New records of Cyanobacterial morphotypes with Leptolyngbya indica sp. nov. from terrestrial biofilms of the Lower Gangetic Plain, India. *Phytotaxa* 4 316 **2** 101-20
[19] Silva CS, Genuario DB, Vaz MG and Fiore MF 2014 *Mar* Phylogeny of culturable cyanobacteria from Brazilian mangroves. *Systematic and applied microbiology* 1 37 **2**100-12
[20] Andreote AP, Vaz MG, Genuário DB, Barbiero L, Rezende-Filho AT and Fiore MF2014 Nonheterocytous cyanobacteria from B razilian saline-alkaline lakes *Journal of phycology* Aug 50 **4** 675-84
[21] Johansen JR, Kovacik L, Casamatta DA, Iková KF and Kastovský J 2011 May Utility of 16S-23S ITS sequence and secondary structure for recognition of intrageneric and intergeneric limits within cyanobacterial taxa: Leptolyngbya corticola sp. nov.(Pseudanabaenaceae, Cyanobacteria). Nova Hedwigia 1 92 **3** 283
[22] Vaz MG, Genuario DB, Andreote AP, Malone CF, Sant'Anna CL, Barbiero L and Fiore MF 2015 *Jan* Pantanalinema gen. nov. and Alkalinema gen. nov.: novel pseudanabaenacean genera (Cyanobacteria) isolated from saline–alkaline lakes. *International Journal of systematic and evolutionary microbiology* 1 65 **1** 298-308
[23] Kelly RP, Port JA, Yamahara KM, Martone RG, Lowell N, Thomsen PF, Mach ME, Bennett M, Prahler E, Caldwell MR, Crowder LB. Harnessing DNA to improve environmental management *Science* 2014 Jun 27 344 **6191** 1455-6
[24] Bálint M, Pfenninger M, Grossart HP, Taberlet P, Vellend M, Leibold MA, Englund G and Bowler D. 2018 Dec Environmental DNA time series in ecology *J.Trends in Ecology & Evolution*. 1 33 **12** 945-57
[25] Deiner K, Bik HM, Mächler E, Seymour M, Lacoursière-Roussel A, Altermatt F, Creer S, Bista I, Lodge DM, De Vere N and Pfrender ME 2017 Environmental DNA metabarcoding: Transforming how we survey animal and plant communities *Molecular ecology* 26 **21** 5872-95
[26] Pawlowski J, Kelly-Quinn M, Altermatt F, Apothéloz-Perret-Gentil L, Beja P, Boggero A, Borja A, Bouchez A, Cordier T, Domaizon I and Feio MJ 2018 Oct The future of biotic indices in the ecogenomic era: Integrating (e) DNA metabarcoding in biological assessment of aquatic ecosystems. *Science of the Total Environment* 1 **637** 1295-310
[27] Altermatt F, Little CJ, Mächler E Wang S, Zhang X and BlackmanRC 2020 Uncovering the complete biodiversity structure in spatial networks: the example of riverine systems. Oikos 129 **15** 607-618

[28] Shogren AJ, Tank JL, Andruszkiewicz E, Olds B, Mahon AR, Jerde CL and Bolster D 2017 *Jul* Controls on eDNA movement in streams: Transport, retention, and resuspension Scientific Reports 11 7 **11**-1

[29] Sansom BJ and Sassoubre LM2017 *Dec* Environmental DNA (eDNA) shedding and decay rates to model freshwater mussel eDNA transport in a river. Environmental Science & Technology 19 51 **24** 14244-53

[30] García-Martínez J, Acinas SG, Antón AI and Rodríguez-Valera F1999 Use of the 16S–23S ribosomal genes spacer region in studies of prokaryotic diversity. *Journal of microbiological methods* 1 36 **1-2** 55-64

[31] Iteman I, Rippka R, de Marsac NT and Herdman M 2000Comparison of conserved structural and regulatory domains within divergent 16S rRNA–23S rRNA spacer sequences of cyanobacteriaThe GenBank accession numbers for the sequences reported in this paper are AF180968 and AF180969 for ITS-L and ITS-S, respectively *Microbiology* 1 146 **6** 1275-86

[32] Boyer Sl, Flechtner VR and Johansen JR 2001 Is the 16S–23S rRNA internal transcribed spacer region a good tool for use in molecular systematics and population genetics? A case study in cyanobacteria. *Molecular biology and evolution* 18 **16** 1057-1069.

[33] Al-Rawi A , Alwash BM, Al-Essa NE and Hassan FM 2018. A New Record Of coelastrella terrestris (reisigl) hegewald & n. Hanagata, 2002 (sphaeropleales, scenedesmaceae) in iraq. *J.Bulletin of the iraq natural history museum (p-issn: 1017-8678, e-issn: 2311-9799) 15* **2** 153-161

[34] Al-Meshhdany Wy, Hassan Fm 2020 *J.* Five diatom species identified by using potential application of next generation dna sequencing. *J.Bulletin of the iraq natural history museum* (p-issn: 1017-8678, e-issn: 2311-9799). 30 16 **1** 39-61

[35] Maulood BK, Hassan FM, Al-Lami AA, Toma JJ and Ismail AM2013 Checklist of algal flora in Iraq Ministry of Environment, Baghdad

[36] Abed IJ, Abdulhasan GA, Moushib LI 2019Molecular and Immunological Methods to Confirm Toxiginicity (Microcystin Production) of Westiellopsis Prolifica Isolated from Tigris River–Iraq. *Baghdad Science Journal* **16** 4

[37] Salman JM, Hassan FM and Abdulameer SH 2018 *Jan* Qualitative and Quantitative study of Epipelion algal community in Euphrates River (Al-Hussainya), Karbala Province-Iraq *Int J of Aquatic Science* 1 9 **1** 30-7

[38] Eaton JW, Moss B. The estimation of numbers and pigment content in epipelic algal populations. Limnology and Oceanography. 1966 Apr;11(4):584-95.

[39] Visco JA, Apothéloz-Perret-Gentil L, Cordonier A, Esling P, Pillet L and Pawlowski J 2015 *Jul* Environmental monitoring: inferring the diatom index from next-generation sequencing data. *Environmental science & technology* 7 49 **13** 7597-605

[40] Stein J 1975 Hand Book Of Phycological Methods. Cambridge Univ. Press. Cambridge **p.**445

[41] Stanier RY, Kunisawa R, Mandel M and Cohen-Bazire G1971 *Jun* Purification and properties of unicellular blue-green algae (order Chroococcales). Bacteriological reviews 35 **2** 171

[42] Sambrook J, Fritsch EF and Maniatis T1989 Molecular cloning: a laboratory manual Cold spring harbor laboratory press

[43] Abed IJ, Abdulhasan GA, Najem AM 2018 Genotype versus phenotype to determine the definitive identification of the genera Chlorella beijerinck, 1890 (chlorellaceae) and Coelastrella chodat, 1922 (scendesmaceae). *J.Bulletin of the Iraq Natural History Museum* (P-ISSN: 1017-8678, E-ISSN: 2311-9799) Jul 1 15 **1** 101-11

[44] Hassan FM, Salih WY and Al-Haideri HH 2020 Next-Generation Sequencing Technologies for Environmental DNA as an Efficient Bio Indicator for Bacterial Biodiversity in Tigris River, Iraq. *Systematic Reviews in Pharmacy* 11 **11** 1107-14

IOP Conf. Series: Earth and Environmental Science **877** (2021) 012025 doi:10.1088/1755-1315/877/1/012025

Flood Control by Weir Design Using HEC-RAS Model: The Case of Al-Musandaq Escape

Tabarak W.Mahdi[1], Ali N.Hillo[1]

[1]Department of Civil Engineering, College of Engineering, University of Wasit, Iraq.

E-mail: twaseem@uowasit.edu.iq, dralinasser@uowasit.edu.iq

Abstract. Proper flood control plays an important part in designing hydraulic structures and environmental safety measures. However, in Iraq, a clear understanding either of the estimation or management of the magnitude of flooding is yet to reach a higher level. This has resulted in grave and frequent damage to much property and life in Maysan town. Therefore, this study was focused on the Flood Hazard contro; in the River Tigris at the downstream of Al-Kut Barrage to the Al-Musandaq Escape, by adopting the HEC RAS model. Here, information related to the hydrological and topographical Digital Elevation Model (DEM) data were used as the input data. The hydrological data enabled the estimation of the flood depth of the river, for April 2019. All the geometric data were prepared by the HEC-RAS model. The unsteady-state model simulation was performed employing the input data. In this study, the best method for flood control and management is to design a weir having an optimal level, the optimum level that does not permit the passage of a flow that exceeds 700 m3/s to the city of Maysan during the flood season and a flow that is not below 250 m3/s during the dry season, is 9.406 m.

Keywords: HEC-RAS; GIS; Flood; Weir.

1. Introduction

The most significant natural disaster might be the flood, with more natural hazards than other hazards that affect more people than all-natural disasters combined. [1],[2]. Moreover, the possibility of floods will rise due to climate change [3]. Social-civil conflicts [4], environmental issues [5], and economic losses [6] are connected to floods. Maysan city in Iraq is an example of flood lands suffering from severe flood events causing ongoing environmental, economic, and social damage. Many hectares of arable land are swamped, crops are destroyed, and houses are destroyed. The inundations need to be recognized and the consequences of the proposed acts to suggest flood control measures evaluated. In situ flood control is a simple solution flood management [7]. Flood observations in situ aren't available always, however. For flood research, other researchers use Data for remote sensing. [8,9]. However, in the cloudy skies, there are typically flood events that limit the use of remote sensing data. Also, observation-based flood research is applicable just for a particular flood occurrence [10]. Therefore, such observation-based maps cannot analyze predicted flooding incidents or the systemic operation of the flood. The use of computational models allows flood events to be simulated, taking into account various scenarios. Therefore, Models numerical are critical resources for the understanding of flood occurrences, flood risk assessment, and flood control planning. Previous research has shown that numeric models are applicable for developing danger maps taking into account different techniques in flood management. [11,12] or to recreate past flood events [13]. There are now numerical models with diverse capabilities and developers available; some are free, while others require a license to be purchased.. The HEC-RAS model Established by the Engineers' Corps of the Army (USACE) is one of the Hydraulic Models most common. HEC-RAS is free software that has been successfully used for flood studies [14,15] with a comfortable graphical user interface. Besides, it is under continuous development and improvement by USACE.

The current study aims to control the April 2019 flood event in Maysan city using the HEC-RAS-v5.0.7 1D and HEC-GeoRAS (Geospatial River Analysis System). A 1D numerical simulation of the April 2019 Maysan flood occurrence was Executed with the HEC-RAS-v 5.0.7 and the daily flow of the Tigris river.

2. Study Area and data

2.1. Study area

The reach of Tigris River is situated between downstream of Kut Barrage and Al-Musandaq Escape, Figure.1, and is about 160 km long. kut Barrage is the primary cross-regulator for the Tigris River flow regulation. At the downstream of Kut Barrage, the elevation of the river levee is as high as 20m. This elevation progressively decreases to 9m at near the Al-Musandaq Escape. The average longitudinal surface slope of the water is around 4cm / km. Within the reach under study, there is one flood escape to protect the reach understudy from flooding, especially since the reach has a mild slope and runs within a flat area. The Al-Musandaq escape is this flood escape. The Al-Musandaq Escape is located about 87 km downstream of the Kut Barrage on the right side of the reach. This escape connects the Tigris River with Uda Marsh [16] with an earth channel of 400 m maximum width. The authorities opened this escape to discharge 1800 m^3 / s during the flood of 1974 to reduce flood pressure on the downstream range, Ministry of Irrigation previously (IMI).

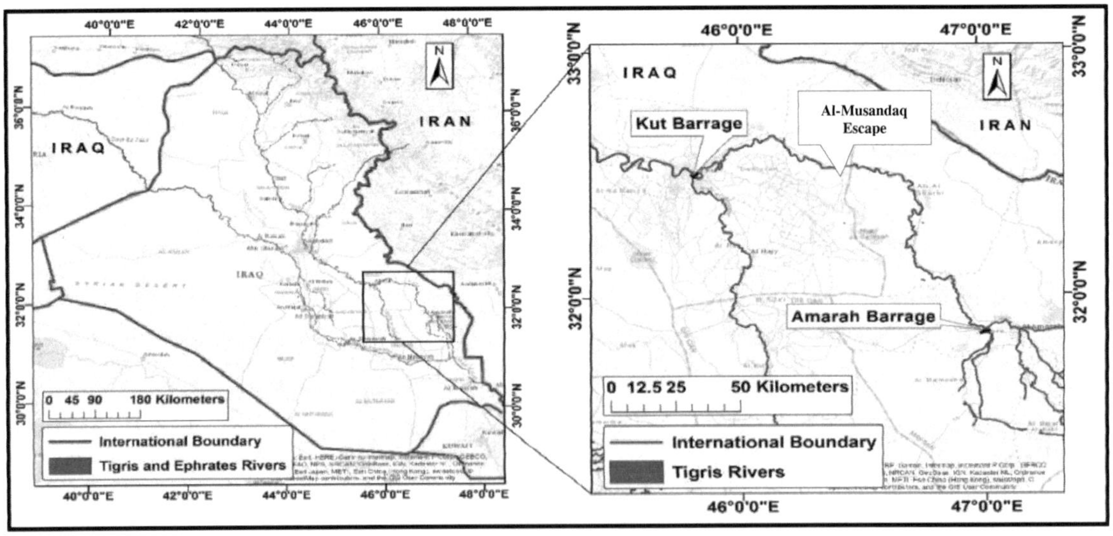

Figure 1. The River Tigris located between the Kut Barrage downstream and Al-Musandaq Escape (GIS software)

2.2. Data

Geographical data providing a geographical overview of the region and flow data providing details on the flow of the Tigris River is the main data forms. The geographic data from the Shuttle radar topography mission (SRTM) with a grid cell size of 90 m was based on the digital elevation model (DEM) figure 2. The research-based on an area between downstream of the Kut Barrage (with the coordinates of 32° 29′ 00″ N latitude, and 45° 50′ 00″ E longitude) up to the Al-Musandaq Escape (with the coordinates of 32° 30′ 00″ N latitude, and 46° 25′ 00″ E longitude). Data for the flow were based on daily data of the Tigris river at Kut Barrage station provided by the IMI. The cross-sections were surveyed, 62 major cross-sections were measured along the Tigris River from D/S Al-Kut Barrage - Al-

Musandaq Escape by GPS (Global Position System). The cross-sections photo is shown in figure 3. The first and last cross-section for the Tigris river and Al-musandaq Escape was shown in figure 4. The study simulated the floods between April 01, 2019, and April 30, 2019.Figure.5 shows the flow discharge of the Tigris river.

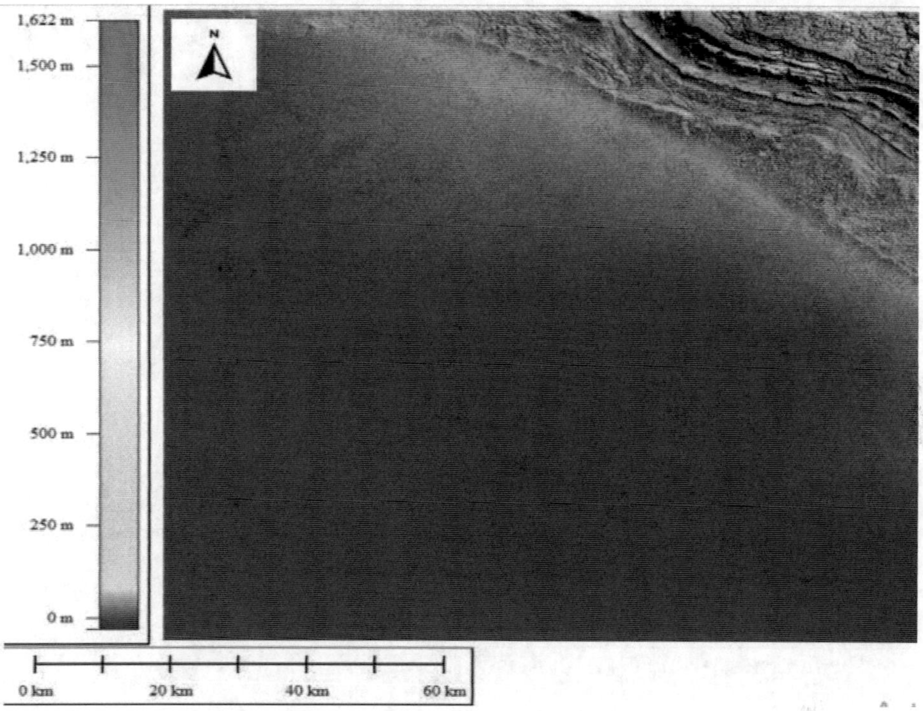

Figure 2. DEM model picture for the study area (global mapper)

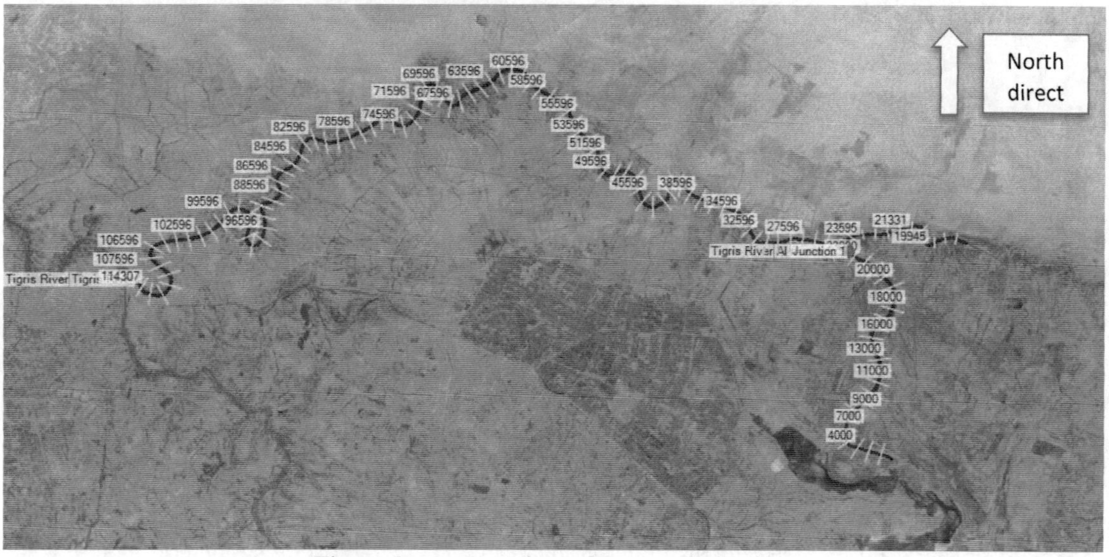

Figure 3. cross-sections along reach the river

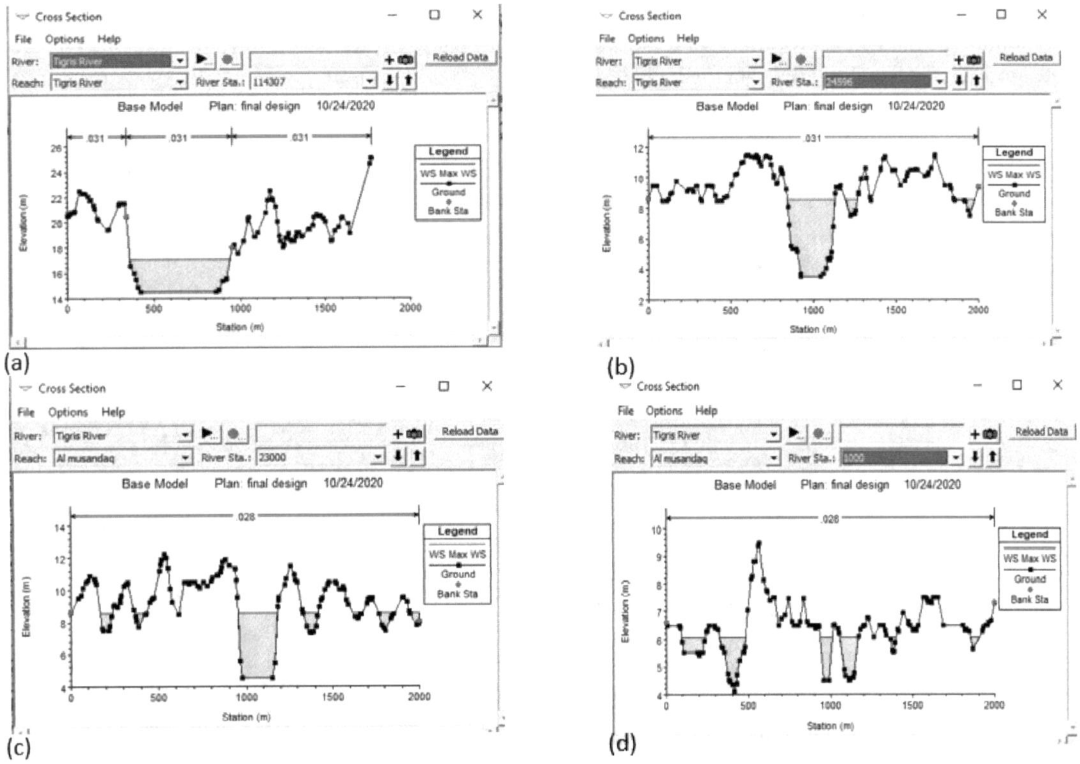

Figure 4. cross-sections : (a) The first cross-section at Tigris river, (b) The final cross-section at Tigris river,(c) The first cross-section at Al-musandaq Escape,(d) The final cross-section at Al-musandaq Escape

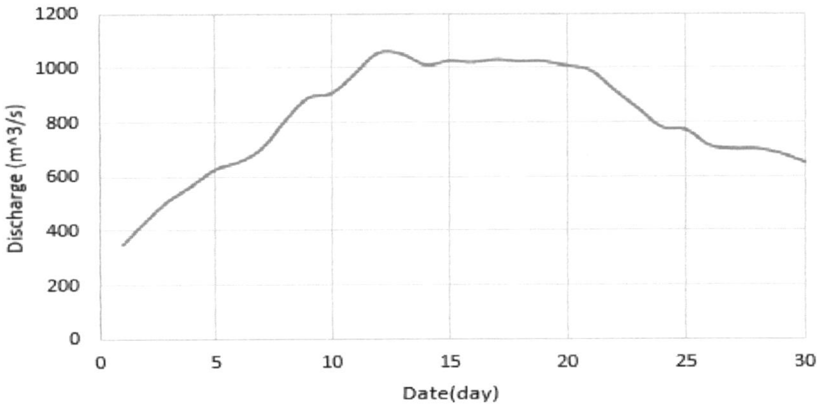

Figure.5 the flow discharge of the Tigris river

3. Theoretical Basis

3.1. One Dimensional unsteady Flow

The study focused on the unsteady state and gradually varying flow of the simulation model, which depends upon the finite difference in the solutions of continuity and momentum. Equations (1) & (2) are used to simulate the River Tigris flow of water. The HEC-RAS 1-D unsteady flow simulator measurement engine [17] is as given,

$$\frac{\delta A_T}{\delta T} + \frac{\delta Q}{\delta x} = q \text{...} (1)$$

$$\frac{\delta Q}{\delta t} + \frac{\delta VQ}{\delta X} + gA\left(\frac{\delta z}{\delta x} + S_o + S_f\right) = 0 \text{......................} (2)$$

Where,

An indicates the cross-section (m^2), V refers to the cross-sectional averaged velocity (m/d), Q represents the flow in cross-section (m^3/d), g is the gravity acceleration (m/s^2), and x shows the distance along the channel (m)

$\frac{\delta z}{\delta x}$= water surface slope implies the water surface slope, So means the bed slope; Sf is the energy slope, t equals the time (day)

3.2. Weir

The HEC-RAS is capable of simulating flow at the hydraulic controls in various ways. Bridges, culverts, inline structures, lateral structures, and SA/2D area connections can all function as hydraulic controls. They effectively break down the conservation equations used between the cross-sections in a 1D reach and/or cells in a 2D area with empirically derived (and usually very stable) equations. The weir equations can be used to define the flow over an obstruction and are available with all the five hydraulic controls identified above. However, several options must be considered when selecting a simulating weir flow in the HEC-RAS. The HEC-RAS approaches weir flow for three different cases: Ungated Inline Weirs, Ungated Lateral Weirs, and Gated Weirs. They all begin with the same standard equation. (3),

But in each of the three cases the weir equation is applied slightly differently.

$$Q = CLH^{2\backslash 3} \text{..} (3)$$

Where, Q is the discharge (m3/s), C refers to the weir coefficient, L represents the weir crest length and H indicates the energy head over the weir crest.

4. Methodology

4.1. HEC -RAS Model Development

To model floods in the Tigris River Basin, HEC-RAS and HEC-GeoRAS (Geospatial River Analysis System), an extension of HEC-RAS, were used. The HEC-RAS model was established by the Hydraulic Engineering Centre, Part of the U.S. Army Corps of Engineers Institute of Water control. [16]. In 1965, Several river modeling programs were developed at the research institute. In 1995, the first version of the HEC-RAS model was released to analyze the rivers, streams, and channels. The HEC-RAS software is an integrated system that can simulate the water flow in rivers and channels by employing a numerical model [18]. as this software is one-dimensional, it implies the lack of any direct modeling of the hydraulic influences of the transverse profiles. [19]. To simulate and produce flood inundation depths, the most important HEC-RAS model is to estimate the water surface profiles of flood events. As per the HEC- RAS manual, the water surface profile measurement is based on the 1D energy equation [20]. In this analysis, the model was built by real values with the help of the GIS environment, three steps were taken to run hydraulic or hydrologic modeling (HEC -RAS). These steps include Data Pre-processing, Data Modelling Phase, and Data Post-processing [21]. The first step in designing the HEC-RAS hydraulic model is to define the directory in which the researcher wants to function and enter a title for the new project. Then it imports the cross-sections from the excel file that were measured by GPS. and edited in the geometry edit window figure.6 [22]. In the new HEC RAS project, flow data from the river, all RAS layers, and the manning coefficients were established step by step according to the HEC-RAS manual to successfully simulate unsteady-state flow. In the current study, when the output data of the program was exported to RASmapper. Results were showed into Rasmapper after running in HEC-RAS[23] The following procedures have been carried out to achieve the purpose of the analysis, as shown in Figure.7 below..

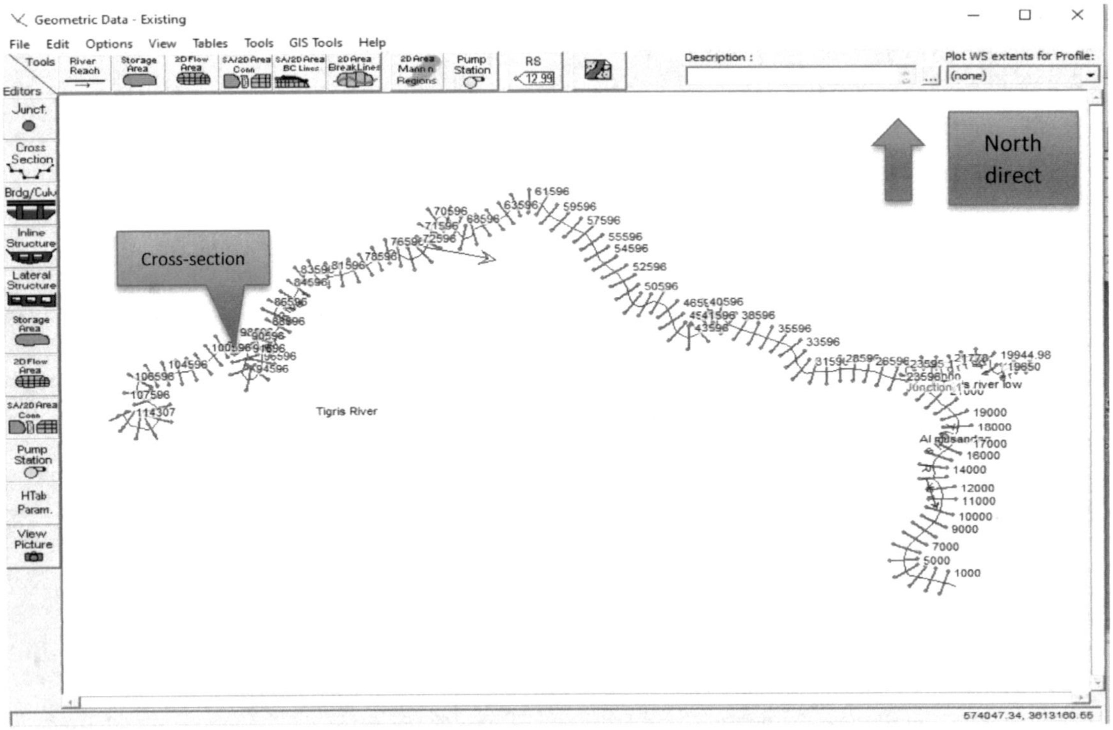

Figure 6. Edit geometry window in the HEC-RAS model

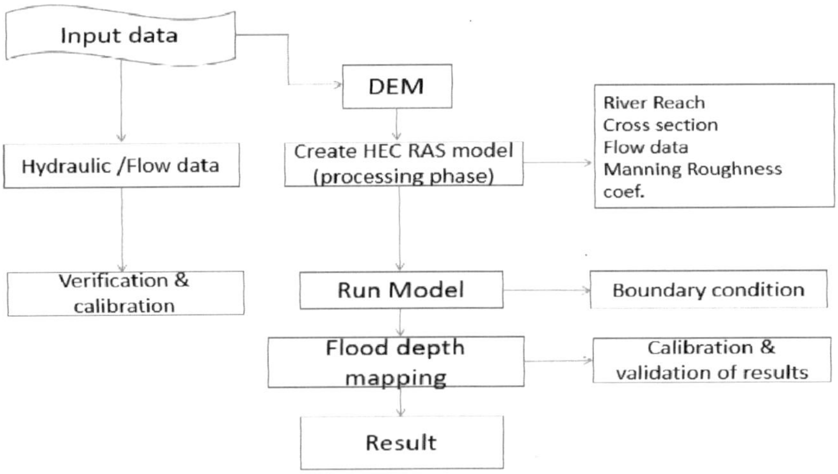

Figure 7. Flow Chart of the Tigris River Flood Modelling

4.2. HEC -RAS Model Calibration

The most sensitive parameter is Manning's roughness coefficient in river simulation. Calibration of Manning's roughness coefficient value of the river must be attempted by simulating these flows using the HEC-RAS model along with the study distance. Calibration is an inverse mathematical problem since it attempts to evaluate the original and local parameters that describe a function from unique variables from a given measurement data set. The calibration of hydrodynamic models has historically been performed manually. By changing the parameter sets, the calibration method minimizes the objective function; showing the river and taking into account the ranges of the parameter values for the flow state. The most widely used objective function is the correlation coefficient (r) Equation (4)[24] between simulated and observed data.

$$r = \frac{n(\sum xy) - (\sum x)(\sum y)}{\sqrt{\left[n\sum x^2 - \sum(x)^2\right]\left[n\sum y^2 - \sum(y)^2\right]}} \quad \ldots\ldots\ldots\ldots\ldots (4)$$

where, n= number of data points, X_i= observed flow, y= simulated flow

5. Results and Discussion

5.1 HEC-RAS Calibration and Validation

For the main channel and its flood plain, different Manning's n values were assumed to investigate and evaluate their impact on water surface elevation. The range of these values tested ranges between (0.021-0.031) for the Tigris River and between (0.02-0.028) for Al-Musandaq escape. To compare the simulated and observed flows, the r test was used. The comparison of simulated and observed flows for values of Manning's n is shown in Figure.8. Manning's roughness coefficient 'n' was 0.031 for the river Tigris and 0.028 for the Al-Musandaq Escape. The values of r, are 0.995 and 0.991 for calibration and validation respectively, which were in the acceptable range. Figure.9 shows a graphic comparison of the correlation between the observed and simulated flow data, indicating that the model demonstrated very good performance in fitting correlation.

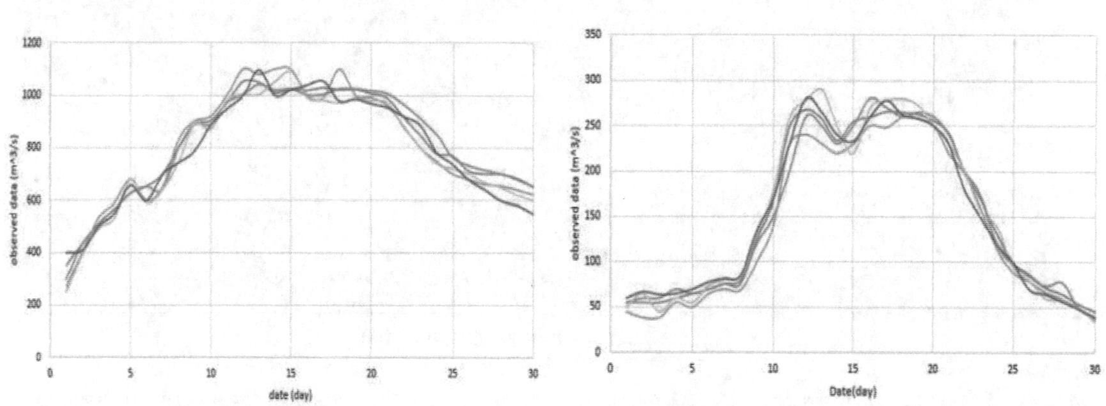

Figure 8. The comparison of simulated and observed flows for values of Manning's n: (a) Tigris river in comparison, (b) Al-Musandaq escape n comparison

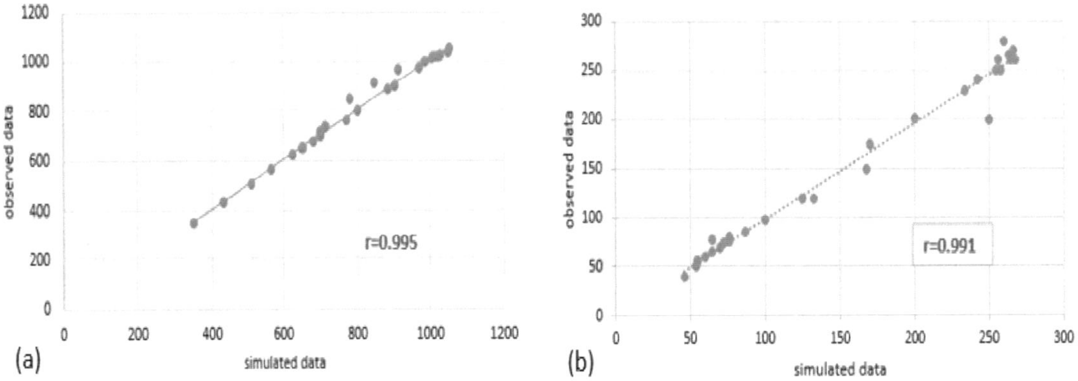

Figure 9. Statistical parameter of unsteady flow calibration: (a) the Tigris river, (b) Al-Musandaq Escape

5.2 Results of one-dimensional mathematical model

To get a clear perception of the inundated flood depth for April 2019, the HEC-RAS was simulated for the discharge values of the River Tigris at the Al-Kut barrage d/s up to the Al-Musandaq Escape. GIS based preliminary: The current study showed the output map of the depths of the flood inundation prepared by simulation of the HEC-RAS based on the Digital Elevation Model (DEM). As shown in Figure 10, the map reveals the flood depths in the area under study, in the event of a flood. Evidently, the maximum height of 7.3 m can be observed on April 13, when the flood peak was achieved., the very high flood risk depths are identified in dark blue.

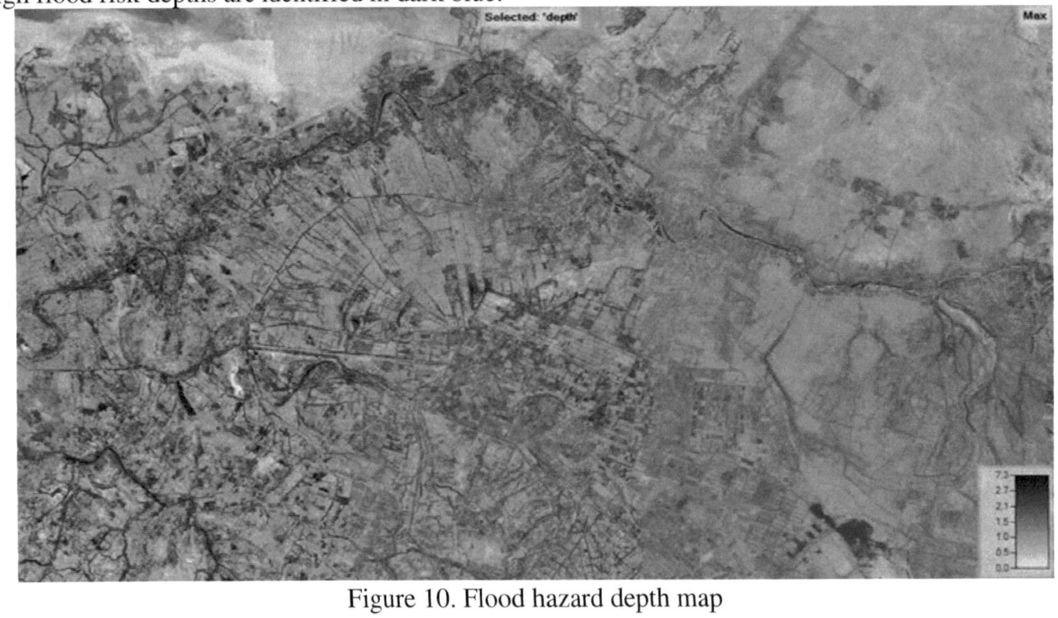

Figure 10. Flood hazard depth map

5.3 Weir design

A weir was designed in front of the Al-Musandaq Escape at station 23+410, as shown in Figure (11). The parameters were entered in the inline structure weir window, where the distance is 20 m, width is 10 m, and the weir coefficient is equal to 3.1. Several scenarios were conducted to achieve the optimum level that does not permit the passage of any flow exceeding 700 m^3/s into the city of Maysan during the flood season and no flow below 250 m3/s during the dry season; the scenarios were performed for four peak flows, namely, (4000, 2500, 1050, 533, and 250) m^3/s. After implementing the many scenarios, the optimum level for designing a weir that does not allow the passage of more than 700 m^3/s drainages to the city of Maysan and no less than 250 m^3/s is 9.406 m.

IOP Conf. Series: Earth and Environmental Science **877** (2021) 012025 doi:10.1088/1755-1315/877/1/012025

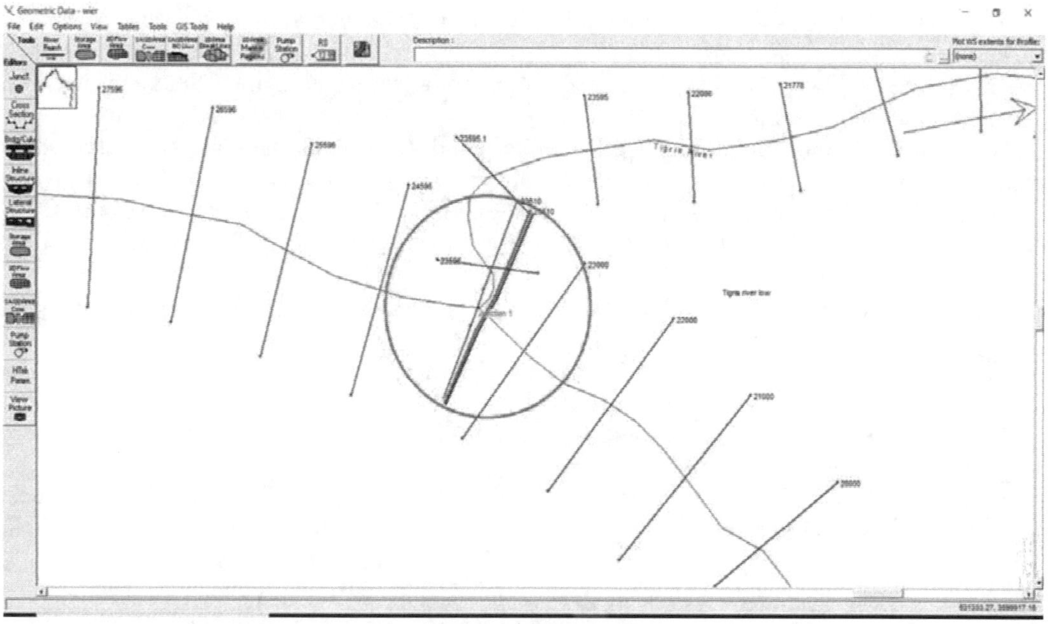

Figure 11. Weir location

6. CONCLUSION

In this study, the 1D HEC-RAS model was used to generate the inundation model in the Tigris River floodplain. For the further generation of the flood hazard depth map from the model's inundation data, the Geographical Information System was used. In April 2019, calibration and validation of the 1D HEC-RAS model were carried out where the roughness coefficient 'n' of Manning was the main calibration parameter. The flow data from the Tigris river was used for calibration and validation. A strong correlation between the observed and simulated flow data was seen as a result. For Manning's roughness coefficient 'n' 0.031 for the River Tigris and 0.028 for the Al-Musandaq Escape, the result showed a strong correlation between the observed and simulated flow results. The correlation coefficient, r is 0.995 and 0.991 for calibration and validation respectively, which were found in the appropriate range. In this analysis, the 1D HES-RAS model was used to generate the Tigris River floodplain flood inundation model.The best method for flood control and management is to design a weir having an optimal level, the optimum level that does not permit the passage of a flow that exceeds 700 m^3/s to the city of Maysan during the flood season and a flow that is not below 250 m^3/s during the dry season, is 9.406 m.

Recommendations

It recommends the following suggestions for future studies:

- Development Al-Musandaq system to perform the flood operation functions for the distribution of water from the Tigris River to the Marshlands of Maysan city.
- Studying the possibility to use the Al-Musandaq escape for irrigation of some areas adjacent to the path for the winter season only to meet the population's needs.

REFERENCES

[1] Aerts, J. C. J. H. and Botzen, W. J. W. 2011 Climate change impacts on pricing long-term flood insurance: A comprehensive study for the Netherlands *Global Environmental Change* **21**1045–1060.

[2] Agrawal, R. and Regulwar, D. G. 2016 Flood analysis of Dhudhana river in upper Godavari basin using HEC-RAS *Int J Eng Res* **1** 188–191.

[3] Akoglu, H. 2018 User's guide to correlation coefficients *Turkish journal of emergency medicine* **18** 91–93.
[4] Al-Rubaie, A. A. K. 2008 Ecological and morphological study of Iraqi Southern marshes. *Mesopotamian Journal of Marine Science* **23** 1-17.
[5] Ban, H.-J. 2017 Flood monitoring using satellite-based RGB composite imagery and refractive index retrieval in visible and near-infrared bands *Remote Sensing* **9** 313.
[6] Cameron, T., Ackerman, P. E. and HEC-GeoRAS, G. I. S. 2012 Tools for Support of HEC-RAS Using ArcGIS User's Manual *US Army Corps of Engineers Institute for Water Resources, Hydrologic Engineering Center (HEC): Davis, CA, USA.*
[7] Center, A. D. R. 2008 Natural Disasters Data Book 2008: An Analytical Overview. Kobe Japan: ADRC.
[8] Daham, M. H. and Abed, B. S. 2020 One and Two-Dimensional Hydraulic Simulation of a Reach in Al-Gharraf River *Journal of Engineering* **26** 28–44.
[9] Das, T.2013 Increases in flood magnitudes in California under warming climates *Journal of Hydrology* **501** 101–110.
[10] Erena, S. H., Worku, H. and De Paola, F. 2018 Flood hazard mapping using FLO-2D and local management strategies of Dire Dawa city, Ethiopia *Journal of Hydrology: Regional Studies* **19** 224–239.
[11] Esper Angillieri, M. Y. and Fernández, O. M. 2017 Morphometric analysis of river basins using GIS and remote sensing of an Andean section of Route 150, Argentina: A comparison between manual and automated delineation of basins *SciELO* **34** 150-156
[12] Getahun, Y. S. and Gebre, S. L. 2015 Flood hazard assessment and mapping of flood inundation area of the Awash River Basin in Ethiopia using GIS and HEC-GeoRAS/HEC-RAS model *Journal of Civil & Environmental Engineering* **5** 1.
[13] Ghimire, R., Ferreira, S. and Dorfman, J. H. 2015 Flood-induced displacement and civil conflict *World Development* **66** 614–628.
[14] Gül, G. O., Harmancıoğlu, N. and Gül, A. 2010 A combined hydrologic and hydraulic modeling approach for testing efficiency of structural flood control measures *Natural hazards* **54** 245–260.
[15] Haq, M. 2012 Techniques of Remote Sensing and GIS for flood monitoring and damage assessment: A case study of Sindh province, Pakistan *The Egyptian Journal of Remote Sensing and Space Science* **15** 135–141.
[16] Knighton, D. 2014 *Fluvial forms and processes: a new perspective*. Routledge.
[17] Li, J. and Shi, W. 2015 Effects of alpine swamp wetland change on rainfall season runoff and flood characteristics in the headwater area of the Yangtze River *Catena* **127** 116–123.
[18] Mohammadi, S. A., Nazariha, M. and Mehrdadi, N. 2014 Flood damage estimate (quantity), using HEC-FDA model. Case study: the Neka river *Procedia Engineering* **70** 1173–1182.
[19] Nigusse, A. G. and Adhanom, O. G. 2019 Flood Hazard and Flood Risk Vulnerability Mapping Using Geo-Spatial and MCDA around Adigrat, Tigray Region, Northern Ethiopia *Momona Ethiopian Journal of Science* **11** 90–107.
[20] Patel, C. G. and Gundaliya, P. J. 2016 Floodplain Delineation Using HECRAS Model—A Case Study of Surat City *Open Journal of Modern Hydrology* **6** 34–42.
[21] Ruiz-Bellet, J. L. 2015 Historical, hydraulic, hydrological and meteorological reconstruction of 1874 Santa Tecla flash floods in Catalonia (NE Iberian Peninsula) *Journal of Hydrology* **524** 279–295.
[22] Samadi, A., Sadrolashrafi, S. S. and Kholghi, M. K. 2019 Development and testing of a rainfall-runoff model for flood simulation in dry mountain catchments: A case study for the Dez River Basin *Physics and Chemistry of the Earth, Parts A/B/C* **109** 9–25.
[23] Schober, B., Hauer, C. and Habersack, H. 2015 A novel assessment of the role of Danube floodplains in flood hazard reduction (FEM method) *Natural Hazards* **75** 33–50.
[24] Sherwood, K. L. M. M. C. R. and Brosnahan, S. M. 2018 Baseline coastal oblique aerial photographs collected US Army Corps of Engineers Field Research Facility, Duck, North Carolina, June 9, 2017 USGS.

Design and Implementation of Sunlight Tracking Based on the Internet of Things

Mohammed Hassan Majhool[1], Haider TH. Salim ALRikabi[1], and Mansour S. Farhan[1]

[1]Electrical Engineering Department, College of Engineering, Wasit University, Wasit, Iraq

E-mail: hdhiyab@uowasit.edu.iq, mhsn32645@gmail.com

Abstract. In this article, for smart management and control of the solar pursuing device, an archetype is designed to evaluate or verify administration system and control. The newly-built prototype is designed for several applications to be a powerful platform for understanding the smart solar energy method. The Internet of Stuff integrates ordinary items using the Internet to enter the physical world. The IoT technology is facilitated to track solar PV (photovoltaic) power generation that can improve plant upkeep, tracking, and efficiency. This would include monitoring the solar panel and pointing it into the sunlight. It's all feasible to utilize LDR sensors. The IoT automatically preserves track of the voltage supply the solar panel detects the towards of sunlight. The suggested system demonstrates the online utilization of solar PV power.

Keywords: ESP32, LDR, SOLAR PANEL, VOLTAGE SENSOR, IOT.

1. Introduction

Electricity considers one of the most important devices which are used in everyone's life. The consumption of energy is increasing daily dramatically while energy resources are decreasing. To equilibrium the shortage of electricity, many sources are used to produce the electricity[1]. There are two methods to generate electricity: the first method is the traditional method and the other is the unconventional method. In addition, fossil and nuclear fuels are used to generate the electrical energy, but they are not unconventional[2]. Sustainable energy sources can be attained by using solar energy as a source. Solar energy is widely available throughout the world. The sun light has been used to produce energy for billions of years. Sunlight can act as an significant source of electricity the produced by transforming it into electrical energy[3]. This application is known as solar thermal energy, and it is conservative. Many ecological sources are presented such as wind, rain, tides, geothermal energy, natural biofuels, and traditional biomass. Solar energy has enormous advantages. Nowadays, frequent power outages are very communal in Iraq. So, it is essential to investigate renewable energy. The Rapid growth of renewable energy requests has been enabled by the steep reduction in cost over the preceding decades and specialized change in its stable productivity, quality, and life[4, 5]. Iraq is one of the countries that has a very hot climate in the summer and it needs continuous energy to overcome the high temperatures. the operating capacity of the electrical stations is about 25GW, but the actual need is about 35GW. Therefore, Iraq suffers from a severe shortage of electricity supplies. some institutions, factories, and agricultural fields have used solar cells to avoid the shutdown. The Internet of Things simply means the network of physical things. The tracking of sunlight can be used to increase the efficiency of generating the electricity by using the internet of things. The collection of the sun's radiation that falls on the ground surface is not constant because it changes depending on the location, time, and climatic conditions. The solar panel can be checked by using the Internet of Things. Several procedures have been considered for the solar panel tracking system by using IoT. solar, wind, hydropower, and geothermal are much eco-friendlier than energy sources than the devices which are used now. Solar energy is more effective than any other green energy, and harmless to the environment. To achieve more productivity from the solar panel, various techniques have been studied, and one of them is a solar tracking device, which regulates the system by straiting it with the light, rendering the solar tracking is important to improve efficiency[6]. Tracking systems were chosen that depend on the Earth's physical features, efficacious and easy to use. In[7-9], the development of a highly efficient dual-axis tracker in solar panel tracking is discussed. The researcher used a light-dependent resistor (LDR), two controlling motors to move the panel directly toward sunlight. The tracker panel showed better performance and efficiency than a fixed panel. In[10], the authors suggested using a single-axis solar tracker to fill the deficit in the production of electrical energy in a city in Egypt. In[11-13], conventional solar energy systems can be classified into two main groups, Active and passive. Active solar tracking systems use electric motors and gears to increase the power of solar panels, while a passive tracking system is used to compress gas with high boiling point gases. This work was carried out by Khatoon and Kumar who created a prototype to realize and manage the system, keeping pace with modern applications and making it a smart system based on the technology of the Internet of things, thus achieving supervision of solar photovoltaic energy[14-17]. The researchers used LDR sensors with an Arduino microcontroller. Vinod Kumar and John, suggested solar tracker setup which prepared from mechanical and electrical construction system. The mechanical structure is used to control the Azimuth and Elevation motor of the solar follower depend on the angle collected from the website, while electrical structure used for control motor and evaluation the device data[18]. The Internet of things is considered as a bridge between two construction, which let all scheme completely automatic and make solar tracker precise, cheap[19, 20]. Diggavi, Thula, et al., suggested a mechanism to show the maximum power generation using a solar tracking system, display its results and monitor using the thing speak platform. The LDR sensor is connected to a Node MCU microcontroller and compares the display results and output of the solar panel in dual-axis condition[21]. In[22], A solar tracking system based on Arduino was proposed to track the maximum radiation incident on the surface of the board and could be obtained at any time during the days year.

2. The Requirements of the Proposed Work

2.1. LDR SENSOR

This is a kind of passive element, such as a resistor with resistance variation by the intensity of light. A photo resistor is a device that alters the conductivity of photo responsive materials. The electrical resistance of the photo resistor is lowered. The number of electrons is proportional to the photon frequency.

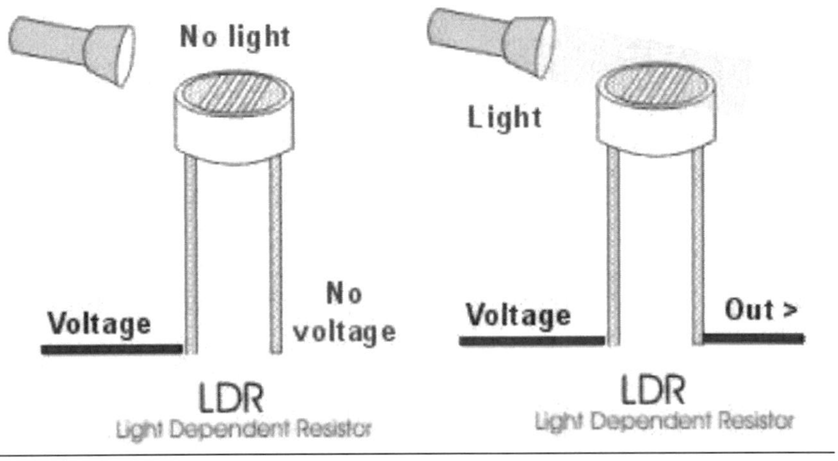

Figure 1. LDR CHIP

2.2. *Microcontroller ESP32S*

The ESP32 is a powerful, integrated Wi-Fi, a full TCP/IP stack, and Bluetooth 4.2. Due to the low cost, great power, and connectivity, the ESP32 is ideal for Internet of Things development. this chip was chosen because it decreases the charge of the checking system with high treating performances. ESP32 board based on Tensilica 32 - bit LX56 microprocessor, CPU Xtensa with 2 - core[23].

Figure 2. ESP32S board used as a controller3.3 Communication

2.3. *HTTP/ HTTPS (Hypertext Transfer Protocol)*

HTTP is a request and comeback protocol between a customer and a server. By using the HTTP server, devices can communicate with the Ask Sensors server. GET is one of the most commonly-used HTTP requests to request data from a specified resource and it enables you to write data to or read data from the Ask Sensors IoT platform.

2.4. *L298N Motor Driver*

Several techniques can be used to control a DC motor. We will use the L298N motor driver which can provide up to 3A of current at 35V. Additionally, it allows to run of two DC motors simultaneously and build a robot. The L298N integrated motor driver is shown in the following figure 4[24-26].

IOP Conf. Series: Earth and Environmental Science **877** (2021) 012026 doi:10.1088/1755-1315/877/1/012026

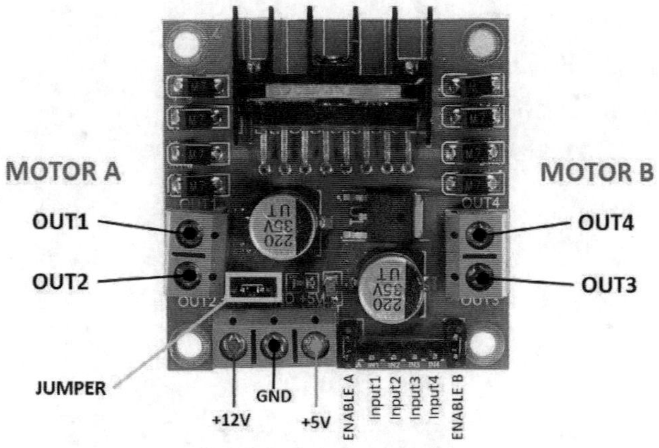

Figure 3. L298N DRIVER

2.5. Voltage Sensor

A voltage sensor is used to monitor the voltage in an item. Voltage sensors can detect both AC voltage or DC voltage. The input of this sensor can be a voltage signal and the output is an electrical switch.

Figure 4. A cross-section of a voltage sensor

2.6. Solar panel

Solar energy is a group of photovoltaic cells that generate electricity from sunlight and convert light into electricity from the sun in an industrial area, domestic and street lights, they are wide applications of solar panels.

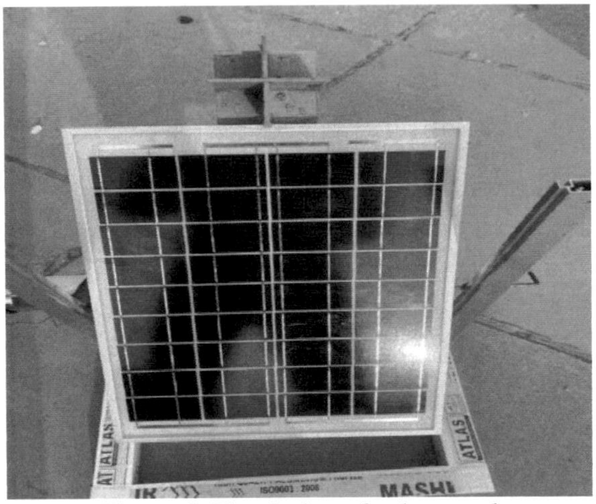

Figure 5. The design of solar panel

2.7. DC Motor Electric Gear Box (Gearbox 12V – 2A)

Table 1. Summary Specification of DC gearbox motor[27]

speed	260 rpm
Brand	chines
Voltage	12vdc
Current	2A
mounted	37mm

Figure 6. Gearbox motor device

2.8. Ask sensor IOT Cloud Platform
It is an IOT platform designed to be easy in the practical application of various IoT system projects, gather and envision your devices data simply, controller your actuators distantly, and achieve linked strategies in real-time. allowing users to communicate, visualize and analyse data coming from their sensors via the cloud. You can connect various controllers like Arduino, ESP32, or any other device in few minutes and start developing IoT applications[28-31].

3. Experiment Setup
This experiment can be used for monitoring the system tracking the output of the solar panel using ESP32S technologies, to optimize the system. It is done by using a 10W prototype shown in figure

Figure 7. prototypes of the sunlight tracker system

3.1 Flowchart

The flowchart of the PV tracker system is assumed in the procedure of the steps below as shown in figure (8) .

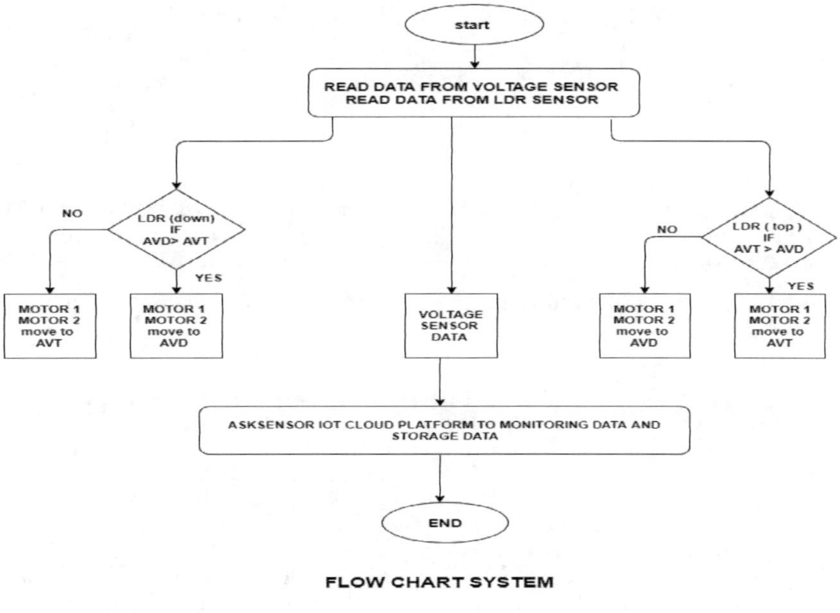

Figure 8. The flow chart of proposed work

The performance of solar panels has been monitored by using IoT. The sensor of the system using in the project, LDR & voltage sensor for detecting the sun location and transmit the data to Node MCU

ESP32. After the microcontroller progression the information from a sensor, L298N motor driver that hold PV, it moves toward the sun. The PV gathers the energy from the sun and send value to ESP32, and ESP32 sends data via the router to IOT shown in the figure 9.

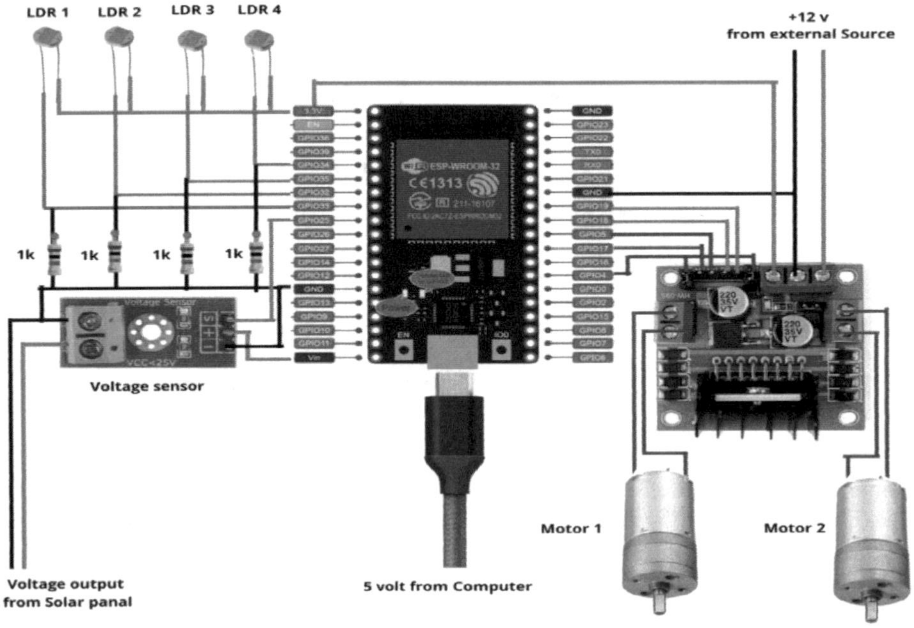

Figure 9. The connection map of the proposed circuit diagram

3.2. Results

After completing and connecting the experiment, the resulting curves are monitored by the cloud, and the 4LDR values are compared with the voltage sensor during real-time. Figure (10,11,12,13,14) show the curves of LDRs values of voltages VS time. Measuring from the solar panel, the parameter may be monitored from the serial monitor in Arduino IDE. The stream data made and saved in the Ask sensor platform.

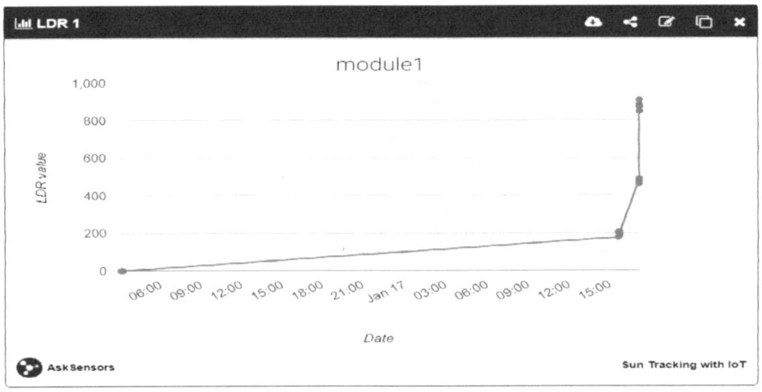

Figure 10. LDR1 vs Time

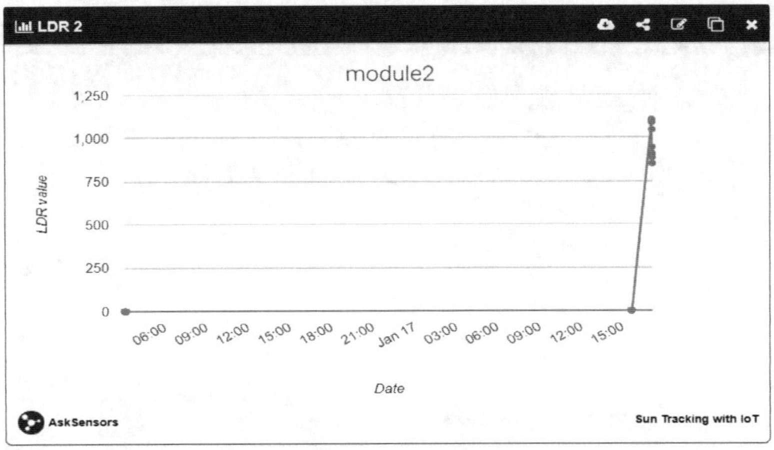

Figure 11. LDR2 vs Time

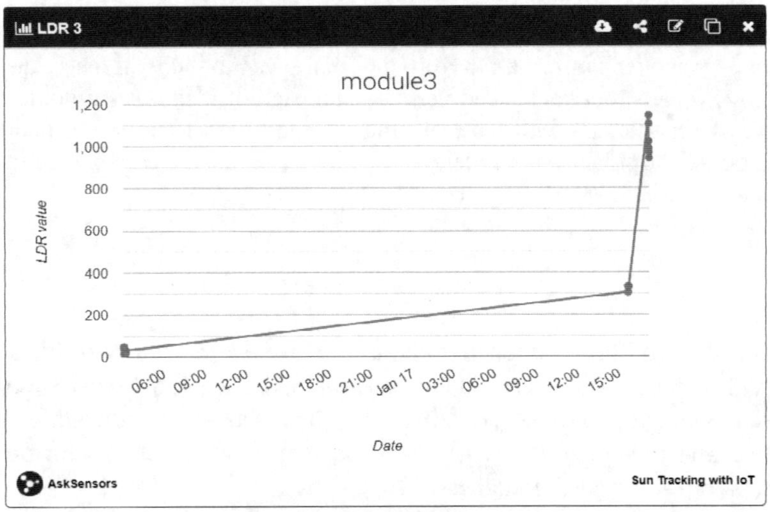

Figure 12. LDR3 vs Time

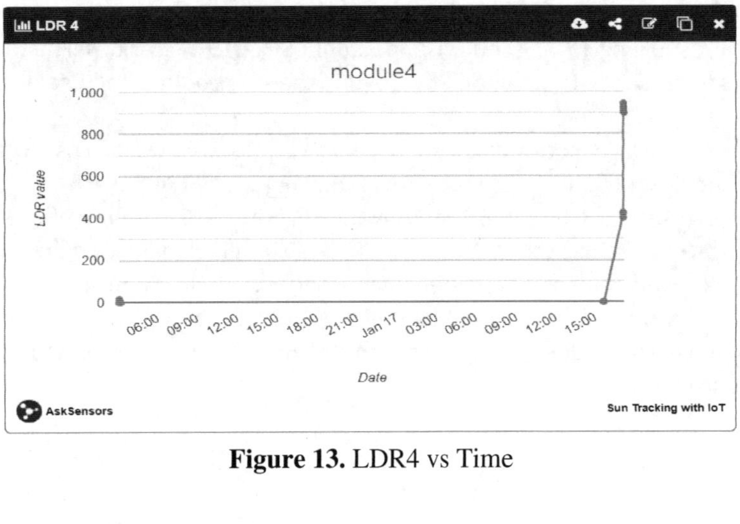

Figure 13. LDR4 vs Time

Figure 14. Voltage vs Time

4. Conclusion

Solar tracker has been successfully developing, and monitoring systems are equipped with a Node MCU ESP32 unit to transfer and read information from the solar panel to the IOT. There are two motors and are classified into two parts, horizontal and vertical, and each of them uses two motors to define the system's functions. Experimented with the real-time reading of the LDR values and compared to the voltages coming out of the solar panel. Finally, the solar tracker is the best choice in producing solar cells, in addition to being environmentally friendly.

References

[1] Infield D and Freris L 2020 *Renewable energy in power systems*: John Wiley & Sons)

[2] Tareen W U K, Anjum Z, Yasin N, Siddiqui L, Farhat I, Malik S A, Mekhilef S, Seyedmahmoudian M, Horan B and Darwish M J E 2018 The prospective non-conventional alternate and renewable energy sources in Pakistan—A focus on biomass energy for power generation, transportation, and industrial fuel **11** 2431

[3] Shaikh M R S 2017 A review paper on electricity generation from solar energy

[4] Herzog A V, Lipman T E, Kammen 2001 Renewable energy sources *Perspectives and overview of life support systems and sustainable development* **76**

[5] Khazaal H F, Alrikabi H T S, Abed F T and Kadhm S I 2019 Water desalination and purification using desalination units powered by solar panels *Periodicals of Engineering and Natural Sciences* **7** 1373-82

[6] Carballo J A, Bonilla J, Roca L and Berenguel M J S E 2018 New low-cost solar tracking system based on open source hardware for educational purposes **174** 826-36

[7] Zolkapli M, Al-Junid S, Othman Z, Manut A and Zulkifli M M 2013 High-efficiency dual-axis solar tracking developement using Arduino. In: *2013 International Conference on Technology, Informatics, Management, Engineering and Environment*: IEEE) pp 43-7

[8] Zubaidi S L, Hashim K, Ethaib S, Al-Bdairi N S S, Al-Bugharbee H and Gharghan S K J J o K S U-E S 2020 A novel methodology to predict monthly municipal water demand based on weather variables scenario

[9] Tuama H, Abbas H, Alseelawi N S and ALRikabi H T S 2020 Bordering a set of energy criteria for the contributing in the transition level to sustainable energy in electrical Iraqi Projects *Periodicals of Engineering and Natural Sciences* **8** 516-25

[10] Ahmed M S I, Saleh S M and Mohamed 2020 Proposed Design for Single Axis Photovoltaic Solar Tracker **45** 1-11

[11] Nadia A-R, Isa N A M, Desa M K 2018 Advances in solar photovoltaic tracking systems: A review, *Renewable and sustainable energy reviews* **82** 2548-69

[12] Li L, Li H, Xu Q and Huang W J R E 2015 Performance analysis of azimuth tracking fixed mirror solar concentrator **75** 722-32

[13] Abed F T, ALRikabi H T S and Ibrahim I A 2020 Efficient Energy of Smart Grid Education Models for Modern Electric Power System Engineering in Iraq. In: *IOP Conference Series: Materials Science and Engineering*: IOP Publishing) p 012049

[14] Khatoon F and Kumar S 2020 *Innovations in Electronics and Communication Engineering*: Springer) pp 299-309

[15] Ethaib S, Omar R, Kamal S M M, Awang Biak D R and Zubaidi S L J P 2020 Microwave-assisted pyrolysis of biomass waste: a mini review **8** 1190

[16] Al-dabag M, ALRikabi H S and Al-Nima R 2021 Anticipating Atrial Fibrillation Signal Using Efficient Algorithm *International Journal of Online and Biomedical Engineering (iJOE)* **17** 106-20

[17] Mohammed B, Chisab R and Alrikabi H 2020 Efficient RTS and CTS Mechanism Which Save Time and System Resources *international Journal of Interactive Mobile Technologies* **14** 204-11

[18] Hoffmann F M, Molz R F, Kothe J V, Nara E O B and Tedesco L P C J R e 2018 Monthly profile analysis based on a two-axis solar tracker proposal for photovoltaic panels **115** 750-9

[19] Vinodhkumar R and John S 2020 IoT based dual-axis solar tracking system. In: *IOP Conference Series: Materials Science and Engineering*: IOP Publishing) p 032024

[20] Alrikabi H, Alaidi A H and Nasser K 2020 The Application of Wireless Communication in IOT for Saving Electrical Energy *International Journal of Interactive Mobile Technologie* **14** 152-60

[21] Diggavi K C, Thula M and Pakkiraiah B 2020 *Innovations in Electrical and Electronics Engineering*: Springer) pp 569-78

[22] Ingole A N 2016 Arduino based solar tracking system. In: *Satellite Conference ICSTSD 2016 International Conference on Science and Technology for Sustainable Development. Kuala Lumpur, Malaysia. Retrieved from* https://vdocuments. *mx/arduino-based-sola r-tracking-system. html,*

[23] Taştan M and Gökozan H J A S 2019 Real-time monitoring of indoor air quality with internet of things-based E-nose **9** 3435

[24] Zhang J, Zhang L, An X, Luo H, Yao K 2016 Adaptive coupled synchronization among three coupled chaos systems and its application to secure communications **2016** 134

[25] Yahya O H, AlRikabi H T S, Al_Airaji R M and Faezipour M 2020 Using internet of things application for disposing of solid waste *International Journal of Interactive Mobile Technologies* **14** 4-18

[26] Naseer A H, Alaa A D, Haider Th.Salim Alrikabi, Faisal Theyab Abed 2021 Monitoring the Consumption of Electrical Energy Based on the Internet of Things Applications *International Journal of Interactive Mobile Technologies (iJIM)* **15**

[27] Legrouri A, Baird T and Fryer J J J o C 1993 Electron optical studies of fresh and reduced vanadium pentoxide-supported rhodium catalysts **140** 173-83

[28] Hala A. Naman N A H, Mohand Lokman Al-dabag, Haider Th.Salim Alrikabi 2021 Encryption System for Hiding Information Based on Internet of Things *International Journal of Interactive Mobile Technologies (iJIM)* **15**

[29] Imandasari T, Sadewo M G, Windarto A P, Wanto A, Wijaya H O L and Kurniawan R 2019 Analysis of the Selection Factor of Online Transportation in the VIKOR Method in Pematangsiantar City. In: *Journal of Physics: Conference Series*: IOP Publishing) p 012008

[30] Zubaidi S L, Abdulkareem I H, Hashim K S, Al-Bugharbee H, Ridha H M, Gharghan S K, Al-Qaim F F, Muradov M, Kot P and Al-Khaddar R J W 2020 Hybridised Artificial Neural Network Model with Slime Mould Algorithm: A Novel Methodology for Prediction of Urban Stochastic Water Demand **12** 2692

[31] Mohammed B K, Mortatha M B, Abdalrada A S, ALRikabi H T S 2021 A comprehensive system for detection of flammable and toxic gases using IoT *Periodicals of Engineering and Natural Sciences* **9** 702-11

Performance Evaluation of Al-Muamirah Wastewater Treatment Plant

Alia Haider Al-Wardy[1*], [2]Riyadh Jasim Mohammed Al-Saadi[1], Saif Salah Alquzweeni[3]

[1]Department of Civil Engineering, College of Engineering, University of Kerbala, Karbala, Iraq
[2]Department of Civil Engineering, College of Engineering, University of Kerbala, Karbala, Iraq
[3]Department of Civil Engineering, College of Engineering, University of Babylon, Babil, Iraq

[*]Corresponding author: m09141212@s.uokerbala.edu.iq

Abstract. Wastewater treatment is a process, which is being done on the wastewater to change its quality to be within the required specifications of water discharged to the surface waterway or water used for different purposes. In the present study, the performance of the oxidation ditch system related to Al-Muamirah wastewater treatment plant in Hilla city, province of Babylon, Iraq, in removing the pollutant of municipal were evaluated. The samples were taken from input wastewater and output treated water of the plant to measure the pollutant parameters. These parameters are the Biochemical Oxygen Demand (BOD_5), Chemical Oxygen Demand (COD), Total Suspended Solids (TSS), Ammonia (NH_3), Phosphorus (PO_4), and the quantitative measure of the acidity or basicity (pH). Data were analyzed by using Excel software. The results show that the efficiency of removal of BOD_5, COD, TSS, NH_3 and PO_4 was 91%, 78%, 93%, 69% and 68%, respectively on a monthly basis. Accordingly, it can be recommended that the Al-Muamirah wastewater treatment plant has an acceptable efficiency of wastewater treatment and producing water in accordance with Iraqi environmental standards of water discharged to the surface water or water of other uses.

Keywords: BOD; COD; Oxidation ditch; Performance evaluation; Removal efficiency; Wastewater treatment plant (WWTP)

1. Introduction

Water scarcity would affect about 9 billion people worldwide by 2075, according to forecasting research **[1]**. Many countries are suffering from the negative effects of climate change, which have resulted in decreasing freshwater resources. Iraq is located in the Middle East's eastern region, which is dry to semi-arid, On the other hand, because of the rising population, the need for freshwater is projected to rise. This adds to the pressure already exerted on water facilities. Furthermore, water contamination has grown substantially as a result of rapidly expanding industrial activities. This region, in general, has a water scarcity, which is likely to worsen in the future due to the effects of numerous causes, including climate

change, rapid population expansion, fast economic development, rises in per capita consumption and contamination. The Tigris and Euphrates Rivers are the major sources of freshwater in Iraq. From 2009 to 2014, these revisers faced severe water shortages, which are likely to worsen as a result of climate change and rising water demand upstream. Furthermore, in Iraq, where all settlements were centered on the Tigris and Euphrates rivers, freshwater supplies were a critical component in the emergence of cities to ensuring the flourishing of both industry and agriculture **[2]**. Additionally, several studies have shown that the continual release of wastewater into the environment is exacerbating the problem of water sacristy by polluting freshwater supplies **[1]**. As a result, municipal water security is critical for achieving a sustainable environment in modern cities, particularly in light of global warming and socioeconomic factors. Furthermore, most cities are near freshwater sources. Freshwater shortage is a typical challenge for policymakers for the reasons stated. Recently, the World Economic Forum confirmed that water scarcity is one of the largest international risks because of the limited amount of accessible freshwater (approximately 0.014% of the total amount of water on Earth). Other important causes that lead to water shortage include inadequate management of freshwater sources. As a result, water utilities should assist and improve municipal water system management **[3]**.
Water makes up 99.9% of wastewater, while the remaining 1% contains organic and inorganic contaminants, both dissolved and suspended, as well as microorganisms, and the wastewater treatment aims to remove these pollutants **[4]** .Several factors affect the wastewater characteristics, including the reason for which the water was used, the environment, the population's social and economic conditions, and others **[5]**. The general concept of biological wastewater treatment is that the water contains a large number of microorganisms that have the potential to reduce organic pollutants and nutrients in wastewater. The characteristics of wastewater and the purpose of the treatment process influence the selection of a suitable biological wastewater treatment technology **[6]**.
Wastewater treatment plants have great environmental importance, which requires maintaining their efficiency so that the concentrations of pollutants that discharge to the rivers after treatment should be within acceptable limits.**[7]** Wastewater treatment plants are traditionally designed using experimental equations and simplified systems specifications using references or guidelines that can be used for designing purposes and sizing treatment units by selecting the wastewater characteristics, operating conditions, and deciding the effluent requirements **[8]**.
Oxidation ditch (extended aeration) is a biological treating technique for activated sludge that removes biodegradable organics by using long solids retention times (SRTs). Usually, oxidation ditches are full mix system, but they can be changed to reach plug flow states **[9]**. A single or multichannel structure in an oval, circle, or horseshoe-shaped tank with mechanical aeration and mixing equipment is the typical design for oxidation ditch treatment systems. Water that has been screened and degritted enters the channel and is mixed with the activated sludge that has been returned. The tank design, as well as the aeration and mixing equipment, support unidirectional channel flow, allowing the energy used for aeration to provide mixing in a system with a long hydraulic retention period. The aeration/mixing system used produces a velocity in the channel of 0.25 to 0.30 m/s (0.8 to 1.0 ft/s), which is sufficient to keep the activated sludge suspended. The mixed liquor completes a tank circulation in 5 to 15 minutes at these channel velocities, and the magnitude of the channel flow can dilute the influent wastewater flowrate by a factor of 20 to 30. As the wastewater leaves the aeration zone, the DO concentration decreases and denitrification may occur downstream from the aeration zone. Brush-type or surface-type mechanical aerators are used for mixing and aeration **[10].** Hydraulic retention periods in the oxidation ditch range between 6 to 30 hours in most municipal wastewater treatment plants.**[9].** Small- and medium-scale wastewater treatment plants often use oxidation ditches. Simple construction, low sludge production, flexible operating mode, and special capability of nitrification and denitrifications in the same tank are all advantages of oxidation ditches. **[11]**. Because of the longer aeration period, the energy cost is higher in

this system than in other activated sludge processes**[12]**. In general, the process is divided into two aeration basins, one anoxic and the other aerobic. The first reactor, which operates under anoxic conditions, receives wastewater and returned activated sludge (RAS). The mixed liquor is then pumped into the second reactor, which is set up to operate under aerobic conditions. The process is then reversed, with the second reactor operating in an anoxic conditions**[9]**. The oxidation ditch is usually preceded by preliminary treatment such as bar screens and grit removal. While primary settling before an oxidation ditch is often used, it is not common in this design. Depending on the effluent conditions, tertiary filters can be needed after clarification. Prior to final discharge, disinfection is required, and reaeration may be required. The flow to the oxidation ditch is aerated and mixed with secondary clarifier return sludge. Figure 1 shows the sketch of the oxidation ditch process.

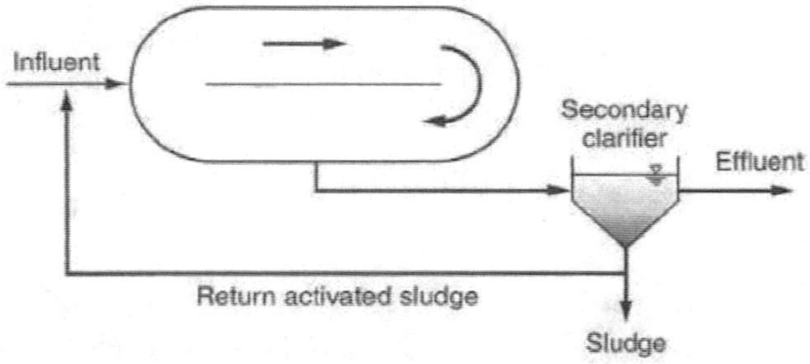

Figure 1. Sketch of Oxidation ditch process **[10]**

[13] Presented a study to evaluate a wastewater treatment plant in India working with a flow rate of about 23MLD as a conventional activated process and the results showed that the efficiency of the plant to remove BOD and TSS are 94.56% and 93.72% respectively. **[14]** Presented a study to evaluate a wastewater treatment plant in Egypt and the results showed that the efficiency of removal is more than 90% for COD, BOD, and TSS. This means that the performance of the plant was acceptable and in agreement with the specifications. **[15]** In a small-scale wastewater treatment facility, conducted an analysis to analyze and evaluate a biological treatment unit. They found that the effluent concentration of TSS, BOD, and COD was less than 10 mg/L,20 mg/L and 98 mg/L respectively, which means that the WWTP has a good level. The extended aeration method, which is commonly utillized to handle the wastewater in small communities, has the highest BOD removal rate among activated sludge systems, with an efficiency of 90-98 percent **[16]**. Because of its advantages, the extended aeration triggered sludge system is the most common method for safe wastewater treatment in a residential tower, resots, palaces, hostels, restaurants, and health centers, infirmaries, labor camps, and offices in manufacturers, companies, and organizations. **[17]**. **[18]** converted a conventional activated sludge system to extended aeration in order to improve removal efficiency. According to **[16]**, the COD removal efficiency of the extended aeration powered sludge system was 73.3 percent, and the linear alkyl benzene sulfonate removal efficiency was 96.7 percent. The average efficiency of removal of pollutants BOD_5, TSS, and COD , according to **[19]** was 57.7%, 70.8 percent, and 61.4 percent, respectively. **[14]** showed that the efficiency of removal is more than 90% for COD, BOD, and TSS. According to **[20]**, the efficiency of removal of COD, TSS , and BOD5 from industrial wastewater was 98.2 percent, 97.6 percent, and 96.66 percent, respectively.

The aim of our research is to evaluate the performance of the oxidation ditch system at eliminating pollutants of municipal wastewater in Al-wastewater Muamirah's treatment plant in the province of

Babylon. Up to our knowledge, this study is considered the first one on the performance evaluation of the oxidation ditch system in wastewater treatment in the region.

2. Case study

Al Muamirah WWTP was established in 2019. The total area of the treatment plant is 57 Acres, and it serves about 970,000 people in the city of Hilla, the capital of Babel Governorate, central Iraq. The plant operates with activated sludge system, oxidation ditch process. The treatment plant's discharge capacity is 107,000 meter cubic per day. The average daily input flow to the treatment plant is about 30000 meter cubic per day, the small input amount related to the incomplete pipelines conveying the sewers of Hilla. Al Muamirah WWTP plant shown in figure 2 is consists of four stages of treatment:

I. Preliminary (physical) treatment:
a) The primary lift station: It contains 4 submersible pumps with a capacity of 2,400 cubic meters for each pump to raise the water to the grit chamber facility.
b) Screens: It contains 3 fine screens and 3 coarse screens use for disposal of coarse and fine solid waste.
c) Grit chamber: It contains 2 sand removal basin, and 2 Archimedes pump used to remove sand and oil by a physical machine

II. Secondary (biological) treatment
a) Anaerobic tanks: A two closed basin each one contains two entry gates, one of which comes from the grit chamber and the other is attached to the return pipe. These ponds represent the first stage of the removal of nitrates and a portion of phosphorus by anaerobic bacteria.
b) Anoxic tanks: An open basin consists of three basins for each large aeration basin and works to reduce sedimentation and add dissolved oxygen to the water and contains devices for measuring dissolved oxygen content. These basins represent the aerobic treatment stage, which is the second stage of the nitrate removal process and at this stage, gas is released Nitrogen.
c) Oxidation Ditch: It consists of two main basins each basin is ovoid with a depth of approximately 6 meters and the capacity of each basin is 50,000 cubic meters. It consists of opposite arcs and separate channels. Each basin equipped with five oxygen addition motors and ten submersible mixers that reduce sedimentation in the basin, add oxygen, develop aerobic bacteria, and create turbulence in the water to increase its speed, i.e. increase sludge.
d) Sedimentation tanks: There are four sedimentation basins each 45 meters in diameter and containing a sweeping bridge. In these ponds, the sludge is disposed of for later treatment and disposal, as well as disposal of floating materials, while the treated water is discharged to the contact basins.
e) Collection and distribution basin: in which alum is added to increase sedimentation and contains three arcs, in which water is distributed to sedimentation basins, treated water is discharged into contact basins, and sludge is drained back.

III. Purification stage (chlorination): The sterilization process takes place in the contact basins, where chlorine is added to the treated water in these basins to kill microorganisms and microbes. There are two lines for the water to come out according to the specifications.

IV. Sludge treatment unit:
a) Holding tank: It is used for sludge collection and storage, it contains a mixer to prevent sedimentation, and there is an outlet to the squeezing stage.
b) Dewatering: The facility contains five squeezing machines, squeezing and condensing sludge, and speeding up fertilizer manufacturing.
c) Drying beds: It consists of 16 drying basins used to separate the sludge with filter media and sun drying.

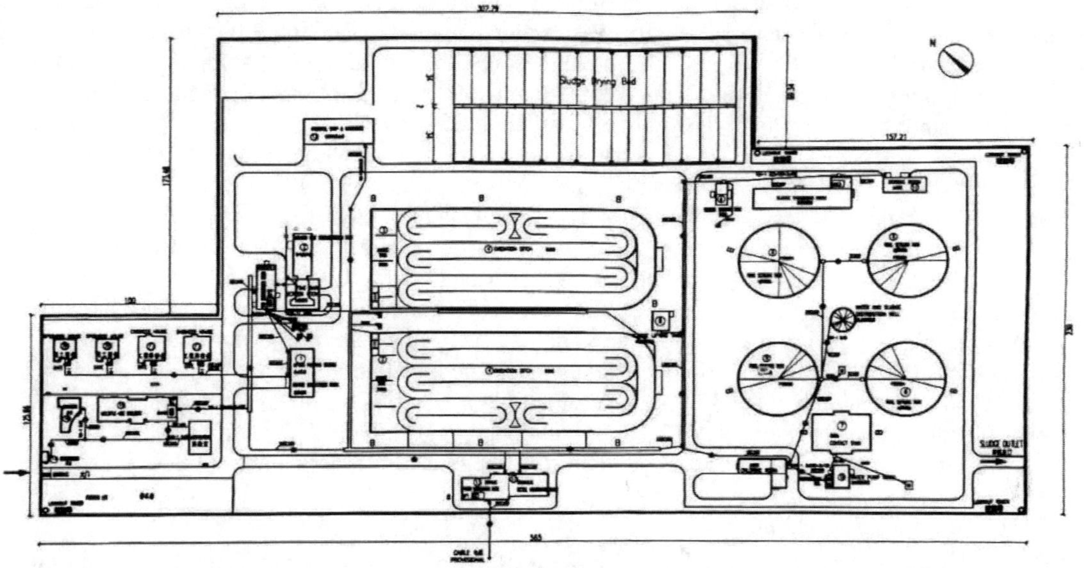

Figure 2. Schematic diagram of Al-Muamirah Wastewater Treatment

3. Data and methods

The purpose of our study was to evaluate the performance of Babylon's Al-Muamiarh WWTP over a one-year period (2020). The characteristics of raw wastewater have been studied to know the strength of pollutants before the treatment. Also, concentrations of effluent pollutants after treatment have been examined and compared with the upper limits of Iraqi standard. In addition, percentage removal of each pollutant should be calculated to know the efficiency of the plant. Samples were taken weekly for one year. A total of Forty-eight samples were taken. Two sampling points were identified are the wastewater treatment plant influent and the wastewater treatment plant effluent. The contamination parameters tested for these samples were the biochemical oxygen demand (BOD5), the chemical oxygen demand (COD), total suspended solids (TSS), Ammonia (NH3), Phosphorus (PO4), and the quantitative measure of the acidity or basicity (PH). The protocols required for the sampling process were considered by selecting the appropriate depth, the time of sampling, the equipment used and the transport to the laboratory. Water and wastewater testing standards **[21]** were followed during the experiments, which were carried out in the plant laboratory. Finally, the values of the tested parameters were analyzed by using Excel 2010 software.

4. Results and discussion

4.1. Raw wastewater assessment

For influent BOD, the average monthly value was 130 mg/L,while the maximum monthly value was 181 mg/L in December. These values are under the design value of Al-Muamirah WWTP (300 mg/L) which means there is no over organic loading during the study year. There is a slight variation in the concentration of influent BOD throughout the year due to the difference in climate and water consumption between months of the year as well as the operational conditions of the lifting stations that pump wastewater to the treatment plant **[22]**. For influent COD, the average monthly concentration was 217 mg/L, while the highest monthly amount was 328 mg/L in December which are all lower than the design value of Al-Muamirah WWTP (400 mg/L). This variation may be related to the characteristics of wastewater according to each season and the influent industrial wastewater from industrial regions that reach Al-Muamirah WWTP across the city wastewater network. **[22]**. For influent TSS, the average monthly concentration was 186 mg/L, while, the highest monthly amount was 344 mg/L in December and these values are less than the design value of the plant (350 mg/L). There is a high variation of TSS concentration among the months of the year and that could be related to the operational conditions of the lifting stations that pump wastewater to the treatment plant **[10]**. For influent NH3, the average concentration was 13 mg/L while the maximum monthly value was 18.5 mg/L in December. These values are less than the maximum design value of Al-Muamirah WWTP which was 20 mg/L. For influent PO4, the average monthly concentration was 6 mg/L which is lower than the design value of Al-Muamirah WWTP (7 mg/L), while the maximum monthly value was 11.6 mg/L in March that is more than the maximum design value of Al-Muamirah WWTP due to illegal industrial wastewater discharge to the sewerage network. In general, there is no high variation due to the stability of phosphorus concentration over the seasons **[23]**. The average influent pH was 7.68. The maximum monthly value of pH was 7.95 in June, these values are less than the design value of the plant which is 8. The average effluent BOD/COD ratio was 0.6, this is within the range (0.3 – 0.8) mentioned in **[10]**, the variation in values may be related to the characteristics of wastewater and the amount of biodegradable organic matter **[22]**.

4.2. Treated wastewater assessment

The average monthly effluent concentration of BOD was 12 mg/L, while the BOD maximum monthly concentration was 16 mg/L in January which are all matches the Iraqi standard (less than 40 mg/L). For effluent COD, the average monthly value was 45 mg/L, and the highest monthly concentration value examined for the COD was 80 mg/L in January which does meet with the Iraqi standard (under 100 mg/L). This indicates an effective performance of plant operation. For effluent TSS, the average monthly concentration value was 11 mg/L, and the maximum monthly concentration value examined for the TSS was 14.2 mg/L in November which are all meet with the Iraqi standard (less than 60 mg/L). This may be related to providing sufficient detention time in the sedimentation tanks and in the oxidation ditch. For effluent PO4, the average concentration was 2 mg/L, while the highest monthly concentration value examined for the PO4 was 2.6 mg/L in April which are all meet with the Iraqi standard (less than 3 mg/L). For effluent concentrations of NH3, the average was 4 mg/L, and the maximum monthly amount value examined for the NH3 was 7.33 mg/L in August which are all meet the Iraqi standard (less than 10 mg/L). There is no high variation in effluent pH and the average monthly value was 7.5, and the highest monthly concentration value examined for the PH was 7.7 in each of May and June It was within the permissible limit (6-9.5). The average effluent the BOD/COD ratio was 0.27 which is within the range (0.1-0.3) mentioned in **[10]**.

4.3. Removal efficiency of the plant

Figure 3 illustrates the average monthly removal efficiency of pollutant parameters in the inflow and outflow of wastewater of the treatment facility. The findings also revealed the efficiency of BOD5, COD, TSS, NH3, and PO4 removal over the course of the study year (2020) as a monthly average. The average monthly value of BOD removal efficiency is 91 % and this good removal may be due to the stability of microorganisms that are working with a good performance due to sufficient transferred oxygen in the aeration tanks **[10]**. The average monthly value of COD removal efficiency is 78% which can be considered stable but less than the requirements, this may be related to the nature of influent wastewater which contains high non-biodegradable organic matter **[23]**. The average monthly value of TSS removal efficiency is 93 %, this refers to that the plant is working with high efficiency probably due to the good operation of the sedimentation tanks and a sufficient hydraulic retention time (HRT) **[10]**. On the other

IOP Conf. Series: Earth and Environmental Science **877** (2021) 012027 doi:10.1088/1755-1315/877/1/012027

hand, the average monthly removal efficiency of nutrients NH3 and PO4 is 69 % and 68%, respectively. These efficiencies are of moderate values that may be related to the characteristics of influent wastewater. Briefly, all the studied pollutant parameters in the treated wastewater of Al-Muamirah WWTP have met the Iraqi standard of water discharged to surface water or water of other uses. This indicates that the oxidation ditche system work efficiently in removing the studied pollutant parameters. Accordingly, it can be concluded that this plant has a good performance in treating wastewater.

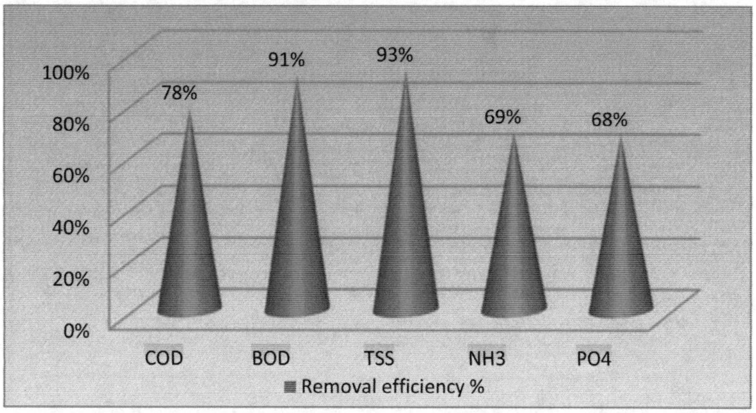

Figure 3. The average monthly efficiency of removal of the pollutant parameters studied in the treatment plant's wastewater.

Figures 4-9 shows the monthly variation of COD, BOD5, TSS, NH3, PO4, and PH, respectively for the inflow and outflow of wastewater of the treatment plant during the study period. Figure 10 shows the monthly variation of the BOD/COD ratio in the inflow and outflow of wastewater of the treatment plant. The average monthly values of the studied pollutant parameters of the inflow and outflow of wastewater of the Al-Muamirah wastewater treatment plant with a comparison of Iraqi standard are shown in Table 1.

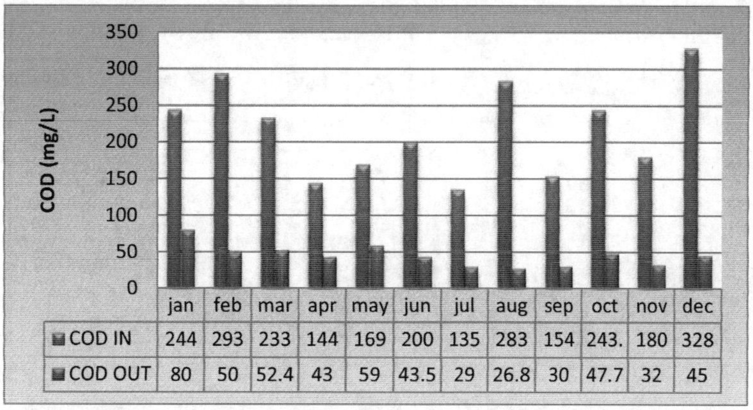

Figure 4. The monthly variation of COD in the inflow and outflow of wastewater of the treatment plant.

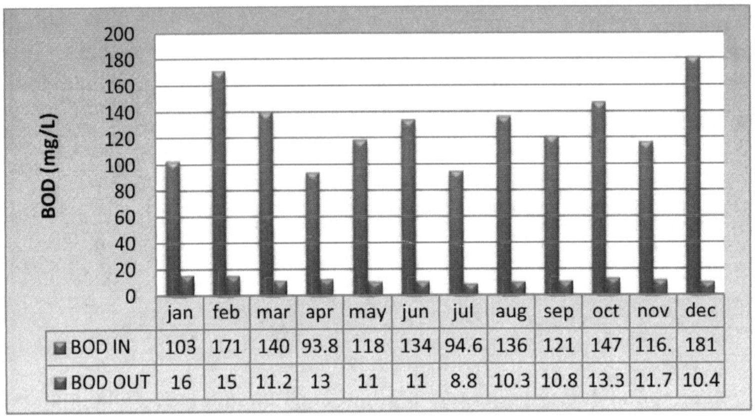

Figure 5. The monthly variation of BOD in the inflow and outflow of wastewater of the treatment plant.

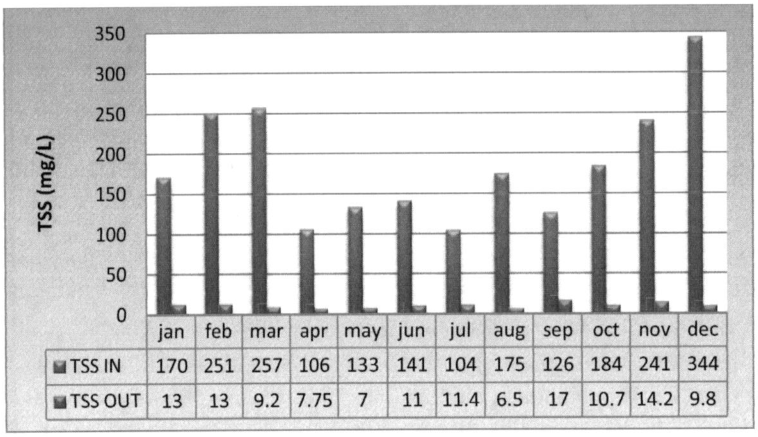

Figure 6. The monthly variation of TSS in the inflow and outflow of wastewater of the treatment plant.

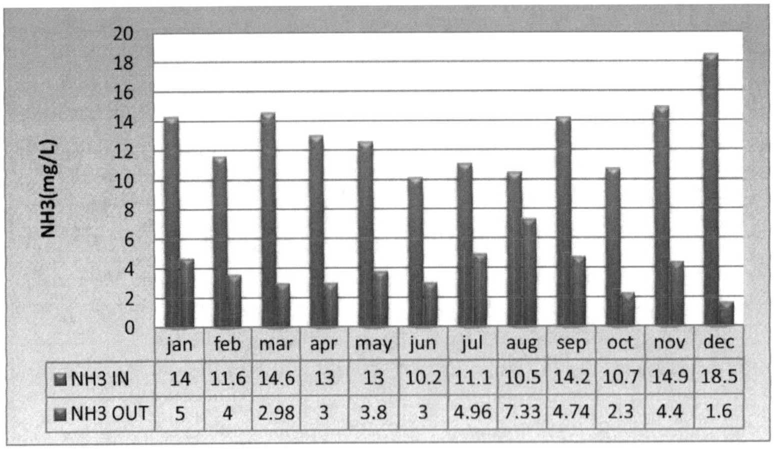

Figure 7. The monthly variation of NH3 in the inflow and outflow of wastewater of the treatment plant.

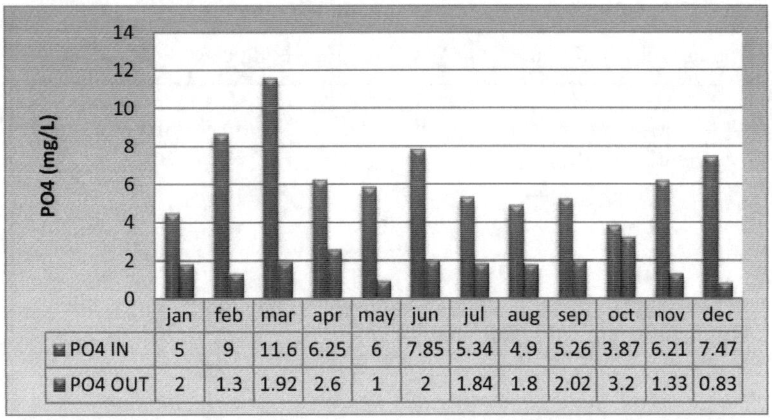

Figure 8. The monthly variation of PO4 in the inflow and outflow of wastewater of the treatment plant.

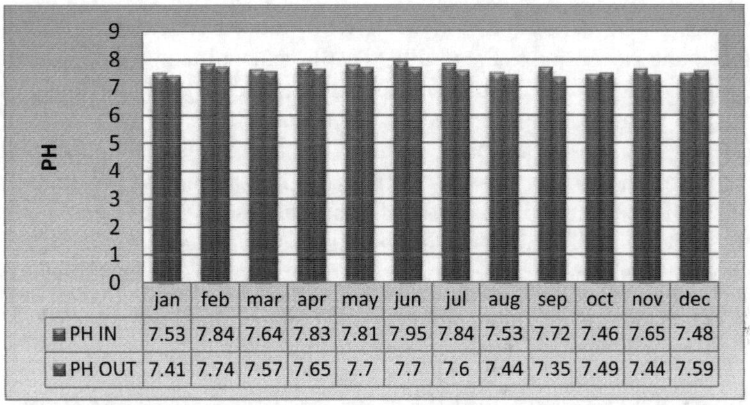

Figure 9. The monthly variation of PH in the inflow and outflow of wastewater of the treatment plant.

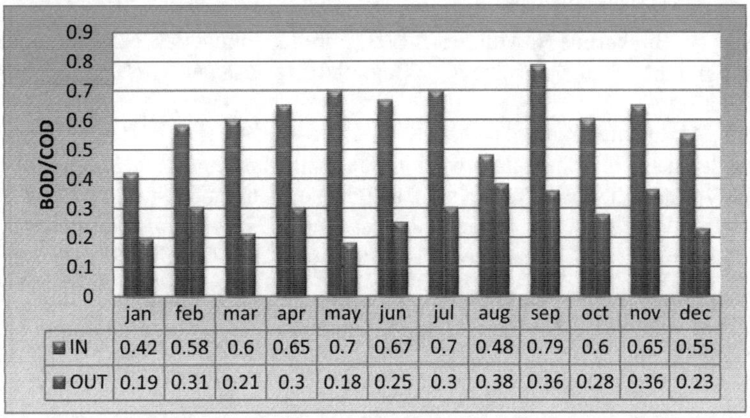

Figure 10. The monthly variation of the BOD/COD ratio in the inflow and outflow of wastewater of the treatment plant.

Table 1. The average monthly value of the studied pollutant parameters in the inflow and outflow of wastewater of the Al-Muamirah WWTP with a comparison of Iraqi standard.

Parameter	Inflow of wastewater (Influent)	Max. design of plant	Outflow of treated wastewater (Effluent)	Iraqi standard
BOD (mg/L)	130	300	12	40
COD (mg/L)	217	400	45	100
TSS (mg/L)	186	350	11	60
NH3 (mg/L)	13	20	4	10
PO4 (mg/L)	6	7	2	3
pH	7.68	8	7.5	6-9.5

5. Conclusion

The present research was conducted to assess the performance of Al-Muamirah WWTP. The pollutant parameters (COD, TSS, BOD5, NH3, and PO4) with pH parameter are examined for inflow and outflow of wastewater of this plant. According to the results, the efficiency of removal of BOD5, TSS, COD, NH3, and PO4 was 91 percent, 93 percent, 78 percent, 69 percent, and 68 percent, respectively, on a monthly basis. All of the studied pollutant parameters were considered to be below the Iraqi standard. Therefore, it can be advised that this plant has an efficient performance to remove the studied pollutant parameters. This means that the water leaving the station is suitable for discharge into the Euphrates River as well as suitable for non-human use such as agriculture. Additional studies should be carried out on evaluating the performance of the Al-Muamirah WWTP by measuring other pollutants and comparing them with the specification. Other studies should also be carried out on evaluating the performance WWTPs to other cities that operate with the same system.

References

[1] Zubaidi, S L, Hashim K, Ethaib S, Al-Bdairi N S S, Al-Bugharbee H, Gharghan S K 2020 A novel methodology to predict monthly municipal water demand based on weather variables scenario, *Journal of King Saud University - Engineering Sciences* 1018-3639

[2] Zubaidi S L, Al-Bugharbee H, Muhsin YR, Hashim K, Alkhaddar R 2020 Forecasting of monthly stochastic signal of urban water demand: Baghdad as a case study. *IOP Conf. Ser.: Mater. Sci. Eng.* **888** 012018

[3] Zubaidi S L, Abdulkareem I H, Hashim K S, Al-Bugharbee H, Ridha H M, Gharghan S K, et al. 2020 Hybridised artificial neural network model with slime mould algorithm: A novel methodology for prediction of urban stochastic water demand. *water MDPI* **12** 2692.

[4] Samer M. 2015 Wastewater treatment engineering: *BoD–Books on Demand*

[5] Henze M, Gujer W, Mino T, Matsuo T, Wentzel M, Marais G. 1995 Wastewater and biomass characterization for the activated sludge model no. 2: biological phosphorus removal. *Water Science and Technology*. **31** 13-23

[6] Rajasulochana P, and Preethy V 2016 Comparison on efficiency of various techniques in treatment of waste and sewage water–*A comprehensive review. Resource-Efficient Technologies,* **2** 175-184

[7] Awad S F and Khudhair B H 2020 Performance assessment of biological trickling filter for treating municipal wastewater using simulation gps-x model [thesis]: *University of Baghdad*

[8] Arif, A U A, Sorour M T, Aly S A 2018 Design and comparison of wastewater treatment plant types (activated sludge and membrane bioreactor), using GPS-X simulation program: case Study of Tikrit WWTP (Middle Iraq). *Journal of Environmental Protection*, **9** 636-651.
[9] US EPA. 2000 Wastewater Technology Fact Sheet: Oxidation Ditches (Doctoral dissertation, *US EPA, Washington, DC, USA*).
[10] Metcalf & Eddy Abu-Orf M, Bowden, G, Burton, F L, Pfrang W, Stensel H D, and AECOM (Firm). 2014 Wastewater engineering: treatment and resource recovery. *McGraw Hill Education.*
[11] Ammary B Y, & Radaideh J A 2005 Simultaneous nitrification and denitrification in an oxidation ditch plant. *Chemical and biochemical engineering quarterly*, **19** 207-212.
[12] Eslami H, Ghelmani S V, Salehi Vaziri, A, Hosseinshahi D, Ghaleaskari S, Talebi-Hemmatabadi P, Merajimoghadam T 2016 Comparing the efficiency of stabilization ponds and subsurface constructed wetland in domestic sewage treatment in city of yazd. *Journal of Water and Wastewater; Ab va Fazilab (in persian)*, **26** 100-106.
[13] Kumar K, Kumar P, Babu M 2010 Performance evaluation of waste water treatment plant. *International Journal of Engineering Science and Technology* **2** 7785-96.
[14] Hegazy M H, Gawad M A 2016 Measuring and Evaluating the Performance of a Wastewater Treatment Plant. *Proceedings of the World Congress on Civil, Structural, and Environmental Engineering, Paper No AWSPT* **111**.
[15] Mažeikienė A, Vaiškūnaitė R. 2018 Analysis and Assessment of Biological Treatment Processes in a Small-Scale Wastewater Treatment Plant. *Polish journal of environmental studies* **27**.
[16] Pirsaheb M, Khamutian, R, Khodadadian M (2014) A comparison between extended aeration sludge and conventional activated sludge treatment for removal of linear alkylbenzene sulfonates (Case study: Kermanshah and Paveh WWTP). Desalination and Water Treatment, *Taylor & Francis* **52** 4673-4680.
[17] Metcalf and Eddy 2003 wastewater engineering: treatment and reuse. *McGraw Hill. New York, NY*, **384**.
[18] Takdastan A, Kordestani B, Nisi A, JalilzadehYangjeh R (2016). Study of operational and maintenance problems and parameters of extended aeration activated sludge process in golestan hospital wastewater treatment plant, ahvaz, and their solutions. *Journal of Environmental Health Enginering*, **3** 270-279.
[19] Nikmanesh M S, Eslami H, Momtaz S M, Biabani R, Mohammadi, A, Shiravand, B, & Zarei Mahmoudabadi T. 2018 Performance evaluation of the extended aeration activated sludge system in the removal of physicochemical and microbial parameters of municipal wastewater: a case study of nowshahr wastewater treatment plant. *Journal of Environmental Health and Sustainable Development*, **3** 509-517.
[20] Zazouli M A, Ghahramani E, Ghorbanian AlahAbad M, Nikouie A, Hashemi M. 2010 Survey of Activated Sludge Process Performance in Treatment of Agghala Industrial TownWastewater in Golestan Province in 2007 *Iranian Journal of Health and Environment*, **3** 59-66.
[21] Association A P H, Association A W W, Federation W P C, Federation W E. 1912 Standard methods for the examination of water and wastewater: *American Public Health Association.***8**
[22] Ramadan A E M M, Abdel-Rahman A, Abdullah A M, Eltawab O A 2017 Evaluation of wastewater treatment plants in El-Gharbia Governorate, Egypt. *Organic Chemistry Current Research*, **6** 1-12.
[23] Pereira N C, Wang L K, Hung Y-T. 2009 Handbook of Environmental Engineering: Biological Treatment Processes **8**

Concrete Properties using Treated Recycled EPS

Hasan Jasim Mohammed[1], and Yasir Gaib Hussein[1]
Tikrit University, Civil Engineering Department, Collage of Engineering, Iraq

E-mail: dr.hassanjassim@tu.edu.iq, yasir.g.hussein42354@st.tu.edu.iq

Abstract. The study explores the mechanical properties of treated recycled extended polystyrene (TEPS) concrete, treated by two methods, one by heating, and the other by immersed recycled EPS in cement neat. By substituting 0 %, 15 %, 25 %, and 35 % of the coarse aggregate volume with treated recycled EPS, (for both method). Treated recycled TEPS concrete ratios are experimentally prepared, while the cement is substituted thru 10 % silica fume (SF). Tests were carried out, like compressive strength, splitting tensile strength, modulus of rupture, and density. The outcomes display the decreasing of the compressive strength, tensile strength and modulus of rupture of TEPS concretes with rise TEPS percentage around 26 %, 17 % and 32 %, respectively (35% TEPS) related to standard concrete. They also show that TEPS concrete density decrease about 30 % of normal concrete. The TEPS is suitable in concrete and meets provisions.

1. Introduction

Concrete materials comprising cement, fine aggregate, and coarse aggregate are popular materials, which, with an average density of 2300 kg /m^3, are considered concrete. Compression is concrete, efficient yet fragile in strain. Concrete has many advantages, comprising great elasticity, struggle to bending, and little creep, retraction, and permeability [1].

Several experts on the mechanical properties of concrete have carried out studies. A variety of influences, like the compressive strength of concrete and the tensile strength and density [2]. TEPS seems to be a very lightweight with excessive pressure resistance and water and excellent properties for cushioning [3-6].

EPS is commonly waste. To reduce the undesirable effects of this waste on the environment, a recycling approach is required, as shown in figure (1) [7].

The first use of EPS in concrete was as an artificial aggregate. Concern regarding the use of EPS concrete for various applications, such as wall panels and blocks [8-10]. In addition to impact resistance [11]; a confined structure, small water absorption and efficient sound as well as thermal isolation characteristics are shown in recycled EPS.

Replacement of coarse aggregate as a part volume without the EPS variation [12]; an experimental investigation is involved. Recycled EPS is an unwanted non-eco-friendly material from the processing of packaging and industrial advances. It poses a storage problem [13-14]. Moreover, causes environmental problems due to large amounts of retention and the long life of the substance and its chemically inactive existence. Using waste, EPS in the manufacture of concrete will solve this problem. It will help to solve the environmental crisis by seeking ways to use it in building.

[1] Hasan Jasim Mohammed.

Figure 1. EPS negative effects on environment

The study objective is to explore the mechanical properties of concrete containing treated recycled EPS. The concrete mixes utilized consist of varying ratios of TEPS portions as a part substitution of coarse aggregate, at quantities of (0%, 15%, 25%, and 35%). The mechanical properties like; density, compressive strength, modulus of rupture, and tensile strength of the TEPS concrete with new piece sizes and forms are studied.

2. Experimental Work and Materials
Type I of an ordinary Portland cement is utilized. Chemical and physical examination of cement in compliance with ASTM C150 / C150M-19a [15]; as shown in tables (1) and (2).

Table 1. Cement Chemical Characteristics.

Specifications	Results	ASTM C150
CaO	53.65	-
Al_2O_3	4.90	≤ 8%
SiO_2	15.2	≤ 21%
Fe_2O_3	3.0	≤ 5%
MgO	3.5	≤ 5 %
C_4	1.5	≤ 2.5 %
SO_3	3.2	≤ 4 %
Loss on Ignition(L.O.I)	0.5	≤ 1.5 %
Unsolvable	0.7	(0.66-1.02)
L.S.F	6.3	-
C_3S	3.2	< 5 %
C_2S	9.2	-
C_3A	50.2	-
C_4AF	53.65	-

A fine river aggregate (sand) was used. Sieve grading results and the chemical and physical properties experiments performed by the requirements of ASTM C778-17 [16]; are shown in tables (3) and (4).

Table 2. Cement Physical Characteristics.

Physical Characteristics	Test Results	ASTM C150
Particular shallow area (m^2/kg)	259 m^2/kg	(230 m^2/kg) lower limit
Setting period		
Initial:	2 hrs. 45 min	$\geq$ 45 min
Final:	8 hrs. 10 min	$\leq$ 10 hrs.
f'_c (MPa)		
at 3-day	22.68 MPa	$\geq$ 15 MPa
at 7-day	39.44 MPa	$\geq$ 23 MPa

Table 3. Sand outcomes of sorting check.

Size of sieve (mm)	Passing Cumulative (%)	Requirements
4.75	100	95-100
2.36	82	80-100
1.18	67	50-85
600	53	25-60
300	28	5-30
150	5	0-10

Table 4. Natural features of sand.

Features	Results	Requirements
Gravity of Specific	2.48	-
Humidity ratio	6.4 %	-
Immersion %	1.2 %	-
Unit Weight, kg/m^3	1520	-
SO_3, %	0.05%	< 0.5
Material Smaller 0.075 mm	4.4%	< 5

A normal coarse aggregate (gravel) is utilized, measuring a dimension of 12.5 mm. The experiment outcomes of the gravel features agreeing with ASTM C33 / C33M-18 [17]; specifications are presented in tables (5-7).

Table 5. Coarse aggregate sieving test.

Size of sieve	Passing Cumulative (%)	Limits
20 mm	100	100
14 mm	95	90 – 100
10 mm	55	50 – 85
5 mm	4	0 – 15
2.36 mm	1	0 – 5

Table 6. Coarse aggregates physical test.

Features	Results	Requirements
Gravity of Specific	2.65	-
Moisture content	5.3%	-
Absorption (%)	0.6%	-
Unit Weight (Dry), kg/m^3	1720	-
SO_3, %	0.05	$\leq$ 0.1

Table 7. Coarse aggregates chemical test.

Features	Results	Requirements
SO_3, %	0.05	-
Material < 75 µm%	0.04	$\leq$0.1
Salts	0.5	-

IOP Conf. Series: Earth and Environmental Science **877** (2021) 012028 doi:10.1088/1755-1315/877/1/012028

Reused TEPS; It's generated in the shape of cuts. In this analysis, cutting of recycled EPS by hand to parts in incremental dimensions to achieve a similar CA element in volume occupation and cement mortar restriction [18-20]. Then cured by two methods, first by warming: Put EPS cubes in an oven at 130 °C for 15 minutes [21]. Secondly by submerging EPS in cement neat (2 hours) earlier by means of in the concrete mixture, as presented in figure (2).
Recycled TEPS's sieve analysis reveals that it coincides with coarse aggregate grading. Following ASTM C33 / C33M-18 [17]; the sieving and unit weight of the TEPS aggregate were determined after cutting. As shown in table (8), a density of 14 kg/m^3 was measured.

Table 8. Sieving of TEPS test.

Size of sieve	Passing Cumulative (%)	Requirements
20 mm	100	100
14 mm	95	90 – 100
10 mm	60	50 – 85
5 mm	4	0 – 15
2.36 mm	2	0 – 5

Figure 2. a) EPS treated by heated

Figure 2. b) EPS immersed in cement neat

For the corresponding cement, SF is utilized as a substitution. SF increases the linking between the cement paste and TEPS, that lead enlarged the strength of concrete [4-6]. SF was applied to the mix (substituting 10 % of the weight of cement).

The amounts of mixture data used herein are shown in the table (9). The mixes contain replacing 0 %, 15 %, 25 %, and 35 % of the volume of the CA thru reused TEPS (two methods of treatment). By means of the similar concrete mixture design approach (1: 1.41: 1.98) suggested by the ACI code necessities [22]; the normal concrete was prepared. The unit weight of the concrete mix was 2321 kg/m^3, (W/C=0.39), and (f'_c= 35 MPa), using a superplasticizer 4.9 kg/m^3.

Table 9. Materials amounts in mixes (kg/m^3).

Mix designation	Cement	SF	Sand	Coarse Aggregate	Water	Treated EPS (kg)		% Replaced TEPS
						Heating	Cement mortar	
NC	490	0	680	960	191	0	0	0
TH15	441	49	680	816	153	18.228	0	15
TH25	441	49	680	720	153	30.38	0	25
TH35	441	49	680	624	153	42.532	0	35
TC15	441	49	680	816	153	0	0.932	15
TC25	441	49	680	720	153	0	1.553	25
TC35	441	49	680	624	153	0	2.174	35

SP : 4.9 kg/m^3

They measured the gravel and sand first. The blending was carried out in a mixer (0.05 m^3 volume). In the mixer, ordinary concrete was mixed by putting fine and coarse aggregate. Whereas the blender was working and the mixing nonstop for one minute, the water contributed about one-third of the amount. Then the cement was applied and, after three minutes of mixing, water was discharged steadily keen on the mixer. At a point where a uniform mixture had been collected, the process was stopped. After that the fresh unit weight and were assessed.

In the mixing of materials, a special sequence was used for TEPS concrete mixes. First, inside the mixer, about 40 % of the water with SP was dispensed. After that, TEPS pieces were extra and then mixing nonstop around two minutes to confirm complete water and superplasticizer wetting. Then, the other materials are putting into the mixer and gradually applied the remaining water. While, the mixing was in motion to achieve, a consistent mixture and smooth disperse TEPS pieces. Three to five minutes of mixing continued, and then the fresh densities and slump were assessed.

For testing the various parameters, standard test specimens were named. Under ASTM C143 / C143M-15a [23]; a slump test was conducted for all forms of mixes. Agreeing to ASTM C39/C39M-18 [24]; at 28 days of curing, six cylinders with a diameter of 150 mm and a height of 300 mm were used to classify the compressive strength. Acceptable to calculate the splitting tensile strength (ASTM C496 / C496M-17) [25]; and to assess the concrete density (ASTM C29 / C29M-17a) [26]; as shown in figure (3), the cylinder size was 150 mm diameter x 300 mm height. To calculate the rupture module, prism specimens of sizes 100 to 100 to 500 mm were cast. The test was performed in compliance with ASTM C78-02 [27].

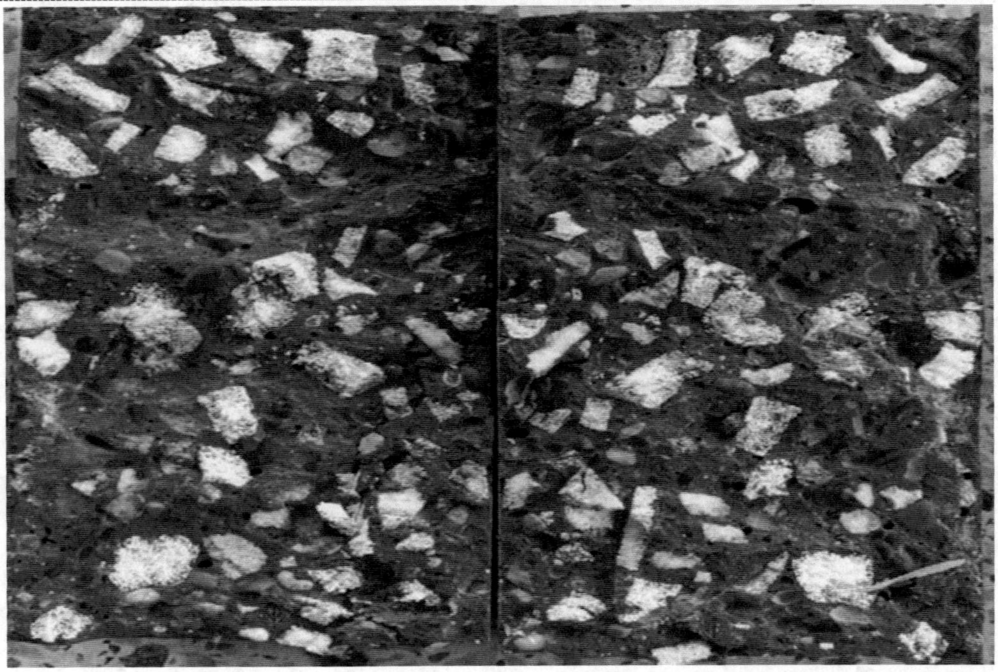

Figure 3. Splitting tensile strength test (sample failure)

3. Results and Discussion

The results of all forms of concrete mixes are shown in table (10). A normal concrete mixture was utilized in this analysis such as a guide intended for comparison. For the other mixtures, a main variable was used, which substituted the usual coarse aggregates with a recycled TEPS material. Moreover, silica fume was utilized to recover the compressive strength. In addition, super-plasticizer was used to improve the workability of TEPS concrete.

The compressive strength decreased when the ratio of TEPS in the mixes enlarged (TH15, TH25, TH35, TC15, TC25 and TH35). The reducing is because of the features of TEPS compared with natural gravel, as it is brittle.

Voids and pores of TEPS mix associated to ordinary concrete that is extra cause lead the reduction in concrete compressive strength. Experimental results are similar to the results getting in some studies [4-5, 18]. ACI code was adopted 17 MPa compressive strength. So, the all types of TEPS concrete could be used for structural considerations [28]. This performance also happened in the tensile strength of concrete. The density as reduced to reach the lightweight limits, due to the little weight of the recycled TEPS particles more than natural aggregate. Moreover, the modulus of rupture was decreased about 32%, at the TH35 concrete.

Table 10. Concrete mixes results.

Concrete Mix designation	Slump (mm)	Fresh density (kg/m³)	Hardened density (kg/m³)	Tensile stress (MPa)	Compressive strength (MPa)	Modulus of rupture (MPa)
NC	40	2430	2382	3.89	39.33	10.3
TH15	55	2234	2120	3.79	38.757	9.9
TH25	76	2150	2032	3.65	33.24	8.8
TH35	94	2075	2008	3.39	30.11	7.0
TC15	62	2200	1920	3.68	38.23	10.0
TC25	85	2112	1820	3.43	31.85	9.4
TC35	100	2033	1665	3.22	29.15	9.3

4. Conclusions

A test of the concrete mechanical properties containing treated recycled EPS was provided for the present experimental work. This represents a recent trend of experiments utilizing waste and rubbish as substitute materials for the concrete's existing gravel. The following findings were discovered:

1. Growing in the ratio of TEPS substitution results in an increase in the results of the slump. This occurred because of the super-plasticizer effect, seven to eight minutes, especially with mix time addition.
2. Up to a 30 %, concrete density decrease of the normal mixture density can be achieved by decrease by considered quantities of coarse aggregate and replacing it with the lighter recycled TEPS.
3. As a partial substitute for gravel, the recycled TEPS content in the concrete mix reductions the unit weight (density) and compressive strength (28 days) about 30 % and 26%, respectively, relative to the normal concrete.
4. SF effects on growing compressive strength and density is the most essential contribution to the reasonably great density. The recycled TEPS compressible formal by casting causes a rise in material quantities to pay for the decrease in concrete volume. Interaction between the materials in the mix during the mixed methods.
5. Recycled TEPS material has a significant influence on the splitting tensile strength of concrete, presenting a reduction of about 17 %, whereas at the maximum substitution ratio with recycled TEPS, the modulus of rupture is decreased by up to 32%.

5. Acknowledgments

The study is conducted in Tikrit University. The writers desire to be thankful the university professionals for their help on the study.

6. References

[1] Djamaluddin R 2013 Flexural behaviour of external reinforced concrete beams *Procedia Engineering* **54** 252-260

[2] Arezoumandi M, Smith A, Volz JS and Khayat KH 2015 An experimental study on flexural strength of reinforced concrete beams with 100% recycled concrete aggregate *Engineering Structures* **88** 154-162

[3] Canakci H, Alak D and Celik F 2016 Evaluation of Shear Strength Properties of Modified Expanded Polystyrene Aggregate *Procedia engineering* **161** 606-10
[4] Babu KG, and Babu DS 2003 Behaviour of lightweight expanded polystyrene concrete containing silica fume *Cement and Concrete Research* **33** (5) 755-62
[5] Sadrmomtazi A, Sobhani J, Mirgozar M and Najimi M 2012 Properties of multi-strength grade EPS concrete containing silica fume and rice husk ash *Construction and Building Materials* **35** 211-19
[6] Chen B and Liu N 2013 A novel lightweight concrete-fabrication and its thermal and mechanical properties *Construction and building materials* **44** 691-98
[7] greenmax-machine.com 2019 ttps://greenmax-machine.com/.
[8] Ling I and Teo D 2011 Properties of EPS RHA lightweight concrete bricks under different curing conditions *Construction and Building Materials* **25** (8) 3648-55
[9] Xu Y, Jiang L, Xu J and Li Y 2012 Mechanical properties of expanded polystyrene lightweight aggregate concrete and brick *Construction and Building Materials* **27** (1) 32-38
[10] Fernando P, Jayasinghe M and Jayasinghe C 2017 Structural feasibility of Expanded Polystyrene (EPS) based lightweight concrete sandwich wall panels *Construction and building materials* **139** 45-51
[11] Mohammed HJ and Zain M 2016 Experimental application of EPS concrete in the new prototype design of the concrete barrier. *Construction and Building Materials* **124** 312-42
[12] Herki BMA and Khatib JM 2016 Structural behaviour of reinforced concrete beams containing a novel lightweight aggregate https://wlv.openrepository.com/handle/2436/609863.
[13] Tamut T, Prabhu R, Venkataramana K and Yaragal SC 2014 Partial replacement of coarse aggregates by expanded polystyrene beads in concrete *Int J Res Eng Technol* **3** (02) 238-41
[14] Mohammed HJ and Omar KA 2020 Flexural behavior of reinforced concrete beams containing recycled expandable polystyrene particles *Journal of Building Engineering* **32** 101805
[15] ASTM, C150M-19a C 2019 *Standard specification for Portland cement* www.astm.org.
[16] ASTM, C778-17 2017 *Specification for standard (sand) fine aggregate* www.astm.org.
[17] ASTM, C33M-18 C 2018 *Specification for standard concrete aggregates* www.astm.org.
[18] Babu DS, Babu KG and Wee T 2005 Properties of lightweight expanded polystyrene aggregate concretes containing fly ash *Cement and Concrete Research* **35** (6) 1218-23
[19] Chen B and Liu J 2004 Properties of lightweight expanded polystyrene concrete reinforced with steel fiber *Cement and Concrete Research* **34** (7) 1259-63
[20] Cadere CA, Barbuta M, Rosca B, Serbanoiu AA, Burlacu A and Oancea I 2018 Engineering properties of concrete with polystyrene granules *Procedia Manufacturing* **22** 288-93
[21] Kan A. and Demirbog˘a R 2009 A new technique of processing for waste-expanded polystyrene foams as aggregates *Journal of Materials Processing Technology* **19** 2994-00
[22] ACI, 211 C 1991 *Standard Practice for Selecting Proportions for Normal, Heavyweight, and Mass Concrete*
[23] ASTM, C143M-15a C 2015 *Standard test method for slump of hydraulic-cement concrete* www.astm.org.
[24] ASTM, C39/C39M-18 2018 *Standard test method for compressive strength of cylindrical concrete specimens* www.astm.org.
[25] ASTM, C496M-17 C 2017 *Standard test method for splitting tensile strength of cylindrical concrete specimens* www.astm.org.
[26] ASTM, C29M-17a C 2017 *Standard test method for bulk density ("unit weight") and voids in aggregate* www.astm.org.
[27] ASTM, C78–02 2002 *Standard test method for flexural strength of concrete (using simple beam with third-point loading)* www.astm.org.
[28] ACI, 318M-14 2014 *Building code requirement for structural concrete*

Properties of Concrete Containing Recycled Demolition Aggregate

Hasan Jasim Mohammed[1], **and Zeina Saad Sabir**[2]
Tikrit University, Civil Engineering Department, Collage of Engineering, Iraq

E-mail: dr.hassanjassim@tu.edu.iq, zeenasaad1986@yahoo.com

Abstract. The purpose of this research is to investigate the properties of Recycled Demolition Aggregate (RDA) concrete. Five RDA concrete ratios are prepared experimentally by substituting, 0%, 25%, 50%, 75% and 100% of the gravel weight with RDA. While, the 10% of cement is substituted by silica fume (Si). Adding steel fibres (SF) (0.5 %, 1.0 % and 1.5 %). Treated RDA with cement mortar and superplasticizer (SP) admixture added to (1%) of total cementitious materials (TCM). The concrete properties exams performed such as; density, compressive strength, splitting tensile strength, and modulus of rupture. The tests concluded that the compressive strength, splitting tensile strength and rupture modulus values of RDA concretes are reduced with an increased RDA ratio relative to normal concrete. Density of RDA concrete reduces around 9% of normal concrete. The RDA is suitable in concrete and meets specifications.

1. Introduction

Natural materials containing coarse aggregate, fine aggregate and cement, with an average density of 2300 kg/m^3, are traditional materials considered to be concrete. Concrete is strong of compression but weak of tension. Concrete has many advantages, including high elasticity, resistance to bend, and low creep, shrinkage and permeability [1]. Several investigators of the properties of concrete have carried out studies. A variety of influences, such as the compressive strength of concrete and the tensile strength and density [2]. Recycled demolition aggregate (RDA) is the aggregate that is produced and processed from demolition waste after collapse. A cheap and widespread material requires a solution for recycling to minimize its negative effects on the environment, as presented in figure (1). For both structural and non-structural construction, RDA concrete may be used. As a means of economic practicality and environmental realization, the use of RDA in construction work as structural grade concrete can occur alongside the greater reduction of waste materials [3].

Due to its weak distribution of particle size, RDA is typically weakly graded. Because of crushing and handling using various kinds of grinders, it could be too coarse or too fine. In RDA, the quantity of finer fractions is greater. Attributed to the existence of the old attached paste or cement mortar surrounding it, it has an old adsorption transition zone [4]. RDA has lower mechanical properties than natural aggregate [4-6]; such as low crushing strength, low impact resistance and lower abrasion resistance. RDA, which is produced by the treatment of demolition of waste construction, has been considered as a probable substitute for natural aggregate. Nonetheless, RDA has a higher porosity, lower density, higher water absorption, lower resistance to abrasion and higher layer content [7-8].

[1] Hasan Jasim Mohammed.

Figure 1. Demolition waste

Large tons of waste concrete are produced by many countries; such a massive amount of waste has a significant effect on the environment [9-10]. Thereby, waste concrete recycling, which can solve this problem and conserve landfill space, protect natural resources and reduce the release of carbon. In Iraq, from 2014 to 2017, several buildings were demolished or exploded during military activities. Due to the very large amount of natural resources used during concrete processing, the accomplishment of waste concrete is certain. It could lead to a major reduction in the inactive waste that is sent to landfills, and natural resources will be safe in effect. The arrangement of recycled aggregate (RA) with mortar on their surface enhanced a significant part of this waste, resulting in high water absorption, decreased strength and increased capacity to penetrate chloride [11-12].

This study purposes to explore the mechanical properties of RDA concrete. As a partial substitute of gravel, the concrete samples used include various RDA ratios, at weights of 0 %, 25 %, 50 %, 75 % and 100 %. Addition of steel fibres (0.5 %, 1.0 % and 1.5 %) and cement mortar treated RDA. Partial silica fume replacement by cement weight (10%) and superplasticizer (SP) mixture added to (1%) of total cement materials (TCM). The RDA concrete properties are studied, for example density, compressive strength, splitting tensile strength and modulus of rupture.

2. Materials and Experimental Work

The cement was used an ordinary Portland (Type I). The cement tests on the (chemical and physical) conducted agreeing to the ASTM C150 / C150M-19a [13]; as demonstrated in tables (1) and (2). A river fine aggregate (sand) was used. The experiments results of the grading, chemical and physical properties, accomplished agreeing to specifications ASTM C778-17 [14]; are displayed in tables (3-4). A normal coarse aggregate (gravel) is utilized, measuring a dimension of (12.5) mm. Tables (5–7) show the experiment outcomes of the gravel features agreeing to reference [15].

Coarse RDA was sourced from a building demolition and replaced natural aggregate at five weight ratios: 0 %, 25 %, 50 %, 75 % and 100 %. RDA was a mix of pieces of brick, ceramic, tile and concrete. The used coarse RDA is illustrated in figure (2). Furthermore, the coarse RDA pass on through several stages before using are collecting, crashing and grading, then the material and sand separation and sieve analysis of particle size (5-12 mm), as figure (3) and table (8).

Silica fume (Si) utilized as replacement for further cement material. The Si was supplementary into the mix (substituting 10% of the weight of cement) to rise the attachment between RDA and the cement paste that improved the concrete strength [10-12]. Steel fibres with two bent ends were used, The steel fibres (SF) added to the concrete composition at three weight contents 0.5 %, 1 % and 1.5 % with a diameter 0.55 mm and length 35 mm. Steel fibre conforms to ASTM-A820 TYPE-1 [16]. The properties of the SF is described in table (9).

Table 1. Cement Chemical Characteristics.

Configuration	Results	ASTM C150
CaO	32	-
Al_2O_3	3.5	$\leq 8\%$
SiO_2	18	$\leq 21\%$
Fe_2O_3	4	$\leq 5\%$
MgO	3	≤ 5 %
C_4	2	≤ 2.5 %
SO_3	1	≤ 4 %
L.O.I	0.75	≤ 1.5 %
Inexplicable Material	0.85	(0.66-1.02)
L.S.F	8.28	-
C_3S	3.5	< 5 %
C_2S	11.16	-
C_3A	73.1	-
C_4AF	32	-

Table 2. Cement Physical Characteristics.

Physical Structures	Test Outcomes	Limits
Definite surface area (m^2/kg)	289 m^2/kg	(230 m^2/kg) lower limit
Setting period		
Initial:	3 hrs. 15	≥ 45 min
Final:	8 hrs. 35	≤ 10 hrs.
Compressive strength (MPa)		
at 3-day	18.32 MPa	≥ 15 MPa
at 7-day	30.22 MPa	≥ 23 MPa

Table 3. Sand results of grading test.

Size of sieve (mm)	Passing Cumulative (%)	Requirements
4.75	98	95-100
2.36	83	80-100
1.18	58	50-85
600	52	25-60
300	17	5-30
150	5	0-10

Table 4. Features of sand.

Features	Results	Requirements
Gravity of Specific	2.58	-
Moisture content	6.8 %	-
Absorption %	1.1 %	-
Unit Weight (Dry), kg/m^3	1520	-
SO_3, %	0.08%	0.5 (max. value)
Solid < 0.075 mm	3.2%	5 (max. value)

IOP Conf. Series: Earth and Environmental Science **877** (2021) 012029 doi:10.1088/1755-1315/877/1/012029

Table 5. Coarse aggregate sieving test.

Size of sieve	Passing Cumulative (%)	Limits
20 mm	100	100
14 mm	95	90 – 100
10 mm	55	50 – 85
5 mm	2	0 – 15
2.36 mm	0	0 – 5

Table 6. Coarse aggregates physical test.

Features	Results	Requirements
Gravity of Specific	2.64	-
Moisture content	6.3%	-
Absorption (%)	0.7%	-
Unit Weight (Dry), kg/m^3	1730	-
SO_3, %	0.06	≤ 0.1

Table 7. Coarse aggregates chemical test.

Features	Results	Requirements
SO_3, %	0.05	-
Material < 75 µm%	0.04	≤ 0.1
Salts	0.5	-

Figure 2. RDA used in concrete mixtures

IOP Conf. Series: Earth and Environmental Science **877** (2021) 012029 doi:10.1088/1755-1315/877/1/012029

Figure 3. Sieve analysis of RDA particles

Table 8. Sieving of RDA test.

Size of sieve	Passing Cumulative (%)	Requirements
20 mm	100	100
14 mm	100	90 – 100
12.5 mm	84	50 – 85
5 mm	0	0 – 15

Table 9. Features of steel fibre.

Commercial name	Configuration	Property	Specifications
BUNDREX	**Hooked**	**Density**	2740 kg/m^3
		Colour	Grey
		Standard	ASTM
		Length	35 mm
		Nominal diameter	0.55 mm
		Aspect ratio (Lf/Df)	65

Superplasticizer used to achieving the required workability, the W/C need be certainly reduce to a minimum. A superplasticizer (SP) (MegaFlow 500) was used. It confirmed ASTM C 494, Type F & G standards [17]. SP recommended quantity is (0.5 %-2 %) of cement weight, so 1% is the used dosage herein. Typical properties for MegaFlow 500 at 25 °C is stated in table (10).

Table 10. Typical properties for MegaFlow 500 (at 25 °c).

Feature	Test Method	Results
Constituent	-	Single
Formula	-	Liquid
Colour	-	Whitish to light brown
Specific gravity	-	1.09 +/- 0.02
Air entrainment	-	Less than 1% over control mix
Chloride content	-	Nil to BSEN 934-2
pH	ASTM C494	5.5 - 8.0

IOP Conf. Series: Earth and Environmental Science **877** (2021) 012029 doi:10.1088/1755-1315/877/1/012029

Table (11) reveals the mixture details quantities used herein. The mixtures contain replacing 0 %, 25 %, 50 %, 75 % and 100 % of the gravel weight with RDA, partial substitution 10 % Si from the cement weight, Steel fibre at three weight contents: 0.5 %, 1 % and 1.5 %. The normal concrete was arranged in the similar concrete mixture method (1: 1.39: 1.96) recommended by reference [18]. The concrete mixture density of 2321 kg/m^3, a (W/C = 0.39), and f'_c= 35 MPa thru a 4.9 kg/m^3 superplasticizer.

Table 11. Quantities of materials in mixture parts **(kg/m^3)**

Mix designation	Cement	Si	SF	SP	Sand	Water	Coarse aggregate	RDA	% Replaced RDA
NC	490	0	0	0	680	191	960	0	0
RDA25	490	0	0	0	680	191	720	240	25
RDA50	490	0	0	0	680	191	480	480	50
RDA75	490	0	0	0	680	191	240	720	75
RDA100	490	0	0	0	680	191	0	960	100
RDA50-M	490	0	0	0	680	191	480	480	50
RDA50-Si	441	49	0	0	680	191	480	480	50
RDA50-SF0.5	490	0	39.5	0	680	191	480	480	50
RDA50-SF1	490	0	78.5	0	680	191	480	480	50
RDA50-SF1.5	490	0	118.6	0	680	191	480	480	50
RDA50-SP	490	0	0	4.9	680	186	480	480	50
RDA50-Si- SP	441	49	0	4.9	680	186	480	480	50
RDA50-SP-SF1	490	0	78.5	4.9	680	186	480	480	50
RDA50-Si-SP-SF1	441	49	78.5	4.9	680	186	480	480	50
RDA50-Si-SF1	441	49	78.5	0	680	191	480	480	50

They first weighted the gravel and sand. The mixing was carried out in mixer (0.05 m^3 volume). In the mixer, standard concrete was mixed by way of putting coarse and fine aggregates. Although the blender was working and the mixing sustained for one minute, the water put about one-third of the amount. Then the cement was added and, after three minutes of mixing, water was dispensed gradually into the blender. At a point that a uniform mixture had been obtained, the process was stopped and then the slump and fresh unit weights were evaluated.

In the mixing of materials, a special sequence was used for the RDA concrete mixes. First, inside the mixer, about 40 % of water was dispensed, and then dry RDA pieces were extra and mixing continuous around two minutes to guarantee complete water moistening. The solid materials are thus put into the mixer and progressively added the last 60 percent of the water whereas the mixing was in development to get a consistent mix. Three to five minutes of mixing proceeded, and then the fresh densities and slump were measured.

A slump test was performed for all types of mixes in agreement with reference [19]. Agreeing to reference [20]; 6 cubes sized (150×150×150) mm used to determine the compressive strength at 28 days of curing. The cylinder dimension was 150 mm diameter x 300 mm height to determine indirect tensile strength (ASTM C496 / C496M-17) [21]; in addition to calculate the concrete density [22]. Prism samples of sizes 100×100×500 mm were made to evaluate the modulus of rupture. The test was performed in agreement to ASTM C78-02 [23]; as displayed in figure (4).

IOP Conf. Series: Earth and Environmental Science **877** (2021) 012029 doi:10.1088/1755-1315/877/1/012029

-a- sample under test -b- sample failure

Figure 4. RDA concrete test for prisms

3. Results and Discussion

Table (12) demonstrations the compressive strength results of all types of concrete mixes. A ordinary concrete mixture was utilized as a guide intended for comparison. For the other mixtures, a main variable was used, which substituted the normal coarse aggregates with the RDA material. In addition, super-plasticizer and Si were utilized in these mixtures to recover a few of the workability and compressive strength lost, minimizing the loss causing from RDA usage. As the amount of RDA content in the mixes increased, the compressive strength decreased (25 %, 50 %, 75 %, and 100 %). This reduction is due to using the RDA that was utilized as a replacement for normal gravels, where the porosity of the recycled aggregate which leads to the absorption of the mixture water and thus less workability. There is an important difference in compressive strength as compared to conventional coarse gravels. Simultaneously, the unit weight of RDA is lesser than the normal gravel because of higher porosity, larger water absorption, lower abrasion resistance and greater content of layers [7-8].

For this, to decrease the opposing effects of RDA on the performance of concrete structures or materials, some technique needed to develop it. Such using of mineral additives (silica fume, steel fibre, et.) as cement replacement, which is one active method. These outcomes are alike several studies [9-11]. ACI code was implemented 17 MPa compressive strength. So, the all kinds of RDA concrete might be used for structural considerations [24]. This performance also happened in the tensile strength of concrete. The density as reduced due to higher porosity of the RDA particles more than natural aggregate. Furthermore, the modulus of rupture was increased up to 34% at (RDA50-Si-SP-SF1) when mixing additives together.

Table 12. Concrete mixes results.

Concrete designation	Slump (mm)	Density (kg/m^3)	Tensile stress (MPa)	Compressive strength (MPa)	Modulus of rupture (MPa)
NC	70	2462	3.39	32.05	6.76
RDA25	65	2387	3.09	31.03	6.72
RDA50	62	2280	3.08	29.56	5.97
RDA75	55	2311	3.05	23.43	5.81
RDA100	45	2235	3.02	22.86	5.71
RDA50-M	55	2340	3.97	36.18	6.84
RDA50-Si	58	2310	4.30	56.87	7.26
RDA50-SF0.5	57	2336	4.23	37.17	7.35
RDA50-SF1	60	2388	5.71	46.46	7.71
RDA50-SF1.5	65	2346	7.92	50.68	8.57
RDA50-SP	60	2361	4.84	37.74	6.83
RDA50-Si- SP	66	2306	5.13	65.61	7.29
RDA50-SP-SF1	54	2362	5.59	60.96	8.52
RDA50-Si-SP-SF1	58	2375	7.60	67.94	9.07
RDA50-Si-SF1	68	2309	5.10	66.04	8.88

4. Conclusions

The experimental work has offered a test of the mechanical properties of concrete comprising RDA. This represents a recent trend of experiments using discarded waste and rubbish as substitute materials for the concrete's normal gravels. The following findings were discovered:

1) An increase in the RDA replacement ratio allows the effects of the slump to decrease. However, the results increased particularly the super-plasticizer effect when using additives, if mix time addition seven to eight minutes.
2) Up to a 9 %, concrete density reduction of the control mixture density can be accomplished by deducting considered quantities of coarse aggregate and substituting it with the RDA in the highest percentage used (RDA100) .
3) RDA in the concrete mix as a partial substitute of gravel decreases the compressive strength about 27 %, 29 % at RDA75, RDA100, respectively, compared with the normal concrete. However, at (RDA50-Si-SP-SF1) (Mixing additives together), the compressive strength is increase up to 112 %.
4) RDA content has a little influence on the concrete tensile strength, displaying a reduction about 11 % at RDA100. But, when Mixing additives together (RDA50-Si-SP-SF1) displaying an increase of up to 124 %, while the modulus of rupture is increase up to 34 % under the same conditions.

5. Acknowledgments

The study is conducted in Tikrit University. The writers hope to be grateful to the college professionals who assisted them during tests.

6. References

[1] Djamaluddin R 2013 Flexural behaviour of external reinforced concrete beams *Procedia Engineering* **54** 252-60
[2] Arezoumandi M, Smith A, Volz JS and Khayat KH 2015 An experimental study on flexural strength of reinforced concrete beams with 100% recycled concrete aggregate *Engineering Structures* **88** 154-62
[3] Behera M, Bhattacharyya SK, Minocha AK, Deoliya R and Maiti S 2014 Recycled aggregate from C & D waste & its use in concrete – A breakthrough towards sustainability in construction sector: A review *Construction and Building Materials* **68**, 501516

[4] McNeil K and Kang TH-K 2013 Recycled concrete aggregates: a review *International Journal of Concrete Structures and Materials* **7** (1) 61–69
[5] Hansen TC 1992 *Recycling of Demolished Concrete and Masonry* 1st edition (Oxfordshire, UK: Taylor and Francis)
[6] Sagoe-Crentsil KK, Brown T and Taylor AH 2001 Performance of concrete made with commercially produced coarse recycled concrete aggregate *Cement and Concrete Research* **31** (5) 707–12
[7] Juan D, and Gutiérrez MS 2009 Study on the Influence of Attached Mortar Content on the Properties of Recycled Concrete Aggregate *Construction and Building Materials* **23** (2) 872-77
[8] Evangelista L, Guedes M, de Brito J, Ferro AC and Pereira MF 2015 Physical, chemical and mineralogical properties of fine recycled aggregates made from concrete waste *Construction and Building Materials* **86** 178–88
[9] Matias D, de Brito J, Rosa A and Pedro D 2013 Mechanical properties of concrete produced with recycled coarse aggregates-influence of the use of superplasticizers *Construction and Building Materials* **44** 101–09
[10] Xiao JZ, Li WG, YH Fan, YH X and Huang X 2012 An overview of study on recycled aggregate concrete in China (1996–2011) *Construction and Building Materials* **31** 364–83
[11] Martin- Morales M, Zamorano M, Ruiz-Moyano A, and Valverde- Espinosa I 2011 Characterization of recycled aggregates construction and demolition waste for concrete production following the Spanish Structural Concrete Code EHE-08 *Construction and Building Materials* **25** (2) 742–48
[12] Mohammed HJ and Zain M 2016 Experimental application of EPS concrete in the new prototype design of the concrete barrier *Construction and Building Materials* **124** 312-42
[13] ASTM, C150M-19a C 2019 *Standard specification for portland cement* www.astm.org
[14] ASTM, C778-17 2017 *Specification for standard (sand) fine aggregate*www.astm.org
[15] ASTM, C33M-18 C 2018 *Specification for standard concrete aggregates* www.astm.org
[16] ASTM A820/A820M - 16 2016 *Standard specification for steel fibers for fiber-reinforced concrete* www.astm.org
[17] ASTM C494/C494M–99a 1999 *Standard specification for chemical admixtures for concrete* www.astm.org
[18] ACI, 211 C 1991 *Standard practice for selecting proportions for normal, heavyweight, and mass concrete*
[19] ASTM, C143M-15a C 2015 *Standard test method for slump of hydraulic-cement concrete* www.astm.org
[20] ASTM, C39/C39M-18 2018 S*tandard test method for compressive strength of cylindrical concrete specimens* www.astm.org
[21] ASTM, C496M-17 C 2017 *Standard test method for splitting tensile strength of cylindrical concrete specimens* www.astm.org
[22] ASTM, C29M-17a C 2017 *Standard test method for bulk density ("unit weight") and voids in aggregate* www.astm.org
[23] ASTM, C78–02 2002 *Standard test method for flexural strength of concrete (using simple beam with third-point loading)* www.astm.org
[24] ACI, 318M-19 2019 *Building code requirement for structural concrete*

IOP Conf. Series: Earth and Environmental Science **877** (2021) 012030 doi:10.1088/1755-1315/877/1/012030

Stratigraphy and Depositional Environment of Mauddud Formation in Ratawi Oil wells of Southern Iraq

Maha Razaq Manhi[1] and Hamid Ali Ahmed Alsultani[2]
[1] MSc Student, Applied Geology Department, College of Science, University of Babylon, Iraq.
[2] Applied Geology Department, College of Science, University of Babylon, Iraq.
Email: sci.hamid.ali@uobabylon.edu.iq

Abstract. The Mauddud Formation is Iraq's most significant and widely distributed Lower Cretaceous formation. This Formation has been investigated at a well -23 and a well-6 within Ratawi oil field southern Iraq. In this work, 75 thin sections were produced and examined. The Mauddud Formation was deposited in a variety of environments within the carbonate platform. According to microfacies analysis studying of the Mauddud Formation contains of twelve microfacies , this microfacies Mudstone to wackestone microfacies , bioclastic mudstone to wackestone microfacies , Miliolids wackestone microfacies ,Orbitolina wackestone microfacies , Bioclastic wackestone microfacies , Orbitolina packstone microfacies , Peloidal packstone microfacies , Bioclastic packstone microfacies , Peloidal to Bioclastic packstone microfacies , Bioclastic grainstone microfacies , Peloidal grainstone microfacies , Rudstone microfacies . Deep sea, Shallow open marine, Restricted, Rudist Biostrome, Mid – Ramp, and Shoals are the six depositional environments in the Mauddud Formation based on these microfacies.

1. Introduction

The Mauddud Formation was originally described by Henson of the Qatar Petroleum Company's subsurface section, well Dukhan No. 1. Ain Mauddud [1] is the source of its name. [2]Mauddud Formation is a subsurface, largely neomorphosed and dolomitized limestone. The Mauddud Formation's lower contact is gradational and conformable with the Nahr Umr, Lower Balambo, and Sarmord Formations. In Iraq's North Central, North, and Northeast, the top contact has a break and is either nonsequential or unconformable; it is an unconformity [3]. Mauddud Formation is 55 meters thick and composed of limestone rich in Orbitolina and Trocholina samples at the type locality where the formation was originally identified. The Mauddud Formation was discovered in the Northern Arabian Gulf, particularly along the Saudi Arabia–Kuwait border, to be made up of limestone with thicknesses varying from 30.5 to 97.6 meters. [4]. [5] utilized the Zubair well no.3 as a supplemental type section for the Mauddud Formation in Iraq. They described the formation as a detrital organic limestone with strata of blue to green shale that was locally pseudo oolitic creamy in color [6]. The Mauddud Formation fades between Wara and Nahr Umr in western Kuwait, as well as on the eastern face of the Hail – Rutbah Arch [7]. The age of the Mauddud Formation is still debated ,figure 1, with some claiming an Albian age and others claiming a Cenomanian date. Abundant fossils cited by [8] support an Albian age. Because of the widespread presence of certain Orbitolina concave group species, the formation was previously thought to stretch into the Cenomanian [8] .In Middle and Southern Iraq, the Mauddud Formation is the most common Lower Cretaceous Formation.The aims of study include microfacies analysis, deposional environment interpretation and determination of the sedimentary cycles of the Mauddud Formation.

IOP Conf. Series: Earth and Environmental Science **877** (2021) 012030 doi:10.1088/1755-1315/877/1/012030

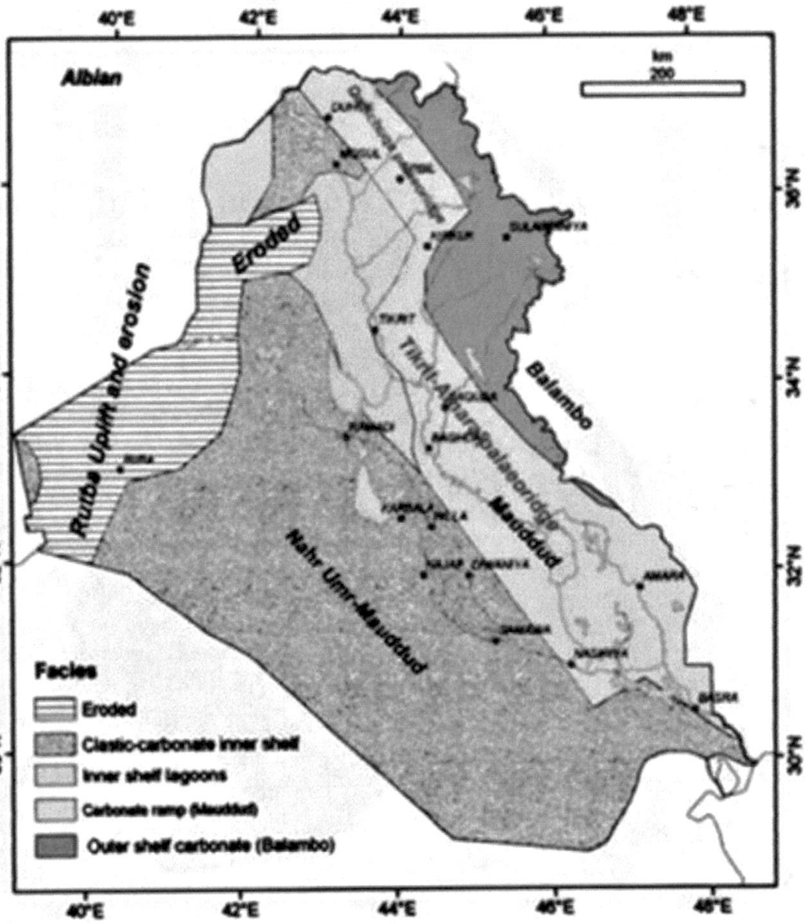

Figure.1 Albian paleogeography map show the depositional basin of Mauddud Formation (Jassim and Goff 2005).

2. Location of study area

The study region lies in the southern portion of Iraq , figure 2, on the stable shelf of the Mesopotamian basin. The Ratawi field lies 50 kilometers west of Basra, and it is a flat, semi-desert terrain positioned inside the tectonic components at the platform edge of the Mesopotamian fore deep. And it progressively drops to sea level as it approaches the north, while its highest point reaches around (120) m. at sea level as it approaches the south, figure 3.

3. Methodology

In this study the Mauddud formation section was reported from the A well -23, A well -6. Table (1).Studying Petrography of these wells by thin sections examination which prepared in the workshop of the Department of Geology, these thin sections have been studied by the aid of binocular microscope in order to determine the fossils content and then Microfacies Analysis, then recognize depositional Environment and stratigraphic analysis and determine the sedimentary of Mauddud formation in studied oil wells.

IOP Conf. Series: Earth and Environmental Science **877** (2021) 012030 doi:10.1088/1755-1315/877/1/012030

Table.1: Sampling for the collected borehole

Well no.	Top(m)	Bottom (m)	Mauddud Core samples	Thin section
A.23	2575	2691	27	**35**
A.6	2458	2596	29	**40**

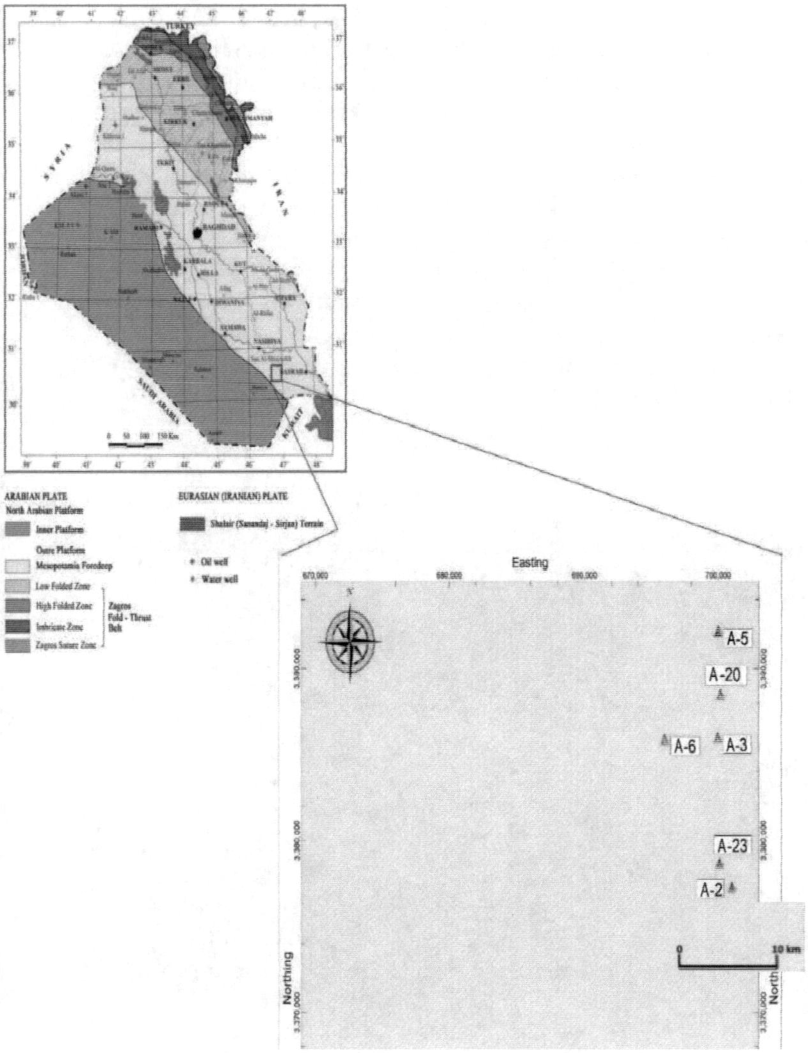

Figure.2 Location map of the study wells

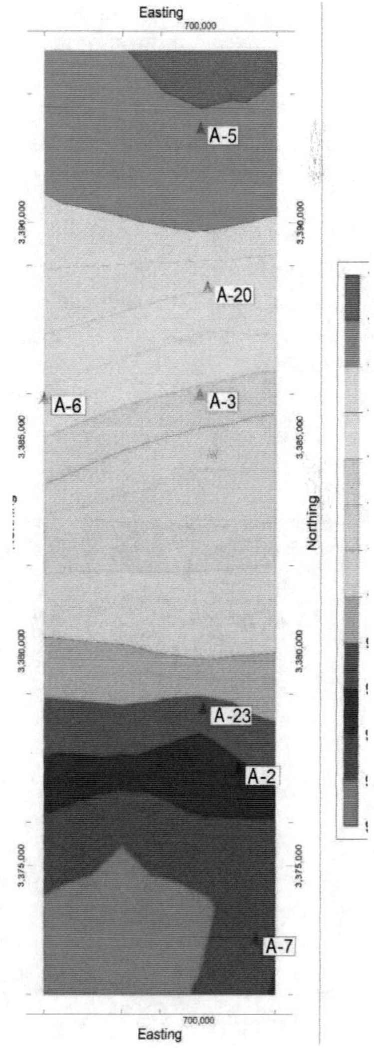

Figure.3 Isopach map of Ratawi Field of Mauddud Formation in the study area.

4. Results and discussion

4.1. Petrography

Petrographic analysis of (75) thin section of samples from the A well -23 and A well - 6, as they were used in this study. On the basis of Dunham's classifications, thin sections were categorized [9].Microfacies are identified and classified into facies associations based on grain, fabric, and bio content differentiation. According to studies, the carbonate grains of the Mauddud Formation are both skeletal and non-skeletal. Skeletal grains are mostly composed of benthonic foraminifera, Mollusca, algae, reef, and rudist, whilst non-skeletal grains are mostly composed of peloids, Ooids, and lithoclastic .The foraminifera in the Mauddud Formation are mostly skeletal grains. [10] Foraminifera are benthic organisms that live in or on sediments on the seafloor. Planktonic foraminifera dwelling in the upper 100 meters of the seas were mentioned in the Mauddud Formation. The Orbitolinidae are enormous benthic foraminifers that dominated the Early to Mid-Cretaceous period, figure-4a, 4b, 4c, 4d, 4e, 4f . Nezzazata less common in this formation. figure 5a). Milionis are also important structural granules, respectively , figure 5b. [11] "Calcareous algae are important in micropaleontology as records of ancient life, and they can be used in the interpretation of paleoenvironments and age determination" of strata. figure 5c . The bioclastic of Mauddud Formation include rudist, echinoderm, coral fragments and Mollusca,figure 5e . Peloids are generally spherical, ellipsoidal rounded or sub rounded, to irregular in shape and commonly has high organic contents. Fecal pellets, produced by deposit-feeding animals,figure 5f .There is a difficulties in the segregation of the two types in thin sections, so the term lithoclastic had been used by many authors [13]. Limeclastic or limestone clast is suitable [12]. [13] the groundmass or matrix denotes the interstitial material between larger grains .In Mauddud Formation the groundmass is composed of micrite and/or Microspar which formed by neomorphism .Micritization, neomorphism, dissolution, cementation, dolomitization, and compaction are some of the diagenetic processes that have been influenced by the Mauddud Formation.

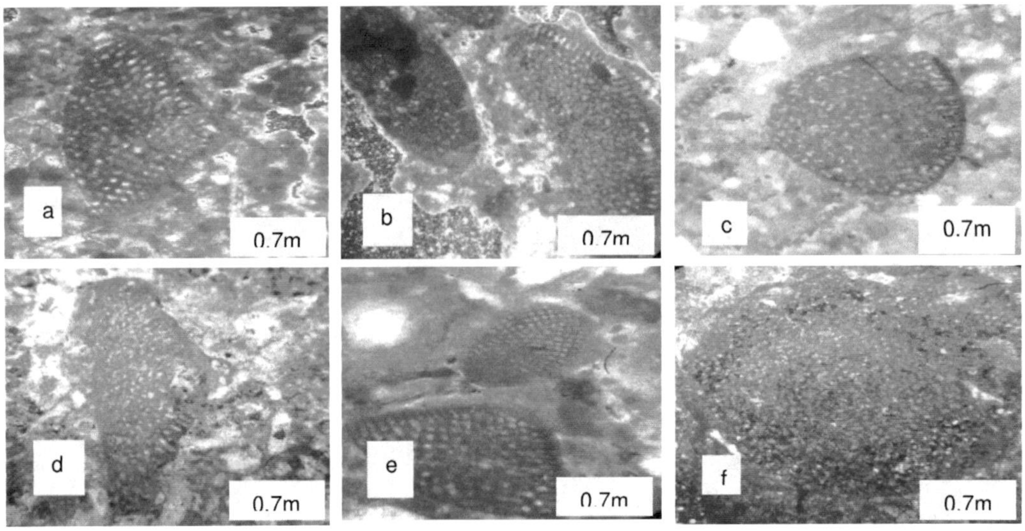

Figure.4: Explain the types Orbitolna, a-*Iraqia simplex(Henson)* sp (A . 23, depth . 2585m),b-*Orbitolna sefini* sp,(A . 23, depth .2515m),c-*Neoiraqia* sp (A .23 , depth . 2520m),d- *Orbitolna concave* sp (A . 6, depth .2655m),e- *Orbitolina conica* sp (A . 6, depth .2674m),f-*Conicorbitolina conica* sp(A . 23, depth .2589m).

IOP Conf. Series: Earth and Environmental Science **877** (2021) 012030 doi:10.1088/1755-1315/877/1/012030

Figure.5 Explain the types skeletal and non-skeletal grains, a-*Nezzazata* sp,(A . 23, depth 2482 m),b-Miliolids, (A . 6, depth.2630 m),c-Calcareous algae(A . 6, depth .2639),d-*Echinoderm spine* sp (A . 6, depth 2618m),e- mollusca,(A . 23, depth.2630M),f-peloidal (A . 6, depth .2651).

4.2. Microfacies

In Mauddud Formation microfacies, there are twelve distinct microfacies include Mudstone to wackestone microfacies , Bioclastic mudstone to wackestone microfacies , Miliolids wackestone microfacies , Orbitolina wackestone microfacies , Bioclastic wackestone microfacies , Orbitolina packstone microfacies , Peloidal packstone microfacies , Bioclastic packstone microfacies, figure 6a,b,c,d,e,f. Peloidal to Bioclastic packstone microfacies , Bioclastic grainstone microfacies , Peloidal grainstone microfacies , Rudstone microfacies , figure 7a, b, c, d, e, and f .

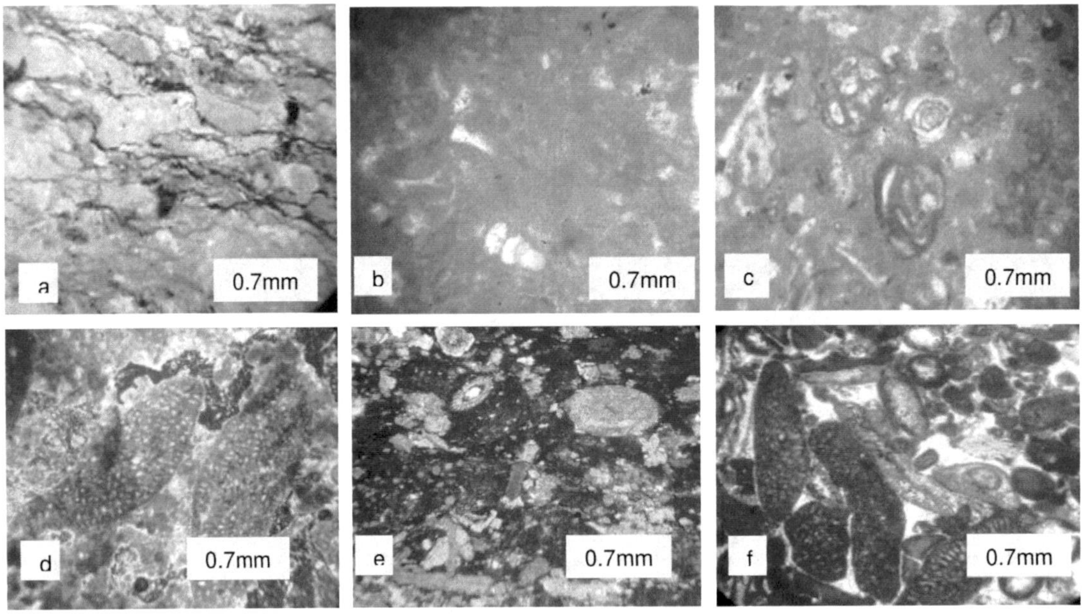

Figure.6 Explain the types microfacies, a,Mudstone to wackestone microfacies (A . 23, depth 2467 m),b- Bioclastic mudstone to wackestone microfacies (A . 6,depth 2642m),c-Miliolids wackestone microfacies (A . 23,, depth 2643m),d- Orbitolina wackestone microfacies (A . 23,, depth .2629 m),e- Bioclastic wackestone microfacies (A . 6, depth 2577),f-Orbitolina packstone microfacies A . 23, depth (2660 m)

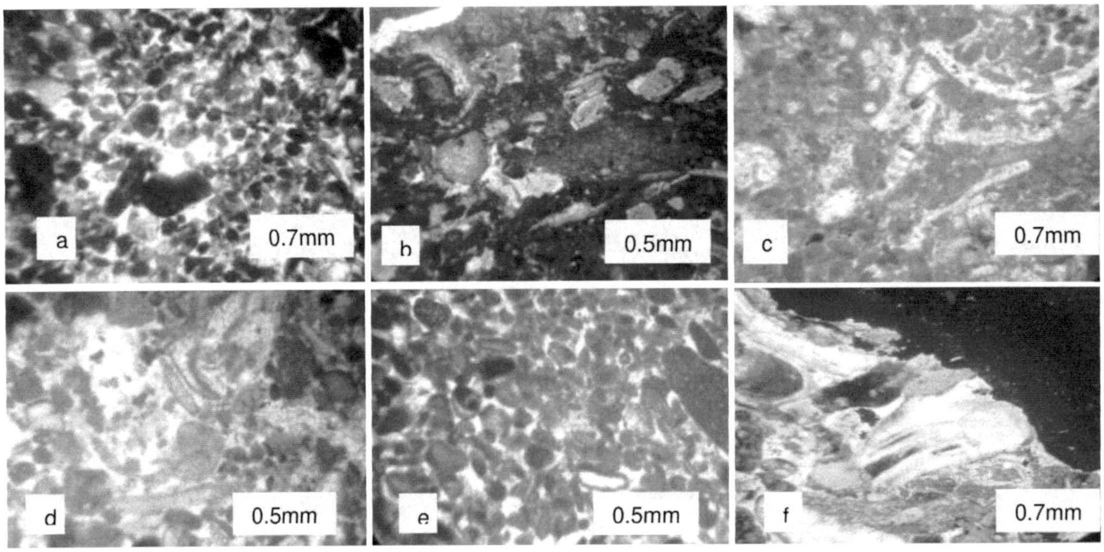

Figure.7 Shows the types microfacies, a-Peloidal packstone microfacies(A . 6, depth (2617 m), b-Bioclastic packstone microfacies (A . 23,. depth (2485 m) ,c-Peloidal to Bioclastic packstone microfacies (A.6, depth (2610 m),d-Bioclastic grainstone microfacies (A.6 ,depth 2517m),e-Peloidal grainstone microfacies(A.23,,depth 2585 m),f-Rudstone microfacies(A.23, depth 2600m).

4.3. Depositional Environment

The kind of microfacies largely affected the Mauddud Formation's paleoenvironment. The depositional texture and grain types (skeletal and non-skeletal) were used to determine the known microfacies. Flugel's suggested models of standard microfacies and carbonate depositional environment belts were compared to facies associations [14].Six major environments were recognized, they are open marine, deep sea, restricted and shoals figures 8, 9.

4.3.1. Deep marine environment

Bioclastic mudstone to wackestone microfacies describes the deep marine association facies in A–6 wells. Planktonic foraminifera, benthonic foraminifera bioclasts, and tiny benthic foraminifera on a backdrop of mainly micrite suggest deposition below typical wave base and under low-energy hydrodynamic circumstances. The presence of planktonic and echinoderm pieces, as well as fine bioclastic, might indicate a deep marine environment on the outer ramp. Dissolution is a critical diagenetic process. A well - 23 the facies of mudstone to wackestone microfacies. This microfacies is compared to Wilson's microfacies [15].

4.3.2. Shoal environment

The shoal environment is a belt of strong tidal current and wave activity that runs along the carbonate platform's seaward edge. In this environment, deposition depths are less than (5-10) meters above wave-base [16].Peloidal grainstone microfacies and Peloidal packstone microfacies characterize this habitat in the Mauddud Formation. Grain supported texture distinguishes these microfacies. Benthic foraminifera, Mollusca, rudist pieces, and peloids make up this collection. The peloids have been sorted and rounded to perfection. This facies is typically seen on high-energy shoals [14].

4.3.3. Shallow open marine environment

One of the most frequent facies in the Mauddud Formation carbonates in the research region is the shallow open marine facies association. This is the formation's most frequent facies connection Microfacies such as orbitolina wackestone, orbitolina packstone, Peloidal to bioclastic packstone, and bioclastic grainstone define it. The bioclasts are silt to sand in size and some cases coarser, and at various levels within the succession, a specific type of bioclasts may locally dominate, especially the rudist bioclasts which occur as well worn (sand- size or coarser) angular or tabular grains. Other important fossils include benthonic foraminifera such as Orbitolinids, Nezzazata, miliolid, Textularia, calcareous green algae, coral, and coralline algae with echinoderms and gastropods. Sponge spicules, bryozoa, brachiopod, and pelagic foraminifera were less common. The Orbitolina flourished in tropical to subtropical water along shallow coastlines where the temperature would have been between 15 and 25C [17]. Orbitolina occurrence with miliolids suggests shallow water, and its association with rudist fragments may indicate deeper more open water [18].

4.3.4. Restricted marine environment

Miliolids wackestone and Mudstone to wackestone make up the majority of this facies relationship. Miliolids, Textularids, Nezzazata, and much more benthonic foraminifera are numerous and varied. Coral pieces, rudist fragments, and mollusks are among the other fossils found. The matrix is basically fine lime mud, with a significant amount of argillaceous material in certain instances. In a lagoonal setting, this pattern may be observed [19].

4.3.5. Mid – Ramp environment

The bioclastic Wackestone and packstone make up most of the Mid-Ramp facies association. Rudistid packstone and Orbitolina wackestone are two other significant microfacies. The main skeletal grains include Orbitolina, Nezzazata, rudist fragments, and pelecypods. The primary diagenetic processes in this facies relationship are dolomitization and dissolution.

4.3.6. Rudist Biostrome environment

Rudstone is the main component of this microfacies. This association is mostly comprised of Rudist fragments and Orbitolina shells. The reef is a carbonate formation with an organism-built wave-resistant structure [16], Wilson's facies belt -5 of organic buildup [15] and Flugel's facies belt -5 of organic buildup [20].

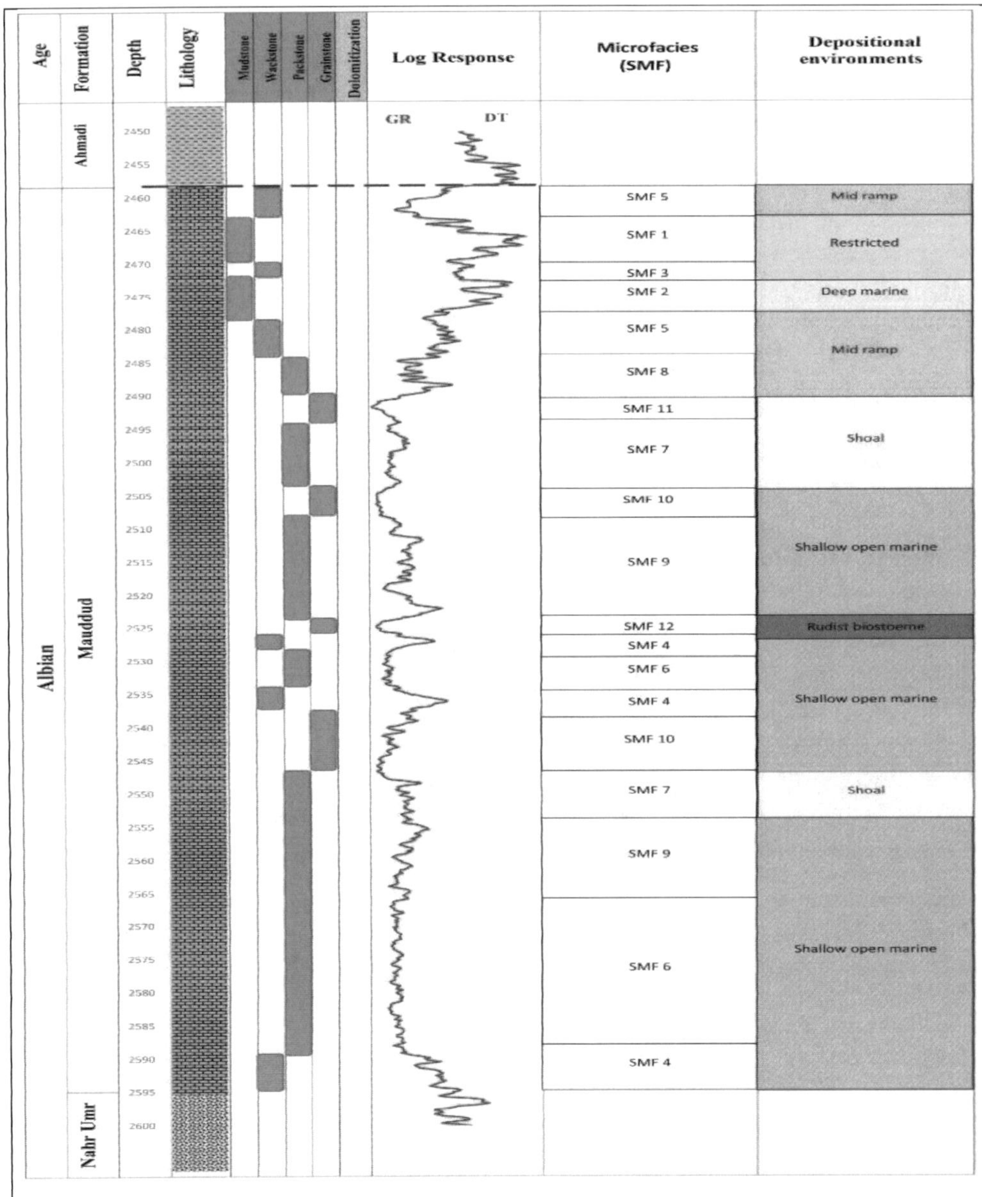

Figure 8. Microfacies succession and depositional environments of Mauddud Formation in A-23 well

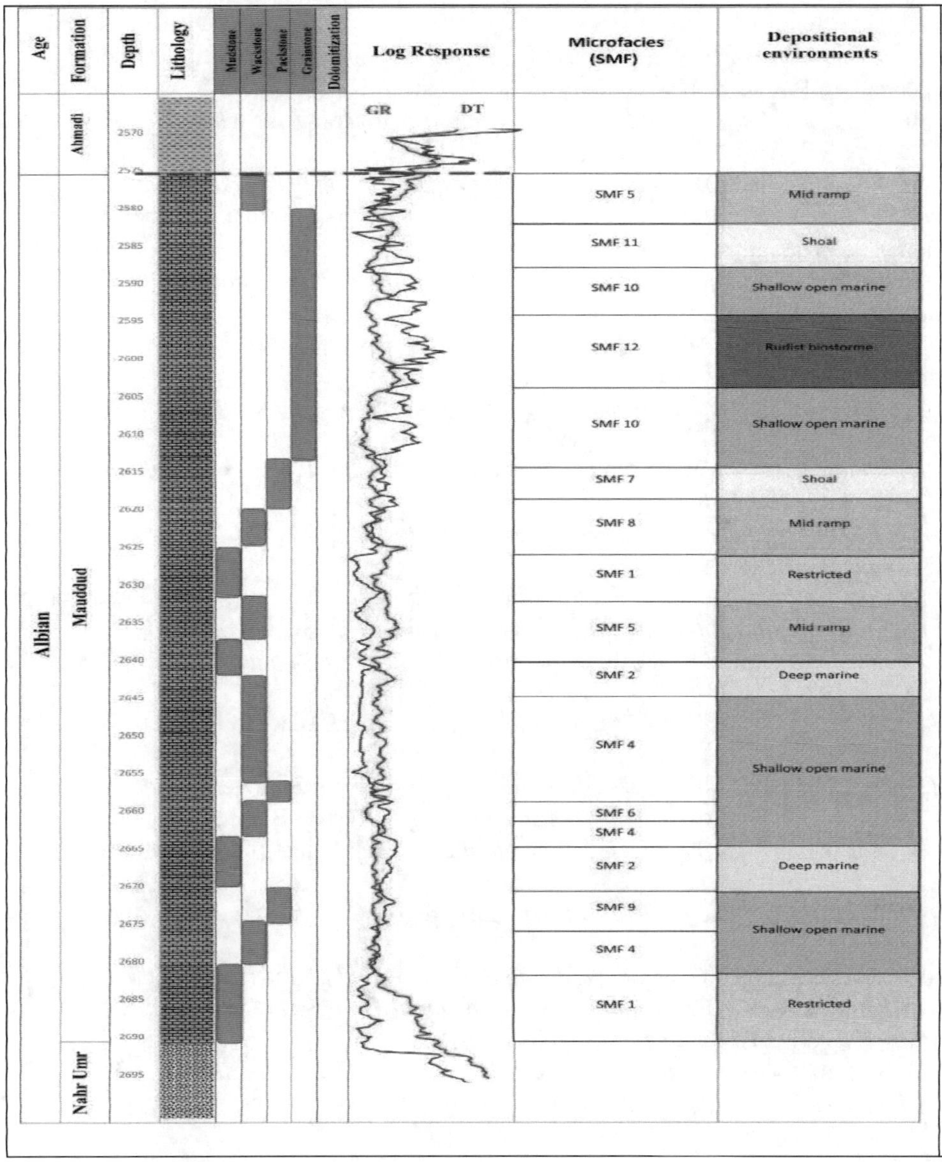

Figure9. Microfacies succession and depositional environments of Mauddud Formation in A - 6 well.

5. Conclusion

The Mauddud Formation in A well - 23 and A well -6 in Ratawi oil field southern Iraq is composed Petrographic study in Mauddud Formation shows types fossils the following:-Iraqia simplex(Henson) sp, Orbitolna sefini sp,Neoiraqia sp ,Orbitolna concave sp,Orbitolina conicasp,Conicorbitolinaconicasp,mesoritolina(subconcava)sp,Orbitolina qatarica(Henson)sp, Nezzazata sp, Rudist fragments, Textularia sp, algae, Echinoderm spine , Mollusca. Twelve microfacies have been identified. , and depending on these microfacies have been distinguished six depositional environments in a Mauddud Formation, including deep marine environment, shoals environment, shallow open marine environments, restricted marine environment ,mid – ramp environment, rudist biostrome environment twelve types of microfacies were identified these microfacies.

6. References

[1] Mohammed, M. U., 1981. *The microfacies of Mauddud Formation*. Unpubl.M.Sc.Thesis,University of Baghdad, 61p.
[2] Shubber, B.A., 1986. *Sedimentology of Mauddud Formation in middle and southernIraq*. Unpubl M.Sc. Thesis, University of Baghdad, 104 p.
[3] Jassim, S.Z., and Goff, J.C., 2006. *Geology of Iraq*, Dolin, Prague and MoravianMuseum, Brno, 337p.
[4] Al-Dabbas, M.A., Jassim, J.A., and Qaradaghi, A.I., 2012: *Sedimentological anddepositional environment studied of Mauddud Formation, centeral and southern Iraq.* Arabian Journal of Geosciences, Volume: 5, Issue: 2, p 297 – 312.
[5] Owen, R.M.S., and Nasr,S.N.,(1958).*The stratigraphy of Kuwait – Basrah area.InHabitat of Oil-A Symposium, Spec.* Publ. AAPG, p1252 – 1278.
[6] Sayyab, A.S., and Mohammed, M.U., 1985. *Biostratigraphic study of some subsurface sections of Mauddud Formation (south of Iraq)*. Iraqi. Sci., V.26, p 45 – 67.
[7] Ibrahim, M.W., 1981. *Lithostratigraphy and subsurface geology of the Albian rocks of south Iraq*, Journal of petroleum geology, 4, 2, p 147 – 162.
[8] Bellen,R.C. van, H. V. Dunnington, R. Wetzel and D. M. Morton, 1959, *Lexique Stratigraphique international*. v.3, Asia, fasc. 10a, Iraq, 333p.
[9] Dunham, R. J., 1962, *Classification of carbonate rocks according to depositionaltexture*. In :Ha, W.E. classification of carbonate rocks.
[10] Heckel, P.H., 1970, *Recognition of ancient shallow marine environments*, In Righy, J.K. and Hamablin, W.K. (eds.), *Recognition of ancient Sedimentry environments*, SEPM, SPEC.
[11] Haq, B. U., Hardenbole, J. and Vail, P.R., 1988. *Chronology of fluctuation sea levels since the Triassic*, Science, V.235, PP. 1156-1167.
[12] Tucker, M.E., and Wright, V.P. 1990. *Carbonate sedimentology*. Blackwell Scientific Publications, Oxford, 482p.
[13] Flugel, E.,2010. *Microfacies of Carbonate Rocks, Verlag Berlin Heidelberg*. Springer,984p
[14] Wilson, J. L. 1975. *Carbonate Facies in Geologic history*.Springer. Verlag, Berlin, 471p
[15] Tucker, M.E., 1985. *Sedimentary Petrology: An introduction*.Blackwell Scientific publications, 252p.
[16] Douglass, R. C., 1960. *Revision of the family Orbitolinidae*. Micropal., V. 6, no. 3, pp. 249 – 270
[17] Sadooni,F.N. and Alsharhan,A.S.,2003. *Stratigraphy, microfacies , and petroleum potential of the Mauddud Formation (Albian – Cenomanian) in the Arabian Gulf basin.* The American Association of petroleum geologists bulletin, V.87, No.10, p1633 – 1680
[18] RODRÍGUEZ, A.J., 2007. Seq*uence stratigraphic analysis of the Cotui Limestone, Upper Cretaceous of Southwestern Puerto Rico*. Unpubl M.Sc. Thesis, University of Puerto Rico.99p.
[19] Flugel, E., 1982. *Microfacies analysis of limestone*, Springer, p.633 .

IOP Conf. Series: Earth and Environmental Science **877** (2021) 012031 doi:10.1088/1755-1315/877/1/012031

Prediction and Forecasting of Maximum Weather Temperature Using a Linear Autoregressive Model

Salah L. Zubaidi[1*], Hussein Al-Bugharbee[2], Khalid Hashim [3,4], Nabeel Saleem Saad Al-Bdairi [1], Sabeeh L. Farhan [5,6], Asad Al Defae[1], Mohammed J. Jameel[7]

[1] Department of Civil Engineering, College of Engineering, Wasit University, Iraq
[2] Department of Mechanical Engineering, College of Engineering, Wasit University, Iraq
[3] Department of Civil Engineering, Liverpool John Moores University, UK
[4] Department of Environmental Engineering/ University of Babylon, Iraq
[5] Department of Architecture, Engineering, College of Engineering, Wasit University, Iraq
[6] Architecture, Design and the Built Environment, Nottingham Trent University, UK
[7] Center of Computer, University of Anbar, Iraq

* Corresponding author: salahlafta@uowasit.edu.iq

Abstract. This paper investigates the autoregressive (AR) model performance in prediction and forecasting the monthly maximum temperature. The temperature recordings are collected over 12 years (i.e., 144 monthly readings). All the data are stationaries, which is converted to be stationary, via obtaining the normal logarithm values. The recordings are then divided into 70% training and 30% testing sample. The training sample is used for determining the structure of the AR model while the testing sample is used for validating the obtained model in forecasting performance. A wide range of model order is selected and the most suitable order is selected in terms of the highest modelling accuracy. The study shows that the monthly maximum temperature can accurately be predicted and forecasted using the AR model.

Keywords: Autoregressive Model; Baghdad City; prediction model; temperature.

1. Introduction

Climate change and climate variability are anticipated to cause significant issues for the ecosystem (i.e., increase the temperature in future) [1, 2]. Industrialisation and the massive use of fossil fuels caused an increase the greenhouse gases that led to an increase in the impact of climate change [3, 4]. It has located a substantial impact on the environment of residential area in various places of the world [5-7]. These influences differ concerning the region, the type, and the importance.

The climatic factors have impacted, directly and indirectly, both individuals and their residential environment along different periods [8]. Temperature is considered the most vital climatic factor, which impacts the growth, development and yield of crops [9]. Additionally, the system of dwellings is developed in response to climatic factors [10, 11].

Different regions face a harmful influence of climate change that led to decreasing the quantity[12-16] and quality [17-22] of freshwater resources. High temperatures (i.e., dry day) lead to increase urban water consumption [23]. Additionally, various studies showed that urban water consumption was driven by maximum temperature [24-28].

In the last few years, precise forecast of temperature is a problem which has attracted the researchers' attention, since it has several various applications in the field such as industry, agriculture or energy. Recently, different models are used in different areas [29-33], and studies forecast the maximum temperature by various techniques [34-36]. The AR model used successfully in different applications [37-39]. In this research, Auto-regressive (AR) model will employ to predict the monthly temperature.

2. Area of study and data set

Iraq is one of the Arab countries that lies in arid to the semi-arid area, and Baghdad is the capital of Iraq and locates in the centre of the country[40]. The weather is wet and cold in winter and dry and hot in summer (i.e., the temperature reaches 45 °C). Iraq faced an acute climate change cased adversely impact the people, residential area, and freshwater sources[41-43]. The historical monthly data if maximum temperature along twelve years (2003-3014) used to build and assess the model.

3. Methodology

The procedure of this research divides into, data pre-processing, and auto-regressive model.

3.1. Data Pre-processing

It has a considerable influence on the accuracy of the forecast techniques. It can be separated here into two phases: normalisation and cleaning. Normalisation time series assistances to decrease the impact of outliers and makes the data to be normal or near-normal distribution [44, 45]. In this research, a natural logarithm is used for normalising the data due to its ability to decrease the influence of multicollinearity among predictor factors[28, 46].

3.2. Auto-Regressive Model (AR)

In autoregressive (AR) model, the output pertaining to a particular variable can be predicted from the past observations of that variable[47]. This model has a linear form. As such, the simplicity of this model coupled with its powerful prediction increases the popularity of this model in different disciplines in which time series data need to be analysed. In water demand forecasting, city engineers and water authorities are working collectively to maintain the balance between the demand and supply of drinking water to residents in their city. Hence, to achieve this goal, a sound statistical method should be used. Accordingly, there is a growing interest in applying autoregressive model in water demand forecasting. The outputs in this model are merely dependent on the previous observations of the same variable [37, 38].
To mathematically formulate autoregressive models, Eq. (1) is used to relate the current observation with the past ones in a linear relationship as illustrated:

$$X_t = \theta_0 + \sum_{i=1}^{p} k_i X_{t-1} + \varepsilon_t \tag{1}$$

Where; X_t and X_{t-1} are the observations in periods *t* and *t-1*, p is the order of the AR model considered, k_i is the autoregressive parameters, θ_0 is the constant term, and ε_t is the disturbance term for period *t*. A least-square algorithm using MATLAB is utilized to accurately predict the unknown coefficients in the AR model.

4. Results and discussion

Data were normalised, cleaned and divided into two sets training (70%) and testing (30%). Figure 1 shows various model goodness of fit for both training and testing sample. These models of fitness are obtained in different model order. As it was mentioned earlier that the training sample was used for obtaining the AR model coefficients while the testing sample is used for evaluating the forecasting capability of the model. The model goodness of fit contains coefficients of regression(R), mean absolute error (MAE), mean square error (MSE), and root mean square error (RMSE). From all the curves of the model goodness of fit, it is clear that the best model order is 10.

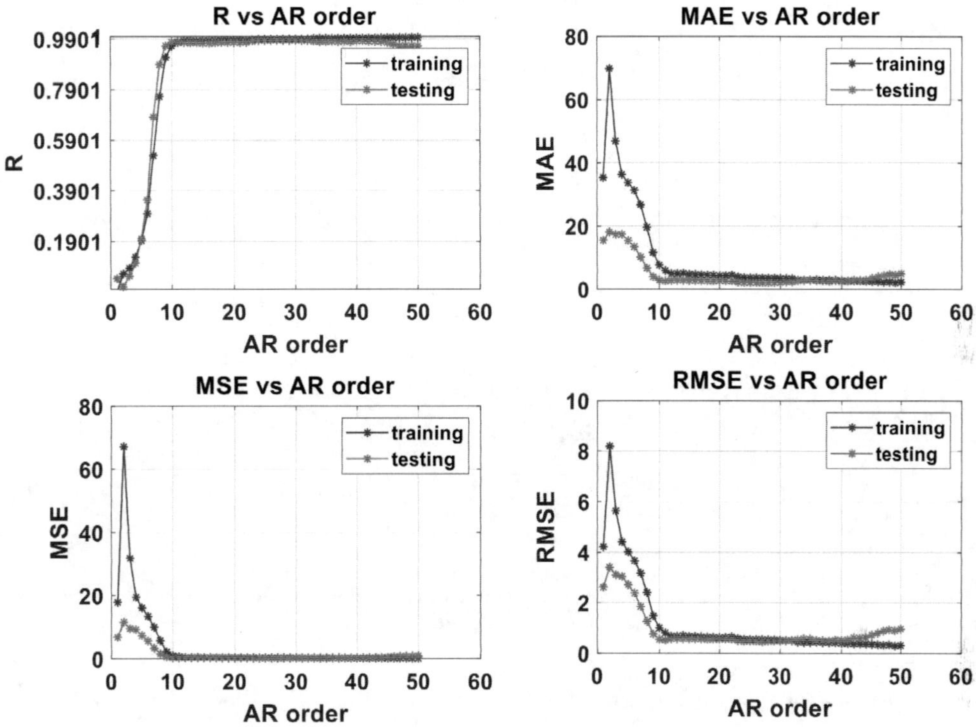

Figure 1. Assess the performance of the AR model by four goodness of fit.

Additionally, the graphical test is used to examine the performance of AR model. Figure 2 visualises the model forecasting performance. It shows the real measure temperature as well as the forecasted ones in the testing stage.

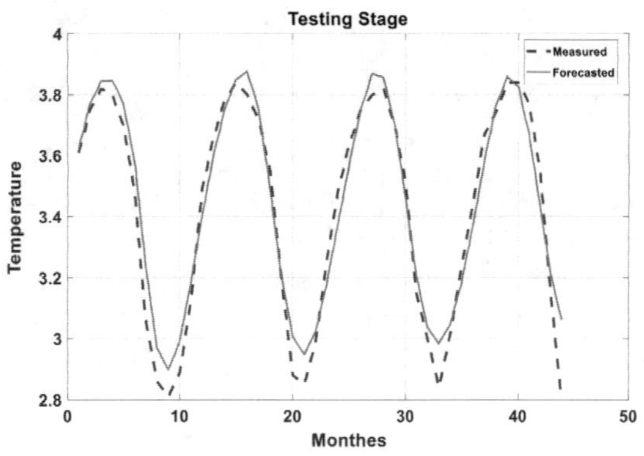

Figure 2. Measured and forecasted temperature.

Moreover, coefficient of determination is employed to assess the performance of AR model to forecast the monthly temperature. Figure 3 shows that the value of R^2 is 0.96, which means that the performance of AR model is good according to the Dawson et al., [48].

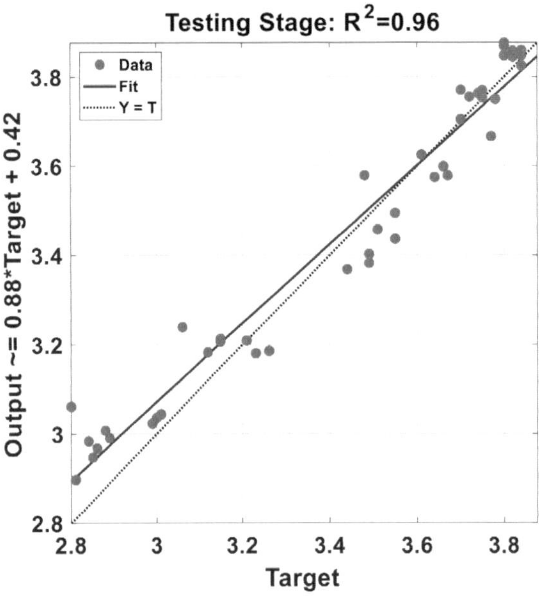

Figure 3. AR model performance.

Based on the above statistical tests, it reveals that this methodology is able to forecast the maximum temperature.

5. Conclusion

This research examines the performance of a method to predict and simulate the monthly maximum temperature. Monthly time series of historical maximum temperature for Baghdad City along twelve years (2003-2014) were used for building and assessing the suggested practice. The method contains data pre-processing and AR model. Data were divided into training (70%) and testing (30%). Different statistical tests were used to examine the methodology. The results show that the methodology forecast the data with good accuracy in the testing stage with R^2 equal to 0.96. Extra research should be conducted concerning forecast maximum temperature for the rest cities in Iraq.

References

[1] Zubaidi S L, Hashim K, Ethaib S, Al-Bdairi N S S, Al-Bugharbee H and Gharghan S K 2020 A novel methodology to predict monthly municipal water demand based on weather variables scenario *Journal of King Saud University - Engineering Sciences* 1-7

[2] Mohammed R and Scholz M 2019 Climate variability impact on the spatiotemporal characteristics of drought and Aridityin arid and semi-arid regions *Water Resources Management* **33** 5015-33

[3] Salman S A, Shahid S, Ismail T, Ahmed K and Wang X-J 2018 Selection of climate models for projection of spatiotemporal changes in temperature of Iraq with uncertainties *Atmospheric Research* **213** 509-22
[4] Zubaidi S L, Abdulkareem I H, Hashim K, Al-Bugharbee H, Ridha H M, Gharghan S K, Al-Qaim F F, Muradov M, Kot P and Al-Khaddar R 2020 Hybridised Artificial Neural Network Model with Slime Mould Algorithm: A Novel Methodology for Prediction of Urban Stochastic Water Demand *Water* **12** 1-18
[5] Jasim I A, Farhan S L, Al-Maliki L A and AL-Mamoori S K 2021 Climatic Treatments for Housing in the Traditional Holy Cities: A Comparison between Najaf and Yazd Cities. In: *IOP Conference Series: Earth and Environmental Science*: IOP Publishing) pp 1-10
[6] Farhan S L and Nasar Z A 2020 Urban identity in the holy cities of Iraq: Analysis of architectural design trends in the city of Karbala *Journal of Urban Regeneration and Renewal* **14** 210-22
[7] Farhan S, Akef V, Antón D, Hashim K and Zubaidi S 2021 Factors influencing the transformation of Iraqi holy cities: the case of Al-Najaf *Przegląd Naukowy Inżynieria i Kształtowanie Środowiska* **30** 365-75
[8] Zubaidi S L, Kot P, Hashim K, Alkhaddar R, Abdellatif M and Muhsin Y R 2019 Using LARS –WG model for prediction of temperature in Columbia City, USA. In: *IOP Conference Series: Materials Science and Engineering,* (Najaf, Iraq: Materials Science and Engineering. IOP) pp 1-9
[9] Kadiyala M D, Nedumaran S, Singh P, S C, Irshad M A and Bantilan M C 2015 An integrated crop model and GIS decision support system for assisting agronomic decision making under climate change *Science of the Total Environment* **521-522** 123-34
[10] Farhan S L, Jasim I A and Al-Mamoori S K 2019 The Transformation of The City of Najaf, Iraq: Analysis, Reality and Future Prospects *Journal of Urban Regeneration and Renewal* **13** 1-12
[11] Farhan S, Akef V and Nasar Z 2020 The transformation of the inherited historical urban and architectural characteristics of Al-Najaf's Old City and possible preservation insights *Frontiers of Architectural Research* 1-17
[12] Mohammed R, Scholz M, Nanekely M, Mokhtari Y and assessment r 2018 Assessment of models predicting anthropogenic interventions and climate variability on surface runoff of the Lower Zab River *Stochastic Environmental Research and Risk Assessment* **32** 223-40
[13] Mohammed R and Scholz M 2018 Flow–duration curve integration into digital filtering algorithms for simulating climate variability based on river baseflow *Hydrological Sciences Journal* **63** 1558-73
[14] Hashim K S, Kot P, Zubaidi S L, Alwash R, Al Khaddar R, Shaw A, Al-Jumeily D and Aljefery M H 2020 Energy Efficient Electrocoagulation Using Baffle-Plates Electrodes for Efficient Escherichia Coli Removal from Wastewater *Journal of Water Process Engineering* **33** 1-7
[15] Hashim K S, Hussein A H, Zubaidi S L, Kot P, Kraidi L, Alkhaddar R, Shaw A and Alwash R 2019 Effect of Initial Ph Value on The Removal of Reactive Black Dye from Water by Electrocoagulation (EC) Method *Journal of Physics: Conference Series* **1294** 1-6
[16] Ethaib S and Zubaidi S L 2020 Removal of Methylene Blue Dye from Aqueous Solution Using Kaolin. In: *IOP Conference Series: Materials Science and Engineering,* (Nasiriyah, Iraq: IOP) pp 1-7
[17] Al-Marri S, AlQuzweeni S S, Hashim K S, AlKhaddar R, Kot P, AlKizwini R S, Zubaidi S L and Al-Khafaji Z S 2020 Ultrasonic-Electrocoagulation method for nitrate removal from water. In: *IOP Conference Series: Materials Science and Engineering,* (Najaf, Iraq: IOP) pp 1-9

[18] Alnaimi H, Idan I J, Al-Janabi A, Hashim K S, Gkantou M, Zubaidi S L, Kot P and Muradov M 2020 Ultrasonic-electrochemical treatment for effluents of concrete plants. In: *IOP Conference Series: Materials Science and Engineering,* (Najaf, Iraq: IOP) pp 1-10

[19] Alyafei A, AlKizwini R S, Hashim K S, Yeboah D, Gkantou M, Al Khaddar R, Al-Faluji D and Zubaidi S L 2020 Treatment of effluents of construction industry using a combined filtration-electrocoagulation method. In: *IOP Conference Series: Materials Science and Engineering,* (Najaf, Iraq: IOP) pp 1-8

[20] Ethaib S, Omar R, Kamal S M M, Awang Biak D R and Zubaidi S L 2020 Microwave-Assisted Pyrolysis of Biomass Waste: A Mini Review *Processes* **8**

[21] Ethaib S, Omar R, Kamal S M M, Awang Biak D R and Zubaidi S L 2020 Toward Sustainable Processes of Pretreatment Technologies of Lignocellulosic Biomass for Enzymatic Production of Biofuels and Chemicals: A Review *BioResources* **15** 10063-88

[22] Hashim K S, Ewadh H M, Muhsin A A, Zubaidi S L, Kot P, Muradov M, Aljefery M and Al-Khaddar R 2020 Phosphate removal from water using bottom ash: Adsorption performance, coexisting anions and modelling studies *Water Science and Technology* **3** 1-17

[23] Zubaidi S L, Ortega-Martorell S, Kot P, Alkhaddar R M, Abdellatif M, Gharghan S K, Ahmed M S and Hashim K 2020 A Method for Predicting Long-Term Municipal Water Demands Under Climate Change *Water Resources Management* **34** 1265-79

[24] Adamowski J, Fung Chan H, Prasher S O, Ozga-Zielinski B and Sliusarieva A 2012 Comparison of multiple linear and nonlinear regression, autoregressive integrated moving average, artificial neural network, and wavelet artificial neural network methods for urban water demand forecasting in Montreal, Canada *Water Resources Research* **48** 1-14

[25] Zubaidi S L, Dooley J, Alkhaddar R M, Abdellatif M, Al-Bugharbee H and Ortega-Martorell S 2018 A Novel approach for predicting monthly water demand by combining singular spectrum analysis with neural networks *Journal of Hydrology* **561** 136-45

[26] Zubaidi S L, Gharghan S K, Dooley J, Alkhaddar R M and Abdellatif M 2018 Short-Term Urban Water Demand Prediction Considering Weather Factors *Water Resources Management* **32** 4527-42

[27] Rasifaghihi N, Li S S and Haghighat F 2020 Forecast of urban water consumption under the impact of climate change *Sustainable Cities and Society* **52**

[28] Zubaidi S L, Ortega-Martorell S, Al-Bugharbee H, Olier I, Hashim K S, Gharghan S K, Kot P and Alkhaddar R M 2020 Urban Water Demand Prediction for a City That Suffers from Climate Change and Population Growth: Gauteng Province Case Study *Water* **12** 1-17

[29] Aljaaf A J, Van Tonder L, Mallucci C, Al-Jumeily D, Hussain A and Alloghani M 2019 Patients Attitude to Technology *Journal of medical systems* **43** 1-7

[30] Aljaaf A J, Mohsin T M, Al-Jumeily D and Alloghani M 2021 A fusion of data science and feed-forward neural network-based modelling of COVID-19 outbreak forecasting in IRAQ *Journal of Biomedical Informatics* **118** 1-8

[31] Al-Bugharbee H, Abolfathi A and Trendafilova I 2018 Vibration-Based Damage Detection of Structural Joints in Presence of Uncertainty *MATEC Web of Conferences* **148** 1-6

[32] Bugharbee H A and Trendafilova I 2018 A New Methodology for Fault Detection in Rolling Element Bearings Using Singular Spectrum Analysis *MATEC Web of Conferences* **148** 1-5

[33] Garcia D, Trendafilova I and Al-Bugharbee H 2014 Vibration-based health monitoring approach for composite structures using multivariate statistical analysis. In: *EWSHM-7th European workshop on structural health monitoring,* (France, Nantes: hal-01022019) pp 1743-50

[34] Cobaner M, Citakoglu H, Kisi O and Haktanir T 2014 Estimation of mean monthly air temperatures in Turkey *Computers and Electronics in Agriculture* **109** 71-9

[35] Appelhans T, Mwangomo E, Hardy D R, Hemp A and Nauss T 2015 Evaluating machine learning approaches for the interpolation of monthly air temperature at Mt. Kilimanjaro, Tanzania *Spatial Statistics* **14** 91-113

[36] Salcedo-Sanz S, Deo R C, Carro-Calvo L and Saavedra-Moreno B 2015 Monthly prediction of air temperature in Australia and New Zealand with machine learning algorithms *Theoretical and Applied Climatology* **125** 13-25
[37] Al-Bugharbee H and Trendafilova I 2015 Autoregressive Modelling for Rolling Element Bearing Fault Diagnosis *Journal of Physics: Conference Series* **628** 1-8
[38] Al-Bugharbee H and Trendafilova I 2016 A Fault Diagnosis Methodology for Rolling Element Bearings Based on Advanced Signal Pretreatment And Autoregressive Modelling *Journal of Sound and Vibration* **369** 246-65
[39] Zubaidi S L, Kot P, Alkhaddar R M, Abdellatif M and Al-Bugharbee H 2018 Short-Term Water Demand Prediction in Residential Complexes: Case Study in Columbia City, USA. In: *11th International Conference on Developments in eSystems Engineering (DeSE),* (Cambridge, United Kingdom: 11th International Conference on Developments in eSystems Engineering (DeSE). IEEE) pp 31-5
[40] Zubaidi S L, Al-Bugharbee H, Muhsen Y R, Hashim K, Alkhaddar R M and Hmeesh W H 2019 The Prediction of Municipal Water Demand in Iraq: A Case Study of Baghdad Governorate. In: *12th International Conference on Developments in eSystems Engineering (DeSE),* (Kazan, Russia: 12th International Conference on Developments in eSystems Engineering (DeSE). IEEE) pp 274-7
[41] Farhan S L, Hashim I A J and Naji A A 2019 The Sustainable House: Comparative Analysis of Houses in Al Kut Neighborhoods-Iraq. In: *2019 12th International Conference on Developments in eSystems Engineering (DeSE),* (Kazan, Russia: IEEE) pp 1031-6
[42] Al-Maliki L A, Farhan S L, Jasim I A, Al-Mamoori S K and Al-Ansari N 2021 Perceptions about water pollution among university students: A case study from Iraq *Cogent Engineering* **8** 1895473
[43] Zubaidi S L, Al-Bugharbee H, Muhsin Y R, Hashim K and Alkhaddar R 2020 Forecasting of monthly stochastic signal of urban water demand: Baghdad as a case study. In: *IOP Conference Series: Materials Science and Engineering,* (Najaf, Iraq: IOP) pp 1-7
[44] Tabachnick B G and Fidell L S 2013 *Using Multivariate Statistics* vol sixth ed (United States of America: Pearson Education, Inc)
[45] Pallant J 2016 *SPSS Survival Manual: A step by step guide to data analysis using IBM SPSS*: Open University Press/McGraw-Hill)
[46] Zubaidi S L, Al-Bugharbee H, Ortega-Martorell S, Gharghan S K, Olier I, Hashim K S, Al-Bdairi N S S and Kot P 2020 A Novel Methodology for Prediction Urban Water Demand by Wavelet Denoising and Adaptive Neuro-Fuzzy Inference System Approach *Water* **12** 1-17
[47] Al-Bugharbee H and Trendafilova I 2014 Fault diagnosis in roller element bearings by using a linear autoregressive model. In: *the 26th International Conference on Noise and Vibration Engineering,* (Belgium, Leuven: Katholieke Universiteit Leuven) pp 2765-76
[48] Dawson C W, Abrahart R J and See L M 2007 HydroTest: A web-based toolbox of evaluation metrics for the standardised assessment of hydrological forecasts *Environmental Modelling & Software* **22** 1034-52

Updated Moving Forecasting Model of Air Maximum Temperature

Khalid Hashim [1,2], Hussein Al-Bugharbee [3], Salah L. Zubaidi [4*], Nabeel Saleem Saad Al-Bdairi [4], Sabeeh L. Farhan [5,6], Saleem Ethaib [7]

[1] Department of Civil Engineering, Liverpool John Moores University, UK
[2] Department of Environmental Engineering/ University of Babylon, Iraq
[3] Department of Mechanical Engineering, College of Engineering, Wasit University, Iraq
[4] Department of Civil Engineering, College of Engineering, Wasit University, Iraq
[5] Department of Architecture, Engineering, College of Engineering, Wasit University, Iraq
[6] School of Architecture, Design and the Built Environment, Nottingham Trent University, UK
[7] Department of Petroleum and Gas engineering, College of Engineering, University of Thi-Qar, Al Nassirya, Iraq

* Corresponding author: k.s.hashim@ljmu.ac.uk

Abstract. In the current study, a moving forecasting model is used for the purpose of forecasting maximum air temperature. A number of recordings are used for building the AR model and next, to forecasting some temperature values ahead. Then the AR model coefficients are updating due to shifting the training sample by adding new temperature values in order to involve the change in temperature time series behaviour. The current work shows a high performance all over the temperature time series, which considered in the analysis.

Keywords

Autoregressive Model; Baghdad City; prediction model; temperature.

1. Introduction

Global warming and the variability of climate are likely to cause a substantial problem for the ecosystem and go worst based on different scenarios of climate change (i.e., the temperature increased)[1, 2]. Massive emission of greenhouse gases resulting from natural disaster (e.g., volcanoes) or human activities (using fossil fuels) led to an increase in the influence of global warming [3, 4]. It has located a considerable impact on the environment of residential area in various places of the world [5-7]. These effects differ in terms of the region, the type, and the importance.

The climatic variables have affected, directly and indirectly, both individuals and their residential environment during various periods [8]. Temperature is the most vital climatic variables that impact the growth, development and yield of crops [9]. Also, the dwellings' system is enhanced in response to climatic variables [10, 11].

Various countries face a harmful effect of global warming that led to reducing the quantity[12-16] and quality [17-22] of potable water resources. High temperatures (i.e., dry day) causes increase municipal water demand [23]. Also, various research revealed that municipal water demand was driven by maximum temperature [24-28].

Recently, precise prediction of temperature is a problem that has attracted the researchers' attention, since it has many various applications in the area such as industry, agriculture or energy. Different techniques and models are applied in several areas [29-33], and research forecast the maximum temperature by different techniques [34-36]. The AR model applied effectively in different applications [37-39]. In this research, Auto-regressive (AR) model will employ to forecast the monthly temperature.

2. Area of study and data set

Iraq is one of the Arab countries that lies in arid to the semi-arid area, and Baghdad is the capital of Iraq and locates in the centre of the country[40]. The weather is wet and cold in winter and dry and hot in summer (i.e., the temperature reaches 45 C). Iraq faced an acute climate change cased adversely impact the people, residential area, and freshwater sources[41-43]. The historical monthly data if maximum temperature along twelve years (2003-3014) used to build and assess the model.

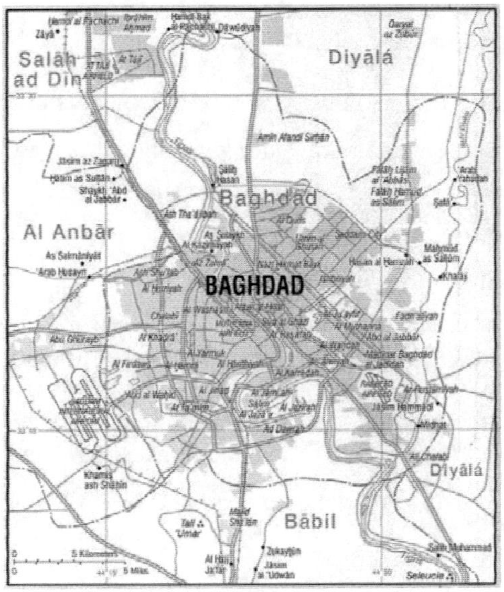

Figure 1. Location map of Baghdad city.

3. Methodology

The procedure of this research divides into, data pre-processing, and auto-regressive model.

3.1 Data Pre-processing

It has a considerable influence on the accuracy of the forecast techniques. It can be separated here into two phases: normalisation and cleaning. Normalisation time series assistances to decrease the impact of outliers and makes the data to be normal or near-normal distribution [44, 45]. In this research, a natural logarithm is used for normalising the data due to its ability to decrease the influence of multicollinearity among predictor factors[28, 46].

3.2 Autoregressive Model (AR)

The autoregression model (AR) is a common approach to process and handle time series data to predict and forecast the variable of interest depending on past observations of the same variable at different stages [47]. This model receives a growing interest in different fields in which time series data are prevalent and the prediction of current and future values corresponding to a certain variable is needed. The valuable insights obtained from this model are the main cause of its popularity compared to other methods. Also, the solid inference of this model enables the city engineers and water authorities to overcome any shortage in drinking water provided to residents in a certain city.

To mathematically formulate autoregressive models, Eq. (1) is used to relate the current observation with the past ones in a linear relationship as illustrated [37, 38]:

$$X_t = \theta_0 + \sum_{i=1}^{p} k_i X_{t-1} + \varepsilon_t \tag{1}$$

Where; X_t and X_{t-1} are the observations in periods *t* and *t-1*, p is the order of the AR model considered, k_i is the autoregressive parameters, θ_0 is the constant term, and ε_t is the disturbance term for period *t*. A least-square algorithm using MATLAB is utilized to accurately predict the unknown coefficients in the AR model.

4. Results and discussion

Firstly, the time series data of maximum temperature are normalised and cleaned. The data then splitted into training set (70%) and testing set (30%). The AR model fitness metrics, i.e., R, MAE, MSE, and RMSE are shown in Figure 2A for the shifting steps. The overlapping of the new and old training sample is taken into consideration in order to avoid the adverse effect of the time series sudden change. The results reveal that the modelling process is quite accurate which makes the forecasting outputs are reliable. From the Figure 2B, it can be seen that the minimum R is 0.9 while the maximum MAR equals 1.5. The greatest MSE is no more than 0.4 and the RMSE equals 0.6 as a maximum.

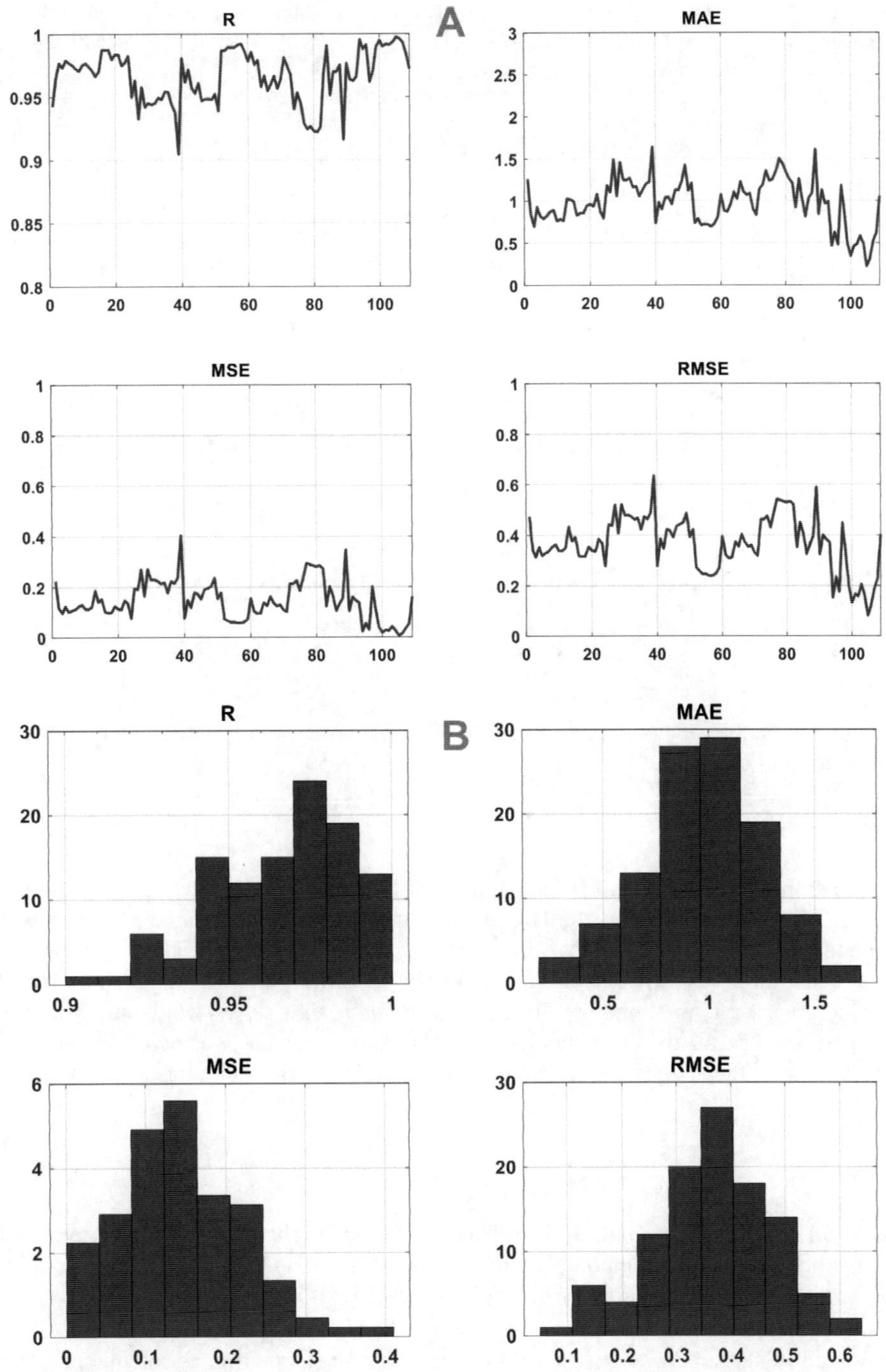

Figure 2. Four fitness metrics along many shifting of data to assess the AR model.

Figure 2 provides a comparable visualisation for measure and forecasted sample, which includes 16 monthly readings. The graph shows very similar values and trend which reflects a reliable forecasting performance.

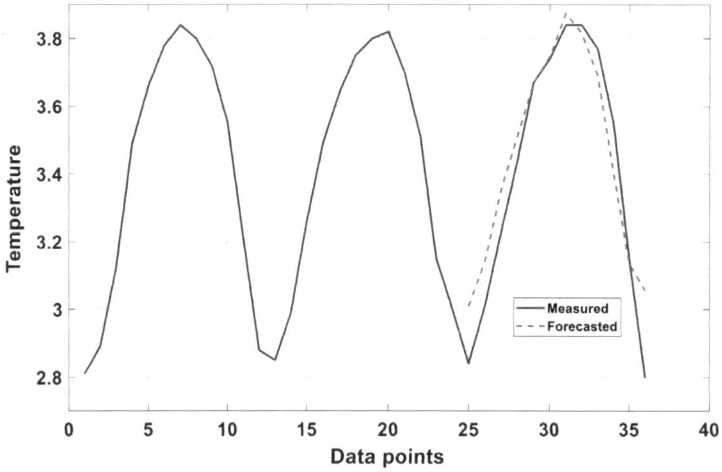

Figure 3. Measured and forecasted temperature.

From all the above tests, the AR model can simulate effectively the monthly maximum temperature along with different shifting of data.

5. Conclusions

Maximum temperature forecast is a significant component in active modern city planning and management due to it can help to find appropriate tools that used in building and industrials. In this study, data preprocessing and AR model used to forecast the monthly time series of temperature in Baghdad City over twelve years. the AR model coefficients are updating due to shifting the training sample by adding new temperature values in order to involving the change in temperature time series behaviour. The current works shows a high performance all over the temperature time series, which considered in the analysis (i.e., minimum R=0.9. Lastly, the suggest technique of this research can be applied as an early ground to base further research.

6. References

[1] Zubaidi S L, Hashim K, Ethaib S, Al-Bdairi N S S, Al-Bugharbee H and Gharghan S K 2020 A novel methodology to predict monthly municipal water demand based on weather variables scenario *Journal of King Saud University - Engineering Sciences* 1-7

[2] Mohammed R and Scholz M 2019 Climate variability impact on the spatiotemporal characteristics of drought and Aridityin arid and semi-arid regions *Water Resources Management* **33** 5015-33

[3] Salman S A, Shahid S, Ismail T, Ahmed K and Wang X-J 2018 Selection of climate models for projection of spatiotemporal changes in temperature of Iraq with uncertainties *Atmospheric Research* **213** 509-22

[4] Zubaidi S L, Abdulkareem I H, Hashim K, Al-Bugharbee H, Ridha H M, Gharghan S K, Al-Qaim F F, Muradov M, Kot P and Al-Khaddar R 2020 Hybridised Artificial Neural Network Model with Slime Mould Algorithm: A Novel Methodology for Prediction of Urban Stochastic Water Demand *Water* **12** 1-18

[5] Jasim I A, Farhan S L, Al-Maliki L A and AL-Mamoori S K 2021 Climatic Treatments for Housing in the Traditional Holy Cities: A Comparison between Najaf and Yazd Cities. In: *IOP Conference Series: Earth and Environmental Science*: IOP Publishing) pp 1-10

[6] Farhan S L and Nasar Z A 2020 Urban identity in the holy cities of Iraq: Analysis of architectural design trends in the city of Karbala *Journal of Urban Regeneration and Renewal* **14** 210-22

[7] Farhan S, Akef V, Antón D, Hashim K and Zubaidi S 2021 Factors influencing the transformation of Iraqi holy cities: the case of Al-Najaf *Przegląd Naukowy Inżynieria i Kształtowanie Środowiska* **30** 365-75

[8] Zubaidi S L, Kot P, Hashim K, Alkhaddar R, Abdellatif M and Muhsin Y R 2019 Using LARS – WG model for prediction of temperature in Columbia City, USA. In: *IOP Conference Series: Materials Science and Engineering,* (Najaf, Iraq: Materials Science and Engineering. IOP) pp 1-9

[9] Kadiyala M D, Nedumaran S, Singh P, S C, Irshad M A and Bantilan M C 2015 An integrated crop model and GIS decision support system for assisting agronomic decision making under climate change *Science of the Total Environment* **521-522** 123-34

[10] Farhan S L, Jasim I A and Al-Mamoori S K 2019 The Transformation of The City of Najaf, Iraq: Analysis, Reality and Future Prospects *Journal of Urban Regeneration and Renewal* **13** 1-12

[11] Farhan S, Akef V and Nasar Z 2020 The transformation of the inherited historical urban and architectural characteristics of Al-Najaf's Old City and possible preservation insights *Frontiers of Architectural Research* 1-17

[12] Mohammed R, Scholz M, Nanekely M, Mokhtari Y and assessment r 2018 Assessment of models predicting anthropogenic interventions and climate variability on surface runoff of the Lower Zab River *Stochastic Environmental Research and Risk Assessment* **32** 223-40

[13] Mohammed R and Scholz M 2018 Flow–duration curve integration into digital filtering algorithms for simulating climate variability based on river baseflow *Hydrological Sciences Journal* **63** 1558-73

[14] Hashim K S, Kot P, Zubaidi S L, Alwash R, Al Khaddar R, Shaw A, Al-Jumeily D and Aljefery M H 2020 Energy Efficient Electrocoagulation Using Baffle-Plates Electrodes for Efficient Escherichia Coli Removal from Wastewater *Journal of Water Process Engineering* **33** 1-7

[15] Hashim K S, Hussein A H, Zubaidi S L, Kot P, Kraidi L, Alkhaddar R, Shaw A and Alwash R 2019 Effect of Initial Ph Value on The Removal of Reactive Black Dye from Water by Electrocoagulation (EC) Method *Journal of Physics: Conference Series* **1294** 1-6

[16] Ethaib S and Zubaidi S L 2020 Removal of Methylene Blue Dye from Aqueous Solution Using Kaolin. In: *IOP Conference Series: Materials Science and Engineering,* (Nasiriyah, Iraq: IOP) pp 1-7

[17] Al-Marri S, AlQuzweeni S S, Hashim K S, AlKhaddar R, Kot P, AlKizwini R S, Zubaidi S L and Al-Khafaji Z S 2020 Ultrasonic-Electrocoagulation method for nitrate removal from water. In: *IOP Conference Series: Materials Science and Engineering,* (Najaf, Iraq: IOP) pp 1-9

[18] Alnaimi H, Idan I J, Al-Janabi A, Hashim K S, Gkantou M, Zubaidi S L, Kot P and Muradov M 2020 Ultrasonic-electrochemical treatment for effluents of concrete plants. In: *IOP Conference Series: Materials Science and Engineering,* (Najaf, Iraq: IOP) pp 1-10

[19] Alyafei A, AlKizwini R S, Hashim K S, Yeboah D, Gkantou M, Al Khaddar R, Al-Faluji D and Zubaidi S L 2020 Treatment of effluents of construction industry using a combined filtration-electrocoagulation method. In: *IOP Conference Series: Materials Science and Engineering,* (Najaf, Iraq: IOP) pp 1-8

[20] Ethaib S, Omar R, Kamal S M M, Awang Biak D R and Zubaidi S L 2020 Microwave-Assisted Pyrolysis of Biomass Waste: A Mini Review *Processes* **8**

[21] Ethaib S, Omar R, Kamal S M M, Awang Biak D R and Zubaidi S L 2020 Toward Sustainable Processes of Pretreatment Technologies of Lignocellulosic Biomass for Enzymatic Production of Biofuels and Chemicals: A Review *BioResources* **15** 10063-88

[22] Hashim K S, Ewadh H M, Muhsin A A, Zubaidi S L, Kot P, Muradov M, Aljefery M and Al-Khaddar R 2020 Phosphate removal from water using bottom ash: Adsorption performance, coexisting anions and modelling studies *Water Science and Technology* **3** 1-17

[23] Zubaidi S L, Ortega-Martorell S, Kot P, Alkhaddar R M, Abdellatif M, Gharghan S K, Ahmed M S and Hashim K 2020 A Method for Predicting Long-Term Municipal Water Demands Under Climate Change *Water Resources Management* **34** 1265-79

[24] Adamowski J, Fung Chan H, Prasher S O, Ozga-Zielinski B and Sliusarieva A 2012 Comparison of multiple linear and nonlinear regression, autoregressive integrated moving average, artificial neural network, and wavelet artificial neural network methods for urban water demand forecasting in Montreal, Canada *Water Resources Research* **48** 1-14

[25] Zubaidi S L, Dooley J, Alkhaddar R M, Abdellatif M, Al-Bugharbee H and Ortega-Martorell S 2018 A Novel approach for predicting monthly water demand by combining singular spectrum analysis with neural networks *Journal of Hydrology* **561** 136-45

[26] Zubaidi S L, Gharghan S K, Dooley J, Alkhaddar R M and Abdellatif M 2018 Short-Term Urban Water Demand Prediction Considering Weather Factors *Water Resources Management* **32** 4527-42

[27] Rasifaghihi N, Li S S and Haghighat F 2020 Forecast of urban water consumption under the impact of climate change *Sustainable Cities and Society* **52**

[28] Zubaidi S L, Ortega-Martorell S, Al-Bugharbee H, Olier I, Hashim K S, Gharghan S K, Kot P and Alkhaddar R M 2020 Urban Water Demand Prediction for a City That Suffers from Climate Change and Population Growth: Gauteng Province Case Study *Water* **12** 1-17

[29] Aljaaf A J, Van Tonder L, Mallucci C, Al-Jumeily D, Hussain A and Alloghani M 2019 Patients Attitude to Technology *Journal of medical systems* **43** 1-7

[30] Aljaaf A J, Mohsin T M, Al-Jumeily D and Alloghani M 2021 A fusion of data science and feed-forward neural network-based modelling of COVID-19 outbreak forecasting in IRAQ *Journal of Biomedical Informatics* **118** 1-8

[31] Al-Bugharbee H, Abolfathi A and Trendafilova I 2018 Vibration-Based Damage Detection of Structural Joints in Presence of Uncertainty *MATEC Web of Conferences* **148** 1-6

[32] Bugharbee H A and Trendafilova I 2018 A New Methodology for Fault Detection in Rolling Element Bearings Using Singular Spectrum Analysis *MATEC Web of Conferences* **148** 1-5

[33] Garcia D, Trendafilova I and Al-Bugharbee H 2014 Vibration-based health monitoring approach for composite structures using multivariate statistical analysis. In: *EWSHM-7th European workshop on structural health monitoring,* (France, Nantes: hal-01022019) pp 1743-50

[34] Cobaner M, Citakoglu H, Kisi O and Haktanir T 2014 Estimation of mean monthly air temperatures in Turkey *Computers and Electronics in Agriculture* **109** 71-9

[35] Appelhans T, Mwangomo E, Hardy D R, Hemp A and Nauss T 2015 Evaluating machine learning approaches for the interpolation of monthly air temperature at Mt. Kilimanjaro, Tanzania *Spatial Statistics* **14** 91-113

[36] Salcedo-Sanz S, Deo R C, Carro-Calvo L and Saavedra-Moreno B 2015 Monthly prediction of air temperature in Australia and New Zealand with machine learning algorithms *Theoretical and Applied Climatology* **125** 13-25

[37] Al-Bugharbee H and Trendafilova I 2015 Autoregressive Modelling for Rolling Element Bearing Fault Diagnosis *Journal of Physics: Conference Series* **628** 1-8

[38] Al-Bugharbee H and Trendafilova I 2016 A Fault Diagnosis Methodology for Rolling Element Bearings Based on Advanced Signal Pretreatment And Autoregressive Modelling *Journal of Sound and Vibration* **369** 246-65

[39] Zubaidi S L, Kot P, Alkhaddar R M, Abdellatif M and Al-Bugharbee H 2018 Short-Term Water Demand Prediction in Residential Complexes: Case Study in Columbia City, USA. In: *11th International Conference on Developments in eSystems Engineering (DeSE),* (Cambridge, United Kingdom: 11th International Conference on Developments in eSystems Engineering (DeSE). IEEE) pp 31-5

[40] Zubaidi S L, Al-Bugharbee H, Muhsen Y R, Hashim K, Alkhaddar R M and Hmeesh W H 2019 The Prediction of Municipal Water Demand in Iraq: A Case Study of Baghdad Governorate. In: *12th International Conference on Developments in eSystems Engineering (DeSE),* (Kazan, Russia: 12th International Conference on Developments in eSystems Engineering (DeSE). IEEE) pp 274-7

[41] Farhan S L, Hashim I A J and Naji A A 2019 The Sustainable House: Comparative Analysis of Houses in Al Kut Neighborhoods-Iraq. In: *2019 12th International Conference on Developments in eSystems Engineering (DeSE),* (Kazan, Russia: IEEE) pp 1031-6

[42] Al-Maliki L A, Farhan S L, Jasim I A, Al-Mamoori S K and Al-Ansari N 2021 Perceptions about water pollution among university students: A case study from Iraq *Cogent Engineering* **8** 1895473

[43] Zubaidi S L, Al-Bugharbee H, Muhsin Y R, Hashim K and Alkhaddar R 2020 Forecasting of monthly stochastic signal of urban water demand: Baghdad as a case study. In: *IOP Conference Series: Materials Science and Engineering,* (Najaf, Iraq: IOP) pp 1-7

[44] Tabachnick B G and Fidell L S 2013 *Using Multivariate Statistics* vol sixth ed (United States of America: Pearson Education, Inc)

[45] Pallant J 2016 *SPSS Survival Manual: A step by step guide to data analysis using IBM SPSS*: Open University Press/McGraw-Hill)

[46] Zubaidi S L, Al-Bugharbee H, Ortega-Martorell S, Gharghan S K, Olier I, Hashim K S, Al-Bdairi N S S and Kot P 2020 A Novel Methodology for Prediction Urban Water Demand by Wavelet Denoising and Adaptive Neuro-Fuzzy Inference System Approach *Water* **12** 1-17

[47] Al-Bugharbee H and Trendafilova I 2014 Fault diagnosis in roller element bearings by using a linear autoregressive model. In: *the 26th International Conference on Noise and Vibration Engineering,* (Belgium, Leuven: Katholieke Universiteit Leuven) pp 2765-76

Forecasting of Air Maximum Temperature on Monthly Basis Using Singular Spectrum Analysis and Linear Autoregressive Model

Nabeel Saleem Saad Al-Bdairi [1], Salah L. Zubaidi [2*], Hussein Al-Bugharbee [1], Khalid Hashim [3,4], Sabeeh L. Farhan [5,6], Asad Al Defae [2]

[1] Department of Mechanical Engineering, College of Engineering, Wasit University, Iraq
[2] Department of Civil Engineering, College of Engineering, Wasit University, Iraq
[3] Department of Civil Engineering, Liverpool John Moores University, UK
[4] Department of Environmental Engineering/ University of Babylon, Iraq
[5] Department of Architecture, Engineering, College of Engineering, Wasit University, Iraq
[6] School of Architecture, Design and the Built Environment, Nottingham Trent University, UK

* Corresponding author: nsaleem@uowasit.edu.iq

Abstract. In this research, the singular spectrum analysis technique is combined with a linear autoregressive model for the purpose of prediction and forecasting of monthly maximum air temperature. The temperature time series is decomposed into three components and the trend component is subjected for modelling. The performance of modelling for both prediction and forecasting is evaluated via various model fitness function. The results show that the current method presents an excellent performance in expecting the maximum air temperature in future based on previous recordings.

Keywords: Autoregressive Model; Baghdad City; prediction model; temperature.

1. Introduction

The ecosystem has faced significant problems resulting from the variability of climate and global warming. These problems are likely to go worst depend on various climate change's scenarios (i.e., the temperature increased)[1, 2]. The natural disaster (e.g., volcanoes), industrialisation, and rapid increase of urbanisation led to Enormous greenhouse gases emission that in the last influence adversely climate change [3, 4]. It has set a considerable influence on the environment of residential areas in many places of the world [5-7]. These impacts differ in relation to the region, the type, and the importance.

The factors of climatic have directly and indirectly impacted both inhabitants and their residential environment along with short-, medium-, and long-terms [8]. Temperature is considered the most vital climatic variable that influences the growth, development and yield of crops [9]. In Addition, the system of houses is advanced in relation to factors of climatic [10, 11].

Different areas have faced a damaging impact of climate change that causes diminishing the quantity [12-16] and quality [17-22] of drinking water resources. High temperatures (i.e., dry days) lead to the growing urban water needed [23]. In addition, various studies presented that urban water needed was driven by maximum temperature [24-28].
Lately, a precise estimate of maximum temperature is a problem that has attracted the attention of researchers, since it has several various usages in the applications such as industry, agriculture or energy. Many approaches and methods are utilised in several fields [29-33], and research simulate the maximum temperature by different techniques [34-36]. The AR technique employed efficiently in several applications [37-39]. In this study, Auto-regressive (AR) approach will be applied to estimate the monthly maximum temperature.

2. Area of study and data set

Iraq is one of the Arab countries that lies in arid to the semi-arid area, and Baghdad is the capital of Iraq and locates in the centre of the country[40]. The weather is wet and cold in winter and dry and hot in summer (i.e., the temperature reaches 45 C). Iraq faced an acute climate change cased adversely impact the people, residential area, and freshwater sources[41-43]. The historical monthly data if maximum temperature along twelve years (2003-3014) used to build and assess the model.

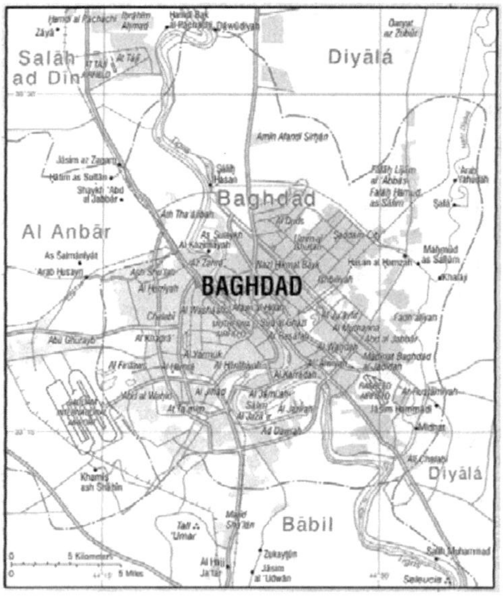

Figure 1. Location map of Baghdad city.

3. Methodology

The technique of this study divides into, data pre-processing, and auto-regressive approach.

3.1 Data Pre-processing

This technique has a significant effect on the precision of the prediction techniques. It can be separated here into two stages: normalisation and cleaning. Normalisation time series assistances to reduction the influence of outliers and makes the data to be normal or near-normal distribution [44, 45]. In this study, a natural logarithm is used for normalising the data due to its capability to reduction the influence of multicollinearity among predictor factors[28, 46].

3.2 Autoregressive Model (AR)

In autoregressive (AR) model, the output pertaining to a particular variable can be predicted from the past observations of that variable [47]. This model has a linear form. As such, the simplicity of this model coupled with its powerful prediction increases the popularity of this model in different disciplines in which time series data need to be analysed. In water demand forecasting, city engineers and water authorities are working collectively to maintain the balance between the demand and supply of drinking water to residents in their city. Hence, to achieve this goal, a sound statistical method should be used. Accordingly, there is a growing interest in applying autoregressive model in water demand forecasting. The outputs in this model are merely dependent on the previous observations of the same variable [37, 38].

To mathematically formulate autoregressive models, Eq. (1) is used to relate the current observation with the past ones in a linear relationship as illustrated [37, 38]:

$$X_t = \theta_0 + \sum_{i=1}^{p} k_i X_{t-1} + \varepsilon_t \tag{1}$$

Where; X_t and X_{t-1} are the observations in periods t and t-1, p is the order of the AR model considered, k_i is the autoregressive parameters, θ_0 is the constant term, and ε_t is the disturbance term for period t. A least-square algorithm using MATLAB is utilised to accurately predict the unknown coefficients in the AR model.

4. Results and discussion

Initially, the maximum temperature data are normalised and cleaned. Afterwards, the time series of maximum temperature was analysis. Figure 2 visualises the temperature time series components obtained from the singular spectrum analysis. The first component represents the trend component, which follows the same fluctuation of the original time series. It is clear that this component, i.e. trend component, has the greatest portion of the variance of the original time series.

The data of trend then categorised into training set (70%, 101 data points) and testing set (30%, 43 data points). The AR approach fitness criteria, i.e., R^2, MAE, MSE, and RMSE are shown in Table 1 for the training and testing stages. The table presents the model fitness at order 10 for both the prediction of the training sample and forecasting the testing sample. The comparison shows that the use of the combined technique outperforms the use of AR alone for both the prediction and forecasting processes. This can be clearly seen from the reduction in the MAE, MSE, and RMSE values when the SSA-AR is used.

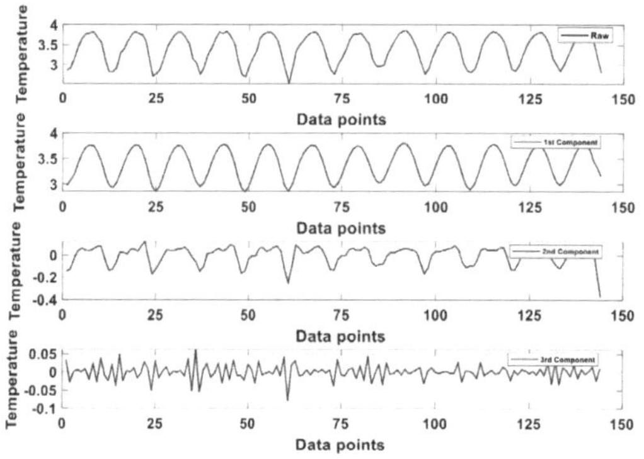

Figure 2. The normalised and the first three signals of maximum temperature time series.

Table 1: Model fitness for AR and SSA-AR methods

	Training sample				Testing sample			
	R^2	MAE	MSE	RMSE	R^2	MAE	MSE	RMSE
AR	0.92	7.67	1.03	1	0.94	2.92	0.3	0.55
SSA-AR	0.98	3.16	0.17	0.41	0.94	2.36	0.19	0.44

Additionally, Figure 3 represents the visual comparison between the measured and forecasted values of the testing sample. As it was mentioned earlier that the testing sample includes 44 recordings. It is clearly seen how precise the model forecasting is.

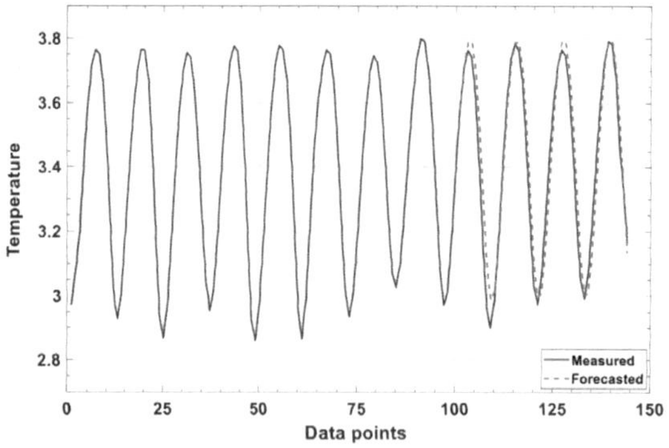

Figure 3. The comparison between measured and forecasted maximum temperature time series.

Moreover, Figure 4 illustrates the model forecasting error histogram. It can be seen that the absolute error value is 0.15 as a maximum while the majority of the errors range in between [-0.1 +0.1].

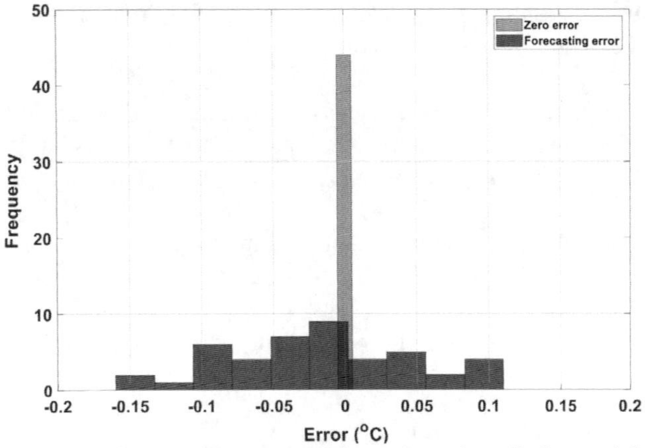

Figure 4. The error histogram of the suggested prediction model.

Based on the above statistical tests, the suggested methodology has the ability for simulating efficiently the monthly maximum temperature considering the pervious data.

5. Conclusions

Maximum temperature estimate is an important component in dynamic modern city planning and management because it helps to find suitable tools that utilised in building and industrials. In this research, the combined model includes the SSA and AR techniques employed for forecasting the monthly time series of temperature in Baghdad City over twelve years. The SSA used to denoise the time series of maximum temperature. The AR technique used to simulate the time series of maximum temperature. The outcomes reveal that the SSA is a suitable technique for noise removal. Also, the suggested methodology is effective to forecast the maximum temperature data in the testing stage (i.e., it is yield coefficient of determination 0.98 and 0.94 for training and testing stage, respectively). These results can be considered as an initial base of combined techniques for additional research in future.

6. References

[1] Zubaidi S L, Hashim K, Ethaib S, Al-Bdairi N S S, Al-Bugharbee H and Gharghan S K 2020 A novel methodology to predict monthly municipal water demand based on weather variables scenario *Journal of King Saud University - Engineering Sciences* 1-7

[2] Mohammed R and Scholz M 2019 Climate variability impact on the spatiotemporal characteristics of drought and Aridityin arid and semi-arid regions *Water Resources Management* **33** 5015-33

[3] Salman S A, Shahid S, Ismail T, Ahmed K and Wang X-J 2018 Selection of climate models for projection of spatiotemporal changes in temperature of Iraq with uncertainties *Atmospheric Research* **213** 509-22

[4] Zubaidi S L, Abdulkareem I H, Hashim K, Al-Bugharbee H, Ridha H M, Gharghan S K, Al-Qaim F F, Muradov M, Kot P and Al-Khaddar R 2020 Hybridised Artificial Neural Network Model

with Slime Mould Algorithm: A Novel Methodology for Prediction of Urban Stochastic Water Demand *Water* **12** 1-18

[5] Jasim I A, Farhan S L, Al-Maliki L A and AL-Mamoori S K 2021 Climatic Treatments for Housing in the Traditional Holy Cities: A Comparison between Najaf and Yazd Cities. In: *IOP Conference Series: Earth and Environmental Science*: IOP Publishing) pp 1-10

[6] Farhan S L and Nasar Z A 2020 Urban identity in the holy cities of Iraq: Analysis of architectural design trends in the city of Karbala *Journal of Urban Regeneration and Renewal* **14** 210-22

[7] Farhan S, Akef V, Antón D, Hashim K and Zubaidi S 2021 Factors influencing the transformation of Iraqi holy cities: the case of Al-Najaf *Przegląd Naukowy Inżynieria i Kształtowanie Środowiska* **30** 365-75

[8] Zubaidi S L, Kot P, Hashim K, Alkhaddar R, Abdellatif M and Muhsin Y R 2019 Using LARS – WG model for prediction of temperature in Columbia City, USA. In: *IOP Conference Series: Materials Science and Engineering,* (Najaf, Iraq: Materials Science and Engineering. IOP) pp 1-9

[9] Kadiyala M D, Nedumaran S, Singh P, S C, Irshad M A and Bantilan M C 2015 An integrated crop model and GIS decision support system for assisting agronomic decision making under climate change *Science of the Total Environment* **521-522** 123-34

[10] Farhan S L, Jasim I A and Al-Mamoori S K 2019 The Transformation of The City of Najaf, Iraq: Analysis, Reality and Future Prospects *Journal of Urban Regeneration and Renewal* **13** 1-12

[11] Farhan S, Akef V and Nasar Z 2020 The transformation of the inherited historical urban and architectural characteristics of Al-Najaf's Old City and possible preservation insights *Frontiers of Architectural Research* 1-17

[12] Mohammed R, Scholz M, Nanekely M, Mokhtari Y and assessment r 2018 Assessment of models predicting anthropogenic interventions and climate variability on surface runoff of the Lower Zab River *Stochastic Environmental Research and Risk Assessment* **32** 223-40

[13] Mohammed R and Scholz M 2018 Flow–duration curve integration into digital filtering algorithms for simulating climate variability based on river baseflow *Hydrological Sciences Journal* **63** 1558-73

[14] Hashim K S, Kot P, Zubaidi S L, Alwash R, Al Khaddar R, Shaw A, Al-Jumeily D and Aljefery M H 2020 Energy Efficient Electrocoagulation Using Baffle-Plates Electrodes for Efficient Escherichia Coli Removal from Wastewater *Journal of Water Process Engineering* **33** 1-7

[15] Hashim K S, Hussein A H, Zubaidi S L, Kot P, Kraidi L, Alkhaddar R, Shaw A and Alwash R 2019 Effect of Initial Ph Value on The Removal of Reactive Black Dye from Water by Electrocoagulation (EC) Method *Journal of Physics: Conference Series* **1294** 1-6

[16] Ethaib S and Zubaidi S L 2020 Removal of Methylene Blue Dye from Aqueous Solution Using Kaolin. In: *IOP Conference Series: Materials Science and Engineering,* (Nasiriyah, Iraq: IOP) pp 1-7

[17] Al-Marri S, AlQuzweeni S S, Hashim K S, AlKhaddar R, Kot P, AlKizwini R S, Zubaidi S L and Al-Khafaji Z S 2020 Ultrasonic-Electrocoagulation method for nitrate removal from water. In: *IOP Conference Series: Materials Science and Engineering,* (Najaf, Iraq: IOP) pp 1-9

[18] Alnaimi H, Idan I J, Al-Janabi A, Hashim K S, Gkantou M, Zubaidi S L, Kot P and Muradov M 2020 Ultrasonic-electrochemical treatment for effluents of concrete plants. In: *IOP Conference Series: Materials Science and Engineering,* (Najaf, Iraq: IOP) pp 1-10

[19] Alyafei A, AlKizwini R S, Hashim K S, Yeboah D, Gkantou M, Al Khaddar R, Al-Faluji D and Zubaidi S L 2020 Treatment of effluents of construction industry using a combined filtration-electrocoagulation method. In: *IOP Conference Series: Materials Science and Engineering,* (Najaf, Iraq: IOP) pp 1-8

[20] Ethaib S, Omar R, Kamal S M M, Awang Biak D R and Zubaidi S L 2020 Microwave-Assisted Pyrolysis of Biomass Waste: A Mini Review *Processes* **8**

[21] Ethaib S, Omar R, Kamal S M M, Awang Biak D R and Zubaidi S L 2020 Toward Sustainable Processes of Pretreatment Technologies of Lignocellulosic Biomass for Enzymatic Production of Biofuels and Chemicals: A Review *BioResources* **15** 10063-88

[22] Hashim K S, Ewadh H M, Muhsin A A, Zubaidi S L, Kot P, Muradov M, Aljefery M and Al-Khaddar R 2020 Phosphate removal from water using bottom ash: Adsorption performance, coexisting anions and modelling studies *Water Science and Technology* **3** 1-17

[23] Zubaidi S L, Ortega-Martorell S, Kot P, Alkhaddar R M, Abdellatif M, Gharghan S K, Ahmed M S and Hashim K 2020 A Method for Predicting Long-Term Municipal Water Demands Under Climate Change *Water Resources Management* **34** 1265-79

[24] Adamowski J, Fung Chan H, Prasher S O, Ozga-Zielinski B and Sliusarieva A 2012 Comparison of multiple linear and nonlinear regression, autoregressive integrated moving average, artificial neural network, and wavelet artificial neural network methods for urban water demand forecasting in Montreal, Canada *Water Resources Research* **48** 1-14

[25] Zubaidi S L, Dooley J, Alkhaddar R M, Abdellatif M, Al-Bugharbee H and Ortega-Martorell S 2018 A Novel approach for predicting monthly water demand by combining singular spectrum analysis with neural networks *Journal of Hydrology* **561** 136-45

[26] Zubaidi S L, Gharghan S K, Dooley J, Alkhaddar R M and Abdellatif M 2018 Short-Term Urban Water Demand Prediction Considering Weather Factors *Water Resources Management* **32** 4527-42

[27] Rasifaghihi N, Li S S and Haghighat F 2020 Forecast of urban water consumption under the impact of climate change *Sustainable Cities and Society* **52**

[28] Zubaidi S L, Ortega-Martorell S, Al-Bugharbee H, Olier I, Hashim K S, Gharghan S K, Kot P and Alkhaddar R M 2020 Urban Water Demand Prediction for a City That Suffers from Climate Change and Population Growth: Gauteng Province Case Study *Water* **12** 1-17

[29] Aljaaf A J, Van Tonder L, Mallucci C, Al-Jumeily D, Hussain A and Alloghani M 2019 Patients Attitude to Technology *Journal of medical systems* **43** 1-7

[30] Aljaaf A J, Mohsin T M, Al-Jumeily D and Alloghani M 2021 A fusion of data science and feed-forward neural network-based modelling of COVID-19 outbreak forecasting in IRAQ *Journal of Biomedical Informatics* **118** 1-8

[31] Al-Bugharbee H, Abolfathi A and Trendafilova I 2018 Vibration-Based Damage Detection of Structural Joints in Presence of Uncertainty *MATEC Web of Conferences* **148** 1-6

[32] Bugharbee H A and Trendafilova I 2018 A New Methodology for Fault Detection in Rolling Element Bearings Using Singular Spectrum Analysis *MATEC Web of Conferences* **148** 1-5

[33] Garcia D, Trendafilova I and Al-Bugharbee H 2014 Vibration-based health monitoring approach for composite structures using multivariate statistical analysis. In: *EWSHM-7th European workshop on structural health monitoring,* (France, Nantes: hal-01022019) pp 1743-50

[34] Cobaner M, Citakoglu H, Kisi O and Haktanir T 2014 Estimation of mean monthly air temperatures in Turkey *Computers and Electronics in Agriculture* **109** 71-9

[35] Appelhans T, Mwangomo E, Hardy D R, Hemp A and Nauss T 2015 Evaluating machine learning approaches for the interpolation of monthly air temperature at Mt. Kilimanjaro, Tanzania *Spatial Statistics* **14** 91-113

[36] Salcedo-Sanz S, Deo R C, Carro-Calvo L and Saavedra-Moreno B 2015 Monthly prediction of air temperature in Australia and New Zealand with machine learning algorithms *Theoretical and Applied Climatology* **125** 13-25

[37] Al-Bugharbee H and Trendafilova I 2015 Autoregressive Modelling for Rolling Element Bearing Fault Diagnosis *Journal of Physics: Conference Series* **628** 1-8

[38] Al-Bugharbee H and Trendafilova I 2016 A Fault Diagnosis Methodology for Rolling Element Bearings Based on Advanced Signal Pretreatment And Autoregressive Modelling *Journal of Sound and Vibration* **369** 246-65

[39] Zubaidi S L, Kot P, Alkhaddar R M, Abdellatif M and Al-Bugharbee H 2018 Short-Term Water Demand Prediction in Residential Complexes: Case Study in Columbia City, USA. In: *11th International Conference on Developments in eSystems Engineering (DeSE),* (Cambridge, United Kingdom: 11th International Conference on Developments in eSystems Engineering (DeSE). IEEE) pp 31-5

[40] Zubaidi S L, Al-Bugharbee H, Muhsen Y R, Hashim K, Alkhaddar R M and Hmeesh W H 2019 The Prediction of Municipal Water Demand in Iraq: A Case Study of Baghdad Governorate. In:

12th International Conference on Developments in eSystems Engineering (DeSE), (Kazan, Russia: 12th International Conference on Developments in eSystems Engineering (DeSE). IEEE) pp 274-7

[41] Farhan S L, Hashim I A J and Naji A A 2019 The Sustainable House: Comparative Analysis of Houses in Al Kut Neighborhoods-Iraq. In: *2019 12th International Conference on Developments in eSystems Engineering (DeSE),* (Kazan, Russia: IEEE) pp 1031-6

[42] Al-Maliki L A, Farhan S L, Jasim I A, Al-Mamoori S K and Al-Ansari N 2021 Perceptions about water pollution among university students: A case study from Iraq *Cogent Engineering* **8** 1895473

[43] Zubaidi S L, Al-Bugharbee H, Muhsin Y R, Hashim K and Alkhaddar R 2020 Forecasting of monthly stochastic signal of urban water demand: Baghdad as a case study. In: *IOP Conference Series: Materials Science and Engineering,* (Najaf, Iraq: IOP) pp 1-7

[44] Tabachnick B G and Fidell L S 2013 *Using Multivariate Statistics* vol sixth ed (United States of America: Pearson Education, Inc)

[45] Pallant J 2016 *SPSS Survival Manual: A step by step guide to data analysis using IBM SPSS*: Open University Press/McGraw-Hill)

[46] Zubaidi S L, Al-Bugharbee H, Ortega-Martorell S, Gharghan S K, Olier I, Hashim K S, Al-Bdairi N S S and Kot P 2020 A Novel Methodology for Prediction Urban Water Demand by Wavelet Denoising and Adaptive Neuro-Fuzzy Inference System Approach *Water* **12** 1-17

[47] Al-Bugharbee H and Trendafilova I 2014 Fault diagnosis in roller element bearings by using a linear autoregressive model. In: *the 26th International Conference on Noise and Vibration Engineering,* (Belgium, Leuven: Katholieke Universiteit Leuven) pp 2765-76

Assessment of Natural Radioactivity for Some Secondary Ceilings Samples in Iraq

Ali Saeed Al Rmahi[1] and Ali Abid Abojassim[1,*]

[1]Department of Physics, Faculty of Science, University of Kufa, Al-Najaf, Iraq.

*Corresponding author's email: ali.alhameedawi@uokufa.edu.iq

Abstract. This research focuses on measuring of specific activity of natural radioactive sources (^{238}U, ^{232}Th and ^{40}K) in some types of secondary ceilings that used as a decorate material and available in Iraqi markets. The measurements were done using gamma ray spectroscopy system that based on NaI(Tl) Detector. Also, the radiological hazard indices have been assessed for all samples in present study. The results show that the specific activity vary from 8.7±0.6 to 32.9±2.3 Bq/kg for ^{238}U, 2.9±0.2 to 40.3±1.5 Bq/kg for ^{232}Th, and 117.4±2.6 to 649.1±7.9Bq/kg for ^{40}K. The latter levels were compared with the world mean values that reported by the UNSCEAR 2008. It was found that all values of ^{238}U and ^{232}Th were below the world wide published values, while the values of only two samples of ^{40}K were above the upper range of the world wide published values. Regarding the average values of radiological hazard risk were found to be within the permissible limit according to the OECD, UNSCEAR 2000, and ICRP. It can be concluded that natural radioactivity levels together with radiological hazard risk studied from the most of the secondary ceilings samples available in local markets of Iraq within natural rates of permissible limits and may not cause any danger to the human when being used.

1. Introduction

Radiological and toxicological contaminants are one the of environmental pollutions that can cause cancerous disease and other changes in the health of humans and the living world in general. This is, therefore, substantially influencing the changes in the environment [1]. The earth and its residents are exposed to radiation emitted from radioactive materials that originally found in the crust of the earth, the sun, and other sources. Radioactivity on the earth can be originated from two types of sources: natural and artificial ones. These sources are currently part of the ecosystem. The natural radioactivity major sources are the nuclides of long half-lives that existed since the formation of the earth and nuclides produced via cosmic rays [2].The natural radionuclides of concern are mainly ^{238}U, ^{235}U ,and ^{232}Th, and their progenies and ^{40}K. Radiation is existed in all parts of our life. It, perhaps, has been found on the earth since it was formed. As a result, life has been developed in an environment that may have a few amounts of ionized radiation [3]. Therefore, one can conclude that radiation is available in everywhere around us (in the atmosphere, the water sources, the buildings and even in the construction materials used in these buildings). Natural materials such as soil, sand, cement and, rock etc., which contain amounts of natural radioactivity of ^{238}U, ^{232}Th and ^{40}K were used as building materials for construction of buildings and houses [4]. Life development has brought the people to spend most of their time (almost ninety percent of life) in enclosed and confined spaces such as residential buildings (houses and offices) and transportation vehicles. Buildings that are used as shelter should protect us from the environmental conditions (Pollutants, heat, cold and noise). Unfortunately, these shelters do not appear to be safe enough due to the many types of pollutants within them. In this regards, the latter pollutants may have detrimental effects on our health, which in turn be increased dramatically with time. On a daily basis, our body is exposing to many natural sources of radiations that are presented in water, food, and building materials. The motivation of the present study is to assess the specific activity of natural radioactivity in some types of secondary ceilings samples that used in most Iraqi building materials as a decorate materials. Exposure to high levels of radiation could be a reason for the significant damage to humans bodies and can lead to death [5] (WHO, 2016). In this study building materials (secondary ceilings) samples were selected because it is in direct touch with human. There are many previous studies for measuring radioactivity levels in building materials using gamma ray spectroscopy [6-8]. The purpose of this study is to evaluate the natural radioactivity raised from (^{238}U, ^{232}Th and ^{40}K) in some types of secondary ceilings that available in Iraqi markets using NaI(Tl) detector with dimensions of "3x3". Moreover, ten radiological hazard parameters using different equations were calculated.

2. Materials and Method

2.1. A *Collection of Samples*

Twenty samples were collected in this study that reflects various types of secondary ceilings. They were taken from the local markets in Al-Najaf governorate , as shown in Tables (1).

Table 1. Information about the secondary ceilings

No.	Sample name	Sample code	Country of Origin
1	Saten	S1	Germany
2	Cardcles	S2	
3	Turbo	S3	
4	Fuga	S4	
5	Akwsatak	S5	Turkey
6	Gypsum Board	S6	
7	Rota trofor	S7	
8	Flogan	S8	
9	Summit	S9	Iran
10	ABS	S10	
11	Azran	S11	
12	Arcopa	S12	
13	Fs.AB	S13	China
14	MDF- KS	S14	
15	CNC - KS	S15	
16	Peld	S16	
17	Against fire	S17	Saudi
18	Anti-humidity	S18	
19	Cement board	S19	
20	Tchno	S20	Bulgaria

2.2. *Preparations of Samples*

The collected samples were crushed into small pieces first, then they were converted into a fine powder, using electric grinder. The fine powder, after that, was sieved to obtain grain size of about 300μm for about 750gm in weight using special sieves. Then, the samples were dried at 100 °C for 2 hours using an oven (Model Memmert GmbH+ Co. KG, Germany). Next, the samples were packed in (1L) polyethylene Marinelli beaker of constant volume ((See Figure 1). Then, All samples were stored for about four weeks before the measurement, to allow secular equilibrium between ^{222}Rn and ^{226}Ra [9].

FIGURE 1. Samples in Marinelli beaker

2.3. Measurement of Samples

Gamma-ray spectrum from each samples was recorded using Sodium iodide doped with thallium NaI(Tl) with "3×3" crystal dimensions (See Figure 2). It was processed using the MAESTRO-32 software. NaI(Tl) detector was calibrated by radioactive standard sources (^{137}Cs, ^{60}Co, ^{22}Na, ^{54}Mn, and ^{152}Eu) of known energies and activity. The specific activity for ^{238}U and ^{232}Th were determined using gamma-lines 1765 keV (^{214}Bi) and the gamma-ray lines 2614 keV (^{208}Tl), respectively [9,10]. The specific activity for ^{40}K was determined directly by its gamma-line of 1460 keV [9,10]. The samples were placed on the detector and measured for a period of 18000 sec.

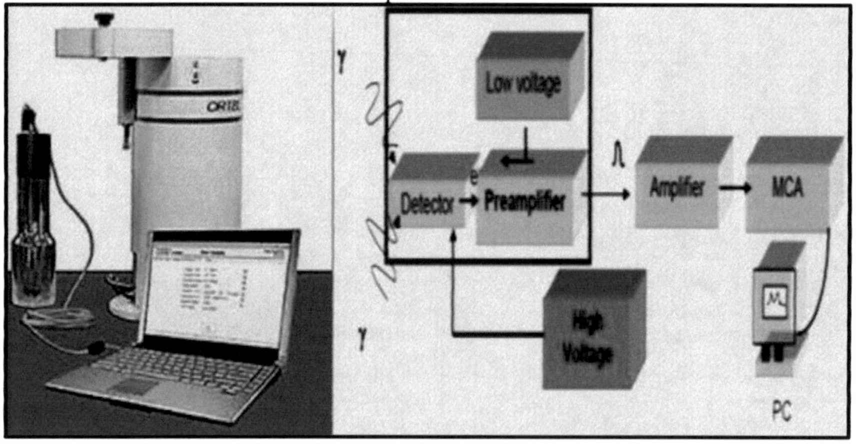

FIGURE 2. Block diagram of NaI(Tl) spectrometer system.

2.4. Theoretical calculations

2.4.1. Specific Activity (A)

The specific activities of ^{238}U, ^{232}Th, and ^{40}K (A_U, A_{Th} and A_k) radionuclides were calculated using following equation [11,12]:

$$A\left(\frac{Bq}{kg}\right) = \frac{N}{I_\gamma\, \varepsilon\, M\, T} \dots\dots\dots (1)$$

where, N is net area under photo peak, I_γ is the probability of gamma decay , ε is the efficiency of detector, M is the mass of sample, and T is time measured.

2.4.2. External Hazard Index (H_{ex})

The external hazard index was calculated using the following equation [13]:

$$H_{ex} = \frac{A_U}{370} + \frac{A_{Th}}{259} + \frac{A_K}{4810} \ldots\ldots\ldots. (2)$$

2.4.3. Internal Hazard Index (H_{in})

The internal hazard index was calculated using the following equation [14]:

$$H_{in} = \frac{A_U}{185} + \frac{A_{Th}}{259} + \frac{A_K}{4810} \ldots\ldots\ldots. (3)$$

2.4.4. Representative Level Index (I_γ)

Representative level index was calculated using the following equation [15].

$$I_r = \left(\frac{1}{150}\right) A_U + \left(\frac{1}{100}\right) A_{Th} + \left(\frac{1}{1500}\right) A_K \ldots\ldots (4)$$

2.4.5. Alpha Index (I_α)

Alpha index was calculated using the following equation [13]:

$$I_\alpha = \frac{A_U}{200\ (\frac{Bq}{kg})} \ldots\ldots\ldots (5)$$

2.4.6. Radium Equivalent Activity (Ra_{eq})

Radium equivalent activity was calculated using the following equation [16]:

$$Ra_{eq}\left(\frac{Bq}{kg}\right) = A_U + 1.43\ A_{Th} + 0.077 A_K \ldots\ldots\ldots (6)$$

2.4.7. Exposure Rate ($\dot{X}$)

The exposure rate was calculated as the following equation [14,17]:

$$\dot{X}\left(\frac{\mu R}{h}\right) = 1.90\ A_U + 2.82\ A_{Th} + 0.197 A_K \ldots\ldots\ldots (7)$$

2.4.8. Absorbed Dose Rate in Air (D_r)

The absorbed dose rate in air 1 meter was calculated using the following equation [18]:

$$D_r\left(\frac{nGy}{h}\right) = 0.462\ A_U + 0.604\ A_{Th} + 0.0417 A_K \ldots\ldots\ldots (8)$$

2.4.9. Annual Gonadal Equivalent Dose (AGED)

Annual gonadal equivalent dose was calculated using the following equation [19- 21] as:

$$AGED\ (\frac{mSv}{y}) = 3.09\ A_U + 4.18\ A_{Th} + 0.314 A_K \ldots\ldots\ldots (9)$$

2.4.10. Annual Effective Dose Equivalent (AEDE)

Annual effective dose equivalent indoor was calculated using the following equation [22].

$$AEDE_{indoor}\left(\frac{mSv}{y}\right) = \left[D_r\left(\frac{mGy}{hr}\right) \times 8760\ hr \times 0.8 \times \frac{0.7Sv}{Gy}\right] \times 10^{-6} \ldots\ldots (10)$$

2.4.11. Excess Lifetime Cancer Risk (ELCR)

Excess lifetime cancer risk indoor according to Duration of Life (DL =70 year) and Risk Factor(RF=0.05 y/Sv) was calculated using the following equation [11, 16]:

$$ELCR = AEDE \times DL \times RF \ldots\ldots (11)$$

3. Results and Discussion

The results of specific activity for ^{238}U, ^{232}Th, and ^{40}K in different types of secondary ceilings samples of the present study were shown in Table (2), all were measured in units of Bq/kg. Examining the results in Table 2, the range of specific activity of ^{238}U with the standard error was 8.7±0.6 Bq/kg in the sample S16 to 32.9±2.3 Bq/kg in the sample S14, with an average of 22.4±1.6 Bq/kg. Also, the specific activity of ^{232}Th was ranged between relatively high values in S14, where it reached 40.3±1.5 Bq/kg, and the lowest value in the S4 which was 2.9±0.2 Bq/kg, with an average of 12.2±1.8 Bq/kg. While, the results of specific activity for ^{40}K were ranged from 117.4±2.6 Bq/kg in sample S4 to 649.1±7.9 Bq/kg in sample S6, with an average of 282.2±26.6 Bq/kg. According to the above results, it was found that the specific activity of ^{238}U in all samples present study are within the permissible limit set by the UNSCEAR 2008 (i.e.33 Bq/kg) [18], as shown in Figure 3. From figure 4, it was found that the specific activity of ^{232}Th in all samples in present study were less than the permissible limit set by the UNSCEAR 2008 (i.e. 45 Bq/kg) [18]. The present results show that the values of the specific activity of ^{40}K in all samples were less than the recommended value of (420) Bq/kg, that given by worldwide UNSCEAR 2008 [18], except samples S6 and S12 (see Figure 5). So, It recommend not to use this type of secondary celling (S6 and S12). The results of the natural radioactivity in secondary ceilings samples as building materials in the present study were varied, because of the different geological nature of original materials and basic components that made of the samples in the present study. The calculation of ten hazard indices such as (Ra_{eq}, H_{ex}, H_{in}, I_γ, I_α), and (Exposure, D_r, AGED, $AEDE_{indoor}$, ELCR) in secondary ceilings samples from Iraq markets are listed in tables 3 and 4, respectively. Form Table 3, it can be seen that the average value of Raeq, H_{ex}, H_{in}, I_γ, I_α were 61.6±4.9 Bq/kg, 0.166±0.013, 0.227±0.016, 0.459±0.036, and 0.112±0.008, respectively. The highest value of Raeq, and (H_{ex}, H_{in}, I_γ I_α) were 107.8 Bq/kg, (0.291, 0.380, 0.772, 0.164) in S14 which were less than the recommended value of 370 Bq/kg [23], and 1[24] (see Figure 6), respectively. Form Table 4, it is found that the average value of Exposure, D_r, AGED, $AEDE_{indoor}$, and ELCR in secondary ceilings samples were 132.6±10.5μR/h, 29.5±2.3 nGy/h, 208.9±16.3 mSv/y, 0.145±0.011 mSv/y, and $(0.500\pm0.039)\times10^{-3}$, respectively. The highest value of D_r, and $AEDE_{indoor}$ were 48.9 nGy/h, and 0.240 mSv/y in S14 which were less than the recommended value of 55 nGy/h [25], and 0.42 [26], respectively. While, the highest value of AGED was 340.7 mSv/y in S14 which were larger than the recommended value of 300 Bq/kg [27]. Comparing the results of the specific activity for ^{238}U, ^{232}Th and ^{40}k that shown in the present study sample that produced in different countries, it can be seen that the highest average value of the specific activities for ^{238}U and ^{232}Th was in Chinese sample, but for ^{40}K was in Turkey sample as shown in Figure 7. While, the radiological hazard index (Ra_{eq}, H_{ex} and H_{in}) were highest values in Chines, as shown in Figure 8. This increase in values may be due to geological nature for the origin of soil that made of secondary ceilings samples. But, the average values for all samples of secondary ceilings of specific activity for (^{238}U, ^{232}Th, and ^{40}K) and radiological hazard index were within the world's average according to UNSCEAR 2008, UNSCEAR2000, OECD and ICRP.

Table 2. Results of specific activity for ^{238}U, ^{232}Th, and ^{40}K in present samples.

No.	Specific activity in Bq/kg

IOP Conf. Series: Earth and Environmental Science **877** (2021) 012034 doi:10.1088/1755-1315/877/1/012034

	Sample code	^{238}U		^{232}Th		^{40}K	
		Average	**±Error**	**Average**	**±Error**	**Average**	**±Error**
1	S1	14.8	0.9	6.7	0.4	234.0	3.8
2	S2	22.9	1.2	9.9	0.5	270.4	4.1
3	S3	22.4	1.4	15.8	0.7	315.6	5.5
4	S4	14.1	0.9	2.9	0.2	117.4	2.6
5	S5	12.1	0.9	11.0	0.5	250.3	4.3
6	S6	23.6	1.4	22.6	0.9	649.1	7.9
7	S7	19.9	1.2	18.8	0.7	348.5	5.4
8	S8	25.2	1.4	7.6	0.5	241.9	4.6
9	S9	32.8	1.6	8.1	0.5	149.7	3.5
10	S10	25.2	1.7	7.0	0.4	300.0	5.0
11	S11	24.1	1.5	9.0	0.5	290.0	4.5
12	S12	31.0	2.0	11.6	0.8	460.9	8.2
13	S13	32.6	2.1	13.0	0.8	393.6	7.6
14	S14	32.9	2.3	40.3	1.5	224.8	6.2
15	S15	26.0	1.8	12.5	0.8	332.8	6.8
16	S16	8.7	0.6	3.9	0.3	143.3	2.7
17	S17	30.6	1.4	13.0	0.6	310.5	4.8
18	S18	18.6	1.4	16.6	0.8	254.5	5.5
19	S19	17.3	1.0	5.8	0.3	178.4	3.3
20	S20	14.0	0.8	7.4	0.3	178.9	2.9
Minimum		8.7±0.6		2.9±0.2		117.4±2.6	
Maximum		32.9±2.3		40.3±1.5		649.1±7.9	
Average±S.E		22.4±1.6		12.2±1.8		282.2±26.6	
Worldwide [18]		33		45		420	

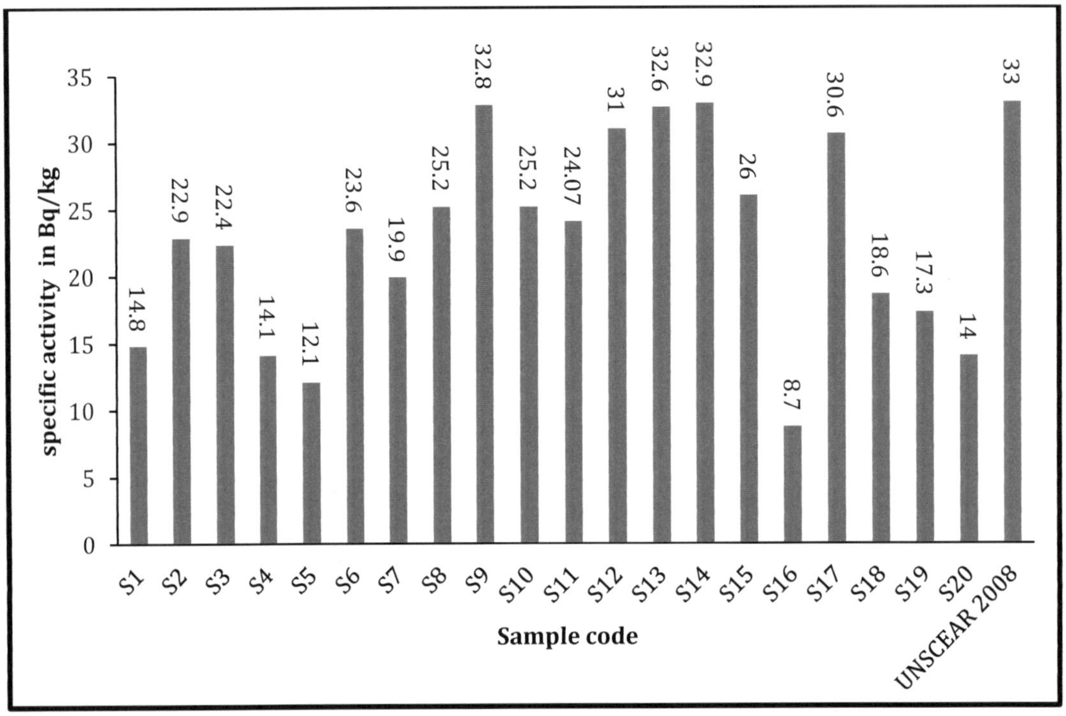

FIGURE 3. The comparison of specific activity between present samples with UNSCEAR limit, for ^{238}U.

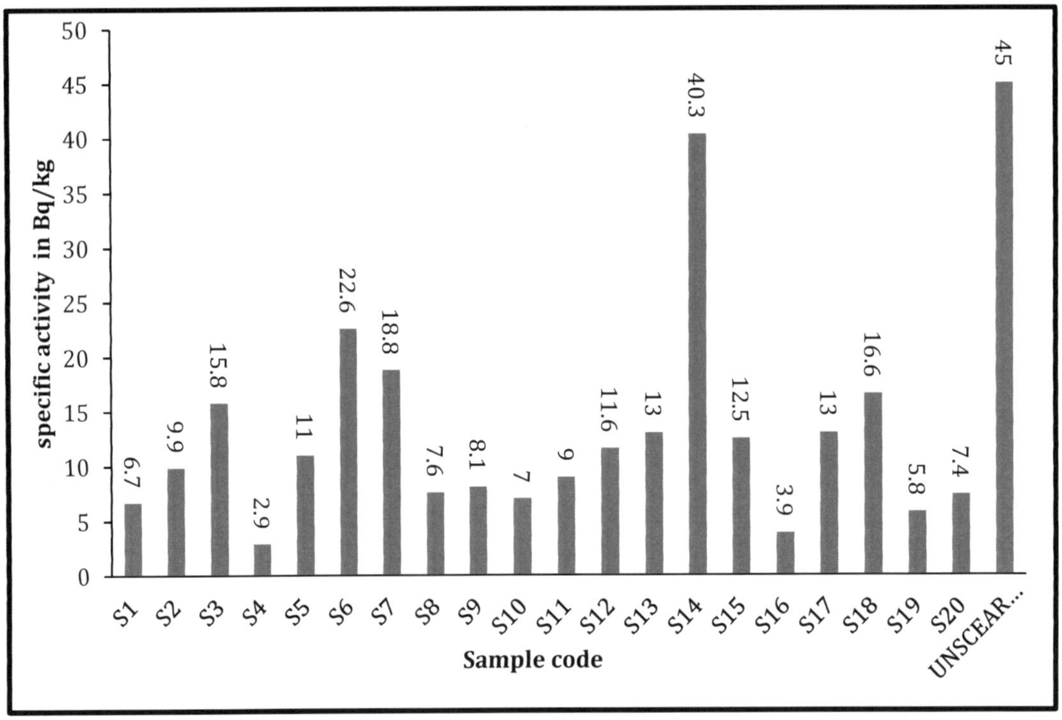

FIGURE 4. The comparison of specific activity between present samples with UNSCEAR limit, for ^{232}Th.

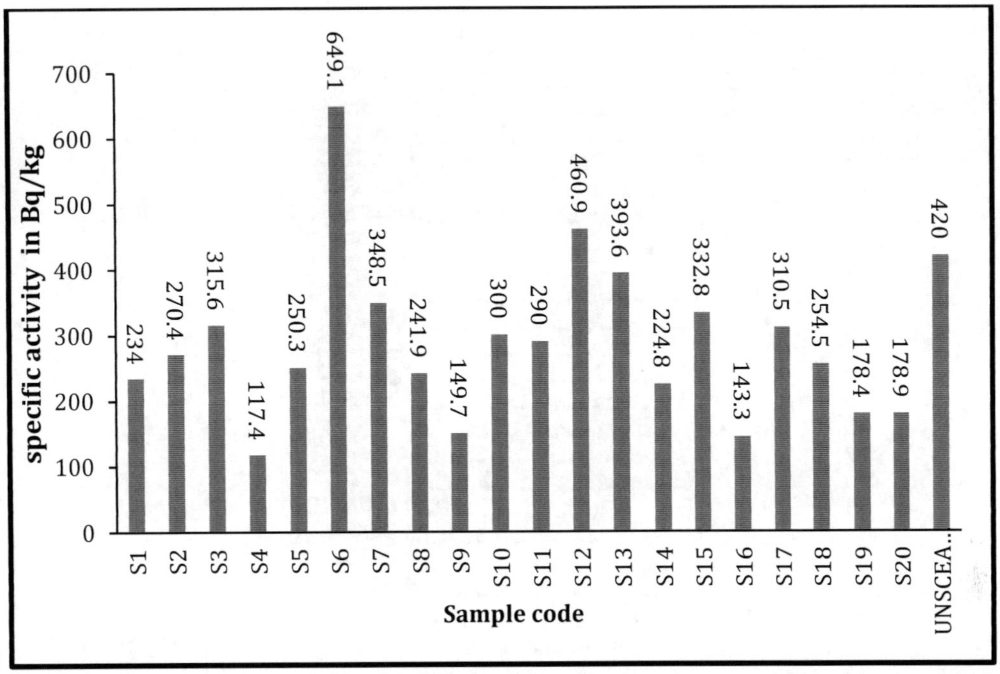

FIGURE 5. The comparison of specific activity between present samples with UNSCEAR limit, for ^{40}K.

Table 3. Results of Ra_{eq}, H_{ex}, H_{in}, I_γ and I_α

No.	Sample code	Ra_{eq} (Bq/kg)	H_{ex}	H_{in}	I_γ	I_α
1	S1	42.4	0.115	0.155	0.322	0.074
2	S2	57.9	0.156	0.218	0.432	0.114
3	S3	69.3	0.187	0.248	0.518	0.112
4	S4	27.3	0.074	0.112	0.201	0.071
5	S5	47.1	0.127	0.160	0.358	0.061
6	S6	105.9	0.286	0.350	0.816	0.118
7	S7	73.6	0.199	0.253	0.553	0.099
8	S8	54.7	0.148	0.216	0.405	0.126
9	S9	55.9	0.151	0.240	0.399	0.164
10	S10	58.3	0.158	0.226	0.438	0.126
11	S11	59.3	0.160	0.225	0.444	0.120
12	S12	83.1	0.224	0.308	0.630	0.155
13	S13	81.5	0.220	0.308	0.610	0.163
14	S14	107.8	0.291	0.380	0.772	0.164
15	S15	69.5	0.188	0.258	0.520	0.13
16	S16	25.3	0.068	0.092	0.193	0.043
17	S17	73.1	0.197	0.280	0.541	0.153
18	S18	61.9	0.167	0.218	0.460	0.093
19	S19	39.3	0.106	0.153	0.292	0.086
20	S20	38.4	0.104	0.141	0.287	0.07
Average± S.E.		61.6±4.9	0.166±0.013	0.227±0.016	0.459±0.036	0.112±0.008
Worldwide		<370[23]	<1[24]	<1[24]	<1[24]	<1[24]

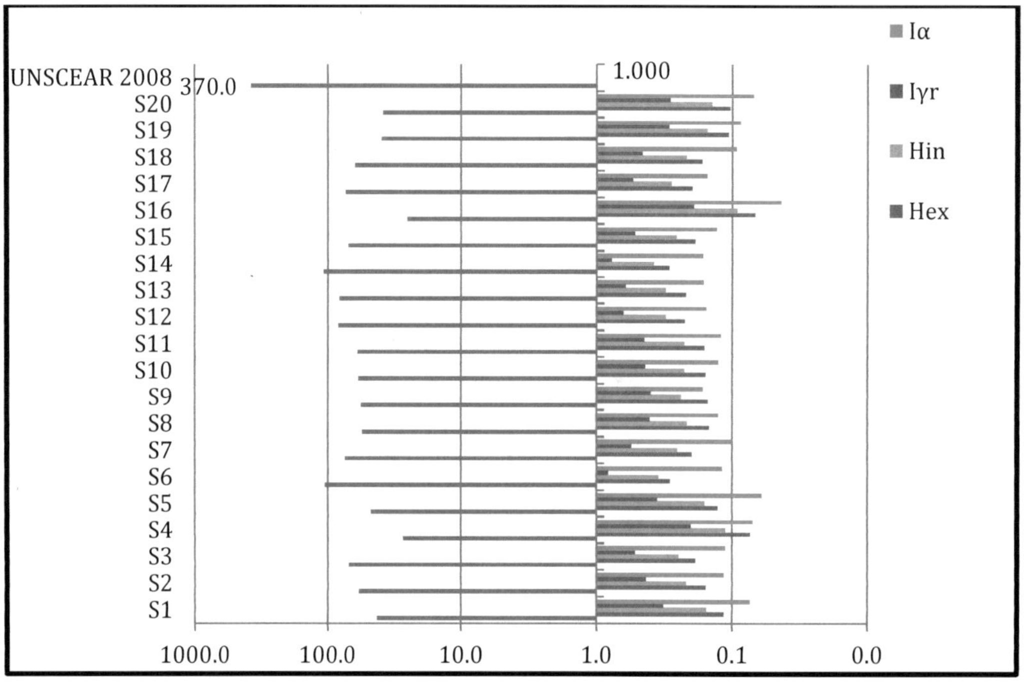

FIGURE 6. The comparison of Ra_{eq}, H_{ex}, H_{in}, I_γ and I_α between present samples with world limit.

Table 4. Results of $\dot{X}$, D_r, AGED, $AEDE_{indoor}$, and ELCR

No.	Sample code	Exposure (μR/h)	D_r (nGy/h)	AGED (mSv/y)	$AEDE_{indoor}$ (mSv/y)	ELCR×10^{-3}
1	S1	93.1	20.6	147.2	0.101	0.354
2	S2	124.7	27.8	197.0	0.137	0.478
3	S3	149.3	33.1	234.4	0.162	0.567
4	S4	58.1	13.2	92.6	0.065	0.226
5	S5	103.3	22.7	162.0	0.111	0.389
6	S6	236.4	51.6	371.2	0.253	0.886
7	S7	159.5	35.1	249.5	0.172	0.602
8	S8	117.0	26.3	185.6	0.129	0.452
9	S9	114.7	26.3	182.2	0.129	0.451
10	S10	126.7	28.4	201.3	0.139	0.487
11	S11	128.2	28.6	203.1	0.141	0.492
12	S12	182.4	40.5	289.0	0.199	0.696
13	S13	176.1	39.3	278.7	0.193	0.675
14	S14	220.4	48.9	340.7	0.240	0.840
15	S15	150.2	33.4	237.1	0.164	0.574
16	S16	55.8	12.4	88.2	0.061	0.212
17	S17	156.0	34.9	246.4	0.171	0.600
18	S18	132.3	29.2	206.8	0.143	0.502
19	S19	84.4	18.9	133.7	0.093	0.325
20	S20	82.7	18.4	130.4	0.090	0.316

IOP Conf. Series: Earth and Environmental Science **877** (2021) 012034 doi:10.1088/1755-1315/877/1/012034

Average ± S.E.	132.6±10.5	29.5±2.3	208.9±16.3	0.145±0.011	0.500±0.039
Worldwide	------	55[25]	≤ 300 [27]	0.42 [26]	------

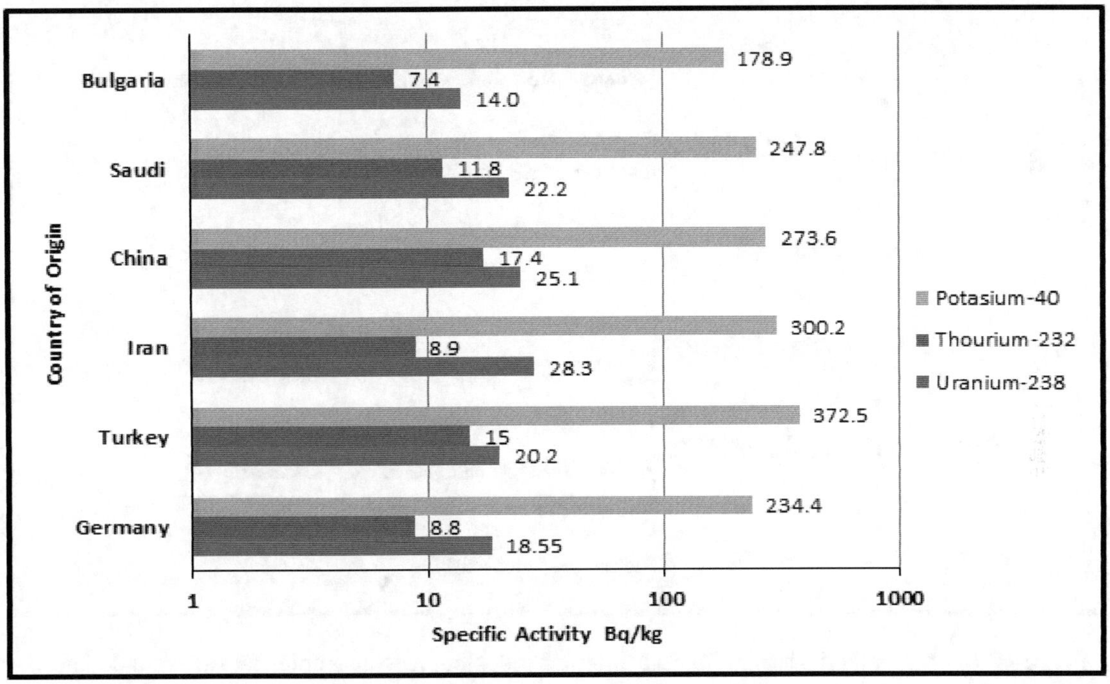

FIGURE 7. The comparison of specific activity in present samples for different countries.

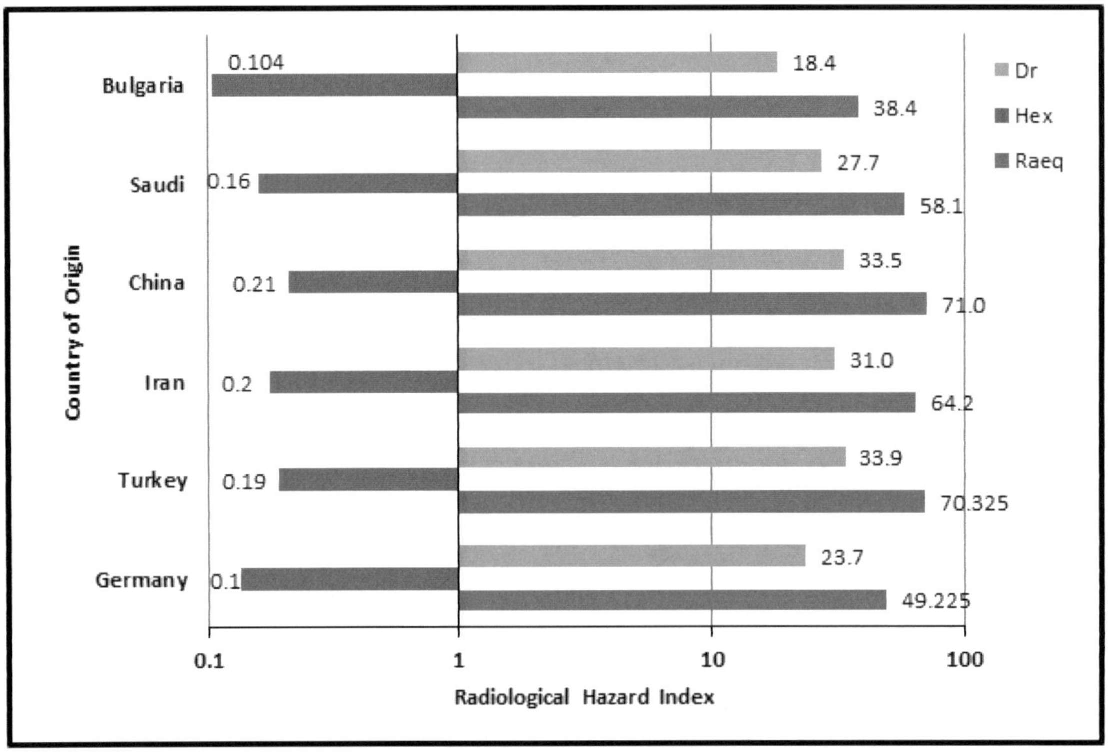

FIGURE 8. The comparison of radiological hazard index in present samples for different countries.

4. Conclusions

Natural radioactivity and radiological hazard index in samples of present study were within the allowed limit according to world limit (UNSCEAR, UNSCEAR, OECD, and ICRP). Therefore, our gamma spectroscopic investigations allow us to confirm that samples of secondary ceilings were safe, except samples S6 and S12.

5. References

[1] Rao, C. S. (2007). *Environmental pollution control engineering*. New Age International.
[2] Tykva, R., & Berg, D. (Eds.). (2004). *Man-made and natural radioactivity in environmental pollution and radiochronology*. Springer Science & Business Media.
[3] Cherry, S. R., Sorenson, J. A., & Phelps, M. E. (2012). *Physics in nuclear medicine e-Book*. Elsevier Health Sciences.
[4] Whittle, K. (2020). *Nuclear materials science*. IOP Publishing.
[5] World Health Organization. (2016). Ionizing radiation, health effects and protective measures. *World Health Organization*.
[6] Trevisi, R., Leonardi, F., Risica, S., & Nuccetelli, C. (2018). Updated database on natural radioactivity in building materials in Europe. *Journal of environmental radioactivity*, *187*, 90-105.
[7] Maxwell, O., Emmanuel, J. S., Olusegun, A. O., Cyril, E. E. O., Ifeanyi, A. T., & Embong, Z. (2019). A study of natural radioactivity in some building materials in Nigeria. *Radiation protection dosimetry*, *183*(3), 332-335.
[8] Rashid, J. M., & Mansor, M. A. (2020). Evaluation of natural radioactivity for building materials samples used in Tall Al Ubaid Archaeologist in Dhi-Qar governorate-Iraq. *Samarra Journal of Pure and Applied Science*, *2*(1), 53-66.

[9] Dhahir, D. M., Mraity, H. A. A., Abojassim, A. A., Najam, L. N., & Al-kazrajy, H. Y. Y. (2020). Natural radioactivity in soil samples of some school in AL-Shatrah at DhiQar governorate, Iraq. *Malaysian Journal of Science*, 104-114.

[10] Abojassim, A. A., & Rasheed, L. H. (2019). Mapping of Terrestrial Gamma Radiation in Soil Samples at Baghdad Governorate (Karakh Side), Using GIS Technology. *Nature Environment and Pollution Technology*, *18*(4), 1095-1106.

[11] Al-Hamidawi, A. (2014). Assessment of radiation hazard indices and excess life time cancer risk due to dust storm for Al-Najaf, Iraq. Wseas Trans. Environ. Dev, 10, 312.

[12] Abojassim, A. A., (2017). Estimation of human radiation exposure from natural radioactivity and radon concentrations in soil samples at green zone in Al-Najaf, Iraq. Iranica journal of energy and environment, 8(3), 239-248.

[13] Krieger, R. (1981). Radioactivity of construction materials. Betonwerk Fertigteil Techn, 47(468).

[14] Venturini, L., & Nisti, M. B. (1997). Natural radioactivity of some Brazilian building materials. Radiation protection dosimetry, 71(3), 227-229.

[15] Mahler, R. L., & Hamid, A. (1994). Evaluation of water potential, fertilizer placement and incubation time on volatilization losses of urea in two northern Idaho soils. Communications in soil science and plant analysis, 25(11-12), 1991-2004.

[16] Abojassim, A. A., Oleiwi, M. H., & Hassan, M. (2016). Natural radioactivity and radiological effects in soil samples of the main electrical stations at babylon governorate. Yaderna fyizika ta energetika, 17(3), 308-315.

[17] Kahn, B., Eichholz, G. G., & Clarke, F. J. (1983). Search for building materials as sources of elevated radiation dose. Health physics, 45(2), 349-361.

[18] UNSCEAR.(2008). Sources and effects of ionizing radiation: Report to the general assembly, (United Nations, New York), with scientific annexes,2,1-219.

[19] United Nations Scientific Committee on the Effects of Atomic Radiation. (1988). Sources, Effects and risks of ionizing radiation. UNSCEAR 1988 report to the general assembly, with scientific annexes, New York.

[20] Arafa, W. (2004). Specific activity and hazards of granite samples collected from the Eastern Desert of Egypt. Journal of environmental radioactivity, 75(3), 315-327.

[21] Okogbue, C., & Nweke, M. (2018). The 226Ra, 232Th and 40K contents in the Abakaliki baked shale construction materials and their potential radiological risk to public health, southeastern Nigeria. Journal of environmental geology, 2(1).

[22] UNSCEAR.(1994). Sources and Effects of Ionizing Radiation. UNSCEAR 1994 Report to the General Assembly, with Scientific Annexes. United Nations publications , 49.

[23] Nuclear Energy Agency. (1979). Exposure to radiation from the natural radioactivity in building materials: report. OECD.

[24] Kaiser, S. (1999). Radiological protection principles concerning the natural radioactivity of building materials. Radiation Protection, 112.

[25] United Nations. Scientific Committee on the Effects of Atomic Radiation. (2000). Sources and effects of ionizing radiation: sources (Vol. 1). United Nations Publications.

[26] Abojassim, A. A., & Rasheed, L. H. (2021). Natural radioactivity of soil in the Baghdad governorate. Environmental Earth Sciences, 80(1), 1-13.

[27] ICRP, International Commission on Radiological Protection ICRP publication 65, Annals of the ICRP 23(2). Pergamon Press, Oxford. 1993.

Preparation and Optimization Green Gel Casting Technique for Manufacturing Near-Net-Shape Ceramics Using Genetic Algorithm

Mohammed A. Ahmed Al-dujaili[1,*], Imad Ali Al-hydary[1] and Montaha Abdalhussien[1]

[1]Department of Ceramics Engineering and Building Materials, Faculty of Materials Engineering, University of Babylon, Babylon, Hilla, Babylon, P.O. Box: 4, Iraq

*Corresponding author's email: adujaili@uobabylon.edu.iq

Abstract. Gel casting technique is a promising technology that has ability to produce near-net shape ceramics via using toxic and non-ecofriendly agents . The current work aim to develop green gel casting technique using water as a solvent, agar as a gelling agent, and the microwave thermal treatment instead of cross linker. 8mo l% Yttria stabilized zirconia was selected as a case study to produce near-net shape ceramics. The experimental work involved the preparation of Yttria stabilized zirconia nanoparticles via chemical precipitation method. The effect study of agar ratio, Yttria stabilized zirconia solid loading percent on the physical, mechanical, surface properties of the prepared ceramics and selecting of suitable casting conditions. The study has been found that the microwave thermal treatment develops thermally activated cross linking in the agar aqueous solution leading to higher glass transition temperature for agar. The green combination (agar aqueous solution and microwave treatment) can be used as alternative to (monomer, solvent, cross linker) Companion. Also, using the ultrasonic treatment can effectively eliminate needs for dispersants, also the vacuum de-airing treatment. Yttria stabilized zirconia ceramic with high dimensional accuracy, low surface roughness (Ra=2. 81 nm) can be obtained using an agar ratio of (0.4%) and solid loading of (65%). The sample can be moulded with complex shape and the green gel, also the pre-sintered body is machineable. The sintered samples have a porosity of (31%) and compressive strength of (234MPa). Regression analysis and genetic algorithm are showed that the obtained microhardness, compressive strength, and surface roughness are predictable.

Keywords: 8YSZ, Near net shape, Gel casting, Genetic algorithm method

1. Introduction

Near Net Shape (NNS) it is an innovative concept of industrial manufacturing. This technicality mains focus on the parts production, as close as possible to their shape or final contour without need for chipping or grinding operations implementation [1]. NNS's may reducing or eliminating the need for the finishing the surface like machining or grinding and converts benefits of bottom line that can reduce your cost to two-thirds [2].The ongoing drive to use easy methods for near shaped advanced ceramic components in almost a grid shape has led to a revolutionary class of forming techniques known for its wet processing techniques [3]. GC is referred to as a one of the common performing techniques developed to date for fabricating complex-shaped ceramic bodies from highly concentrated and low-viscosity slurries that can be solidified with gelation once the mold is filled. So ; Facilitate mold removal before drying and sintering [4].

Monomer, catalyst, dispersant, solvent, binder, plasticizer, cross linker, and powder of ceramic in gel casting method are mixed thoroughly to produce a high solid loading and low viscosity homogenous suspension [5]. The current work is an attempt to develop green gel casting method that has an ability to manufacture near net shape ceramic. Yttria stabilized zirconia was chosen as a case study to obtain the work. The prediction process was defined as the process of detecting the next steps or results on the basis of any proof or confirmation of these steps [6]. Regression is one of the statistical techniques for the determining of linear relationship between more than one variable. Regression is primarily used for the prediction and causal inference [6]. There are many studies that are focused on the employment of the genetic algorthim for optimization of problems, in the turning processes and joining. In addition, it is present in the field of metals. As, for ceramics the GA is not used commonly and has rarely been documented.

The genetic algorthim method starts with a set of randomly generated individuals in a named population[7]. Each individual is called a chromosome, and every chromosome is evaluated using a certain function is called the fitness function. After that, a new population is generated by employing the genetic algorthim operator (selection, crossover and mutation) which replaces the old population with the new one. Algorthim factors (selection, crossover and mutation) which replaces the outdated population with the new one. This operation is repeated until the optimal solution it was received[8].

2. Materials and methods

2.1Preparation of (8 Yttria stabilized zirconia) Nanoparticles

Yttria stabilized cubic Zirconia Nanoparticles was prepared via the chemical precipitation method through using zirconyl chloride octahydrate and yttrium chloride hexahydrous with concentrated ammonium hydroxide. The suitable amounts of the two chemicals were dissolved into the 10 L of distilled water and magnetically stirred at room temperature for 1hr 250 ml of ammonia solution (NH_4OH, 25 NH_3%) was then added drop-wise to the reaction mixed solutions up to pH value of 9 to promote sedimentation. The resulting suspension shown in Figure 1 (a) was magnetically stirred at room temperature for another 1hr. The precipitate was separated by high speed refrigerated centrifuge (GL-12, China). The gel like material, shown in Figure 1 (b), was washed with distilled water until the PH value of 7 was obtained at 100°C for 48hr in the air. Finally, dry powders, shown in Figure 1 (c), drying was performed were grounded in an agate mortar and then calcined in a furnace at temperature 850°C for 2hr with a heating rate of 8°C/min. Figure 2 shows the flowchart of the preparation procedure.

Figure 1. The steppes of sample preparation by chemical precipitat

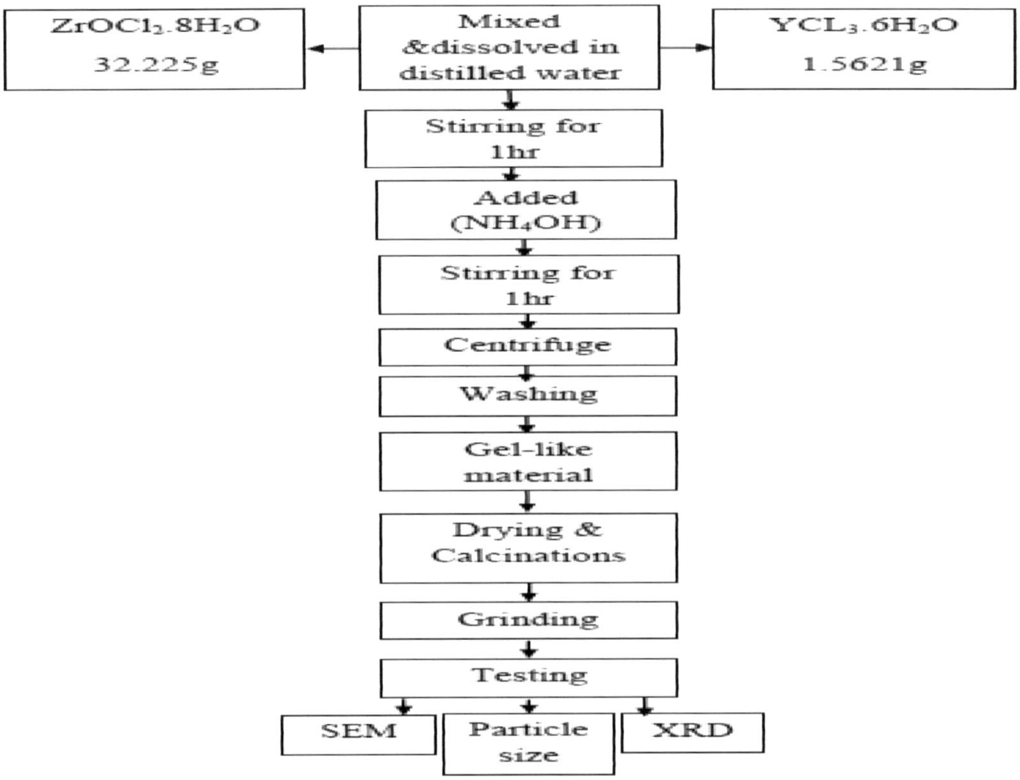

Figure 2. Shows the flow chart of 8YSZ nano particles procedure

2.2Preparation of agar gel

Prior to use the agar in gel casting, the main characteristics of the agar gel were identified. These include the gelation temperature and the minimum agar to water ratio. The gelation temperature is necessary to find out the casting temperature of the slurry in the gel casting process, while, the minimum agar to water ratio is important to get green sample that can be safely handled after the gelation. The agar solution has been prepared via dissolving agar powder in distilled water using microwave treatment for 2min with power of (900W) and frequency of (2450 MHz). The solution was cooled down to get the gel and dried to obtain the solid agar. After many experiments, the minimum agar to water ratio was found to be 0.4wt%. This ratio produces gel that can maintain its shape and consistency up on handling. Based on that, the agar to water ratio of (0.4, 0.5, 0.6, 0.7, 0.8) was specified to be used throughout the study.

2.3Preparation of dense(8YSZ) samples

8YSZ sample were prepared by gel casting method using solid loading percents of (50, 53, 56, 59, 62, 65 wt. %). These percents were calculated based on the weight percent of the 8 Yttria stabilized zirconia powder to the agar solution containing different amount of agar. The upper limit of solid loading was specified, after many experiments, to get castable gel at the specified casting temperature. For preparing the gel casting slurries the agar solution was prepared by dissolving the desired amount of agar in distilled water with the help of microwave treatment for 1min using power of (900W) and frequency of (2450 MHz). The solution was then transferred to water bath at 55°C. The desired amount of 8 Yttria stabilized zirconia nano particles was added to the agar solution at 55°C under sonication and mixing with speed of (1600 rpm) for 15min.

IOP Conf. Series: Earth and Environmental Science **877** (2021) 012035 doi:10.1088/1755-1315/877/1/012035

After that the slurry was poured in PVC molds and then cooled down to 10°C in refrigerator. The 8 Yttria stabilized zirconia gel as demolded and aged at room temperature for 24hr and dried in a convection oven at 55°C. The green samples calcined and sintered in two steps. The green samples were calcined at 650◦C, After that the samples were subjected to pre-sintering at 1100◦C, The final sintering was done at 1500 °C with 8 °C/min and holding time for 2 hr. The experiment population consists of 30 samples as shown in Tables 1,2, and 3, each sample is considered a chromosome and each chromosome consists of a number of genes. These genes have been represented by input parameters, in the present work. The regression equation or the fitness function was formed, using Minitab 17software for microhardness (y1), it can be expressed as follows :

$$y_1 = -3.95 + 0.1432\,(X_1) - 1.58\,(X_2) - 0.0239\,(X_1)*(X_2) \quad (1)$$

Where: y_1: microhardness, x_1: solid loading, x_2: agar ratio. The obtained regression equation for compression test (y_2) can be expressed as follows:

$$y_2 = 1025 - 27.9\,X(1) - 1284\,X(2) + 0.315\,(X1)*(X1) + 902\,(X2)*(X2) + 2.03\,(X1)* \quad (2)$$

Where: y_2: compression, x_1: solid loading, x_2: agar ratio

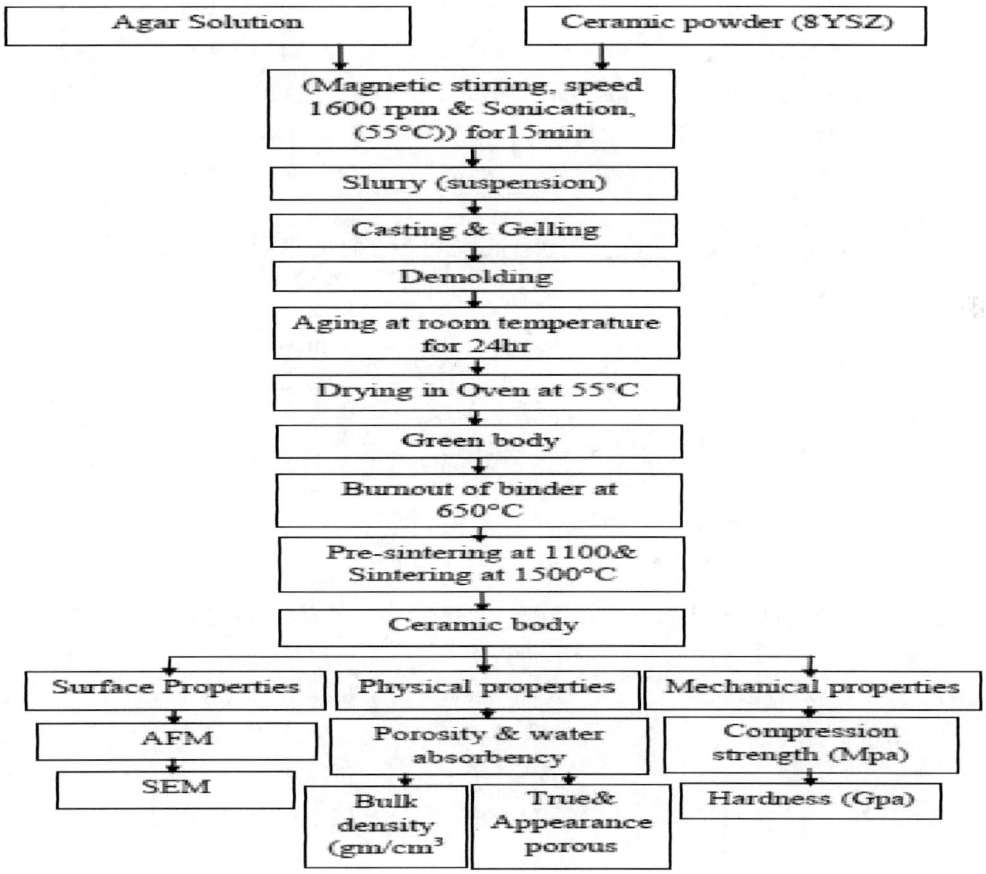

Figure 3. Flow chart of the experimental work

Table 1. Regression equations for the roughness of the surface of the sample (Ra)

	Model summary				
Agar ratios	S	R-sq	R-sq(adj)	R-sq(pred)	Formulas
0.4	0.195654	98.70%	97.83%	91.61%	f(x) = 46.8 - 1.273 x + 0.00906 x*x
0.5	0.0399648	99.13%	98.55%	96.36%	f(x) = 8.21 - 0.1053 x + 0.000403 x*x
0.6	0.216105	92.91%	88.19%	52.69%	f(x) = 8.5 - 0.095 x - 0.00011 x*x
0.7	0.162006	96.21%	93.69%	75.83%	f(x) = 23.76 - 0.635 x + 0.00456 x*x
0.8	0.0638663	98.60%	97.67%	93.54%	f(x) = 7.09 - 0.069 x - 0.00004 x*x

Table 2. Regression equations for the roughness of the side of the sample (Ra)

	Model summary				
Agar ratios	S	R-sq	R-sq(adj)	R-sq(pred)	Formulas
0.4	0.126837	98.82%	98.03%	87.98%	f(x)= 39.99-1.126x+0.00843x*x
0.5	0.159831	96.80%	94.67%	87.86%	f(x) = -11.83 + 0.646 x - 0.00664 x*x
0.6	0.158672	97.14%	95.23%	90.72%	f(x) = 7.70 - 0.012 x - 0.00101 x*x
0.7	0.456331	95.34%	92.23%	80.52%	f(x) = 16.9 - 0.191 x - 0.00081 x*x
0.8	0.230109	98.20%	97.00%	93.28%	f(x) = 9.0 + 0.053 x - 0.00250 x*x

Table 3. The physical properties of 8YSZ ceramic body

Agar ratio	Solid loading (wt. %)	Bulk density (gm/cm^3)	Porosity%	Real porosity (cm^3)	Apparent porosity%	Water absorption%
0.4	50	3.3342	40.23129921	0.3844	0.3832	9.9034
	53	3.4221	39.37327034	0.355	0.3739	11.5668
	56	3.68025	37.7602875	0.31	0.3637	10.9918
	59	3.7224	31.70870472	0.2532	0.3021	8.6439
	62	3.8787	35.23517052	0.2869	0.3506	9.59154
	65	4.09694	31.08719579	0.26	0.3086	7.98472
0.5	50	3.0442	41.80051874	0.2862	0.4309	12.76009
	53	3.1357	40.41475384	0.2818	0.3935	11.5375
	56	3.1941	39.90488845	0.2413	0.3839	10.574
	59	3.2893	36.85413088	0.2911	0.3264	10.1755
	62	3.4278	33.74696991	0.3501	0.3251	9.166
	65	3.54766	36.31452581	0.3436	0.346	9.9818
0.6	50	2.9997	41.76988509	0.6132	0.3951	12.9329

	53	3.1015	40.65626	0.4011	0.3508	11.5334
	56	3.2432	36.10864669	0.4558	0.3406	10.8923
	59	3.4274	36.19815125	0.2932	0.3351	10.641
	62	3.5755	35.214377	0.2842	0.3309	8.81089
	65	3.4525	34.19375065	0.341	0.3358	10.1309
0.7	50	2.96886	41.86305732	0.49	0.4352	13.9266
	53	3.11188	40.76754985	0.455	0.3896	12.5252
	56	3.33999	38.4732253	0.3583	0.3468	12.0453
	59	3.59293	36.27571009	0.3209	0.3521	11.307
	62	3.55956	35.22558255	0.2924	0.3519	9.16955
	65	3.61977	27.86040378	0.268	0.2636	8.2315
0.8	50	2.81265	45.82384871	0.6161	0.4573	15.918089
	53	2.98322	42.79678814	0.2672	0.3907	13.4907
	56	3.15291	38.23802882	0.5129	0.3664	11.3046
	59	3.22382	39.99250575	0.4296	0.3871	12.0093
	62	3.53632	38.95207668	0.3369	0.3742	10.5917
	65	3.67044	30.2552509	0.3296	0.3014	10.0553

3. Results and discussion

3.1Result of particle size

Figure 4 Show the particle size distribution of the calcined 8YSCZ nanoparticles were shown in Figure 3. The powder has unmoral particle size distribution with main value of the particle size of D_{50}= 8.631 μm.

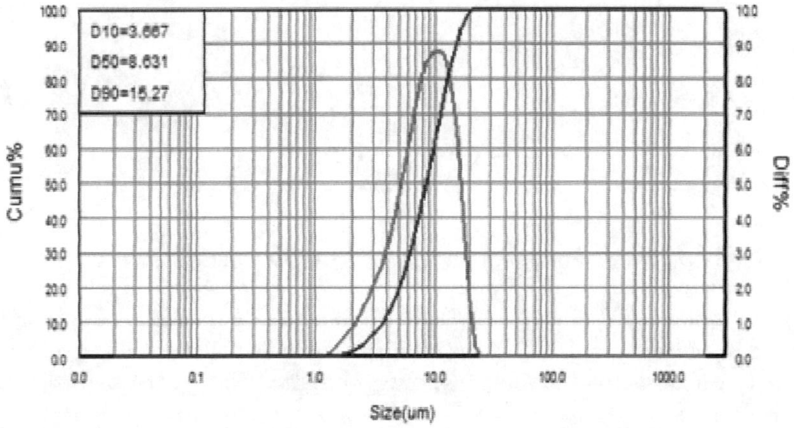

Figure 4. Show the particle size distribution of the calcined($Y_{0.08}$ $Zr_{0.92}$ O_2) powder.

3.2Result of XRD

Figures 5 and 6 show the x-ray diffraction pattern of 8YSZ sample sintered at 850˚C 1300˚C, respectively. The diffraction data is in full agreement with JCPDS (00-030-1468). The sharp peaks and the high intensity indicate the high crystallinity of the sample.

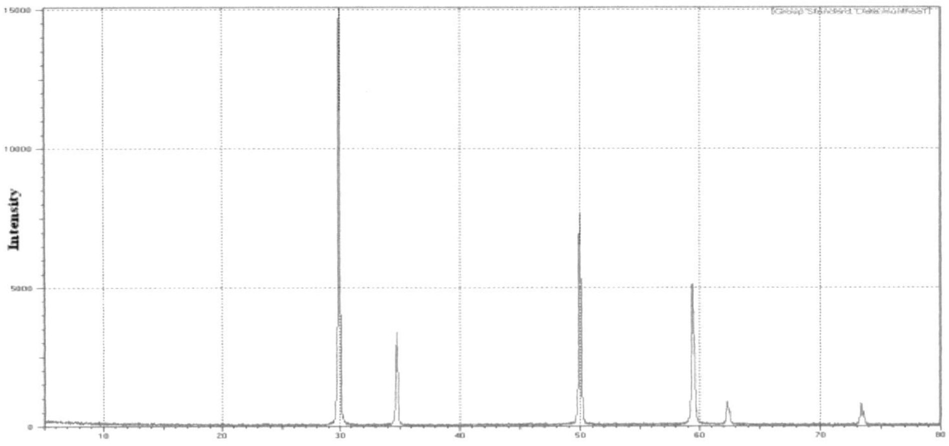

Figure 5. X-ray diffraction analysis of 8YSZ powder calcined at 850˚C

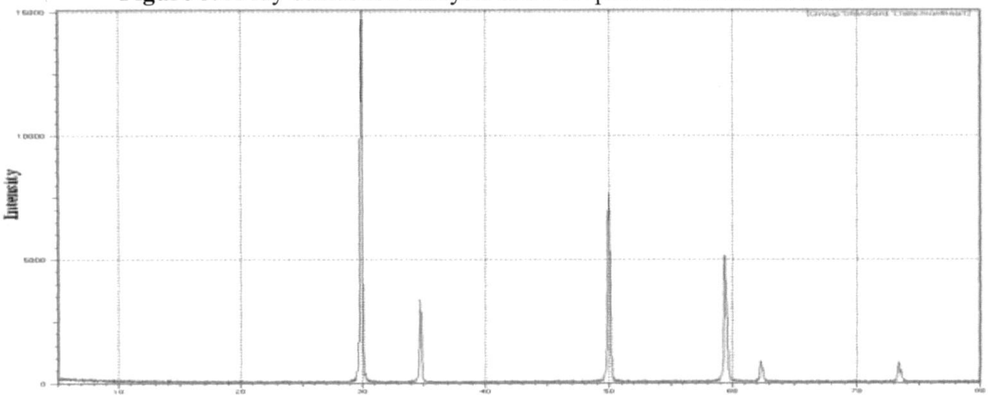

Figure 6. X-ray diffraction analysis of 8YSZ powder calcined at 1300˚C.

3.3Result of SEM

Figure 7 show the SEM micrograph of 8 Yttria stabilized zirconia powder at different magnification. SEM observation indicated that the powder consists of agglomerated particles in the form near to spheres of diameter less than 100nm.

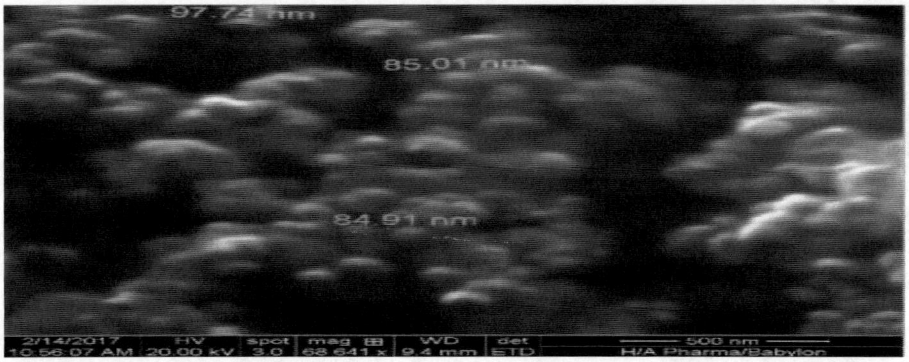

Figure 7. SEM micrograph of 8YSCZ powder prepared by chemical precipitation

3.4Result of DSC

Figure 8 Shows the results of DSC patterns for agar powder and the agar gel after drying. This test was performed to find out the glass transition temperature (Tg) of the agar and confirm that thermal activation is developed during the microwave heat treatment. The Tg can be found from the endothermic peak below 100˚C for agar. It represents the midpoint of that peak. It has been found that the agar powder has Tg of 81.88˚C which is comparable to that reported for agar in the literature, The agar gel exhibited abroad endothermic peak at many locations ranging from about (75-84) °C [9], as well as; reported endothermic peaks of three types of agar gel at many locations ranging from (75- 90) °C [10]. While the Tg of the dried Agar gel is 93.46˚C. The higher Tg value indicates that the choice of the agar crossed linked during the heat treatment due to thermal activation.

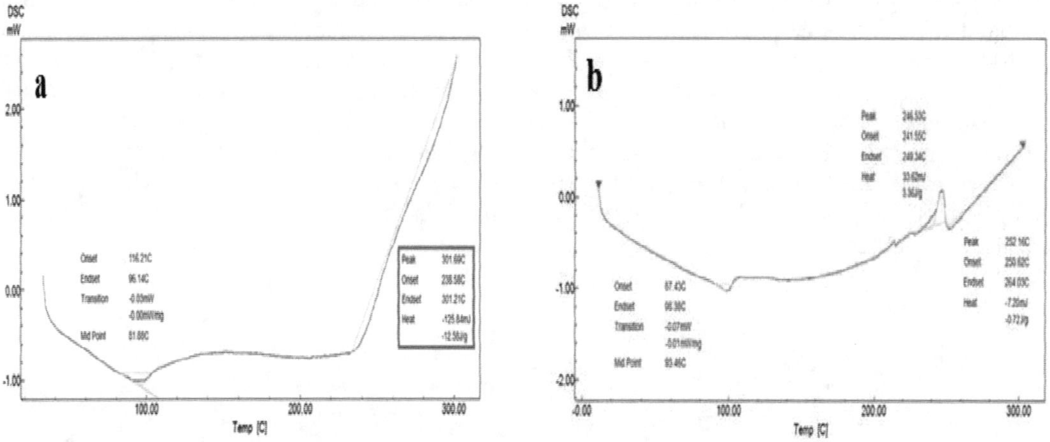

Figure 8. DSC analysis for Agar agar type 1(a) Agar powder and (b) Agar gel after drying

3.5Result of viscosity

Figure 9 shows the results of the cone-on-plate viscosity test for the agar solution. It was observed that here is a change in the profile of viscosity-temperature diagram. This change indicates the transformation of the solution to gel and hence the viscosity increase. However, the change was occurred at temperature 59°C. The pouring temperature was chosen to be 55°C which is accompanied with moderate viscosity value of (17 mpa.s). It is necessary to has high viscosity enough to move the ceramic particles suspended in the gel and, at the same time, low enough to be poured at molds with complex shapes.

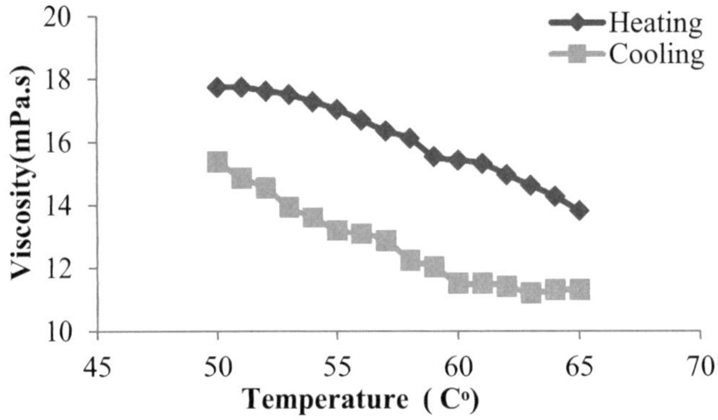

Figure 9. The effect of temperature on the viscosity of Agar gel for heating and cooling.

3.6Results of physical properties

3.6.1 Drying and firing shrinkage.

Figure 10 shows the dependence of drying shrinkage on the solid loading and agar ratio. It has been observed that the drying shrinkage decreased with increased solid loading and decreased agar ratio. On the other hand; when the solid loading increases the friction between the particles increases. This can hinder the settle of the particles during the casting process. Thus, the compaction can be reduced leading to less values of firing shrinkage. The counter balance between these two effects produced the net firing shrinkage. The net firing shrinkage of 8 Yttria stabilized zirconia ceramic body is shown in Figure 11. It can be noticed that, as general trend, the firing shrinkage decreases with increasing the solid loading. This indicates that the friction effect generally dominated. However, many exceptions are here. This may belong to homogenous distribution of the particles.

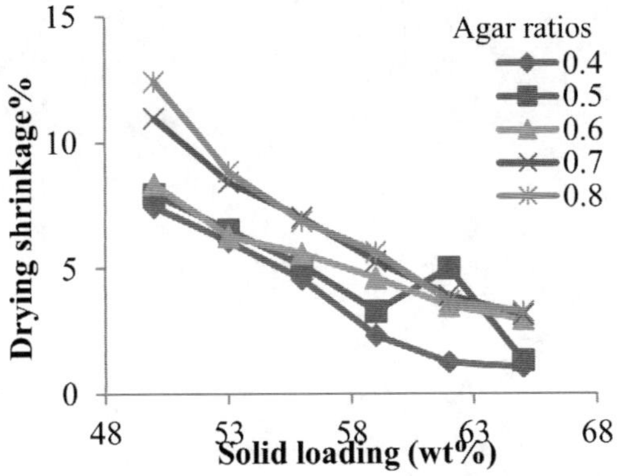

Figure 10. Drying shrinkage of 8YSZ ceramic samples

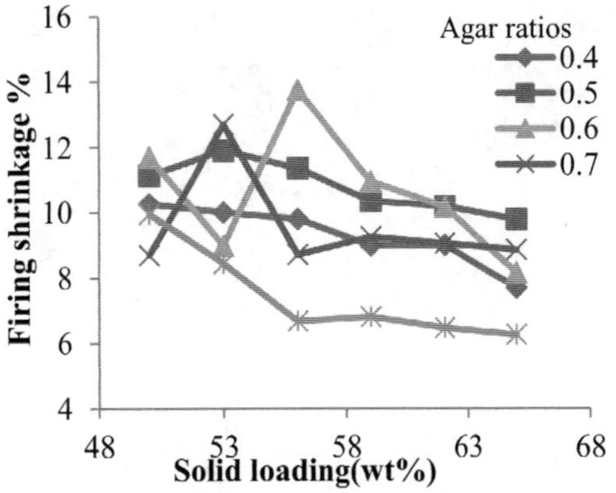

Figure 11. Firing shrinkage of 8YSZ ceramic samples sintered at 1500°C

3.6.2 Result of porosity.

Figures 12 and 13 show that the apparent porosity and real porosity decrease along with increasing solid loading. This is because of two reasons. The first is whenever more quantity of powder is added; the area of contact between the powder particles is considerably increased. Subsequently, the increase in the contact will increase compaction between the particles and this decrease the pore size. The second reason is related to the ratio of agar. The higher ratio of agar leads to higher porosity due to the higher volume left behind the agar during the heat treatment. Also, the affect is analoge to that noticed for apparent porosity. These two properties influenced similarity by the same factors and similar to water absorption results as shown in Figure 14.

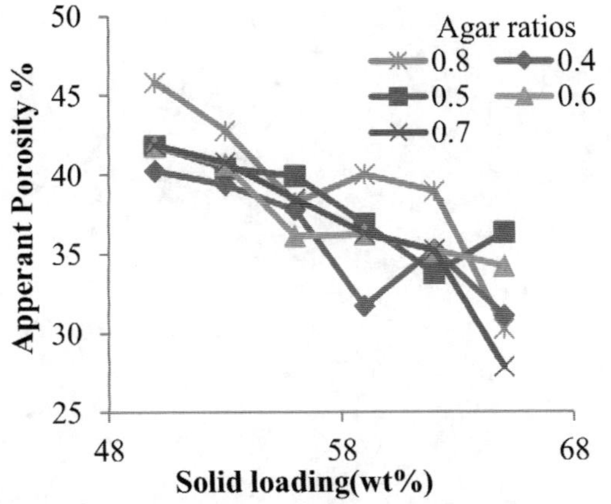

Figure 12. Shows the effect of solid loading and agar ratio on apparent porosity of 8YSZceramic

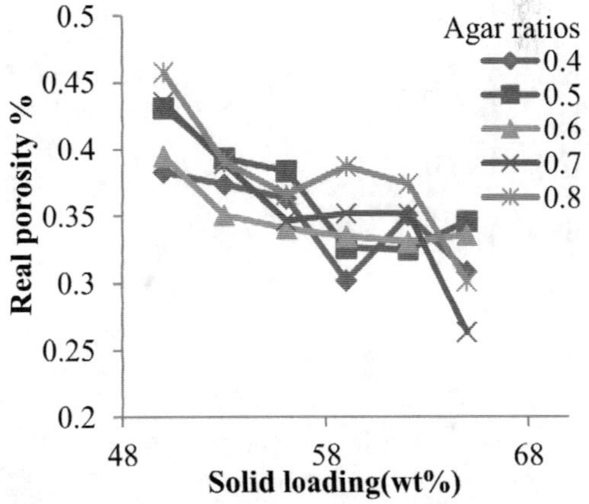

Figure 13. The effect of solid loading and agar ratio on the real porosity of 8YSZ ceramic

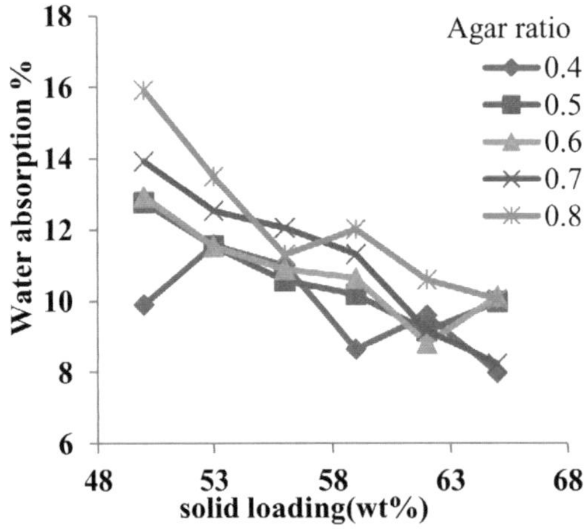

Figure 14. Shows the effect of solid loading and agar ratio on water absorption of 8YSZ ceramic

3.6.3 Green and bulk density.

Figures 15 and 16 show the increment in density as the solid loading increase for green and bulk density. Also, the density increase when lower amount of agar is used. This is because of the interlink between the atoms, resulting in an increases in efficient compactness, packing and change dimensionality in the 8 Yttria stabilized zirconia structure. The increase in density is ascribed to the reduction in average interatomic spacing in the 8 Yttria stabilized zirconia structure during the heat treatments.

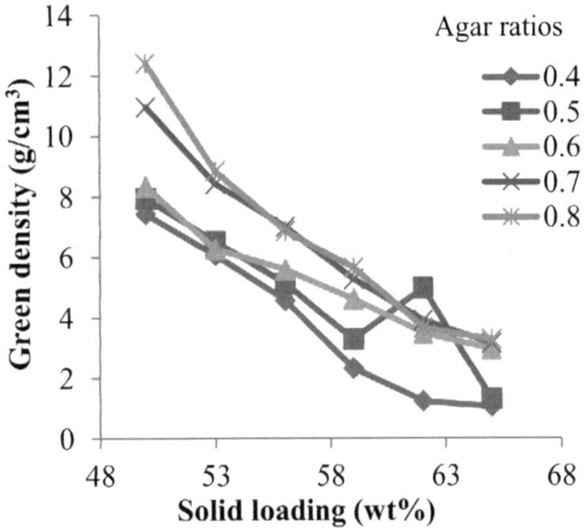

Figure 15. The effect of solid loading and agar ratio on the green density of 8YSZ ceramic

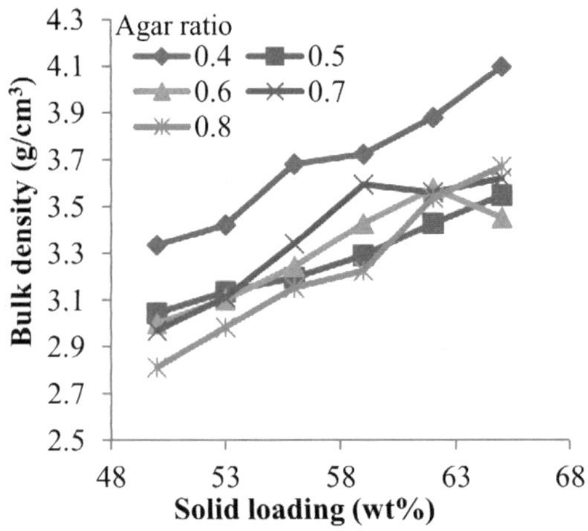

Figure 16. The effect of solid loading and agar ratio on the bulk density of 8YSZ ceramic

3.7Result of compressive strength

Figure 17 shows that with increasing solid loading content from 50 to 65wt % leads to increase the compressive strength from 34 to 234 Mpa. Also, decreasing agar ratio increases the compressive strength especially at high solid loading. The enhancement of the compressive strength is related to the reduction of porosity that leads to improvement in the density and compression strength as well.

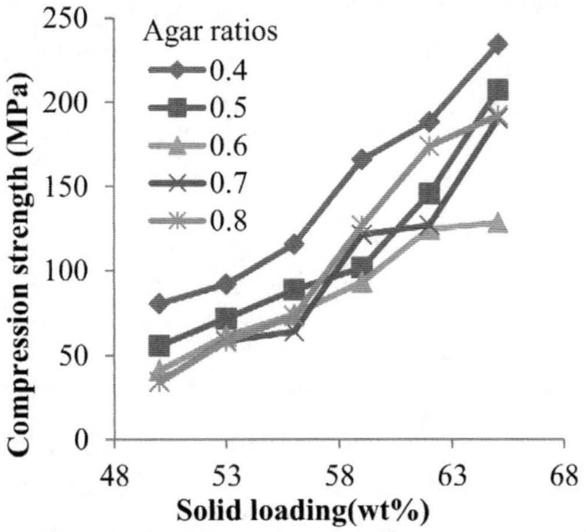

Figure 17. The effect of solid loading and agar ratio on the compression strength of 8YSZ ceramic

3.8 Results of near net shape testing

3.8.1 Mechanical profile.

Figures 18, 19, 20, 21, and 22 showed the roughness values of the prepared samples using Ra and Rz measurements. The roughness was tested for the free surfaces (upper surface of the sample) as well as the interface between the sample and the mold (side). It can be noticed that the roughness, for the surface and the side, decreases with increasing the solid loading for specific agar ratio. This is due to the reduction of porosity which affects the roughness in this level. However, it has been observed that the roughness decreases by increasing agar ratio. This is might be due to the higher dispersion of 8 Yttria stabilized zirconia nano particles when high ratio of agar is used. It is important to note that:-

1- The Rz roughness measurement is more sensitive to the solid loading than the Ra measurement.
2- The effect the mold material of the surface roughness is obtained at high agar ratio. The roughness of (side) is higher than that of the (surface) when the agar ratio is higher than 0.6%.

IOP Conf. Series: Earth and Environmental Science **877** (2021) 012035 doi:10.1088/1755-1315/877/1/012035

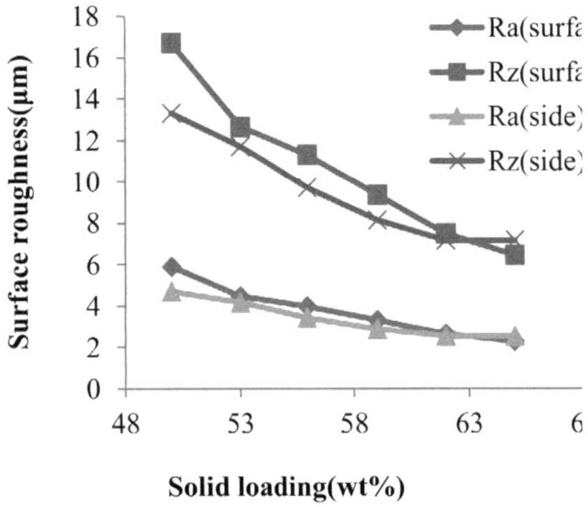

Figure 18. Surface roughness Ra and Rz of 8YSZ samples prepared by gel casting in the agar ratio of 0.4 wt%.

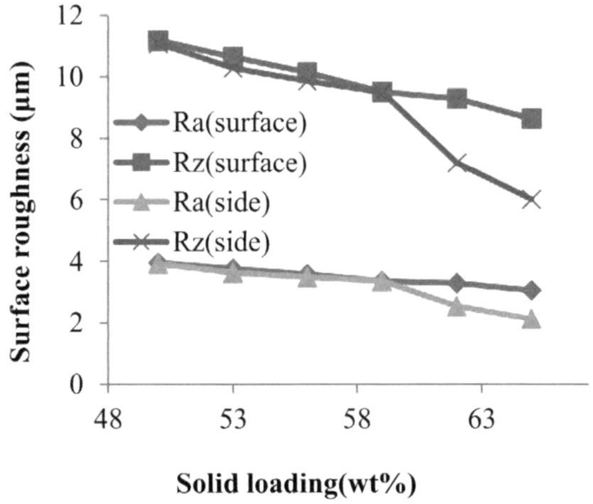

Figure 19. Surface roughness Ra and Rz of 8YSZ samples prepared by gel casting in agar ratio of 0.5 wt%

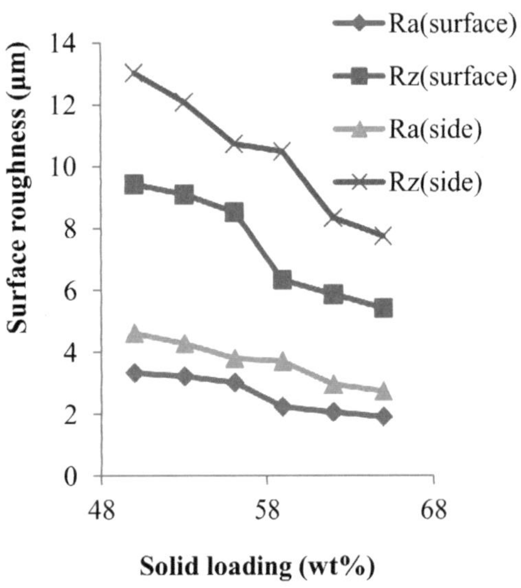

Figure 20. Surface roughness Ra and Rz of 8YSZ samples prepared by gel casting in the agar ratio of 0.6 %

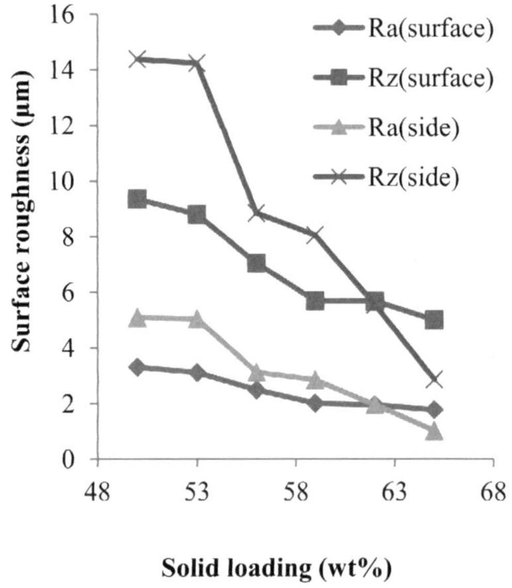

Figure 21. Surface roughness Ra and Rz of 8YSZ samples prepared by gel casting in agar ratio of 0.7 wt%

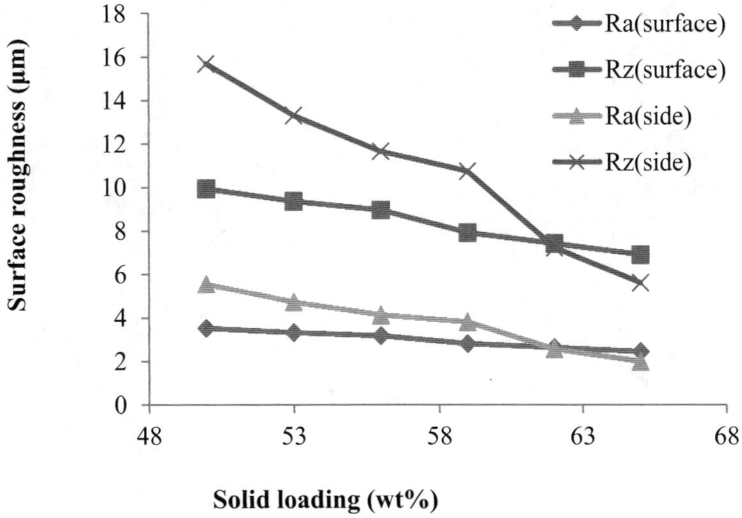

Figure 22. Surface roughness Ra and Rz of 8YSZ samples prepared by gel casting in agar ratio of 0.8 wt%

3.8.2 Results of AFM.

Figure 23 Shows surface roughness in nano level obtained by AFM test for a sample with 65% solid loading with different agar ratios. It can be noticed that the roughness increases with increasing agar ratio in the samples as shown in Figure 24. Comprising this result with that obtained for micro level roughness indicates that the removing of agar during the heat treatment creates voids in nano level. Thus; agar is responsible on the nano level roughness while the dispersion of the particles and the sintering process are responsible on the micro level porosity. The AFM images of the surface of 8 Yttria stabilized zirconia samples with different agar ratios are shown in Figure 25. It obvious that increasing the agar ratio.

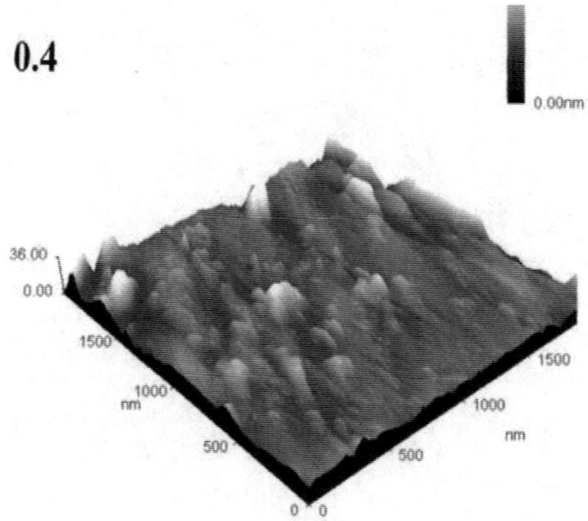

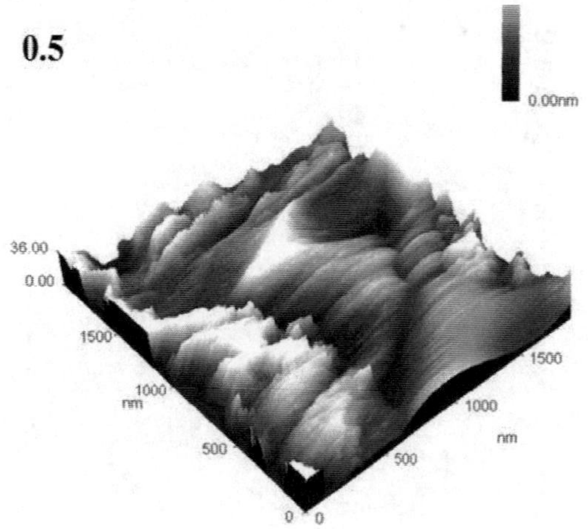

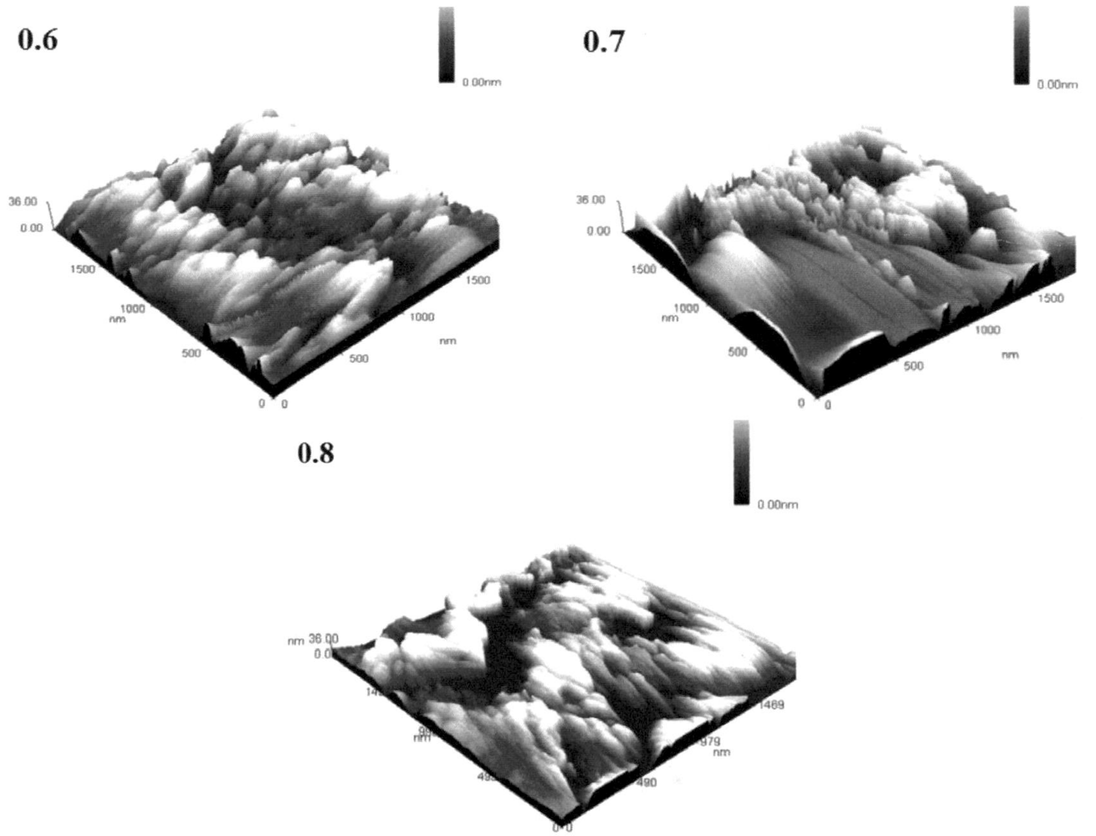

Figure 23. AFM images for the surface of 8YSZ samples with solid loading 65% and different agar ratio

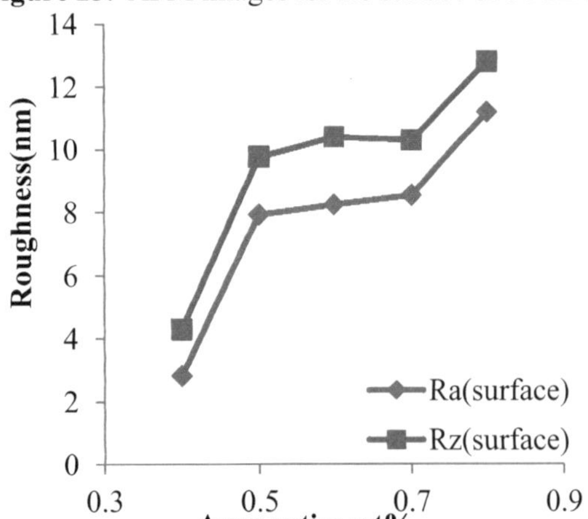

Figure 24. AFM images for the surface of 8YSZ samples with solid loading 65% and different agar ratio

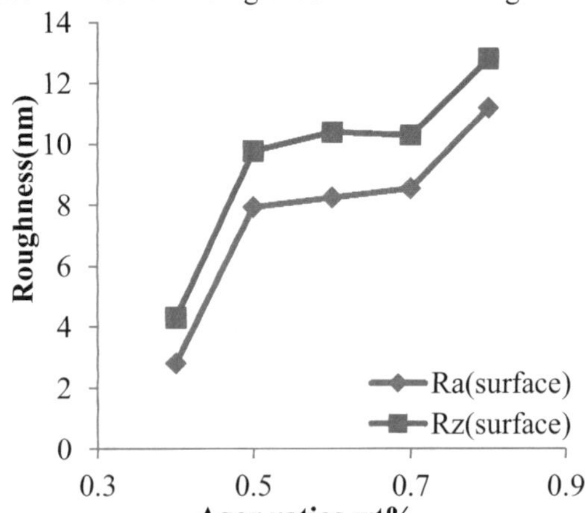

Figure 25. Nano level roughness of 8YSZ samples prepared by gel casting with solid loading 65wt% and different agar ratios

3.9 Results of the genetic algorithm

The results of a genetic algorithm can be representation as a sketch of the fitness value and the number of the generation, generally as the generation's progress, the fitness value begins to stabilize, at a certain value that represents the optimum value.

3.10 Results of the GA for optimum microhardness

The GA shows that the best fit for the maximization of the microhardness at a population size of 0.4 agar, is found to be 418HV which is equal to the theoretical hardness of 8YSZ ceramic body. As shown in the Figure 26.

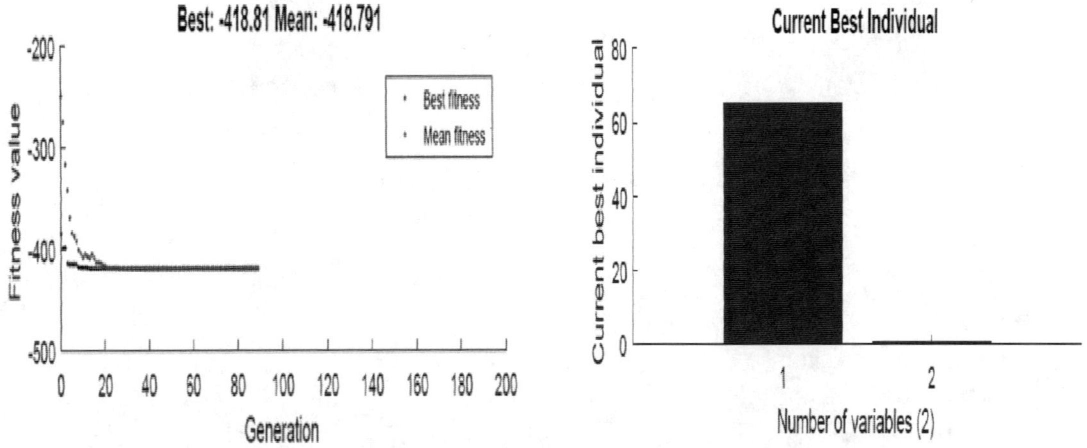

Figure 26. The generation verses the fitness value for microhardness

3.11 Results of the GA for optimum compressive strength

For the compressive strength, the boundaries were taken as the same as that for the microhardness. As shown in the Figure 27, As well as for roughness and AFM results, as shown in the Figures 28, 29, 30, 31, 32 and 33.

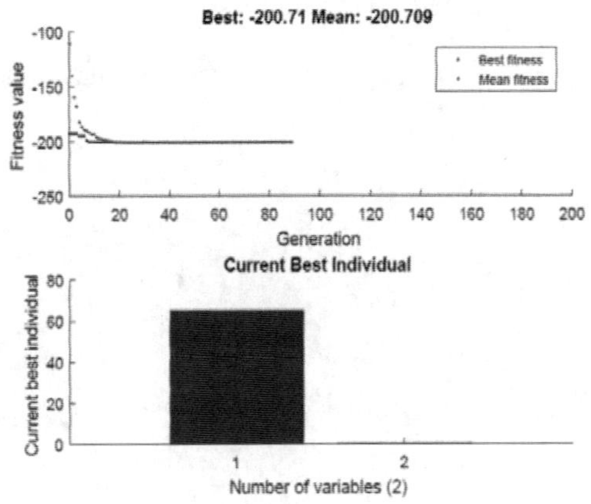

Figure 27. The generation verses the fitness value for compression test

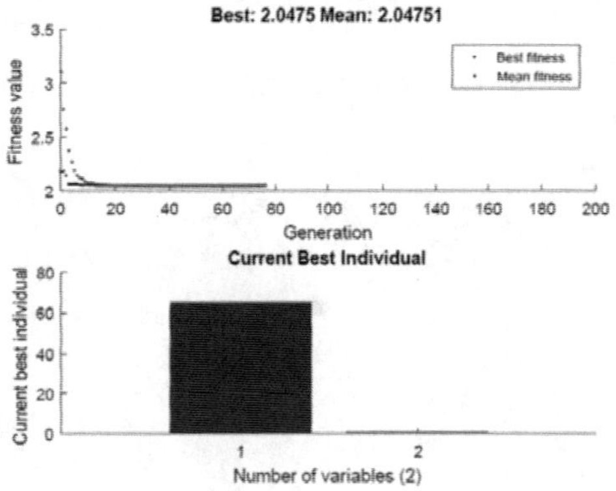

Figure 28. The generation verses the fitness value for (Ra) surface

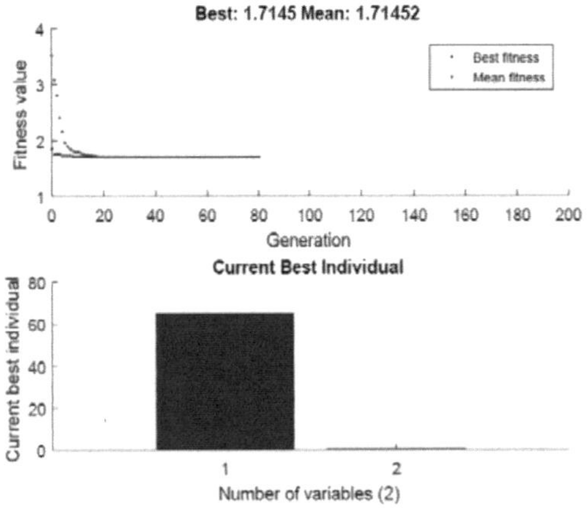

Figure 29. The generation verses the fitness value for (Ra) side

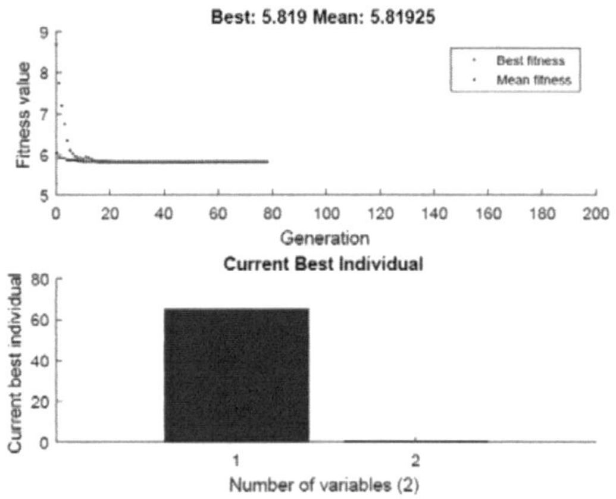

Figure 30. The generation verses the fitness value (Rz) surface

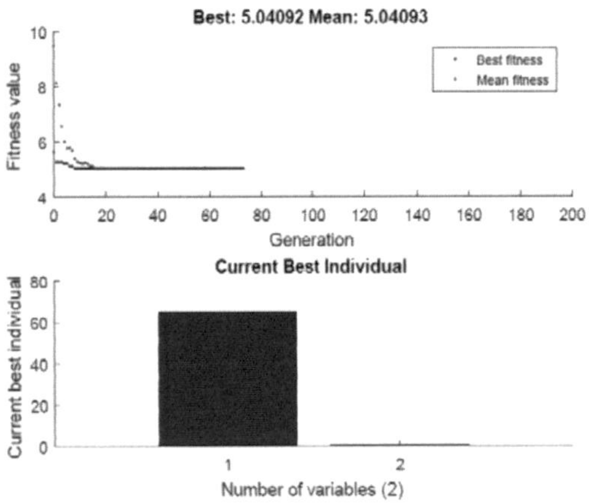

Figure 31. The generation verses the fitness value (Rz) side

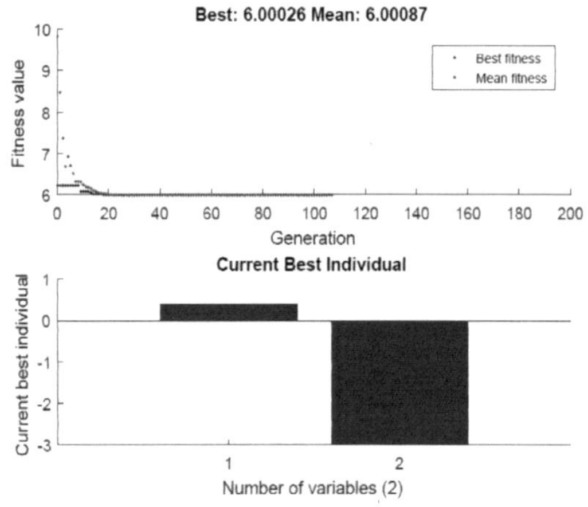

Figure 32. The generation verses the fitness value for Ra

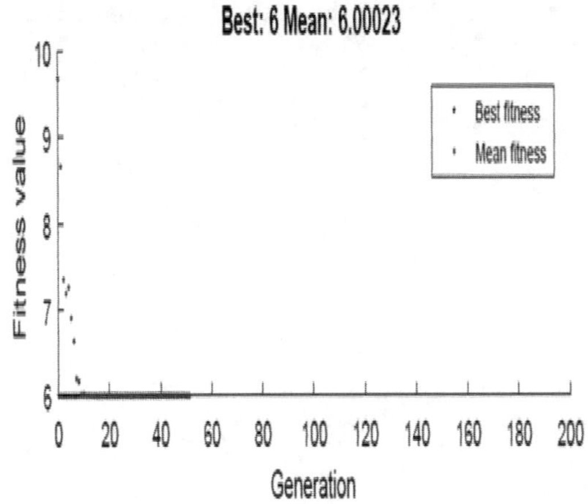

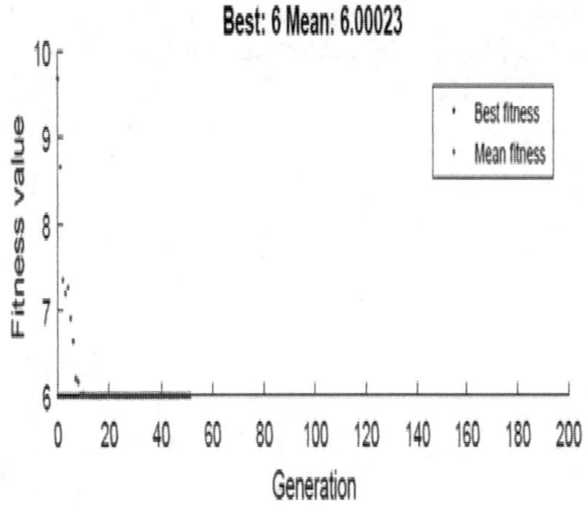

Figure 33. The generation verses the fitness value for Rz

4. Conclusions

High crystalline yttria stabilized zirconia with a cubic phase is prepared successfully using the chemical precipitation method. The results displayed that the particle size and the particle size distribution of 8YSZ can be controlled by adjusting the method. The new synthesized, Agar, as natural agent, can be used as a gelling agent in gel casting method without using any toxic or harmful agent. Thermal activation is a appropriate activation process for the agar solution; it can be used effectively rather of the using cross linking agents. Usage of ultrasonic treatment is a possible substitute for the vacuum treatment to accomplish the de-airing step. Increasing the solid loading with decrease agar ratio improves the surface roughness, dimension accuracy, microhardness, and compressive strength, demands the use of extreme values of the processing parameters to obtain highly near net shape ceramics according to the results of the genetic algorithm.

References

[1] Kramar D and Kopac J, Machining of near-net shape forged pinions 2012 *Manuf. Eng.* **55** 716-720.

[2] The Quadrant group of companies, Guidelines to identify Near-Net-Shape opportunities. 2016 quadrantplastics.com. 800-366-0300.

[3] Leo S, Tallon C, Stone N and Franks G V, Near-Net-Shaping Methods for Ceramic Elements of (Body) Armor Systems 2014 *J. Amer. Cera. Soci.* **97** 3013-3033

[4] `Olhero S M, Garcia-Gancedo L, Button T W. Alves F J, Ferreira J M, Innovative fabrication of PZT pillar arrays by a colloidal approach 2012 *J. Euro. Cera. Soci.* **32** 1067-1075

[5] Huang Y, Yang J, The New Methods and Techniques Based on Gel-Casting 2010 *Nov. Collo.Form. Cera*, 325-370

[6] Campbell D, Campbell S, Introduction to regression and data analysis 2008 *StatLab Workshop Series*

[7] Al-dujaili M A A, Jaheel S, Abbas H N, Preparation of HA/β-TCP scaffold and mechanical strength optimization using a genetic algorithm method 2017 *J. Aust. Cera. Soci.* **53** 41-48

[8] Al-dujaili M A A, Al-hydary I A D, Al-Jabar A J A, 2017 Optimization of the physical properties of barium titanates using a genetic algorithm approach *J. Aust. Cera. Soci.* **53** 673-686

[9] Sharifi L, Ghanbarnezhad S, Ghofrani S, Mirhosseini S H, High porous alumina bodies: production and properties via gel-casting technique 2014 *Inter. J. Adva. Scie. and Techn.* **65** 59-70

[10] Abbas S, Maleksaeedi S, Kolos E, Ruys A J, 2015 Processing and properties of zirconia-toughened alumina prepared by gelcasting *Mater.* **8.7** 4344-4362

Physical Characteristics and Compressive Strength of Na-Geopolymer Paste Designed by a Taguchi Method

Mohammad A. Ahmed Al-dujaili[1,*], Imad A. Disher Al-hydary[1], Zainab Zayer Hassan[1]

[1] Department of Ceramics Engineering and Building Materials, Faculty of Materials Engineering, University of Babylon, Babylon, P.O.Box:4, Iraq

*Corresponding author's email: adujaili@uobabylon.edu.iq

Abstract. Geopolymer paste is a revolutionary building material that the chemical activity of inorganic molecules will create. It is an alternative to traditional Portland cement and is more Eco-friendly. This analysis aimed to classify the mixtures and their process parameters suitable for the development of Geo-polymer paste with one of the ultimate compressive powers, the highest-lowest porosity, and the lowest-lowest final and initial setting time. In the experimental design of the Geo-polymer-based-metakaolin, a Taguchi methodology has been utilized. Five variables parameters were chosen that are mostly to influence the properties of the geopolymer. These are the quantity of Si, alkali, the proportion of alkali reagents, duration of blending, and water amount. These variables' influence has been calculated at 7 and 28 days on compressive strengths, porosity, density, and setting time. The analysis indicates that the strong compressive strength (115MPa) of Geopolymer paste could be achieved with the formula (1Na2O. Al2O3. 3.8SiO2.xH2O) utilizing suitable processing conditions under which the molar ratio of alkali silicate to alkali hydroxide must be held within the range of 3.25-3.02.

Keywords: Geopolymer paste, Metakaolin-based Geopolymer, Taguchi method, Effect Alkali activator

1-INTRODUCTION

Geopolymers are relatively recent products that were first produced and patented in the 70s by Joseph Davidovits. These are inorganic polymeric products with a zeolite-like chemical structure, but their composition is either semi-crystalline or amorphous. They are formulated along with an alkaline-activating solvent (mainly Na+ hydroxides and silicates) from a combination of many aluminosilicate materials (with high Si and Al content).

Because of polycondensation, the geopolymer reaction process, geopolymerization, is exothermic. Alkali-activation, also called geopolymerization, was stated by Palomo et al. [1]. It is a chemical method that enables vitreous materials (partially or entirely amorphous and/or metastable) to be transformed into very compact composites that lute. Van Jaarsveld et al. [2] proposed that a highly alkaline medium is needed for geopolymerization to happen. This is essential for the breakdown of silica and alumina and the hydrolyzation of the raw material particles' surfaces. Such an atmosphere can be accomplished by utilizing alkaline solutions for activation in a clear or combination form.

To achieve the optimal processing parameters[3,4], a genetic algorithm (GA) technique was implemented. The Taguchi technique is one of the statistical techniques utilized to optimize the process parameters and enhance the efficiency of the components generated. This since achieves the procedure's desired outcomes; the Taguchi approach requires determining the right control factors. To perform a series of tests, Orthogonal Arrays (OA) have been utilized. The effects of these tests are utilized to evaluate the data and predict the generated components' consistency. [5].

The study of the effect these parameters on Geopolymers' properties by full factorial design requires carry out many experiments. It becomes complex and difficult when the factor number increases. To overcome this issue, experimental design methods, such as the Taguchi method, may be used. This method proves its efficiency in many fields, including biotechnology, environmental engineering, software testing, service system, education, and production planning. The Taguchi technique is a methodical technique for extending different variables concerning efficiency, cost, and quality. [6].

A measure of robustness utilized in Taguchi designs to define control variables that minimize uncertainty in a product or method by mitigating the impact of uncontrollable variables (noise factors). (noise factors). Throughout manufacturing or the procedure used, noise factors may not be managed but could be controlled throughout experimentation. A temperature could range to a low and high value between 40 °C and 80°C . The number of tests to be done rises by raising the number of levels to adjust a parameter.

The signal-to-noise (S/N) proportion is computed for every factor level combination. The equation for the smaller-is-better S/N proportion utilizing base 10 logs is:(5)

$$S/N = -10*\log(\Sigma(Y^2)/n) \quad (1)$$

Whereas: Y = responses for the given factor level combination and n = number of responses in the factor level combination.

However, in the nominal is best (I), the signal-to-noise (S/N) proportion is computed for each factor level combination. The equation for the nominal-is-best (I) S/N proportion utilizing base 10 log is:

$$S/N = -10*\log(s^2). \quad (2)$$

Whereas s = standard deviation of the responses for all noise factors for the given factor level combination. The nominal is the best (II) of the signal-to-noise (S/N) proportion is computed for every factor level combination. The equation for the nominal-is-best (II) S/N proportion utilizing base 10 log is:

$$\frac{S}{N} = 10^* \log\left(\frac{\bar{Y}^2}{s^2}\right) \quad (3)$$

Also, it could be a choice to utilize the adjusted equation for the nominal-is-best S/N proportion. The equation is:

$$\frac{S}{N} = 10^* log \left(\frac{\bar{Y}^2 - \frac{s^2}{n}}{s^2}\right) \quad (4)$$

Whereas: $\bar{Y}$ = mean of responses for the given factor level combination. While, S = standard deviation of the given factor level combination responses, n = responses' number in the factor level combination.

Theoretically, a solid aluminosilicate source of geopolymerization could be any substance comprising aluminum and silicon. Nevertheless, blast furnace slags and ash from the burning of fossil fuels and calcinated clay are the most widely utilized components. The most widely utilized clay is kaolin that converts into metakaolin upon thermal activation. The resulting geopolymer is purer when metakaolin is utilized as the aluminosilicate source, which could have more easily defined characteristics than the Geopolymers derived from agro-industrial waste. Without the complications generated by utilizing fly ash, slags, and other alternate raw materials that include many hard-to-characterize amorphous phases, the geopolymers generated with metakaolin are deemed a " model method.' For instance, fly ash is not a well-defined substance, but it involves many crystalline and vitreous stages. [7].

In the aqueous form, an alkaline compound is required by the activator. Thus, sodium hydroxide, sodium silicate, and silica gel are traditionally used substances. A combination of silicate (solid dissolved in water) and hydroxide (solid dissolved in water) with silica gel (solid dissolved in water) is usually utilized. SiO2 waste is the silicate of the solution formed, while the hydroxide ensures high alkalinity in the solution[8]. The alkaline solution has a significant function throughout geopolymerization and impacts the production of mechanical strength. Part of its selection relies on its reactivity and the expense of the products used. Geopolymers' mechanical performance is strongly impacted by the Si/Al proportion, hydroxide amount [9], geopolymerization reaction efficacy [10], water amount [11], and curing situation [12]. Depending on the resources used or based on their molar proportion, geopolymers may exhibit various thermal, chemical, and mechanical behaviors. The current research work aims to find the mixes, and their processing parameters, which are suitable to produce a Geopolymer paste with one of the following features: Highest compressive strength, highest/ lowest porosity, highest / lowest initial and final setting time.

2- Materials and methods

2.1 The Starting Materials

Metakaolin, sodium silicate, sodium hydroxide, and silica gel have been utilized as initial materials to synthesize Geopolymer cement. Metakaolin, which was used in this study, was obtained by the calcination of kaolin clay, supplied from the local area (Dwaikhla, Western Iraqi Desert). The kaolin was calcined at 750 °C for three hours in an air atmosphere using a heating rate of 5°C/ min. Equation (5) was used to describe the composition of the prepared geopolymer.

$$\mathbf{1Na_2O.\ Al_2O_3.\ n\ SiO_2.\ xH_2O} \quad \textbf{(5)}$$

where (n) is the number of moles of SiO_2 in the formula of the geopolymer. For each mix formula, the effect of (n) and (x) was studied, and the mixing time and the alkali reactants ratio using a Taguchi method to design the experiments. The Taguchi method suggested 25 experiments for the Geopolymer mix (GP1).
Four parameters were selected based on the previous studies, including the parameters that strongly affect the geopolymer's manufacture. The upper and lower limits of each parameter were selected depending on the finding of the prior studies and many primary experiments[13]. The criteria used to select these limits' values include: (1) The paste of the geopolymer should be easily mixable. (2) The setting time should be neither too short nor too long. 3) The resulting Geopolymer body should be free of macro-cracks. The parameters magnitudes are demonstrated in Table (1).

Table 1. The affected parameters the manufacture of the geopolymer

N	H_2O (ml)/10.73 MK-750	SS/SH (wt/wt)	Mixing time (min)
3.2	8	1.5	5.0
3.4	9	2.26	7.5
3.6	10	3.02	10.0
3.8	11	3.78	12.5
4	12	4.54	15.0

2.2. Preparation of Geopolymer paste

The alkaline liquid utilized in this investigation was a mixture of sodium hydroxide (SH), and sodium silicate (SS). At first, the water is placed in the beaker, and then the sodium hydroxide is weighed and added to the water to reach the desired molarity(M). The sodium hydroxide pellet release heat as they dissolved in the water. When the sodium hydroxide addition is complete, the sodium silicate is added to the solution. At this time, they placed the solution on stirrers, a magnetic rod in the solution, and stirred at 80°C and 600 rpm. After all, the sodium silicate dissolved. Silica gel was added to the solution. Approximately one hour later, all the silica dissolved completely. After that, the desired water volume was added to compensate for the water lost due to evaporation, and the solution was cooled naturally to room temperature. After the alkaline solution is cooled to room temperature, the metakaolin (MK- 750) was added to the solution and mixed using a mechanical mixer at a fixed speed (3550rpm) desired to mix time. Molds made of plastic with diameter = 2.1 cm and height = 4.2 cm have been utilized for molding the pastes of geopolymer. The specimens should be kept in a specific condition at a lab with temperature ± 2 (23 °C) for one day and demolded. Then, these samples have been cured under the specific condition for 7 or 28 days.

2.3. Characterizations

The chemical composition of the kaolin was found using a wet chemical method. The kaolin wet chemical analysis test is currently being used to analyze the content levels of heavy metals in kaolin. This analysis was performed at the Iraqi geological and mining survey. The starting materials' main oxides were declared in terms of their weight percentages (Table 2). The crystalline phases of the starting materials have been recognized by the X-ray diffraction (XRD) technique (Shimatzu 6000 diffractometer, Japan). Cukα X-ray as source radiation was used with 30 mA and 40 kV at a 5°/min scanning speed. Nickel was used as a filter. Particle size analysis tests have been made to identify the distribution of kaolin and metakaolin powder's particle size by using a laser particle size analyzer. The synthesized samples were investigated with the Microvickers Hardness device model (TH-717) with a load of 9.8N and a holding time of 15 seconds. The diagonals distance was measured with the device of Vickers hardness.

The SEM was utilized to examine the features of the Geopolymer paste's microstructure; these features include the pores shapes, micro-cracks, and the presence of crystal. Geopolymer paste setting times are calculated utilizing the Vicat needle in compliance with ASTM C191. The penetration was recorded for each 15min period until a 5 cm or less penetration was achieved. Interpolation at a depth of 5 cm calculated the initial setting. The final setting time was also calculated, which is described as the time when the needle did not sink into the paste. Differential thermal analysis (DTA) was utilized to observe transition phases, crystallizations, and decomposition temperatures. According to the Archimedes method, the density, porosity, and water absorption of the synthesized samples were measured. Simultaneously, the compressive strength has been determined utilizing the formula ($\sigma c = P/A$). The physical, mechanical, and thermal properties were measured as an average of two-time measurements.

3. RESULTS AND DISCUSSION

3.1 XRD Analysis of Kaolin

Figure 1 demonstrates the kaolin powder XRD form. The pattern confirms the crystalline structure of the powder utilized. The common characteristic peaks of kaolin were observed according to (ICCD=00-001-0527) and (ICCD=00-033-1161) for kaolinite and SiO_2 minerals, respectively. It is well- known that the heat treatment of kaolin at moderate temperature leads to the formation of metakaolin; this was confirmed for the kaolin heat treated a 750°C for 3h (MK-750) as demonstrated in Figure 2. The pattern asserts the amorphous structure of metakaolin powder. XRD analysis of metakaolin demonstrates a sharp peak that refers to free quartz in the kaolin powder.

IOP Conf. Series: Earth and Environmental Science **877** (2021) 012036 doi:10.1088/1755-1315/877/1/012036

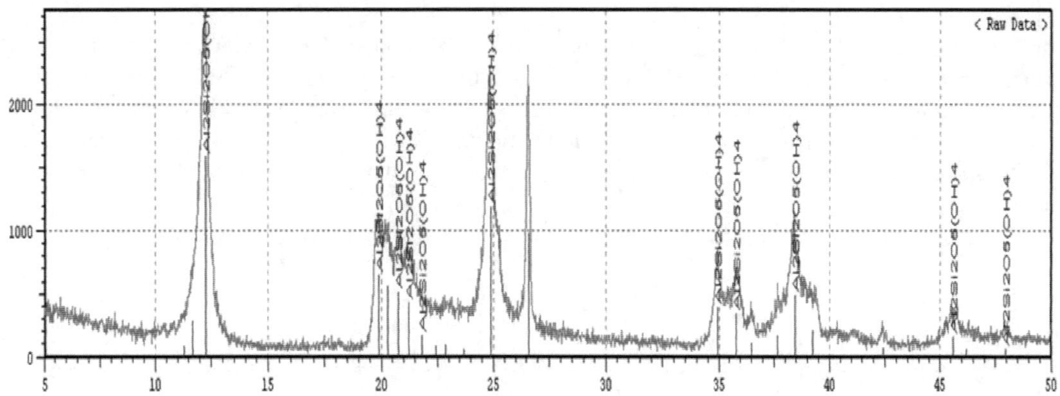

Figure 1. Kaolin XRD form

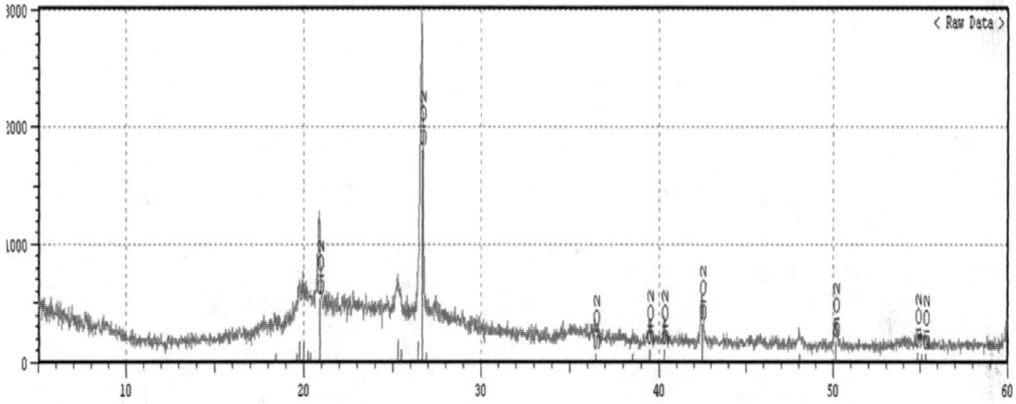

Figure 2. Metakaolin (MK 750) XRD form

3.2 Results of The Chemical Analysis of Kaolin

Table (2) demonstrates the outcomes of the wet chemical analysis of the kaolin. It can be seen that the SiO2 amount in the clay is higher than the stoichiometric amount in the kaolinite, which is (48.77%). Furthermore, that confirms the XRD result regarding the presence of free quartz and helps calculate the amount of free quartz. Approximately, that should be excluded when the geopolymer's composition is calculated because of quartz's inertness, which makes quartz an inert component during the synthesis of the geopolymer.

Table 2. Demonstrates the kaolin wet chemical analysis results

SiO_2 %	Fe_2O_3 %	Al_2O_3 %	TiO_2 %	CaO %	MgO %	SO_3 %	P_2O %	K_2O %	Na_2O %	Cl %	LOI %
48.77	1.76	34.27	1.47	0.43	0.08	0.11	0.02	0.43	0.17	0.03	12.46

3.3 Analysis of Kaolin Particle Size

Figure 3 demonstrates the distribution of kaolin powder particle size. The results demonstrated that the kaolin is composed of micro-sized particles mainly below 20μm. The particle size distribution is multimodal with a D_{50} of 3.8μm. The distribution of MK-750 particle size is demonstrated in Fig (4); it is well known that metakaolin has fine particle size as a compound with kaolin as it is produced due to the breaking of the kaolinite structure. However, the agglomeration and aggregation lead to large secondary particles, as confirmed in Figure 4.

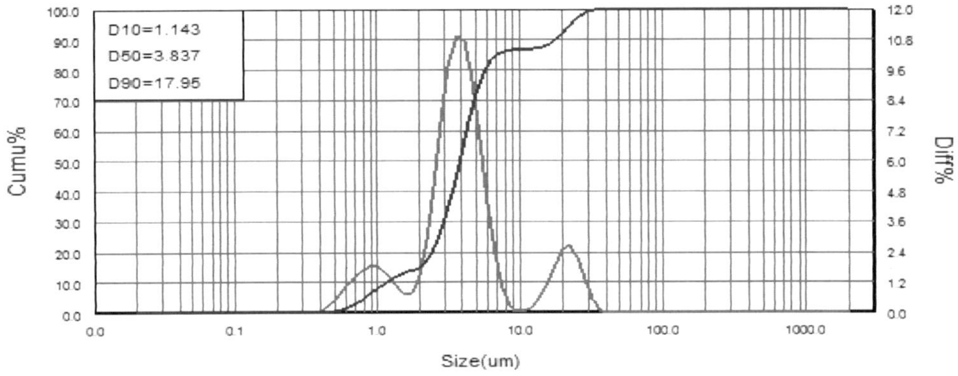

Figure 3. The analysis of kaolin powder particle size

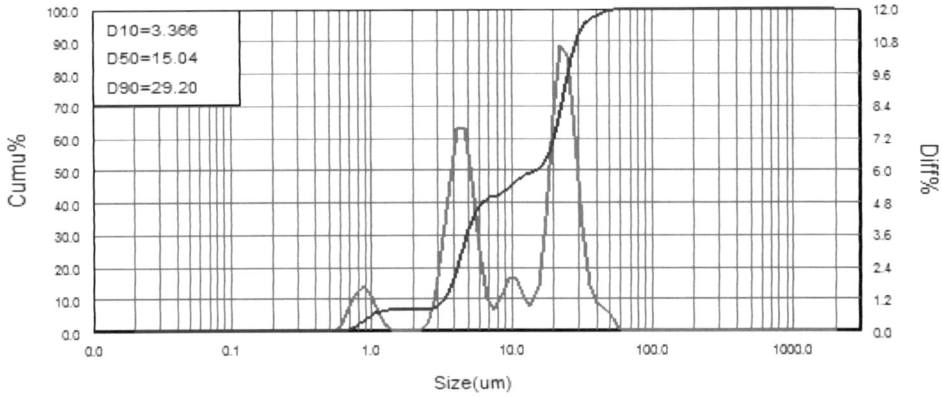

Figure 4. The analysis of Metakaolin powder particle size.

3.4 DTA Analysis of Kaolin

Figure 5 demonstrates the differential thermal analysis (DTA) for kaolin. The endothermic peak at 530 °C was recorded for the kaolin calcination. To form metakaolin, this event is finished at 575°C, and no other thermal event could be observed, indicating that 750°C is a suitable choice for the formation of metakaolin.

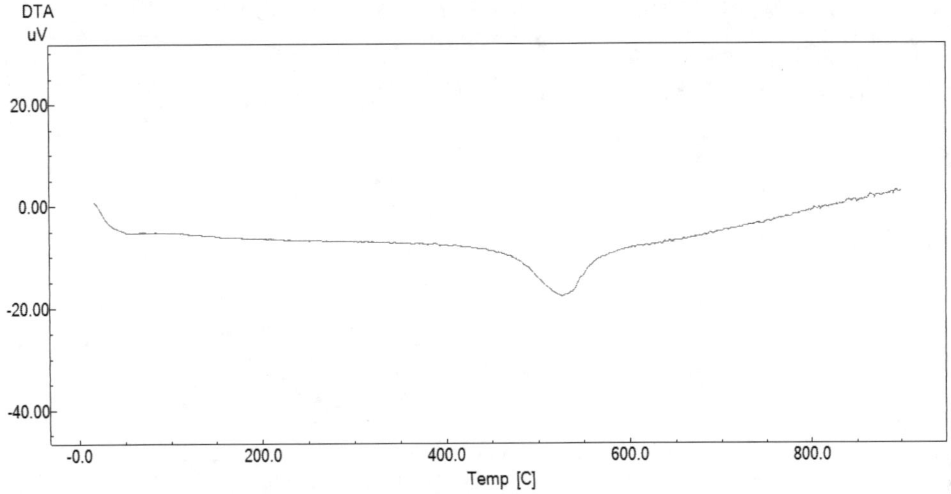

Figure 5. Different thermal analysis (DTA) for kaolin

3.5 Geopolymer paste Result

3.5.1 Compressive Strength

3.5.1.1 Compression Strength at 7 Days

Table (3) illustrates the Taguchi analysis's findings for the compression strength after seven days since the casing date and gives the rank for the process parameters utilized in the current study. The results indicate that the most influential factors on the compressive strength are silica content mixing time water content SS/SH ratio. It can be observed that the compression strength after seven days since the casing date for all the tested sample is higher than 30MPa and reaches up to (61MPa). These values are too high as compared with that reported in pieces of literature as demonstrated in Table (4)

Table 3. Taguchi Analysis of response for 7-days compressive strength for GP1

Level	SiO_2	H_2O/10.73 MK-750	SS/SH	Mixing time
1	33.42	42.60	32.86	41.82
2	39.22	32.60	35.00	35.62
3	45.46	33.22	42.64	35.68
4	31.60	39.70	36.80	30.00
5	32.82	34.40	35.22	39.40
Delta	13.86	10.00	9.78	11.82
Rank	1	3	4	2

These results indicate the advantage of the current study over the previous studies.

Table 4. Values of compressive strength at 7-days compared with other studies

Compressive strength at 7-days (MPa)	Reference
61	Current study
8.61	[14]
31	[15]
28. 53	[16]
18.94	[16]
32	[17]
21.33	[18]

Figure 6 demonstrates the major impact plot for the compression strength at 7 days of Geopolymer paste of batch GP1which satisfies the formula (Na_2O. Al_2O_3. $nSiO_2$). In general, the strength increases by increasing the silica content up to 3.6, where the best strength can be obtained. The silica content controls the Si/Al proportion that impacts hydrolysis, dissolution, and Geopolymer condensation reaction. Depending on prior studies, many academics stated that the optimal proportions of SiO_2/Al_2O_3 were between 3 and 3.8 that was somewhat less compared to the ranges that Davidovits stated them [19]. Moreover, De Silva et al. proved that [20] the Geopolymer system condensation response with little proportion of Si/Al tends to occur between types of silicate and aluminate, producing primarily structures of poly(sialate). In the system of geopolymer that has a great Si/ Al proportion, Geopolymer structures of poly(sialate-disiloxo) and poly(sialate-siloxo) are made due to condensation response of predominant between silicate types, making silicates (oligomeric), which reacted with $Al(OH_4)^{4-}$. Greater numbers of siloxo units have been contributing to improving strength.

Rahier, et al. [21] report that low and high amounts of water in producing Geopolymers will negatively affect the rate of reaction. This was supposed as a result of the reduced concentration of OH at a high amount of water. In other words, at low amounts of water, despite increasing OH concentration, the reactive types ($H_3SiO_4^-$ monomer monodeprotonated) for reacting between oligomers' silicate reductions and reducing the reaction rate of geopolymerization.

For the concentration of SS/SH, the strength also generally increases by increasing the SS/SH ratio due to the acceleration of polymerization; the highest strength was obtained at the ratio of 3.02. Mixing time is one of the variables that affect the Geopolymer strength where it works to homogenize the alkali solution with the powder; the result demonstrates the best compressive strength at a time of 5 min.

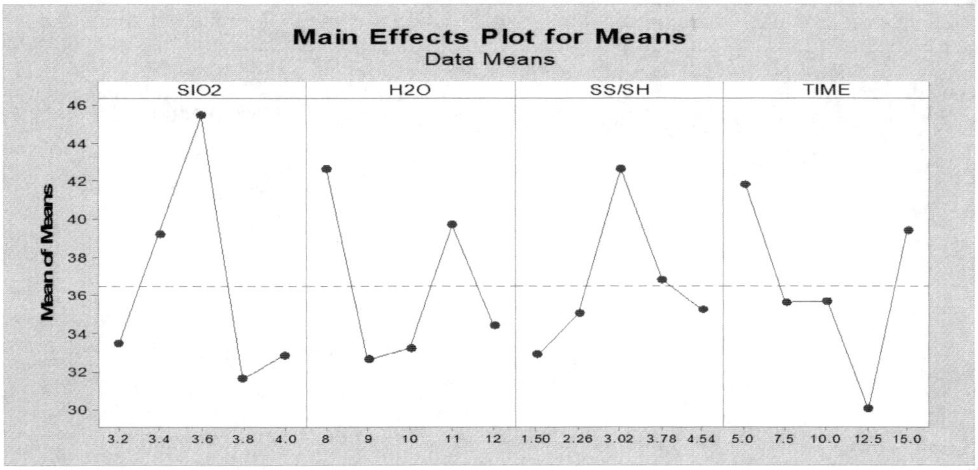

Figure 6. The major impacts plot for the compressive strength at 7 days of batch GP1

3.5.1.2 Compression Strength after 28 Days

The Taguchi analysis outcomes for the 28-days compressive strength are shown in table (5) for batch GP1. It could be observed that the silica amount was still the most effective factor in the strength of the sample under compression. However, compared with the 7-days strength under the same batch compression, the water content gets the second-order among the other effective factors; this might be connected to the amount of water on the porosity of the resulting Geopolymer body. On the order hand, the mixing time and SS/SH proportion's influence is limited to 10-16MPa only.

As shown in Figure 7, the 28-days compression strength increases when the silica content increases up to 3.8. Similarly, it increases with increasing the water content up to 11ml; this is maybe due to facilitating the mixing process as reported by Ghosh, et al. [22]. The achieved 28-day compressive strength was as high as (114MPa), which is a very high value compared with the results reported by many researchers, as demonstrated in Table (6).

Table 5. Taguchi Analysis of response for 28-days compressive strength for GP1

Level	SiO_2	H_2O/10.73 MK-750	SS/SH	Mixing time
1	48.56	58.56	64.00	76.62
2	71.62	66.84	69.32	63.42
3	68.70	65.00	73.14	60.44
4	77.60	78.42	62.70	71.18
5	75.28	72.94	72.60	70.10
Delta	29.04	19.86	10.44	16.18
Rank	1	2	4	3

Table 6. Values of compressive strength at 28-days compared with other studies

Compressive strength at 28-days (MPa)	Reference
114	Current study
48.1	[23]
79	[24]
33-42	[25]
32	[26]
50	[27]
62	[15]
48.45	[18]
60-85	[13]

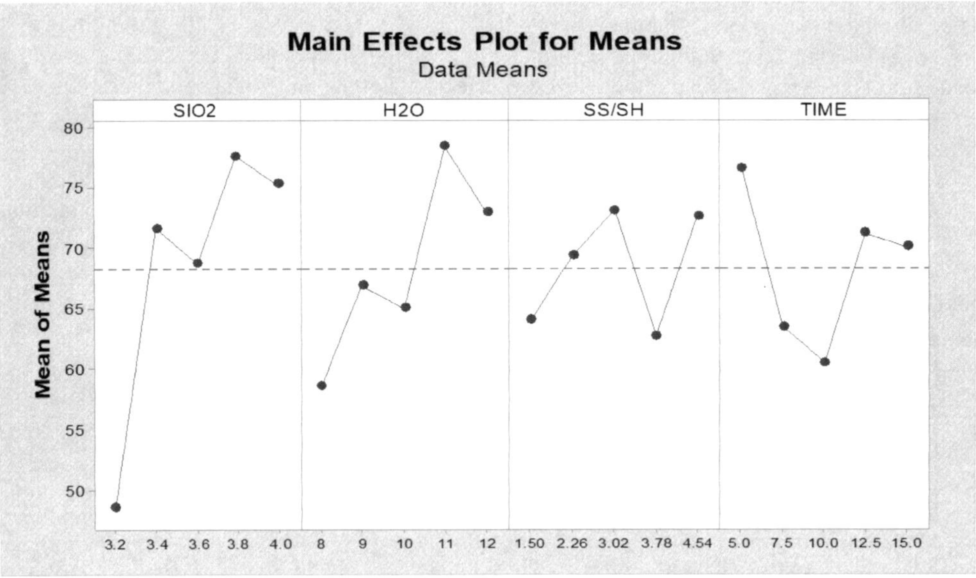

Figure 7. The major impacts plot for the compressive strength at 28 days of batch GP1

3.5.2 Setting Time

Figure 8 demonstrates the major impact plot for the setting time of GP1. In batch GP1, silica content is the most influential factor in the initial and final setting time. When geopolymer contains sodium ions only, as in GP1batch, the setting time decreases by increasing the silica content up to 3.8. This indicates that the polymerization process rate, in the presence of Na ions only, increases when the silica content is increased.

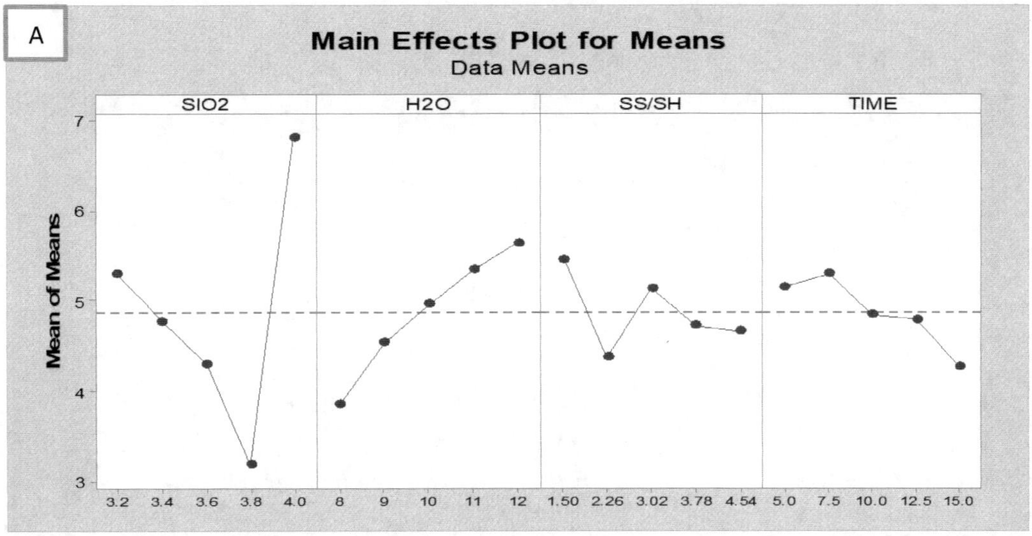

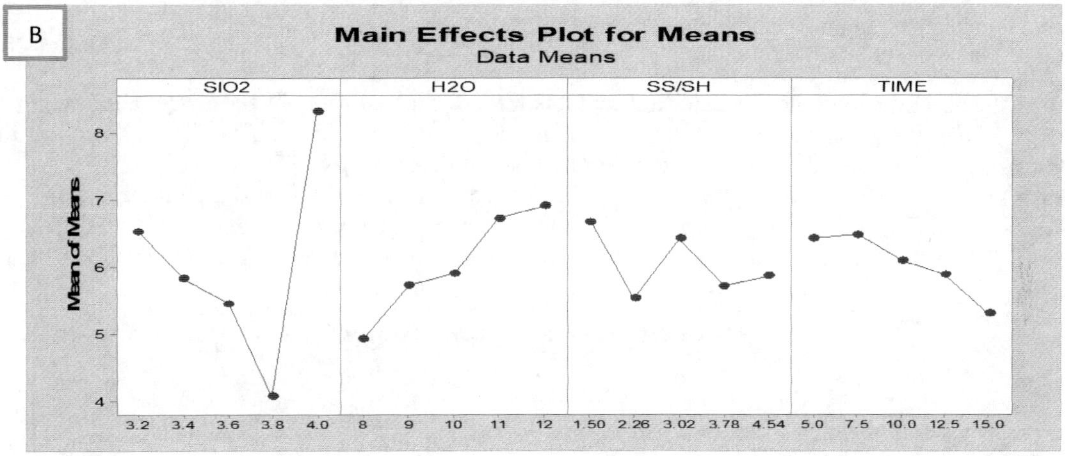

Figure 8. (A) initial setting time and (B) the final setting time of Geopolymer batch GP1

3.5.3 Bulk Density

Figure 9 show the Taguchi major impact plot for the bulk density of the geopolymer patches. The results indicate that the density is impacted mainly by the water content, which decides the geopolymer's porosity. The SS/SH ratio seems to have a minor impact on the bulk density of the prepared geopolymer. The impact of the silica amount on the bulk density of Na-geopolymer is high. It has been found that the bulk density of Na-based geopolymer is high. This result is in contrast with the previous study Lizcano, et al. [28], where it has been specified that the bulk density magnitude note down for metakaolin Geopolymers' Na-based are (1.25–1.72 g/cm^3) and K-based are (1.39–1.82 g/cm^3).

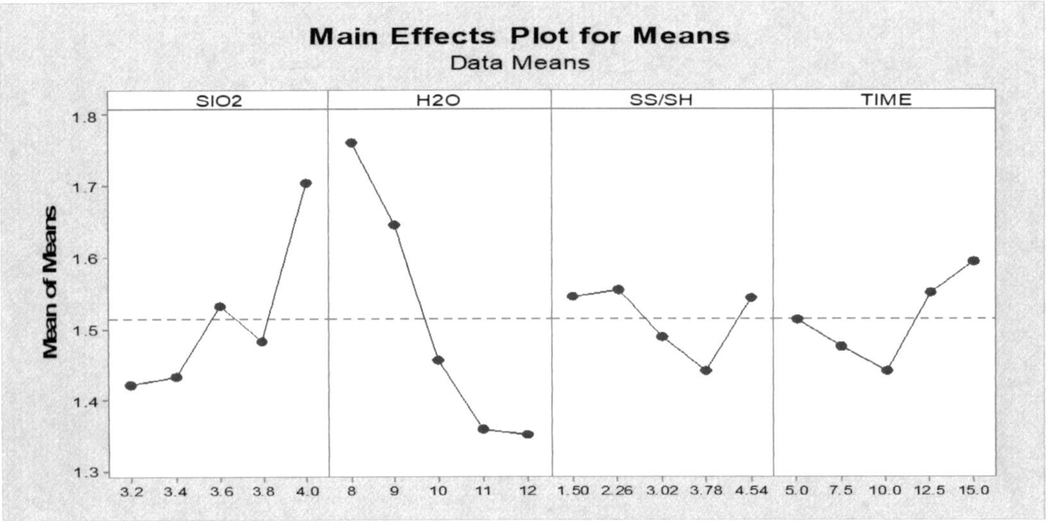

Figure 9. The major impacts plot for the bulk density of GP1

3.5.4 Porosity and Water Adsorption

Figures 10 and 11 demonstrate the major impact plots for the prepared geopolymer's water absorption and porosity, respectively. As expected, the water content was found to be the most effective factor on the porosity; hence, the water absorption caused the formation of pores due to:

1-The removal of the extra water, which is commonly added to enhance the workability of the paste.

2-The removal of the water which is produced during the condensation polymerization.

Tables (7) and (8) list the value of the process parameters, suggested by Taguchi analysis, to obtain geopolymer with the highest or lowest porosity, respectively, from each batch.

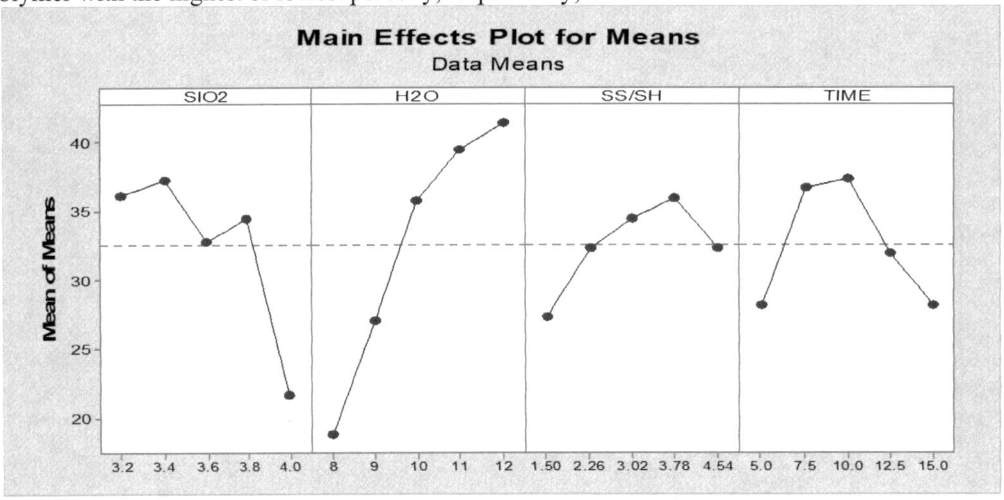

Figure 10. The major impact plot for the porosity of GP1 batch

Table 7. the process parameters required to prepare geopolymer with the highest porosity

Batch	SiO_2	H_2O/10.73MK-750	SS/SH	Mixing time
GP1	**3.4**	**12**	**3.78**	**10**

Table 8. the process parameter required to prepare geopolymer with the lowest porosity.

Batch	SiO_2	H_2O/10.73MK-750	SS/SH	Mixing time
GP1	4.0	8	1.5	15

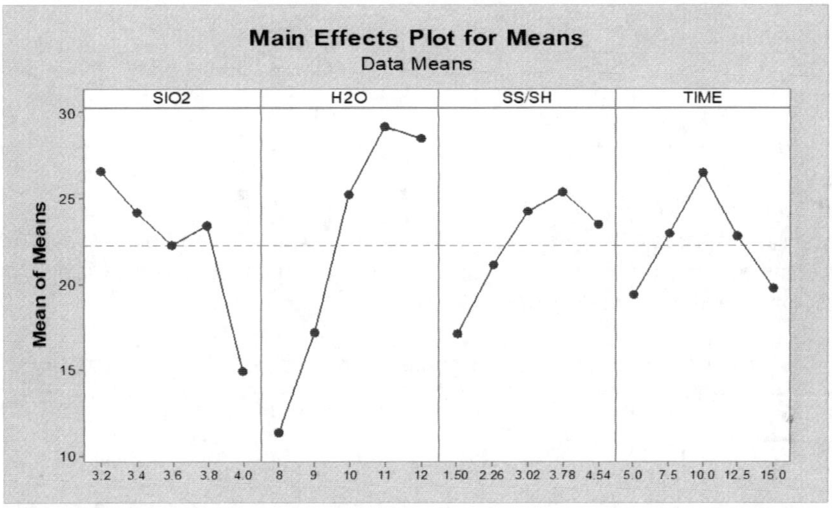

Figure 11. The major impact plot for the water absorption of GP1batch

3.6 Results of Optimum Mixes

3.6.1Compressive Strength of Geopolymer paste

Table (9) illustrates the value of the processing variables required to obtain the optimal compressive strength according to Taguchi analysis.

Table 9. Value of the processing variables to obtain the optimal compressive strength according to Taguchi analysis

Batch	SiO_2	H_2O/10.73MK-750	SS/SH	Mixing time
GP1	3.8	11	3.02	5

Table (10) demonstrates the compressive strength of Geopolymer cement at 7 days and 28 days. The result revels that the suggested optimum mix achieved compressive strength higher than that obtained from the experimental works; this confirms that the Taguchi method can suggest the optimum conditions to get optimum results.

Table 10. Compressive strength of geopolymer paste prepared according to the optimum mix

Batch	Compressive strength at 7 days (MPa)	Compressive strength at 28 days (MPa)
GP1	96	115

3.6.2 Vickers Hardness of Geopolymer Paste

Table (11) shows the Vickers hardness of Geopolymer cement past, prepared according to the optimum mixes.

Table 11. Vickers Micro-hardness of Geopolymer cement

Batch	Vickers hardness (HV)
GP1	28.83

3.6.3 Physical properties of Geopolymer Paste

Table (12) demonstrates the result of the density, porosity, and water absorption of Geopolymer paste. The consequences show that batch GP1 has the greatest density magnitude and lowest porosity. This result explains the reason behind the hardness result as well as the compressive strength.

Table 12. Density, water absorption, and porosity of Geopolymer paste

Batch	Density (g/cm^3)	Water absorption %	Porosity %
GP1	1.4	2.9	4

3.6.4 Setting Time of Geopolymer Paste

Table (13) shows the geopolymer paste's final and initial setting time. The results indicate that the batch GP1 setting times within the range accepted for cement applications.

Table13. The initial and final setting time of the Geopolymer paste

Batch	Setting time (min)	
	Initial	Final
GP1	130	270

3.6.5 Microstructure of Geopolymer Paste

A few spherical voids and micro-cracks that can be traced to air bubbles throughout the geopolymerization process and micro-cracks attributable to fracture are seen in Figure 12 for batch GP1. It is stressed that the large structure observed tends to validate the amorphous properties of these geopolymers. These bubbles added to the overall porosity. In the tests, the unreacted metakaolin particles can not be detected; the unreacted raw metakaolin particles could serve as structural faults decreasing the Geopolymer paste's compressive power [29]. Slower and more uniform reaction kinetics may be identified with completely reacted matrices with homogenous distribution and a dense gel process. The slower method involved in

Geopolymer paste's strength production helps the precursor to melt continuously, resulting in high strength of a more homogeneous composition at later ages [30].

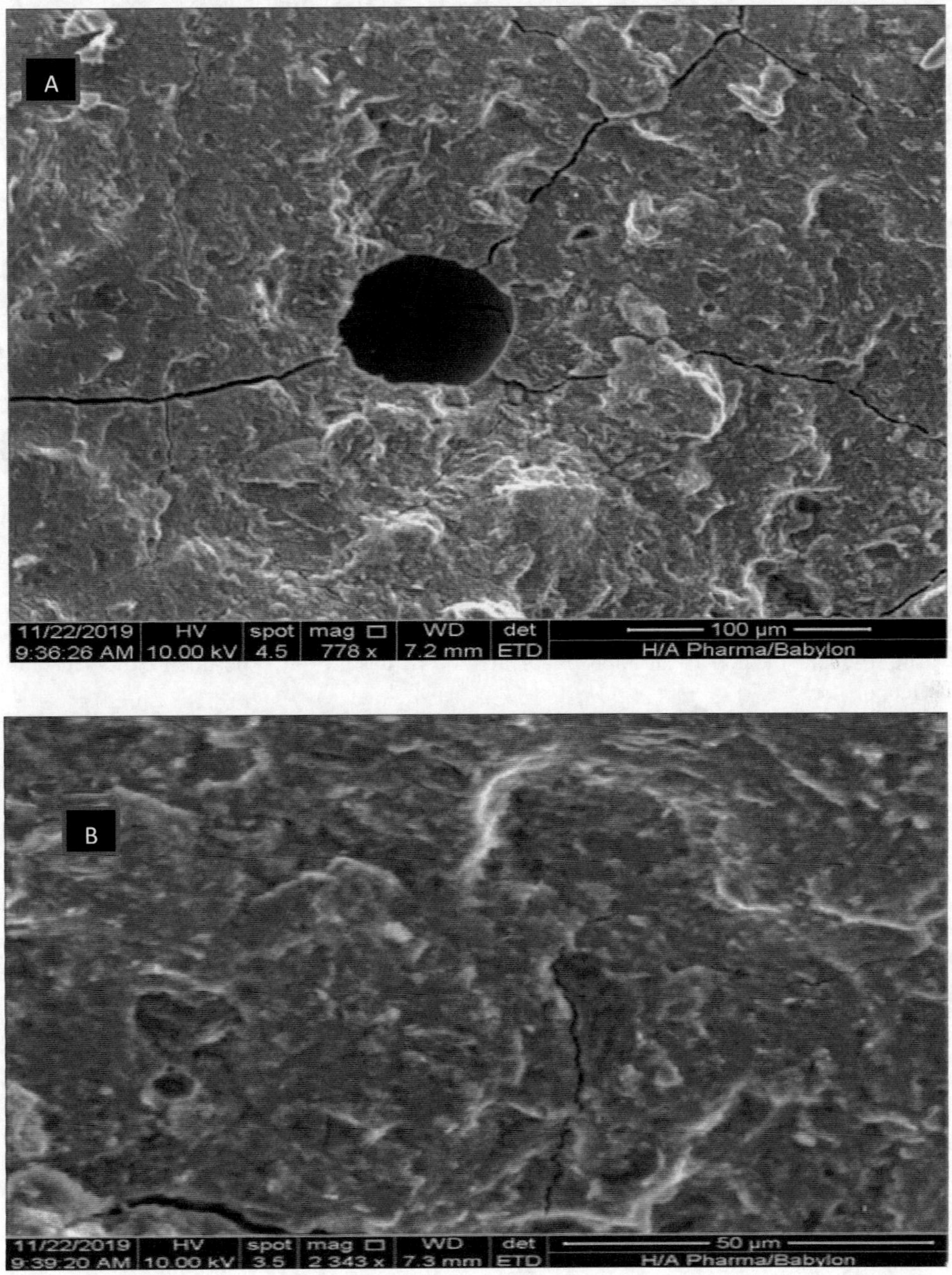

Figure 12. SEM images of geopolymer paste for batch GP1

4. CONCLUSIONS

From the findings of the current study, the following can be concluded:

1. The Taguchi method is an appropriate method to design the experiments of metakaolin-based Geopolymers.
2. Controlling the silica content, water content, SS/SH ratio, and the mixing time is required to produce a geopolymer with a compressive strength that exceeds 100MPa.
3. The porosity of geopolymer can be low as 3% or high as 40%, depending on the mix composition and processing conditions.
4. The alkali hydroxides to silicates molar proportion must be kept between 3.02-3.25 in order to obtain high compressive strength.

Acknowledgment

We hereby acknowledge that no financial support from the Department of Ceramics Engineering and Building Materials-College of Materials Engineering- the University of Babylon or other organizations regarding this manuscript.

Reference

[1] A. Palomo, M.W. Grutzeck, M.T. Blanco, Alkali-activated fly ashes. A cement for future,1999 Cem. Concr. Res. 29, 1323–1329.

[2] JGS van Jaarsveld, J.S.J. van Deventer, L. Lorenzen, The potential use of geopolymeric materials to immobilise toxic metals: part I. Theory and applications, 1997 Miner. Eng. 10 ,659–669.

[3] Al-Jabar, A. J. A., Al-Dujaili, M. A. A., & Al-Hydary, I. A. D. Prediction of the physical properties of barium titanates using an artificial neural network. 2017Applied Physics A, 123(4), 274.

[4] Al-dujaili, M. A. A., Al-hydary, I. A. D., & Al Jabar, A. J. A. Optimization of the physical properties of barium titanates using a genetic algorithm approach.2017 Journal of the Australian Ceramic Society, 53(2), 673-686.

[5] Athreya, S., & Venkatesh, Y. D. Application of Taguchi method for optimization of process parameters in improving the surface roughness of lathe facing operation. International Refereed 2012 Journal of Engineering and Science, 1(3), 13-19.

[6] Amir Ahmad Dar, B S Abdur Rahman. An Application of Taguchi L9 Method in Black Scholes Model for European Call Option.2018 International Journal of Entrepreneurship (Print ISSN: 1099-9264; Online ISSN: 1939-4675)

[7] P. Duxson, J.L. Provis, G.C. Lukey, S.W. Mallicoat, W.M. Kriven, J.S.J. Van Deventer, Understanding the relationship between geopolymer composition, microstructure and mechanical properties,2005 J. Colloid Interface Sci. A. 269 ,47–58.

[8] F.T. Lima, F.J. Silva, C. Thaumaturgo, Xanes an borda-k do Si do sistema caulimmetaculim-geopolímero, 2005 Rev. Brasil. Apl. Vácuo 24 (2) , 90–92.

[9] Yao, X., Zhang, Z., Zhu, H., & Chen, Y. Geopolymerization process of alkali–metakaolinite characterized by isothermal calorimetry.2009 Thermochimica Acta, 493(1-2), 49-54.

[10] Pacheco-Torgal, F., Castro-Gomes, J., & Jalali, S. Alkali-activated binders: A review. Part 2. About materials and binders manufacture. 2008 Construction and building materials, 22(7), 1315-1322.

[11] Zuhua, Z., Xiao, Y., Huajun, Z., & Yue, C. Role of water in the synthesis of calcined kaolin-based geopolymer. 2009 Applied Clay Science, 43(2), 218-223.

[12] Davidovits, J. (1982). US Patent No. 4,349,386. Washington, DC: US Patent and Trademark Office.

[13] Malkawi, A. B., Nuruddin, M. F., Fauzi, A., Almattarneh, H., & Mohammed, B. S. Effects of alkaline solution on properties of the HCFA geopolymer mortars.2016 Procedia Engineering, 148, 710-717.

[14] Mustafa Al Bakri, A. M., Abdulkareem, O. A., Kamarudin, H., Khairul Nizar, IRafiza, A. R., Zarina, Y., & Alida, A. Microstructure Studies on the Effect of the Alkaline Activators of Fly Ash-Based Geopolymer at Elevated Heat Treatment Temperature. 2013 In Applied Mechanics and Materials (Vol. 421, pp. 342-348). Trans Tech Publications.

[15] Albitar, M., Ali, M. M., Visintin, P., & Drechsler, M.. Effect of granulated lead smelter slag on strength of fly ash-based geopolymer concrete. 2015 Construction and Building Materials, 83, 128-135.

[16] Al-Shathr, B. S., & Al-Attar, T. S. Effect of curing system on metakaolin based geopolymer concrete. 2016 Journal of University of Babylon, 24(3), 569-576.

[17] Wijaya, A. L., & Ekaputri, J. J. Factors influencing strength and setting time of fly ash based-geopolymer paste. 2017 In MATEC Web of Conferences (Vol. 138, p. 01010). EDP Sciences.

[18] Risdanareni, P., Puspitasari, P., Kartika, D., & Djatmika, B.. Mechanical properties of geopolymer paste with fly ash variation. 2016, October In AIP Conference Proceedings (Vol. 1778, No. 1, p. 030025). AIP Publishing.

[19] Davidovits, J. (1982). US Patent No. 4,349,386. Washington, DC: US Patent and Trademark Office.

[20] Silva, P., Sagoe-Crenstil, K.,& Sirivivatnanon, V. Kinetics of geopolymerization: role of Al2O3 and SiO2.2007 Cement and Concrete Research, 37(4), 512-518.

[21] Rahier, H., Wastiels, J., Biesemans, M., Willlem, R., Van Assche, G., & Van Mele, B. Reaction mechanism, kinetics and high temperature transformations of geopolymers. 2007Journal of Materials Science, 42(9), 2982-2996.

[22] Ghosh, K., & Ghosh, P. Effect of synthesizing parameters on compressive strength of fly ash based geopolymer paste. 2012 Int J Struct Civ Eng, 1(8), 1-11.

[23] Barbosa, V.F., MacKenzie, K. J., & Thaumaturgo, C. Synthesis and characterisation of materials based on inorganic polymers of alumina and silica: sodium polysialate polymers.2000 International Journal of Inorganic Materials, 2(4), 309-317.

[24] Cheng,T. W., &Chiu, J. P.. Fire-resistant geopolymer produced by granulated blast furnace slag. 2003 Minerals Engineering, 16(3), 205-210.

[25] Song, X. J., Marosszeky, M., Brungs, M., & Munn, R. Durability of fly ash-based geopolymer concrete against sulphuric acid attack. 2005, April In Proceedings of the international conference on durability of building materials and components, Lyon, France (pp. 369-375).

[26] Reddy, B. S. K., Varaprasad, J., & Reddy, K. N. K. Strength and workability of low lime fly-ash based geopolymer concrete. 2010 Indian Journal of Science and Technology, 3(12), 1188-1189.

[27] Nath, P., & Sarker, P. K. Effect of GGBFS on setting, workability, and early strength properties of fly ash geopolymer concrete cured in ambient condition.2014 Construction and Building Materials, 66, 163-171.

[28] Lizcano, M., Kim,H. S., Basu, S., & Radovic, M. Mechanical properties of sodium and potassium activated metakaolin-based geopolymers. 2012 Journal of Materials Science, 47(6), 2607-2616.

[29] Zhang, F., Zhang, L., Liu, M., Mu, C., Liang, Y. N., & Hu, X. Role of alkali cation in compressive strength of metakaolin based geopolymers. 2017 Ceramics International, 43(4), 3811-3817.

[30] da Silva Rocha, T., Dias, D. P., França, F. C. C., de Salles Guerra, R. R., & de Oliveira, L. R. D. C.. Metakaolin-based geopolymer mortars with different alkaline activators (Na+ and K+). 2018 Construction and Building Materials, 178, 453-461.

Rechargeable Flexible Paper Battery using PAV, PSSPEDOT Polymer

Thakir H. Abed[1], Meethaq. M.Abed[2], Burak Y. Kadem[3], Ahmad T. Jaiad[4]

[1]Ministry of Education, Directorate of Education-Babylon city, Babil, Iraq
[2]Polymer engineering, Al-Furat Al-Awsat Technical University, Technical Institute of Kerbala, 56001, Karbala, Iraq
[3]College of Renewable Energies and Environmental Science, Al-Karkh University for Science, Baghdad, Iraq
[4]Ministry of Electricity, Babil, Iraq

*Email: mee1sa2012@gmail.com

Abstract. The main idea of this research is to design a rechargeable paper battery from local cheap and available materials .The practical part is represented by adding Polyvinyl alcohol (PVA) to conductive polymer (PSS PEDOT) with adding different mineral salts for then study the quantum of electrical conductivity and heat influence on electrical conductivity and acidity factors of the Electrolyte solution.The next step was to produce a rechargeable, flexible battery manufactured from regular cellulose paper, sulfone, and ionic solution.The measurements were made using modern laboratory devices to study the electrical properties, conductivity, resistance (Hall Effect), and voltage quantum of the ionic solution. Through the results we obtained, we noticed an increase in the conductivity of the ionic solution when adding mineral salts. The voltage quantum of one battery ranged between 0.3-0.3 volts, also several batteries were connected in series, and the result was 1.8 volts. Manufactured paper batteries can be recharged with a direct current source and recover their efficiency marked up 98-99%.

Keywords: paper battery, metal salts, polymer PVA, PEDOT: PSS.

1. Introduction

Polymeric materials are considered one of the most used materials at present. Polymeric materials are cheap, easy to manufacture, with specifications that often reach very good. It is known that most of the polymeric materials are insulating materials and that these materials have been used in the electrical and electronic industries for a long time as an auxiliary material, but the good properties of these materials led to their inclusion in the list of candidate materials for study to be used as a basic material in the electrical and electronic industries. These materials have very good electrical insulation specifications, as well as some polymeric materials have good thermal insulation characteristics as well as good mechanical and optical properties and properties. Therefore, the process of changing the electrical properties of the polymeric insulating materials is one of the important matters with good positive returns in the electrical and electronic industries [1]. On the other hand, some metals have been replaced by polymers in many uses because these polymers possess good conductivity properties, including their use in batteries. Like the case that the polymeric materials can change their electrical properties using several techniques, four of which are popular techniques, which are [2].

- Ion implantation technique.
- Creating defects by using the nuclear radiation technique.
- Creating defects by using alloys of different materials technique.
- Creating defects by adding impurity materials technique.

Each technology has advantages and disadvantages that differ from other technologies. The ion implantation technique is one of the limited feasibility techniques due to the inability of this technique to provide defects (charge carriers) with an appropriate density that significantly changes the electrical properties. The technique of changing the electrical properties is by using nuclear radiation, the polymeric materials showed that their effect is limited to changing the behavior of charge carriers with a slight increase in the conductivity value of the polymeric materials (this increase in conductivity is less than the amount required to convert the material from an insulating material to a conductive) [3]. The technique of changing the electrical properties is by using polymeric alloys, this technique often leads to the emergence of new stages in the electrical properties of these phases, which lead to a reduction in the efficiency of the performance of these materials, as well as that the change that occurs in the conductivity of these materials is often within a range Alteration unable to convert material from dielectric to conductive [4]. As for the conversion of the polymeric material by deformation from an insulating material to a conductive material, this method is one of the methods that have great chances of success becauseThe ability of polymeric materials to contain deformation materials (due to the possibility of converting these materials into liquid materials).

The likelihood of impurity binding with polymeric chains is great due to the availability of a large number of binding sites with these gigantic compounds.The availability of many elements that can be homogeneous with polymeric materials in the liquid state.The normal battery mechanisms work.Electricity in general is the process of movement of charges through a conductive path such as metal wire, which is commonly called an electrical circuit. The battery has three main parts. Two electrodes are the cathode (-), the anode (+), and the electrolyte solution. The cathode and the anode are the two terminals in a standard battery that are usually connected in the external electrical circuit. The chemical reaction process present in the electrolyte solution leads to the building of negative charges on the anode (electrons), which produces an electric potential difference between the cathode and the anode (this voltage difference is unstable at the beginning until the process of collecting negative charges at the anode continues) [5]. Where the process of collecting the electrons at the anode in order to get rid of the potential difference generated by them. This process is done as follows, the electrons perform a repulsive process with each other to salvage to a place where the electrons are less abundant, which is the cathode, but the electrolyte solution prevents this from happening directly from the anode to the cathode. When the circuit is closed, the electrons can move to the cathode. This process is usually limited in batteries due to the electrochemical reaction present that will change the chemistry of the anode and cathode and thus stop supplying the electrons over time. Therefore, rechargeable batteries are represented by reversing this process by connecting the battery with an external source such as a solar cell [6].

2. Experimental Part

2.1 Samples Preparation

In this research, various mineral salts were used i.e, lithium chloride (LiCl2, copper chloride (CuCl2), nickel chloride (NiCl2), zinc chloride (ZnCl2), cadmium chloride (CdCl2), and tin chloride (SnCl2). separately, 1 mg of these salts dissolved in Methanol, the solution was placed in the vibrator using ultrasound waves (Ultrasonic bath) for about half an hour to obtain a homogeneous solution and good solubility. On the other hand, polyethylene dioxythiophene polystyrene sulfonate (PEDOT: PSS) was used, which is one of the most popular conductive polymers that are available either. In the form of dry pills or the form of a liquid dissolved in water.Ratios of salts solutions mixed with 1 milligram of (PEDOT: PSS) then the mixture was placed in the vibrator using ultrasound for about half an hour to ensure complete homogeneity between the two materials. Then 5 mg of polyvinyl alcohol was dissolved in distilled water, after that the mixture placed again in a vibrator using ultrasound. The work was done according to certain percentages of mixing the solutions to obtain a homogeneous ionic solution in proportions as follows (5%, 10%, 15%, 20%, and 25%), the ratio of (PEDOT: PSS) solution with salts to the ratio of polyvinyl alcohol solution. To ensure the stability of the solutions for a sufficient period, the solutions were stored in suitable small containers tightly sealed, where the solutions recorded fairly good stability after being shaken with ultrasound for 20 minutes to ensure the homogeneity of the final solution. The stability of the solutions was laboratory after keeping the samples for 4 consecutive days, and the stability of the solutions was very good.

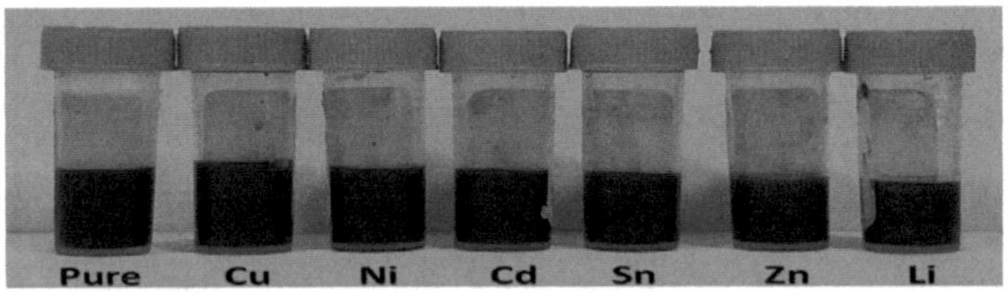

Figure.1: The solutions used after shaking with ultrasound

The electrolyte solution was prepared in the form of a compound of (PEDOT: PSS) with PVA with the used salts, a normal cellulose paper is coated with this prepared material in various proportions to obtain a semi-dry and flexible electrolyte solution with the flexibility of the paper.Various samples were prepared to be suitable for the type of measurements made during this research as figure.1.

3. Results and Discussion

3.1. The effect of temperature on the conductivity and the (PH):

factor of the prepared liquids, the electrical conductivity of the prepared liquids was calculated using (DDS307W-micro processer conductivity meter of British origin), the acidity (PH) ratio of the prepared liquids was calculated using a (PH meter) from the British fisher scientific company with the change of temperature. This device contains two electrodes, one to measure the temperature of the liquid and the second to measure the percentage of acidity in the liquid. The conductive polymer used in this research was vaccinated with several different salts separately. The effect of mineral salts on the conductivity of the ionic solution at various temperatures was studied, as in Figure .2.

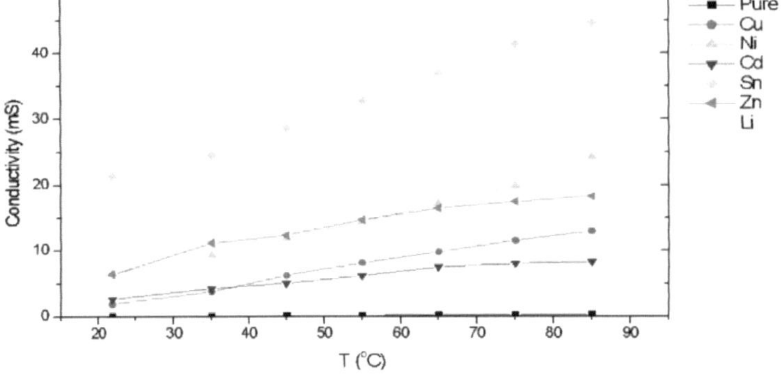

Figure .2: The effect of temperature on the electrical conductivity of ionic polymeric solutions using various salts

It can be recognized from the figure that the electrical conductivity using tin chloride recorded the highest readings with the conductivity of (22mS), followed by lithium chloride with the conductivity of (9mS) compared to the electrical conductivity of the polymeric solution without salts with the conductivity of (1 mS). This increase was also evident with the increase in temperatures. The figure shows the gradual increase in the electrical conductivity of the electrolytic polymeric solutions with increasing temperature. Where we deduce from this relationship the possibility of using these solutions in regions with high temperatures such as Iraq and the Arab Gulf states. The effect of the acidity and base ratio of these solutions was also studied with the change of temperature, as it was found that when the temperature increased, the PH factor decreased and the electrical conductivity of the ionic liquid increased because the increase in the concentration of salts inside the solution would increase the number of reduced electrons and break the bonds between the molecules of the substance, and increase the temperature working On the increase in the movement of free charges inside the liquid. Figure .3 shows the effect of temperature on the PH factor of solutions. Where acids are defined as chemical substances that contain an empty electron field capable of acquiring a pair of electrons, while bases are considered to have electronic orbits in which the electrons are free [7]. It is noticeable that all prepared electrolytic solutions are within the acid range, where their value is less than 7, and these values are preferred in preparing batteries in general, as in Figure (2)

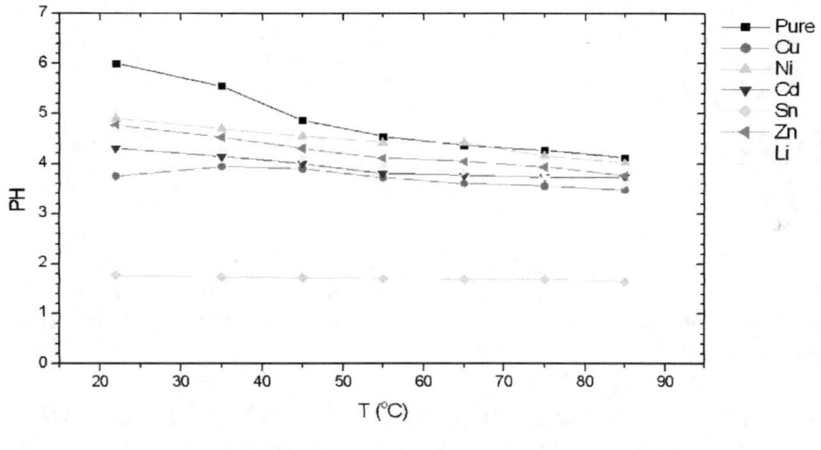

Figure .3: The effect of temperature on the PH factor of ionic polymeric solutions using various salts

4. Hall Measurements

Some electrical properties of the prepared membrane were calculated on a standard sheet of liquids prepared using a Hall effect measurement device from Ecopia. Where the paper used was in the form of a square with dimensions (1.5 x 1.5) cm to be suitable for measurement within the special base of the device shown in Figure (5). The sample contains four electrodes prepared from the silver paste. The electrical properties of the compositions prepared using a Hall phenomenon were calculated. Table (1) shows the values of the electrical conductivity, resistivity, and surface resistance of the models, while Table (2) shows the values of the Hall coefficient, Bulk concentration of the carriers, and the concentration of the carriers on the sheet concentration, a small paper was coated by placing some drops of the solution on it and letting the paper dry to form a membrane, then making measurements of Hall for the membrane.

IOP Conf. Series: Earth and Environmental Science **877** (2021) 012037 doi:10.1088/1755-1315/877/1/012037

Table .1: some electrical properties extracted using Hall's phenomenon

	Sheet Resistivity (Ohm/seq)	Resistivity (Ohm.cm)	Conductivity(1/Ohm.cm)
Pure	5.33E+08	1.60E+05	6.25E-06
CuCl2	3.40E+09	7.20E+05	1.39E-06
ZnCl2	7.30E+09	2.19E+06	4.57E-06
SnCl2	6.91E+09	2.07E+06	4.82E-06
NiCl2	6.91E+09	2.08E+06	4.81E-06
LiCl	2.63E+08	7.90E+04	1.27E-05
CdCl2	6.50E+09	1.95E+06	5.13E-06

Table .2: Concentration of carriers on the sheet, inside the material and the Hall parameter for the studied materials

	Bulk concentration (cm3) x10^{11}	sheet concentration (cm2) x10^{6}	Hall coefficient (cm3/C)
Pure	-135	-4060	-4.62E+05
CuCl2	-0.278	-8.35	-2.44E+08
ZnCl2	-0.277	-8.32	-2.52E+08
SnCl2	-0.132	-3.97	-4.72E+08
NiCl2	-3.71	-111	-1.69E+07
LiCl	5.38	161	1.16E+07
CdCl2	-0.278	-8.35	-2.25E+08

The results of Hall phenomenon measurements show the extent to which the composites are affected by the quality of salts added to polymers, as we note the value of the electrical conductivity of the polymer solution without salts (pure), the lowest values of which were recorded compared to polymeric solutions with salts.

The highest value of electrical conductivity recorded was $-11.27\text{x}10^{-5}\Omega^{-1}$ cm^{-1} when adding lithium chloride (LiCl) to the polymeric solution compared to the polymeric solution without salts (pure) $6.25\text{x}10^{-6}\Omega^{-1}cm^{-1}$. A remarkable decrease was also observed in the values of resistivity and surface resistance of the membranes coated on the paper for the same solution, as shown in Table (1).On the other hand, there was a noticeable change in the quality of the majority charge present in the membranes coated on paper, and this difference came according to the different types of salts used. Where we notice that some values contain the negative sign present in the values of the carrier concentration, whether surface or total or Hall coefficient, while there are positive values for others. When a voltage is applied to a semiconductor, the charge carriers try to travel in one straight direction, but the collision with the particles of the material continues, and thus it continuously deviates from its path, but if the applied voltage is high, it will force it to walk in a straight path. When a magnetic field is shed in another direction, the electrons will be forced to change course. Consequently, the electric field will force the electrons to gather to the bottom of the material and the gaps to collect at the top of the material, and the voltage difference between the two ends of the material will generate a voltage known as the Hall voltage. From Table (2) it can be seen that the majority of charge carriers are gaps in nickel chloride and lithium chloride, while the rest of the salts have negative values, which means that electrons are the majority charge carriers. For example, lithium chloride has an ionic bond between the metal (lithium) and the non-metal (chlorine) and when they separate, the positively charged lithium-ion is formed, which is responsible for transporting charges in the compound, and since the electron affinity of the element of lithium is very small, the possibility of its separation from the chloride is easy Compared to the rest of the salts, in which the electron affinity is higher than the lithium salt-forming minerals.

5. Optical Microscope

The pre-prepared liquids were poured into a plastic container (Petre dish) and the drying process was carried out for 48 hours at room temperature to obtain a thick membrane suitable for the process of measuring surface properties and knowing the homogeneity of the composite materials used. Images were taken of the prepared samples surface using an optical microscope (Hot selling 8LED 1000xUSB digital microscope) from KALINU TECHNOLOGY CO., LTD, China, which works through a computer program to capture images with a magnification of up to 1000 times. The surface of the dry ion solution membrane was scanned using a light microscope as shown in Figure .4. The observed change between the electrolytic polymeric solutions can be seen using various salts.
Membrane images give a clear reflection of the nature of the films and the homogeneity between the membrane -forming materials of conductive and insulating polymers and salts.
Homogeneity is important in various applications, including the application of the battery, as this homogeneity in the surface indicates a good mixture of materials and thus no unwanted agglomerations. Also, this homogeneity facilitates the process of moving the charge easily between components in a better way than if it were in the form of clusters.

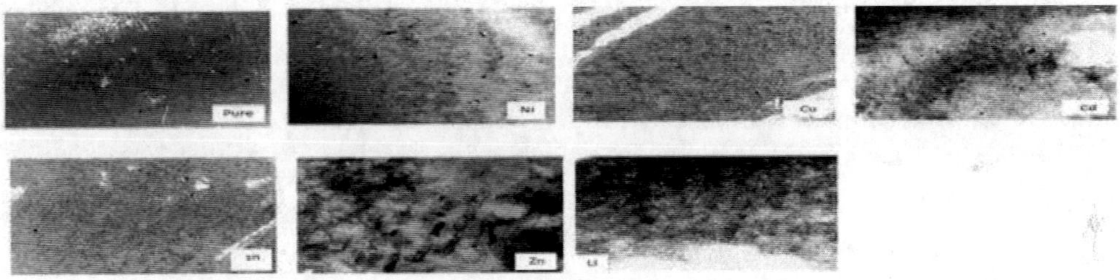

Figure.4: Optical microscopy images of membranes prepared from polymeric solutions

6. Battery Measurements

The voltages generated for the prepared samples calculated using a time-dependent system, which records the voltages that vary with time, as the model is linked to the Avometer device from (Uni-T), which is linked through a special program with the computer.

6.1. Electrolytic solution measurements (voltage with time)

measuring the voltage change with time for electrolytic solutions prepared using a graphite electrode with a titanium electrode at room temperature and as shown in Figure .4.

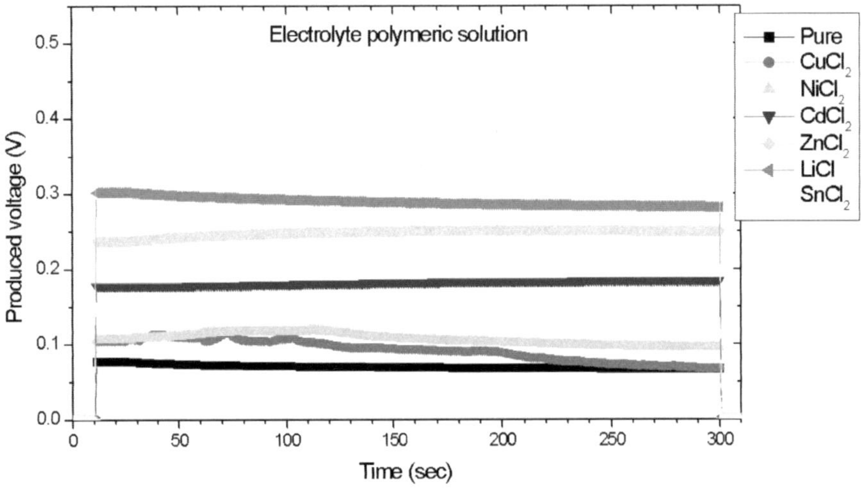

Figure .5: Voltage change with time of the electrolyte solution

Figure 5 shows that the values of the generated voltages change with the change of the added salts. The highest values of voltages were recorded when using tin chloride salts, about 0.5 volts, but this value began to gradually decrease until it reached the stability limit of about 0.4 volts. These values are consistent with the results of the electrical conductivity of SnCl2 solution, which recorded the lowest PH values and the highest conductivity values. Whereas, the use of lithium chloride recorded relatively stable voltages at 0.32 volts. As for the voltages of the rest of the salts used, they ranged at values less than 0.2 volts.

6.2. Paper Battery Measurements (Voltage with Time)

A battery was manufactured using the electrolyte solution, as was previously explained, and this battery was foldable without any effect on the recorded voltage values. Figure 6 shows the shape of the paper battery before and after folding it.

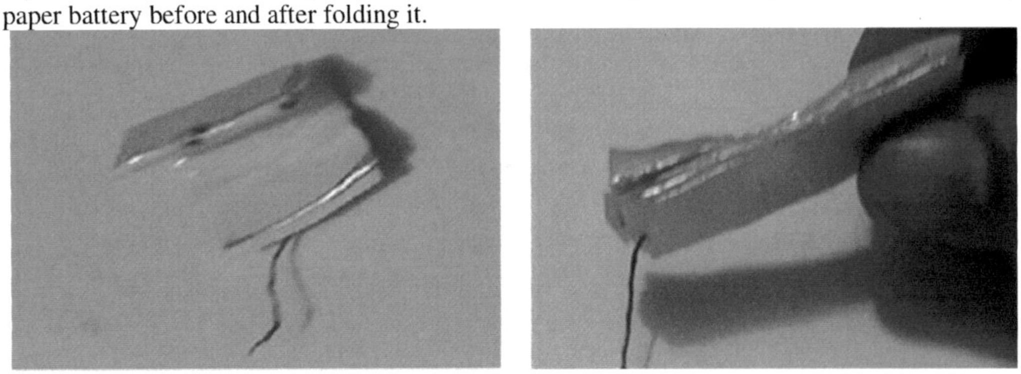

After - Folding **Before - Folding**

Figure.6: The paper battery before and after folding

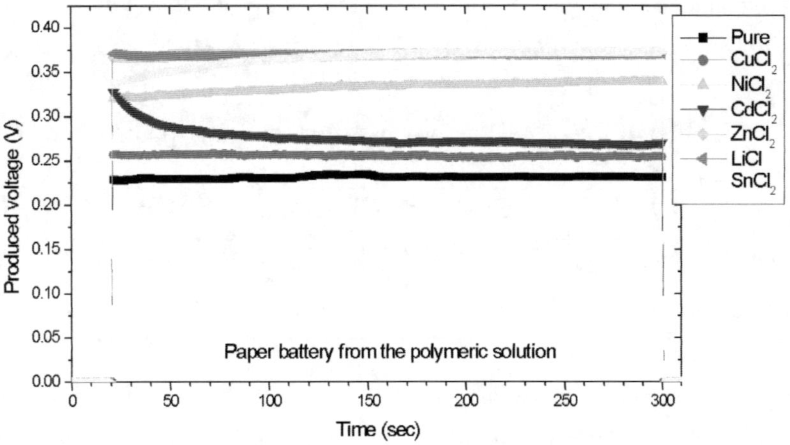

Figure. 7: The change in voltage over time of the paper battery

At the beginning of the battery manufacture, its voltage was unstable due to the chemical reaction taking place between its components, then after about half an hour, its voltage began to stabilize. After that, a measurement of the voltage change over time at room temperature was performed. The results showed a clear change in the values of the voltages generated using different solutions with relatively regular voltages with time, more regular than the voltages generated by the electrolyte solution, and more stable and preserving their charge, as shown in Figure.7 The highest recorded voltages were approximately 0.37 volts for the electrolyte solution, using lithium, tin, and zinc chlorides separately. After that, the battery was left without charging, and the process of losing its charge was monitored over time for a continuous week, and the results were as shown in table (3).

Table .3: shows the results of a voltage change for a week.

Time	Pure	$CuCl_2$	$NiCl_2$	$CdCl_2$	$ZnCl_2$	LiCl	$SnCl_2$
1day	0.231	0.256	0.337	0.271	0.371	0.371	0.373
2day	0.221	0.231	0.273	0.251	0.343	0.363	0.361
3day	0.214	0.221	0.247	0.232	0.312	0.333	0.345
4day	0.201	0.210	0.212	0.205	0.287	0.297	0.303
5day	0.184	0.185	0.186	0.176	0.265	0.281	0.292
6day	0.157	0.157	0.167	0.142	0.236	0.264	0.267
7day	0.104	0.132	0.143	0.110	0.212	0.249	0.246

The results showed that the battery loses a small portion of its charge during one day due to the continuation of the chemical reaction between the components of the battery, in addition to the temperature factor of the surrounding battery, the loss of a characteristic inherent in any type of battery where it loses part of its charge during time. Charging under normal conditions, it was recharged to a DC source with a voltage of 9 volts, and the test results showed that the battery was received from the source and the charger can be recharged with an increase in its voltage, as shown in table .4.

Table .4: illustrates the process of recharging the battery.

	Pure	$CuCl_2$	$NiCl_2$	$CdCl_2$	$ZnCl_2$	LiCl	$SnCl_2$
Before charging	0.104	0.132	0.143	0.110	0.212	0.249	0.246
After charging	0.226	0.242	0.315	0.264	0.353	0.362	0.361

6.3. Connect a group of cells to increase the voltage:

A number of cells were connected in series as in Figure. 8 and the resulting voltages were calculated, where the voltages were the product of the collection of cell voltages in the case of series connection. It was also noticed that the effect of polarity in these cells on linking the series was clear. Figure 8.a shows linking four types of thin paper cells with each other in succession, where the output voltage is 1.819 volts. While Figure 8.b shows the same cells when combined and isolated with a sheet of paper between every two cells to avoid leakage between cells as the product of these cells was approximately the same voltage of 1.503 volts. Figure 7.c shows the mechanism of increasing a second cell to connect the series (five cells in a series) and how the voltage increases with the increase in the number of cells, but to avoid the final cell becoming somewhat thick, using a small number of cells is preferred in this case.

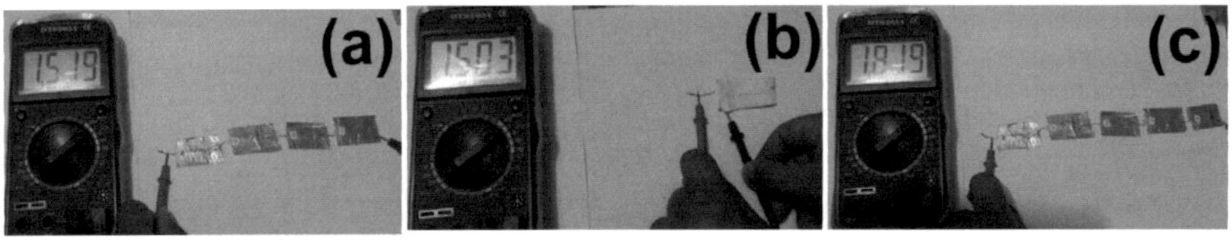

Figure.8: four cells linked in series, b) four cells linked and linked in a row, and c) five cells linked in a series

6.4 Ion dry solution

A quantity of liquid polymeric ionic solution was placed in a clean glass container and left to dry relatively, not completely, so we obtained a dry ionic solution (in the form of a membrane). A battery was made using this membrane, this solution helped the battery to give it an adequate amount of moisture and flexibility because it is a polymeric material and it helped the battery maintain some moisture that helps to maintain the battery's work and maintain its charge. Table (5) shows the results of the voltages generated from these cells manufactured using the polymeric solution as an electrolytic gelatinous solution before and after shipment after one week. As the results showed that the voltage values increase significantly after charging, as it is known that when charging a battery, the electrode connected to the negative pole of the charger will show a negative charge as a result of the accumulation of electrons on it. Between them, due to the repulsion of these electrons with the accumulated electrons on the electrode.

Table .5: shows the voltage results for paper batteries using gelatinous solution before and after charging the battery

	Pure	$CuCl_2$	$NiCl_2$	$CdCl_2$	$ZnCl_2$	LiCl	$SnCl_2$
Before charging	0.29	0.3	0.37	0.3	0.5	0.37	0.37
After charging	0.4	0.41	0.44	0.31	0.55	0.58	0.65

7. Conclusions

The possibility of obtaining a flexible rechargeable battery and its production from available, low-cost, and non-toxic materials, it can be used in portable electrical devices, especially mobile phone chargers, laptops, as well as operating remote tracking devices. That's where conventional batteries were often used. Polyvinyl alcohol (PVA) and (PEDOT: PSS) are non-toxic materials and do not affect the health of users. The use of polymeric compounds as an electrolyte solution enriched with salts gave good results in generating voltages suitable for daily use. Therefore, it is possible to develop this experiment by using other polymeric materials and to study the extent to which the paper battery performance improves.

References

[1] Nicholson, J., 2017. The chemistry of polymers. Royal Society of Chemistry .

[2] Nuri Al-Jamal, Solid State Physics, 2000 edition.

[3] N.S. Hudak, Chloroaluminate-Doped Conducting Polymers as Positive Electrodes in Rechargeable Aluminium Batteries, J. Phys. Chem. C. 118 (2013) 5203-5215.Q.F. Li, N.J . Bjerrum, Aluminium as anode for energy storage and conversion: A review, J. Power sources. 110 (2002) 7-8.

[4] Utracki, L.A. and Weiss, R.A. eds., 1989. Multiphase polymers: blends and ionomers. American Chemical Society.

[5] Sharrow, J.F., Solomon, K.W., Peterkin, F.E., Bernardes, J.S. and Hankla, B.J., US Secretary of Navy, 2006. Battery mechanism. U.S. Patent 7,108,940.

[6] Goodenough, J.B. and Park, K.S., 2013. The Li-ion rechargeable battery: a perspective. Journal of the American Chemical Society, 135(4), pp.1167-1176.

[7] Lewis, Gilbert (1923). Valence and the Structure of Atoms and Molecules.

A Review of New Solar Still Design Comprising a Five-Sided Glass Cover and Equipped with an External Tank for PCM

Abbas Sahi Shareef[1], Hayder Jabbar Kurji, Hassan Abdulameer Matrood

Mechanical Engineering Department, Engineering College, Kerbala University, Kerbala, Iraq

[1] E-mail: abbas.al-rakabi@uokerbala.edu.iq

Abstract: Various human activities have led to the consumption of large quantities of pure water, which has led researchers to find efficient and economical methods for desalinating seawater and water containing impurities. In this review paper, solar energy where it is permanent, abundant and environmentally friendly, to produce pure water was discussed using a new solar distillation device, representing the paper's novelty. The distillation was designed and used in the way led to increase efficiency and improve productivity by adding a solar collector to the system and equipped with a tank containing phase change material (PCM). It has a low melting point and can change the phase by absorbing the system's latent heat to maintain the system's temperature. Which contributes to increasing the distillation period even after sunset, thus increasing the daily productivity of freshwater. Using phase change materials will increase distillation hours from (3-4) hours after sunset, increasing the amount of production between (75 - 90) %.

Keywords: Desalination technology, Solar still, Phase change material, evacuated tube collectors

1. Introduction

The development in all areas of life, besides the quantity of population growth, has led to an increase in pure water utilization, which makes up only 2.5% of the water on the earth surface [1]. It has become necessary to find effective methods for desalination of salty water. All manners of water desalination that people use that consume energy, whether electrical or thermal, and this energy may be expensive and damage the environment, figure 1 shows the energy consumption for the period from 1992 to 2016, prompting researchers to find abundant and sustainable energy [2]. The most crucial type of renewable energy is solar energy for its abundance. It has no negative impact on the environment as it has been used to desalinate water in solar distillation systems [2] [3].

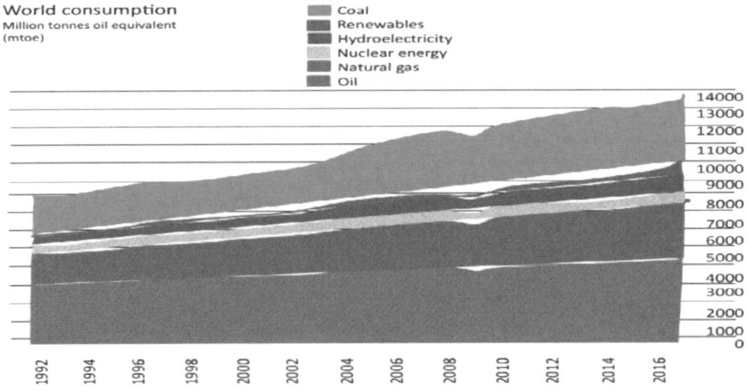

Figure 1. World energy consumption

This research aims to review the performance of the solar distillation system and increase its efficiency through the design of a new solar distiller and the effect of the evacuated tube solar collector and phase change materials on the distiller's daily productivity.

2. Desalination technology

Desalination techniques are a series of procedures that remove salts and other impurities from unsuitable water and produce clean water [2]. The water desalination processes are divided into two main parts (as shown in figure 2.): membrane processes (in which the salty water passes through porous membranes to separate unwanted materials and branches from them). Moreover several types are branched out from them, the most important of which is reverse osmosis [2]. The second section is the thermal processes (in which the water is heated to the temperature of evaporation to separate the salts and impurities, and then the steam is condensed to produce freshwater). Then these processes are divided into several types (vapor compression, multi-stage, multi-effect boiling and solar still) [2]. There is another way called electrolysis that relies on the electrical potential difference principle between two joined films [3]

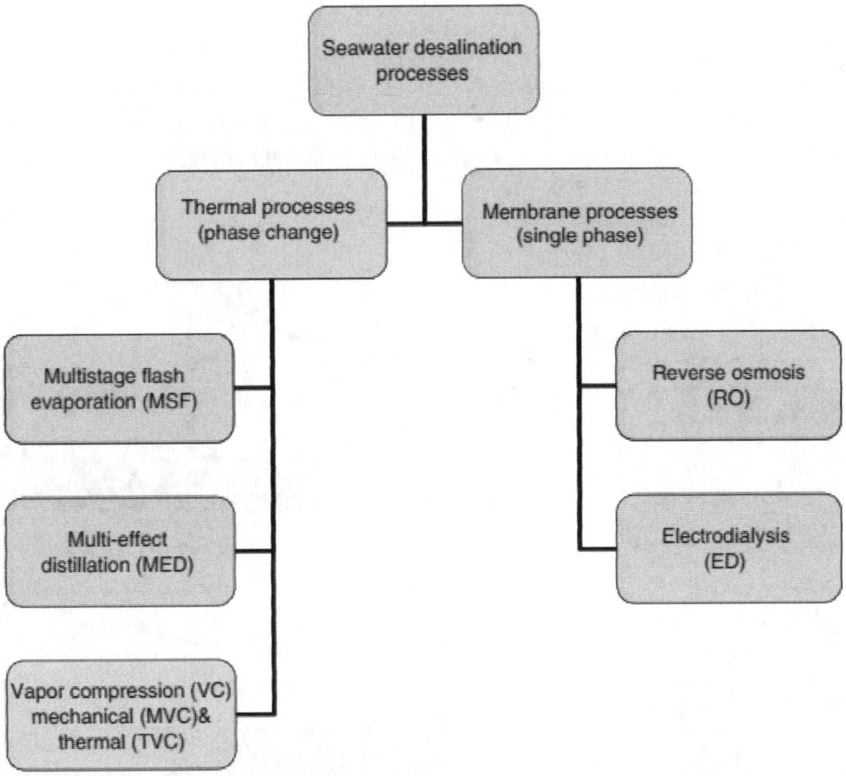

Figure 2. Classification of Desalination processes

2.1. *Solar energy desalination:*

The work of the solar distillation principle depends on the utilization of solar radiation to heat the saltwater to the point of evaporation to separate unwanted materials and then condense the vapour on a glass or plastic surface [3] [4]. According to the thermal power source, the solar distillates are classified to Passive stills and Active stills, where each class has several types, as shown in figure 3.

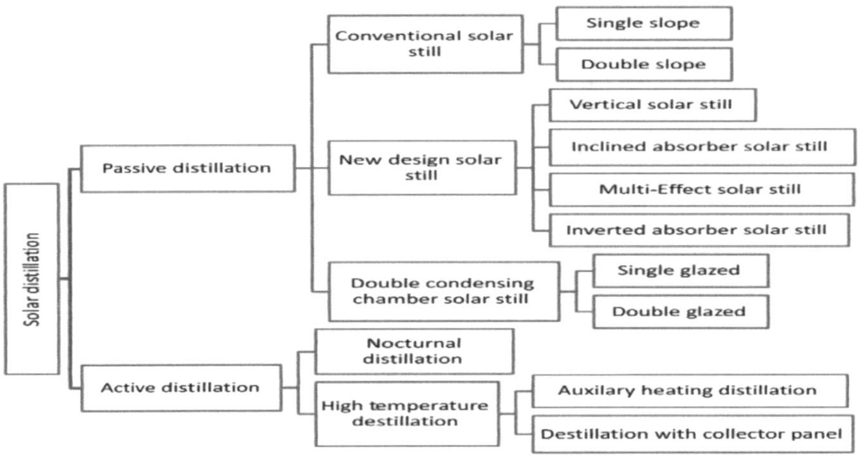

Figure 3. Classification of solar still

2.1.1. Solar still

The conventional solar distillation consists of a saltwater basin with a glass or transparent plastic cover above it that allows solar rays to enter the basin, and tubes for the entry of water and a stream inside the distiller to collect pure water [3]. There are several traditional distillates as shown in figure 4, including (single slope, Double slope glazing cover, V-shape and Spherical solar still).

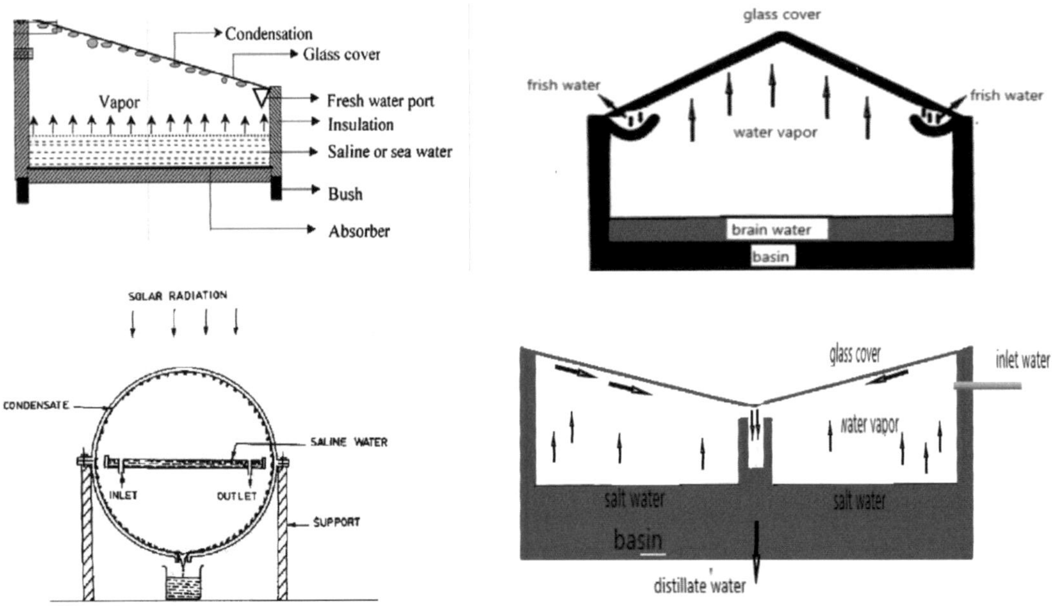

Figure 4. (Single slop, double slop, spherical and v type solar still)

There are other types of solar distillates, Hashim [4] designed a single slop distillate with a double basin (SSDBS) and conducted an experimental study to compare the new distillate and the traditional distillate daily. Shareef and others [6], studied the performance of a new solar distiller with a glass cover with four faces experimentally, and the experiment was in the city of Karbala, Iraq, the outcomes of the investigation, presented that the daily work efficiency of the distillate was 48% at a temperature of 24.9 C° and the daily production amount was 6 liters/day.

2.1.2. Factors affecting solar stills productivity:

1. Glass– water temperature difference.
2. Free surface area and deepness of water.
3. Inlet water temperature.
4. Environmental air temperature.
5. Angle, thickness and material for glass cover.
6. Wind velocity.
7. Insulation solar still.

3. Solar collector

Solar collectors of various kinds are used to absorb solar rays and convert them into heat [8][9] that can be used in several applications such as water heating, building heating, electrical energy production, water desalination, and others [10]. There are many shapes and types of solar collectors, as shown in figure 5.

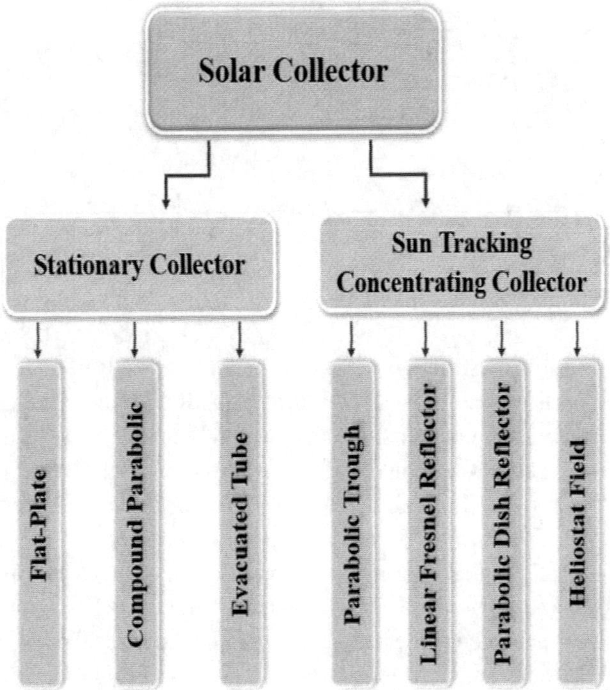

Figure 5. Classification of solar collectors [11]

3.1. Concentrating collectors

To get the maximum amount of solar radiation, it is necessary to track the sun's position, and there are several types of these complexes such as (Parabolic trough collectors, Linear Fresnel reflectors, Central tower receivers, parabolic dish reflectors) [12].

3.2. Stationary collectors

These solar collectors remain stationary and directed to the sun at a specific inclination angle, having different types such as (Flat plate collectors, compound parabolic collector, and Evacuated tube collectors) [12] as shown in figure 6.

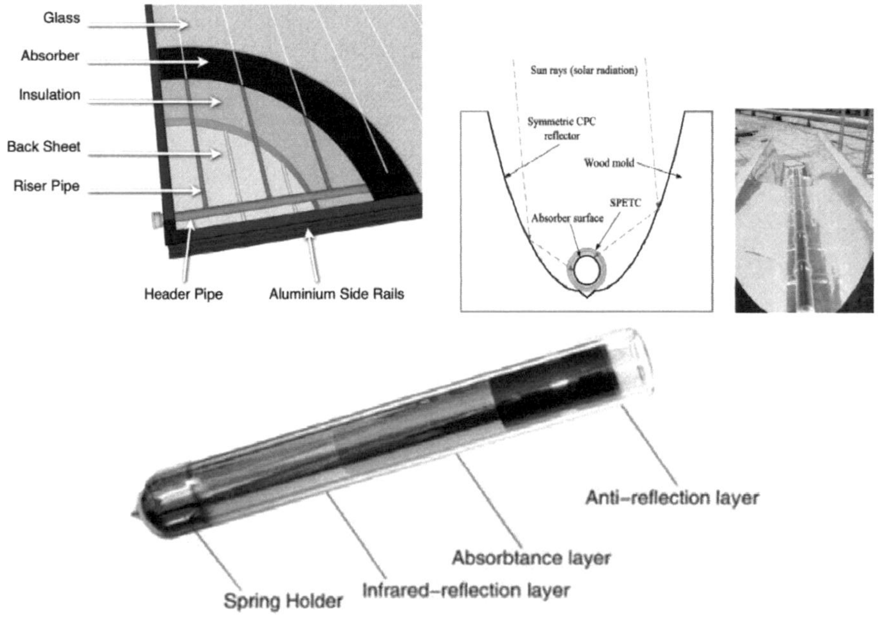

Figure 6. Stationary collectors (Flat plate, Compound parabolic, and Evacuated tube collectors)

3.3. Solar still with solar collector

Providing pure water using solar distillers is one of the best methods used to desalinate water, especially in the absence of equipment for other means. To improve distillation productivity, solar collectors are a couple to the system to increase solar radiation absorbed to heat the water entering the distillate [23]. Al-Masha and others [14] experimentally investigate a traditional solar distiller's performance by adding an evacuated tube collector to it. They found it through experience that the distillate's productivity increased by 102%, and the distillate's efficiency increased from 36% to 71%. The ocean's temperature and the intensity of solar radiation are the most important climatic factors affecting the production of the distillery. Pramod B V and others [15] studied the performance of a pyramidal solar distiller with a vacuum tube and a black stone placed at the base of the basin to increase the absorption of solar radiation. They concluded from this work that the solar collector's addition contributed to an increase of productivity 40% over the traditional distillate. S. Mamouri and others [16] introduced a solar-powered desalination system consisting of distillate and connected with a vacuum tube. The results showed an increase in the system’s production when using the solar collector and increased the distiller's efficiency’s work. The production rate reached 1.02 kg / m^2 per hour, and the efficiency was 22.9%.

4. Phase change material

Phase change material can absorb and release an enormous amount of heat. When enough heat applies to these materials, their phase changes from solid to liquid, and they keep heat in the form of latent heat

[17][18], and when the ambient temperature decreases, the previous process reverses and turns from liquid to solid, in which heat is lost to the surrounding[19].

4.1. Classification of phase change materials

There are three main types of phase change materials, and several classes are branched out of them, as shown in figure 7. [3].

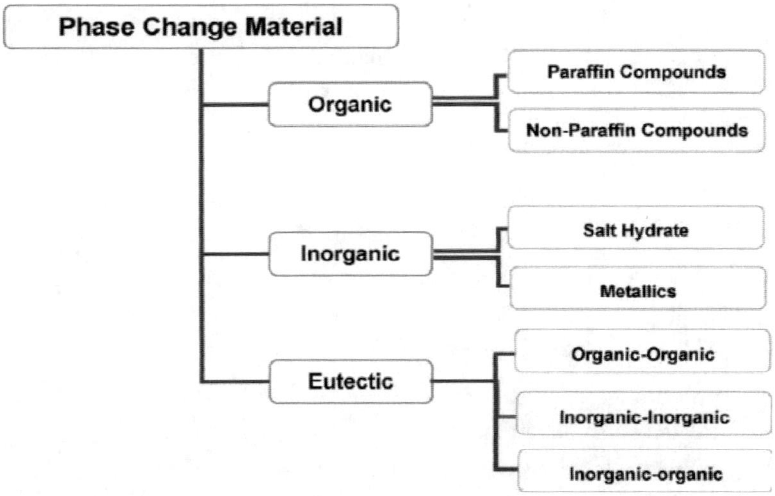

Figure 7. Classification of phase change material

4.1.1. Organic PCM

They are materials that contain carbon in their composition, such as paraffin wax and others. These materials are characterized by the ability to melt and solidify several times without affecting their properties during this process, as well as resistance to corrosive conditions [20].

4.1.2. Inorganic PCM:

This type comprises salts, minerals, and salt hydrates, which have a high density, which leads to large enthalpy and covers a wide range of temperatures. This type's disadvantage is thermal instability and can be subject to phase separation or corrosion conditions [20].

4.1.3. Eutectics

A eutectic is defined as a combination that includes two or more substances divided into organic, inorganic, and organic-inorganic, for a low melting point called a melting point. Each component melts and solidifies at the same time to form a homogeneous mixture of dissolved crystals [3].

4.2 Properties of Phase Change Materials

There are many properties that must be considered for choosing the appropriate PCM type, the most important of which are explained below [21].

Table 1. Properties of phase change materials

Thermal properties	Physical properties	Chemical properties	Economics properties	Kinetic properties.
suitable phase change temperature	high density	long-term chemical stability	Cheap	no supercoiling
high latent heat of transition	small volume change	compatibility with materials of construction	available	sufficient crystallization rate
suitable thermal conductivity	low vapor pressure	no toxicity		
		no fire hazard		

4.3. Application Phase Change Materials

Many applications use phase change materials [21] :

1. Buildings Applications.
2. Waste Heat Recovery Application.
3. Mobilized Thermal Energy Storage.
4. Medical applications.

Phase change materials have been used in solar distillation applications as they store heat to take advantage of it to continuity the distillation process in the absence of solar radiation. Shafii [22] designed a new desalination system consisting of four main parts: a solar collector, an evaporator, a basin with variable phase materials and an air condenser. The experiment showed that varying phase materials contributed to increasing the daily output by 30%. The presence of an air condenser is essential because the radiative cooling when using variable phase materials alone. Wen-Long Cheng, Yan-Kai Huo, Yong-Le Nian [23] Studied experimentally a new (shape-stabilized phase change material (SSPCM))used in solar water desalination system with simulation model design. The conclusions got from the study: SSPCM stability has avoided leakage and packaging problems within the solar distillation basin. The simulation model showed that increasing the thermal conductivity from (0.2 to 4) kg / m .k and increasing the melting temperature from 34 to 50 degrees Celsius contributed to increasing the daily productivity from 21% to 57%. Shareef and others [3] Studied developing the solar distillation system and enhancing the productivity of freshwater by adding phase change materials to the system.Notice in this paper that the use of these materials led to increasing the distillation period to several hours at night, which leads to an increase in the daily productivity and the efficiency of the system's work. The cost per litre of the system uses phase change materials compared to the conventional distillation. Rashid and others [24] studied experimentally to improve the four-sided solar distiller's performance by adding a tank containing phase change materials to the system. New phase change materials were tested (Polyvinyl pyrrolidine (PVP K-30), polyethylene glycol (PEG 6000) and carboxymethylcellulose sodium salt. (CMC), they found through experience that the system's operating time increased between 3 to 5 hours, and the amount of production increased between 20 to 110% compared to the distillate that does not contain the reservoir of phase change materials. Shafii [25] designed a new distillation system comprising a conventional distillation with connecting evacuated tubes and an external condensation tank that contains phase change materials used to store thermal energy in the form of latent heat and also used to condense the steam produced in the distillate. They concluded from the experiment that the daily production rate increased by 86%, reaching 6.555 kg / m^2 day, with work efficiency reaching 50%.

5. New solar still

The new system consists of three main parts: a solar distiller, a phase change material tank and a solar collector. The solar distillery is a metal basin with dimensions (75 * 75 * 10) cm and a five-sided glass cover (which contributes to increasing the area of condensation and allowing the entry of solar rays from all directions and for the longest possible period into the distillery). The system is supplied with water through copper tubes that pass inside the collection from the air's evacuated tubes (number 7). These solar radiators increase the water temperature by absorbing solar rays and converting them into thermal energy that heats the water inside. Part of the steam formed inside the distillate is directed through copper tubes to an insulated external tank (75*75*25) cm containing phase change materials that act as a condenser for the steam passing through it, as well as storing the heat energy absorbed from the steam in the form of latent heat that is used to continue heating the water inside the distillate after the sun passes away. The heat is transferred from the phase change material to the distillate bed by a set of heat pipes immersed inside the phase change material and connected from the top to the distiller base. The table below shows the dimensions and materials used in the manufacture of the novel system.

Table 2. Dimensions and materials of new solar still

Component	Dimensions	Remarks
Basin	(75*75*10)cm	Material: Plate stainless steel
Glass cover	Thick 4mm	Material: Widow glass
	(75*30*40*44) cm Left and right side	
	(75 * 40*40*46) cm back side	
	(75 * 30*30*37) cm Front side	
	(44 * 44 * 46*37) cm top side	
PCM tank	(75*75*25)cm	Material: Plate stainless steel
Heat pipe	Numbers (70)	Material: copper pipe
	33cm length	(Acetone) working fluid
	1 cm in diameter	
ETC	Numbers (7)	Material: Evacuated glass tubes
	180 cm length	
	5.8 cm Do, 4.6 cm Di	
Insulation	5cm thick	Material: glass wool

Figure 8. Five-sided glass cover solar still

The efficiency of the active SS [26]

$$\eta = \frac{(\dot{m}_{ew})L}{\dot{q}_u + I_S(t)A_g} \quad (1)$$

Where (m_{ew}) denotes the hourly yield through solar still (kg/h), (L) designates the latent heat of distillation measured in J/kg,(Is (t)) solar radiation on SS (W/m^2), (A_g) denotes area of the glass cover (m^2),(q_u) heat gain from the ETC tank through flowing water (W).

The hourly yield is given by [27]:

$$\dot{m}_{ew} = \frac{h_{ew}[T_w - Tg]}{L}(A_s)(3600) \quad (2)$$

(A_s) the basin liner still area, (L) is latent heat, (h_{ew}) heat loss coefficient by evaporation from water surface (W /m2°C), (T_w) water temperature (°C), (T_g) glass cover temperature (°C).

Useful thermal energy supplied to the still through evacuated tubes [27]

$$Q_u = A_{ET}\, F_R\, [(\alpha\tau)_e\, I_{effe} - U_{LE}\, [\frac{AL}{AET}]\, (Tw - T_a)] \quad (3)$$

(A_{ET}) is a diameter of absorber glass tube × total length of the tubes, (A_L) = πA_{ET},

(F_R) heat removal factor, $(\alpha\tau)_e$ effective absorptance – transmittance product,

(U_{LE}) overall heat transfer coefficient, $(I)_{effe}$ Intensity of solar radiation, (T_w) Water temperature, (Ta) Ambient temperature

6. Conclusion

Solar collectors have a significant role in raising the water's temperature entering the distillate, which helps increase the evaporation process and thus increase daily production. To increase the distillation hours until after sunset, phase change materials are used to keep thermal energy in the form of latent heat. Through the research, the following conclusions were obtained:

1. Desalination processes are varied, but the best method is solar distillation, as solar energy is available, renewable, and environmentally friendly.
2. Distillation productivity is affected by several factors such as wind speed, incoming water temperature, insulation of the distillation basin, and the glass cover angle. Still, the most important influencing factors are the intensity of solar radiation and the ambient temperature.
3. The new design of the distillation cover helped to increase the surface area to condense the vapor formed inside the distillation device, and because of its five-sided shape that makes it face solar radiation throughout the daylight hours, allowing the largest amount of solar energy to enter the distillation basin, thus increasing the amount of steam produced.
4. The efficiency of the distillation process increased when solar collectors used.
5. Used phase change materials contribute to increasing the distiller's daily productivity between (75 - 90) %, because it increases the hours of daily distillation from (3-4) hours.

Reference

1. Alkaisi, A., Mossad, R., & Sharifian-Barforoush, A. (2017). A review of the water desalination systems integrated with renewable energy. *Energy Procedia, 110, 268-274.*

2. Islam, M. S., Sultana, A., Saadat, A. H. M., Shammi, M., & Uddin, M. K. (2018). Desalination technologies for developing countries: A review. *Journal of Scientific Research, 10(1), 77-97.*

3. Rashid, F. L., Shareef, A. S., & Alwan, H. F (April 2020). Enhancement of Fresh Water Production in Solar Still Using New Phase Change Materials. *Journal of Advanced Research in Fluid Mechanics and Thermal Sciences*, 61, 63-72.

4. Rashid, F. L., Shareef, A. S., & Alwan, H. F (2020). Performance Enhancement of a New Passive Solar Still Design for Water Desalination. *Journal of Mechanical Engineering Research and Developments*

5. Hashim, A. Y. (2011). A new design of a single slop double-basin solar still (SSDBS). *Journal of Basrah Researches (Sciences), 37(2A), 42-49.*

6. Shareef, A. S., Rashid, F. L., & Alwan, H. F. (2018). Experimental study of new design solar still in Karbala-Iraqi weathers. *International Journal of Mechanical Engineering and Technology*, 9(13), 1465-1472.

7. Sharshir, S. W., Yang, N., Peng, G., & Kabeel, A. E. (2016). Factors affecting solar stills productivity and improvement techniques: a detailed review. *Applied Thermal Engineering, 100, 267-284.*

8. Pramod, B. V. N., Raj, J. P., Krishnan, S. H., & Kotebavi, V. (2018, February). Performance analysis of a solar still coupled with evacuated heat pipes. *In IOP Conf. Ser. Mater. Sci. Eng. (Vol. 310).*

9. Shareef, A. S., Abbod, M. H., & Kadhim, S. Q. (2016). Experimental investigation for flow rate effect on a flat plate solar collector with the using of Al2O3 nanofluids as a heat transfer fluid. *Int J Mech Mechatronics Eng, 16, 42-8.*

10. Singh, A. K. (2020). A Review Study of Solar Desalting Units with Evacuated Tube Collectors. *Journal of Cleaner Production, 123542.*

11. Olfian, H., Ajarostaghi, S. S. M., & Ebrahimnataj, M. (2020). Development on evacuated tube solar collectors: A review of the last decade results of using nanofluids. *Solar Energy, 211, 265-282.*

12. Suman, S., Khan, M. K., & Pathak, M. (2015). Performance enhancement of solar collectors—A review. *Renewable and Sustainable Energy Reviews, 49, 192-210.*

13. Shareef, A. S., Rashid, F. L., & Alwan, H. F. (2020, January). Water solar distiller productivity enhancement using solar collector and phase change material (PCM). In IOP Conference Series: *Materials Science and Engineering (Vol. 671, No. 1, p. 012150). IOP Publishing.*

14. Mamouri, S. J., Derami, H. G., Ghiasi, M., Shafii, M. B., & Shiee, Z. (2014). Experimental investigation of the effect of using thermosyphon heat pipes and vacuum glass on the performance of solar still. *Energy, 75, 501-507.*

15. Chaturvedi, A. K. (2015). A Review of Solar Flat Plate Liquid Collector" s Components. *International Journal, 3(1), 138-141.*

16. Al-Mashat, S. M., & Ibrahim, M (2017). Experimental Investigation of Single Basin Solar Still Coupled with Evacuated Tube Heat Pipe.

17. Shareef, A. S., Al-Mousawi, F. N., & Sachit, H. S. (2020, January). Experimental study of a PCM storage system integrated with a thermal solar collector. In IOP Conference Series: *Materials Science and Engineering (Vol. 671, No. 1, p. 012018). IOP Publishing.*

18. Shareef, A. S., Rashid, F. L., Hadi, A., & Hashim, A. (2019). Water–polyethylene glycol/ (SiC–WC) and (CeO2–WC) nanofluids for saving solar energy. *International Journal of Scientific & Technology Research, 8(11).*

19. Boda, M. A., Phand, R. V., & Kotali, A. C. (2017). Various applications of phase change materials: thermal energy-storing materials. *International Journal of Emerging Research in Management and Technology, 6(4), 167-171.*

20. Iqbal, K., Khan, A., Sun, D., Ashraf, M., Rehman, A., Safdar, F, & Maqsood, H. S. (2019). Phase change materials, their synthesis and application in textiles—a review. *The Journal of the Textile Institute, 110(4), 625-638.*

21. Zarma, Ismaila. (2017). Thermal Energy Storage in Phase Change Materials:-Applications, Advantages and Disadvantages

22. Amarloo, A., & Shafii, M. B. (2019). Enhanced solar still condensation by using a radiative cooling system and phase change material. *Desalination, 467, 43-50.*

23. Cheng, W. L., Huo, Y. K., & Nian, Y. L. (2019). Performance of solar still using shape-stabilized PCM: an experimental and theoretical investigation. *Desalination, 455, 89-99.*

24. Rashid, F. L., Shareef, A. S., & Alwan, H. F (2019). Enhancement of Fresh Water Production in Solar Still Using New Phase Change Materials. *Journal of Advanced Research in Fluid Mechanics and Thermal Sciences, 61, 63-72.*

25. Faegh, M., & Shafii, M. B. (2017). Experimental investigation of a solar still equipped with an external heat storage system using phase change materials and heat pipes. *Desalination, 409, 128-135.*

26. *Dev, R., & Tiwari, G. N. (2012).* Annual performance of evacuated tubular collector integrated solar still. *Desalination and water Treatment, 41(1-3), 204-223.*

27. Sampathkumar, K., Arjunan, T. V., & Senthilkumar, P. (2012). Single Basin Solar Still Coupled with Evacuated Tubes-Thermal Modeling and Experimental Validation. *International Energy Journal*, *12*(1).

IOP Conf. Series: Earth and Environmental Science **877** (2021) 012039 doi:10.1088/1755-1315/877/1/012039

Influence Eggshells powder additive on thermal stress of fiberglass/polyester composite tubes

Fadhel Abbas Abdulla*, M S Qasim, Ahmed Ali Farhan Ogaili

Mechanical Engineering Department, College of Engineering,Mustansiriyah University,Baghdad-Iraq

[1]E-mail: a_ogaili@uomustansiriyah.edu.iq

Abstract. In this study, the influences of eggshell powder application on random fiberglass/polyester composite pipes thermal stresses were investigated experimentally and numerically. The experimental part involved industrializing tube samples which are Egg Shell Powder with Polyester at 50% volume fraction.(E50), Random Fiberglass with plyester at 50% volume fraction (F50) and Egg Shell Powder with Random Fiberglass with Polyester at 50% volume fraction (E25F25). Resin molding procedure and experimental rig design to study how tube samples are faced with thermal loads. Pipes are made with (95 mm) inner diameter, (400 mm) circumference, and (5 mm) wall thicknesses. Computational Fluid Dynamic, ANSYS software package version 11, is simulated with product form SHELL63. Results show that, composite samples (F50) had an average longitudinal strain, while composite samples (E50) had a low longitudinal strain. The composite (F50) had average longitudinal and hoop stress (62.2 MPa, and 61.1 MPa) respectively, while the composite (E50) had low longitudinal hoop stress (28.5 MPa, ad 30.8 MPa) respectively.

1.Introduction

Composite materials which pursue an established field of gain in composites using naturalistic and polymeric template materials. Natural materials are very attentive due to the many advantages such as frugal and environmental benefits obtained from natural composites. When temperatures are modified, thermal stresses and deformations in pipes are produced. There are many thermal stress applications, such as the nuclear sector, the martial industry, the shooter tube, and turbine engineering. The increase in emissions means that naturalistic materials are used instead of artificial materials[1],[2]. As a result, several research items are being used on natural materials, such as the modulation of the lamination and the corresponding axis [3],[4]. Chia-Wei Su [5] theorize the thermal stress in the composite material ring. 3-D finite element software used with ANSYS version 10 to create thermal stress. Results showed that composite materials play a significant thermal stress role. F A Abdullah [6] theoretically and experimentally tested thermal insulators were natural composite materials of fibers (and eggshells, black feather, white features, Jute). Lee's theorem disk has been used to calculate different forms of thermal conductivity. Study revealed composite material (Jute) was superior to other naturalistic materials as a composite thermal insulator. The eggshell powder from the egg manufacturing sector was used as a part of the stand up for Portland Common Cement in cement lute [7]. Cement lute 1:3 is a mixed ratio in which cement is partially replaced by 5 to 30 percent eggshell powder with a 5 percent rise in cement weight. O M Sanusi et, al, [8] advanced polymer mold composite (P.M.C.) to incorporate the body shield by using the handling method for producing wood ash with epoxy. Findings ash of wood ratio that the variability was 2.3 percent and that tensile strength and hardness relative to enhanced resin were improved. The mechanical and thermal characteristics of the composite material Polypropylene (P.P.) with various naturalistic fibers were studied by A L de Lemos et al [9]. The findings showed that composites consolidated the naturalistic fibers' performance to a minimum that the glass fiber consolidated composite.

In this work,the eggshell powder group's effects on stress of composite pipes [polyester with fiberglass, random] have been studied experimentally and numerically via the molding process. Test the pipeline characteristics when the thermal load is in use due to thermal stresses with building experience to study thermal stresses' effects. For the analysis of numerical and experimental tests, the ANSYS program was used.

2. Experimental work

For this study, three composite material samples with various fibers(natural and glass fiber), volume fraction, and density analyze each composite's mechanical properties. The composites were characterized using the tensile method [10]. A composite tube was manufactured using the resin molding process with specific thermal loads. In this quest, tensile industrialization experimental sample tests are carried out on traction equipment and thermal-mechanical test samples, which are automated and tested by a thermal-mechanical device.

2.1. Samples research

For the three composite forms, various types of natural fibers were used as a reinforcement material. The composite materials mold at a 50 percent ratio with various volume fractions and a fixed volume fraction. In the table, (1) composite material properties were shown.

Table 1. Material properties

symbol	Reinforcement	Matrix	Vf (%)
E25F25	Egg Shell Powder + Glassfiber		25 & 25
F50	Glassfiber	Polyester	50
E50	Egg Shell Powder		50

2.2. Tensile test

Samples characteristics properites were calculated in compliance with the ASTM D 638 standard[11]. The form of the composite dimension was (20cm*30cm*0.5 cm) modeled by the process of infusion. The tensile sample was analyzed for the samples cut by a cutting machine shown in figure (1-a). Nine samples used for each composite material were cut in the tensile test sample, three samples. Dimensions of the samples are 50 mm long, 13 mm wide, and 5 mm thick. The device used for the tensile test was the WDW-200E traction device speed was set to 2 mm/min [12,13], as shown in figure (1-b).

2.3. Thermal expansion coefficient (α)

Thermo Mechanical Analyzer (T.M.A.) is the equipment used to determine the thermal expansion coefficient, as shown in figure (2-a). Thermal, mechanical analysis (T.M.A.) is a difference in size and automatic variance in characteristics when the specimen fills a thermal device with or without the sample [14],[15]. Test containers, sample holders, oven, tracers, and observatory are the T.M.A. facilities. T.M.A. is mounted onto a screen monitor to display the results and linked to a printer for the results printed. The sample sizes were (2cm*0.5cm*0.5 cm), as shown in figure (2-b) displays the T.M.A. test samples [16],[17].

Figure 1. a) the tensile sample according to ASTMD638, b) WDW-200E traction device.

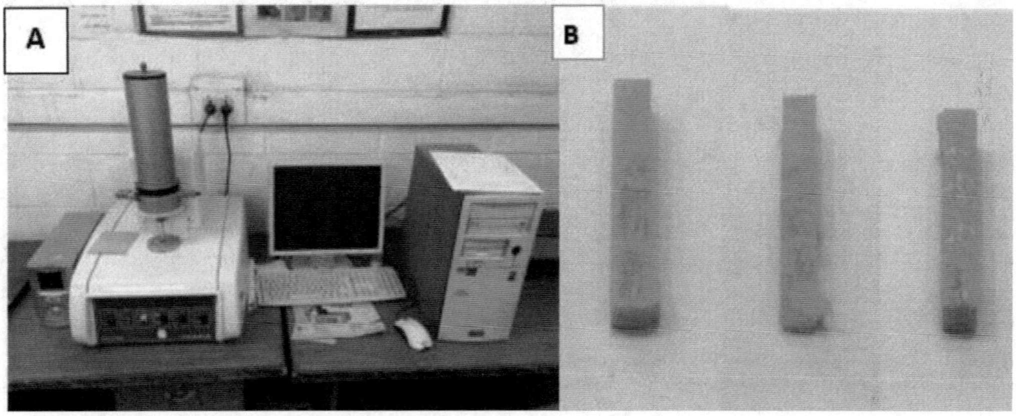

Figure 2. A) Thermo-Mechanical Analyzer (T.M.A.)B) T.M.A. exam samples.

2.4. Tube industrialization

Many technologies are used to industrialize plastic pipes in a major commercial application [18]. There are three tube samples with normal parameters, as seen in the table (1). Test tube with dimensions (400 mm) circumference, (90 mm) internal diameter and (5 mm) width. Two plastic tubes molds were used in resin casting. The outer conduit is the husk of length (400 mm) and ID (100 mm). However, the internal tube is the core that has (90 mm) O.D. and (500 mm) length. The essence tube is inserted into the husk tube as seen in figure (3). Tube diameter difference between husk, essence allows a blank tube frame bore with suitable dimensions [19]. Design lubrication is important in the first step of the molding process to prevent the production tube from adhering to mold. The eggshell powder was disordered with polyester resin and hardener handle until it entered a homogenous state. Then the mix was filled with a surgical syringe. The tube sample was solidified after five hours and ready to be collected. Repeated this method with all composite tube samples. Fig (4) indicates that resin molding creates the (E50) pipe sample [20],[21].

2.5. Experimental research rig

Heat air flows into a sample of composite tubes, causing thermal stress on a composite tube. The measuring equipment consists of three parts: Boundary conditions, part of the temperature, and finally, the date of conquest. The restricting conditions for the hunt were to verify X and Y position

IOP Conf. Series: Earth and Environmental Science **877** (2021) 012039 doi:10.1088/1755-1315/877/1/012039

at either e' of the tube and to determine the path of the Z down the line. There are two iron jowls found on all ends of the tubing in the X and Y directions. There was a strong line in the Z- direction with twain iron caps, where the twain tubes opened with the tightening of the rope at the two ends of the pipe [22].

Moreover, the tube sample, chops, and bar have been held within its hobs on double iron platforms. Temperature section in contagious experimental inventory that the thermal capacity of the tube specimen can be provided. Heat weapons, which supply hot air inside temperature up to (600 KBC) and attached to an iron jowl, were used to specify the temperature component of the experimental experiments. Four K thermocouples are used at the top and bottom of the outer tube for a specified temperature, which gives a minimum of eight thermocouples. Temperature recorder with a 12-channel thermocouple reader linked to the eight inaugurating thermocouples on the tube [23]. The thermocouples' temperature sensitivity and recorded with timing on the S.D. mark opened in the temperature recorder, when hot air flows through the tube. The date conquest scheme consists of two tubular strain gages wired to tensile gages and then linked to a device to monitor each test's readings. The data logger is wired with a device. The hiding point of the strain gauge ensures a relative difference in resistance [24]. The variation in the readings of resistance is small, resulting in mV or mA value. The strain data logger was tested in (mA). Equations (1), (2), and (3) were used to obtain stress interest. Fig (5) demonstrates the experimental rig [25].

$$\sigma_L = \frac{E}{(1-\nu^2)} (\varepsilon_L + \varepsilon_H . \nu) \quad (1)$$

$$\sigma_H = \frac{E}{(1-\nu^2)} (\varepsilon_L . \nu + \varepsilon_H) \quad (2)$$

$$\sigma_{VON} = \sqrt{(\sigma_H)^2 + (\sigma_L)^2 - \sigma_H\, \sigma_L} \quad (3)$$

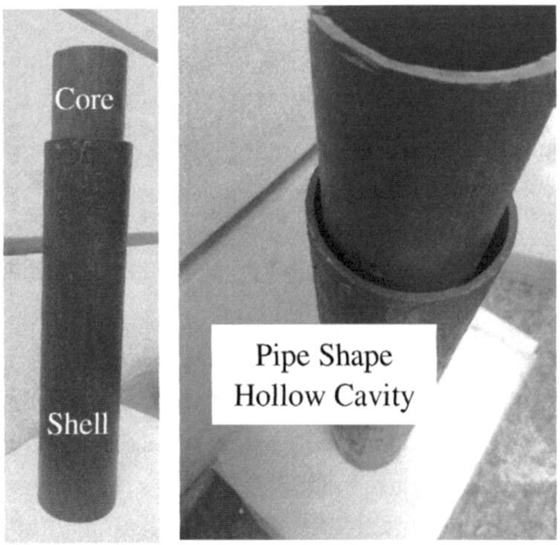

Figure 3. Composite Tube Casting

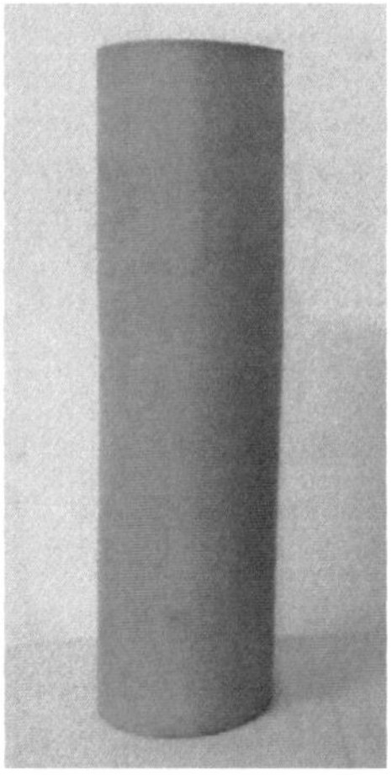

Figure 4. E50 tube sample

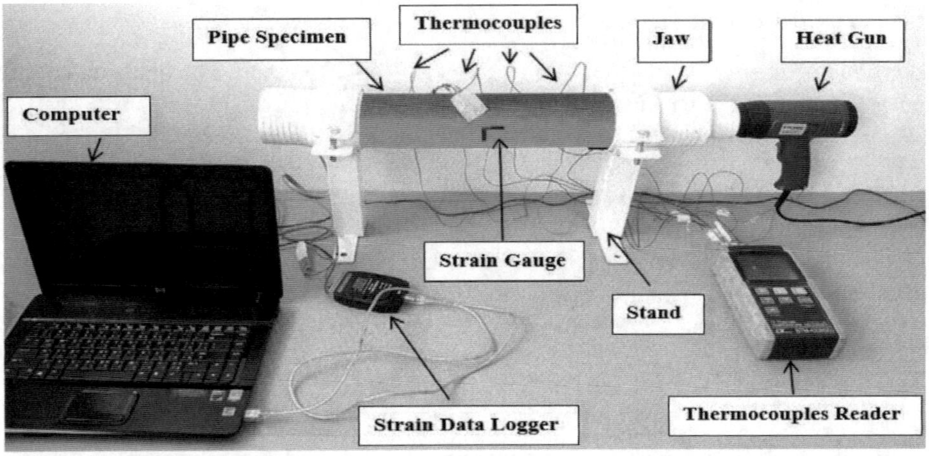

Figure 5. Experimental Work (Rig)

3. Numeric Analysis

Exhaustive Finite Element Approach (F.E.A.) operations from various studies were explored to demonstrate stresses and strains of specific loading status. Mechanical tension measurement of the tube is of greater significance in this research. The ANSYS version 11 software kit was used to test stresses and pressures in three nations. Form of feature, the sort (SHELL63) factor has been used to enable loads in the standard and plane direction. The SHELL63 dimension has six degrees of freedom (D.O.F.) in each node: operation and rotation in the contractual axes x, y, and z. Meshing and boundary states of the F.E.A. grid is very important in the C.F.D. analysis that is needed for the F.E.A. research. The number of grid elements shown in Figure (6), for the tube used in this search is (2000), and the nodes number (1209). The F.E.A. model's boundary condition is provided with a specific method for the model during regular day-to-day operations to clarify the boundary condition related to the usage of the tube concept in this search. Boundary state (seen in figure 7) is used by taking the diameter nodes on both sides of the tubes and adding all the D.O.F. Thermal load is placed inside pipeline[26].

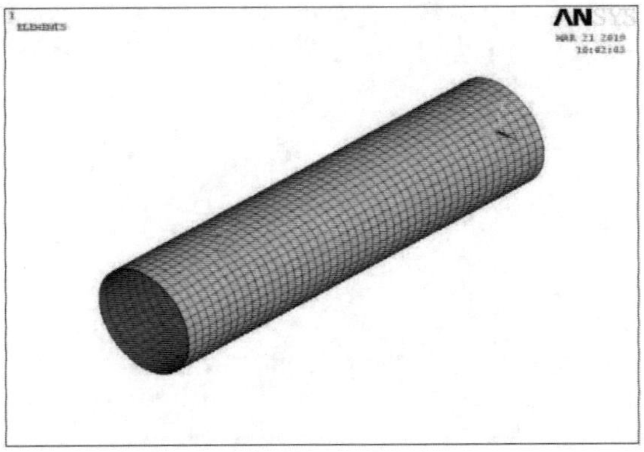

Figure 6. Tube meshing

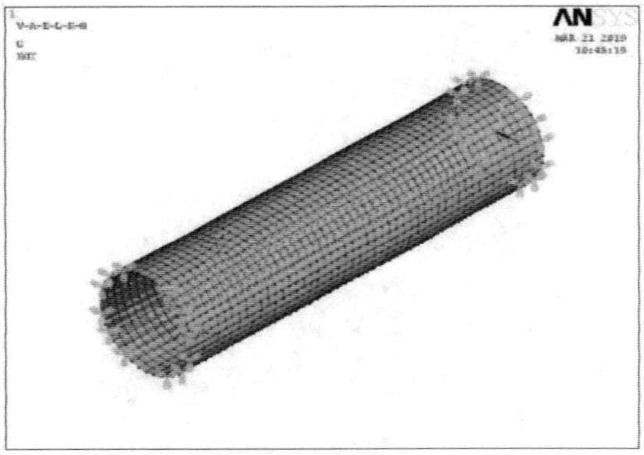

Figure 7. Boundary Conditions of the tube

4. Results and discussion

The effect of eggshell powder's addition to the fiberglass composite tube was also demonstrated by the behavioral variations of temperature, stress, and pressure. The substance performance characteristics from the tensile analysis and thermal, mechanical review for all three samples are increasing.

IOP Conf. Series: Earth and Environmental Science **877** (2021) 012039 doi:10.1088/1755-1315/877/1/012039

4.1. Tensile test and electronic thermal analysis tests

Component properties, such as elasticity panel, the fish ratio, the thermal expansion coefficient, the production pressure, and the tensile strength production and the thermal, mechanical test seen in the table (2). Table (2) reveals that (F50) reveals the average elasticity values (21GPa), however (E50) implies minimal elasticity values (15.42 Gpa) and the Poisson ratio values vary from (0.42) to (0.47). The thermal expansion coefficient (α), the performance stress (Ty) and the ultimate pressure (5-0) can be seen from statistics indicating that (F50) indicates the final stress and the thermal expansion coefficient (206 MPa) and (8,6 to 6 m / mk) respectively [27,28]. The highest yield pressure (E25F25) is (120 MPa). Nevertheless, the minimum ultimate stress, yield stress, and thermal expansion coefficient of (41,2 MPa), (36,7 MPa) and (2,90E-06 m / mk) are provided.

Table 2. Characteristics of the tube sample utilized

symbol	v	σult (MPa)	α (m/mk)	E (Gpa)	σy (MPa)
E50	0.42	41.2	2.90E-06	15.42	36.7
F50	0.47	205	8.50E-06	21	101
E25F25	0.46	200	6.70E-06	18.8	120

4.2. Specimens strain and stress effected by temperature rise.

Figure (10) to Figure (14) illustrates the temperature effect rise on tension and strain for E50, F50, and E25F25 samples. Fig (10) offers the importance of temperature with historical pressure. Observed from the figure (10) the maximums temperature of (225 ° C) by (E25F25) and (125 ° C), the (E50), min. temperature. Additionally, (E50) shows maximums strain when a match to the (F50) and (E25F25) with different (36 percent) and (12 percent). Figure (11) demonstrates importance hoop strain with temperature, and recognize findings mentioned figure (10) about strain. pressure gap between (E5o) and (F5o) and (E25F25) is (19 percent) and (30 percent), respectively. Figure (12) indicates the importance longitudinal stress with temperature. Figure (13) indicates the importance hoop stress with temperature. Figure (14) views the importance von mises stress with temperature.

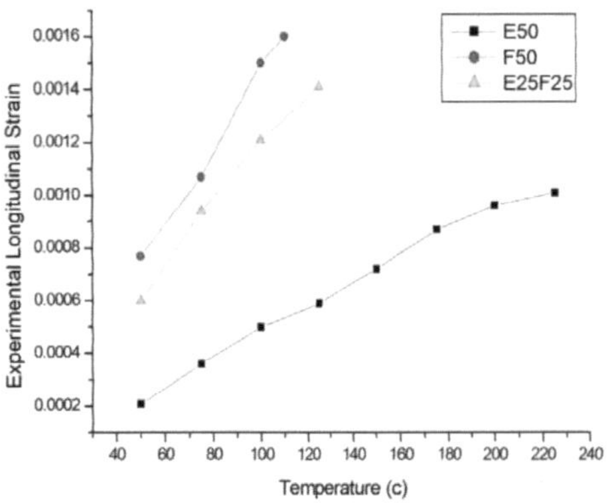

Figure 10. Experimental Longitudinal Strain with Temperature

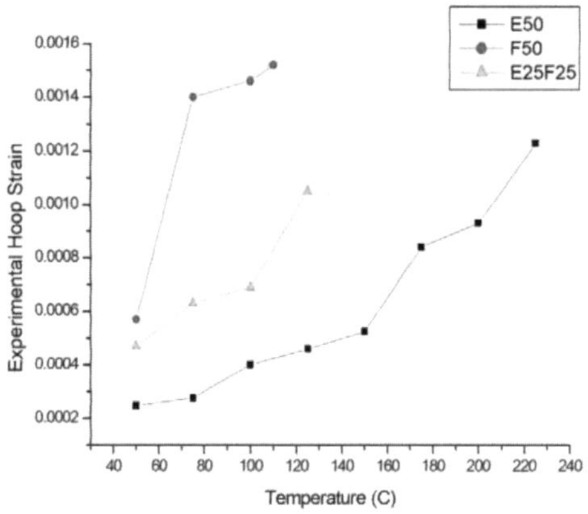

Figure 11. Experimental Hoop Strain with Temperature

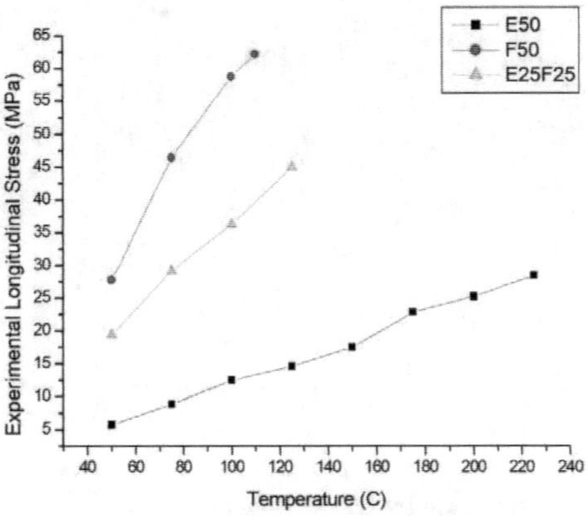

Figure 12. Experimental Longitudinal Stress with Temperature

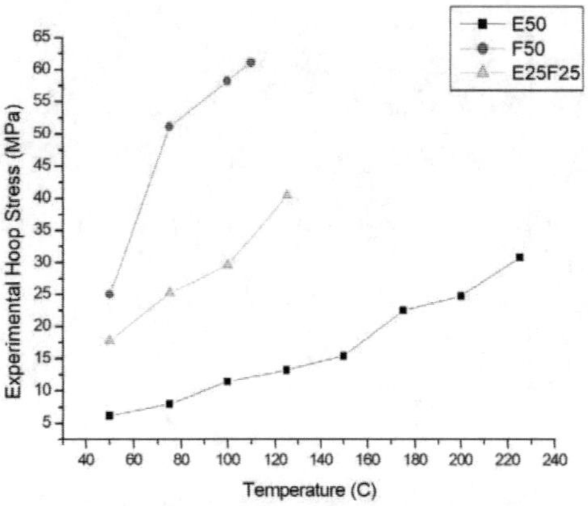

Figure 13. Experimental Hoop Stress with Temperature

4.3. Comparison of empirical and laboratory tests of pressure

Figures (15) to (19) give the relation between numerical von mises tension, numerical longitudinal strain, experimental hoop strain temperature rise and experimental longitudinal strain. It can be seen, a significant change in heights between numerical tension and experimental (14.1%) and (18%) .

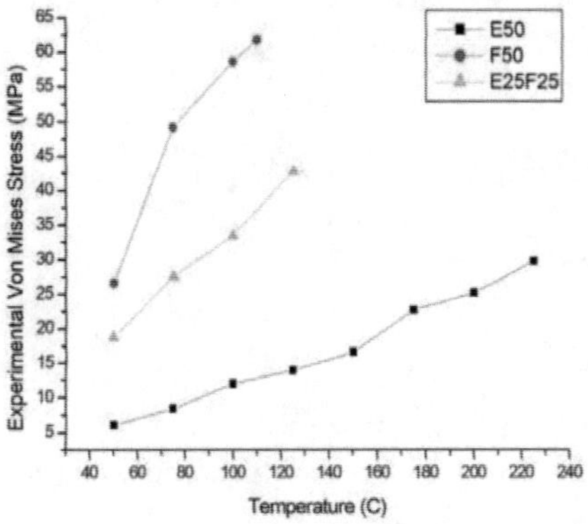

Figure 14. Temperature and Experimental Stress

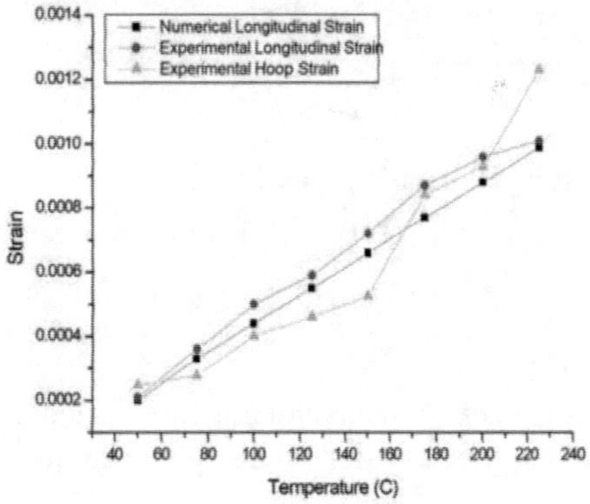

Figure 15. Experimental Strain and Numerical for E50

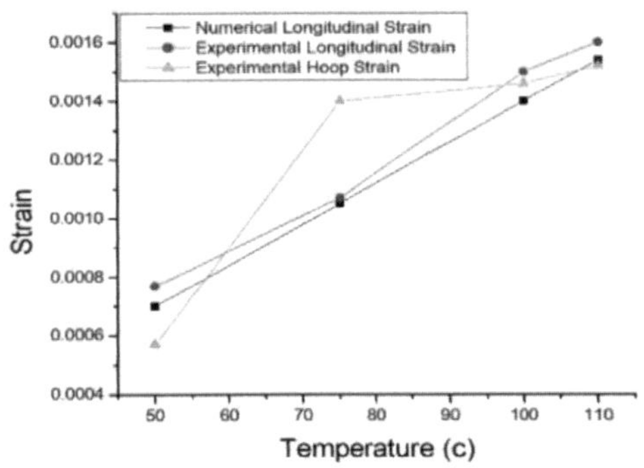

Figure 16. Experimental Strain and Numerical for F50

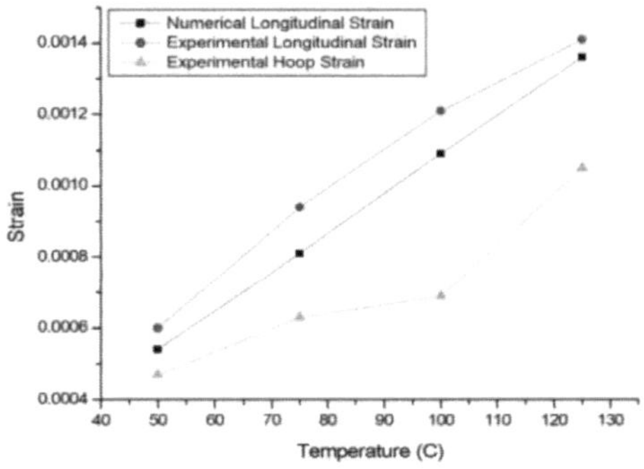

Figure 17. Experimental Strain and Numerical for E25F25

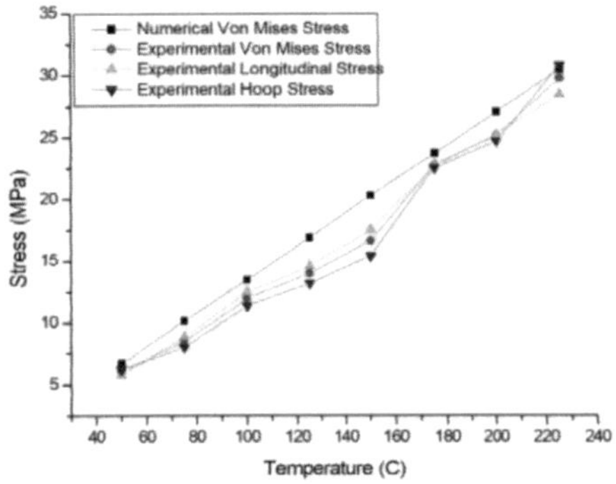

Figure 18. Experimental Stress and Numerical for E50

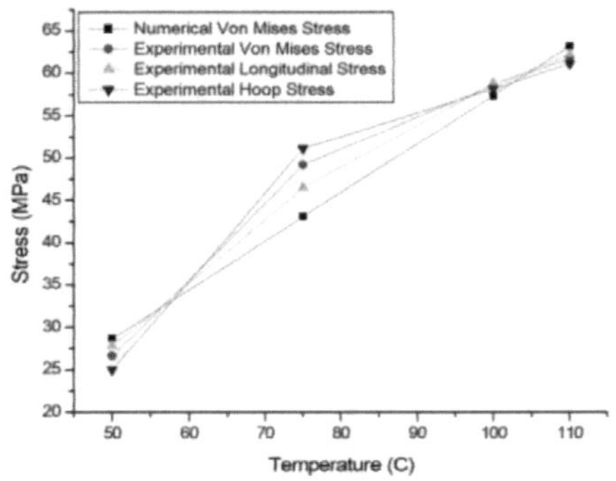

Figure 19. Experimental Stress and Numerical for F50

5. Conclusion

This study can back up many conclusions which constant volume fraction (50%):

1. Concerning the temperature keeper, the composite (E50) gives (225 C.) and the composite (F50) gives (110 C).
2. Supplement eggshell powder gives changes in temperature (110 C) to (125 C).
3. Composite (F50) had an average longitudinal strain, while composite (E50) had a low longitudinal strain.
4. Hoop stress (F50) decreased after eggshell powder fraction.
5. The composite (F50) had 62.2 MPa average longitudinal stress , and (E50) had 8.5 MPa stress .
6. The composite (F50) had average hoop stress (61.1 MPa), and (E50) had 30.8 MPa stress .
7. The (F50) composite had the average stress of (61,7 MPa), and (E50) had 29,8 MPa stress.

ACKNOWLEDGMENTS

The authors (Fadhel Abbas Abdulla ,M S Qasim and Ahmed Ali Farhan Ogaili) would like to thank Mustansiriyah University (www.uomustansiriyah.edu.iq) Baghdad-Iraq for its support in the present work.

References

1. Qasim S M, Mohammed F A and Hashim R 2015 Numerical investigation of the thermal behavior of heated natural composite materials IOP Conf. Ser. Mater. Sci. Eng. 95
2. Abdullah, F. A., & Khalaf, W. A. (2018). Experimental investigation of composite materials subjected to torsional stresses at high shear strain rate. Int. J. Mech. Mechatronics Eng., 18(1), 64-75.
3. Abdulla, F. A., Fadhil, H. A., & Abdulwahid, J. N. (2018, December). Experimental Study of the Creep Behaviour of Nano-Composites Carbon Fibres. In IOP Conference Series: Materials Science and Engineering (Vol. 454, No. 1, p. 012126). IOP Publishing.
4. khalid Hamdan, Z., & Abdullah, F. A. (2018). Investigation of the Adding Nano Particles to Composite Material under High Strain Rate Torsion with Hygrothermal Effect. Technology, 9(6), 1098-1114.
5. Chia-Wei Su, "Thermal stresses of composite beams with rectangular and tubular cross-sections", MSc thesis, The University of Texas at Arlington in Partial Fulfillment, 2007.
6. Jassim, R. F., & Abdulla, F. A. (2020, November). Investigation Experimentally the Effect of Thermal Stresses on the Straight and Curved Natural Composite Material Pipes. In IOP Conference Series: Materials Science and Engineering (Vol. 928, No. 2, p. 022065). IOP Publishing.
7. Abdulla, F. A., & Abdullah, A. H. (2020, November). Effect of Shot Penning on Wear rate of Eggshell natural composite Materials. In IOP Conference Series: Materials Science and Engineering (Vol. 928, No. 2, p. 022091). IOP Publishing.
8. Olawale M. Sanusi, Olufemi D. Komolafe, Tunde O. Ogundana, Mesach O. Olaleke, and Yunusa Y. Sanni, "Development of Wood-ash/Resin Polymer Matrix Composite for Body Armour Application, "FUOYE Journal of Engineering and Technology, Volume 1, Issue 1, 2016.
9. Alessandra Luiza de Lemos, Pamela Galera Prestes Pires, Marcelo Lopes de Albuquerque, Vagner Roberto Botaro, Jane Maria Faulstich de Paiva, Nei Sebastiao Domingues Junior, " Biocomposites reinforced with natural fibers: thermal, morphological and mechanical characterization", Materials Journal, Vol.22, No.2, 2017.
10. American society for testing and materials international, "Standard test method for tensile properties of plastics," D638, 2000.
11. Abdulla, F. A., Moustafa, N. M., & Hussein, A. F. (2018). Effect of uv-radiation on fatigue behaviour of natural composite materials. Int. J. Mech. Prod. Eng. Res. Dev., 8, 727-40.
12. Moustafa N M, Abdulla F A and Nori A F 2018 PID control system for a variable speed horizontal axis wind mill Int. J. Mech. Eng. Technol. 9 1080–7.
13. Jebur, N. A., Abdulla, F. A., & Hussein, A. F. (2018). Experimental and numerical analysis of below knee prosthetic socket. Int. J. Mech. Eng. Technol., 9, 1-8.
14. Abdulla, F. A., Fadhil, H. A., & Abdulwahid, J. N. (2018, December). Experimental Study of the Creep Behaviour of Nano-Composites Carbon Fibres. In IOP Conference Series: Materials Science and Engineering (Vol. 454, No. 1, p. 012126). IOP Publishing.
15. Abdul-Kareem, H. S., Abdulla, F. A., & Abdulrazzaq, M. A. (2019, May). Effect of Shot Peening and Solidification on Fatigue Properties of Epoxy Base Composite Material. In IOP Conference Series: Materials Science and Engineering (Vol. 518, No. 3, p. 032017). IOP Publishing.
16. Al-Ameen E S, Abdulhameed J J, Abdulla F A, Ogaili A A F and Al-Sabbagh M N M 2020 Strength characteristics of polyester filled with recycled GFRP waste J. Mech. Eng. Res. Dev. 43 178–85
17. Tariq, S. Z., & Abdullah, F. A. (2020, February). Effect of wood ash additive on the thermal stresses of random fiberglass/polyester composite pipes. In IOP Conference Series: Materials Science and Engineering (Vol. 745, No. 1, p. 012062). IOP Publishing.
18. Metteb, Z. W., Abdalla, F. A., & Al-Ameen, E. S. (2020, March). Mechanical properties of recycled plastic waste with the polyester. In AIP Conference Proceedings (Vol. 2213, No. 1, p.

020067). AIP Publishing LLC
19. Hamdan, Z. K., Abdalla, F. A., & Metteb, Z. W. (2020, March). Effect of acids salts and water on natural composite materials. In AIP Conference Proceedings (Vol. 2213, No. 1, p. 020075). AIP Publishing LLC
20. Abdulla, F. A., Hamid, K. L., Ogaili, A. A. F., & Abdulrazzaq, M. A. (2020, November). Experimental study of Wear Rate Behavior for Composite Materials under Hygrothermal Effect. In IOP Conference Series: Materials Science and Engineering (Vol. 928, No. 2, p. 022009). IOP Publishing.
21. Al-Ameen, E. S., Abdulla, F. A., & Ogaili, A. A. F. (2020, June). Effect of Nano TiO2 on Static Fracture Toughness of Fiberglass/Epoxy Composite Materials in Hot Climate regions. In IOP Conference Series: Materials Science and Engineering (Vol. 870, No. 1, p. 012170). IOP Publishing.
22. Abdulla, F. A. (2020, July). Experimental and Numerical Investigation of Shot-Peening and Solidification Effects on the Endurance Limit of Composite Material. In IOP Conference Series: Materials Science and Engineering (Vol. 881, No. 1, p. 012058). IOP Publishing.
23. Qasim, M. S., Fadhel, A. A., & Mohammed, H. R. (2020, September). Experimental study of thermal behavior of heated Natural Composite Materials. In IOP Conference Series: Materials Science and Engineering (Vol. 916, No. 1, p. 012091). IOP Publishing.
24. Hussein, A. F., Abdulla, F. A., Hamdan, Z. K., & Ali, W. A. (2020, June). Investigation of stress concentration factor for natural composite material. In IOP Conference Series: Materials Science and Engineering (Vol. 870, No. 1, p. 012155). IOP Publishing.
25. Ogaili, A. A. F., Abdulla, F. A., Al-Sabbagh, M. N. M., & Waheeb, R. R. (2020, November). Prediction of Mechanical, Thermal and Electrical Properties of Wool/Glass Fiber based Hybrid Composites. In IOP Conference Series: Materials Science and Engineering (Vol. 928, No. 2, p. 022004). IOP Publishing.
26. Bader, S. T. (2020, November). Investigation of Natural Composite Materials Pipe Under Thermal Load. In IOP Conference Series: Materials Science and Engineering (Vol. 928, No. 2, p. 022070). IOP Publishing.
27. Abdulla, F. A., Hamid, K. L., Ogaili, A. A. F., & Abdulrazzaq, M. A. (2020, November). Experimental study of Wear Rate Behavior for Composite Materials under Hygrothermal Effect. In IOP Conference Series: Materials Science and Engineering (Vol. 928, No. 2, p. 022009). IOP Publishing.
28. Shyaa, A. K., & abbas Abdulla, F. (2020, November). Enhancement Thermal Conductivity of PCM in Thermal Energy storage. In IOP Conference Series: Materials Science and Engineering (Vol. 928, No. 2, p. 022090). IOP Publishing.

Improvement of Thermal Performance of Coiled Tube Heat Exchanger Utilizing Air Bubble Injection Technique

Saif Salah Hasan[1*], Ali Shakir Baqir[1], Hameed B Mahood[2]

[1]Engineering Technical College, Al-Furat Al-Awsat Technical University, Al-Najaf 31001, Iraq
[2]Department of Chemical and Process Engineering, Faculty of Engineering and Physical Sciences, University of Surrey, Guildford GU2 7XH, UK

E-mail: saif.salah@student.atu.edu.iq

Abstract. The heat transfer enhancement in terms of temperature of a vertical helically coiled tube heat exchanger is carried out experimentally. The experiments were achieved in a heat exchanger with a 50 cm height and 15 cm internal diameter under four different cold and three hot water mass flow rates and four different airflow rates. At the same time, the temperature difference was taken invariant ($\Delta T=20°C$). To avoid some uncertainties, the hot side temperature of the heat exchanger was measured via k-type thermocouples. The results showed that the increase of air injection flow rate improved heat transfer from the hot stream flowing in the coil to the shell's cold stream. An intimate thermal mixing when air injected is clearly observed, which could be responsible for the heat exchanger's thermal enhancement. Finally, the injected air pressure was noticed to be having only a minor effect on thermal performance improvement.

Keywords: heat exchanger, coil tube, air injection bubbles, and sparger.

1. Introduction

Heat exchangers are equipment utilized to transfer thermal energy between two different thermal energy content streams. Hence, they are widely implemented in various industrial applications, such as in the power plants, chemical industries, refinery, food industry, and pharmaceuticals. Accordingly, a huge effort has been directed to improve the heat exchanger performance efficiency utilizing various techniques[1]. However, all enhancement techniques inevitably involve a significant increase in the pressure drop, cost, energy, materials, and heat exchangers weight. Minimizing the pressure drop while improving heat exchanger thermal efficiency increases the target of many investigators.

Among various enhancement techniques, such as using a twisted tape, corrugated tubing, wavy stripes, barbed wires, and internal and external fins, air injection as bubbles seems the most promising approach. This improvement technique's merit is in its capability to significantly improve the thermal efficiency of heat exchangers with a relatively low increase in the pressure drop. However, Dizaji et al. [2] has experimentally studied the thermal enhancement of a helically coiled tube heat exchanger by injecting sub-millimeter bubbles of air inside the side shell. Their results showed a significant increase in the NTU as a result to the injection of air as bubbles. Moosavi et al. [3] used the same experimental setup and technique of [2] with only changing the coil pitch and studying the increment in the overall heat transfer coefficient of the heat exchanger's, and its pressure drop. Panahi [4] Experimentally tested the impact of tiny bubbles injecting on the effectiveness of a coiled tube heat exchanger.

Khorasani and Dadvand [5] researched experimentally the enhancement of effectiveness and the NTU due to the injection of air bubbles in the horizontal helical tube heat exchanger. The results showed that the increase of air flow rate leads to an increment in number heat transfer unit. The NTU was raised to about 1.3 to 4.3 times that without air bubble injection. El-Said and Alsood [6] tested the possibility of improving the thermal efficiency of a multiple tubes heat exchanger by injecting air as small bubbles using a cross flow and parallel flow air injection technique. The results revealed that the injection of air bubble was significantly enhanced the heat exchanger efficiency, and the cross-flow injection method was having more advantage. Baqir et al. [7] similarly investigated the optimal operating conditions of a micrometer air bubble injection in a spiral tube heat exchanger's side shell. The obtained results indicated that the heat transfer coefficient's improvement due to air bubble injection was about 130%. Kreem et al. [8] observed an intimate thermal mixing in the shell side of heat exchanger by bubbles and they claimed that could be the main reason of the improvement of the thermal performance of heat exchanger. Two different air bubble injection techniques (parallel and cross-flow) into the shell side of a shell and tube heat exchanger were tested by Subesh et al. [9]. The possible improvement of the heat exchanger's thermal performance, including Nu, and the overall heat transfer coefficient due to the bubble injection, were measured under a wide range of the shell side and injected airflow rates. They observed that the cross-injection air bubble technique was more efficient than the parallel but with a higher pressure drop. Ghashim et al. [10] studied experimentally the effect of injecting small air bubbles on the heat transfer and pressure drop of a helical coiled tube heat exchanger working under turbulent flow conditions. The flow rate of air was 1.5, 2.5 and 3.5 l/min and the hot fluid Re was from 9000 to 50000 and Nu, effectiveness, NTU and exergy losses were evaluated. The results showed, for the entire range of the tested Re, that the air injected led to improve Nu by 64 – 126 %, whilst increased the pressure drop from 66 – 85%. Sokhal et al. [11] evaluated the influence of bubbles injection on the efficiency and exergy of a shell and tube heat exchanger. The results showed that the increase of the injected air bubble flow rate improves the heat exchanger's performance while the exergy loss increases with increasing the flow rate. Compared with no bubble injection, the minimum dimensionless exergy loss and Nu were found to be about 27.49% and 13 %.

It was evident from the literature review that the temperature distribution and thermal mixing of the shell side of a heat exchanger are still unclear. Therefore, the present short paper, which represents primary experimental results, aims to evaluate a coiled tube heat exchanger's temperature distribution and observe the thermal mixing that the thermal enhancement is based on.

2. Experimental setup and test procedures

Figure 1 illustrates the diagram of the experimental rig. The testing rig comprises a heating unit, cooling unit, and heat exchanger. The heating unit, which is used to provide the hot water to the system, consists of a copper helical coil, a hot water storage tank, variac electrical heaters, and a wattmeter. The cooling unit supplies the shell or cold water during the experiments, includes a cold-water storage tank and a refrigeration cooling system. Finally, the heat exchanger unit consists of a cylindrical PVC shell, coiled copper tube, other fittings, valves, and connections parts. Table 1 gives a summary of the heat exchanger geometry. Besides, the column was insulated with 3 mm EPE foam and a sparger to generate air bubbles with a hole number and size of 1400 and 0.1mm, respectively, placed in the shell's bottom, as shown in fig 2. Furthermore, 8 K-thermocouples are used to measure the temperature distribution of the system. The cold and hot water and air flow rates were measured via rotameters. Air was provided to the system using a compressor type HAILEA ACO-318. The system's flow configuration was counterflow, as clarified in fig. 3. All thermocouples were connected to a PICOLOG 6 TC- 08 data logger with 0.1° C a resolution, which is connected to read the thermocouples reading directly on a PC.

The experiments start with circulating the hot and cold water in the coil side and shell side of the heat exchanger until it reaches a steady-state condition where the temperature measurements were taken. After that, the air is injected at a specific flow rate directly into the heat exchanger's side shell via the sparger. The thermocouples' readings were taken when the new steady-state condition is reached. This procedure was repeated for different operational conditions shown by Table (2).

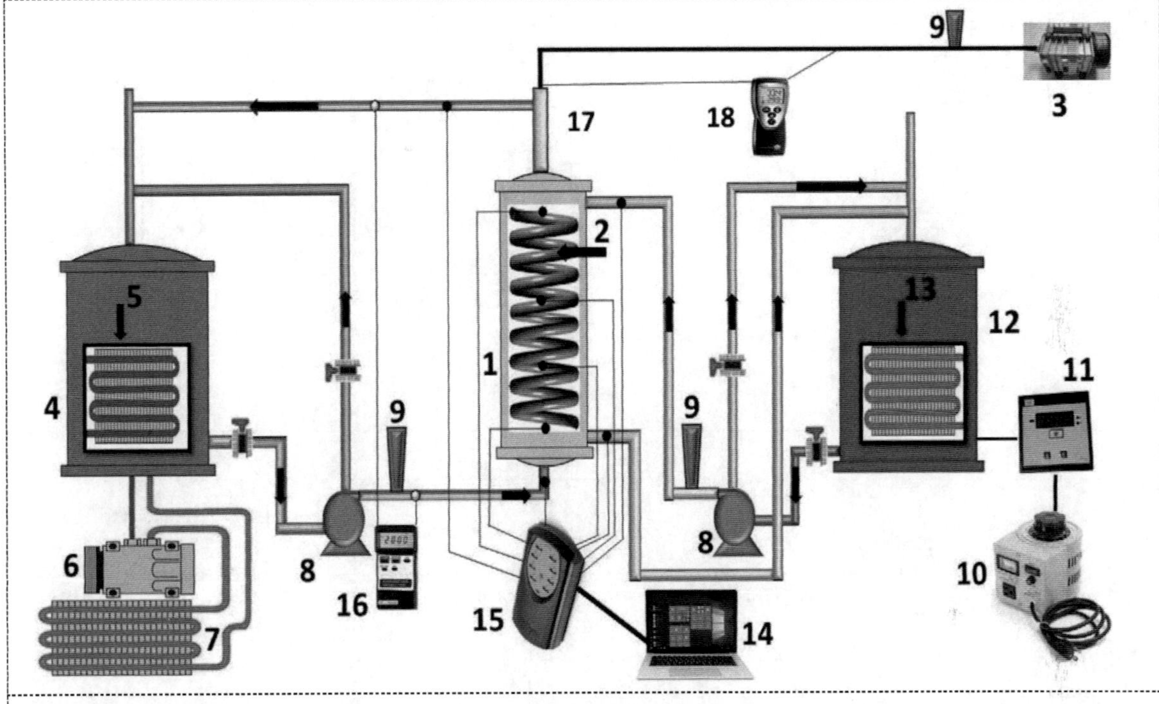

Figure 1. A schematic view of the setup of experiments: 1. Shell helical tube, 2. Helically coiled tube, 3. Air Pump, 4. A tank of cold water, 5. evaporator, 6. Compressor, 7. Condensor,8. Water pump, 9. rotameter, 10. Variac, 11. wattmeter, 12. A Tank with hot water, 13. heater. 14. PC, 15. Logger of Data, 16. manometer, 17. A vent of air, 18. Air temperature thermometer.

Figure 2. sparger

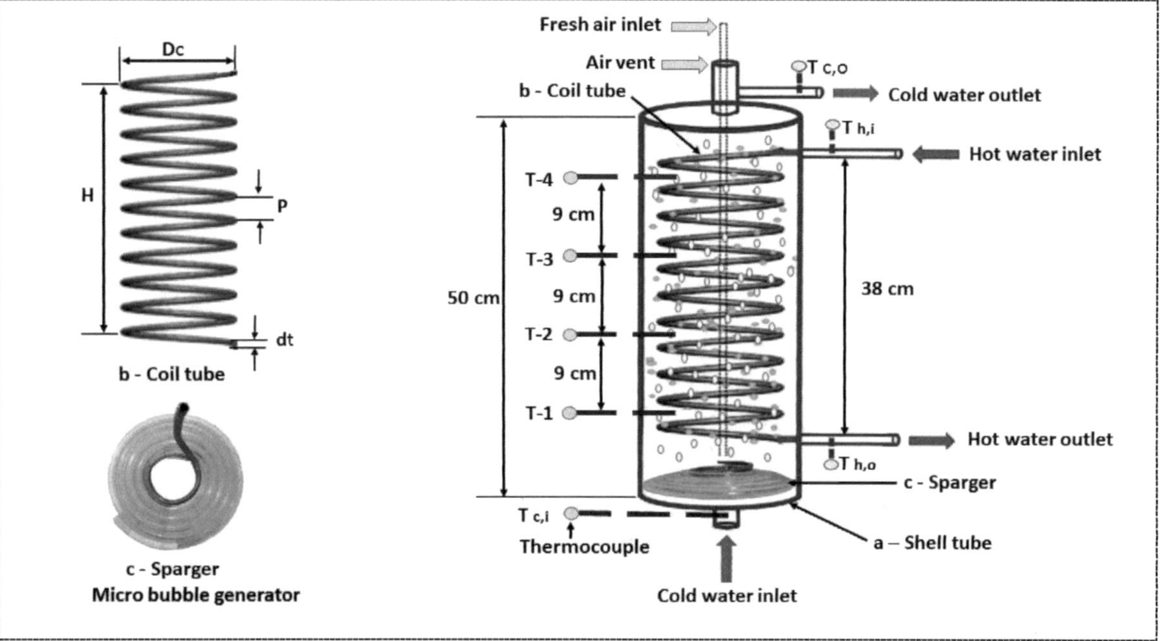

Figure 3. Test section

Table 1. Summary of heat exchanger geometry

Tube	Do (mm)	t (mm)	H (mm)	L (mm)	Dc (mm)	di (mm)	N	P (mm)
Shell	152.4	3	500	-	-	-	-	-
Coiled tube	-	1.6	380	3939 (bco)	114	4.4	11	30

Table 2. Different operational condition

shell and coiled tube inlet temperature (C°)	The coil-side water flow rate of the coil (LPM)	The water flow rate on the side of the shell (LPM)	Airflow rate (LPM)
T coil = 35 ° C T shell = 15 ° C	1	4,6,8	0.5,1,1.5,2
	1.5	4,6,8	0.5,1,1.5,2
	2	4,6,8	0.5,1,1.5,2

3. Results and discussion

In the present experimental work, the enhancement of coil and shell tube heat exchangers' thermal performance is investigated. However, different side-shell flow rates and side coil flow rates were tested with an invariant temperature difference between them. Figures 4, 5, and 6 show the variation of the outlet coil (hot side) temperature with the side shell flowrate at five various injecting air flow rates, and injection pressure (P =2 and 4 bar) and temperature difference (ΔT=20°C). The figures clearly show that the influence of the flow rate of shell side and the injected air bubble flow rate on the coil side outlet temperature. Obviously, the outlet temperature of the hot fluid in the coil is reduced with the increase of the flow rate of the fluid in the shell. Meanwhile, the injected air bubble within the shell side led to a significant decrease in the outlet hot fluid temperature. This indicates that the air bubble has improved the exchange of heat between the hot and the cold sides.

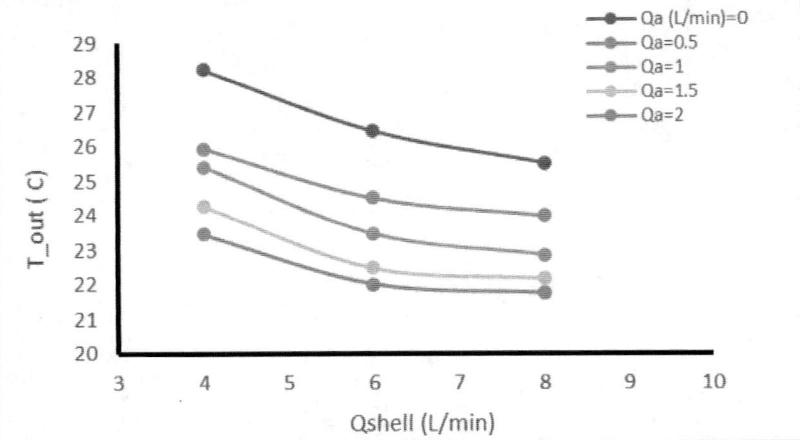

Figure 4. The hot water outlet temperature of the coil side with a shell-side flow rate at varying airflow rate of injection and continuous injection pressure

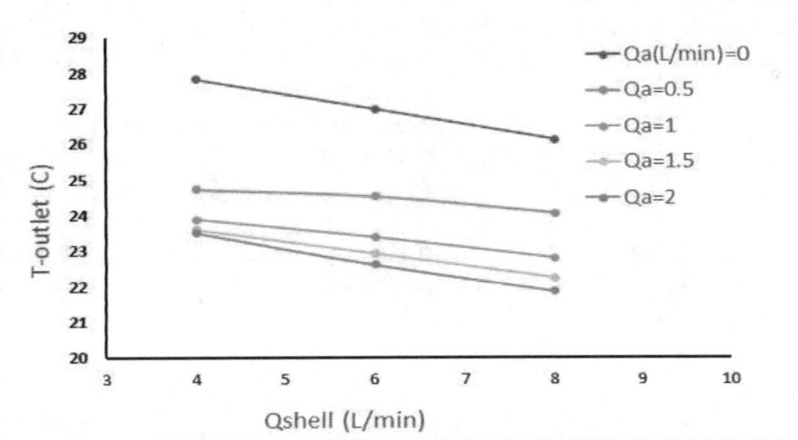

Figure 5. Hot water outlet temperature on the coil side with a shell-side flow rate at varying flow rates of injection air and steady injection pressure

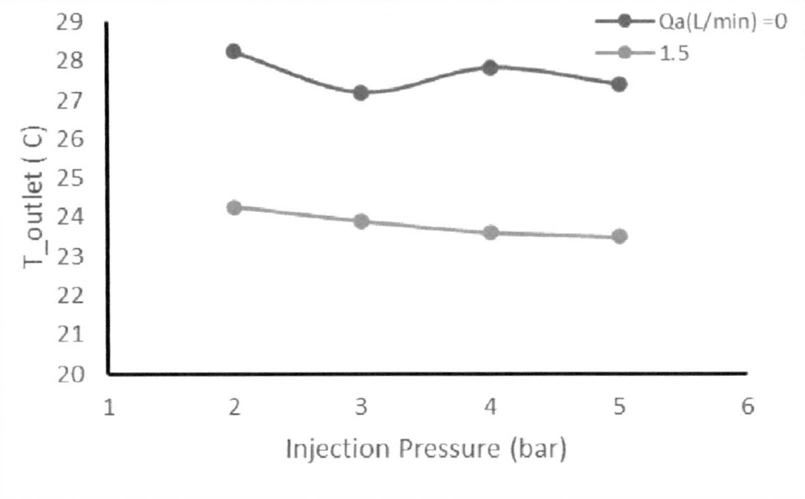

Figure 6. Impact of bubble injection pressure at constant and airflow rates of 1.5 and 0 L/min on coil side outlet temperature

Figure 7 shows the temperature of the heat exchanger's shell side with and without air bubble injection with time for side - shell flow rate, side coil flow rate, and an injection pressure of 4 L/min, 2 L/min, and 3 bar, respectively. The observed results in the figure are new and give, as the first time, evidence that the improvement of the thermal efficiency is achieved mainly by the thermal blending of the water in the shell. However, the coil side (hot side) temperature is significantly decreased with an increasing flow rate of the injected air. Meanwhile, the temperature distribution along the axial dimension (shell side) of the heat exchanger starts with different temperature values and is in convergence with air injection and became nearly at the same temperature at an injected airflow rate of 1.5 L/min. After that, the temperature of water in the shell increases steadly. These results can give a clear idea of what is going on for the water in the shell as aresult to the air bubble injection.

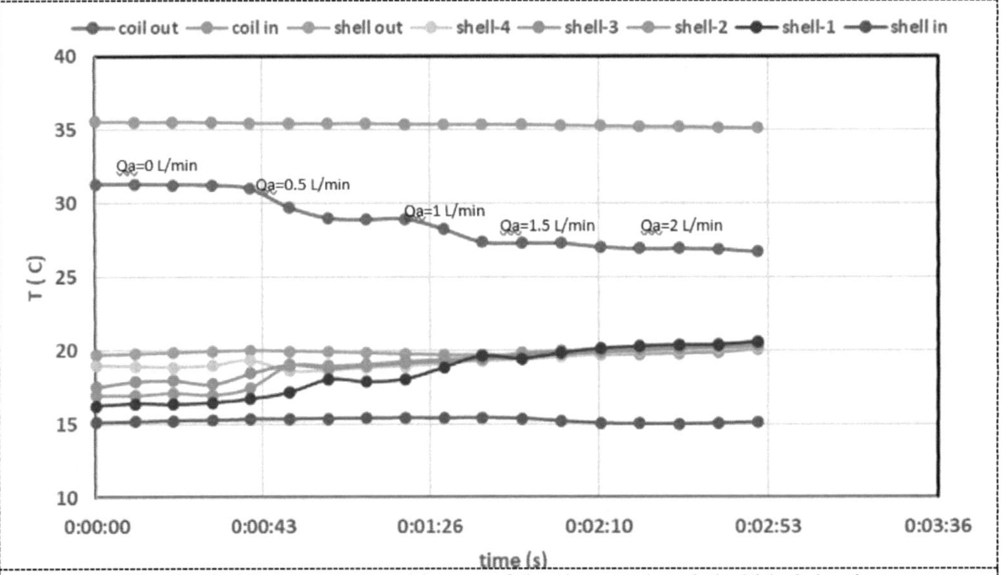

Figure 7. The shell side temperature history of the time under air bubble injection

4. Conclusion

The temperature measurement of the coil type heat exchanger working with and without air bubble injection is carried out experimentally. Three different shells or cold side flow rates, four different air injection pressure, four airflow rates, three hot water flow rates, and a steady temperature difference were examined. The effect of the flow rate of bubbles on the coil and shell heat exchanger's thermal performance was investigated. Accordingly, the following conclusions can be drawn:

1- The temperature of the coil side is decreased with an increase in the injected air flow rate. The significant decrease in the temperature seems to occur at the beginning of air injection (air flow rate = 0.5 L/min).
2- The effect of injected air is more apparent at a low shell-side flow rate.
3- The air injection pressure seems to have only a slight effect on the heat transfer enhancement.
4- The water shell's intimate thermal mixing due to the air injection could be responsible for the significant enhancement in the heat exchanger's thermal efficiency.

References

[1] Kareem, Z. S., Abdullah, S., Lazim, T. M., Jaafar, M. M., & Wahid, A. F. A. (2015). Heat transfer enhancement in three-start spirally corrugated tube Experimental a.pdf.".

[2] Dizaji, H. S., Jafarmadar, S., Abbasalizadeh, M., & Khorasani, S. (2015). Experiments on air bubbles injection into a vertical shell and coiled tube heat exchanger; exergy and NTU analysis. Energy Conversion and Management, 103, 973-980.".

[3] "Moosavi, A., Abbasalizadeh, M., & Dizaji, H. S. (2016). Optimization of heat transfer and pressure drop characteristics via air bubble injection inside a shell and coiled tube heat exchanger. Experimental Thermal and Fluid Science, 78, 1-9.".

[4] D. Panahi, "Evaluation of Nusselt number and effectiveness for a vertical shell-coiled tube heat exchanger with air bubble injection into shell side," Experimental Heat Transfer, vol. 30, no. 3. pp. 179–191, 2017, doi: 10.1080/08916152.2016.1233145.

[5] S. Khorasani and A. Dadvand, "Effect of air bubble injection on the performance of a horizontal helical shell and coiled tube heat exchanger: An experimental study," Applied Thermal Engineering, vol. 111. pp. 676–683, 2017,

[6] E. M. S. El-Said and M. M. A. Alsood, 'Experimental investigation of air injection effect on the performance of horizontal shell and multitube heat exchanger with baffles,' Appl. Therm. Eng., vol. 134, no. September 2017, pp. 238–247, 2018.".

[7] A. S. Baqir, H. B. Mahood, and A. R. Kareem, "Optimisation and evaluation of NTU and effectiveness of a helical coil tube heat exchanger with air injection," Thermal Science and Engineering Progress, vol. 14. 2019, doi: 10.1016/j.tsep.2019.100420.

[8] Tishk International University, Erbil Polytechnic University, Institute of Electrical and Electronics Engineers, and Institute of Electrical and Electronics Engineers. Iraq Section, "Proceedings of the 5th International Engineering Conference (IEC2019) : June 23-25, 2019, Erbil, Kurdistan Region-Iraq.".

[9] Subesh, T., Dilip Raja, N., Logesh, K., Ramesh, V., Venkatasudhahar, M., & Surrya Prakash, D. (2020). Study on performance of horizontal single pass shell and multi-tube heat exchan.pdf.".

[10] S. L. Ghashim and A. M. Flayh, "Experimental investigation of heat transfer enhancement in heat exchanger due to air bubbles injection," Journal of King Saud University - Engineering Sciences. 2020, doi: 10.1016/j.jksues.2020.06.006.

[11] G. S. Sokhal, G. S. Dhindsa, K. S. Sokhal, M. Ghazvini, M. Sharifpur, and M. Sadeghzadeh, "Experimental investigation of heat transfer and exergy loss in heat exchanger with air bubble injection technique," Journal of Thermal Analysis and Calorimetry. 2020, doi: 10.1007/s10973-020-10192-1.

Experimental Investigation of a Single Slope Solar Still Performance- Evaporation Process Enhancement Using Evacuated Pipes

Zahraa Abdulkareem Jaafar[1], Hassanain Ghani Hameed
Mechanical Power Department, AL-Furat Al-Awsat Technical University, Engineering Technical College/Najaf, Najaf-Iraq

[1]E-mail: zahraa.abdulkareem@student.atu.edu.iq

Abstract. The single slope solar still productivity strongly depends on the amount of energy absorbed by the solar still basin plate. Therefore, increasing either the basin plate's absorption or enhancing the heat transfer with raw water will increase the pure water production rates. To improve the evaporation and the solar still thermal performance, custom-designed evacuated copper pipes with different diameters and water filling rates are experimentally investigated in this paper. Moreover, it has been noticed that pure water productivity is significantly affected as it improved by 90.09% when using a 15mm diameter pipe with a 50% filling ratio.
Keywords: solar still, distillation, evacuated tubes, solar still evaporation enhancement, solar still productivity

1. Introduction

The solar stills depend on solar radiation to convert the salty and contaminated water into drinking water based on the evaporation and condensation processes. Therefore enhancing any of these processes will improve the solar still productivity.Many attempts to evaluate the solar still performance in several methodologies. Meysam [1] used an external condenser filled with phase change material (PCM)for improving the condensation rates and simultaneously store the latent heat of condense vapor within the PCM for using it after the day time. The performance of the solar still was enhanced by 86% , and productivity records 6.555 kg/m^2 with a total efficiency of 50%. Adding internal mirrors for increasing the incidence of radiation and adding black painted metal balls was also investigated by Hassanain Ghani[2].The study focuses on the effect of these parameters on different water depths. The results show that solar still productivity could be enhanced by 22.9% using internal mirrors, as well as productivity improved when using metal balls and decreasing water level. Pankaj Dumka et al. [3] investigate numerically and experimentally the effect of adding a magnetized ring of ferrite inside the still basin. the addition has a great effect on the evaporation process as well as the pure water productivity which improved by 49.22%. In contrast, Mohit Bhargva [4] studied multiple configurations to improve both the evaporation and condensation. Using an internal reflector to gain more energy, evacuated tube collector, heat exchanger, and external condenser. Different configurations were used to improve solar still performance. The experimental results show that the optimum configurations were combining the solar with an evacuated tube, internal reflector, and the external condenser. The improved configuration was directly compared simultaneously with a conventional solar still with the same dimensions. The enhanced solar still productivity was 2259 ml/m^{2,} with an efficiency of 33.4%. Also, active ways are investigated using equipment that requires an external power source. Gnanaraj et al [5] improve the solar still by adding an external mirror for reflecting more radiation into the solar still basin. the increased radiation enhances the still performance by increasing the freshwater productivity by 41%. While, Z. Jaafar [6] study the effect of adding galvanized wick and using a special design solar collecting tank. Using the special tanks helps preheated the input water, which significantly affected the solar still productivity with 48.83%. While using the galvanized wick, improving the productivity by 86.65% using 25*25 mm grid. A. S. Abdullah [7] studied the effect of using a rotating drum inside the solar still basin to improve the evaporation process by increasing evaporation's surface. The drum was equipped with a heater and Nano-particles. An external condenser was also used to increase the condensation rates at the same time. A computational fluid dynamic (CFD) model was developed for estimating the results. The numerical and experimental results give excellent agreement for the solar still productivity. The drum was rotated with different speeds varying from 0.1 up to 4 revolutions per minute (RPM). The productivity increases as the drum speed increases, and the solar still productivity was enhanced by 350% when using an external condenser and drum rotating with 4 (RPM). Zakaria Haddada [8] Investigate experimentally the effect of using vertical rotating wick (VRW) or improve the solar still evaporation as well as the still productivity. The wick was black painted to increase the absorption, and it had a direct effect on the still performance, and the productivity increased by 14.72% and 51% in summer and winter respectively. A.E. Kabeel [9] Investigate experimentally the effect of using Nano-particles to improve the basin water thermal performance. Also, he used an external condenser to enhance the condensation rates in the solar still. The experimental results show that using Nano-particles is more efficient than using an external condenser. Nano-particle use was enhanced by 116%, while it improved by 53.2% by using an external condenser. Other researchers' studied enhancing the evaporation processes of conventional solar still. AK, Kaushal [10] suggested adding a floating wick and a heat exchanger for recycling the wasted heat. The design approaches productivity of 21% more than the conventional, while Mohit bhargva[11] studied combining the solar still with an evacuated tube solar collector to collect more energy. A heat exchanger was equipped in this experiment. The modified solar still performance was enhanced by 160% compared to the conventional solar still. Other methods are used to collect more energy like using an external reflector in the study presented by Lei Mu [12] and other Colleagues. The study investigated combining the reflector with a Fresnel lens for concentrating solar radiation on the absorbing plate. The evaporation presses and the productivity was significantly affected, and it recorded pure water productivity of 467% more than the conventional still. Corrugated basin solar still was also investigated by Z. M. Omara et al. [13] experimentally. The experiment conducted double metal wicks with internal reflectors together, and the freshwater productivity enhanced by 145% with an efficiency of 59%. Ramanathan [14] study the effect of enhancing the evaporation inside the solar still by using a flat plate absorber. Absorbing more energy increases the evaporation rates and improving pure water production by 25%.

2. Experiment set-up

The experiments included three separated single slope solar stills, each one contains a basin fabricated from a galvanized iron sheet of 1.5 mm thickness with an area of 1 m². All sides of the basin were 5 cm height. That iron basin mounted in a wooden frame, insolated with a 3 cm thick white cork for isolating the system and maintaining the collected energy within the solar still and minimizing heat losses. All the solar stills were covered with a 4 mm window glass with a transparency of 88% tilted with 32.1° and facing the south direction to receive the maximum solar radiation possible across the year, as shown in Figure 1.

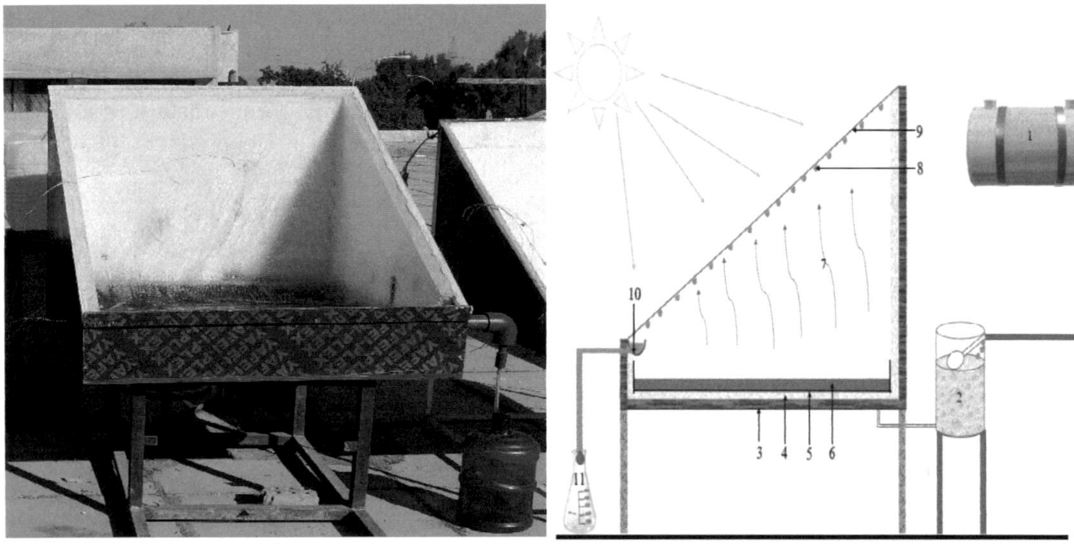

Figure 1. Pictorial and Schematic view for the solar still 1. Raw water tank. 2. Level tank. 3. Wooden frame. 4.insulation cork. 5.basin liner. 6. Raw water. 7.water vapor8. Condensed water drops. 9. Glass cover. 10.pure water collecting tray. 11.pure water collecting tank

3. Experimental procedures

All the experiments were conducted in May 2020 at the technical college's rooftop- Najaf /al-forat Al-Awsat technical university (32.5665 41.6256). A direct comparison was used to obtain accurate results by carrying out the experiment with three stills simultaneously and location to ensure that the three stills run at the same environmental conditions. Using 19 K-type thermocouples and a digital data logger are used to record the thermal data each hour from the beginning until the end of the experiment. The solar radiation and the produced pure water were also measured hourly using a digital solar meter and scaled cylinder.

The three solar stills thermal performance were experimentally investigated for 100% and 50% filling ratio under three cases, 15 mm copper pipes, 7mm copper pipes (as in figure 2), and conventional still without adding pipes.

The first comparison was between the solar stills with 15 & 7 mm pipes with the conventional still. The copper tubes were evacuated using two-stage evacuated pumps. After that, it filled partially with 50% of its original volume with water using a custom design feeding mechanism, as in figure 3.

Another experiment was also conducted using the same pre-mentioned configuration. With 100% pipes filing to study changing the pipes contained water on the solar still thermal performance and freshwater productivity.

IOP Conf. Series: Earth and Environmental Science **877** (2021) 012041 doi:10.1088/1755-1315/877/1/012041

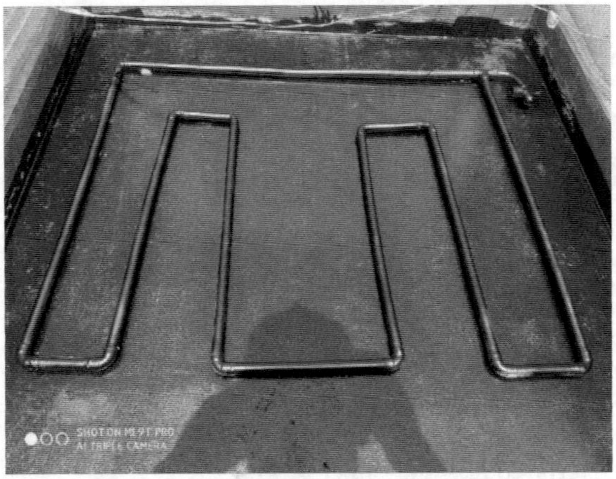

Figure 2. evacuated copper pipe

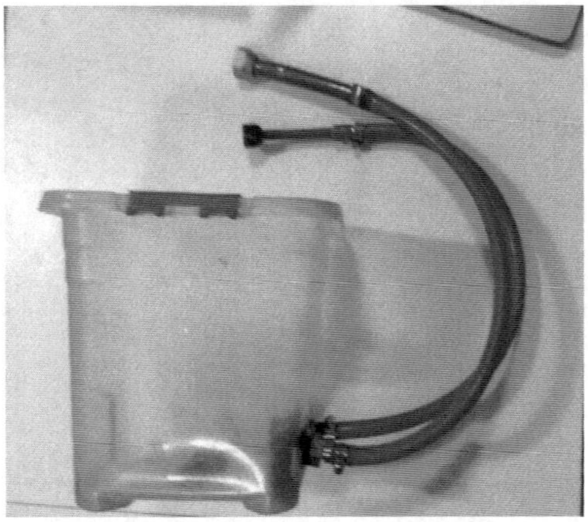

Figure 3. feeding mechanism

4. Results and discussion

4.1. Effect of solar radiation

The experimental Investigation covers several parameters that directly affected the solar still performance, such as solar radiation, ambient temperature, basin water, vapor, and glass cover temperatures. The first experiment was conducted using a 100% water filling ratio, and as it was noticed that the temperatures of the basin water and vapor rise as the solar radiation rises and continue rising until reaching its beak value at 01:30 PM, while the solar radiation reaches its maximum value at 11:30 AM. As in figure (4-6)

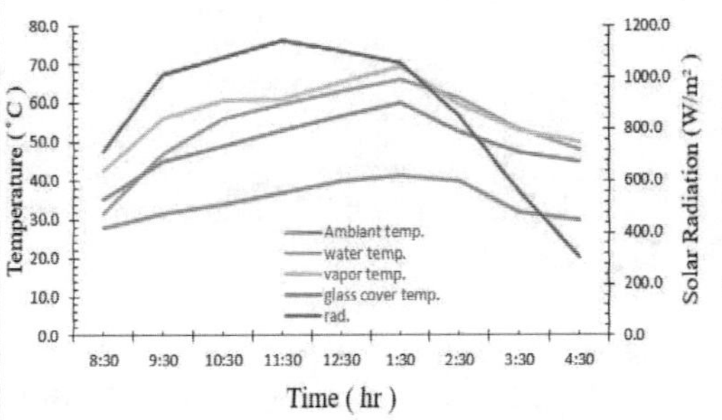

Figure 4. variation of temperature and solar radiation with time for conventional solar still

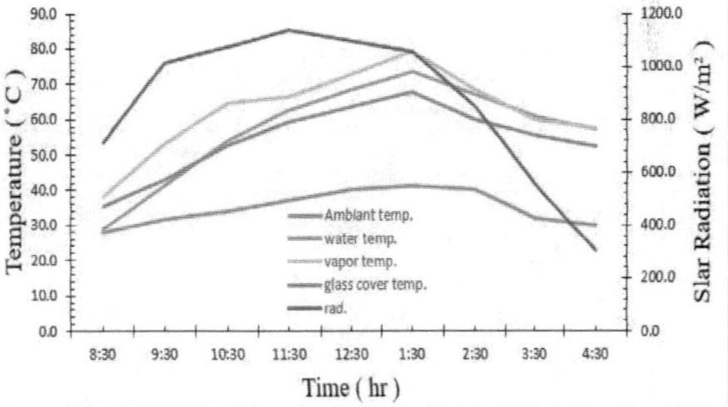

Figure 5. variation of temperature and solar radiation with time for solar still equipped with 15 mm diameter copper pipe

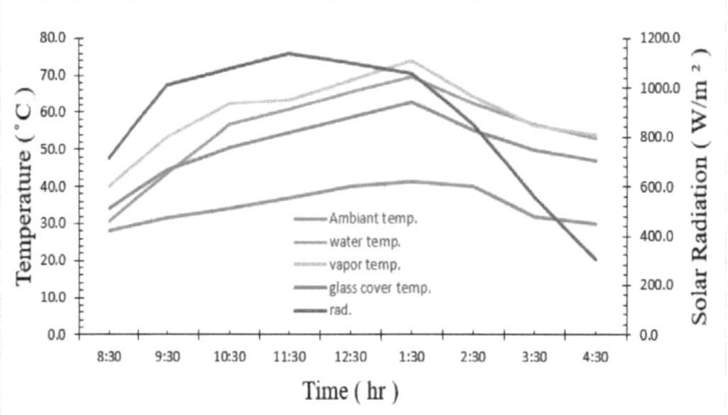

Figure 6. variation of temperature and solar radiation with time for solar still with 7 mm diameter copper pipe

Another experiment was also conducted for the same copper pipes with 50% water filling to investigate reducing the trapped water and increasing the evacuated area on the thermal performance. As shown in figures 7 and 8, both the vapor, water, and glass cover temperatures started to rise with the rise of radiation incidence, reaching its maximum value at 12:30 PM.

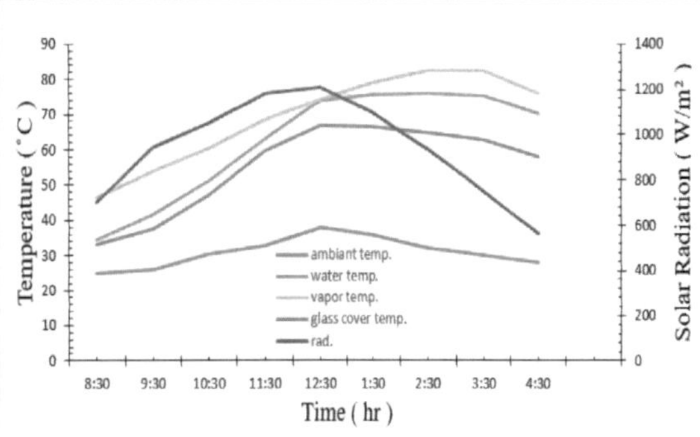

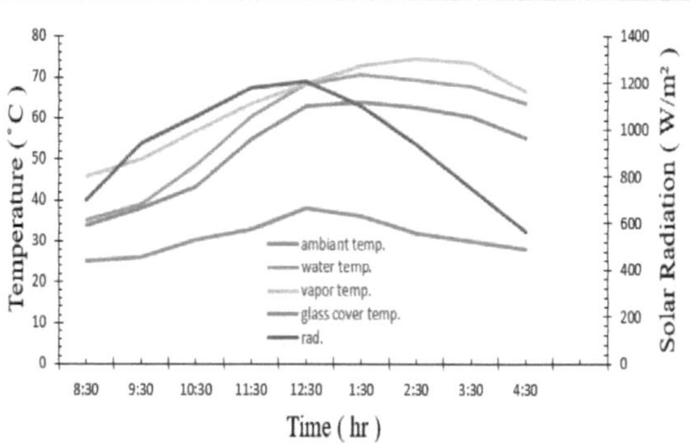

Figure 7. variation of temperature and solar radiation with measurement time for solar still with 15 mm diameter copper pipe filled with 50% water

Figure 8. variation of temperature and solar radiation with measurement time for solar still with 7 mm diameter copper pipe filled with 50% water

4.2. Effect of using evacuated pipes

Using evacuated copper pipes filled with 100% positively affects the thermal performance inside the solar still. It is shown in Figures 9 and 10 the temperatures for both vapor basin water of conventional solar still is slightly more in the first hour, due to the needed energy to be absorbed by the basin plate and the black painted copper pipe. After that, the temperatures of them both start to overcome vapor and water temperatures in the conventional solar still and reach the maximum value at 1:30 PM with a temperature difference of 9.8 °C and 4.6 °C for 15 mm, 7 mm pipes respectively. Simultaneously, basin water temperature differences were recorded as 7.5 °C and 3.7 °C for 15 and 7 mm pipes respectively. After that, the temperature starts to drop until the end of the experiment.

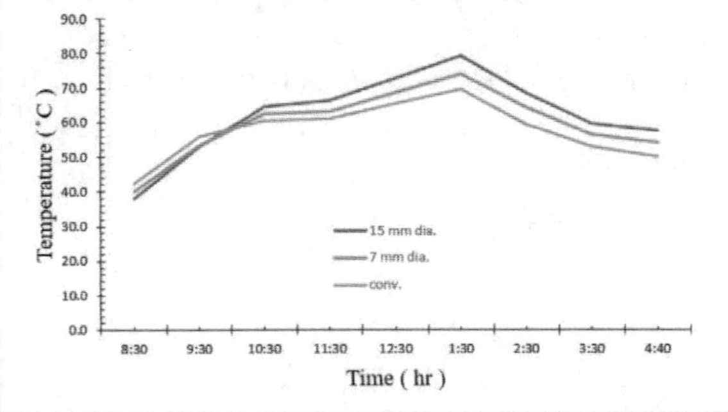

Figure 9. variation of vapor temperature with time of the experiment

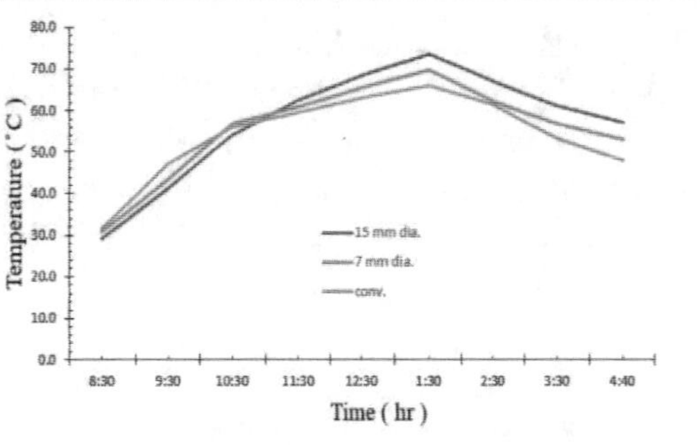

Figure 10. variation of basin water temperature with time of experiments

As for the 50% filing ratio, an experiment was conducted for the three solar stills at the same environmental conditions. Adding evacuated pipes was also positively affect the thermal performance of the solar still. Both vapor and basin water temperatures start to rise with the rise of solar radiation. The difference between stills equipped with 15 or 7 mm diameter with conventional solar still was quite a few at the beginning of the experiment. The temperature difference starts to grow with time due to the rise of incidence radiation and the partially filled pipes' additional effect. As shown in Figures 11 and 12.

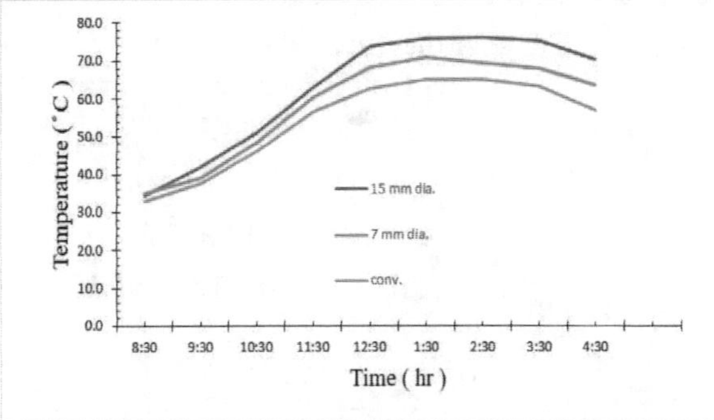

Figure 11. variation of basin water temperature with the time of the experiment

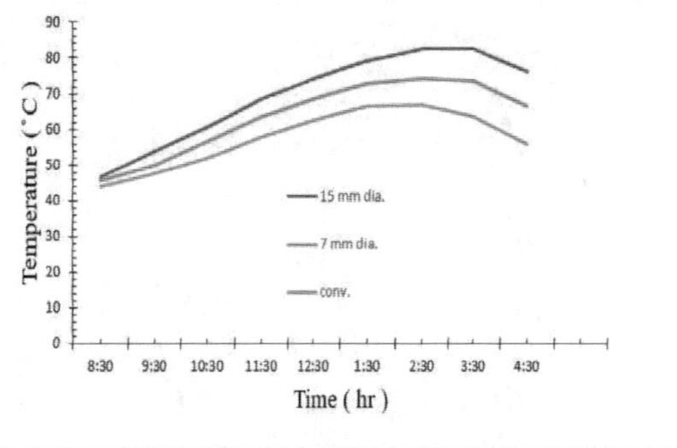

Figure 12. variation of vapor temperature with time of the experiment

4.3. Daily productivity

As expected, the accumulative freshwater productivity in both solar stills improved by 15 and 7 mm evacuated pipes was significantly enhanced during the experiment. The accumulative productivity in the still with 100% filling ratio was not at its best values at the first hours of the experiment due to the large amount of energy needed to raise the trapped water temperature inside the copper pipes. In comparison, it requires less energy when the filling rate is 50%. The accumulative pure water productivity rises along with the experiment time until reaching its maximum value at 1:30 and 2:30 for 100% and 50% filling ratio, respectively, as shown in figures 12 and 13.

Using evacuated pipes with a diameter of 15mm has more effect than using 7 mm pipes due to the increase in tarped water amount. The productivity was enhanced by 46.77% and 19.93% when using 15 and 7 mm pipes respectively. Moreover, as it is obverse in figures 13 and 14, using copper pipes with a 50% filing ratio has more effectiveness than the 100% filling ratio. Moreover, the accumulative pure water productivity records enhance by 90.09% and 27.83% for 15 and 7 mm pipes respectively, compared to the conventional solar still.

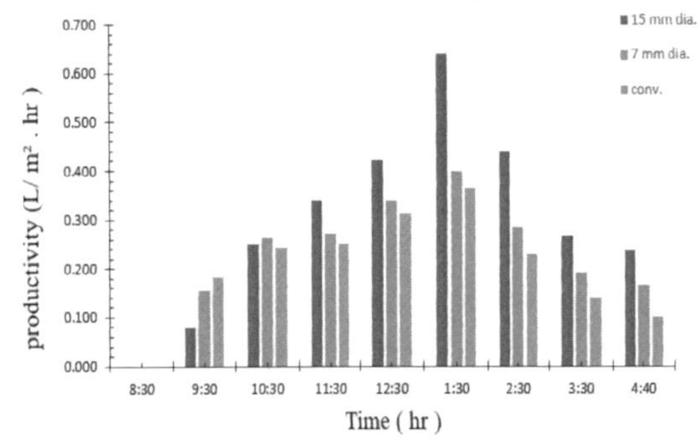

Figure 13. Productivity variation with time for the three types of stills at 100% filling ratio

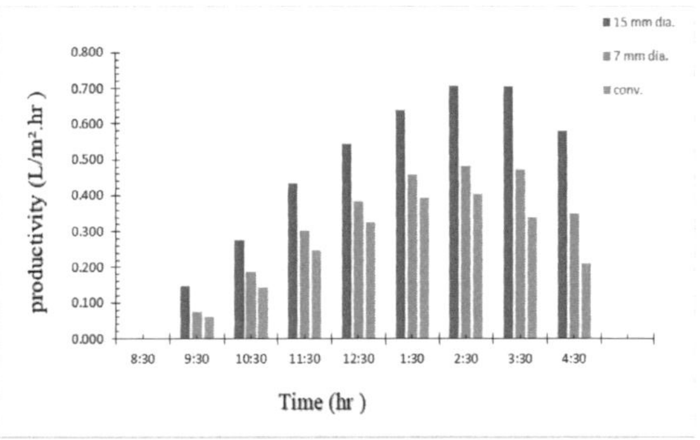

Figure 14. Productivity variation with time for the three types of stills at 50% filling ratio

Table 1: The accumulative pure water productivity for the three solar stills

Solar still with copper pipe	15 mm pipe	7 mm pipe	Convectional
100% filling	2.68 L/m^2.day	2.19 L/m^2.day	1.826 L/m^2.day
50% filling	4.03 L/m^2.day	2.71 L/m^2.day	2.12 L/m^2.day

5. Conclusion

The main parameter of this study focuses on improving the solar still thermal performance. To achieve that particular purpose, two different diameters of evacuated pipes filled different water levels. Two improved solar stills compared to a reference conventional still working under the same conditions. The experimental results that added evacuated pipes significantly affected the solar still thermal performance and pure water productivity. Based on the above discussion, some conclusions could be conducted from this study and as below.

1- Adding black painted evacuated pipes significantly affect the solar still productivity.
2- Increasing the evacuated pipe diameter enhancing the solar still productivity by 27.8% when using 7 mm pipe, while it was 90.09% using 15 mm pipe.

3- Reducing the water filling ratio, enhancing the solar still thermal performance and productivity, and the enhancement was improved from 46.77% up to 90.09% when using 100% and 50% filling ratios.

6. References

[1] Faegh M, Shafii MB. Experimental investigation of a solar still equipped with an external heat storage system using phase change materials and heat pipes. *Desalination* 2017; 409: 128–135.

[2] Hassanain Ghani Hameed. Enhancement of a Single – Slope Solar Still Productivity For a Different Water Depths. *Eng Sci Ref Sci J*; 20–2.

[3] Dumka P, Kushwah Y, Sharma A, et al. Comparative analysis and experimental evaluation of single slope solar still augmented with permanent magnets and conventional solar still. *Desalination* 2019; 459: 34–45.

[4] Bhargva M, Yadav A. Effect of shading and evaporative cooling of glass cover on the performance of evacuated tube - augmented solar still. *Environ Dev Sustain*. Epub ahead of print 2019. DOI: 10.1007/s10668-019-00375-8.

[5] Gnanaraj SJP, Ramachandran S, Christopher DS. Enhancing the productivity of double-slope single-basin solar still with internal and external modifications. *Int J Ambient Energy* 2018; 39: 777–782.

[6] Zahraa Abdulkareem Jaafar, Hassanain Ghani Hameed RHH. Jaafar_2020_IOP_Conf._Ser.__Mater._Sci._Eng._928_022096.pdf. *IOP Conf Ser Mater Sci Eng* 2020; 928 022096: 11.

[7] Abdullah AS, Essa FA, Omara ZM, et al. Rotating-drum solar still with enhanced evaporation and condensation techniques : Comprehensive study. *Energy Convers Manag* 2019; 199: 112024.

[8] Haddad Z, Chaker A, Rahmani A. Improving the basin type solar still performances using a vertical rotating wick. *Desalination* 2017; 418: 71–78.

[9] Kabeel AE, Omara ZM, Essa FA. Enhancement of modified solar still integrated with external condenser using nanofluids: An experimental approach. *Energy Convers Manag* 2014; 78: 493–498.

[10] Kaushal AK, Mittal MK, Gangacharyulu D. An experimental study of floating wick basin type vertical multiple effect diffusion solar still with waste heat recovery. *Desalination* 2017; 414: 35–45.

[11] Bhargva M, Yadav A. Productivity augmentation of single-slope solar still using evacuated tubes, heat exchanger, internal reflectors and external condenser. *Energy Sources, Part A Recover Util Environ Eff* 2019; 00: 1–21.

[12] Mu L, Xu X, Williams T, et al. Enhancing the performance of a single-basin single-slope solar still by using Fresnel lens: Experimental study. *J Clean Prod* 2019; 239: 118094.

[13] Omara ZM, Kabeel AE, Abdullah AS, et al. Experimental investigation of corrugated absorber solar still with wick and re fl ectors. *DES* 2016; 381: 111–116.

[14] Ramanathan V, Kanimozhi B, Bhojwani VK. Experimental study on productivity of modified single-basin solar still with a flat plate absorber. *IOP Conf Ser Mater Sci Eng*; 197. Epub ahead of print 2017. DOI: 10.1088/1757-899X/197/1/012032.

Experimental and Numerical Study of Thermal Performance for Flat Plate Solar Water Heater in Najaf

Mohammed J Alghurabe[1], Dhafer M Al-Shamkhee[2], Assaad A Alsahlani[2]
[1] Construction committee- Najaf Governorate, Iraq
[2] Engineering Technical College / Al-Furat Al-Awsat Technical University, Najaf, Iraq.
[1] E-mail : Mhd81@atu.edu.iq

Abstract. This work presented the numerical and experimental study of the thermal performance of a flat plate collector (FPC). This study focuses on analyzing the performance of (FPC) in the climatic conditions of Najaf and calculating the thermal energy produced by the collector for domestic use, which reduces the electricity consumption that Iraq is witnessing a severe shortage in its supply. Also, various working fluids (water, oil engine, ethylene glycol-water mixture) were tested to determine the best working fluid that improves the collector's efficiency. The experiments were performed in Najaf, Iraq (32° 2' N / 44° 18' E) on January 9, 2019. The simulation study of the (FPC) is performed by COMSOL Multiphasic 5.3 software. The numerical results were validated with experimental results and there was good convergence between them. The results showed that the average daily efficiency of the solar collector (FPC) was 37.17%, and the highest outlet water temperature of the collector was 57.1C°. The collector achieved a useful cumulative useful heat during the day of about 3.3557 MW, this contributes to reducing the use of electricity and achieving the required economic feasibility of use (FPC). Finally, the engine oil gave better results in improving efficiency compared to other working fluids.

1. Introduction

A solar water heater is a device that uses the thermal energy generated from the sun's rays in heating water for homes and commercial purposes. Solar heaters produce hot water with temperatures ranging from (40-97) C^{O}. And it has an advantage that they reduce dependence on electrical energy and reduce environmental pollution, as well as produce more energy than photovoltaic panels and achieved desired economic feasibility. The solar collector is designed to convert the solar radiation falling on the aperture area into heat in the absorber part and transfer heat from it by the working fluid [1,2]. The thermal solar system consists mainly of a collector, heat transfer system including (working fluid and circulation pump), and a storage tank.

There are several types of solar water heaters, but the most popular type is flat plate collector (FPC), Because it is cheap, can be manufactured locally, and is easy to install and maintain. It is used for both domestic and industrial applications that require low and medium temperatures [3][4]. The solar collector is affected by collector design, type of working fluid, and weather conditions. Therefore, many efforts are currently focused on improving (FPC) performance through various technologies. These studies include increasing the absorber's ability to capture solar radiation, reducing heat loss from the collector, enhancing the thermal properties of working fluids, and some studies focus on affecting the weather conditions in different locations on the collectors. Finally, many researchers studying the economic feasibility of using these collectors in the long term. In recent decades, There are several studies done on the thermal performance of (FPC), among them: Harrabi, et al. [5] studied numerical and experimental exergetic and energetic performances of (FPC) according to EN 12975 procedure. The suggested method allows measuring the effects of uncertainty in both thermophysical properties of the working fluid and operation conditions on estimating the performance of the (FPC). The analysis gave good agreement between the numerical and experimental results. The results indicated that the ambient temperature is the parameter most influencing the performance estimate and that a deviation of approximately ± 1C° may affect the energy efficiency of 13.7% and exergy efficiency of 3.9%. Rashid et al. [6] investigated numerically the effect of the use of rectangular channels and fins in (FPC). The study was conducted with flow rates ranging between (9-13) kg/s. the thermal performance analysis and flow characteristics show that a solar water heater with a fin was more efficient when it was without a fin. The results showed the heat transfer coefficient increases when the mass flow rate increase. It was proved that the performance of the collector improved with the increase in turbulence flow in the channel. And the use of the rectangular channel with the fins in the flow path of the collector improved the efficiency by 28% compared to without using the fins. Morrison and Braun [7] Searched in a comparative study of the solar water heaters working by thermosyphon using two models of tank-type: vertical and horizontal. The numerical simulation was inconsistent with the results obtained from two different places. This study showed that the thermosiphon system works well when it can prepare sufficient hot water for daily use. It has been proven that the horizontal tank system gave better results than the vertical tank system when they are working under the same weather conditions. Chow et al. [8] investigated the daily production of the solar water heater (SWH) from domestic hot water (DHW) in a virtual building in Hong Kong. by fixing central (SWH) in the west and south vertical facades of the building. The economic feasibility study focused on technical and cost terms. The simulation results showed that the thermal efficiency of (SWH) was 38.4%, the average annual temperature of the hot water was 41.4°C, The annual thermal energy extracted is estimated at 904GJ, and the payback period was 9.2 years, The payback period could be shorter if taking into account the reduction of energy consumed in cooling as a result of the solar collector blocking the solar radiation on the facades of the buildings. Hobbi and Siddiqui [9] studied the performance of flat plate collectors working to provide domestic hot water to a single family home in Montreal, Canada. The study was conducted by simulation using the TRNSYS software tool to analyze all major design parameters of the collector to determine its optimum value. Considering the optimization parameter was the solar fraction. The numerical results showed that when increasing the circulation flow leads to a rapid increase in solar fraction and collector efficiency. The system can produce 30–62% in winter and 83–97% in summer from hot water needs. Finally, the locally coated non-selective collector can save 54% of the annual water heating needs. Serale et al. [10] Studied improving the thermal performance of (FPC) produced limited temperature. The thermal improvement requires using latent heat stored from the slurry phase change material (PCS) Besides, the water, surfactants. The study suggested the integration of the solar thermal system with slurry

PCMs where the existing model system was introduced on n-eicosane PCS. The results showed the thermal and physical properties of the system increased compared to using a regular solar collector. Sami et al. [11] Evaluated the economic and energetic possibility of the integration of the (FPC) with houses that have high energy performance. In this study, four houses are chosen in different locations that have different weather in Algeria. The investigation of this study depends on using the F-Chart method and meteorological data for specific rejoin. The study focused on minimizing the area for (FPC) to reduces the installation cost versus energetic and economic aspects. The results explained the solar water heater contributed to saving energy about 57% and 46%in southern and northern regions respectively. And the saving in the annual operating cost of (FPC) reached 69% southern and 51% in northern regions respectively. Yousefi et al. [12] Studied improving the efficiency of (FPC) by using nanofluid (Al2O3-water) with working fluid. Nanofluid used in experiments has a mass concentration of 0.2% and 0.4 %. The nanoparticle has a diameter of 15 nm and the working fluid flow value is variable (1-3) L/min. Results explained the enhancement in the collector efficiency was 28.3% when using 0.2% wt Al2O3 nanofluid compared to if a water-based fluid was used, the maximum enhancement in efficiency with a surfactant was 15.63%. Catalin-george and Sebastian [13] studied numerically investigation efficiency of (FPC) in distinct weather in Romania. The effect of the slope angle on the efficiency of a collector in all locations is examined with flow rate (24, 36, 48) kg/h. The numerical results revealed that the better slope angle of (FPC) was 45° in different values of flow rate in all locations, and the difference between inlet and outlet water temperature of (FPC) ranged from 10 ^oC to 15 ^oC. Ranjith and Karim [14] investigated experimentally and numerically impact of testing propylene glycol-water mixture in (FPC) as working fluid in different volume concentrations (25, 50, 75, 100%) of propylene glycol (PG) with varied flow rates values (0.008, 0.0167,0.024) kg/s. the results indicated the collector efficiency decrease when using propylene glycol–water in a mass flow rate of 0.0167 kg/s at 25% PG, and efficiency of collector increase when (PG) volume concentration was 25% to 50%, While at volume concentration from 50% to 100% at a mass flow rate of 0.024 kg /s, the efficiency increased by 5% and 14%, respectively. Saw and Owolabi [15] evaluated experimentally the efficiency enhancement of (FPC) by use phase change material PCM and nanoparticle. The collector was installed with a tilt angle of 10° and had a mass flow rate of 0.5 kg/min. The experimental result showed when using PCM and (Nano-PCM) the outlet water temperature of the collector was 40.2ºC and 40.8ºC respectively, and without the PCM state, the temperature of outlet water of the collector was 35.2ºC. The enhancement in solar collector efficiency reached 6.9% and 8.4% when used PCM and nano-PCM respectively.

The purpose of the research is to study the experimental and numerical performance of (FPC) in Najaf city and calculate the efficiency and thermal energy produced To reduce dependence on electrical energy, which is witnessing a severe shortage of supply in Iraq. Also, several types of working fluids have been tested to improve efficiency.

2. The experimental Work:

2.1. Description of the system

The thermal performance of solar heaters is tested in several ways, depending on the purpose of using the hot water produced. The study will focus only on evaluating the thermal performance of the collector within the heating system. Figure (1) shows the solar heating system to heat the room. The system consists of several parts and the main part is the solar collector (FPC) and it has the specifications shown in Table 1. (FPC) is installed above the college building in the south direction and a tilt angle of 30° as shown in Figure 2.

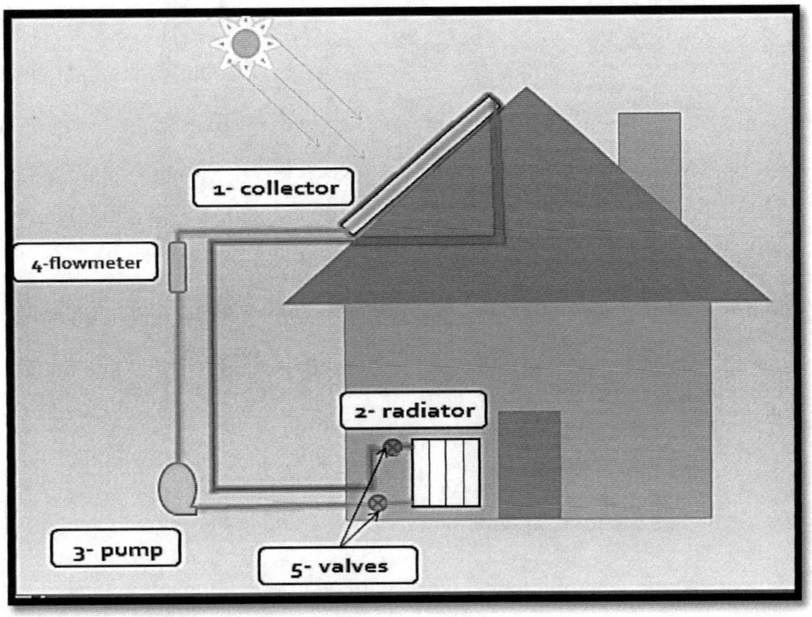

Figure 1. Scheme of the experimental rig

Table 1. The solar collector specifications

Specification	details
Dimension of collector	(2000x1000x80) mm
The thickness of the glass cover	3.2 mm
Glass transmittance	95%
The thickness of the aluminum Absorber plate	0.4 mm
Absorber plate absorptivity	95%
The emissivity of absorber	5%
Copper header tube diameter	22mm / 2 headers
Copper riser tube diameter	8mm / 7 risers
Glass wool insulation thickness	50 mm
The air gap between absorber and glass	25mm

Figure 2. Flat plate collector

The indoor radiator is fixed on the wall inside the laboratory with dimensions (100x60) cm, as shown in Figure (3). The (FPC) is connected to the radiator by insulated tubes. A flowmeter is installed to measure the flow of working fluids entering the collector. The flow rate is controlled by a set of valves and the heating system operating by forced convection in a closed system.

Figure 3. Indoor radiator

A data logger has (7) thermocouples type (T) fixed in different places on the collector: two thermocouples fixed at inlet and outlet pipes of a collector, two thermocouples fixed on the back and edge of the (FPC), and three thermocouples fixed on different locations of the glass cover. The solar irradiance is measured by the solar radiation sensor and ambient temperature and wind speed measured from the Meteorological station in the laboratory. as shown in Figure (4).

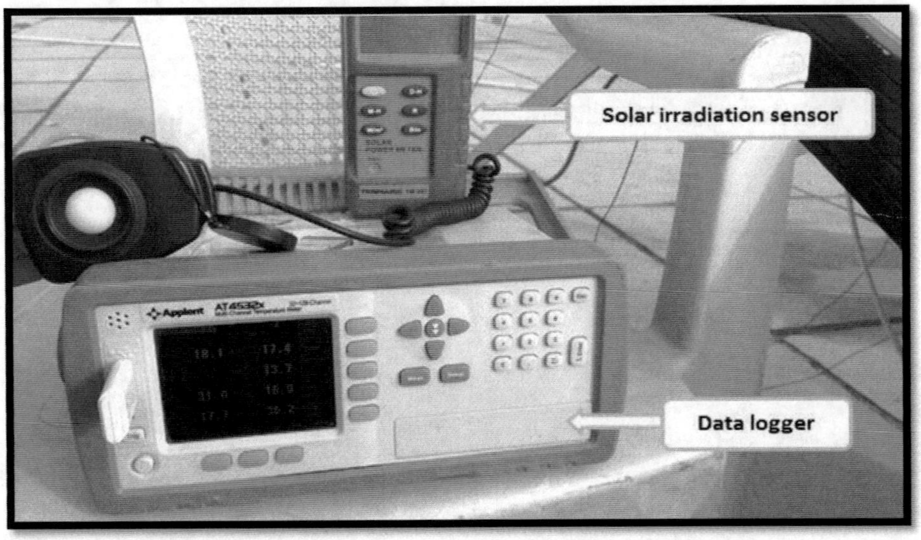

Figure 4. Showed the solar irradiance sensor and data logger

2.2. Test procedures

The solar system is investigated empirically at Engineering Technical college/ Al-Furat Al-Awsat University in Najaf, Iraq (32° 2' N / 44° 18' E). The solar collector (FPC) is tilted at an angle of 30^0 facing the south direction. Tests were conducted on January 9, 2019, For the period 9:00 AM to 5:00 PM. After sunrise, the (FPC) gradually heats the working fluid due to sunlight falling on it. Then the electric pump circulates the working fluid from the collector to the radiator inside the room, causing the radiator temperature to rise. Then heat transfer occurs between the radiator surface and the room air by free convection, this will warm the room gradually. Then the working fluid returns from the radiator to the solar collector to be heated again. The working fluid flows inside the system at a constant volume flow rate of 60 L/ h.

3. System Modeling and Simulation:

Flat plate collector studied by 3-D numerical investigations conducted by (CFD) approach and governing equations were analyzed by COMSOL Multiphasic 5.3 software tool. And the numerical analysis was done by the finite element method. The study took into account many processes: the model exposed to the solar irradiance causing heat generation in the absorber plate, convection heat transfer and radiation occur between side walls and glass cover with the surrounding, radiation heat transfer between the absorber plate and glass cover is considered, and convection heat transfer between an absorber plate and air gab. The solar energy absorbed was simulated by taking data of weather conditions for Najaf city on January 9, 2019, which was recorded by a weather station in the laboratory. Data recorded included, wind speed, solar irradiance, and ambient temperature. The mass flow rate inside the system was 60L/hr. An inlet water temperature of collector data taken from experimental data.

3.1. The domain of the computational simulation

Transient conditions simulations are performed on a flat plate collector that has an area of a (2 m^2). Due to the large difference in the dimensions of the pipe diameter (0.008 m) and the length of the absorber tube (1.85 m), an excellent mesh distribution is required in the tube cross-section. There are assumptions that must be taken into account for the simplified simulation model as follows:

- The heat loss by radiation in the margins of the collector is ignored.

- Neglecting heat loss at the bottom base of the collector.

- The working fluid flow remains constant in the collector and is divided equally in all riser tubes.

- The field of the flow is symmetric to the (y – z) plane, And a half-section of elements (tube, absorber plate, air gap, and glass cover) is considered for the numerical study. The domain of the computational simulation is shown in Figure (5).

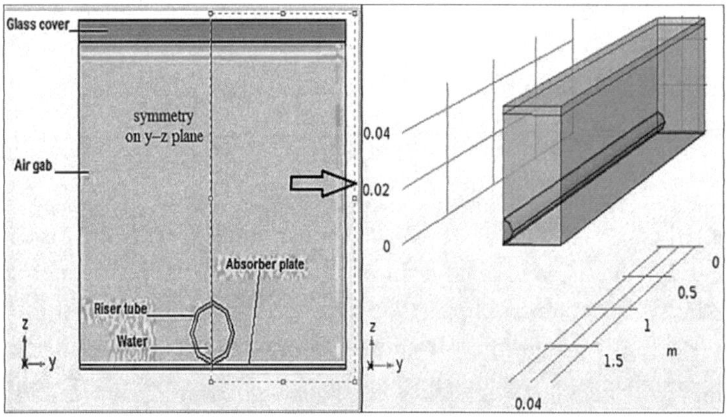

Figure 5. Showed a schematic of the CFD model

One of the requirements for a numerical solution is to know the boundary conditions of the model, which are:

1- In the inlet tube:

a) $\int \rho . U_{in} dA$ (1)
When m˙=(0.016667/7)kg/s

b) $T_{in} = T_{in(t)}$ as shown in Fig (10) (2)

2- In the outlet tube:

a) P_0=0 (gauge pressure) (3)

b) $\frac{\partial T}{\partial x} = 0$ (4)

3- At the wall, a non-slip condition was imposed on the walls U = 0

4- The symmetry is taken on the plan in the middle of the tube.

$u_n = 0$, normal component of velocity equal zero (5)

5- Initial temperature $T_i = T_{am}$ (9) at 9:00 am (6)

6- Convective heat flux on outer glass cover

$$Q_c = h_c\left(T_g - T_a(t)\right) \quad (7)$$

$$h_c = 5.7 + 3.8V_{out}(t) \quad (8)$$

Where $V_{out}(t)$ wind speed which is changing with time taking values from the weather station.

7- Heat absorber by glass

$$Q_g = A_c.G(t).\alpha \quad \text{for glass } \alpha = 0.05 \quad (9)$$

$A_{c=Aparture}$ collector area

8- Heat absorbed by the absorber plate

$$Q_p = A_c.G(t).\alpha_p \quad \text{for plate } \alpha_p = 0.95 \quad (10)$$

9- Insulation wall $Q_n=0$ (11)

3.2. The Governing Differential Equations

Governing differential equations describe the thermo-physical behavior of many heat and mass transfer processes that occurs at (FPC). The equations of the fluid flow and heat transfer in the system are considered. The assumptions are taken into account: incompressible, laminar, steady flow, And three-dimensional, The physical properties will be considered constant, viscous Newtonian flow within the (FPC) are governed by the equations of continuity, momentum, and energy [16]. the continuity equation for flow inside the tube can be written as:

$$\frac{\partial u}{\partial x} + \frac{\partial v}{\partial y} + \frac{\partial w}{\partial z} = 0 \quad (12)$$

The momentum equations can be written in the x, y, and z directions as follows:

$$\frac{\partial u}{\partial t} + \frac{\rho u \partial u}{\partial x} + \frac{\rho v \partial u}{\partial y} + \frac{\rho w \partial u}{\partial z} = -\frac{\partial P}{\partial x} + \mu\left[\frac{\partial^2 u}{\partial x^2} + \frac{\partial^2 u}{\partial y^2} + \frac{\partial^2 u}{\partial z^2}\right] \quad (13)$$

$$\frac{\partial v}{\partial t} + \frac{\rho u \partial v}{\partial x} + \frac{\rho v \partial v}{\partial y} + \frac{\rho w \partial v}{\partial z} = -\frac{\partial P}{\partial y} + \mu\left[\frac{\partial^2 v}{\partial x^2} + \frac{\partial^2 v}{\partial y^2} + \frac{\partial^2 v}{\partial z^2}\right] \quad (14)$$

$$\frac{\partial w}{\partial t}+\frac{\rho u \partial w}{\partial x}+\frac{\rho v \partial w}{\partial y}+\frac{\rho w \partial w}{\partial z}=-\frac{\partial P}{\partial z}+\mu\left[\frac{\partial^2 w}{\partial x^2}+\frac{\partial^2 w}{\partial y^2}+\frac{\partial^2 w}{\partial z^2}\right]-\rho g z \tag{15}$$

The energy equations for fluid and the structure of (FPC) are as following:

$$\rho Cp\left(\frac{\partial T}{\partial t}+u\frac{\partial T}{\partial x}+v\frac{\partial T}{\partial y}+w\frac{\partial T}{\partial z}\right)=k\left[\frac{\partial^2 T}{\partial x^2}+\frac{\partial^2 T}{\partial y^2}+\frac{\partial^2 T}{\partial z^2}\right] \tag{16}$$

The efficiency of the collector is the useful energy of (FPC) to the solar energy incident on the aperture area of a collector as follows.

$$\eta=\frac{Q_u}{I_T A_c} \tag{17}$$

$$Q_u=A_c F_R\left[S-U_L\left(T_{pm}-T_a\right)\right] \quad \text{A. Kalogirou [17]} \tag{18}$$

3.3. Meshing

The process of creating the mesh is critical in the numerical simulation. In this work, the grid was generated by using COMSOL Multiphysics 5.3 software. The type of the mesh works directly on the accuracy of the numerical result. The grid was made a successively finer type and another fine type at locations where it was needed, as shown in Figure (6).

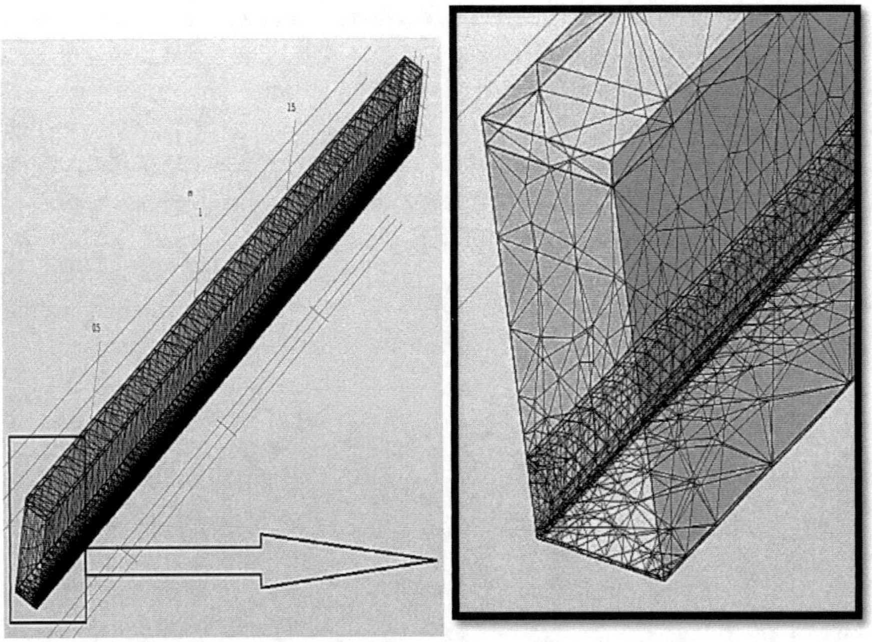

Figure 6. The domain of the computational simulation and grid setup

To choose the appropriate mesh type in the solution of the model, eight numbers of grid points were selected to check the values change of the temperature differences between the outlet and the inlet. Figure (7) shows that the region of meshes starting from 150,000 to 450,000 has no significant effect on the temperature difference. Therefore grids generation with 150, 245 was considered an acceptable computational field.

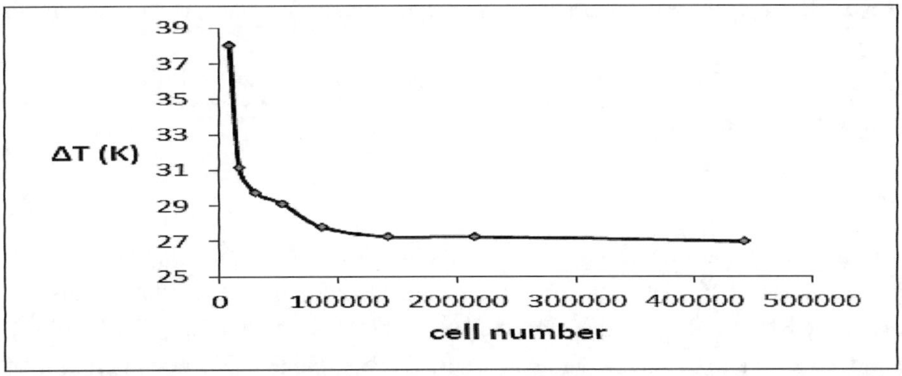

Figure 7. Showed temperature difference of water between inlet and outlet and grid number

4. Validation Model:

Ekramian et al. [18] investigated a numerical study of the same solar collector model that Cruz-Peragon et al. [19] used in his experimental work for the same parameters and conditions. The study performed on the (FPC) contains 15 riser tubes. The thickness of the absorber plate and glass cover are

2 mm and 4 mm, respectively. The distance between one tube and another was 30 mm and the inner diameter of the tube was 10 mm, and the tube's length of 450 mm. A flat plate collector examined under constant operation condition: Solar irradiance was 936.8 W/m^2, ambient temperature, Inlet water temperature, and water flow rate are, 23.2 ˚C31˚C and 6.42 kg/h respectively. So a single riser was considered and the average temperature of the fluid in four sections was taken through the simulation. The numerical results of E. Ekramian explained that the average relative error between the experimental and numerical results is about 5.5%. While in our work, a simulation study was performed on (FPC) and under the same operating conditions in the work of Cruz-Peragon et al. The simulation results are highly accurate and close to the experimental results, and the mean relative error ratio of the experiment to the numerical results was about 0.062%.

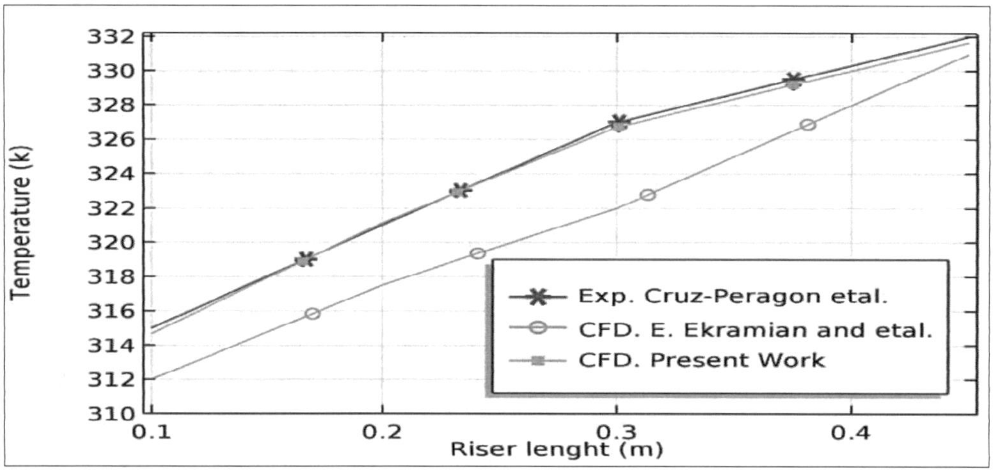

Figure 8. showed the comparison between numerical and experimental temperature

5. The results and Discussions

5.1 The experimental Results:

The experimental work was conducted in period 9:00 AM to 5:00 PM on January 9, 2019, in Najaf, the weather was clear sky, relatively cold, and solar irradiance in intermediate limits. The ambient temperature ranged from 9.8C° in the morning to reach a maximum value 17C° at 4:00 pm. Solar radiation starts at low rates at sunrise and reaches the highest value at the noon period then returns to decrease in the afternoon· as shown in Figure (9). The inlet water temperature of (FPC) was constantly changing due to the flow of water in a closed system as it increased by a certain amount in each cycle, The wind speed was high most of the time, ranging between (1.8 to 4) m/s, as shown in Figure (10).

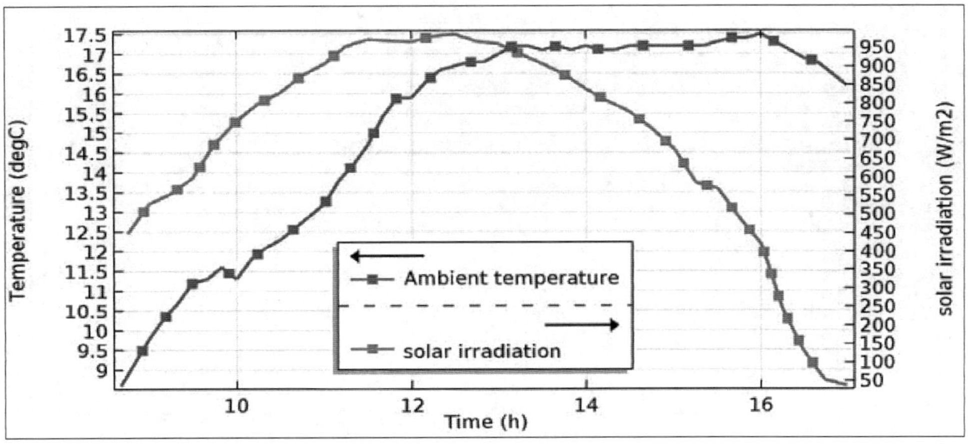

Figure 9. showed hourly variation in ambient temperature and solar irradiance

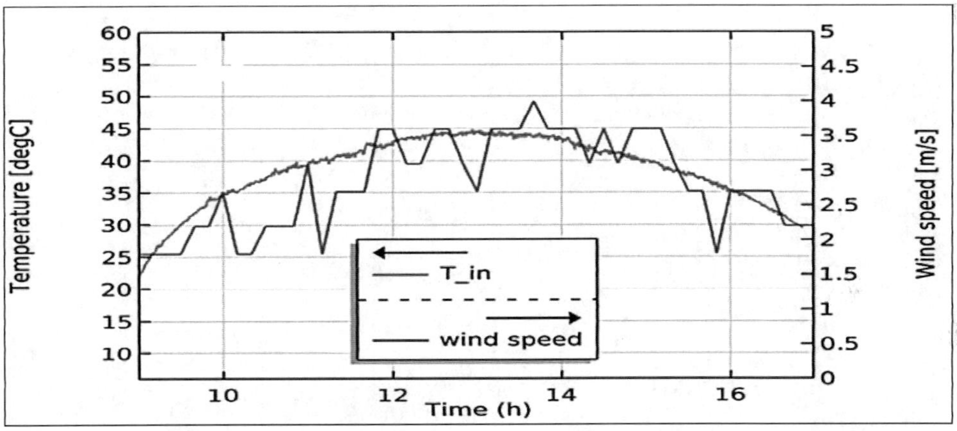

Figure 10. showed the inlet fluid temperature and wind speed

The experiments were conducted at a flow rate of 60 L/hr. Figure(11) showed the change in the outlet water temperature (T_{out}) during the period 9:00 am to 5:00 pm, An analogy is observed between the change in (T_{out}) and the solar irradiance and noted (T_{out}) increases as the intensity of solar radiation increases. (T_{out}) starts with low values at the beginning of sunrise and then gradually increases to reach the highest value of 57.1 ^{o}C at 12:58 pm. In the afternoon the values of (T_{out}) begin to decrease, reaching the lowest value at sunset. Figure (12) showed the temperature difference between the outlet and inlet water of the collector (ΔT). Temperature difference indicates the ability of the collector to heat water and allows comparison between other types of collectors under the same conditions and specifications. The collector efficiency increases as the temperature difference increases. It noted the change in (ΔT) values is almost similar to the change in (T_{out}) and the maximum of (ΔT) reached 12.5 ^{o}C.

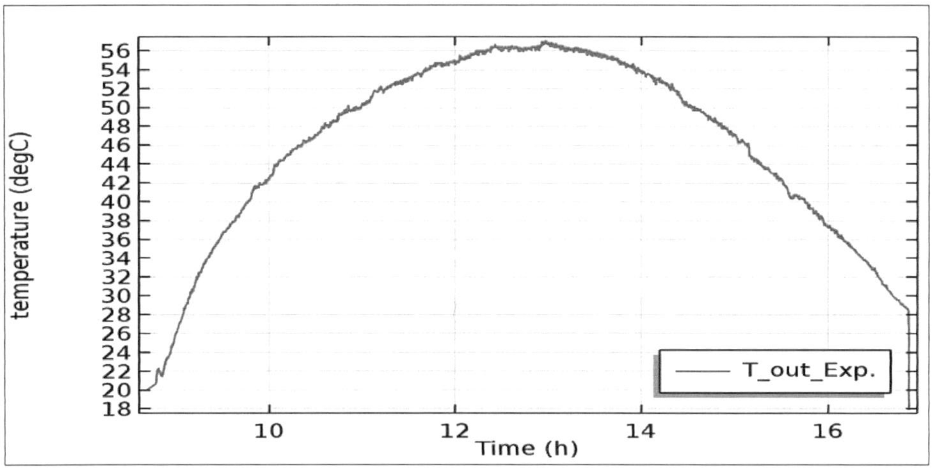

Figure 11. showed the time variation for the outlet temperature of the collector(T_{out})

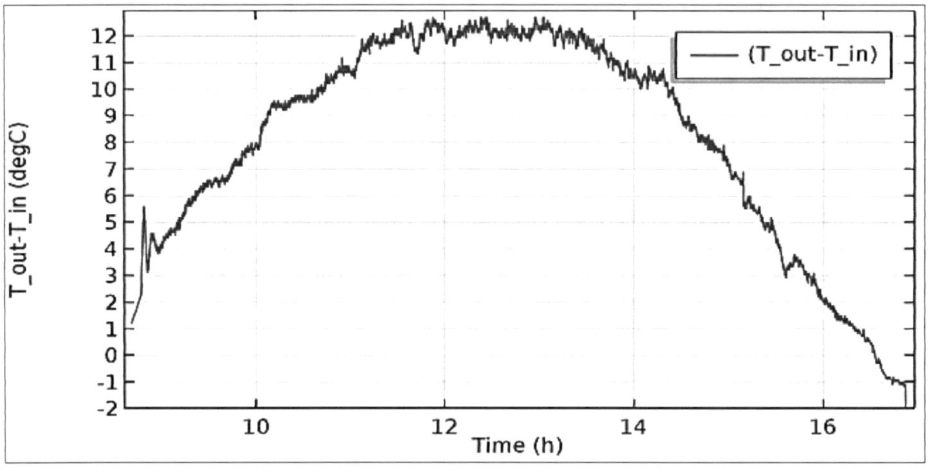

Figure 12. showed the time variation for the difference between outlet and inlet temperature of the collector (ΔT)

The heat obtained from (FPC) is called useful heat (Q_u) which is the heat that is extracted from the collector by passing the working fluid through the absorbent part. The useful heat values are low at the beginning of the sunrise and gradually increase to reach the peak value of 886W at midday then its values gradually decrease in the afternoon until reach the lowest value at sunset as shown in Figure (13). The results indicate that in the period from 10 am to 3 pm, the solar collector can cover all or a large part of the required heat while at other times the values are much lower than the required heat.

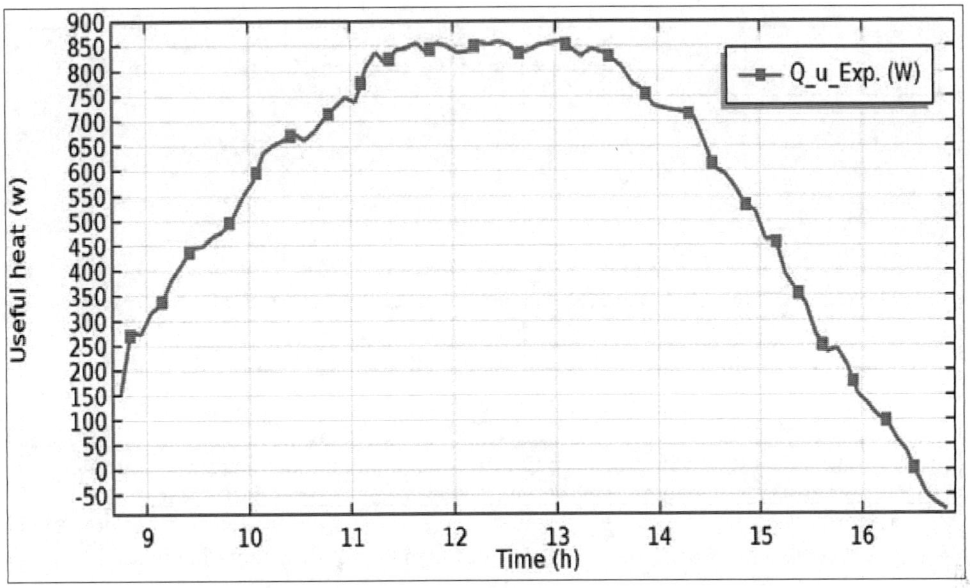

Figure 13. showed the time variation of the useful heat of the collector (Q_u).

5.1.1 The effect of testing different working fluids on collector efficiency:

The type of working fluid is one of the influencing factors to improve collector efficiency because they have different thermal properties. Experiments were carried out on the water on January 9, 2019, (ethylene glycol –water) mixture (50-50)% on January 10, 2019, and oil engine grade (10W-30) on January 12, 2019. The results showed that the engine oil gave better results in enhancement efficiency than other liquids as shown in Figure (14). The overall daily efficiency of water, (ethylene glycol-water) mixture, and oil engine were 42.73%, 48.31%, and 64% respectively. The working fluids tests showed variation in collector efficiency is due to the difference in physical properties (density, viscosity, thermal conductivity, and specific heat) and noted that the most important physical property affecting the increase in heat transfer is specific heat. As the engine oil, has the lowest specific heat of 1.99[kJ/kg.c] while the specific heat of ethylene-glycol-water mixture and water were 3.686[kJ/kg.c] and 4180[kJ/kg.c] respectively. The heat transfer increases as the specific heat of the working fluid are low and vice versa, This gives preference to engine oil in increasing heat exchange and extracting the maximum amount of useful heat from the collector. Although the engine oil gave the highest efficiency of the collector, it has disadvantages represented by its high viscosity, which causes the pressure drop and needs more energy to pump it than other liquids.

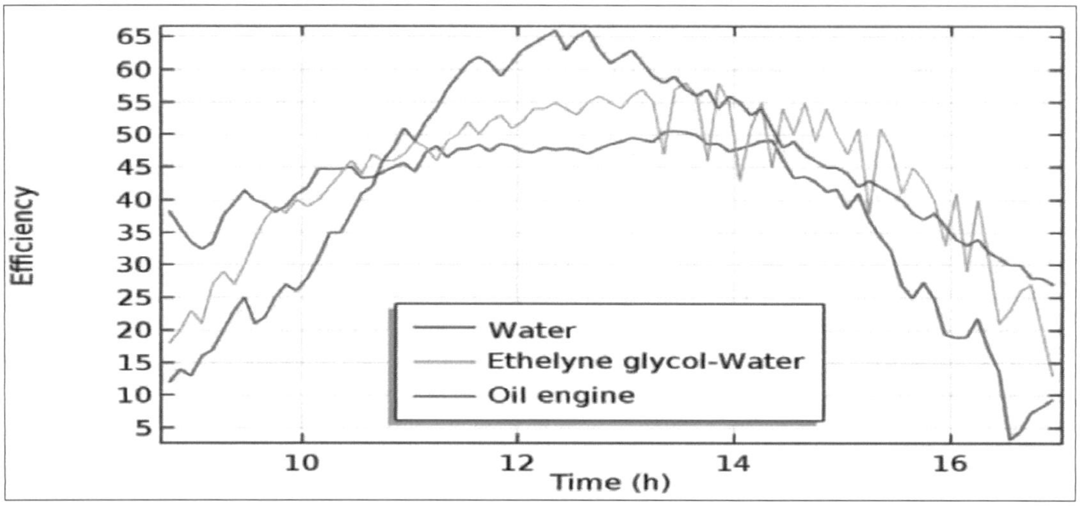

Figure 14. showed the time variation of the collector efficiency for using different types of working fluid

5.3 Numerical Results:

The performance of a solar collector is investigated numerically and the analysis of the simulation model is presented. Figure(15) Shows the experimental and numerical outlet water temperatures of the collector (T_{out}). It was observed that the variation of the experimental and numerical values of (T_{out}) was similar during most of the test period, and the two curves were almost identical. and this indicates the high accuracy in solving the numerical study and the relative error of experimental to numerical temperatures was less than 0.42% as shown in Figure (16). The temperature contour of the section from a flat plate collector contains one riser tube shown in Figure (17).

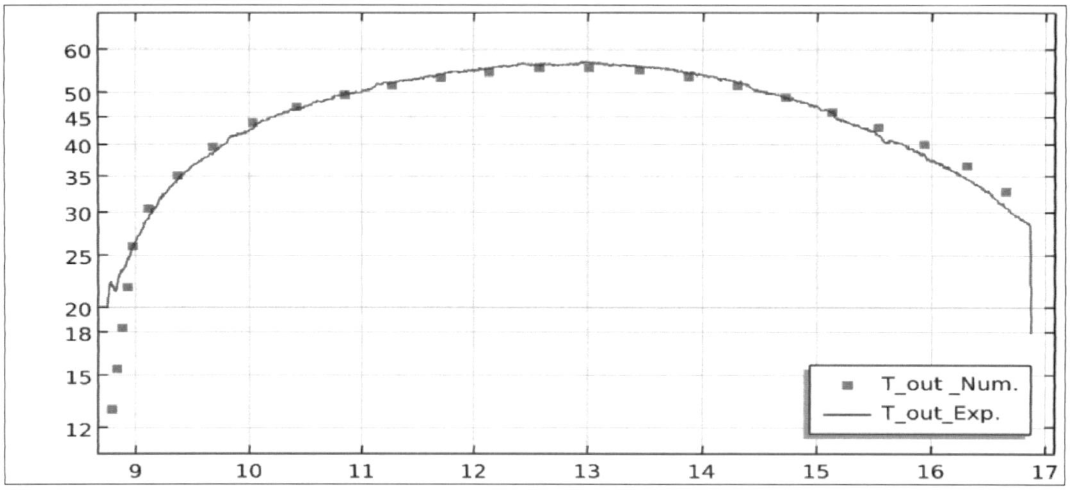

Figure 15. hourly variation for numerical an experimental (T_{out}) of the collector

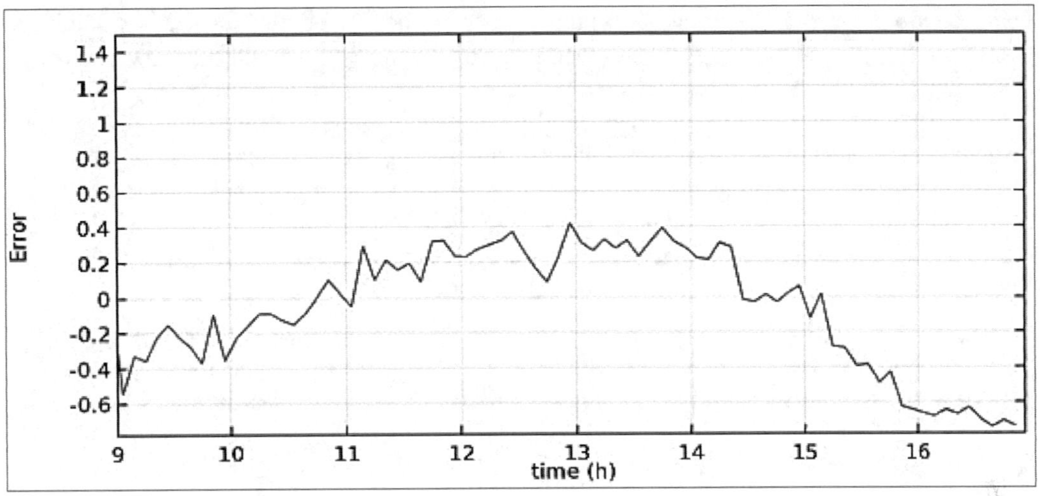

Figure 16. show relative error between numerical and experimental temperatures readings

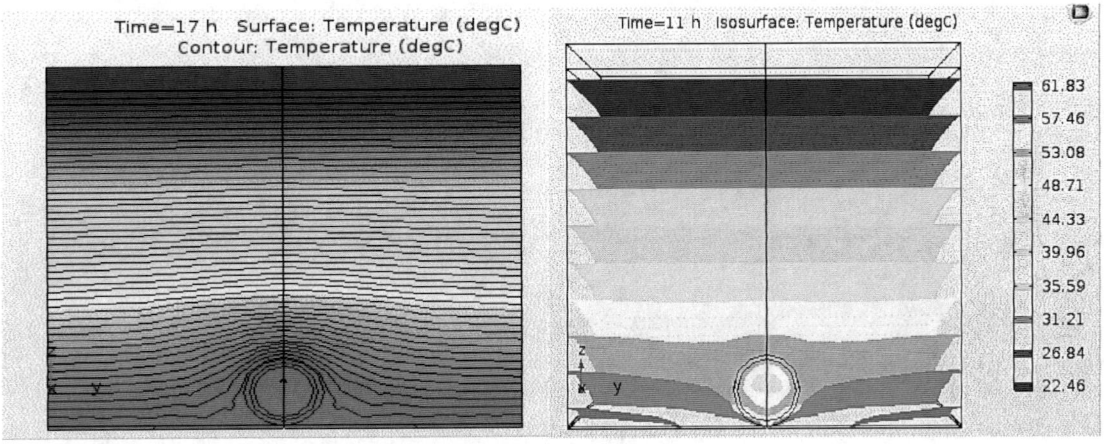

Figure 17. Temperature contour

6. Conclusion:

A flat plate collector (FPC) is investigated numerically and experimentally in cold weather conditions of Najaf city, Iraq. The heat generated by (FPC) used to heat buildings. The results indicated a good convergence between the experimental and numerical results. And the maximum relative error ratio between numerical experimental results was 0.42%. The maximum (T_{out}) of the collector was 57.1 °C, the maximum value for the instantaneous efficiency of the collector reached 50.44%, While the overall daily collector efficiency was 37.17%. It was concluded that a flat plate collector test achieved good results in Najaf city, which has relatively high solar radiation rates and achieving a daily cumulative amount of useful heat of 3.3557MW. This energy contributes to reducing electricity consumption and reducing environmental pollution and working to provide hot water used in heating buildings and domestic use. Finally, the engine oil gave better results in improving efficiency compared to other working fluids.

Keywords: Experimental work, Flat plate collector, Numerical model, Working fluids.

Nomenclature

Q_u *Useful work* (W/m^2).

I_T *Solar irradiation* (W/m^2).

η *Efficiency of the collector.*

A_c *Aperture area of the collector* (m^2).

T_{in} *Temperature of inlet water (°C).*

ρ *Working fluid density* (kg/m^3).

T_{out} *Temperature of outlet water (°C).*

S *Solar radiation absorbed by (FPC)* (W/m^2)

U_L *The overall coefficient of heat loss on the collector area* $(W/(m^2\,K))$.

U_{in} *Inlet working fluid velocity (m/s)*

Q_c *Convection heat transfer* (W)

T_{pm} *Temperature of the plate (°C).*

T_a *Ambient temperature (°C).*

T_g *Glass cover temperature (°C).*

α *Absorption coefficient*

h_c *Convection heat transfer coefficient* $(\frac{w}{m^2.°C})$

V_{out} *Wind speed (m/s)*

F_R *the ratio of the actual useful energy gain that would result if the collector-absorbing surface had been at the local fluid temperature*

Acknowledgments

The authors thank Najaf Governorate / Reconstruction Committee and Engineering Technical College Najaf /Al-Furat Al-Awsat Technical University, for their support in our efforts in accomplishing this research.

References

[1] M. Mwanza, J. Chachak, N. S. Çetin, and K. Ülgen, "Assessment of Solar Energy Source Distribution and Potential in," vol. 5, no. 2, pp. 103–116, 2017.

[2] A. M. Alsayah, M. Hatf, K. Aboaltabooq, M. H. Majeed, and B. A. Saheb, "CFD study to improve PV cell performance by forced air : Modern design," vol. 7, no. 3, pp. 1468–1477, 2019.

[3] I. E. Maree and A. G. Ismaeel, "Experimental and Theoretical Calculation of Efficiency for Flat Plate Solar Collectors in Erbil City," vol. 7, no. 2, pp. 786–793, 2019.

[4] A. B. O, S. W. D, S. T. M, B. A. A, and S. A. O, "Adapting Television Satellite Dish for Low-Temperature Solar Thermal Application : Experimental Evaluation," vol. 6, no. 2, pp. 168–178, 2018.

[5] I. Harrabi, M. Hamdi, and M. Hazami, "Assessment of Uncertainties in Energetic and Exergetic Performances of a Flat Plate Solar Water Heater Assessment of Uncertainties in Energetic and Exergetic Performances of a Flat Plate Solar Water Heater," no. December 2020.

[6] F. Rashid, M. R. I. Sarker, S. S. Tuly, J. Ferdous, and R. A. Beg, "Numerical Study of a Stand-alone Flat Plate Solar Water Heater using Rectangular Flow Channel with Fin Numerical Study of a Stand-alone Flat Plate Solar Water Heater using Rectangular Flow Channel with Fin," no. August 2019.

[7] G. L. Morrison and J. E. Braun, "System modeling and operation characteristics of thermosyphon solar water heaters," *Sol. Energy*, vol. 34, no. 4–5, pp. 389–405, 1985.

[8] T. T. Chow, K. F. Fong, A. L. S. Chan, and Z. Lin, "Potential application of a centralized solar water-heating system for a high-rise residential building in Hong Kong," *Appl. Energy*, vol. 83, no. 1, pp. 42–54, 2006.

[9] A. Hobbi and K. Siddiqui, "Optimal design of a forced circulation solar water heating system for a residential unit in cold climate using TRNSYS," *Sol. Energy*, vol. 83, no. 5, pp. 700–714, 2009.

[10] G. Serale, Y. Cascone, A. Capozzoli, E. Fabrizio, and M. Perino, "Potentialities of a low temperature solar heating system based on slurry phase change materials (PCS)," *Energy Procedia*, vol. 62, pp. 355–363, 2014.

[11] S. Sami, D. Semmar, A. Hamid, R. Mecheri, and M. Yaiche, "Viability of integrating Solar Water Heating systems into High Energy Performance housing in Algeria," *Energy*, vol. 149, pp. 354–363, 2018.

[12] T. Yousefi, F. Veysi, E. Shojaeizadeh, and S. Zinadini, "An experimental investigation on the effect of Al2O3-H2O nanofluid on the efficiency of flat-plate solar collectors," *Renew. Energy*, vol. 39, no. 1, pp. 293–298, 2012.

[13] P. Catalin-George and H. Sebastian, "Numerical Study of the Efficiency of Flat Plate Solar Collectors in Different External Conditions," vol. 659, pp. 435–439, 2014.

[14] P. V Ranjith and A. A. Karim, "A Comparative Study on the Experimental and Computational Analysis of Solar Flat Plate Collector using an Alternate Working Fluid," *Procedia Technol.*, vol. 24, pp. 546–553, 2016.

[15] C. L. Saw and A. L. Owolabi, "Experimental investigation on the effect of PCM and nano-enhanced PCM of integrated solar collector performance," vol. 179, pp. 899–909.

[16] M. Potter and D. C. Wiggert, *Fluid Mechanics - Schaum's Outlines*. 2008.

[17] S. Edition, *Solar Energy Engineering Processes and Systems Second Edition.*

[18] E. Ekramian, S. G. Etemad, and M. Haghshenasfard, "Numerical Analysis of Heat Transfer Performance of Flat Plate Solar Collectors," vol. 1, 2014.

[19] F. Cruz-Peragon, J. M. Palomar, P. J. Casanova, M. P. Dorado, and F. Manzano-Agugliaro, "Characterization of solar flat plate collectors," *Renew. Sustain. Energy Rev.*, vol. 16, no. 3, pp. 1709–1720, 2012.

Use of Thermostone Waste Aggregates for Internal Curing of Reactive Powder Concrete

Mena A. Gawad*[1], Nada M. Fawzi [1]
[1] Civil Engineering department, College of Engineering, University of Baghdad, Baghdad, Iraq
nada.aljalawi@coeng.uobaghdad.edu.iq

Abstract. The concrete need curing for cement hydration that is a chemical reaction in each step require water supply throughout the time period. The traditional concrete cured by external method that prevents the concrete surface dry so that keeping the concrete mixture wet and warm. The internal curing was adopted in normal and high strength concrete such as reactive powder concrete. In present paper, experimental approach is to study the mechanical properties of reactive powder concrete cured internally with thermostone material. The materials that adopted to evaluate and find out the influences of the internal curing on the mechanical properties of reactive powder concrete is focused with different curing methods such as in water, air and combined water and air. Thermostone aggregate are used as partial sand replacement by volume with different percentages to explore the percentage that effects of the concrete mechanical properties. Test results showed that the best partial replacement by thermostone is 5% gave enhancement and increase in compressive strength and flexural resistance strength (modulus of rupture) and concrete density. Highest increasing of compressive strength is 10.07in case of 5% partial replacement at 90 days. In case of cured the specimens up to 90 days, the increase in modulus of rupture is 4.53%

Keywords: Internal curing, Reactive powder concrete, Mechanical concrete properties, Partial replacement, Waste material, Thermostone.

1. Introduction

Different methodology of concrete curing was adapted to reach the concrete to the required strength. The curing of concrete was classified as internal and external curing based on the curing methodology and procedure. In general, the internal curing (IC) was defined as providing water in every part of a freshly placed cementitious mixture using continuous water source so that the concrete become pre-wetted that lead to complete the processes of hydration. The mechanical properties that produced from IC differ that in magnitude than in case of traditional curing. In case of IC in which this methodology required a lot of sources of water that store and easy to supply that pass through the cement paste during the cementitious reactions. The best materials that can be adopting are cells and gels. A cell membrane provides a boundary to water. Such containers allow water to be stored as an entity. The internal curing of concrete was adapted during recent years because of gave benefits to the concrete before and after casting. The internal curing gave in practical more supporting and additional water to the concrete

[1] To whom any correspondence should be addressed.

mixture. The internal curing adapted since 1950s and some researchers applied this technic in pre – stresses concrete in 1967s and this methodology can be adopted for all structural concrete elements. Reactive powder concrete has higher mechanical properties as compared with normal concrete. Rami J,2014 [1], investigated the influenced of thermostone as waste material that added to the concrete mixture as partial replacements in presence of Nano. Different ratio of waste thermostone by weight replacements were considered such as (5, 10, 15 and 20%) and applied destructive and non-destructive tests to investigated the compressive strength and tensile strength of all specimens. Test results showed that the (15%) of partial replacement gave good results while the quality of concrete as mechanical properties that contented normal weight concrete was decreased. J. Justs et al,2015 [2], investigated the impact of internal curing on the absorption capacity and mechanical properties of high performance concrete. Low water to cement ratio was adapted as (0.25) with and without superabsorbent polymers. Test results showed that the mechanical properties are influenced by superabsorbent polymers addition. Jianhui Liu et al, 2017 [3] reviewed the concrete mechanical properties cured by internal curing method. The materials classified into two parts according to water absorbing mechanism. Concluded that the addition of internal curing materials releases internal curing water that lead to increase the later hydration of cement. Kevern and Nowasell, 2018 [4], improved the mechanical properties of concrete by replaced fine aggregates and followed the internal curing for all specimens. Test results pointed out increased in compressive strength at different test ages as compared with the control mixture without replacements due to the additional water supplied by internal curing which promotes a higher degree of hydration and leads to filling the pores with hydration products and increases the strength of cement paste. D. Pytlik et al.,2018 [5], studied the effects of internal curing of high-performance concrete that mixed with hydrogel on the concrete properties. Different percentages of hydrogel were adapted such as (0, 2.5, 5, 7.5 and 10%) in which the specimens were casted based on selected percentages. Test results indicated that the percentages of hydrogel were more effective than the internal curing on the concrete properties. Dale P. et al ,2018 [6], studied the effects of internal curing on the mechanical properties of higher performance concrete. The micro structural specimens with water to cement ratio 0.3was casted and treatments with and without internal curing. The concrete mixer contained slag, silica fume and fly ash. The specimens that cured by used internal curing gave more strength after cured time (120 days). Xuan Chen et al ,2019 [7] studied and analysed the microstructure behaviour of reactive powder concrete. Different parameters that effects on the performance of reactive powder concrete were adopted curing, the analysis gradations of fine aggregate and steel fiber type by used the technic of scanning electron microscope in addition to the mechanical tests. Based on the test and analysis results an optimal mix proportion of reactive powder concrete was developed by considered the economic benefits. Also, it was found that the gradation of reactive powder concrete fine aggregate achieved the densest stacking state gradation after the optimization of mix proportion. Doha M. et al, 2019 [8], reviewed the behaviour of higher performance concrete with different parameters such as density, mechanical properties and microstructure of hydrated cement paste. Concluded from researchers results that the internal curing was more effective at a later age on flexural strength than on compressive strength. Internal curing has enabled the interfacial transition zone increasingly compact and dense that lead to enhancing strength. Shatha D. Mohammed et al ,2020 [9], looked out on the behaviour of the reactive powder concrete compressive strength by adopted different materials as partial replacements of fine aggregate. Three different materials were used as partial replacements as fine aggregate such as porcelain aggregate, glass waste and granular activated carbon with different percentages of (0, 10, 15 and 20%). The specimens were tested at different ages as (7, 28, 60 and 90 days). Test results showed that the glass waste causes an increase in the overall values of the compressive strength.

The aim of present works is to investigate the internal curing impacts on the mechanical properties of reactive powder concrete with and without partial replacement of fine aggregate by using sustainable material to maintain environment that become health and reducing cost. Different specimens are prepared and specified for each tests types. All specimens with different parameters are considered such as curing type, replacements percentages of sustainable materials and the test methodology for each specimen.The sand was replaced in different percentage volume with sustainable materials such as thermostone, the sustainable material used is absorbent and pre-soaked fine thermostone aggregate in water for suitable period as an internal curing.

2. Experimental Program

The materials investigations that adopted to make all specimens to explore the mechanical properties of reactive powder concrete for all curing conditions such as in water, in air and combined curing as water and air are presented here. The mechanical properties of the reactive powder concrete with and without partial replacement of fine aggregate consists of cube and prism for compressive strength, modulus of rupture and densityfor each percentage ratios of partial replacement. All selected materials such as cement, fine aggregate, silica fume, micro steel fiber and thermostone as waste material are tested to specify the chemical and physical properties for each.The specimens are tested at ages of 7, 28 and 90 days after curing under different conditions (water, air and combined). The curing method that adopted is placed all specimens in water tank with up to required time, in case of air curing all specimens placed in laboratory environments up to specified curing ages, also in compound curing (water and air) the specimens placed in water 7 days and the reaming period in air. Fig. 1 shows the schematic representation of test specimens,

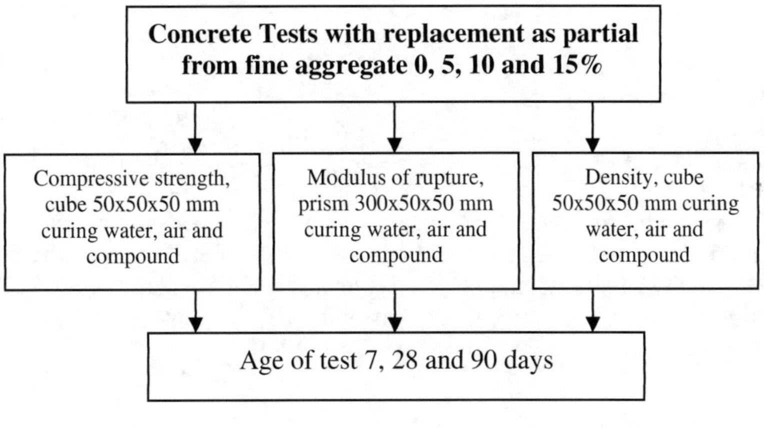

Figure 1. Schematic representation of test specimens

2.1. Materials

The materials that adopted in present study are summarized and discuss below as chemical and physical properties. All mortar components mixed with water matching the requirements of the Iraqi Specification, No. 1703,1992 [10].

2.1.1 Cement

The brand of the cement that popular used in Iraq is Ordinary Portland Cement (OPC) as Tasluja factory was adopted in present study and used to cast all the specimens such as cubes and prisms. The cement package was kept in good conditions to minimize the effect of humidity. Test results indicate that the adopted cement conforms to the Iraqi Specifications No. 5,2019 [11].

2.1.2 Fine Aggregate for Reactive Powder Concrete

The local natural sand Al-Ekhaider has been used in concrete mix for all tested specimen. It was tested to determine the grading and other physical and chemical properties. Test result shows after is within the requirements of the Iraqi Specification No.45,1984. Fig. 2 shows the sieve analysis with lower and upper limits based on the adopted specification matching the requirements of the Iraqi Specification No.45,1984[12].

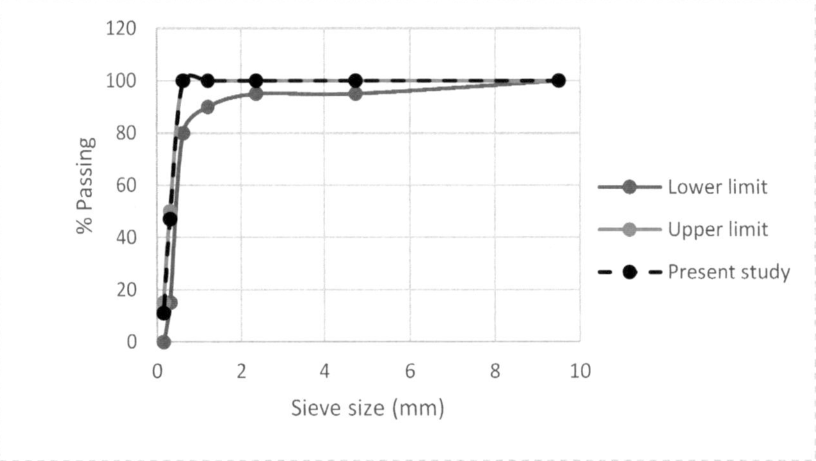

Figure 2. Sieve analysis - fine aggregate.

2.1.3 Micro Silica Fume (SF)

Densified micro silica fume from SIKA company Iraq branch under commercial name (Mega Add MSD) has been used as a mineral admixture. Many trails were applied to reach the best and optimize percentage of the quantity of silica fume as additive material. The replacement of 25% by weight of cement was considered as the best percent. The chemical requirements and physical composition of silica fume is conforming to the requirements of ASTM C 1240-05,2005[13].

2.1.4 Micro Steel Fibers (V_f)

A straight steel fiber manufactured by Ganzhou Daye Metallic Fibers Co., Ltd, China was adopted in all concrete mixtures.

2.1.5 High Range Water Reducing Admixture (HRWRA)

High Range Water Reducing Admixture (HRWRA) was adopted to produce mix with minimum water content. SikaViscocrete-5930 high performance as super plasticizer is used as high range water reducing admixture for concrete, flow ability, very high-water reduction, with high workability retention and high strength development. It is free from chlorides and matching with ASTM C494- type F&G,2005 [14].

2.1.6 Thermostone Aggregate – Partial Replacements

Thermostone aggregate is adopted as partial replacements from fine aggregate that collected from industrial process as waste material (Karbala Thermostone Factory). The main processes on the thermostone is broken into smaller pieces until become same as sand in size by using iron hammer in order to facilitate the insertion of thermostone through the opening of the standard set of sieves of sand. The crushed thermostone is washed and cleaned with water so that no dust resulting from crushing and then dried by spread in air. Sieve analysis for the thermostone crushed particles and then each thermostone aggregate size is replaced by volume with the same size of sand with a certain percentage to have the same grading as original aggregate based on Iraqi Specification No.45,1984 [12]. After all processes complete, the thermostone aggregate is soaked in water for (24) hours to bring the aggregate particles to saturated condition. Fig. 3 shows the prepared processes of thermostone as crushed, sieved, washed and saturated (from left to right).

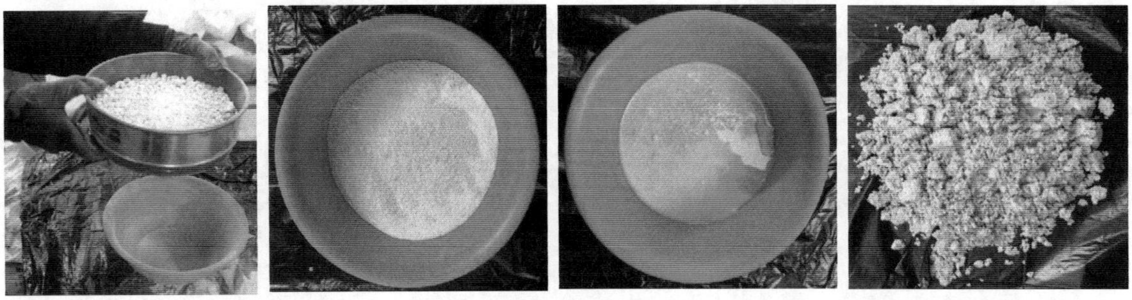

Figure 3. Prepared processes of thermostone as crushed, sieved, washed and saturated

2.2. Concrete Mixing

Many trails are tested to reach the required mixes proportions that gave maximum strength and good flow based on ASTM C-109/C109M,2016[15] and ASTM C-1437,2005[16]. Table 1 lists the mix proportion materials.

Table 1. Mix proportion materials.

W/Cm ratio	Sand [kg/m^3]	%Silica fume*	cement content [kg/m^3]	%Micro steel fiber**	%Super stabilizer***
0.172	960	25	880	0.75	1.50

*by weight of cement, **by volume of mix, ***by weight of binder(cement and silica fume)

Four groups marked as Group A, B, C and D of RPC mixes are adopted. Each group covers certain recognized parameters to investigate the effects of each other in which the average of three specimens for each test is considered. Group A represent the reference specimens that is mean the specimen casted without partial replacement of fine aggregate by thermostone as waste material. Group B represent the specimens that caste with 5% of sand partial replacement while group C and D same as group B but the partial replacement is 10 and 15% respectively. Each group, the specimens are tested under different curing conditions such as water, air and combine curing (water and air). Compressive strength, modulus of rupture and densityare investigating for each group. The specimen groups are lists in Table 2 and Table 3 lists the mix proportions of the concrete, Fig. 4 shows the casted of specimens as cubes with dimensions of 50x50x50 mm and prisms with dimensions of 300x50x50 mm.

Table 2. Details of all the series and its curing conditions.

Specimen mark	Curing method	Group mark	Curing type	% Thermostone replacement
A	1	A	water	0
	2	A1	Air	0
	3	A2	Water+ Air	0
B	1	B	water	5
	2	B1	Air	5
	3	B2	Water +Air	5
C	1	C	water	10
	2	C1	Air	10
	3	C2	Water+ Air	10
D	1	D	water	15
	2	D1	Air	15
	3	D2	Water+ Air	15

Table 3. Mix proportions of RPC mixes

Materials	(A)	(B)	(C)	(D)
Cement(kg/m^3)	880	880	880	880
Sand(kg/m^3)	960	912	864	816
Water(kg/m^3)	189.2	189.2	189.2	189.2
Silica Fume(kg/m^3)	220	220	220	220
SP(kg/m^3)	16.5	16.5	16.5	16.5
%Steel fibres	0.75	0.75	0.75	0.75
Thermostone(kg/m^3)	---	20.9	41.8	62.8

(a) Cubes and prism molds

(b) Cubes and prism casted specimens withinmold

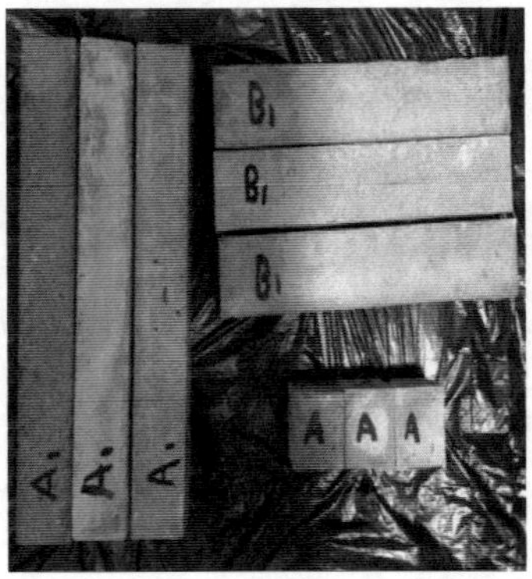

(c) Specimen moulds and

after remove the curing

Figure 4. Casted of specimens as cubes and prisms with curing

3. Test Results

To evaluate the mechanical properties of hardened concrete, different tests are applied such as compressive strength, modulus of rupture and densityunder different curing conditions. Followings are test results and discussions for each experimental investigation.

3.1 Compressive Strength

Compressive strength was classified as the basic parameter to evaluate the strength capacity of concrete. Compressive strength test based on ASTMC109/C109M,2016 [15]. In general, the tested specimens show that increase in compressive strength due to internal curing with increasing of curing age. The enhancement in compressive strength due to continue cement hydration with the progress of curing age that is match with other concluded by Lam ,2005 [17], Bentz, 2007 [18] and Sato et al. ,2011 [19], Amar and Nada [20]. The increase in compressive strength percentage in case of 5% as compared with that 10 and 15%. Also, the increase in percentages of compressive strength of specimens decrease as increased in partial thermostone replacement as waste material. A Fig.5 to 7 shows the variations of compressive strength against the cured method that symbols mentioned above. The compressive strength in case of cured the specimens in the water more than that cured the specimens in air and combined curing. This is due to in water curing the hydration process complete without discrete that lead to increase in the compressive strength. The presences of thermostone as partial replacement from fine aggregate in case of 5% gave more strength of specimen than the control and other percentages specimens. The partial replacements of fine aggregates by 10 and 15% also gave higher compressive strength than the control specimens. The reasons that caused increased in compressive strength because of increase the degree of hydration that lead to increase in the hydration products which fill the voids of concrete, in additions to the decrease the concrete porosity and reduced in the developed stresses.

IOP Conf. Series: Earth and Environmental Science **877** (2021) 012043 doi:10.1088/1755-1315/877/1/012043

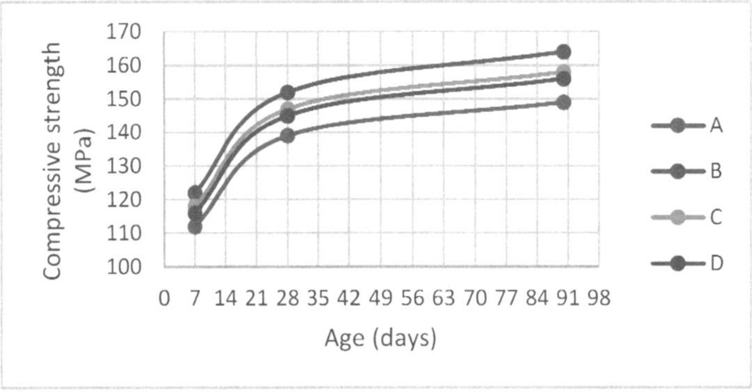

Figure 5. Relationship between compressive strength and curing ages externally cured with water

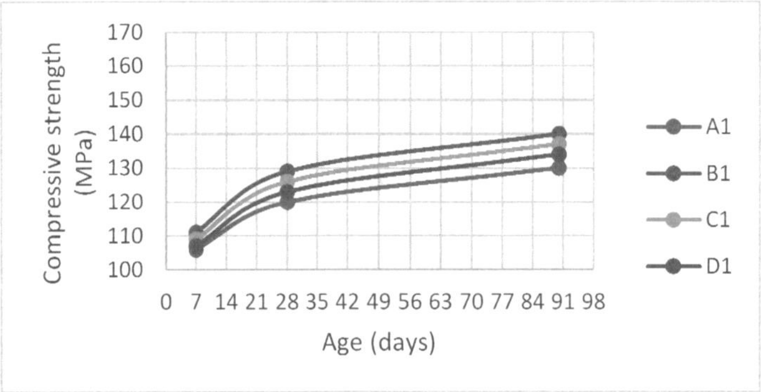

Figure 6. Relationship between compressive strength and curing ages externally cured with air

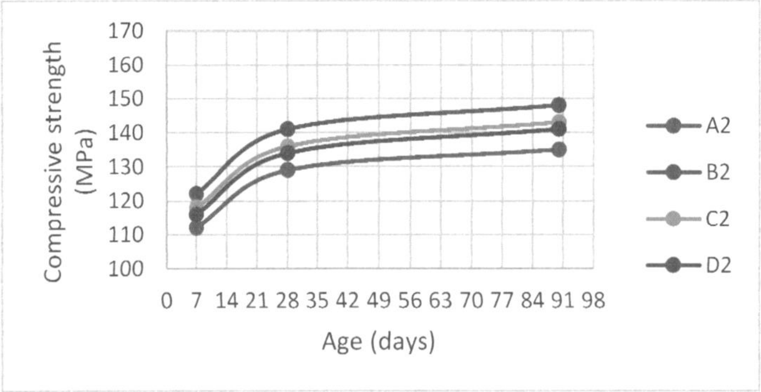

Figure 7. Relationship between compressive strength and curing ages externally cured with water and air

3.2 Modulus of rupture

Modulus of rupture of concrete tests under a 3-point load based on ASTM C293-02, 2002 [21]. The modulus of rupture measures the bond strength of the specimens (cement matrix). Increase in concrete compressive strength lead to increase in modulus of rupture because of the concrete resistance to the applied load. Higher compressive strength of reactive powder makes the concrete more resistance to applied load that cause tension stress due to bending that is mean increase in modulus of rupture of concrete. The increase in modulus of rupture percentage in case of 5% as compared with that 10 and 15%. Also, the increase in percentages of modulus of rupture of specimens decrease as increased in partial thermostone replacement as waste material. Increase in modulus of rupture related to increase in compressive strength of specimen that make the concrete more resistance to flexural under the effect of applied load. Fig. 8 to 10 shows the modulus of rupture variations with ages for different percentages of thermostone partial replacements percentages. In all ages, the modulus of rupture with partial replacement 5% more than the other specimens and all specimens with partial replacements by thermostone are greater than the control specimens

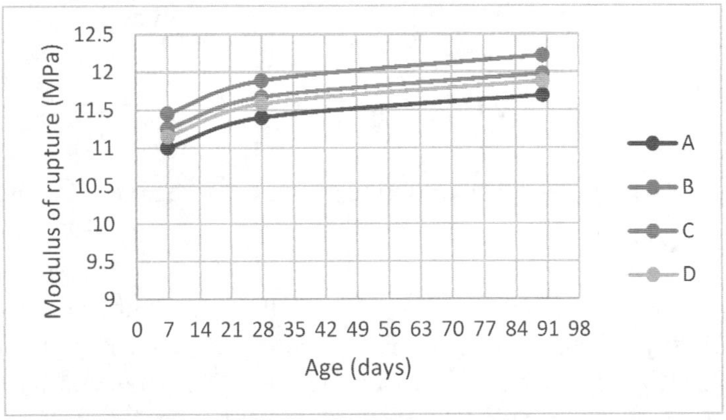

Figure 8. Relationship between modulus of rupture and curing ages externally cured with water

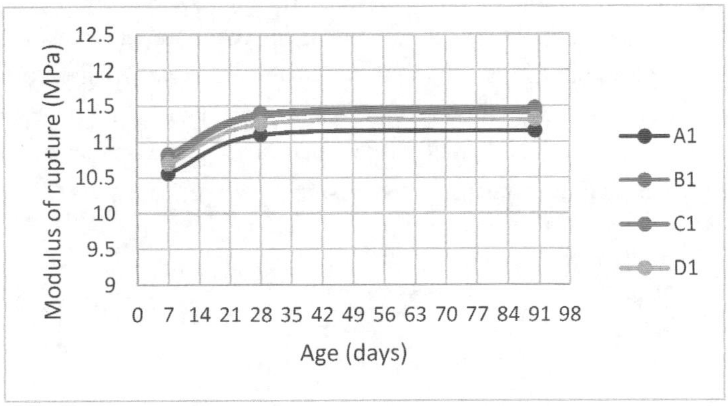

Figure 9. Relationship between modulus of rupture and curing ages externally cured with air

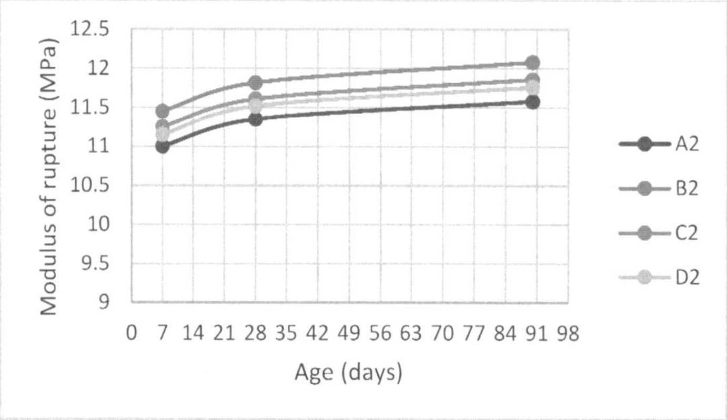

Figure 10. Relationship between modulus of rupture and curing ages externally cured with water and air

3.3 Density

Concrete density is the mass per unit volume in which the throughout the adopted test found out the actual concrete density for all adopted specimens with the considered parameters based on Iraqi Guidelines No. 274, 1992 [22]. The density of concrete differs from concrete to others, it is vastly relying on the mix design and other characteristics such as specific gravity of the aggregates. The increase in the concrete density percentage in case of 5% as compared with that 10 and 15%. Also, the increase in percentages of the concrete density of specimens decrease as in creased in partial thermostone replacement as waste material. Increase in the concrete density due to increase in compressive strength of specimen. Figs. 11 to 13 shows the concrete density variations with ages for different percentages of thermostone partial replacements percentages. In all ages, the concrete density with partial replacement 5% more than the other specimens and all specimens with partial replacements by thermostone are greater than the control specimens.

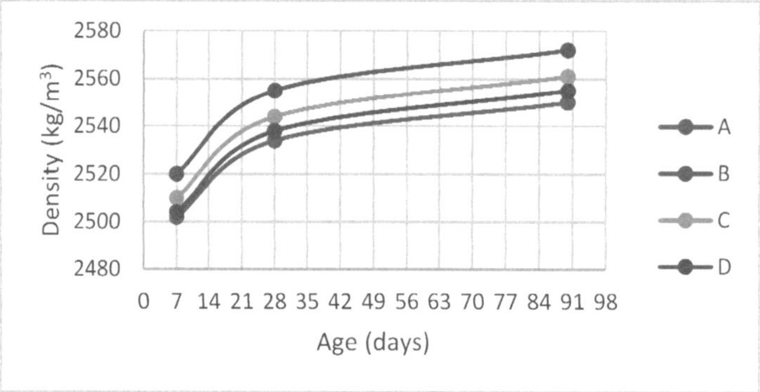

Figure 11. Relationship between density and curing ages externally cured with water

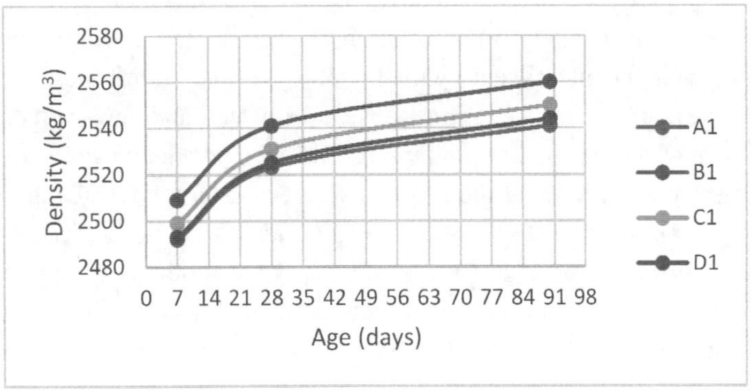

Figure 12. Relationship between density and curing ages externally cured with air

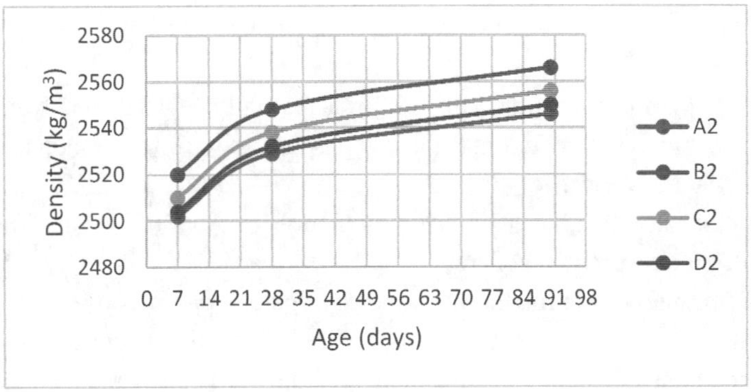

Figure 13. Relationship between density and curing ages externally cured with water and air

4. Discussions

Test results pointed out the mixes are internally cured gave mechanical properties of reactive powder concrete with thermostone as waste material partially replacements more than reference specimens with all partial replacements percentages. These improvements in concrete because of increase degree of cement hydration from internal curing water that make increase the hydration products.

Based on the test results the best partial replacement by thermostone is 5% gave enhancement and increase in compressive strength and modulus of rupture and concrete density due to this percentage is the best for this compositions of RPC with internal curing method that gave improvements in RPC mechanical properties. The increased in the quantity of thermostone aggregate as partial sand replacement causes the formation of large pores in the concrete matrix and in turn leads to decrease the density and other RPC properties as compared with 5% but still more than reference specimens for all cured ages. IC that the water supplied internally that cured the mixture (thermostone aggregate) promote high degree of hydration which leads to fill the pores with hydration product and increase the density of cement paste.

In case of cured the specimens up to 90 days, the increase in compressive strength and modulus of rupture when the specimens cured in water rather than when the specimens cured in air.

Test results indicated that internal curing methodology form the water slowly consumed over time for hydration of concrete matrix. Internal curing allows more effects and improvement concrete modulus of rupture when cured in water than in air for all mixes. Reactive powder concrete mechanical properties and density at age of 90 days with 5% of thermostone aggregate cured in water show increase compare with cured in air and combine curried specimens (water and air) due to increase of cement hydration degree. The combine curried specimens (water and air) gave modulus of rupture greater than the specimens when cured in air only and less than that cured in water due to increase in the hydration cement degree in presence of water so that the concrete strength increase as compare with specimens cured only in air.

The enhancements of compressive strengths, modulus of ruptures and densitiesfor all casted specimens showed and mentioned above with respect to cured methodologies and tested ages with partial replacements percentages by thermostone.

5. Conclusions

The important conclusions based on the test results points as follow:

1. Using of local thermostone waste material that partially replacement from fine aggregate as internal curing materials to internal curing purpose gave improvements in reactive powder concrete mechanical properties.
2. Test results indicated that internal curing methodology from the water slowly consumed over time for hydration of adopted concrete.
3. The internal curing by adopting thermostone aggregate as partial sand replacement is significant enhancing the compressive strength of concrete at all ages of curing as compared with the reference specimens. Results show that there are improvements in the compressive strength with partially replacements as compared with the control specimens for each group and for all cured conditions.
4. There are increase in compressive strength based on test results when percentage of 5% as compared with that 10 and 15%. Also, the increase in percentages of compressive strength of specimens decrease as increased in partial thermostone replacement as waste material. Compressive strength of reactive powder concrete in cases of 10 and 15% gave more than reference but less than 5% replacement.
5. Reactive powder concrete compressive strength cured in water was improved because internal curing with thermostone waste material with different percentages as compared with reference concrete. Highest increasing of compressive strength is10.07% in case of 5% partial replacement at 90 days. In case of cured the specimens up to 90 days, the increase in compressive strength 10.07, 6.04and 4.70% for B, C and D respectively when the specimens cured in water. When the specimens cured in air the increase percentages as compare with the reference within same group are 7.69, 5.38and 3.08% for B1, C1 and D1 respectively, also 9.63, 5.93and 4.44% for B2, C2 and D2 respectively when the specimens cured in water and air.
6. Higher compressive strength of reactive powder makes the concrete more resistance to applied load that cause tension stress due to bending that is mean increase in modulus of rupture of concrete.

7. The presence of super - stabilizer in the concrete mixture and absence of micro steel fiber that lead to increase the concrete workability even the less of water to cement ratio that adopted.

8. Modulus of rupture of reactive powder concrete with partial replacement by thermostone as waste material shows higher values as compared with the reference concrete at all ages and all cured methodologies. Test results indicated that the progress of cured age, the reactive powder concrete specimens show a continuous gain in modulus of rupture that lead to enhancing in the modulus of rupture that make improving the properties of concrete by the additional internal curing water supplied by internal curing materials.

9. The increase in modulus of rupture percentage in case of 5% as compared with that 10 and 15%. Increase in modulus of rupture related to increase in compressive strength of specimen that make the concrete more resistance in tension. In case of cured the specimens up to 90 days, the increase in modulus of rupture 4.53, 2.48and 1.63% for B, C and D respectively when the specimens cured in water. When the specimens cured in air the increase percentages as compare with the reference within same group are 2.87, 2.33and 1.43% for B1, C1 and D1 respectively, also 4.32, 2.42and 1.55% for B2, C2 and D2 respectively when the specimens cured in water and air.
10. Internal curing allows more effects and improvement concrete modulus of rupture when cured in water than in air for all mixes. The combine curried specimens (water and air) gave modulus of rupture greater than the specimens when cured in air only and less than that cured in water. In case of curing initially by water then the specimen's transferee to cured in air make the specimens to reduce in modulus of rupture as compare with the specimen cured total 90 days in water.

11. Increased in the percentage of replacement concrete more or equal to 10% make the density of reactive powder concrete decreased, and at the percentage of replacement of 15% concrete showed lower density but still more than the reference specimen. Increasing reactive powder concrete density of internally cured concrete with replacement of sand by 5% by 0.86% in water.

12. The internal curing method can be used in the construction field because of its ease of inclusive in the concrete mixture.

6. Summary

Three different methodologies with different ages (7, 28 and 90 days) and three percentages of sand aggregate by thermostone as waste material partially(5, 10 and 15%) of reactive powder concrete specimens are investigated. Hardened concrete specimens are tested such as cube and prismatic to find out the reactive powder concrete properties. Density, compressive strength and modulus of rupture are the main test are investigating. Internal curing of tested specimens leading to increase of reactive powder concrete properties due to continue cement hydration with the progress of internal curing age.

7. References

[1] Rami Joseph AghajanSldozian, Effect of Waste Thermostone as an Aggregate in Concrete Containing Nano-SiO2, International Journal of Scientific & Engineering Research, Volume 5, Issue 7, July, (2014).

[2] J. Justs, M.,Wyrzykowski, D. Bajare , P. Lura , Internal curing by superabsorbent polymers in ultra-high performance concrete,Cement and Concrete ResearchVolume 76, October (2015) 82-90.

[3] Jianhui Liu, CaijunShi·Xianwei Ma, Kamal. Khayat, Jian Zhang , Dehui Wang, An overview on the effect of internal curing on shrinkage of high performance cement-based materials, USA Construction and Building MaterialsVolume 146 (2017) 702-712.

[4] J.T. Kevern, Q.C. Nowasell, Internal curing of pervious concrete using lightweight aggregates, Constr. Build. Mater. 161 (2018) 229–235.

[5] D Pytlik, V Bilek and M Bambuchova, Internal curing of HPC with hydrogel, IOP Conf. Series: Materials Science and Engineering 385 (2018).

[6] Dale P. Bentz and Paul E. Stutzma, Internal curing and microstructure of high-performance mortars, National Institute of Standards and Technology, (2018).

[7] Xuan Chen, Dong-wei Wan, Ling-zhiJin, Kai Qian1 Feng Fu1.. Experimental studies and Microstructure Analysis for Ultra High-Performance Reactive Powder Concrete, Construction and Building Materials 229, September (2019).

[8] Doha M. Al Saffara, , Aymen J.K. Al Saadb , Bassam A. Tayehc , Effect of internal curing on behavior of high performance concrete, Case Studies in Construction MaterialsVolume 10, June (2019).

[9] Shatha Mohammed, Rawaa Khalid, HadeelAwad. Compressive Strength Performance of Reactive Powder Concrete Using Different Types of Materials as a Partial Replacement of Fine Aggregate,Article in Key Engineering Materials , August, (2020).

[10] Iraqi Specification, No.1703, water used in concrete, (1992).

[11] Iraqi Specification, No.5, Portland Cement, (2019).

[12] Iraqi Specification No.45, Aggregate from natural sources for concrete and construction, Ministry of Planning, Central Organization for Standardization and Quality Control, (1984).

[13] ASTM C 1240, Standard specification for silica fume used in cementitious mixtures, (2005).

[14] ASTM C 494, Standard specification for chemical admixtures for concrete, Annual Book of ASTM Standards, Vol. 04-02, (2005).

[15]ASTM C-109/C109M,2016, Test method for compressive strength of hydraulic cement mortars using (50mm) cube specimens, Annual Book of ASTM Standards, Vol. (04-02), (2016).

[16] ASTM C 1437-01., Standard test method for flow of hydraulic cement mortar, Vol. (04.01), PP. 1-3, (2005).

[17] Lam, H., Effect of Internal Curing Methods on Restrained Shrinkage and Permeability, Ph.D. thesis, University of Toronto, (2005).

[18] Bentz, D.P., Internal Curing of High-Performance Blended Cement Mortars, ACI Materials Journal, Vol. (104), No. (4) (2007) 408-414.

[19] Sato, R., Shigematsu, A.,Nukushina, T., and Kimura, M., Improvement of Properties of Portland Blast Furnace Cement-Type B Concrete by Internal Curing Using Ceramic Roof Material Waste, ASCE Journal of Materials in Civil Engineering, Vol. (23), No. (6) (2011) 777-783.

[20] Amar Yahia and Nada Mahdi, Enhancing Performance of Self–Compacting Concrete with Internal Curing Using Thermostone Chips, Journal of Engineering, Volume 23, Number (7) (2017) 1-13.

[21] ASTM C293-02, Flexural Strength of Concrete (Using Simple Beam with Center – Point Load),Annual Book of ASTM Standards C293 – 02, (2002) 1-3.

[22] Guidelines No. 274, The Concrete Tests and Test Methods to Measure Density of Hardened Concrete, Central Apparatus for Standardization and Quality Control, Iraq, (1992).

Experimental and Mathematical Research of Convection Heat Transfer for a High Integral Finned Tube Heat Exchanger

M. S. Kassim, H. S. Jaber and H. S. Hashim[1]
Department of Mechanical Engineering, Engineering College, Mustansiriyah University, Baghdad, Iraq
E-mail: munakassim@uomustasnsiriyah.edu.iq

Abstract. This research exhibits an experimental and numerical investigation for the characteristics of heat transfer by utilizing an integral and smooth high finned tube and shows how the integral high fins affect in improving the transfer of heat. In the experimental work, the experimental rig consists of a cold water loop, a hot air loop and the trial part which is a concentric parallel flow double pipe heat exchanger. The first test smooth tube made of brass has external and internal diameter (32 mm) and (22 mm), respectively, the second test section made of the same material has external fins, the third test section possesses internal fins, and the fourth test section has internal and external fins. The dimensions of the fin are (2 mm) in thickness, (2 mm) in height and pitch (2.5 mm) center to center. And, the water flow rates are (6, 8, 10, 12 and 14 L/min). The inlet water to the trial tube was at temperatures (15, 25, 35, and 45oC). The investigational results manifested that the air side heat transfer coefficient of the smooth tube was lesser than the integral high finned tube. The ratio of improvement when using the integral high finned tube was (47.5%, 60.5% and 67.5 %). Numerical simulation was applied on the current heat exchanger to study both the transfer of heat and the field of flow by using ANSYS, FLUENT15 package. Steady state, Newtonian flow, incompressible and three dimensional analyses were assumed. The comparison between experimental work and numerical results elucidated a good agreement.

1. Introduction

Attention of the saving of energy and materials in addition to thrifty desires has prefaced to a work to generate more effective instrument as a heat exchanger (H.E). The objectives of the public thermal hydraulic are to decrease the heat exchanger size wanted for a definite heat job, to fall the agreed temperature difference for the streams of procedure or to decrease the driving power and develop the ability of current H.Es. The investigation of the modified heat transfer interpretation is indicated as heat transfer improvement or strengthening. Generally, this explains an increase in the coefficient of heat transfer [1]. The heat exchanger of water to air is considered as one of the significant types of a dual pipe concentric tube. This kind of H.E has different implementation, such as apartment buildings and condominiums, hybrid schemes, domestic heating, systems of dehumidification, and air conditioning.

2. Literature Survey

Sparrow and Ramsey [2] described an outstanding investigational work on the effect of tip permission for a surprised wall-fixed array of cylinders. They acquired data on the coefficients of heat transfer via carrying out the similarity amid mass and heat transfer by the method of naphthalene sublimation. It was found that when the length of the cylinder rises, the coefficient of heat transfer rises

temperately and the tip permission is amid the pin and the covering drops. Oppositely, the array pressure drop rises noticeably with the rising length of cylinder. This performance was clarified with the inter-cylinder velocities for short pins which were fewer than the mean velocity, while for the longer cylinders, the inter-cylinder velocities tended to approximate the mean value.
Rabas et al. [3] offered a significant thoughtfulness that the thermal presentation is practically not relied on the Re number for the bigger density of fin (0.980 fins/mm). It was noted that the density of fin has much resilient effect on the performance for the bigger diameter of tube (d = 3.175 cm). The low fins, which are lower than 0.635 cm at the variety of Re as 13 ≤ Re ≤ 25*103, were studied.
X. Hu and A. M. Jacobi [4] stated investigational researches of the local features of mass transfer of annularly finned tubes in cross flow. The differences because of boundary layer improvement, forward-edge parting, horseshoe vortices, tube wake, and tip vortices were explained. Furthermore, the commonly situated local maximum in the mass transfer rates related to the horseshoe vortex system was found, and speculation as to their mechanism was obtainable. From the concluding heat transfer performance from the outcomes of mass transfer, it was find that the true fin effectiveness is continuously fewer than that achieved with an expected constant convective heat transfer coefficient. The variance is (3–7%) for the high-conductivity substances like the alloys of aluminum and (9–17%) for the low-conductivity substances such as slight steels.
Ayad et al. [5] carried out an analogous work of eight finned–tube heat exchangers. It was found that there was no influence of fin pitch on the heat transfer presentation for four row coils at Re< 103. On the other hand, the performance of heat transfer is extremely reliant on the fin pitch at Re > 103. Reducing the fin pitch for a two-row conformation leads to a rise in the performance of heat transfer.
C. V. M. Braga et al. [6] studied the parameters of friction and heat mean transfer coefficients for a turbulent flow in annular duct areas with continued longitudinal rectangular fins fixed to the internal wall. The cold air was flowing over the annulus, whereas the warm water was delivered over the internal pipe. The thermal boundary settings were remained fixed temperature through the inner tube surface, the external tube being isolated. There are twenty continued longitudinal fins installed to the inside wall of the annular duct. The tests were accomplished with the operating of H.E in the parallel flow model. The variety of the air temperature difference amid the parts of leaving and entrance was (10 to 25°C). The total tests and determinations were prepared below settings at a steady state. The entrance temperature of water in the experimental rig part was (98°C), and the pressure was upper than (1 atm). The variation of the air flow Re was amid (104 < 5x104). The experimental outcomes for the mean Nu number of the finned annular duct section, in terms of the air flow Re, can be stated according to the subsequent equation:

$$Nu = 0.00529Re^{0.8680}$$

The mean Nu number of the air flowing over the plain annulus is signified as Nuu:

$$Nu_u = 0.00314Re^{0.9474}$$

Ali et al. [7] tested the heat transfer improvement along the exterior surface of a pipe which is made of copper with circular fins joined on the external surface heat exchanger in a rectangular passage with the flow of air intermittently. Their attempts were applied by utilizing four kinds of circular fins with (9.2 cm) external diameter, (3.2 cm) internal diameter, and (0.1 cm) thick tied on pipe of copper. Every kind possesses (5) circular fins. The first kind has 5 fins devoid of inclined blades, the second kind possesses 5 inclined vanes per one fin, the third kind has 7 inclined vanes for each 1 fin and fourth kind possesses 9 inclined vanes for each fin. The results showed that Nu is around (11.80%), (20.25%) and (27.50%) for the 2nd, 3rd and 4th kind, respectively, and it is greater than that for the 1st kind. Furthermore, the development of the transfer of heat and the performance of fin produced a considerable decrease in thermal resistance that can be got from the 4th kind in comparison with the1st kind.
Bashar [8] studied the increase of the convective heat transfer in a single-phase turbulent flow via utilizing triangular fins experimentally and mathematically. In the investigational work, the influence of

space distance (2.2, 2.7 and 3.2 cm) amid each two fins formed, and the volumetric flow rate on the coefficient of heat transfer was observed.

A number of triangular fins are factory-made from Cu. They are connected to the exterior surface of tube. They have (1 cm) length of base, (0.1 cm) thickness, (1 cm) height, (2.2, 2.7, and 3.2) cm space distance amid two fins sequential and (1.5 cm) pitch amid every two of fins. Thermometers and pressure gauges are connected to the tubes and annular sections. The results obtained from the tubes with triangular fins were compared with the smooth tube heat exchanger. It was found that the triangular fins have a significant effect on the heat transfer augmentations. Also, the result presented that the development of heat debauchery for the triangular finned tube is (3.8 to 5.4) times that of smooth tube at space (2.2 cm) distance amid every two sequential fins. While, the spaces 27 and 32 mm will be (3.5 to 4.9) and (3.2 to 4.5).

Ayad [9] investigated the characteristics of the heat transfer of a cross flow air cooled single tube multi passes (integral and smooth low finned tube) and its influence on the heat transfer enhancement. Two test sections from Perspex duct were designated. Every trial part has a test tube (single aluminum tube multi passes) with passes (4 or 8). The outcomes from the experimental work revealed that the air side coefficient of the heat transfer of the smooth tube was lesser than that of the integral low finned tube.

3. Numerical Simulation

Numerical simulation via utilizing ANSYS, FLUENT15 computational fluid dynamic (CFD) set prototypical was accompanied for executing a mathematical simulation across a H.E by utilizing a three-dimensional model. The solution of the equations of momentum, continuity, and energy was used to study the flow domain into H.E. The evaluation of heat transfer was accomplished for a (triangular, smooth) finned tube H.E. Comprising the triangular-finned tube H.E with the parts of entrance and leaving, the internal and external flow was limited via isolating the tube possessing the parts of entrance and leaving as depicted in the figures below. See the figures 1 to 4.

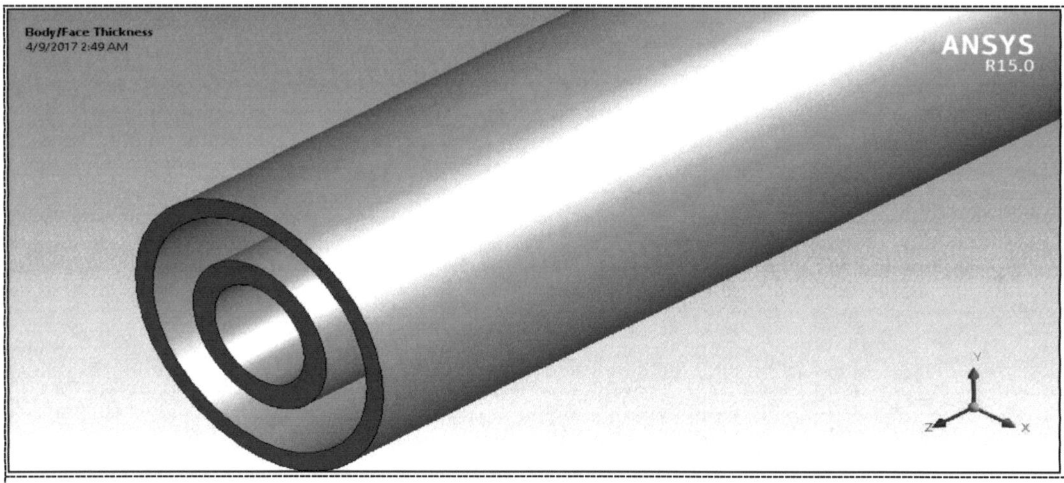

Figure 1. CFD field of the smooth tube H.E

Figure 2. CFD field of the exterior triangular-finned tube H.E

Figure 3. CFD field of the interior triangular finned tube H.E

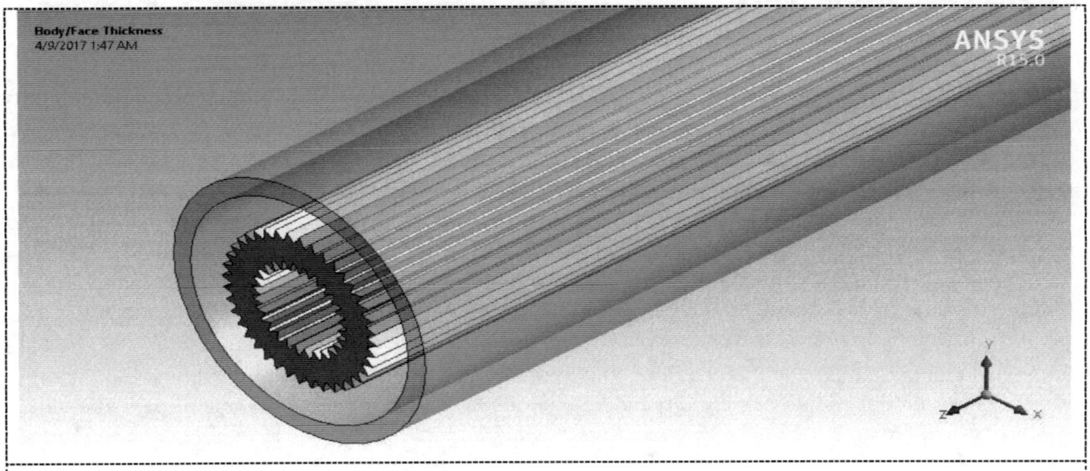

Figure 4. CFD field of the assorted triangular finned tube H.E

4. The Prevailing Equations

The main differential equation for the fluid flow is specified as: Mass conservation or continuity equation, Momentum conservation or Navier Stokes equation, and equation of Energy conservation [11].

4.1. Equation of mass conservation:

$$\frac{\partial(\rho u)}{\partial x}+\frac{\partial(\rho V)}{\partial y}+\frac{\partial(\rho W)}{\partial z}=0$$

4.2. Equation of momentum conservation:

$$\rho\left(u\frac{\partial u}{\partial x}+V\frac{\partial u}{\partial y}+W\frac{\partial u}{\partial z}\right)=\rho X-\frac{\partial p}{\partial x}+\frac{1}{3}\mu\frac{\partial}{\partial x}\left(\frac{\partial u}{\partial x}+\frac{\partial V}{\partial y}+\frac{\partial W}{\partial z}\right)+\mu\nabla^2 u$$

$$\rho\left(u\frac{\partial V}{\partial x}+V\frac{\partial V}{\partial y}+W\frac{\partial V}{\partial z}\right)=\rho Y-\frac{\partial p}{\partial y}+\frac{1}{3}\mu\frac{\partial}{\partial y}\left(\frac{\partial u}{\partial x}+\frac{\partial V}{\partial y}+\frac{\partial W}{\partial z}\right)+\mu\nabla^2 V$$

$$\rho\left(u\frac{\partial w}{\partial x}+V\frac{\partial w}{\partial y}+W\frac{\partial w}{\partial z}\right)=\rho Z-\frac{\partial p}{\partial z}+\frac{1}{3}\mu\frac{\partial}{\partial z}\left(\frac{\partial u}{\partial x}+\frac{\partial V}{\partial y}+\frac{\partial W}{\partial z}\right)+\mu\nabla^2 W$$

4.3. Equation of Energy:

$$\rho c_p\left(u\frac{\partial T}{\partial x}+V\frac{\partial T}{\partial y}+W\frac{\partial T}{\partial z}\right)=\left(u\frac{\partial P}{\partial x}+V\frac{\partial P}{\partial y}+W\frac{\partial P}{\partial z}\right)+K\nabla^2 T+\mu\emptyset$$

Where:

$$\emptyset=2\left[\left(\frac{\partial u}{\partial X}\right)^2+\left(\frac{\partial V}{\partial y}\right)^2+\left(\frac{\partial w}{\partial z}\right)^2\right]+\left[\left(\frac{\partial u}{\partial y}+\frac{\partial V}{\partial X}\right)^2+\left(\frac{\partial V}{\partial z}+\frac{\partial W}{\partial y}\right)^2+\left(\frac{\partial W}{\partial X}+\frac{\partial u}{\partial z}\right)^2\right]-\frac{2}{3}\left[\frac{\partial u}{\partial x}+\frac{\partial V}{\partial y}+\frac{\partial W}{\partial z}\right]$$

5. Boundary Conditions

The velocity at the entrance was defined for the internal and annuli sides throughout this research. Conversely, the temperature of the entrance of the internal tube was (15, 25, 35, 45°C), whereas in annuli it was (70) °C.

6. Mesh Resolution

In the current search, a free mesh was utilized to approximate the computational field inside a limited number of control volumes via utilizing the pattern or the model of finite–volume. The organized mesh is governed unavailable as it is advantageous for simple states and it converts to inadequate and expended time for difficult geometries. The pattern was meshed via utilizing **ANSYS, FLUENT15** (CFD) set. The production and of the modification of mesh scheme are so vital to expect the H.T in the advanced geometries. See the figures 5 to 8.

IOP Conf. Series: Earth and Environmental Science **877** (2021) 012044 doi:10.1088/1755-1315/877/1/012044

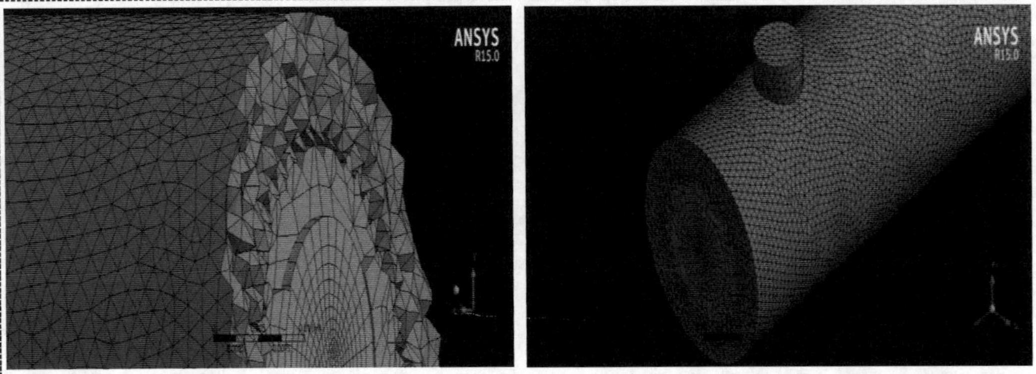

Figure 5. CFD meshing-part of the external finned-tube heat exchanger (pitch 2.5 mm)

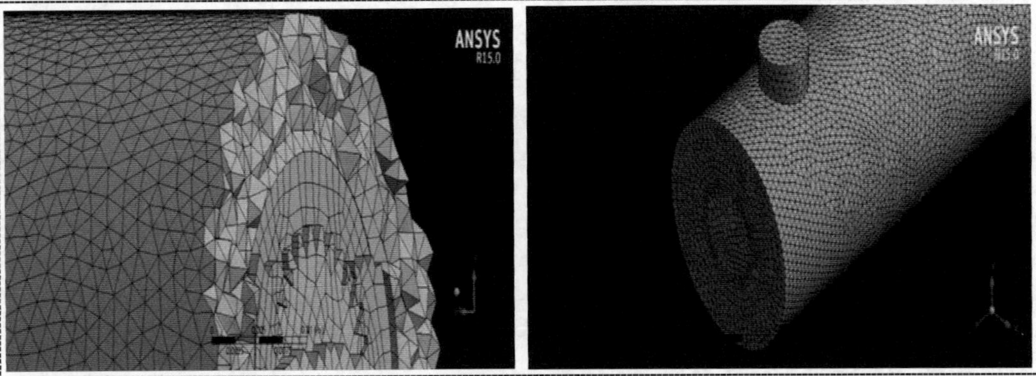

Figure 6. CFD meshing-part of internal finned tube heat exchanger(pitch 2.5mm)

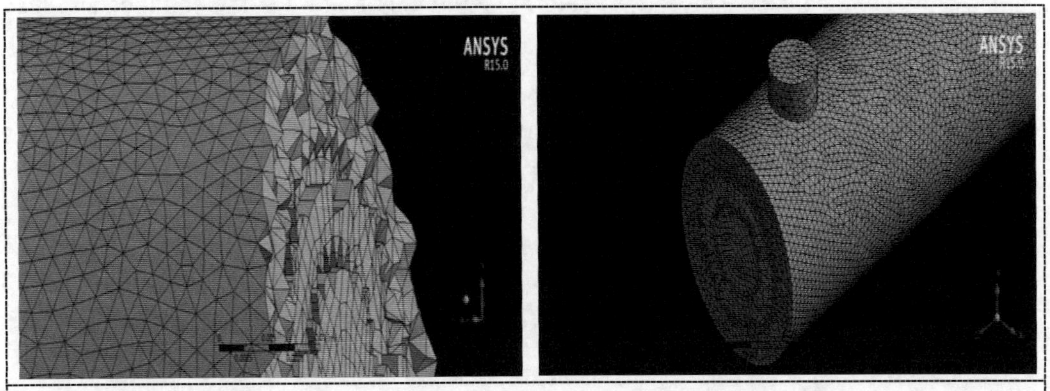

Figure 7. CFD meshing-part of mixed(internal and external) finned tube heat exchanger(pitch 2.5m)

IOP Conf. Series: Earth and Environmental Science **877** (2021) 012044 doi:10.1088/1755-1315/877/1/012044

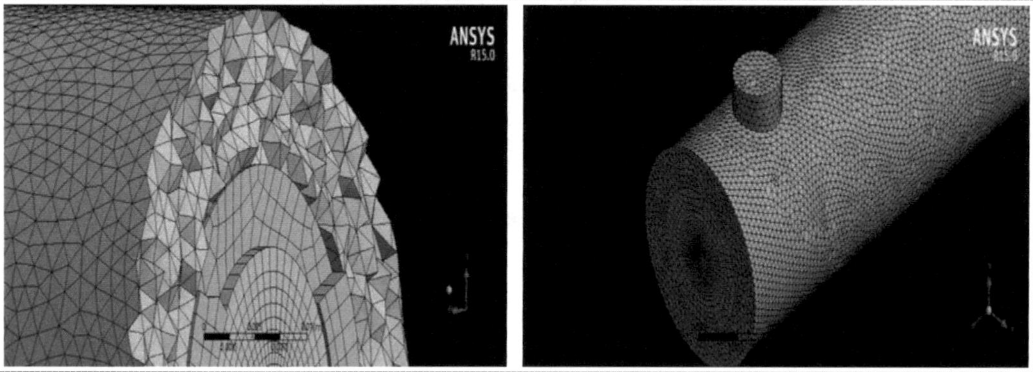

Figure 8. CFD meshing of smooth tube heat exchanger

7. Apparatuses of the Experimental Trial

Figure 1 evinces the graphical drawing of the experimental system. The test section consists of two portions. The first portion is an isolated tube which has factory-made from (PVC) of (5.08 cm) internal diameter, (10 cm) length and (0.5 cm) thickness. The second portion is an interior brass tube devoid of or with triangular brass fins. The integral fins tube consists of three types. The first type is an internal fins tube, the second type is an external fins tube and the third type is a mix (external and internal) fins tube, the all types of fins tube have a pitch of (2.5 mm). The smooth copper tube is (10 cm) long and (2.2 cm, 3.2 cm) internal and external diameter, respectively.
There were outlet and input section in the top and bottom section of the shell respectively suited at a (20 cm) distance of the edge annular, each nozzle has inside diameter and thickness. Nozzle was made from the same material of the annular. The annular was drilled at a distance (20 cm) from the edge ends, on the top and bottom to weld the nozzle.
Several of triangular fins are factory-made from copper, having (2 mm) length of base, (2 mm) height, and (2.5, 3, and 3.5 mm) pitch between each two of fins. Thermometers and pressure gauges are connected to the tubes and annular sections. Thermocouples and pressure gages were mounted on the two ends of the shell, to determine the temperature at the entrance and leaving on the two exchanger sides and the pressure on both shell sides. The tried heat exchanger comprises a single-pass kind concentric tube. See figures 9 and 10.

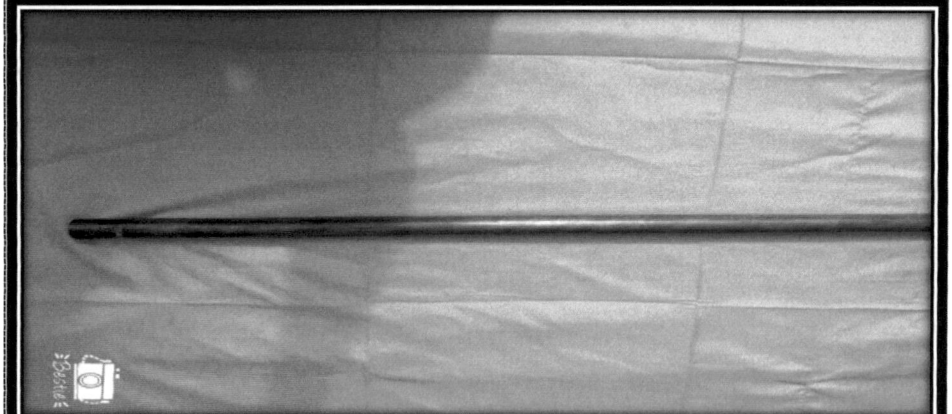

Figure 9. Smooth tube of the enhanced of heat exchanger

IOP Conf. Series: Earth and Environmental Science **877** (2021) 012044 doi:10.1088/1755-1315/877/1/012044

Figure 10. Triangular fins of the enhanced tube side of heat exchanger

The annular and the tube heat exchanger are planned for a parallel flow conformation, in which the cold water passes in same path in the tubes to the hot air which passes in the annular side.
These portions have installed on the border construction. The frame structure is planned for repairs during the test and making a tranquil for operation.

8. Validation

For the assessment of Nu, the investigational information are in a good acceptance with the current relationships, which are the relationship of Dittus–Boelter. The dependability and accurateness of investigational determinations were examined by utilizing a smooth tube alone. The investigational outcomes were related to the expectation of the relationship of Dittus-Boelter:

$$Nu = 0.023Re^{0.8}Pr^{n}$$

Where:
n = 0.33 for cooling
Nu: The Nusselt number
Pr: The Prandial number
Re: The Reynolds number

The Nusselt number was found to increase by increasing the Re number. A good agreement was obtained amid the experimental data and the outcomes acquired from the Dittus-Boelter correlation. The maximum deviation between the experimental data and the Dittus-Boelter correlation was about 5%. See figures 11, 12 and Table 1.

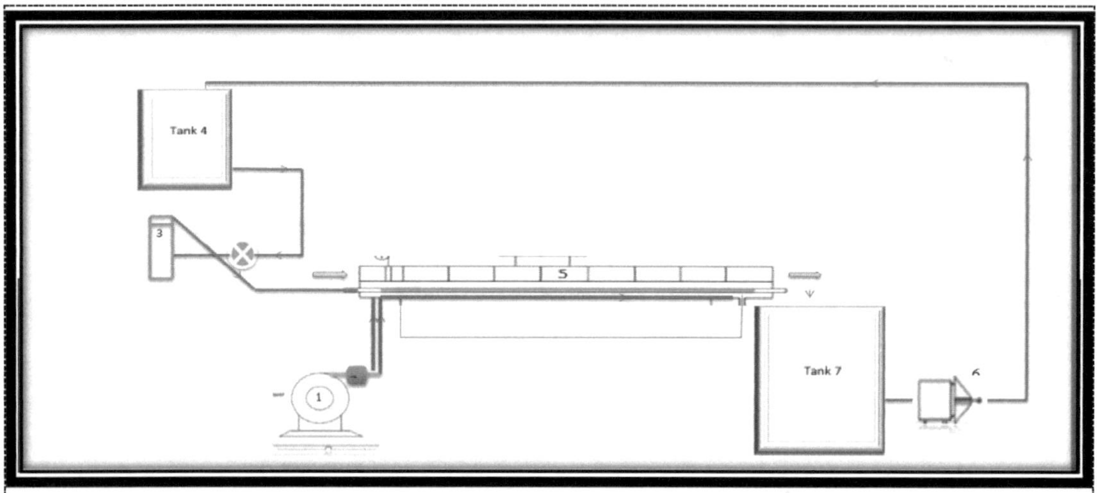

Figure 11. Representation of the experimental apparatuses

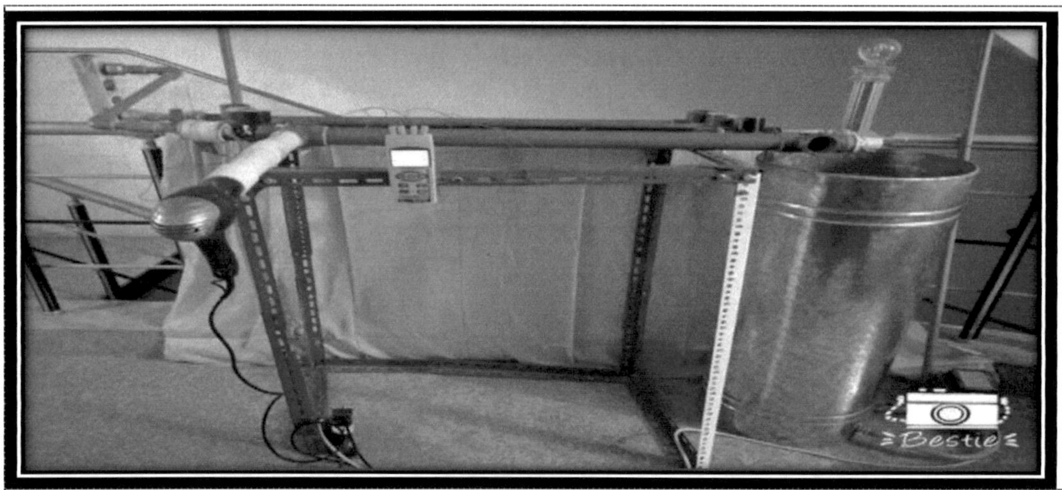

Figure 12. Picture of the experimental apparatuses

$$Q_h = \dot{m}_h cp_h (T_{ho} - T_{hi})$$

Where:
m˙: The mass flow rate (kg.s^{-1})
Q: The heat dissipation (w)
T: The temperature (°C)
Cp: The specific heat

The Subscripts i, o and h is referring to the internal, external and hot, respectively.

$$LMTD = \frac{\Delta T_2 - \Delta T_1}{ln\left(\frac{\Delta T_2}{\Delta T_1}\right)} = \frac{\Delta T_1 - T_2}{ln\left(\frac{\Delta T_1}{\Delta T_2}\right)}$$

Where:

$$\Delta T_1 = T_{hi} - T_{co}$$
$$\Delta T_2 = T_{ho} - T_{ci}$$

Where: ΔTLMTD is the log mean temperature variance (°C).

Table 1. Dittus-Boelter and CFD outcomes for various values of Reynolds number

Re	*Nu Dittus-Boelter*	*Nu CFD*
3019.892	17.217	27.071
4026.522	21.668	31.7113
5033.152	25.907	35.8527
6039.733	39.6343	29.976
8035.044	33.906	46.4237

$$h_i = \frac{Q_h}{A_i(T_m - T_s)}$$

Where:

A: The tube surface area (m^{-2})

hi: The heat transfer coefficient ($W.m^{-2}.\ K^{-1}$) of the tube

$$Nu = \frac{h_i d_i}{K_h}$$

Where:

k: The thermal conductivity ($W.m^{-1}.\ K^{-1}$)

D: The tube diameter (m)

9. Discussions and Results

The finned tube with a base smooth tube as three different cases (external, internal and mixed) integral finned tube at five different volume flow rates of water (6, 8, 10, 12 and 14 liter/min) and one volume flow rate of air under the turbulent flow regime were investigated in a concentric parallel flow heat exchanger to study the improvement of heat transfer owing to finned tube, as shown in figure (13).
The Nu number was found to increase by increasing the Re number. A good acceptance was obtained amid the experimental data and the outcomes attained from the Dittus-Boelter correlation. The maximum deviation between the experimental data and the Dittus-Boelter correlation was about (5%). The assessment among the Nu numbers for the water side of finned tube at three various states (internal, external and mixed) hinge integral finned tube with smooth tube alone is shown in figures (14) to (16).

The investigational outcomes refer to that the Nu number in finned tube at three different states (case 2 to case 4) is greater by (48.7%, 61.7% and 68.1%) than that of case 1 (smooth tube), respectively. The maximum increment in Nu numbers for the triangular fins is roughly (68.1%) higher than those for the plain tube (case 1). At Reynolds number in the range (212769 - 496391) in mixed (internal and external) integral finned tube at the temperature of water is (15°C). The enhancement in the coefficient of heat transfer is owing to decrease the distance between two fins which increase the grossness of the surface of tube, resulting in destroying the boundaries laminar sub layer which coated the tube. The increasing of turbulence causes a better heat transfer between the tube wall and air. These factors caused the higher values of heat transfer coefficient and as a result the higher Nu number values. The Nu number of air side for the finned tube is bigger than that for smooth tube because of the rise in the area of surface via fins and due to the turbulence generated.

The increase in Nu number might be offered via computing the total ratio of improvement $(Nu_{fr}/Nu_s)/(f_{fr}/f_s)^{1/3}$ as manifested in figure (17). When utilizing the states from case 2 to case 4 (external, internal and mixed), the improvement ratio $(Nu_{fr}/Nu_s)/(f_{fr}/f_s)^{1/3}$ is reduced from (2.684) to (1.806) at (6 L/min) from mixed, external to internal from (2.585) to (1.677) at 8 L/min, from (2.681) to (1.696) at 10 L/min, from (2.937) to (1.849) at 12 L/min, and from (2.716) to (1.701) at 14 L/min. See also figures 18-23.

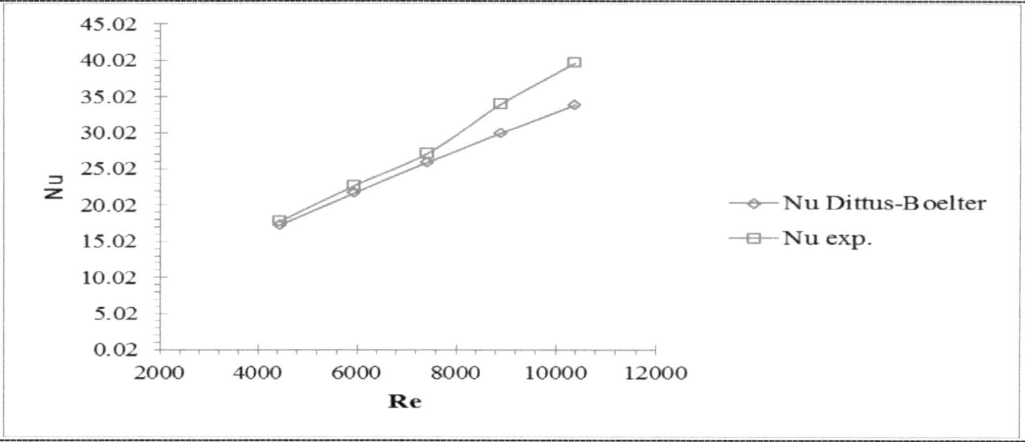

Figure 13. The comparison between experimental data and empirical correlations

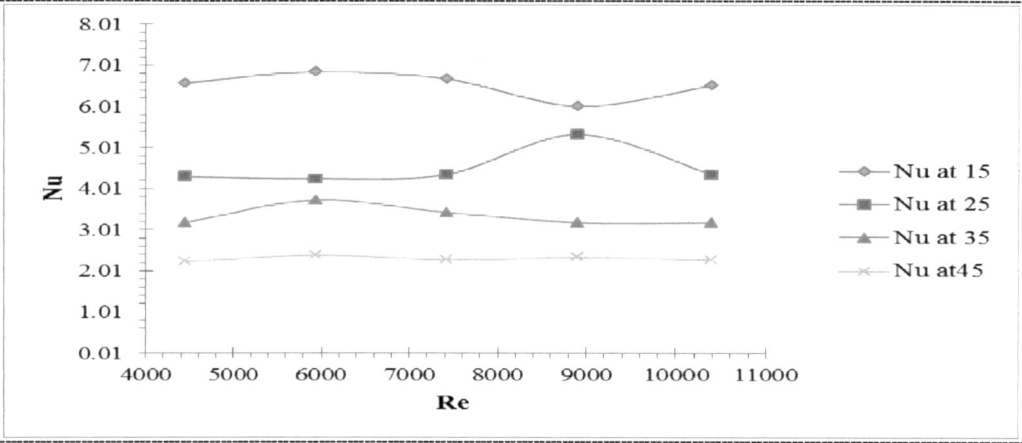

Figure 14. Influence of the water Re number with the air side Nu number at smooth tube.

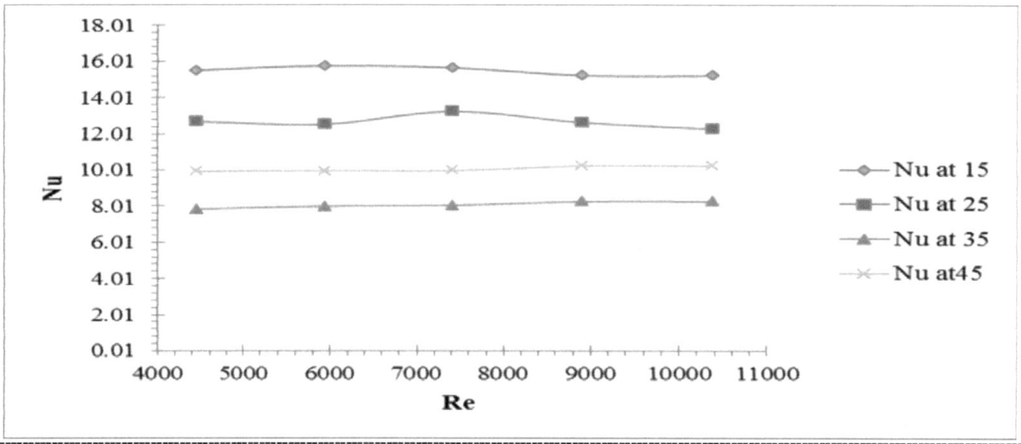

Figure 15. Influence of air side Nu number with the water Re number at the external integral finned tube.

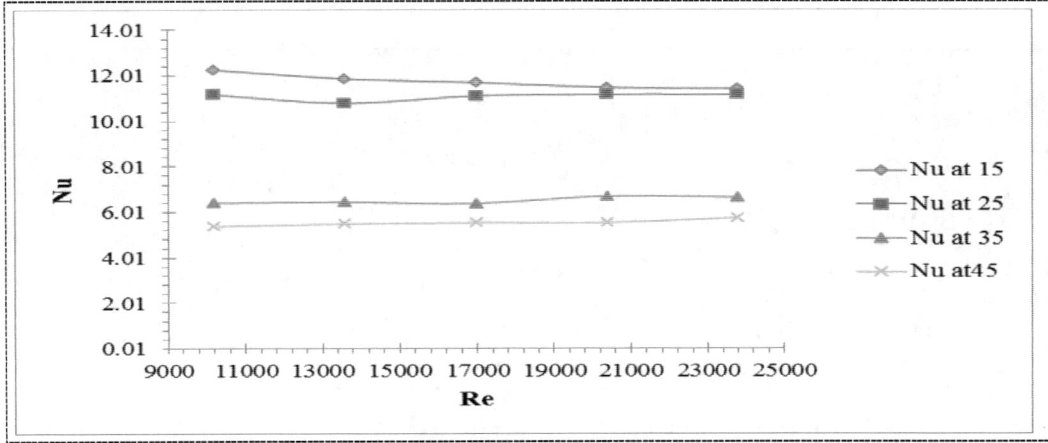

Figure 16. Influence of water Re number with the air side Nu number at the internal integral finned tube.

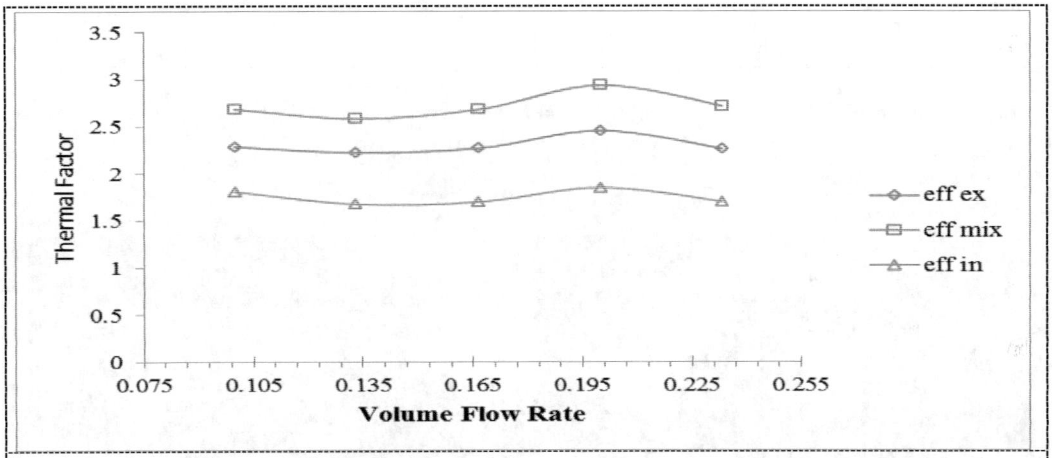

Figure 17. Influence of flow rate on the total ratio of improvement for air side.

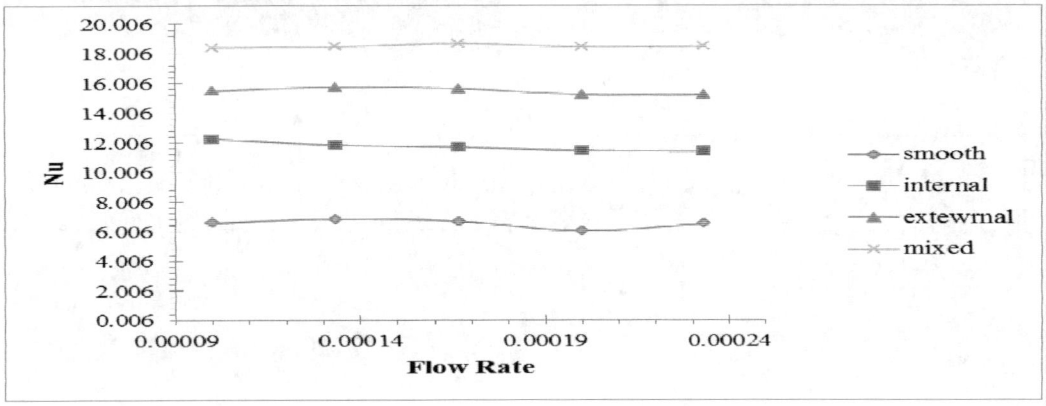

Figure 18. Influence of integral fins on the Nu number for different flow rates

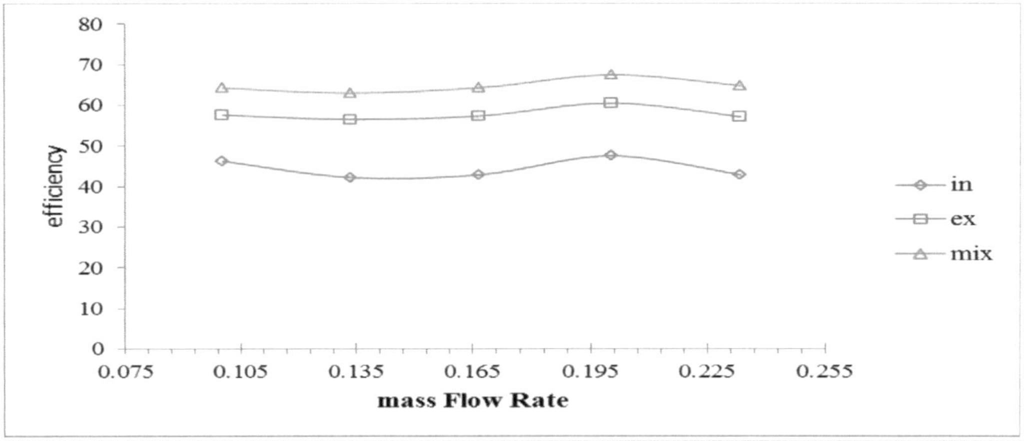

Figure 19. Variation of mass flow rate with the efficiency for finned tube

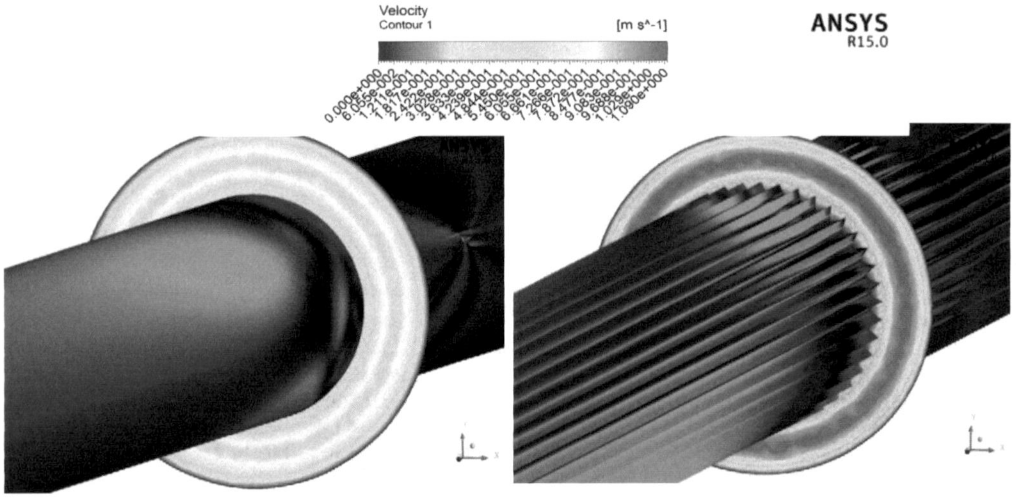

Figure 20. Velocity contours for shell side with air flow over; (a) smooth tube, (b) finned tube with pitch (2.5 mm).

IOP Conf. Series: Earth and Environmental Science **877** (2021) 012044 doi:10.1088/1755-1315/877/1/012044

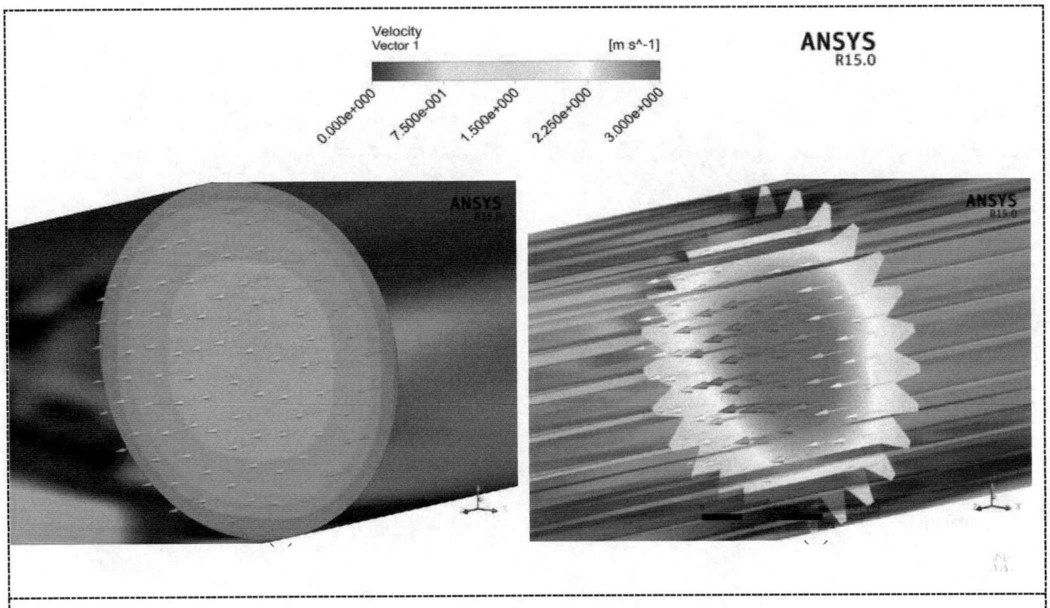

Figure 21. Velocity contours for tube side with water flow within; (a) smooth tube, (b) internal finned tube with pitch (2.5 mm).

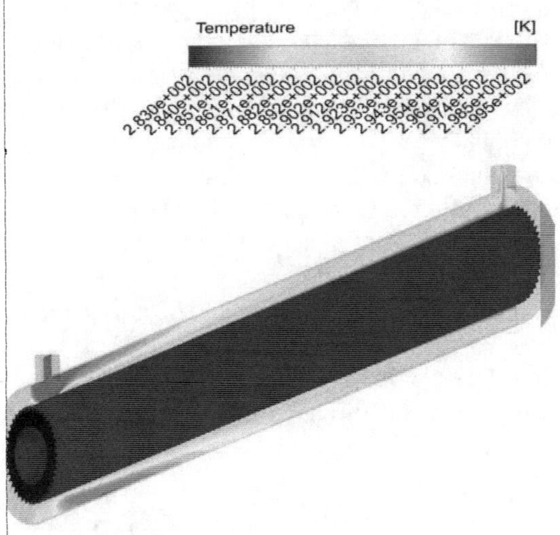

(a) Flow over smooth tube [G04]. (b) Flow over external finned tube with pitch 2.5 mm

[G01-P1].

Figure 22. Temperature contours for the shell side with air flow at (4 m/s at inlet).

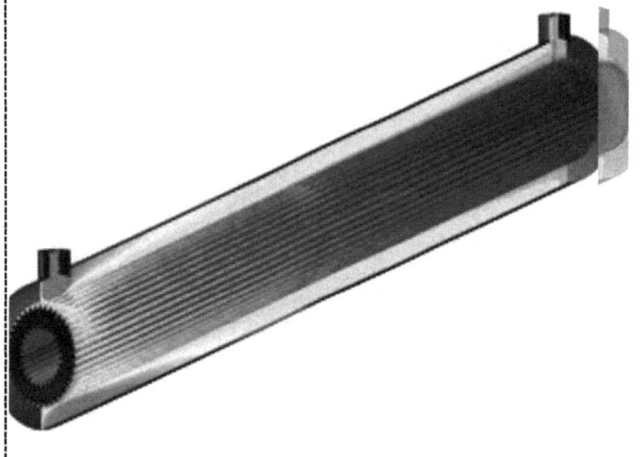

(c) Flow over internal finned surfaces tube
(d) Flow over finned surfaces tube
with pitch 2.5 mm [G03-P1].
with pitch 2.5 mm [G03-P1].

Figure 23. Temperature contours for the shell side with air flow at (4 m/s at inlet).

10. Conclusion

The increasing of H.T is obvious when using fins on the surface of (outer and inner) of internal tube in the current H.E. The appeared improvement is evident about (67.5%) than that of smooth tube. The use of the integral finned tube increases the Nu with the decreased water temperature. The smooth tube offers lesser Nu than the finned tubes. The less pitch (p = 2.5 mm) matched with additional pitch. The outcomes of the numerical simulation elucidated that the addition of fins will improve the dispersal of heat over the heat exchanger, and give numerical results in a good agreement with the results from the experimental work, with a maximum deviation of (10%). The outcomes from the experimental work matched with results obtained from Dittus-Boelter. Good agreements were got with an extreme deviation about (5%). The significant improvement is found in H.E via accepting fins on the mixed (internal and external) integral fins of inner tube. This enhancement appeared clearly in the finned tube with three cases (internal, external and mixed) is (47.5%, 60.5% and 67.5%) than that of smooth tube (case 1), respectively.

11. Future Works

1. Study the effect of another design of an integral finned tube on the flow and heat transfer enhancement.
2. Investigating the effect of using different materials tubes on the heat transfer rate.
3. Study the effect of utilizing various types of fluid on the heat transfer rate.

Acknowledgment

The main author would like to express this gratitude to the Iraqi Ministry of Higher Education and Scientific Research for providing a studentship to conduct this study, as well as the writers would like to show their gratitude to the University of Mustansiriyah (www. uomustansiriyah.edu.iq) in Baghdad-Iraq for the support in conducting this current work.

References

[1] Bergles A., 1999, The Imperative to Enhance Heat Transfer, *Ren. Poly. Inst.*, pp. 13-29.
[2] Sparrow E. and Ramsey J., 1978, Heat Transfer and Pressure Drop for a Staggered Wall-Attached Array of Cylinders with Tip Clearance, *Int. J. Heat Mass Transfer*., **21,** pp. 1369-1377.
[3] Rabas T., Eckels P. and Sabatino R., 1981, The Effect of Fin Density on the Heat Transfer and Pressure Drop Performance of Low-Finned Tube Banks. *Che. Eng. Comm*, **10,** pp. 127-

147.
[4] Hu X. and Jacobi A., 1993, Local Heat Transfer Behavior and its Impact on a Single-Row Annularly Finned Tube Heat Exchanger, *ASME J. of H. T.*, **115,** pp. 66-74.
[5] Rahmah A., 2011, Experimental Study of an Integral Finned–Tube Heat Exchanger, *Uni. of Tech., Mech. Eng. Dep., M.Sc. Thesis.*
[6] Braga C. and Saboya F., 1999**,** Turbulent Heat Transfer, Pressure Drop and Fin Efficiency in Annular Regions with Continuous Longitudinal Rectangular Fins, *Exp. Ther. and flu. Sci.,* ***20***, pp. 55-65.
[7] Al-Jaberi A., Majid H. and Saheb B., 2014*,* Circular Fins with Slanted Blades: Uniform Heat Flux and Isothermal Processes, *Iint. J. of Mech. Eng. and Tech.*, **5,** pp. 133-143.
[8] Essa B., 2016, Experimental and Numerical Investigation to Evaluate the Performance of Triangular finned Tube Heat exchanger, *Uni. of Nah., Mech. Eng. Dep., M.Sc. Thesis.*
[9] Rahmah A., 2011, Experimental Study of an Integral Finned–Tube Heat Exchanger, *Uni. of Tech., Mech. Eng. Dep., M.Sc. Thesis.*
[10] Davidson L., 2009*,* An Introduction to Turbulence Models, *Dep. of Ther. and Flu. Dyn., Chalmers Uni. of Tech.*, Sweden.
[11] Sukhatme S., Jagadish B. and Prabhakaran P., 1990, Film Condensation of R-11 Vapor on Single Horizontal Enhanced Condenser Tubes, *Transactions of the ASME, J. of H. T.*, **112**, pp. 229-234.

Use of Cement Kiln Dust and Silica Fume as partial replacement for cement in concrete

Raid Hussian[1]
[1] Department of Environment, Ministry of Oil, Iraq
Email: r.hussain@yahoo.com

Abstract. Cement is amongst the most polluting materials utilized in the building sector, contributing to a variety of hazardous pollutants, including greenhouse gas emissions. This raises health impacts related to the manufacture of cement. As a result, a substitute substance for conventional cement with low environmental effects and better building characteristics is required. The purpose of the study would be to look at the consequences of using supplementary cementitious materials (SCMS) to substitute cement in a concrete mix partially. This study employed silica fume (SF) and cement kiln dust (CKD) as supplementary cementitious materials. Several concrete mixtures were created by substituting cement by a combination of SF and CKD in three proportions which that 25%, 35%, and 45% within curing periods of (one week and four weeks); the concrete mixtures were tested. The ultrasonic pulse velocity (UPV) test has been used to investigate the concrete mixture's strength in this study. The findings show that the optimal proportion of SF replacement cement and CKD involvement ranged from 25% to 35%. The pulse velocity of specimens improves when the proportion of CKD and SF increases to the optimal percentage, while the larger amounts of these by-products begin to lower the pulse velocity of specimens.

1. Introduction

Cement is an essential part of several forms of the construction industry and is needed in large quantities to create new buildings [1, 2]. Despite the fact that cement has a vital role in the upkeep of civilization and the global financial system, its role in environmental deterioration is well-known [3-5]. Cement factories are estimated to emit 7 to 10% of total carbon dioxide in the air, leading to catastrophic environmental effects such as weather change [6-10]. Therefore, the pollution and consumption of water have risen substantially. The released effluents from concrete plants and casting activities are highly particular due to the chemical composition of conventional concrete [11, 12]. Also, the production of solid waste from the cement industry had resin remarkably, especially from demolishing old concrete structures in the cities (municipal areas) [13-15]. As a result, devastating impacts on the overall quality of water and the degradation of living beings have occurred [16, 17]. Different treatments techniques have been developed to remove many pollutants found in cement plant effluent and other industrial wastewaters [18-20], including filtrations [21-24], coagulation [25-29], electrocoagulation [30-32], sonication-assisted [33, 34], electro-chemical [35-38], electro-physical [39-42], and hybridised methods [43-46]. However, recent studies show an increase in water pollution [47] and freshwater consumption [48-50]. Additionally, these techniques are inadequate to control all anticipated pollutants from cement factories. Because of the huge volumes of carbon dioxide released into the air from the operations related to conventional cement, which is responsible for climate change [51, 52] and global warming [53, 54], as well as contaminants discharged into water bodies, cement manufacture has become a growing subject of attention [1, 55]. As a consequence, reviews show that the development of replacements for cement components, such as silica fume (SF) and cement kiln dust (CKD), is the most viable solution to minimize cement production [9].

Although the utilization of SF and CKD can improve the properties of concrete, Previous studies have proven that the use of these materials also has detrimental effects on the characteristics of concrete. As the percentage of CKD to cement increased, the workability of fresh mixes dropped. Furthermore, as the percentage of CKD increases, the strength of concrete lowers. SCMS materials like SF and CKD are extremely effective and widely employed as cementitious materials because they have large surface surfaces and considerable silica oxides. Earlier research has shown that replacing Cement with SF at a proportion of (0.22-.30) is an effective way to maintain the strength of concrete. The ultrasonic pulse velocity (UPV) is a non-destructive testing method used to assess the quality of concrete buildings. This technique, which involves measuring the velocity of the ultrasonic pulse velocity travelling through concrete members, is used to analyze several attributes of concrete, such as quality and strength [56, 57].
This research assessed the effectiveness of SF and CKD as cement substitutes. The main goal of this study is to investigate if these chemicals impact cement characteristics like strength at different curing periods using an ultrasonic pulse velocity test (one and 4) weeks. The ultrasonic pulse technology was used in this study because it is cheap and accurate [58-61], and also it could be used on the surface of the concrete sample or by embedding it in the body of the concrete and connect it to the receiver using wireless technologies [62-66].

2. Experimentation curriculum

In a number of lab tests, the ultrasonic pulse velocity of the concrete mix produced by partially substituting cement with SF and CKD was measured.

2.1. Materials

Cement kiln dust (CKD) is a by-product of the cement industry. It's a finely powdered substance that looks like Portland cement. Usually, CKD is made up of micron-sized grains recovered by combustion processes during the cement clinker manufacturing process [4, 9]. Cement kiln dust CKD is a fine powdery substance that ranges in hue from grey to brown and is relatively homogeneous in dimension. The manufacturing process, dust collecting technique, chemical properties of CKD, and alkali concentration all influence the gradation of CKD. With fly ash and GGBS in various percentages up to 16%, this product could be used as a cementitious substance. If CKD is utilized separately, the resultant combination may have decreased workability, weight, and setting time due to the high alkali concentration [9, 56].

Silica fume is regarded as a by-product of the silicon and ferrosilicon alloy manufacturing industries, and it is produced at extreme temps from quartz reduction. Because of its properties that promote the cementitious reaction, silica fume is widely utilized as a cementitious material in concrete. It is an ultrafine powder consist of 84-96 non-crystalline silica and about 76 percent silicon [4]. The quantity of Silicon dioxide in silica fume is proportional to the kind of alloy generated in the manufacturing facility.
SF partials are extremely tiny and round, roughly 100 times finer than ordinary cement particles. Previous researches show that the SF concrete has reduced bleeding, porosity, and permeability Because SF oxides react with and consume Calcium hydroxide, which is CH generated during cement hydration. The main binding ingredient in this experiment was Portland Cement, which has strong mechanical characteristics that help the combination to remain coherent. The cement properties used in this study were measured according to BS EN 196-2:2013.
Diagrams 1, 2 and 3 show the chemical composition of SF, CKD, and Portland cement. These features meet the requirements of the BS-EN-197-1(2011) and BS-EN-450-1 standards (2012).
The particle size of the grains, and even the chloride and sulfate concentrations, were checked using the BS EN 12620:2002+A1 standard (2008).
Concrete was prepared and treated using impure and organic-free portable water.

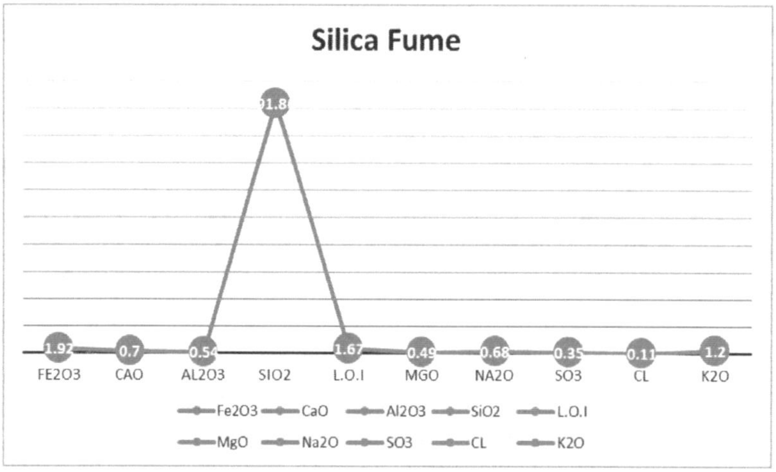

Figure 1: The chemical structure of silica fume.

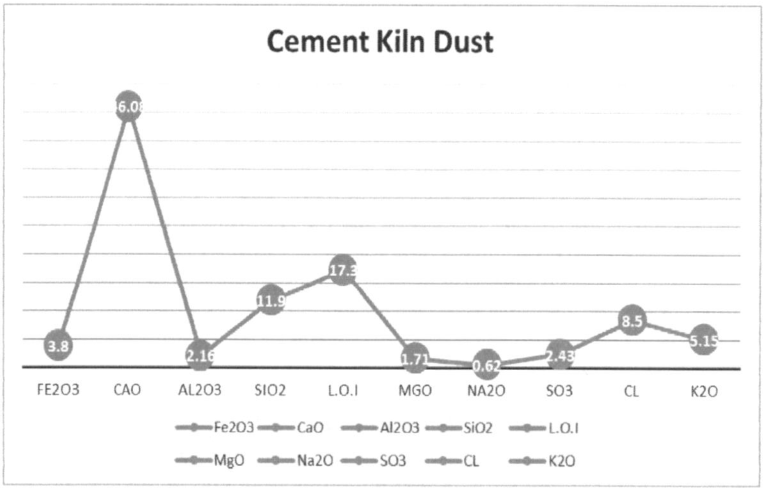

Figure 2: The chemical structure of CKD.

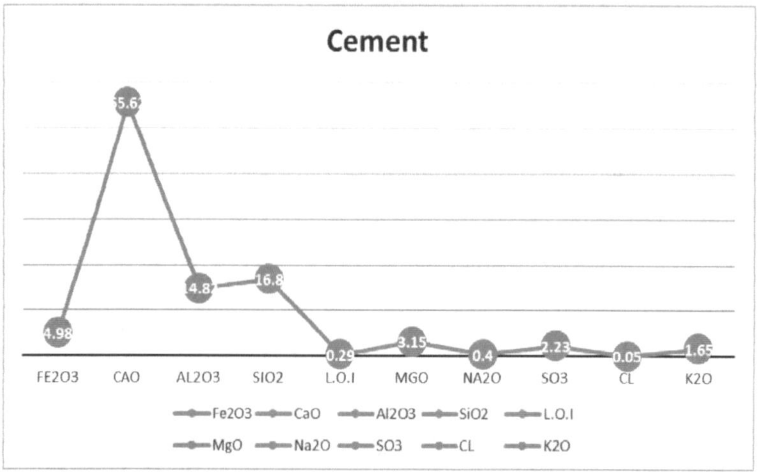

Figure 3: The chemical structure of Portland cement.

2.2. Testing Techniques

For every mix, three prisms (160x40x40 mm) were cast to see how substituting cement with CKD and SF affected the quality of the mortar mixture. These tests are limited to examining items poured from cement mortar. The investigated specimens are then built up of cement in three different concentrations of CKD and SF. Upon completion of the initial setting time of the mixtures, all samples were maintained in good condition, moulded, and placed in water for the cure. BS EN 12504-4:2004 was used to conduct ultrasonic tests on hardening specimens at one week and 4 weeks.

2.3. Design of Mixture

In this study, part of the design process includes determining the proportions of fine aggregate, water, cement, and materials additive ingredients for the control concrete mixes. To match conventional rating curves, fine aggregates have been utilized in the mixture design. The water to binder ratio was 0.4 in all of the mixes. The ratio of sand to a binder that used in this study was 2.4 in all of the mixes. The percentage of each component of the mix is shown in figure (4).

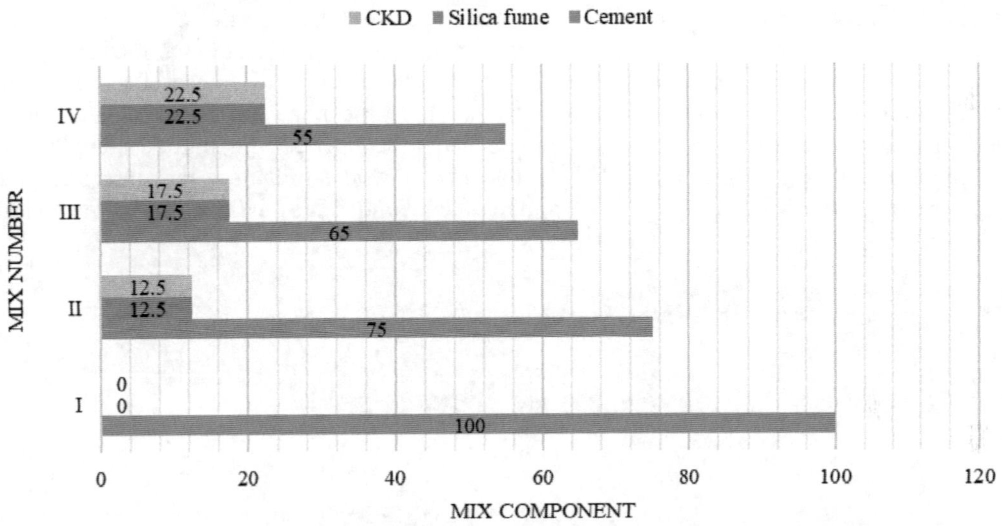

Figure 4: The Mix design.

The ultrasonic pulses are sent through the sample to be examined, and the time required for the pulse to permeate the specimen is measured. A high speed implies that the examined structure is of top condition, whereas a low velocity suggests that the examined structure is of bad condition. Pulse producers, a transducer for converting electronic pulses into mechanical pulses with vibrations of 40 to 50 kHz, and a pulses detector are all used in UPV assessment. The velocities of pulses are determined as follows:

$$\text{Pulse velosity} = \frac{\text{The specimen's thickness}}{\text{The required time for the pulse to penetrate the sample}} \quad (1)$$

3. Results

The results of examining a control concrete mix with substitute material of cement with varying quantities of SF and CKD at different curing periods are explained in Table 1 and Figure 5.

Table 1. One week and 4 weeks ultrasonic pulse velocity testing.

Number of tests	Cement (%)	SF (%)	CKD (%)	UPV test (m/s)	
				One week	Four weeks
MIX1	100	0	0	3828	4008
MIX2	75	12.5	12.5	3850	4102
MIX3	65	17.5	17.5	3785	4055
MIX4	55	22.5	22.5	3578	3863

The main conclusions that may be drawn from the findings of this study are listed below:
The use of a partial replacement of SF and CKD in mixes has been demonstrated to lower concrete pulse velocity values by a tiny portion. In comparison to the control mixture, the pulse velocity of mixtures two and three have been enhanced by 2.3% and 11% after 4 weeks of curing, respectively. Whereas mixture number four decrease the velocity by 3.6 %. This would be based on the view that just a little amount of cementitious materials is required to fill empty fields in the mortar, therefore improving its mechanical properties. Previous studies showed that CKD and SF are ineffectual substances at first and require time to connect with cement components. Therefore, it can be noted that After one week of curing, using 45 percent of a partial substitute decreases the pulse velocity measurements of the mix by around 6.5 percent. At the same time, this value fell to 3.6 after curing for 4 weeks. This is because extra cementitious ingredients reduce concrete compressive strength, which is a key component of the manufacture gel (C-S-H) in concrete. After sitting time, they interact and utilize the moisture components, Ca (OH)2, to allow and initiate hydration of silica fume and cement kiln dust.

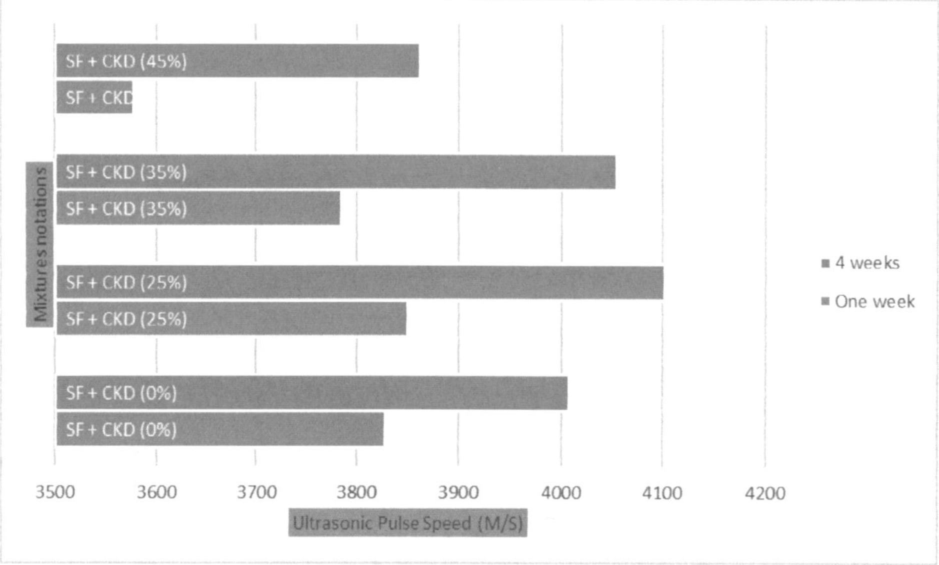

Figure 5. Ultrasonic Pulse Velocity Monitoring with one week and 4weeks curing day mixtures.

According to the observations, specimens that were treated for four weeks had greater pulse velocity readings. That was based on the fact that the curing time improves C-S-H, which leads to a decline in the number of interior gaps or porosity in the conventional concrete, which impacts the properties of concrete and enhances its capacity to withstand compressive stresses.
The Ultrasonic Pulse Velocity method, a non-destructive test methodology, was used in this investigation. As a result, more sophisticated procedures for verifying concrete properties are now accessible. Sensors were used to monitor microcracks, concrete humidity, and other applications in the past. Additional research might utilize the same approach.

4. Conclusions

According to the results of the analysis, it could be stated that as the fraction of cement substituted by silica fume and CKD in concrete increases, the pulse velocity values decrease. The material quality, on the other hand, shows a little improvement with a restricted replacement rate. Whenever the cement in a combination is replaced with extra material, a longer curing period results in a higher-quality specimen.

The use of 25 ~ 35 percent additional cementitious material as a cement substitution for cement could be appropriate proportions, with an increment in this ratio resulting in a slight improvement in quality assessment. Moreover, the used strategy in this research was Ultrasonic Pulse Velocity, which is a traditional instrument; consequently, more current approaches, such as Laser Scanners, are recommended to evaluate the mechanical properties of concrete.

References

[1] Majdi H S, Shubbar A, Nasr M S, Al-Khafaji Z S, Jafer H, Abdulredha M, Masoodi Z A, Sadique M and Hashim K 2020. Experimental data on compressive strength and ultrasonic pulse velocity properties of sustainable mortar made with high content of GGBFS and CKD combinations. *Data in Brief*, **31** 105961-72.

[2] Jasim I, Farhan S and AL-Mamoori S 2021. Climatic Treatments for Housing in the Traditional Holy Cities: A Comparison between Najaf and Yazd Cities. *IOP Earth and Environmental Science,* 012017.

[3] Shubbar A A, Sadique M, Nasr M S, Al-Khafaji Z S and Hashim K S 2020. The impact of grinding time on properties of cement mortar incorporated high volume waste paper sludge ash. *Karbala International Journal of Modern Science*, **6** 1-23.

[4] Shubbar A A, Sadique M, Shanbara H K and Hashim K 2020 *The Development of a New Low Carbon Binder for Construction as an Alternative to Cement. In Advances in Sustainable Construction Materials and Geotechnical Engineering*, Berlin: Springer.

[5] Farhan S, Antón D, Akef V and Hashim K S 2021. Factors influencing the transformation of Iraqi holy cities: the case of Al-Najaf. *Scientific Review Engineering and Environmental Sciences*, **30** 365-75.

[6] Obaid M, Nasr M, Ali I, Ali A and Hashim K S 2021. Performance of green mortar made from locally available waste tiles and silica fume. *Journal of Engineering Science and Technology*, **16** 136-51.

[7] Al-Jumeily D, Hashim K, Alkaddar R, Al-Tufaily M and Lunn J 2019. Sustainable and Environmental Friendly Ancient Reed Houses (Inspired by the Past to Motivate the Future). *11th International Conference on Developments in eSystems Engineering (DeSE),* Cambridge, UK 214-9.

[8] Grmasha R, Al-sareji O, Salman J and Jasim I 2020. Polycyclic Aromatic Hydrocarbons (PAHs) in Urban Street Dust Within Three Land-Uses of Babylon Governorate, Iraq: Distribution, Sources, and Health Risk Assessment. *Journal of King Saud University - Engineering Sciences*, **33** 1-18.

[9] Kadhim A, Sadique M, Al-Mufti R and Hashim K 2020. Long-term performance of novel high-calcium one-part alkali-activated cement developed from thermally activated lime kiln dust. *Journal of Building Engineering*, **32** 1-17.

[10] Al-Sareji O, Grmasha R, Salman J, Idowu I and Hashim K S 2021. Street dust contamination by heavy metals in Babylon governorate, Iraq. *Journal of Engineering Science and Technology*, **16** 3528 - 46.

[11] Alyafei A, AlKizwini R S, Hashim K S, Yeboah D, Gkantou M, Al Khaddar R, Al-Faluji D and Zubaidi S L 2020. Treatment of effluents of construction industry using a combined filtration-electrocoagulation method. *IOP Conference Series: Materials Science and Engineering,* 012032.

[12] Alnaimi H, Idan I J, Al-Janabi A, Hashim K, Gkantou M, Zubaidi S L, Kot P and Muradov M 2020. Ultrasonic-electrochemical treatment for effluents of concrete plants. **888** 1-9.

[13] Abdulredha M, Rafid A, Jordan D and Alattabi A 2017. Facing up to waste: how can hotel managers in Kerbala, Iraq, help the city deal with its waste problem? *Procedia engineering*, **196** 771-8.

[14] Abdulredha M, Ali A and Jordan D 2020. Estimating municipal solid waste generation from service processions during the Ashura religious event. *IOP Materials Science and Engineering,* 012075.

[15] Abdulredha M, Kot P, Al Khaddar R, Jordan D and Abdulridha A 2020. Investigating municipal solid waste management system performance during the Arba'een event in the city of Kerbala, Iraq. *Environment, Development and Sustainability*, **22** 1431-54.

[16] Alwan H, Saleh L and Abdulredha M 2020. Experimental prediction of the discharge coefficients for rectangular weir with bottom orifices. *Journal of Engineering Science and Technology*, **15** 3265-80.

[17] Abdulredha M, Jordan D and Abdulridha A 2018. Benchmarking of the Current Solid Waste Management System in Karbala, Iraq, Using Wasteaware Benchmark Indicators. *World Environmental*

and Water Resources Congress 2018: Groundwater, Sustainability, and Hydro-Climate/Climate Change, 40-8.

[18] Hashim K, Al-Saati N, Hussein A and Al-Saati Z 2018. An investigation into the level of heavy metals leaching from canal-dreged sediment: a case study metals leaching from dreged sediment. *1st Int. Conference on Materials Engineering & Science,* Istanbul Aydın University, Turkey 12-22.

[19] Hashim K, Al-Saati N, Alquzweeni S, Kraidi L, Hussein A and Alwash R 2019. Decolourization of dye solutions by electrocoagulation: an investigation of the effect of operational parameters. *First International Conference on Civil and Environmental Engineering Technologies,* University of Kufa, Iraq 25-32.

[20] Alattabi A, Harris C, Alkhaddar R, Alzeyadi A and Abdulredha M 2017. Online Monitoring of a sequencing batch reactor treating domestic wastewater. *Procedia engineering*, **196** 800-7.

[21] Abdulraheem F S, Al-Khafaji Z, Muradov M, Kot P and Shubbar A A 2020. Natural filtration unit for removal of heavy metals from water. *IOP Materials Science and Engineering,* 012034.

[22] Alenazi M, , Hassan A A, Muradov M, Kot P and Abdulhadi B 2020. Turbidity removal using natural coagulants derived from the seeds of strychnos potatorum: statistical and experimental approach. *IOP Conference Series: Materials Science and Engineering,* 012064.

[23] Hashim K S, Ewadh H M, Muhsin A A, Zubaidi S L, Kot P, Muradov M, Aljefery M and Al-Khaddar R 2020. Phosphate removal from water using bottom ash: Adsorption performance, coexisting anions and modelling studies. *Water Science and Technology*, **83** 1-17.

[24] Al-Saati N H, Hussein T K, Abbas M H, Hashim K, Al-Saati Z N, Kot P, Sadique M, Aljefery M H and Carnacina I 2019. Statistical modelling of turbidity removal applied to non-toxic natural coagulants in water treatment: a case study. *Desalination and Water Treatment*, **150** 406-12.

[25] Omran I I, Al-Saati N H, Al-Saati H H, Hashim K S and Al-Saati Z N 2021. Sustainability assessment of wastewater treatment techniques in urban areas of Iraq using multi-criteria decision analysis (MCDA). *Water Practice and Technology*, **16** 648-60.

[26] Omran I I, Al-Saati N H, Hashim K S, Al-Saati Z N, Patryk K, Khaddar R A, Al-Jumeily D, Shaw A, Ruddock F and Aljefery M 2019. Assessment of heavy metal pollution in the Great Al-Mussaib irrigation channel. *Desalination and Water Treatment*, **168** 165-74.

[27] Ryecroft S P, shaw A, Fergus P, Kot P, Hashim K and Conway L 2019. A Novel Gesomin Detection Method Based on Microwave Spectroscopy. *12th International Conference on Developments in eSystems Engineering (DeSE),* Kazan, Russia 429-33.

[28] Zubaidi S L, Al-Bugharbee H, Muhsen Y R, Hashim K, Alkhaddar R M, Al-Jumeily D and Aljaaf A J 2019. The Prediction of Municipal Water Demand in Iraq: A Case Study of Baghdad Governorate. *12th International Conference on Developments in eSystems Engineering (DeSE),* Kazan, Russia 274-7.

[29] Mousazadeh M, Paital B, Naghdali Z, Mortezania Z, Hashemi M, Aghababaei M, Ghorbankhani M, Lichtfouse E, Sillanpää M and Emamjomeh M M 2021. Positive environmental effects of the coronavirus 2020 episode: a review. *Environment, Development and Sustainability*, **21** 1-23.

[30] Hashim K S, Hussein A H, Zubaidi S L, Kot P, Kraidi L, Alkhaddar R, Shaw A and Alwash R 2019. Effect of initial pH value on the removal of reactive black dye from water by electrocoagulation (EC) method. *2nd International Scientific Conference,* Al-Qadisiyah University, Iraq 12-22.

[31] Hashim K, Kot P, Zubaid S, Alwash R, Al Khaddar R, Shaw A, Al-Jumeily D and Aljefery M 2020. Energy efficient electrocoagulation using baffle-plates electrodes for efficient Escherichia Coli removal from Wastewater. *Journal of Water Process Engineering*, **33** 101079-86.

[32] Hashim K S, AlKhaddar R, Shaw A, Kot P, Al-Jumeily D, Alwash R and Aljefery M H 2020 *Electrocoagulation as an eco-friendly River water treatment method. In Advances in Water Resources Engineering and Management,* Berline: Springer.

[33] Zanki A K, Mohammad F H, Hashim K S, Muradov M, Kot P, Kareem M M and Abdulhadi B 2020. Removal of organic matter from water using ultrasonic-assisted electrocoagulation method. *IOP Conference Series: Materials Science and Engineering,* 012033.

[34] Khalid K, Andy S and Shamma'a A 2021. Water purification from metal ions in the presence of organic matter using electromagnetic radiation-assisted treatment. *Journal of Cleaner Production*, **280** 1-17.

[35] Aqeel K, Mubarak H A, Amoako-Attah J, Abdul-Rahaim L A, Al Khaddar R, Abdellatif M, Al-Janabi A and Hashim K S 2020. Electrochemical removal of brilliant green dye from wastewater. *IOP Conference Series: Materials Science and Engineering,* 012036.

[36] Mohammed A, Hussein A, Yeboah D, Abdulhadi B, Ali A A and Hashim K S 2020. Electrochemical removal of nitrate from wastewater. *IOP Materials Science and Engineering,* 012037.

[37] Abdulhadi B, Kot P, Hashim K, Shaw A, Muradov M and Al-Khaddar R 2021. Continuous-flow electrocoagulation (EC) process for iron removal from water: Experimental, statistical and economic study. *Science of The Total Environment*, **760** 1-16.

[38] Al-Saati N H, Omran I I, Salman A A, Al-Saati Z and Hashim K S 2021. Statistical modeling of monthly streamflow using time series and artificial neural network models: Hindiya Barrage as a case study. *Water Practice and Technology*, **16** 681-91.

[39] Emamjomeh M, Kakavand S, Jamali H, Alizadeh S, Safdari M, Mousavi S and Mousazade M 2020. The treatment of printing and packaging wastewater by electrocoagulation–flotation: the simultaneous efficacy of critical parameters and economics. *Desalination and water treatment*, **205** 161-74.

[40] Emamjomeh M M, Mousazadeh M, Mokhtari N, Jamali H A, Makkiabadi M, Naghdali Z, Hashim K S and Ghanbari R 2020. Simultaneous removal of phenol and linear alkylbenzene sulfonate from automotive service station wastewater: Optimization of coupled electrochemical and physical processes. *Separation Science and Technology*, **55** 3184-94.

[41] Hashim S, Ali S, AlRifaie J Al Khaddar R, Idowu I and Gkantou M 2020. Escherichia coli inactivation using a hybrid ultrasonic–electrocoagulation reactor. *Chemosphere*, **247** 125868-75.

[42] Bareq B, Andy A and Rafid R 2019. Influence of current density and electrodes spacing on reactive red 120 dye removal from dyed water using electrocoagulation/electroflotation (EC/EF) process. *1st Int. Conference on Civil and Environmental Engineering Technologies,* University of Kufa, Iraq 12-22.

[43] Abdulla G, Kareem M, Muradov M, Kot P, Mubarak H, Abdellatif M and Abdulhadi B 2020. Removal of iron from wastewater using a hybrid filter. *IOP Materials Science and Engineering,* 012035.

[44] Alenezi A K, Hasan H A, Hashim K S, Amoako-Attah J, Gkantou M, Muradov M, Kot P and Abdulhadi B 2020. Zeolite-assisted electrocoagulation for remediation of phosphate from calcium-phosphate solution. *IOP Conference Series: Materials Science and Engineering,* 012031.

[45] Alhendal M, Nasir M, Amoako J, Al-Faluji D, Muradov M and Abdulhadi B 2020. Cost-effective hybrid filter for remediation of water from fluoride. *IOP Materials Science and Engineering,* 012038.

[46] Al-Marri S, AlQuzweeni S S, Hashim K S, AlKhaddar R, Kot P, AlKizwini R S, Zubaidi S L and Al-Khafaji Z S 2020. Ultrasonic-Electrocoagulation method for nitrate removal from water. *IOP Conference Series: Materials Science and Engineering,* 012073.

[47] Idowu I A, Hashim K, Shaw A and Nunes L J 2021. Enhancing the fuel properties of beverage wastes as non-edible feedstock for biofuel production. *Biofuels*, **14** 1-8.

[48] Salah Z, Abdulkareem I, Al-Bugharbee H, Ridha H, Gharghan S, Al-Qaim F, Muradov M, Kot P and Alkhaddar R 2020. Hybridised Artificial Neural Network model with Slime Mould Algorithm: A novel methodology for prediction urban stochastic water demand. *Water*, **12** 1-18.

[49] Salah Z, Hashim K, Ethaib S, Al-Bdairi N, Al-Bugharbee H and Gharghan S 2020. A novel methodology to predict monthly municipal water demand based on weather variables scenario. *Journal of King Saud University-Engineering Sciences*, **32** 1-18.

[50] Salah Z, Ortega-Martorell S, Kot P, Alkhaddar R M, Abdellatif M, Gharghan S K, Ahmed M S and Hashim K 2020. A Method for Predicting Long-Term Municipal Water Demands Under Climate Change. *Water Resources Management*, **34** 1265-79.

[51] Zubaidi S, Al-Bugharbee H, and Alkhaddar R 2020. Forecasting of monthly stochastic signal of urban water demand: Baghdad as a case study. *IOP Materials Science and Engineering,* 012018.

[52] Zubaidi S, Ortega-Martorell S, Al-Bugharbee H, Olier I, Hashim K S, Gharghan S K, Kot P and Al-Khaddar R 2020. Urban Water Demand Prediction for a City that Suffers from Climate Change and Population Growth: Gauteng Province case study. *Water*, **12** 1-18.

[53] Zubaidi Salah L, Al-Bugharbee H, Ortega Martorell S, Gharghan S, Olier I, Hashim K, Al-Bdairi N and Kot P 2020. A Novel Methodology for Prediction Urban Water Demand by Wavelet Denoising and Adaptive Neuro-Fuzzy Inference System Approach. *Water*, **12** 1-17.

[54] Zubaidi S, Alkhaddar R, Abdellatif M and Muhsin Y R 2019. Using LARS–WG model for prediction of temperature in Columbia City, USA. *IOP Materials Science and Engineering,* 012026.

[55] Ali A, Al-Shaer A, Hawesah H and Sadique M 2019. Investigating the influence of cement replacement by high volume of GGBS and PFA on the mechanical performance of cement mortar. *1st Int. Conference on Civil and Environmental Engineering Technologies,* University of Kufa, Iraq 31-8.

[56] Kadhim A, Sadique M, Al-Mufti R and Hashim K 2020. Developing One-Part Alkali-Activated metakaolin/natural pozzolan Binders using Lime Waste as activation Agent. *Advances in Cement Research*, **32** 1-38.

[57] Al-Maliki L A, Farhan S L, Jasim I A, Al-Mamoori S K and Al-Ansari N 2021. Perceptions about water pollution among university students: A case study from Iraq. *Cogent Engineering*, **8** 1895473.

[58] Kot P, Hashim K S, Muradov M and Al-Khaddar R 2021 *How can sensors be used for sustainability improvement?. In Methods in Sustainability Science*, Elsevier, United Kingdom: Joe Hayton, p 426.
[59] Kot P, Muradov M, Gkantou M, Kamaris G S, Hashim K and Yeboah D 2021. Recent Advancements in Non-Destructive Testing Techniques for Structural Health Monitoring. *Applied Sciences*, **11** 1-28.
[60] Omer G, Kot P, Atherton W, Muradov M, Gkantou M, Shaw A, Riley M, Hashim K and Al-Shamma'a A 2021. A Non-Destructive Electromagnetic Sensing Technique to Determine Chloride Level in Maritime Concrete. *Karbala International Journal of Modern Science*, **7** 61-71.
[61] Ryecroft S, Shaw A, Fergus P, Moody A and Conway L 2019. A First Implementation of Underwater Communications in Raw Water Using the 433 MHz Frequency Combined with a Bowtie Antenna. *Sensors*, **19** 1813-23.
[62] Ryecroft S, Fergus P, Tang A, Moody A and Conway L 2021. An Implementation of a Multi-Hop Underwater Wireless Sensor Network using Bowtie Antenna. *Karbala International Journal of Modern Science*, **7** 113-29.
[63] Gkantou M, Muradov M, Kamaris G S, Hashim K, Atherton W and Kot P 2019. Novel Electromagnetic Sensors Embedded in Reinforced Concrete Beams for Crack Detection. *Sensors*, **19** 5175-89.
[64] Khalaf M, Hussain A J, Keight R, Al-Jumeily D, Fergus P, Keenan R and Tso P 2017. Machine learning approaches to the application of disease modifying therapy for sickle cell using classification models. *Neurocomputing*, **228** 154-64.
[65] Ghazali R, Hussain A J, Al-Jumeily D and Merabti M 2007. Dynamic ridge polynomial neural networks in exchange rates time series forecasting. *International Conference on Adaptive and Natural Computing Algorithms,* 123-32.
[66] Fergus P, Hussain A, Al-Jumeily D, Huang D-S and Bouguila N 2017. Classification of caesarean section and normal vaginal deliveries using foetal heart rate signals and advanced machine learning algorithms. *Biomedical engineering online*, **16** 1-26.

Removal of COD from petroleum refinery wastewater using electrocoagulation method

Amal H. Khalil[1], Mohammed A. Naji[2,*], Salam M. Naser[3]
[1] Department of Civil Environmental, College of Engineering, University of Babylon, Babylon, Iraq
[2] Karbala refinery Project, Ministry of oil, Kerbala, Iraq
[3]Al-Mustaqbal University College, Babylon, Iraq.
* Corresponding author, Email: Mohammed.naji@krpscop.org

Abstract. This research assessed the removability of chemical oxygen demand (COD) from petroleum effluent using aluminum-based electrocoagulation reactor. A series of batch flow studies have been conducted to evaluate the impact of current density, electrodes separation, and duration of treatment on the removal of COD from the refinery effluent. The COD levels were determined employing the remaining concentrations using spectrophotometer namely Hach-Lang and standard cuvette test (LCC 514, LCK 314, or APC 400). The findings of the current investigation indicate the capacity of the electrocoagulation technique in a relatively short processing time to reduce the COD levels. The greatest efficiency in removing COD has been determined to be 80.0%. After 100 minutes of electrolysis, a current density of 8 mA/cm^2 and electrodes separation of 20 mm achieved the highest percentage removal.

1. Introduction

The oil sector is essential in the financial system mostly because it constitutes the primary component of the country's national revenue and also because the industrialised civilisation in its existing form must be maintained [1]. Nevertheless, the operations of this sector of the economy pose severe ecological consequences since it uses large volumes of water, particularly for distillation, water treatment, desalination, and cooling operations [2, 3]. Around 300 L of water must be generated to produce 1 barrel of oil, meaning the worldwide oil industry creates approximately 34 million barrels of polluted water each day predicated upon on present decade's average global oil output [1]. Moreover, the fast rise in global oil consumption suggests that the oil sector would continue to produce and release contaminated sewage. Based on its source, the chemical makeup of oil effluent is different. The effects of the oil wastewater are increasing because the amount of freshwater is on the Earth is decreasing gradually due to global warming [4-6], uneven distribution of water [7-10], and rapid urbanisation [11-13]. Oil sewage, for illustration, might include metals such as iron, aromas, sulfates, fats, phenols, cadmium, toxic substances, nitrogen, and radiation contaminants [14-16]. Throughout all oil sewage nevertheless, COD is present. The researches indicated the high content of COD which could reach 5.3 g/l in refineries sewage. It should be noted that the allowable quantity of COD is 125 mg/l in the treated effluent. Additionally, the researchers would like to mention the COD come from other industries, like concrete plants [17-23], cement production [24-29], and contact with the polluted air [30-33]. Such elevated contaminant content is a severe danger to the environment, biological diversity, and public health when it substantially pollutes the surface of water and groundwater and also kills fish populations. These contaminants also contaminate river/lake bottoms and lands surrounding oil facilities and factories that need costly clean-up methods. Such contaminants result in severe illness or toxic consequences on public health. As mentioned before, since global warming continues to raise the water demand and rivers in many places of the globe, the necessity for effective treatment solutions is becoming even more pressing. The current rising temperatures of our planet, resulting from the emission of hazardous gasses to the Earth's atmosphere, have worsened these consequences. Accordingly, a number of researches have been done with the purpose of finding an appropriate treatment technique or a series of treatment procedures for treating the sewage of oil before releasing it into the ecosystem. The removal of one or a set of contaminants that originate in oil effluents, like nitrate, organic substance, phosphate, toxic metals, turbidity and bacteria adsorbents, was done using various removal technologies like adsorbents [34, 35], coagulants [36, 37], electrocoagulation [38-40], and combined methods [41-46]. Nevertheless, almost all of the techniques are expensive and not effective enough to satisfy the required limitations variability in oil effluent chemical structure.

Simulation of a wide variety of contaminants, comprising key contaminants in the sewage sludge, like toxic substances, phosphates, ammonia, dyes, and organic compounds, was stated to be effectively removed by the electrocoagulation (EC) process by a large number of academics [47, 48]. The scientists also said that the EC technique is distinct from previous treatment techniques in terms of low cost, unnecessary to qualified personnel, unnecessary petrochemicals, small size, and selectivity [49-51]. Several researchers also highlighted that the EC technique sludge has many electrodes, thus increasing the possibility of recycling it in concrete slurry production and reducing the requirement for waste disposal [52, 53].

The present study investigated how the EC approach can be feasible for the removal of the chemical oxygen demand (COD) from oil sewage. Due to its appealing benefits, the EC technique was used in this investigation. For example, it requires no chemicals additions, cheap running expenses, and simple operation. It allows workers to monitor both electrical and operational parameters to minimise pollution.

2. Methods

Specimens of oil sewage with 450 mg/l COD were processed utilising EC cells with various operational variables for treatment of the reflationary oil effluent. Table 1 shows the key features of the oil sewage samples.

Table 1. Properties of the sewage of the oil.

Parameter	Value	Unit
pH	6.5	--
Conductivity	9.52	ms/cm
TSS	132	Mg/l
Salinity	0.18	%
COD	450	Mg/l

A DC rectifier was used to apply to the desired current density value ranging from 3 to 7 mA/cm2. A manual adjustment has been adopted to change the spacing between the electrodes to have the necessary value ranging between 20 and 30 mm. The electrolysation procedure was done by pumping the samples of sewage until the reactor has been full. The samples have been stored in the tank for durations of 0 to 120 minutes to complete the treatment of the wastewater.

15 mL samples of water were taken frequently from the EC reactor in order to track the elimination process. The specimens were purified at a filter of 0.45 μm. A spectrophotometer namely Hach Lange DR 2800) has

been used to quantify the remaining COD concentration in the filter using the acceptable estimation cuvette test namely LCK 514, LCK 314, or APC 400, based on residual concentrations. Hach Lang, UK, supplied the standard cuvettes. All tests were completed at room temperature.
The removal efficiency is calculated by the following equation [38]:

$$COD\ removla\ \% = \frac{r_0 - r}{r_0} \times 100\% \qquad (1)$$

where r_0 is the influent COD concentration in mg/L and r is the effluent COD concentration in mg/L.
A rectangular EC reactor was used to remove the COD from the oil refinery effluent. The EC reactor is made of a 200 mm-wide, 75 mm-wide, 120 mm-depth rectangular Perspex container and four aluminum electrodes. The electrode is made of aluminum which is pierced plates measuring 70 mm in width and 120 mm in height. There are 35 circular holes in each electrode. During the treatment process, the electrodes mix sewage which improves elimination performance by improving the collision rate between coagulants and contaminants.

3. Results and discussion

3.1. Spacing of electrodes influence

To assess the effect of spacing on COD removal effectiveness, waste samples having 400 mg/l COD concentration with varying distances between 20 and 30 mm were electrolysed. At a constant current of 4 mA/cm2, and a retention duration of 60 min, the electrocoagulation tests were placed. The results showed that the wider the distance, the lower the removal rate. For example, the removal effectiveness of COD declined from around 48% to 36% after the 60 minutes of electrolysing as the distance grew from 20 mm to 30 mm, Figure 1. There are two explanations for this conduct. Increasing distance between electrodes increases the development of anodising metal foil, which decreases the anode melting rate, slowing down the removal rate. Widening the distance between electrodes lessens the electrostatic attraction force that lowers the production of flocs and hence reduces the effectiveness of removal.

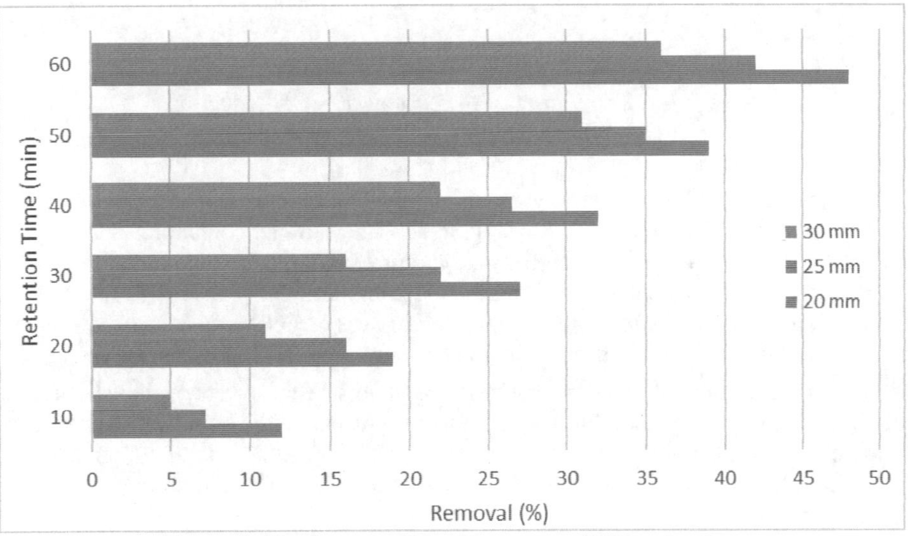

Figure 1: The impact of the distance on COD reduction.

3.2. Current density influence

The formation of the coagulants, flock expansion as well as the sizes and rates of air bubbles are determined by current densities. The effect of COD removal based on current was thus examined using the electrolysis of 400 mg/l COD sewage samples at various currents ranging from 4 to 8 mA/cm2. Spacing and detention duration at 20 mm and 60 min were held constant, respectively.

The results achieved, Figure 2, indicated that the greater the flow, the faster the elimination of COD. Where it can be seen from such achievements that the elimination of COD rose from 49 percent to 62 percent after 60 minutes of electrolysation, as the current increased correspondingly from 4 to 8 mA/cm2. This might be attributed to the improvement of the melted coagulant. As the number of coagulants improved, COD removal was steadily increased and the number of sites in operation improved. But raising the current applied will increase electricity usage.

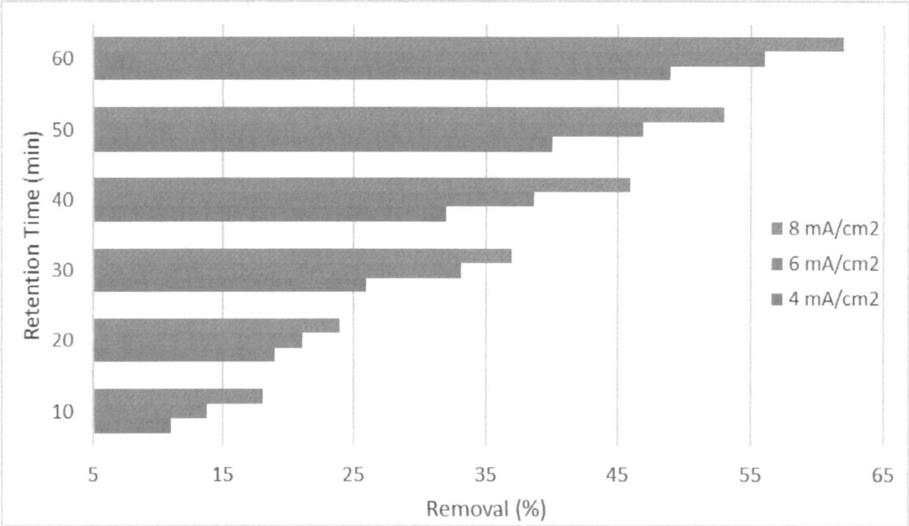

Figure 2: The impact of the current on COD reduction.

3.3. Retention time Influence

The study reveals linear relationships between retention time and coagulant production, which enhances the clearance of pollutants. Therefore, electrolysing a sewage sample containing 400 mg/l COD was investigated in order to eliminate COD for 100 min at current 6 mA/cm2 and a distance of 20 mm. The findings clearly, figure 3, indicate that extending the retention period increases the reduction effectiveness of COD. Where, it has been noted that the removal effectiveness of COD rose from 15 percent to around 81 percent as the retention durations extended from 10 to 100 minutes, correspondingly.

The release of consistent quantities of the anodes' coagulants to the constant voltage and electrolysing periods could clarify this. The quantity of melted aluminum oxides in the liquids, therefore, grows with the growth of the electrolysing times. Consequently, the reduction of COD is also increasing. Longer EC durations also provide improved contact periods between impurities and aluminum oxides, which enhance the extraction of pollutants.

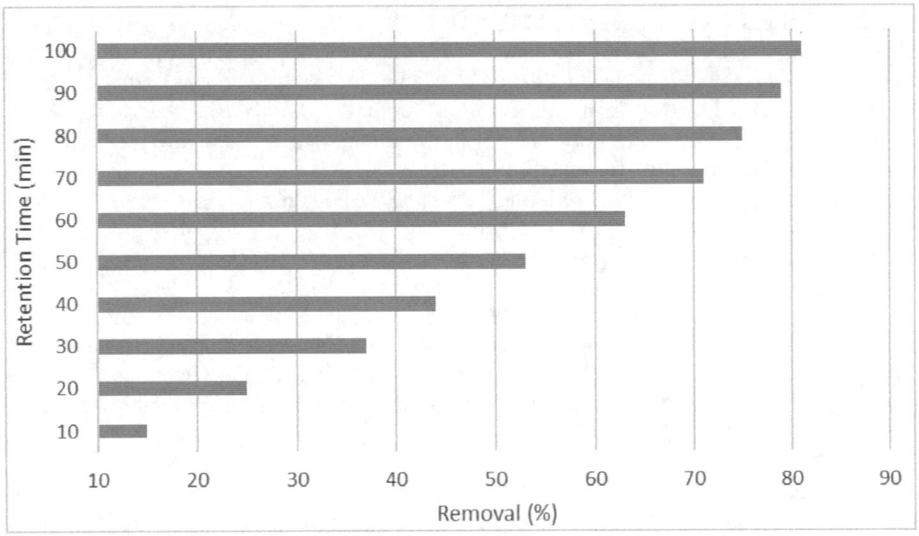

Figure 3: The impact of the Time on COD reduction.

Figure 3, nevertheless, shows that the investigation's last 20 minutes improvement in the removal effectiveness was approximately 6 percent, were not particularly important. This might be related to the formation of a passive layer on the conductors' surfaces that reduces coagulant synthesis and hence the effectiveness of removal. It might thus be fair to maintain the detention duration at 80 minutes, decreasing the operating cost of this procedure in the present research. In conclusion, the findings showed that an important quantity of COD could be removed from sewage from oil refinery effluent, but the full elimination of such contaminants is not sufficiently possible. The author believes thus that this approach may be utilised to increase the performance of other treatment facilities as a pre-processing facility for oil effluent.

Finally, the researchers in this work highlight the recent developments in wireless technologies [54-57], sensing methods [58-60], and IoT [61, 62] could be applied in the development of the electrocoagulation reactor

4. Conclusion

In order to explore the viability of the EC technique as a method for oil sewage, a series of batch flow tests were conducted. The results showed that the COD content in oil refinery effluence might be reduced successfully by this technique. Furthermore, it has been shown that with regards to COD reductions, the efficiency of this technique may be enhanced by increasing the current user or the processing periods or by reducing the distance among poles. Finally, the author suggested that the EC unit could be integrated with other conventional processes to obtain total removal effectiveness at acceptable costs within a reasonably short processing period.

References

[1] Shonza N, Andreatta D, Muniz E, de Freitas R and Porto P 2020. Crude oil wastewater treatment by electrocoagulation in a continuous process with polarity switch. *Environmental Technology*, 1-9.

[2] Hashim K, Al-Saati N, Hussein A and Al-Saati Z 2018. An investigation into the level of heavy metals leaching from canal-dreged sediment: a case study metals leaching from dreged sediment. *First Int. Conference on Materials Engineering & Science,* Istanbul Aydın University, Turkey 12-22.

[3] Hashim K S, AlKhaddar R, Shaw A, Kot P, Al-Jumeily D, Alwash R and Aljefery M H 2020 *Electrocoagulation as an eco-friendly River water treatment method. In Advances in Water Resources Engineering and Management,* Berline: Springer.

[4] Zubaidi S L, Al-Bugharbee H, Muhsen Y R, Hashim K, Alkhaddar R M, Al-Jumeily D and Aljaaf A J 2019. The Prediction of Municipal Water Demand in Iraq: A Case Study of Baghdad Governorate. *12th International Conference on Developments in eSystems Engineering (DeSE),* Kazan, Russia 274-7.

[5] Salah Z, Abdulkareem I H, Hashim K S, Al-Bugharbee H, Ridha H M, Gharghan S K, Al-Qaim F F, Muradov M, Kot P and Alkhaddar R 2020. Hybridised Artificial Neural Network model with Slime Mould Algorithm: A novel methodology for prediction urban stochastic water demand. *Water*, **12** 1-18.

[6] Salah Z, Hashim K, Ethaib S, Al-Bdairi N S S, Al-Bugharbee H and Gharghan S K 2020. A novel methodology to predict monthly municipal water demand based on weather variables scenario. *Journal of King Saud University-Engineering Sciences*, **32** 1-18.

[7] Zubaidi S, Hashim K, Alkhaddar R, Abdellatif M and Muhsin Y R 2019. Using LARS–WG model for prediction of temperature in Columbia City, USA. *IOP Materials Science and Engineering,* 012026.

[8] Salah Z, Ortega-Martorell S, Kot P, Alkhaddar R M, Abdellatif M, Gharghan S K, Ahmed M S and Hashim K 2020. A Method for Predicting Long-Term Municipal Water Demands Under Climate Change. *Water Resources Management*, **34** 1265-79.

[9] Zubaidi S, Muhsin Y, Hashim K and Alkhaddar R 2020. Forecasting of monthly stochastic signal of urban water demand: Baghdad as a case study. *IOP Materials Science and Engineering,* 012018.

[10] Zubaidi S, Ortega-Martorell S, Al-Bugharbee H, Olier I, Hashim K S, Gharghan S K, Kot P and Al-Khaddar R 2020. Urban Water Demand Prediction for a City that Suffers from Climate Change and Population Growth: Gauteng Province case study. *Water*, **12** 1-18.

[11] Farhan S, Antón D, Akef V, Zubaidi S and Hashim K 2021. Factors influencing the transformation of Iraqi holy cities: the case of Al-Najaf. *Scientific Review Engineering and Environmental Sciences*, **30** 365-75.

[12] Al-Maliki L A, Farhan S L, Jasim I A, Al-Mamoori S K and Al-Ansari N 2021. Perceptions about water pollution among university students: A case study from Iraq. *Cogent Engineering*, **8** 1895473.

[13] Jasim I A, Farhan S L, Al-Maliki L A and AL-Mamoori S K 2021. Climatic Treatments for Housing in the Traditional Holy Cities: A Comparison between Najaf and Yazd Cities. *IOP Conference Series: Earth and Environmental Science,* 012017.

[14] Abdulla G, Kareem M M, Hashim K S, Muradov M, Kot P, Mubarak H A, Abdellatif M and Abdulhadi B 2020. Removal of iron from wastewater using a hybrid filter. *IOP Conference Series: Materials Science and Engineering,* 012035.

[15] Abdulraheem F, Al-Khafaji Z, Hashim K S, Muradov M, Kot P and Shubbar A 2020. Natural filtration unit for removal of heavy metals from water. *IOP Materials Science and Engineering,* 012034.

[16] Emamjomeh M M, Mousazadeh M, Mokhtari N, Jamali H A, Makkiabadi M, Naghdali Z, Hashim K S and Ghanbari R 2020. Simultaneous removal of phenol and linear alkylbenzene sulfonate from automotive service station wastewater: Optimization of coupled electrochemical and physical processes. *Separation Science and Technology*, **55** 3184-94.

[17] Al-Jumeily D, Hashim K, Alkaddar R, Al-Tufaily M and Lunn J 2019. Sustainable and Environmental Friendly Ancient Reed Houses (Inspired by the Past to Motivate the Future). *11th International Conference on Developments in eSystems Engineering (DeSE),* Cambridge, UK 214-9.

[18] Omran I I, Al-Saati N H, Hashim K S, Al-Saati Z N, Patryk K, Khaddar R A, Al-Jumeily D, Shaw A, Ruddock F and Aljefery M 2019. Assessment of heavy metal pollution in the Great Al-Mussaib irrigation channel. *Desalination and Water Treatment*, **168** 165-74.

[19] Kadhim A, Sadique M, Al-Mufti R and Hashim K 2020. Developing One-Part Alkali-Activated metakaolin/natural pozzolan Binders using Lime Waste as activation Agent. *Advances in Cement Research*, **32** 1-38.
[20] Kadhim A, Sadique M, Al-Mufti R and Hashim K 2020. Long-term performance of novel high-calcium one-part alkali-activated cement developed from thermally activated lime kiln dust. *Journal of Building Engineering*, **32** 1-17.
[21] Ali A, Al-Shaer A, Hashim K, Hawesah H and Sadique M 2019. Investigating the influence of cement replacement by high volume of GGBS and PFA on the mechanical performance of cement mortar. *First Int. Conference on Civil and Environmental Engineering Technologies,* University of Kufa, Iraq 31-8.
[22] Shubbar A A, Sadique M, Shanbara H K and Hashim K 2020 *The Development of a New Low Carbon Binder for Construction as an Alternative to Cement. In Advances in Sustainable Construction Materials and Geotechnical Engineering*, Berlin: Springer.
[23] Idowu I A, Hashim K, Shaw A and Nunes L J 2021. Enhancing the fuel properties of beverage wastes as non-edible feedstock for biofuel production. *Biofuels*, **14** 1-8.
[24] Majdi H, Ali A, Nasr M, Al-Khafaji Z, Jafer H, Abdulredha M, Masoodi Z, Sadique M and Hashim K 2020. Experimental data on compressive strength and ultrasonic pulse velocity properties of sustainable mortar made with high content of GGBFS and CKD combinations. *Data in Brief*, **31** 105961-72.
[25] Omer G, Kot P, Atherton W, Muradov M, Gkantou M, Shaw A, Riley M, Hashim K and Al-Shamma'a A 2020. A Non-Destructive Electromagnetic Sensing Technique to Determine Chloride Level in Maritime Concrete. *Karbala International Journal of Modern Science*, **6** 1-14.
[26] Shubbar A, Sadique M, Nasr M, Al-Khafaji Z and Hashim K 2020. The impact of grinding time on properties of cement mortar incorporated high volume waste paper sludge ash. *Karbala International Journal of Modern Science*, **6** 1-23.
[27] Obaid M, Nasr M S, Ali I, Ali A and Hashim K 2021. Performance of green mortar made from locally available waste tiles and silica fume. *Journal of Engineering Science and Technology*, **16** 136-51.
[28] Al-Saati N H, Omran I I, Salman A A, Al-Saati Z and Hashim K S 2021. Statistical modeling of monthly streamflow using time series and artificial neural network models: Hindiya Barrage as a case study. *Water Practice and Technology*, **16** 681-91.
[29] Chabuk A, Hammood Z, Abed S, Kadhim M, Hashim K, Al-Ansari N and Laue J 2021. Noise Level in Textile Industries: Case Study Al-Hillah Textile Factory-Company for Textile Industries, Al-Hillah-Babylon-Iraq. *IOP Conference Series: Earth and Environmental Science,* 012048.
[30] Grmasha R, Al-sareji O, Salman J and Jasim I 2020. Polycyclic Aromatic Hydrocarbons (PAHs) in Urban Street Dust Within Three Land-Uses of Babylon Governorate, Iraq: Distribution, Sources, and Health Risk Assessment. *Journal of King Saud University - Engineering Sciences*, **33** 1-18.
[31] Zubaidi Salah L, Al-Bugharbee H, Ortega Martorell S, Gharghan S, Olier I, Hashim K, Al-Bdairi N and Kot P 2020. A Novel Methodology for Prediction Urban Water Demand by Wavelet Denoising and Adaptive Neuro-Fuzzy Inference System Approach. *Water*, **12** 1-17.
[32] Al-Sareji O, Grmasha R, Salman J, Idowu I and Hashim K 2021. Street dust contamination by heavy metals in Babylon governorate, Iraq. *Journal of Engineering Science and Technology*, **16** 3528 - 46.
[33] Omran I I, Al-Saati N H, Al-Saati H H, Hashim K S and Al-Saati Z N 2021. Sustainability assessment of wastewater treatment techniques in urban areas of Iraq using multi-criteria decision analysis (MCDA). *Water Practice and Technology*, **16** 648-60.
[34] Al-Saati N H, Hussein T K, Abbas M H, Hashim K, Al-Saati Z N, Kot P, Sadique M, Aljefery M H and Carnacina I 2019. Statistical modelling of turbidity removal applied to non-toxic natural coagulants in water treatment: a case study. *Desalination and Water Treatment*, **150** 406-12.
[35] Alyafei A, AlKizwini R S, Hashim K S, Yeboah D, Gkantou M, Al Khaddar R, Al-Faluji D and Zubaidi S L 2020. Treatment of effluents of construction industry using a combined filtration-electrocoagulation method. *IOP Conference Series: Materials Science and Engineering,* 012032.

[36] Alenazi M, Hashim K S, Hassan A A, Muradov M, Kot P and Abdulhadi B 2020. Turbidity removal using natural coagulants derived from the seeds of strychnos potatorum: statistical and experimental approach. *IOP Conference Series: Materials Science and Engineering,* 012064.

[37] Alhendal M, Nasir M J, Hashim K S, Amoako-Attah J, Al-Faluji D, Muradov M, Kot P and Abdulhadi B 2020. Cost-effective hybrid filter for remediation of water from fluoride. *IOP Conference Series: Materials Science and Engineering,* 012038.

[38] Hashim K S, Hussein A H, Zubaidi S L, Kot P, Kraidi L, Alkhaddar R, Shaw A and Alwash R 2019. Effect of initial pH value on the removal of reactive black dye from water by electrocoagulation (EC) method. *2nd International Scientific Conference,* Al-Qadisiyah University, Iraq 12-22.

[39] Aqeel K, Mubarak H A, Amoako-Attah J, Abdul-Rahaim L A, Al Khaddar R, Abdellatif M, Al-Janabi A and Hashim K S 2020. Electrochemical removal of brilliant green dye from wastewater. *IOP Conference Series: Materials Science and Engineering,* 012036.

[40] Mohammed A, Hussein A, Yeboah D, Al Khaddar R, Abdulhadi B, Ali A and Hashim K S 2020. Electrochemical removal of nitrate from wastewater. *IOP Materials Science and Engineering,* 012037.

[41] Hashim K, Ali S, AlRifaie J K, Kot P, Shaw A, Al Khaddar R, Idowu I and Gkantou M 2020. Escherichia coli inactivation using a hybrid ultrasonic–electrocoagulation reactor. *Chemosphere*, **247** 125868-75.

[42] Alenezi A K, Hasan H A, Hashim K S, Amoako-Attah J, Gkantou M, Muradov M, Kot P and Abdulhadi B 2020. Zeolite-assisted electrocoagulation for remediation of phosphate from calcium-phosphate solution. *IOP Conference Series: Materials Science and Engineering,* 012031.

[43] Al-Marri S, AlQuzweeni S S, Hashim K S, AlKhaddar R, Kot P, AlKizwini R S, Zubaidi S L and Al-Khafaji Z S 2020. Ultrasonic-Electrocoagulation method for nitrate removal from water. *IOP Conference Series: Materials Science and Engineering,* 012073.

[44] Hashim K S, Shaw A, AlKhaddar R, Kot P and Al-Shamma'a A 2021. Water purification from metal ions in the presence of organic matter using electromagnetic radiation-assisted treatment. *Journal of Cleaner Production*, **280** 1-17.

[45] Abdulhadi B, Kot P, Hashim K, Shaw A, Muradov M and Al-Khaddar R 2021. Continuous-flow electrocoagulation (EC) process for iron removal from water: Experimental, statistical and economic study. *Science of The Total Environment*, **760** 1-16.

[46] Alnaimi H, Idan I J, Al-Janabi A, Hashim K, Gkantou M, Zubaidi S L, Kot P and Muradov M 2020. Ultrasonic-electrochemical treatment for effluents of concrete plants. **888** 1-9.

[47] Abdulhadi B A, Kot P, Hashim K S, Shaw A and Khaddar R A 2019. Influence of current density and electrodes spacing on reactive red 120 dye removal from dyed water using electrocoagulation/electroflotation (EC/EF) process. *First International Conference on Civil and Environmental Engineering Technologies (ICCEET),* University of Kufa, Iraq 12-22.

[48] Hashim K S, Al-Saati N H, Alquzweeni S S, Zubaidi S L, Kot P, Kraidi L, Hussein A H, Alkhaddar R, Shaw A and Alwash R 2019. Decolourization of dye solutions by electrocoagulation: an investigation of the effect of operational parameters. *First International Conference on Civil and Environmental Engineering Technologies (ICCEET),* University of Kufa, Iraq 25-32.

[49] Hashim K S, Khaddar R A, Jasim N, Shaw A, Phipps D, Kot P, Pedrola M O, Alattabi A W, Abdulredha M and Alawsh R 2019. Electrocoagulation as a green technology for phosphate removal from River water. *Separation and Purification Technology*, **210** 135-44.

[50] Emamjomeh M, Kakavand S, Jamali H, Alizadeh S, Mousavi S, Hashim K and Mousazade M 2020. The treatment of printing and packaging wastewater by electrocoagulation–flotation: the simultaneous efficacy of critical parameters and economics. *Desalination and water treatment*, **205** 161-74.

[51] Hashim K, Kot P, Zubaid S, Alwash R, Al Khaddar R, Shaw A, Al-Jumeily D and Aljefery M 2020. Energy efficient electrocoagulation using baffle-plates electrodes for efficient Escherichia Coli removal from Wastewater. *Journal of Water Process Engineering*, **33** 101079-86.

[52] Hashim K, Shaw A and Phipps D, 2019. *Treatment reactor and method of treating a liquid.* WIPO, PCT/GB2019/052493, L J M University, United Kingdom.

[53] Zanki A K, Mohammad F H, Hashim K S, Muradov M, Kot P, Kareem M M and Abdulhadi B 2020. Removal of organic matter from water using ultrasonic-assisted electrocoagulation method. *IOP Conference Series: Materials Science and Engineering,* 012033.

[54] Gkantou M, Muradov M, Kamaris G S, Hashim K, Atherton W and Kot P 2019. Novel Electromagnetic Sensors Embedded in Reinforced Concrete Beams for Crack Detection. *Sensors*, **19** 5175-89.

[55] Ryecroft S P, shaw A, Fergus P, Kot P, Hashim K and Conway L 2019. A Novel Gesomin Detection Method Based on Microwave Spectroscopy. *12th International Conference on Developments in eSystems Engineering (DeSE),* Kazan, Russia 429-33.

[56] Ryecroft S, Shaw A, Fergus P, Kot P, Hashim K, Moody A and Conway L 2019. A First Implementation of Underwater Communications in Raw Water Using the 433 MHz Frequency Combined with a Bowtie Antenna. *Sensors*, **19** 1813-23.

[57] Kot P, Hashim K S, Muradov M and Al-Khaddar R 2021 *How can sensors be used for sustainability improvement?. In Methods in Sustainability Science*, Elsevier, United Kingdom: Joe Hayton, p 426.

[58] Omer G, Kot P, Atherton W, Muradov M, Gkantou M, Shaw A, Riley M, Hashim K and Al-Shamma'a A 2021. A Non-Destructive Electromagnetic Sensing Technique to Determine Chloride Level in Maritime Concrete. *Karbala International Journal of Modern Science*, **7** 61-71.

[59] Kot P, Muradov M, Gkantou M, Kamaris G S, Hashim K and Yeboah D 2021. Recent Advancements in Non-Destructive Testing Techniques for Structural Health Monitoring. *Applied Sciences*, **11** 1-28.

[60] Ryecroft S, Shaw A, Fergus P, Kot P, Hashim K, Tang A, Moody A and Conway L 2021. An Implementation of a Multi-Hop Underwater Wireless Sensor Network using Bowtie Antenna. *Karbala International Journal of Modern Science*, **7** 113-29.

[61] Ghazali R, Hussain A, Al-Jumeily D and Merabti M 2007. Dynamic ridge polynomial neural networks in exchange rates time series forecasting. *International Conference on Adaptive and Natural Computing Algorithms,* 123-32.

[62] Fergus P, Hussain A, Al-Jumeily D, Huang D-S and Bouguila N 2017. Classification of caesarean section and normal vaginal deliveries using foetal heart rate signals and advanced machine learning algorithms. *Biomedical engineering online*, **16** 1-26.

Fluoride removal using electrocoagulation technique

Hind M. Ewadh[1,*], Mustafa J. Al Imari[2], Sabrean F. Jawad[3], Hayfaa A. Mubarak[4]
[1] Environmental Research and Studies Center, University of Babylon, Hilla, Iraq
[2] Department of Medical Laboratory Technique, Al-Mustaqbal University College, Hilla, Iraq
[3] Department of Pharmacy, Al-Mustaqbal University College, Babylon, Iraq
[4] Chemical Engineering Department, College of Engineering, University of Babylon, Iraq
*Corresponding author; Email: hindewadh@uobabylon.edu.iq

Abstract. A modest quantity of fluoride can increase the mineralization of teeth and reduce their cavities. But the presomerence of fluoride in excess in water can lead to severe disease infertility. In the past few decades, scientists have thus been preoccupied with developing ways to reduce sewage fluoride concentrations and reduce their effects on human health. The present study is aimed at using the technology of electrocoagulation to remove fluoride from polluted water. Tests have been done to examine the elimination of fluoride with a rectangular electrocoagulation cell and examine the impact of the experimental aspects on fluoride extraction, specifically electrical current, electrode spacing, and pH. The authors found that 93% of the fluoride has been extracted using 5mm spaced electrodes with a current density of 2 mA/cm^2 and a level of pH of 7 from the polluted water after 20 min of processing. Experimental factors considerably impact the efficacy of fluoride removal. In the acidic environment, greater effectiveness of fluoride removal is being attained. The elimination effectiveness depends directly on the electric current, whereas the distance between poles is adversely linked to fluoride elimination.

1. Introduction

Water is a crucial part of the planet earth's exit forms of life. Nevertheless, just 1% fewer is fresh and suited for consumption. Suitable amounts of drinking water are regarded as one of humankind's significant difficulties due to the increasing population [1, 2]. Additionally, authors have proved by the 2050s, half of the worldwide people may have no access to potable water [3]. This trend is subjected to a rapid increase due to the extensive discharge of industrial wastewaters [4, 5] and consumption of fresh water during the production processes [6-8], urbanisation [9-12], global warming [13-15] and air pollution that is in contact with surface water [16-19]. Other researchers have pointed out about 33.6 million barrels of severely contaminated water are released annually into water sources (surfaces or underground) [20] because of rapid industrialization; thus, the situation is much more complex [21].
Fluoride is among the most prevalent elements where it constitutes around 5% of the earth's surface [22, 23]. Consequently, many watercourses naturally contain fluoride at a low quantity. However, expanding industrialization increased fluoride concentration [24, 25]. For instance, high-temperature chemicals deposit enormous amounts of polluted effluent of fluorides in the manufacturing of the electronics like aluminium and steel [26-28]. The small level of fluoride in water prevents tooth cavitation; nevertheless, large levels can contribute to severe illnesses like skeletal paralysis. Eventually, fluoride levels in drinkable water were restricted to 1,2 mg/l by the World Health Organization [22]. Scholars are now using several processing procedures to eliminate water and wastewater pollutants [29-32]. Electrocoagulation is an appealing treatment method owing to low cost [33-35], flexibility as it can be

IOP Conf. Series: Earth and Environmental Science **877** (2021) 012047 doi:10.1088/1755-1315/877/1/012047

combined with other methods [36-38], produce few sludges [39, 40] and operate using sensors [41-44]. The sludge is full of metal ions which improve the recycling potential of sludge instead of disposal in expensive landfills [45-49]. The electrocoagulation method has been utilized to remove a range of contaminants from the aquatic environment, including toxic substances and microbiological contaminants. For example, this technique was cleaned to reduce arsenic from contaminated water by Fe-Al electrodes, where approximately 100% of arsenic has been eliminated in 0.5 hours [22]. Aluminium electrodes in an acidic environment are also useful for reducing the contamination of other metals like nickel and copper. This method has also been applied in order to eliminate organic contaminants from industrial sewage, with effectiveness of around 85 percent over 45 min [50-52].
Fluoride is also removed from polluted water using electrocoagulation. In artificial industrial effluents employing aluminium electrodes, for example, the fluoride content was reduced. The influence on the wastewater treatment utilizing the electrocoagulation method of many factors, including the current density, and pH have been investigated by several academics [53-55]. The removal of fluoride from drinking water was also achieved via electrocoagulation. The literature utilised this approach to reduce fluoride from artificial drinking water at the starting dose of 2-10 mg/l [22].
The water contamination crisis affects humankind for generations, and safe techniques of treatment of water are necessary for the expected growth in demand for drinking water. Therefore, investigators are now concerned with using safe ways to eliminate water contaminants. Electrocoagulation offers a strong treatment alternative compared to many other standard treatment approaches for water and wastewater because it is cheap, compact, and produces good sludge. Therefore, the project intends to use the electrocoagulation approach to reduce fluoride concentration in wastewater using a perforated plate of aluminium electrodes.

2. Methods

2.1. *Chemicals and electrocoagulation unit*
All reagents included in this experiment were supplied by Sigma-Aldrich without adjustment or decontamination. The Artificial liquid was created to achieve a 500 mg/l of fluoride level throughout this stock solution by melting a certain quantity of fluoride (NaF) into deionized water based on Hashim *et al.,* [56] method. Dilution has been applied to specimens of 1 liter in size to achieve lower levels of fluoride concentrations. Furthermore, the pH-values for the liquid ranging from 5 to 9 were controlled by HCl and NaOH, and NACl was employed to keep contaminated water conductive at 0.3 mS/cm. In this research, a reactor of 1 liter in size having inner dimensions of 50, 200 and 100 mm for wide, length and height correspondingly was used to remove the fluoride from Artificially polluted water. Inside the reactor, four perforated aluminium sheets were vertically placed on improving liquid circulation and enhancing pollutant removal.

2.2. *Methods for experimentation*
The reactor was filled with 500 ml of polluted water. The reactor input and output were linked to accomplishing the electrolyzing procedure, and a peristaltic pump was used for circulating water. The reactor was connected to reliable electricity. In order to test the acidity, conductivity and temperature of the solution, a pH-meter, conductivity meters and thermometer were employed. In all the experiments, the ambient temperature of 20 ±1 °C was maintained. The removal efficiency was determined after the equation reported by as follow [57]:

$$E\ (\%) = \frac{x - x1}{x1} * 100\% \qquad (1)$$

Where x represents the ultimate solution fluoride concentration, whereas x1 is the initial mg/l fluoride concentration, and E is Florida's removal effectiveness.

2.3. *The impact of the test settings*

IOP Conf. Series: Earth and Environmental Science **877** (2021) 012047 doi:10.1088/1755-1315/877/1/012047

In research, the impact on the efficiency of removal of fluid Ions from the water was examined by 3-parameter that are electric current strength, pH-level and distance among plates [53]. The pH influence was examined in this study by maintaining 2 mA/cm^2 of electric current, 5 mm of electrodes separate and an initial 10 mg/l fluoride concentration for 20 minutes, while the pH level changed from 5 to 9. The impact of electric current was investigated utilizing 1, 2 and 3 mA/cm^2 while keeping a pH of 7, spacing of 5mm, fluoride of 10mg/L. Finally, the effect of plate separation was investigated by changes of the plate separation from 5 to 10 with the rest of the parameters maintained.

3. Results and discussions

3.1. pH impact

The pH value has a key role in treating any contaminant [57]. A 600 ml of water samples were treated for 20 minutes using varied pH-values of 5, 7, and 9 with 5 mm distance between electrodes and 2 mA/cm^2 electrical currents to examine the impact of pH-value on the removing efficiency of fluoride. The effect of pH on fluoride removal effectiveness at different pH values is presented in Figure 1. With a reduction in wastewater acidity, the extraction efficiency of the contaminant improved. Acidic levels achieved the best removal effectiveness. However, the removal effectiveness reduces considerably as the wastewater turns alkaline. The difference in fluoride removal efficiency suggests an alkaline condition is not suitable for fluoride by electrocoagulation. Studies claimed that the preference for better fluoride removal is the acidic environment [58].

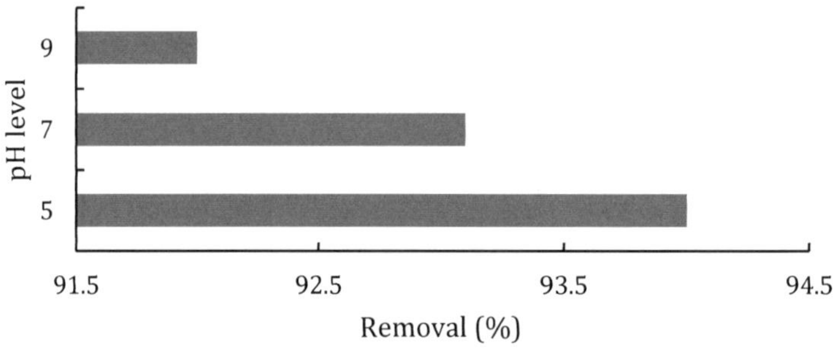

Figure 1: The pH impact on fluoride elimination.

3.2. Electrical current impact

The current significantly affect fluoride removal by controlling bubble sizes and floc formation. In this study, three distinct currents range from 1 mA to 3 mA/cm2 with pH at 7, initial pollutant content at 10 mg/l and spacing at 5 mm, were investigated to determine the impact of the currents on the removal. The reduction of fluoride enhanced greater current as shown in figure 2. for current of 1, 2 and 3 mA/cm2 after 5 min of operation, the removal effectiveness range amounted to around 65%, 74.3% and 77.1%, correspondingly. This verifies the fact that delivering more energy immediately increases the reaction resulting in coagulation-controlled species and accelerates the generation of the flock to increase the efficiency of removal.

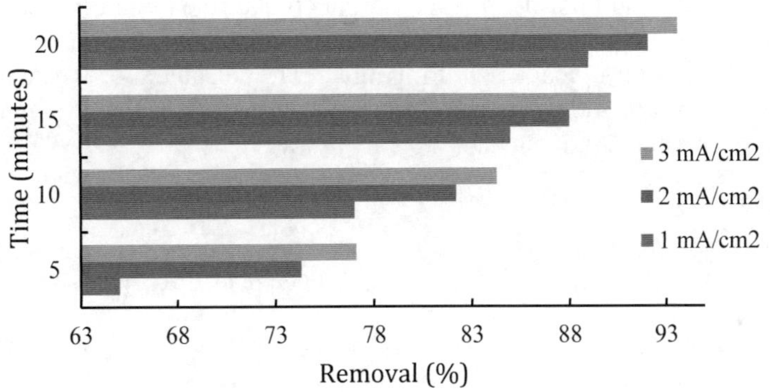

Figure 2: The current impact on fluoride elimination.

3.3. *Spacing impact*

Scholars indicated that electrocoagulation removing efficacy is strongly affected by electrode separation. Therefore, 600 ml of polluted water having initial fluoride content of 10 mg/l, pH-level 7 using an electrical current of 2 mA/cm2, to study the impact of the electrodes separating on the removal performance. The intervals between electrodes were changed from 5 to 10 mm. The results presented in Figure 3 reveal that the separating distance adversely affects the removal efficiency. The fluoride concentration decreased from 93% to 84% after 20 minutes of treatment when the electrode distance increased from 5 mm to 10 mm. Studies link the electrical impedance between electrodes to the changes in pollutant removal performance [31, 56].

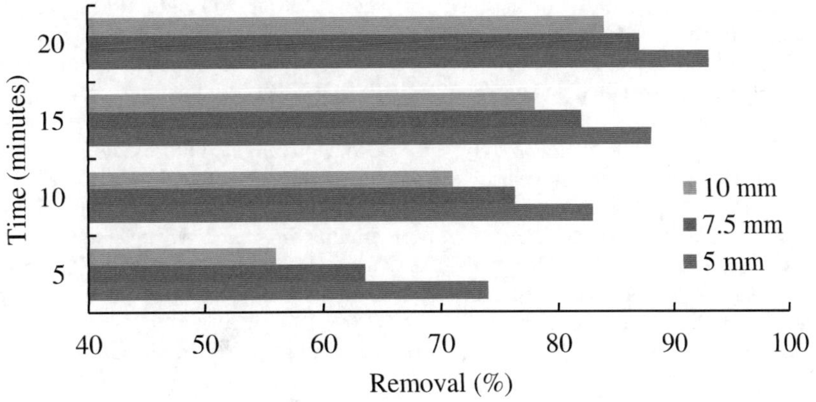

Figure 3: The spacing impact on fluoride elimination.

The improvement of the electrocoagulation in terms of fluoride removal, as was noticed in the results, could be achieved via controlling the operating parameters, such as the current density and pH of the water. Such parameters, especially the pH, can be monitored using sensors that provide real-time measurements, such as the ultrasonic sensors that are in use in the concrete industry [59, 60], or microwave sensors that are used in communications [61, 62] and pollution monitoring [63, 64].

4. Conclusion

The present investigation examined the efficacy of the electrocoagulation method using aluminium electrodes and the influence of the major operational elements (pH, electric current, and distance between electrodes) in terms of fluoride removal from polluted synthetic water. It can be claimed that electrocoagulation could be utilized to extract fluoride from polluted water, based on the current results of this experimental investigation. The operational elements also have a strong impact on fluoride removal efficiency. In removing fluoride, the pH plays a crucial role as a higher removal effectiveness can be obtained in the acid environment. The effectiveness of removal is connected favourably to the electric current as it promotes coagulants formation. The distance between electrodes adversely affects the effectiveness of fluoride removal from polluted water. After 20 minutes, the removal efficiency using 2 mA/cm^2 and 5 mm gap at a 7 pH level was 93%.

References

[1] Shubbar A A, Jafer H, Abdulredha M, Al-Khafaji Z S, Nasr M S, Al Masoodi Z and Sadique M 2020. Properties of cement mortar incorporated high volume fraction of GGBFS and CKD from 1 day to 550 days. *Journal of Building Engineering*, **30** 101327.

[2] Alwan H H, Saleh L A, Al-Mohammed F M and Abdulredha M A 2020. Experimental prediction of the discharge coefficients for rectangular weir with bottom orifices. *Journal of Engineering Science and Technology*, **15** 3265-80.

[3] Zubaidi S L, Al-Bugharbee H, Muhsen Y R, Hashim K, Alkhaddar R M, Al-Jumeily D and Aljaaf A J 2019. The Prediction of Municipal Water Demand in Iraq: A Case Study of Baghdad Governorate. *12th International Conference on Developments in eSystems Engineering (DeSE),* Kazan, Russia 274-7.

[4] Shubbar A A, Al-Shaer A, AlKizwini R S, Hashim K, Hawesah H A and Sadique M 2019. Investigating the influence of cement replacement by high volume of GGBS and PFA on the mechanical performance of cement mortar. *First International Conference on Civil and Environmental Engineering Technologies (ICCEET),* University of Kufa, Iraq 31-8.

[5] Alyafei A, AlKizwini R S, Hashim K S, Yeboah D, Gkantou M, Al Khaddar R, Al-Faluji D and Zubaidi S L 2020. Treatment of effluents of construction industry using a combined filtration-electrocoagulation method. *IOP Conference Series: Materials Science and Engineering,* 012032.

[6] Kadhim A, Sadique M, Al-Mufti R and Hashim K 2020. Long-term performance of novel high-calcium one-part alkali-activated cement developed from thermally activated lime kiln dust. *Journal of Building Engineering*, **32** 1-17.

[7] Kadhim A, Sadique M and Khalid K 2020. Developing One-Part Alkali-Activated metakaolin/natural pozzolan Binders using Lime Waste as activation Agent. *Advances in Cement Research*, **32** 1-38.

[8] Majdi H S, Shubbar A, Nasr M S, Al-Khafaji Z S, Jafer H, Abdulredha M, Masoodi Z A, Sadique M and Hashim K 2020. Experimental data on compressive strength and ultrasonic pulse velocity properties of sustainable mortar made with high content of GGBFS and CKD combinations. *Data in Brief*, **31** 105961-72.

[9] Al-Jumeily D, Hashim K, Alkaddar R, Al-Tufaily M and Lunn J 2019. Sustainable and Environmental Friendly Ancient Reed Houses (Inspired by the Past to Motivate the Future). *11th International Conference on Developments in eSystems Engineering (DeSE),* Cambridge, UK 214-9.

[10] Farhan S, Antón D, Akef V, Zubaidi S and Hashim K 2021. Factors influencing the transformation of Iraqi holy cities: the case of Al-Najaf. *Scientific Review Engineering and Environmental Sciences*, **30** 365-75.

[11] Zubaidi S, Ortega-Martorell S, Al-Bugharbee H, Olier I, Hashim K S, Gharghan S K, Kot P and Al-Khaddar R 2020. Urban Water Demand Prediction for a City that Suffers from Climate Change and Population Growth: Gauteng Province case study. *Water*, **12** 1-18.

[12] Zubaidi Salah L, Al-Bugharbee H, Ortega Martorell S, Gharghan S, Olier I, Hashim K, Al-Bdairi N and Kot P 2020. A Novel Methodology for Prediction Urban Water Demand by Wavelet Denoising and Adaptive Neuro-Fuzzy Inference System Approach. *Water*, **12** 1-17.

[13] Salah Z, Abdulkareem I H, Hashim K S, Al-Bugharbee H, Ridha H M, Gharghan S K, Al-Qaim F F, Muradov M, Kot P and Alkhaddar R 2020. Hybridised Artificial Neural Network model with Slime Mould Algorithm: A novel methodology for prediction urban stochastic water demand. *Water*, **12** 1-18.

[14] Salah Z, Hashim K, Ethaib S, Al-Bdairi N S S, Al-Bugharbee H and Gharghan S K 2020. A novel methodology to predict monthly municipal water demand based on weather variables scenario. *Journal of King Saud University-Engineering Sciences*, **32** 1-18.

[15] Salah Z, OrtegaAbdellatif M, Gharghan S K, Ahmed M S and Hashim K 2020. A Method for Predicting Long-Term Municipal Water Demands Under Climate Change. *Water Resources Management*, **34** 1265-79.

[16] Al-Saati N, Omran I, Salman A, Al-Saati Z and Hashim K 2021. Statistical modeling of monthly streamflow using time series and artificial neural network models: Hindiya Barrage as a case study. *Water Practice and Technology*, **16** 681-91.

[17] Al-Sareji O J, Grmasha R A, Salman J M, Idowu I and Hashim K S 2021. Street dust contamination by heavy metals in Babylon governorate, Iraq. *Journal of Engineering Science and Technology*, **16** 3528 - 46.

[18] Shubbar A, Sadique M, Nasr M, Al-Khafaji Z and Hashim K S 2020. The impact of grinding time on properties of cement mortar incorporated high volume waste paper sludge ash. *Karbala International Journal of Modern Science*, **6** 1-23.
[19] Ali A, Sadique M, Shanbara H and Hashim K 2020 *The Development of a New Low Carbon Binder for Construction as an Alternative to Cement. In Advances in Sustainable Construction Materials and Geotechnical Engineering*, Berlin: Springer.
[20] Zubaidi S, Abdellatif M and Muhsin Y R 2019. Using LARS–WG model for prediction of temperature in Columbia City, USA. *IOP Materials Science and Engineering,* 012026.
[21] Al-Anbari R, Alnakeeb A, Abdulredha M J E and Journal T 2013. Landfill site selection for Kerbala municipal solid wastes by using geographical information system techniques. **32** 13.
[22] Alhendal M, Nasir M, Hashim K, Amoako-Attah J and Abdulhadi B 2020. Cost-effective hybrid filter for remediation of water from fluoride. *IOP Materials Science and Engineering,* 012038.
[23] Grmasha R, Al-sareji O, Salman J, Hashim K and Jasim I A 2020. Polycyclic Aromatic Hydrocarbons (PAHs) in Urban Street Dust Within Three Land-Uses of Babylon Governorate, Iraq: Distribution, Sources, and Health Risk Assessment. *Journal of King Saud University - Engineering Sciences*, **33** 1-18.
[24] Hashim K, Al-Jumeily D, Alwash R and Aljefery M 2020 *Electrocoagulation as an eco-friendly River water treatment method. In Advances in Water Resources Engineering and Management*, Berline: Springer.
[25] Alnaimi H, Idan I, Al-Janabi A, Hashim K, Gkantou M and Muradov M 2020. Ultrasonic-electrochemical treatment for effluents of concrete plants. **888** 1-9.
[26] Hashim K, Al-Saati N, Hussein A and Al-Saati Z 2018. An investigation into the level of heavy metals leaching from canal-dreged sediment: a case study metals leaching from dreged sediment. *First International Conference on Materials Engineering & Science,* Istanbul Aydın University (IAU), Turkey 12-22.
[27] Omran I I, Al-Saati N H, Hashim K S, Al-Saati Z N, Patryk K, Khaddar R A, Al-Jumeily D, Shaw A, Ruddock F and Aljefery M 2019. Assessment of heavy metal pollution in the Great Al-Mussaib irrigation channel. *Desalination and Water Treatment*, **168** 165-74.
[28] Abdulla G, Kareem M, Hashim K, Muradov M, Mubarak H A, Abdellatif M and Abdulhadi B 2020. Removal of iron from wastewater using a hybrid filter. *IOP Materials Science and Engineering,* 012035.
[29] Abdulredha M, Muhsin A, Al-Janabi A, Alajmi B, Gkantou M, Amoako J, Al-Jumeily D, Mustafina J and AlKhayyat A 2021. Using SF and CKD as cement replacement materials for producing cement mortar. *IOP Conference Series: Materials Science and Engineering,* 012007.
[30] Hashim K, Shaw A and Phipps D, 2019. *Treatment reactor and method of treating a liquid*. WIPO, PCT/GB2019/052493, L J M University, United Kingdom.
[31] Hashim K, Al-Saati N, Alquzweeni S, Kraidi L, Hussein A and Alwash R 2019. Decolourization of dye solutions by electrocoagulation: an investigation of the effect of operational parameters. *1st Int. Conference on Civil and Environmental Engineering Technologies,* University of Kufa, Iraq 25-32.
[32] Abdulraheem F, Al-Khafaji Z, Hashim K, Muradov M, Kot P and Shubbar A 2020. Natural filtration unit for removal of heavy metals from water. *IOP Materials Science and Engineering,* 012034.
[33] Alenezi A K, Hasan H, Amoako, Gkantou M, Muradov M and Abdulhadi B 2020. Zeolite-assisted electrocoagulation for remediation of phosphate from calcium-phosphate solution. *IOP Materials Science and Engineering,* 012031.
[34] Al-Marri S, AlQuzweeni S, AlKizwini R, Zubaidi L and Al-Khafaji Z S 2020. Ultrasonic-Electrocoagulation method for nitrate removal from water. *IOP Conference Series: Materials Science and Engineering,* 012073.
[35] Abdulhadi B, Kot P, Hashim K, Shaw A, Muradov M and Al-Khaddar R 2021. Continuous-flow electrocoagulation (EC) process for iron removal from water: Experimental, statistical and economic study. *Science of The Total Environment*, **760** 1-16.
[36] Abdulhadi B, Shaw A and Khaddar R A 2019. Influence of current density and electrodes spacing on reactive red 120 dye removal from dyed water using electrocoagulation/electroflotation (EC/EF) process. *1st Int. Conference on Civil and Environmental Engineering Technologies,* University of Kufa, Iraq 12-22.
[37] Al-Saati N, Hussein T, Abbas M, Hashim K, Al-Saati Z, Sadique M, Aljefery M and Carnacina I 2019. Statistical modelling of turbidity removal applied to non-toxic natural coagulants in water treatment: a case study. *Desalination and Water Treatment*, **150** 406-12.
[38] Alenazi M, Hashim K, Hassan A, Muradov M and Abdulhadi B 2020. Turbidity removal using natural coagulants derived from the seeds of strychnos potatorum: statistical and experimental approach. *IOP Conference Series: Materials Science and Engineering,* 012064.
[39] Mohammed A, Hussein A, Yeboah D, Abdulhadi B, Ali A and Hashim K 2020. Electrochemical removal of nitrate from wastewater. *IOP Materials Science and Engineering,* 012037.
[40] Zanki A, Mohammad F, Muradov M, Kareem M and Abdulhadi B 2020. Removal of organic matter from water using ultrasonic-assisted electrocoagulation method. *IOP Materials Science and Engineering,* 012033.

[41] Gkantou M, Muradov M, Kamaris G S, Hashim K, Atherton W and Kot P 2019. Novel Electromagnetic Sensors Embedded in Reinforced Concrete Beams for Crack Detection. *Sensors*, **19** 5175-89.
[42] Hashim K, Andy A, Rafid R and Shamma'a A 2021. Water purification from metal ions in the presence of organic matter using electromagnetic radiation-assisted treatment. *Journal of Cleaner Production*, **280** 1-17.
[43] Omran I, Al-Saati N, Al-Saati H, Hashim K and Al-Saati Z 2021. Sustainability assessment of wastewater treatment techniques in urban areas of Iraq using multi-criteria decision analysis (MCDA). *Water Practice and Technology*, **16** 648-60.
[44] Idowu I A, Hashim K, Shaw A and Nunes L J 2021. Enhancing the fuel properties of beverage wastes as non-edible feedstock for biofuel production. *Biofuels*, **14** 1-8.
[45] Abdulredha M, Abdulridha A and Jordan D 2020. Estimating municipal solid waste generation from service processions during the Ashura religious event. *IOP Materials Science and Engineering,* 012075.
[46] Abdulredha M, Jordan D and Abdulridha A 2018. Benchmarking of the Current Solid Waste Management System in Karbala, Iraq, Using Wasteaware Benchmark Indicators. *World Environmental and Water Resources Congress 2018: Groundwater, Sustainability, and Hydro-Climate/Climate Change,* 40-8.
[47] Abdulredha M, Kot P, Al Khaddar R, Jordan D and Abdulridha A 2020. Investigating municipal solid waste management system performance during the Arba'een event in the city of Kerbala, Iraq. *Environment, Development and Sustainability*, **22** 1431-54.
[48] Abdulredha M, Rafid A, Jordan D and Alattabi A 2017. Facing up to waste: how can hotel managers in Kerbala, Iraq, help the city deal with its waste problem? *Procedia engineering*, **196** 771-8.
[49] Alattabi A W, Harris C, Alkhaddar R, Alzeyadi A and Abdulredha M 2017. Online Monitoring of a sequencing batch reactor treating domestic wastewater. *Procedia engineering*, **196** 800-7.
[50] Hashim K, Kot P, Zubaid S, Alwash R, Al Khaddar R, Shaw A, Al-Jumeily D and Aljefery M 2020. Energy efficient electrocoagulation using baffle-plates electrodes for efficient Escherichia Coli removal from Wastewater. *Journal of Water Process Engineering*, **33** 101079-86.
[51] Hashim K, Ali S, AlRifaie J, Al Khaddar R, Idowu I and Gkantou M 2020. Escherichia coli inactivation using a hybrid ultrasonic–electrocoagulation reactor. *Chemosphere*, **247** 125868-75.
[52] Hashim K S, Ewadh H M, Muhsin A A, Zubaidi S L, Kot P, Muradov M, Aljefery M and Al-Khaddar R 2020. Phosphate removal from water using bottom ash: Adsorption performance, coexisting anions and modelling studies. *Water Science and Technology*, **83** 1-17.
[53] Aqeel K, Mubarak H A, Amoako-Attah J, Abdul-Rahaim L A, Al Khaddar R, Abdellatif M, Al-Janabi A and Hashim K S 2020. Electrochemical removal of brilliant green dye from wastewater. *IOP Conference Series: Materials Science and Engineering,* 012036.
[54] Emamjomeh M, Kakavand S, Jamali H, Alizadeh S, Safdari M, Mousavi S, Hashim K and Mousazade M 2020. The treatment of printing and packaging wastewater by electrocoagulation–flotation: the simultaneous efficacy of critical parameters and economics. *Desalination and water treatment*, **205** 161-74.
[55] Emamjomeh M, Mousazadeh M, Mokhtari N, Jamali H, Makkiabadi M, Naghdali Z, Hashim K S and Ghanbari R 2020. Simultaneous removal of phenol and linear alkylbenzene sulfonate from automotive service station wastewater: Optimization of coupled electrochemical and physical processes. *Separation Science and Technology*, **55** 3184-94.
[56] Hashim K S, Shaw A, Al Khaddar R, Ortoneda Pedrola M and Phipps D 2017. Defluoridation of drinking water using a new flow column-electrocoagulation reactor (FCER) - Experimental, statistical, and economic approach. *Journal of Environmental Management*, **197** 80-8.
[57] Hashim K, Hussein A, Zubaidi S, Kraidi L and Alwash R 2019. Effect of initial pH value on the removal of reactive black dye from water by electrocoagulation (EC) method. *2nd International Scientific Conference,* Al-Qadisiyah University, Iraq 12-22.
[58] Abdulredha M, Kadhim N, Hussein A, Yeboah D and Hashim K 2021. Zeolite as a natural adsorbent for nitrogenous compounds being removed from water. *IOP Materials Science and Engineering,* 012082.
[59] Kot P, Muradov M, Gkantou M, Kamaris G S, Hashim K and Yeboah D 2021. Recent Advancements in Non-Destructive Testing Techniques for Structural Health Monitoring. *Applied Sciences*, **11** 1-28.
[60] Omer G, Kot P, Atherton W, Muradov M, Gkantou M, Shaw A, Riley M, Hashim K and Al-Shamma'a A 2021. A Non-Destructive Electromagnetic Sensing Technique to Determine Chloride Level in Maritime Concrete. *Karbala International Journal of Modern Science*, **7** 61-71.
[61] Ryecroft S, Shaw A, Fergus P, Kot P, Hashim K, Tang A, Moody A and Conway L 2021. An Implementation of a Multi-Hop Underwater Wireless Sensor Network using Bowtie Antenna. *Karbala International Journal of Modern Science*, **7** 113-29.
[62] Ryecroft S, Shaw A, Fergus P, Kot P, Hashim K, Moody A and Conway L 2019. A First Implementation of Underwater Communications in Raw Water Using the 433 MHz Frequency Combined with a Bowtie Antenna. *Sensors*, **19** 1813-23.

[63] Ryecroft S, Fergus P, Kot P, Hashim K and Conway L 2019. A Novel Gesomin Detection Method Based on Microwave Spectroscopy. *12th Int. Conference on Developments in eSystems Engineering,* Kazan, Russia.
[64] Kot P, Hashim K S, Muradov M and Al-Khaddar R 2021 *How can sensors be used for sustainability improvement?. In Methods in Sustainability Science*, Elsevier, United Kingdom, p 426.

Engineered electrocoagulation reactor for the removal of *E. coli* from wastewaters

Eman A. Al- Imara[1], Rand L. Al-Jaryan[2], Sabrean F. Jawad[3], Mohanad M. Kareem[4], Hayfaa A. Mubarak[5]
[1] Department of Biotic Evolution, Marine Science Centre, Basra University, Basra, Iraq
[2] Agriculture College, Al-Qasim Green University, Iraq
[3] Department of Pharmacy, Al-Mustaqbal University College, Babylon, Iraq
[4] Department of Chemistry, College of Science, University of Babylon, Iraq
[5] Chemical Engineering Department, University of Babylon, Babel, Iraq
*Corresponding author; Email: eman.abdalali@uobasrah.edu.iq

Abstract. As a consequence of natural pollution, water and sewage are polluted in many nations across the globe. Especially in poorer countries, sewage treatment and disposal practices are often substandard. Throughout many limited-income nations, the poor economic condition and absence of resources assistance severely hamper the planning and application of novel water and sewage systems. This has resulted in a rise in the number of bio-contaminants in the environment. The objective of this study is to use electrocoagulation as a low-cost method to remove or lower the amount of bio contaminant in sewage. By transmitting a voltage between the two conductors, disinfectants are generated in place. Sewerage samples were obtained at the Kerbala wastewater treatment plant, which is situated south of Kerbala, Iraq. In this work, steel plates were utilized to create coagulants. Furthermore, the effect of many factors on the performance of the electrolysis device was studied, namely spacing among electrodes and current density. The analysis indicates after 40 min of irradiation employing electrodes spaced 5mm apart and a current intensity of 2 mA/cm2, the E. coli bacteria as a biocontrol agent were killed. Furthermore, the results demonstrated that an initial pH value of 6.0 is appropriate for bio-contaminants removal utilizing electrocoagulation.

Keywords: Electrocoagulation, *E.coli*, wastewater, low-cost.

1. Introduction

Scientists worldwide believe that the planet is presently facing severe and catastrophic water shortages that represent a real threat to the existence of human beings [1, 2]. Scientists have identified many variables that contribute to the scarcity of water supplies, including the continuous global warming (especially the increase in temperature) [3, 4], increase in the population [5, 6] and the uneven distribution of rains [7, 8]. Moreover, other factors, such as the increase in the urban areas (urbanization) to accommodate the increase in the population, resulted in an increase in the generation of wastewater [9-11], and also in the generation of landfilling process that intern pollutes the groundwater [12-14]. The industry plays a serious role in the shortage of water and pollution of the small amount of fresh water on the planet's crust that is available to humans due to the discharging of huge volumes of polluted effluents into the surface water and groundwater [15-17], for instance, cement industry not only discharges polluted wastewater into the environment [18-20] that contain high concentrations of metals [21, 22], dyes [23, 24], organic matter [25, 26], nutrients [27-29] and other pollutants [30-32] but also pollutes the air that intern pollutes surface water [33, 34]. According to the experts, just 1% of the water contained in the planet is fresh and edible by humans and other organisms [35]. Moreover, the dumping of huge amounts of sewage to surface waterways, especially freshwater bodies, has significantly decreased the amount and quality of water globally [36, 37]. Furthermore, changes in the planet climate are cited as major factors influencing the availability of water across the planet [38-40]. For instance, as a result of the dryness that is affecting many regions, freshwater consumption has grown substantially. Furthermore, global warming affects the natural distribution of water across the planet, resulting in a greater scarcity of drinkable water [3]. Although the scientists developed many solutions, such as the development of new treatment methods [41, 42], the development of recycling options to minimise the volume of the industrial by-products [43-45], and the development of eco-friendly alternatives for the traditional industrial products [46-48], water pollution problem still represents the hardest chalnge for scientists [49, 50].

The dumping of sewage, like industrial and municipal sewage, has significantly diminished the availability of freshwater globally [12, 51]. The discharge of sewage to water sources is regarded as among the most pressing problems among water researchers owing to the large amounts of sewage disposed of to waterways and the many sorts of contaminants present in sewage that wound up in freshwater streams. Surface waters in many regions of China, for instance, are heavily contaminated with insecticides since more than 300 million tonnes of insecticides are being used in agricultural operations annually, with the majority of them ending up in rivers and lakes. Aside from insecticides, numerous other contaminants, such as toxic substances, pigments, and organic compounds, can make their way into waterways [52]. The quantity of such contaminants in waterways has risen dramatically in recent decades as a result of industrialization and urbanization. Scientists determined that the percentage of phosphate pollution in rivers and other bodies of freshwater has increased by over 70% in a previous couple of decades [53, 54]. Bio contaminants are the most common waterway contaminants, and their quality of life and the ecosystem has aroused the interest of academics [55, 56].

Microbial pollutants have received a lot of attention in recent years. Scientists began to investigate and increase our understanding of the incidence and transmission of many types of diseases, notably via waterways, in the 1850s. Studies concentrate on hazardous illnesses such as digestive ailments and their transmission via waterways. Although these attempts to understand the spreading of bio-contamination, bio-contamination is still on the rise due to various factors such as urbanization. For example, according to scientists, over 30 reservoirs in a rural part of Bangladesh were discovered to be polluted with E. coli. This issue is exacerbated by a lack of or inadequacy in modern treatment methods, especially in underdeveloped and developing nations. As a result, several studies have been undertaken to decrease biological contamination through the use of low-cost and efficient treatment methods such as disinfection and coagulation. Nonetheless, chemical techniques produce a variety of by-products that have a damaging effect, limiting their use in sewage treatment.

This project aims to use a steel electrocoagulation device to reduce or eradicate contamination from sewage. To obtain a better E. coli extraction efficiency, the effects of electrical charge intensity and electrodes separation on the operation of the steel electrocoagulation device were examined. The author chose the electrocoagulation technique for biotreatment because of its low price, convenient implementation, potential improvement, and minimal waste generation [57, 58]. This eliminates the necessity of sludge treatment and eventual disposal in dumps.

2. Research Methods

2.1. Electrocoagulation device

The device utilized in this study was designed for a variety of substances. The electrocoagulation vessel is a rectangle made of clear plastic with dimensions of 130 mm length, 90 mm depth, and 120 mm width. The electrocoagulation reactor vessel size and dimensions are based on prior investigations that used the electrocoagulation technique for sewage treatment. Eight steel electrodes are used to ensure appropriate water mixture and better sewage treatment. The steel electrodes were constructed in such a way that the water could be stirred as it passed via the device. Each utilized electrode is 110 mm long and 81 mm deep, and it is positioned vertically in the electrocoagulation device. Steel was chosen for use as electrode material in this study since it is relatively inexpensive when compared to other options. Furthermore, the dimensions and location of the electrodes in the device were determined using prior scientific literature.

2.2. Sewer water samples

Real wastewater samples have been utilized in this study to meet the objective of sterilizing sewage utilizing the electrocoagulation technique. The sewage specimens were taken from the Kerbala wastewater treatment facility, which is situated south of Kerbala, Iraq. Before performing the tests, the collected samples were diluted with denoised water. The E. coli organisms were employed as an indication to assess the efficacy of the electrocoagulation technique in terms of sewer water disinfection. To perform the seeding and development of the E. coli bacterium, the American Public Health Association technique was used. The count of E. coli colonies identified in the water samples was contrasted to the number of E. coli colonies found in the treated sewer water.
It is critical to determine the initial amount of E. coli colonies in the sewage before calculating the removal effectiveness of the electrocoagulation device. The initial number of E. coli colonies in the sewage is designated by J1 in this study. The effluent was then processed in an electrocoagulation device with iron electrodes. To obtain the best sanitation efficiency, the current and the spacing were varied. The current densities utilized in this study were 2, 4, and 6 mA/cm2, with electrodes separation ranging from 10 to 30 mm. It is important to note that the pH level of the sewage water is maintained at 6, as this is the usual pH level of greywater. Following each treatment attempt, a specimen of sterilized sewage was seeded and cultured to determine the remaining numbers of E. coli colonies. The reduction effect of the electrocoagulation device was calculated as follows [23]:

$$Removal\ \% = \frac{J1-J2}{J2} \times 100 \quad (1)$$

J1 and *J2* represent the numbers of *E. coli* colonies before and after the electrocoagulation process per water sample.

3. Outcomes

3.1. Current density

Several studies have shown that power density is a critical factor in water purification. The current regulates the production of compounds from the poles, which has a major impact on the extraction efficiency of wastewaters. Furthermore, the current governs the formation of bubbles, which aid in the movement of pollutants to the sewer water top. The impacts of power density on the removal effectiveness of bio-contaminants from greywater have therefore been investigated. The current was increased from 2 to 6 mA/cm2, while all other parameters remained unchanged. The acidity, treatment duration, and electrodes separation were all kept constant at 6, 50, and 10mm, correspondingly. The findings of the impact of altering the current intensity on the extraction efficiency of bio-contaminants from sewage are shown in Figure 1. The E. coli elimination effectiveness increased from 68% at 2 mA/cm2 to 84% at 6 mA/cm2. This could be due to an increase in the oxidation process and generating bobbles as power density increases. Nevertheless, an increase in current density has a significant impact on treatment costs, rendering the treatment technique impractical.

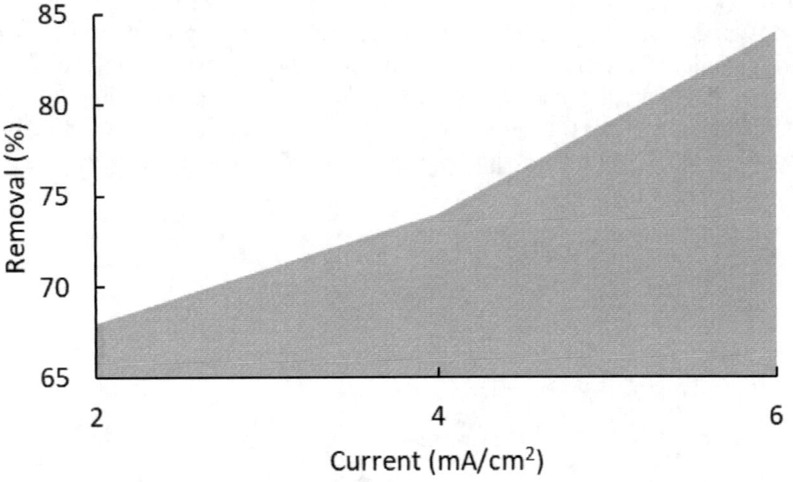

Figure 1: The influence of the current on the effectiveness of bio-contaminant extraction.

Figure 2 shows the impacts of energy use current density. The graph shown that increasing the current density considerably increases power consumption, improving the financial integrity of the electrocoagulation device utilizing steel electrodes. Whenever the current density rose from 1 to 3 mA/cm2, the power consumption increased considerably from 1.6 kW.h/m3 to 14.2 kW.h/m3. According to the foregoing, a current of 4 mA/cm2 offers superior elimination at a much cheaper price than 6 mA/cm2.

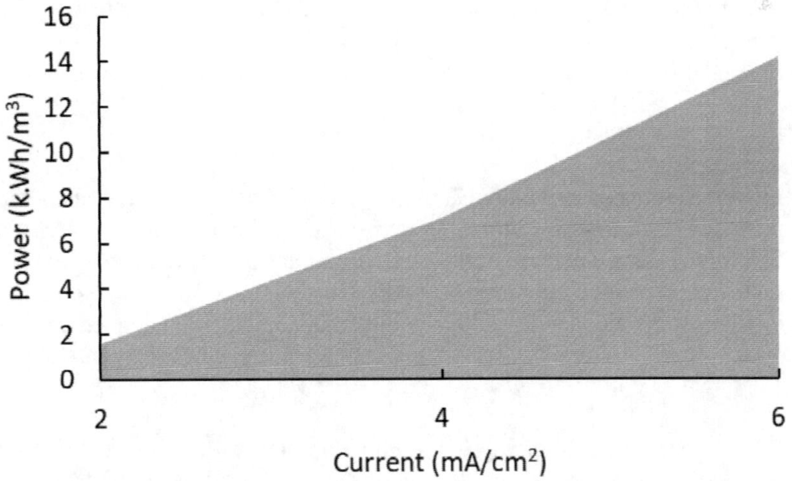

Figure 2: The impact of current density on power consumption.

3.2. Electrodes spacing

According to research on the use of electrolysis in water purification, the distance among poles has a significant impact on the technique's effectiveness. The resistivity is affected by the distance between electrodes, with shorter distances having lower resistivity. As a result, the electrocoagulation unit's removal rate may suffer greatly. Depending on the foregoing, the effect of electrode spacing has indeed been investigated in the present study. The Current, duration of treatment and acidity level was held constant at 4 mA/cm2, 50, and 6.0, accordingly. The electrodes spacing was increased from 10 to 30 mm. Figure 3 depicts the influence of distance on E. coli removal effectiveness from sewage. As could be observed, the longer the distance between poles, the lesser the maximum removal. The figure illustrates that whenever the gap was reduced from 30 to 10 mm, the removal effectiveness rose from 75 to 86 percent. The resistivity rose as the distance between the electrodes grew, resulting in a shift in the E. coli elimination effectiveness.

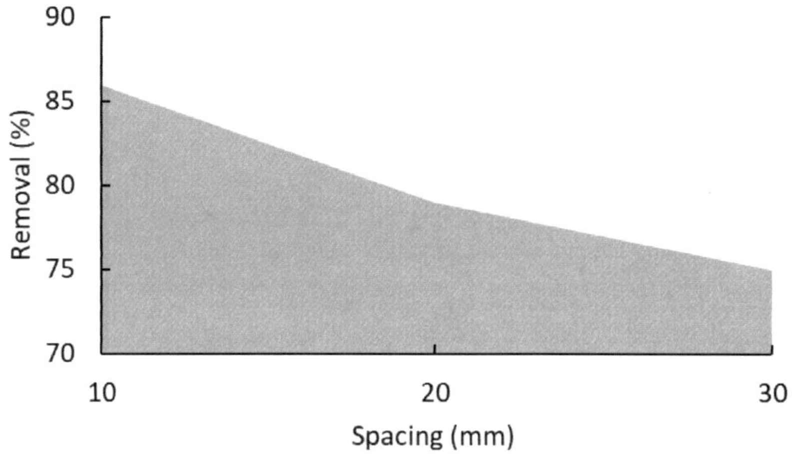

Figure 3: The influence of spacing on the effectiveness of bio contaminant removal.

The authors of the present work recommend investigating the recent sensing technologies; examples of these technologies is the electro-magnetic sensors [59-61] and sonic sensors [62-64], in water treatment (in the electrocoagulation plants) to achieve an efficient monitoring process for the removal of pollutants.

4. Conclusion

The applicability of the electrocoagulation technique as a sterilizer for sewage treatment plant discharge utilizing steel plates with varying separation and current is explored in this research. This study found that steel may be utilized as an electrode in the electrocoagulation technique to remove the existence of E. coli microorganisms in sewage treatment plant discharge. It took 50 minutes to eliminate the E. coli germs from the effluents discharged. Based on findings, it is possible to deduce that the power density and plate separation impact the removal effectiveness of E. coli in sewage treatment plant discharge.

References

[1] Zubaidi S, Al-Bugharbee H, Muhsen Y, Hashim K, Alkhaddar R, Al-Jumeily D and Aljaaf A J 2019. The Prediction of Municipal Water Demand in Iraq: A Case Study of Baghdad Governorate. *12th International Conference on Developments in eSystems Engineering (DeSE),* Kazan, Russia 274-7.

[2] Zubaidi S, Alkhaddar R, Abdellatif M and Muhsin Y R 2019. Using LARS–WG model for prediction of temperature in Columbia City, USA. *IOP Materials Science and Engineering,* 012026.

[3] Zubaidi S, Ortega-Martorell S, Al-Bugharbee H, Olier I, Hashim K S, Gharghan S K, Kot P and Al-Khaddar R 2020. Urban Water Demand Prediction for a City that Suffers from Climate Change and Population Growth: Gauteng Province case study. *Water,* **12** 1-18.

[4] Zubaidi Salah L, Al-Bugharbee H, Ortega Martorell S, Gharghan S, Olier I, Hashim K, Al-Bdairi N and Kot P 2020. A Novel Methodology for Prediction Urban Water Demand by Wavelet Denoising and Adaptive Neuro-Fuzzy Inference System Approach. *Water*, **12** 1-17.

[5] Salah Z, Hashim K, Ethaib S, Al-Bdairi N S S, Al-Bugharbee H and Gharghan S K 2020. A novel methodology to predict monthly municipal water demand based on weather variables scenario. *Journal of King Saud University-Engineering Sciences*, **32** 1-18.

[6] Salah Z, Ortega-Martorell S, Kot P, Alkhaddar R M, Abdellatif M, Gharghan S K, Ahmed M S and Hashim K 2020. A Method for Predicting Long-Term Municipal Water Demands Under Climate Change. *Water Resources Management*, **34** 1265-79.

[7] Abdulhadi B, Kot P, Hashim K, Shaw A, Muradov M and Al-Khaddar R 2021. Continuous-flow electrocoagulation (EC) process for iron removal from water: Experimental, statistical and economic study. *Science of The Total Environment*, **760** 1-16.

[8] Salah Z, Abdulkareem I, Al-Bugharbee H, Ridha H M, Gharghan S K, Al-Qaim F F, Muradov M, Kot P and Alkhaddar R 2020. Hybridised Artificial Neural Network model with Slime Mould Algorithm: A novel methodology for prediction urban stochastic water demand. *Water*, **12** 1-18.

[9] Al-Jumeily D, Hashim K, Alkaddar R, Al-Tufaily M and Lunn J 2019. Sustainable and Environmental Friendly Ancient Reed Houses (Inspired by the Past to Motivate the Future). *11th International Conference on Developments in eSystems Engineering (DeSE),* Cambridge, UK 214-9.

[10] Farhan S L, Antón D and Hashim K 2021. Factors influencing the transformation of Iraqi holy cities: the case of Al-Najaf. *Scientific Review Engineering and Environmental Sciences*, **30** 365-75.

[11] Omran I I, Al-Saati N H, Hashim K S, Al-Saati Z N, Patryk K, Khaddar R A, Al-Jumeily D, Shaw A, Ruddock F and Aljefery M 2019. Assessment of heavy metal pollution in the Great Al-Mussaib irrigation channel. *Desalination and Water Treatment*, **168** 165-74.

[12] Abdulredha M, Abdulridha A, Shubbar A, Alkhaddar R, Kot P and Jordan D 2020. Estimating municipal solid waste generation from service processions during the Ashura religious event. *IOP Conference Series: Materials Science and Engineering,* 012075.

[13] Abdulredha M, Rafid A, Jordan D and Alattabi A 2017. Facing up to waste: how can hotel managers in Kerbala, Iraq, help the city deal with its waste problem? *Procedia engineering*, **196** 771-8.

[14] Idowu I A, Hashim K, Shaw A and Nunes L J 2021. Enhancing the fuel properties of beverage wastes as non-edible feedstock for biofuel production. *Biofuels*, **14** 1-8.

[15] Hashim K, Rafid M, Kot P and Shamma'a A 2021. Water purification from metal ions in the presence of organic matter using electromagnetic radiation-assisted treatment. *Journal of Cleaner Production*, **280** 1-17.

[16] Mohammed A, Hussein A, Yeboah D, Abdulhadi B, Ali A and Hashim K 2020. Electrochemical removal of nitrate from wastewater. *IOP Materials Science and Engineering,* 012037.

[17] Omran I I, Al-Saati N H, Al-Saati H H, Hashim K S and Al-Saati Z N 2021. Sustainability assessment of wastewater treatment techniques in urban areas of Iraq using multi-criteria decision analysis (MCDA). *Water Practice and Technology*, **16** 648-60.

[18] Alyafei A, AlKizwini R S, Hashim K S, Yeboah D, Gkantou M, Al Khaddar R, Al-Faluji D and Zubaidi S L 2020. Treatment of effluents of construction industry using a combined filtration-electrocoagulation method. *IOP Conference Series: Materials Science and Engineering,* 012032.

[19] Alnaimi H, Idan I J, Al-Janabi A, Hashim K, Gkantou M, Zubaidi S L, Kot P and Muradov M 2020. Ultrasonic-electrochemical treatment for effluents of concrete plants. **888** 1-9.

[20] Hashim K S, Al-Saati N H, Alquzweeni S S, Zubaidi S L, Kot P, Kraidi L, Hussein A H, Alkhaddar R, Shaw A and Alwash R 2019. Decolourization of dye solutions by electrocoagulation: an investigation of the effect of operational parameters. *First International Conference on Civil and Environmental Engineering Technologies (ICCEET),* University of Kufa, Iraq 25-32.

[21] Abdulla G, Kareem M, Muradov M, Kot P, Mubarak H, Abdellatif M and Abdulhadi B 2020. Removal of iron from wastewater using a hybrid filter. *IOP Materials Science and Engineering,* 012035.

[22] Abdulraheem F, Al-Khafaji Z, Muradov M, Kot P and Ali A 2020. Natural filtration unit for removal of heavy metals from water. *IOPMaterials Science and Engineering,* 012034.

[23] Abdulhadi B, Kot P, Andy A and Khaddar R 2019. Influence of current density and electrodes spacing on reactive red 120 dye removal from dyed water using electrocoagulation/electroflotation (EC/EF)

process. *1st Int. Conference on Civil and Environmental Engineering Technologies,* University of Kufa, Iraq 12-22.

[24] Al-Saati N H, Hussein T K, Abbas M H, Hashim K, Al-Saati Z N, Kot P, Sadique M, Aljefery M H and Carnacina I 2019. Statistical modelling of turbidity removal applied to non-toxic natural coagulants in water treatment: a case study. *Desalination and Water Treatment*, **150** 406-12.

[25] Zanki A, Mo F, Muradov M and Abdulhadi B 2020. Removal of organic matter from water using ultrasonic-assisted electrocoagulation method. *IOP Materials Science and Engineering,* 012033.

[26] Hashim K, Al-Saati N, Hussein A and Al-Saati Z 2018. An investigation into the level of heavy metals leaching from canal-dreged sediment: a case study metals leaching from dreged sediment. *1st Int. Conference on Materials Engineering & Science,* Istanbul Aydın University, Turkey 12-22.

[27] Hashim K, Ewadh H, Muhsin A, Zubaidi S, Kot P, Muradov M, Aljefery M and Al-Khaddar R 2020. Phosphate removal from water using bottom ash: Adsorption performance, coexisting anions and modelling studies. *Water Science and Technology*, **83** 1-17.

[28] Al-Marri S, AlQuzweeni S, Zubaidi S and Al-Khafaji Z 2020. Ultrasonic-Electrocoagulation method for nitrate removal from water. *IOP Materials Science and Engineering,* 012073.

[29] Alenezi A K, Hasan H A, Hashim K S, Amoako-Attah J, Gkantou M, Muradov M, Kot P and Abdulhadi B 2020. Zeolite-assisted electrocoagulation for remediation of phosphate from calcium-phosphate solution. *IOP Materials Science and Engineering,* 012031.

[30] Emamjomeh M, Kakavand S, Jamali H, Alizadeh S, Safdari M and Mousazade M 2020. The treatment of printing and packaging wastewater by electrocoagulation–flotation: the simultaneous efficacy of critical parameters and economics. *Desalination and water treatment*, **205** 161-74.

[31] Emamjomeh M, Mousazadeh M, Mokhtari N, Jamali H, Naghdali Z and Ghanbari R 2020. Simultaneous removal of phenol and linear alkylbenzene sulfonate from automotive service station wastewater: Optimization of coupled electrochemical and physical processes. *Separation Science and Technology*, **55** 3184-94.

[32] Hashim K, Ali S, Kot P, Andy A, Al Khaddar R, Idowu I and Gkantou M 2020. Escherichia coli inactivation using a hybrid ultrasonic–electrocoagulation reactor. *Chemosphere*, **247** 125868-75.

[33] Al-Sareji O, Grmasha R, Salman J, Idowu I and Hashim K 2021. Street dust contamination by heavy metals in Babylon governorate, Iraq. *Journal of Engineering Science and Technology*, **16** 3528 - 46.

[34] Grmasha R, Al-sareji O, Salman J and Jasim I A 2020. Polycyclic Aromatic Hydrocarbons (PAHs) in Urban Street Dust Within Three Land-Uses of Babylon Governorate, Iraq: Distribution, Sources, and Health Risk Assessment. *Journal of King Saud University - Engineering Sciences*, **33** 1-18.

[35] Abdulredha M, Al Khaddar R, Jordan D, Al-Attabi A and Alzeyadi A 2017. Public participation in solid waste management during mega festivals: A pilot study. *WCST World Congress on Sustainable Technologies Proceedings 2017,* 38-41.

[36] Al-Maliki L A, Farhan S L, Jasim I A, Al-Mamoori S K and Al-Ansari N 2021. Perceptions about water pollution among university students: A case study from Iraq. *Cogent Engineering*, **8** 1895473.

[37] Jasim I A, Farhan S L, Al-Maliki L A and AL-Mamoori S K 2021. Climatic Treatments for Housing in the Traditional Holy Cities: A Comparison between Najaf and Yazd Cities. *IOP Conference Series: Earth and Environmental Science,* 012017.

[38] Abdulredha M, Jordan D and Hashim K 2017. The development of a waste management system in Kerbala during major pilgrimage events: determination of solid waste composition. **196** 779-84.

[39] Abdulredha M, Muhsin A, Al-Janabi A, Alajmi B, Gkantou M, Amoako J, Al-Jumeily D, Mustafina J and AlKhayyat A 2021. Using SF and CKD as cement replacement materials for producing cement mortar. *IOP Materials Science and Engineering,* 012007.

[40] Abdulredha M, Kadhim N, Hussein A, Yeboah D and Hashim K 2021. Zeolite as a natural adsorbent for nitrogenous compounds being removed from water. *IOP Conference Series: Materials Science and Engineering,* 012082.

[41] Aqeel K, Mubarak H, Amoako J, Abdellatif M, Al-Janabi A and Hashim K 2020. Electrochemical removal of brilliant green dye from wastewater. *IOP Materials Science and Engineering,* 012036.

[42] Hashim K, Kot P, Zubaid S, Alwash R, Al Khaddar R, Shaw A, Al-Jumeily D and Aljefery M 2020. Energy efficient electrocoagulation using baffle-plates electrodes for efficient Escherichia Coli removal from Wastewater. *Journal of Water Process Engineering*, **33** 101079-86.

[43] Kadhim A, Sadique M, Al-Mufti R and Hashim K 2020. Long-term performance of novel high-calcium one-part alkali-activated cement developed from thermally activated lime kiln dust. *Journal of Building Engineering*, **32** 1-17.

[44] Kadhim A, Sadique M, Al-Mufti R and Hashim K 2020. Developing One-Part Alkali-Activated metakaolin/natural pozzolan Binders using Lime Waste as activation Agent. *Advances in Cement Research*, **32** 1-38.

[45] Majdi H, Shubbar A, Nasr M, Al-Khafaji Z, Jafer H, Masoodi Z, Sadique M and Hashim K 2020. Experimental data on compressive strength and ultrasonic pulse velocity properties of sustainable mortar made with high content of GGBFS and CKD combinations. *Data in Brief*, **31** 105961-72.

[46] Ali A, Al-Shaer A, Hawesah H and Sadique M 2019. Investigating the influence of cement replacement by high volume of GGBS and PFA on the mechanical performance of cement mortar. *1st Int. Conference on Civil and Environmental Engineering Technologies,* University of Kufa, Iraq.

[47] Shubbar A, Sadique M, Nasr M, Al-Khafaji Z and Hashim K 2020. The impact of grinding time on properties of cement mortar incorporated high volume waste paper sludge ash. *Karbala International Journal of Modern Science*, **6** 1-23.

[48] Shubbar A, Sadique M, Shanbara H and Hashim K 2020 *The Development of a New Low Carbon Binder for Construction as an Alternative to Cement. In Advances in Sustainable Construction Materials and Geotechnical Engineering*, Berlin: Springer.

[49] Alhendal M, Nasir M, Amoako J, , Muradov M, Kot P and Abdulhadi B 2020. Cost-effective hybrid filter for remediation of water from fluoride. *IOP Materials Science and Engineering,* 012038.

[50] Al-Saati N H, Omran I I, Salman A A, Al-Saati Z and Hashim K S 2021. Statistical modeling of monthly streamflow using time series and artificial neural network models: Hindiya Barrage as a case study. *Water Practice and Technology*, **16** 681-91.

[51] Idowu I A, Atherton W, Hashim K, Kot P, Alkhaddar R, Alo B I and Shaw A 2019. An analyses of the status of landfill classification systems in developing countries: Sub Saharan Africa landfill experiences. *Waste Management*, **87** 761-71.

[52] Zhang Y, Han S, Liang D, Shi X, Wang F, Liu W, Zhang L, Chen L, Gu Y and Tian Y 2014. Prenatal exposure to organophosphate pesticides and neurobehavioral development of neonates: a birth cohort study in Shenyang, China. *Egyptian journal of petroleum*, **27** 1275-90.

[53] Abdulredha M, al-Khaddar R, Kot P, Jordan D and Abdulridha A 2018. Benchmarking of the Current Solid Waste Management System in Karbala, Iraq, Using Wasteaware Benchmark Indicators. *World Environmental and Water Resources Congress 2018: Groundwater, Sustainability, and Hydro-Climate/Climate Change,* 40-8.

[54] Alattabi A W, Harris C, Alkhaddar R, Alzeyadi A and Abdulredha M 2017. Online Monitoring of a sequencing batch reactor treating domestic wastewater. *Procedia engineering*, **196** 800-7.

[55] Abdulredha M, Kot P, Al Khaddar R, Jordan D and Abdulridha A 2020. Investigating municipal solid waste management system performance during the Arba'een event in the city of Kerbala, Iraq. *Environment, Development and Sustainability*, **22** 1431-54.

[56] Alwan H, Saleh L, Al-Mohammed F and Abdulredha M 2020. Experimental prediction of the discharge coefficients for rectangular weir with bottom orifices. *Journal of Engineering Science and Technology*, **15** 3265-80.

[57] Hashim K, Al-Jumeily D, Alwash R and Aljefery M H 2020 *Electrocoagulation as an eco-friendly River water treatment method. In Advances in Water Resources Engineering and Management*, Berline: Springer.

[58] Hashim K, Hussein A, Zubaidi S, Kraidi L and Alwash R 2019. Effect of initial pH value on the removal of reactive black dye from water by electrocoagulation (EC) method. *2nd International Scientific Conference,* Al-Qadisiyah University, Iraq 12-22.

[59] Ryecroft S, Andy S, Fergus P, Moody A and Conway L 2019. A First Implementation of Underwater Communications in Raw Water Using the 433 MHz Frequency Combined with a Bowtie Antenna. *Sensors*, **19** 1813-23.

[60] Ryecroft S, Andy S, Fergus P, Tang A, Moody A and Conway L 2021. An Implementation of a Multi-Hop Underwater Wireless Sensor Network using Bowtie Antenna. *Karbala International Journal of Modern Science*, **7** 113-29.

[61] Ryecroft S, Andy S, Fergus P, Patrick K, Hashim K and Conway L 2019. A Novel Gesomin Detection Method Based on Microwave Spectroscopy. *12th International Conference on Developments in eSystems Engineering (DeSE),* Kazan, Russia 429-33.

[62] Kot P, Muradov M, Gkantou M, Kamaris G and Yeboah D 2021. Recent Advancements in Non-Destructive Testing Techniques for Structural Health Monitoring. *Applied Sciences*, **11** 1-28.

[63] Omer G, Kot P, Atherton W, Muradov M, Gkantou M, Shaw A, Riley M, Hashim K and Al-Shamma'a A 2021. A Non-Destructive Electromagnetic Sensing Technique to Determine Chloride Level in Maritime Concrete. *Karbala International Journal of Modern Science*, **7** 61-71.

[64] Kot P, Hashim K S, Muradov M and Al-Khaddar R 2021 *How can sensors be used for sustainability improvement?. In Methods in Sustainability Science*, Elsevier, United Kingdom, p 426.

Evaluating the Durability of Green Cement Mortar Using Ultrasonic Pulse Velocity

Ali Abdulridha[1,*], Saif S. AlQuzweeni[2], Rasha S. AlKizwini[3], Zahra A. Saleh[2], K. S. Hashem[4]
[1] Department of Civil Engineering, University of Warith AL-Anbiya'a, Kerbala, Iraq
[2] Department of Civil Engineering, University of Babylon, Babylon, Iraq
[3] Department of Environmental Engineering, University of Babylon, Babylon, Iraq
[4] Department of Environmental Engineering, LJM University, UK
* Corresponding author, Email: ali.abd@g.uowa.edu.iq

Abstract. Various experimental studies have highlighted the negative consequences of Portland cement on health and the environment, such as toxic emissions and alkaline sewage. The development of environmentally acceptable substitutes for cement is thus one of the objectives of current investigations. The proposed environmental alternatives to cement, nevertheless, might have detrimental impacts on the concrete's characteristics. This investigation intends to study the suitability as alternatives to cement in cement mortar, using industrial wastes like silica fume and cement kiln dust. As a replacement for cement, the cement mortars developed in this research continue from 0% to 60% silica fume and cement kiln dust. Ultrasonic pulse velocity tests at 1 to 4 weeks of age were conducted on hardened specimens. The findings showed that a low reduction in the pulse velocity resulted from high proportions of silica fume and cement kiln dust replacements, whereas an improvement in the characteristics of the mortars with low replacement ratios. Using low kiln dust and silica fume of 20 to 40%, the durability of mortars may increase.

Keywords: Green mortar, by-products, Silica fume, mechanical properties.

1. Introduction

Concrete is of fundamental element in the building sector in our age; the essential element in the concrete mix is Portland Cement (PC) [1, 2]. Although its essential importance for preserving civilization and the financial system, its function in environmental damage is well established [3, 4]. The findings suggest that cement manufacturers release 6 to 8 percent carbon dioxide in the air [5, 6], resulting in catastrophic environmental impacts such as climate change [7, 8]. Climate change has led to enormous changes in water use [9, 10] and contamination of freshwater sources [9, 11]. Furthermore, the disposal of sewer water from concrete factories and moulding operations is of extremely fundamental importance owing to the chemical composition of PC [12, 13], which has devastating consequences for the quality of water supply and mass killing in that species of living organisms [14, 15]. Several contaminants have been detected in concrete plant sewage like phenols [16, 17], turbidity [18, 19], Phosphate [20, 21], nitrates [22, 23] and metals [24-26]. These pollutants that resulted from the cement industries played a significant role in the global warming phenomenon [27-29] that threatens the existence of human beings due to the severe water shortage [30-32]. Therefore, various types of technologies and strategies have been designed to eliminate the effects of this industry on the environment, such as recycling of by-products in cement or concrete [33] and treatment of the effluents using different methods, including but limited to electrocoagulation reactors [34-38], electro-chemical units [39-42], filtration methods [43-47] and combined methods [48-51].

These approaches remain unable to manage the entire anticipated concrete production pollution [52, 53]. Concrete manufacturing has become an increasing problem because of considerable greenhouse gases and sewage emitted from concrete production activities to atmospheric and aquatic bodies. Thus, researchers suggested that developing (cementitious material) substitutes that substitute for cement, like silica fume (SF) and kiln dust (KD) are the sustainable answer to the cement industry difficulties [54]. The use of SF and KD has negative impacts on concrete mortar, although it is also recognized that these ingredients can increase their performance. In general, when the ratio of KD to cement increases, the working ability of fresh mixes decreases. Furthermore, the mortar mechanical properties similarly decrease by raising the KD [54, 55]. The pozzolanic substances, like SF, are highly reactive and extensively used in concrete because of the large surface area and the high silica oxide content. The materials are ultrafine powdered. It is observed that SF is suitable for substituting the PC with about 0.25 to 0.33 (weight proportions) while maintaining the performance [56].

An in situ, non-destructive test method for measuring the quality of concrete buildings is the ultrasonic pulse velocity test (UPVT). In this approach, concrete strengths and characteristics are measured by measuring the speed of the pulse transmitted through concrete [52, 57]. The effectiveness of SF and KD (in varied proportions) as PC substitutes was evaluated in this study. The main objective of the study is to assess the impacts of the usage of these components for cement mortar durability at various treatment ages between 7 days and 28 days with the use of the UPVT.

2. Experimentation

Several experiments to evaluate the ultrasonic pulse velocities of the concrete mortar by part replacement of the KD and the SF were undertaken in these experimental studies. Mix proportions and KD and SF proportions employed in this study are subsequently discussed.

2.1. Test Constituents

The KD is a cement industry's by-product, where it is disposed of through waste disposal. The KD is a finely powdered substance that includes active Calcium oxide, and its primary characteristics rely on the collecting system's locations, the kind of works, the dust collecting plant, and the sorts of fuel used during the manufacturing process. KD comprises raw, partly calcinated feed, clinch dust, free calcareous residue, and salt alkaline, halide, and other volatile chemicals. KD contains unresponsive raw feed. The resulting mortar might suffer a decrease in workability, setting time, and strengths due to the high alkalis content if only KD is used as cementitious materials.

SF is a by-product in the manufacture of silicon alloys. It is frequently used as a material additive in concrete since its physical and chemical characteristics encourage cement reactions. The strength of the SF is significantly increased when added to concrete, and the lifespan of the concrete can be improved. The smoke released from burners is gathered as SF in the silicon and alloys industries rather than disposed of in sites of slurry. The SF nanoparticles with a mean size of approximately 1/1000th of the typical cement particle sizes are very fine. Hence, it is a particularly reactive cementitious material because of its micro-sizes, substantial surface, and silica oxides concentration.

Due to the strong adhesive qualities that link it with other elements in the mix, the ordinary Portland cement was selected as the primary binding material for this research. The BS EN 196-2:2013 was flowed to test the characteristics of the concrete.

Table 1 shows the chemical characteristics of the used cementitious materials (PC and SF). The characteristics of these materials comply with BS-EN-450-1(2012), and BS-EN-197-1(2011).

Local river sands (<5 mm) have been employed as fine aggregates in this investigation. The characteristics of the fine aggregate (particle size distribution, chloride, and sulfate content) were measured in accordance with the BS EN 12620:2002+A1 (2008). freshwater free of organic particles has also been used to mix and cure concrete.

Table 1. PC, KD, and SF chemicals components.

Component (%)	Sample		
	PC	KD	SF
Al_2O_3	14.92	2.31	0.37
CaO	62.81	45.32	0.31
Cl	0.02	7.98	0.04
Fe_2O_3	5.18	3.69	1.5
K_2O	1.7	5.7	0.6
L.O.I	0.31	17.4	1.59
MgO	3.39	1.51	0.42
Na_2O	0.2	0.29	0.39
SiO_2	20.58	12.29	93.81
SO_3	2.09	2.62	0.22

2.2. *Mixture Design*

The design method in this experimental investigation covers the casting of concrete mixtures by selecting the quantity of cement, moisture, aggregates, and material additive ingredients. The fine aggregates were chosen to fit the conventional gradation curve. For all mixtures, 0.4 is chosen for the water to binder ratio (Cement + CKD +SF) while the sand-to-binding ratio was between one part and 2.5 parts of sand. Table 2 shows the quantity of every element and the percentage of the design of the mixture.

Table 2. Experimental mixture design.

Mixture No.	PC (%)	KD (%)	SF (%)
1	100	0	0
2	60	20	20
3	50	25	25
4	40	30	30

The UPVT is started by delivering an ultrasonic pulse through the specimen and by estimating the time necessary to pierce the sample by the pulse. High speed means high concrete sample quality, whereas slower speed suggests a bad concrete sample quality. The pump generator, the transducer for the transformation of the electronic pulses into a mechanical pulse with a range of 40 to 50 kHz, together with the pulse receiver, comprises UPVT equipment/tools. The speeds of the pulse are determined accordingly:

$$\text{UPV} = \frac{\text{Sample thickness}}{\text{Travel time}} \qquad (1)$$

2.3. *Test approaches*

For each case, three prisms (160x40x40mm) were prepared to investigate the impact of KD and SF substitution on the strength of the mortar mixture. In these instances, materials are tested exclusively using cement mortar. The items examined are then constructed with three KD and SF ratios. All specimens were maintained and demoulded and put in water for cure after one day of casting. Ultrasonic testing was done at 7, 14, and 28 days in accordance with BS EN 12504-4:2004 on hardened specimens.

3. Results

Table 3 lists the findings of experiments carried out with respect to the controlled cement mortar and specimens with partially replacing PC of varying KD and SF percentages at various curing times.

Table 3. The outcome of UVPT at 7, 14, and 28 days old.

Mixture No.	PC (%)	KD (%)	SF (%)	UPVT m/s		
				7 days	14 days	28 days
1	100	0	0	3798	3891	3974
2	60	20	20	3818	3952	4100
3	50	25	25	3458	3668	3789
4	40	30	30	3028	3281	3441

The results revealed that a minor drop of the pulse velocity of cement mortar typically results in a cement substitution of KD and SF in mixtures. It is nonetheless clear that contrasted with others including the second mixture exhibits an increase in UPVT. For instance, the variation in UPVT values of the three mixtures was +3%, -6%, and -14% for the first mixture, at 28 days of curing as shown in Figures 1. This explains why the small quantity of cemented materials may fill the gap in the mortar and improve its mechanical characteristics.

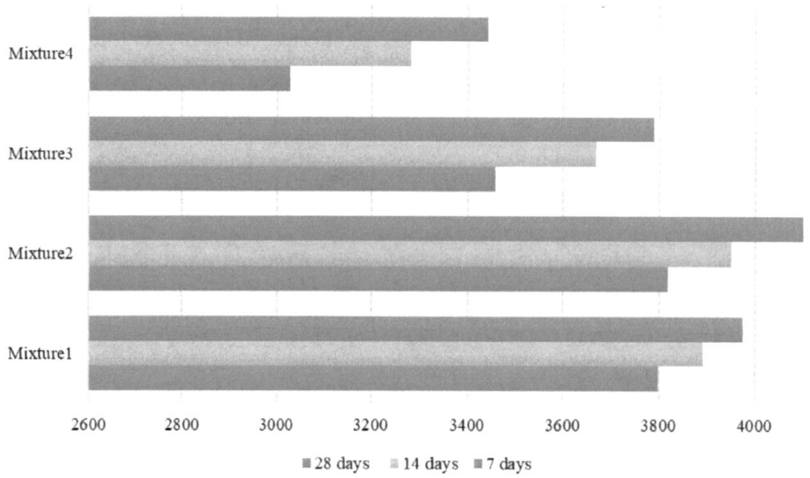

Figure 1: The UPVTs at ages 7,14 and 28 days for all mixtures

However, it can be shown that the pulse velocity of mortar falls by about 20% at a time of seven days when 60% of the parts replaced component is used. But on the other hand, the KD and SF inactive material in the early years was 11 percent lower at 28 curing times, and it took time to interact with the cement components. To summarise, approximately 20 percent can be replaced partially by KD and SF to achieve good sustainability and reduce the environment's pollution. This could be because cementitious materials decrease the cementitious mortar compression capacity, which represents a major factor in mortar manufacturing gel. Due to these effects, Ca(OH)2 activates and initiates hydration of KD and SF after hydration of the cement.

The findings for specimens indicated greater pulse velocities that cured for 28 days. This may also be supported by the fact that the curing ages impact the gain and growth of the gel, which in turn increases the density of the concrete and the toughness of the mixes by lowering the volume of interior voids or permeability of the mortar construction. To illustrate, 7% and 14% increased pulse velocity values for 14 and 28 days for mixture 4, compared to 7 days for the identical samples.

It was noticed that the mechanical properties of the prepared mortar were checked here using ultrasonic pulse velocity method that could have some drawbacks; therefore, it is noteworthy to mention that there are new technologies are currently in use (but in limited scales) to measure the mechanical properties of the concrete and cracking process, such as the embedded sensors [58, 59], surface sensors [59-61], and even wireless methods [62, 63]. These methods could be used in future studies. Also, the treatment plant has been modified in some studies to achieve cost-effectiveness [64, 65].

4. Conclusions

Depending on the outcomes of this investigation, the combined PC-KD and -SF substitution in mortars reduces pulse speed and, in turn, the durability of the mortars as the proportion increases. The durability of the samples, however, is somewhat improved in small volume fractions. Whenever the cement is replaced with more materials in the mixture, an increase in the treatment time will provide greater pulse velocity values. 20 percent can be considered appropriate as supplementary cementitious material; when this increases, the pulse velocity of the mortar slightly decreases.

For future studies, new technologies, such as the embedded sensors, surface sensors, and even wireless methods, are recommended to evaluate the mechanical properties of concrete or mortar. Also, the treatment plant could be modified (such as using baffle plates) to achieve cost-effectiveness.

Acknowledgement

I would want to thank all those who assisted me in carrying out this project.

References

[1] Shubbar A A, Sadique M, Nasr M S, Al-Khafaji Z S and Hashim K S 2020. The impact of grinding time on properties of cement mortar incorporated high volume waste paper sludge ash. *Karbala International Journal of Modern Science*, **6** 1-23.

[2] Shubbar A A, Sadique M, Shanbara H K and Hashim K 2020 *The Development of a New Low Carbon Binder for Construction as an Alternative to Cement. In Advances in Sustainable Construction Materials and Geotechnical Engineering*, Berlin: Springer.

[3] Majdi H S, Shubbar A, Nasr M S, Al-Khafaji Z S, Jafer H, Abdulredha M, Masoodi Z A, Sadique M and Hashim K 2020. Experimental data on compressive strength and ultrasonic pulse velocity properties of sustainable mortar made with high content of GGBFS and CKD combinations. *Data in Brief*, **31** 105961-72.

[4] Grmasha R, Al-sareji O, Salman J and Jasim I 2020. Polycyclic Aromatic Hydrocarbons (PAHs) in Urban Street Dust Within Three Land-Uses of Babylon Governorate, Iraq: Distribution, Sources, and Health Risk Assessment. *Journal of King Saud University - Engineering Sciences*, **33** 1-18.

[5] Al-Sareji O, Grmasha R, Idowu I and Hashim K 2021. Street dust contamination by heavy metals in Babylon governorate, Iraq. *Journal of Engineering Science and Technology*, **16** 3528 - 46.

[6] Shubbar A A, Al-Shaer A, AlKizwini R S, Hashim K, Hawesah H A and Sadique M 2019. Investigating the influence of cement replacement by high volume of GGBS and PFA on the mechanical performance of cement mortar. *First International Conference on Civil and Environmental Engineering Technologies (ICCEET),* University of Kufa, Iraq 31-8.

[7] Zubaidi S, Kot P, Hashim K, Abdellatif M and Muhsin Y 2019. Using LARS–WG model for prediction of temperature in Columbia City, USA. *IOP Materials Science and Engineering,* 012026.

[8] Salah Z, Ortega-Martorell S, Kot P, Alkhaddar R M, Abdellatif M, Gharghan S K, Ahmed M S and Hashim K 2020. A Method for Predicting Long-Term Municipal Water Demands Under Climate Change. *Water Resources Management*, **34** 1265-79.

[9] Alwan H H, Saleh L A, Al-Mohammed F M and Abdulredha M A 2020. Experimental prediction of the discharge coefficients for rectangular weir with bottom orifices. *Journal of Engineering Science and Technology*, **15** 3265-80.

[10] Al-Maliki L A, Farhan S L, Jasim I A, Al-Mamoori S K and Al-Ansari N 2021. Perceptions about water pollution among university students: A case study from Iraq. *Cogent Engineering*, **8** 1895473.

[11] Alattabi A W, Harris C, Alkhaddar R, Alzeyadi A and Abdulredha M 2017. Online Monitoring of a sequencing batch reactor treating domestic wastewater. *Procedia engineering*, **196** 800-7.

[12] Al-Jumeily D, Hashim K, Alkaddar R, Al-Tufaily M and Lunn J 2019. Sustainable and Environmental Friendly Ancient Reed Houses (Inspired by the Past to Motivate the Future). *11th International Conference on Developments in eSystems Engineering (DeSE),* Cambridge, UK 214-9.

[13] Kadhim A, Sadique M, Al-Mufti R and Hashim K 2020. Long-term performance of novel high-calcium one-part alkali-activated cement developed from thermally activated lime kiln dust. *Journal of Building Engineering*, **32** 1-17.

[14] Hashim K S, Ali S S M, AlRifaie J K, Kot P, Shaw A, Al Khaddar R, Idowu I and Gkantou M 2020. Escherichia coli inactivation using a hybrid ultrasonic–electrocoagulation reactor. *Chemosphere*, **247** 125868-75.

[15] Omran I I, Al-Saati N H, Al-Saati H H, Hashim K S and Al-Saati Z N 2021. Sustainability assessment of wastewater treatment techniques in urban areas of Iraq using multi-criteria decision analysis (MCDA). *Water Practice and Technology*, **16** 648-60.

[16] Emamjomeh M M, Mousazadeh M, Mokhtari N, Jamali H A, Makkiabadi M, Naghdali Z, Hashim K S and Ghanbari R 2020. Simultaneous removal of phenol and linear alkylbenzene sulfonate from automotive service station wastewater: Optimization of coupled electrochemical and physical processes. *Separation Science and Technology*, **55** 3184-94.

[17] Alyafei A, AlKizwini R S, Hashim K S, Yeboah D, Gkantou M, Al Khaddar R, Al-Faluji D and Zubaidi S L 2020. Treatment of effluents of construction industry using a combined filtration-electrocoagulation method. *IOP Conference Series: Materials Science and Engineering,* 012032.

[18] Alenazi M, Hashim K S, Hassan A A, Muradov M, Kot P and Abdulhadi B 2020. Turbidity removal using natural coagulants derived from the seeds of strychnos potatorum: statistical and experimental approach. *IOP Conference Series: Materials Science and Engineering,* 012064.

[19] Alenezi A K, Hasan H A, Hashim K S, Amoako-Attah J, Gkantou M, Muradov M, Kot P and Abdulhadi B 2020. Zeolite-assisted electrocoagulation for remediation of phosphate from calcium-phosphate solution. *IOP Conference Series: Materials Science and Engineering,* 012031.

[20] Hashim K S, Ewadh H M, Muhsin A A, Zubaidi S L, Kot P, Muradov M, Aljefery M and Al-Khaddar R 2020. Phosphate removal from water using bottom ash: Adsorption performance, coexisting anions and modelling studies. *Water Science and Technology*, **83** 1-17.

[21] Al-Saati N H, Hussein T K, Abbas M H, Hashim K, Al-Saati Z N, Kot P, Sadique M, Aljefery M H and Carnacina I 2019. Statistical modelling of turbidity removal applied to non-toxic natural coagulants in water treatment: a case study. *Desalination and Water Treatment*, **150** 406-12.

[22] Al-Marri S, AlQuzweeni S S, Hashim K S, AlKhaddar R, Kot P, AlKizwini R S, Zubaidi S L and Al-Khafaji Z S 2020. Ultrasonic-Electrocoagulation method for nitrate removal from water. *IOP Conference Series: Materials Science and Engineering,* 012073.

[23] Mohammed A-H, Hussein A H, Yeboah D, Al Khaddar R, Abdulhadi B, Shubbar A A and Hashim K S 2020. Electrochemical removal of nitrate from wastewater. *IOP Conference Series: Materials Science and Engineering,* 012037.

[24] Abdulhadi B, Kot P, Hashim K, Shaw A, Muradov M and Al-Khaddar R 2021. Continuous-flow electrocoagulation (EC) process for iron removal from water: Experimental, statistical and economic study. *Science of The Total Environment*, **760** 1-16.

[25] Abdulla G, Kareem M M, Hashim K S, Muradov M, Kot P, Mubarak H A, Abdellatif M and Abdulhadi B 2020. Removal of iron from wastewater using a hybrid filter. *IOP Conference Series: Materials Science and Engineering,* 012035.

[26] Hashim K S, Al-Saati N H, Hussein A H and Al-Saati Z N 2018. An investigation into the level of heavy metals leaching from canal-dreged sediment: a case study metals leaching from dreged sediment. *First International Conference on Materials Engineering & Science,* Istanbul Aydın University (IAU), Turkey 12-22.

[27] Zubaidi S, Ortega-Martorell S, Al-Bugharbee H, Olier I, Hashim K S, Gharghan S K, Kot P and Al-Khaddar R 2020. Urban Water Demand Prediction for a City that Suffers from Climate Change and Population Growth: Gauteng Province case study. *Water*, **12** 1-18.

[28] Zubaidi Salah L, Al-Bugharbee H, Ortega Martorell S, Gharghan S, Olier I, Hashim K, Al-Bdairi N and Kot P 2020. A Novel Methodology for Prediction Urban Water Demand by Wavelet Denoising and Adaptive Neuro-Fuzzy Inference System Approach. *Water*, **12** 1-17.

[29] Zubaidi S, Al-Bugharbee H, Muhsen Y, Hashim K, Alkhaddar R, Al-Jumeily D and Aljaaf A J 2019. The Prediction of Municipal Water Demand in Iraq: A Case Study of Baghdad Governorate. *12th International Conference on Developments in eSystems Engineering (DeSE),* Kazan, Russia 274-7.

[30] Salah Z, Abdulkareem I H, Hashim K, Al-Bugharbee H, Ridha H, Gharghan S, Al-Qaim F F, Muradov M, Kot P and Alkhaddar R 2020. Hybridised Artificial Neural Network model with Slime Mould Algorithm: A novel methodology for prediction urban stochastic water demand. *Water*, **12** 1-18.

[31] Salah Z, Hashim K, Ethaib S, Al-Bdairi N S S, Al-Bugharbee H and Gharghan S K 2020. A novel methodology to predict monthly municipal water demand based on weather variables scenario. *Journal of King Saud University-Engineering Sciences*, **32** 1-18.

[32] Zubaidi S, Al-Bugharbee H, Muhsin Y and Rafid M 2020. Forecasting of monthly stochastic signal of urban water demand: Baghdad as a case study. *IOP Materials Science and Engineering,* 012018.

[33] Kadhim A, Sadique M, Al-Mufti R and Hashim K 2020. Developing One-Part Alkali-Activated metakaolin/natural pozzolan Binders using Lime Waste as activation Agent. *Advances in Cement Research*, **32** 1-38.
[34] Abdulraheem F, Al-Khafaji Z, Muradov M, Kot P and Ali A 2020. Natural filtration unit for removal of heavy metals from water. *IOP Materials Science and Engineering,* 012034.
[35] Hashim K S, AlKhaddar R, Shaw A, Kot P, Al-Jumeily D, Alwash R and Aljefery M H 2020 *Electrocoagulation as an eco-friendly River water treatment method. In Advances in Water Resources Engineering and Management*, Berline: Springer.
[36] Hashim K, Al-Saati N, Zubaidi S, Kot P, Kraidi L, Hussein A H, Alkhaddar R, Shaw A and Alwash R 2019. Decolourization of dye solutions by electrocoagulation: an investigation of the effect of operational parameters. *First International Conference on Civil and Environmental Engineering Technologies (ICCEET),* University of Kufa, Iraq 25-32.
[37] Hashim K S, Hussein A H, Zubaidi S L, Kot P, Kraidi L, Alkhaddar R, Shaw A and Alwash R 2019. Effect of initial pH value on the removal of reactive black dye from water by electrocoagulation (EC) method. *2nd International Scientific Conference,* Al-Qadisiyah University, Iraq 12-22.
[38] Hashim K S, Shaw A, AlKhaddar R, Kot P and Al-Shamma'a A 2021. Water purification from metal ions in the presence of organic matter using electromagnetic radiation-assisted treatment. *Journal of Cleaner Production*, **280** 1-17.
[39] Abdulhadi B A, Kot P, Hashim K S, Shaw A and Khaddar R A 2019. Influence of current density and electrodes spacing on reactive red 120 dye removal from dyed water using electrocoagulation/electroflotation (EC/EF) process. *First International Conference on Civil and Environmental Engineering Technologies (ICCEET),* University of Kufa, Iraq 12-22.
[40] Aqeel K, Mubarak H A, Amoako-Attah J, Abdul-Rahaim L A, Al Khaddar R, Abdellatif M, Al-Janabi A and Hashim K S 2020. Electrochemical removal of brilliant green dye from wastewater. *IOP Conference Series: Materials Science and Engineering,* 012036.
[41] Abdulredha M, Kot P, Al Khaddar R, Jordan D and Abdulridha A 2020. Investigating municipal solid waste management system performance during the Arba'een event in the city of Kerbala, Iraq. *Environment, Development and Sustainability*, **22** 1431-54.
[42] Abdulredha M, Rafid A, Jordan D and Alattabi A 2017. Facing up to waste: how can hotel managers in Kerbala, Iraq, help the city deal with its waste problem? *Procedia engineering*, **196** 771-8.
[43] Alhendal M, Nasir M J, Hashim K S, Amoako-Attah J, Al-Faluji D, Muradov M, Kot P and Abdulhadi B 2020. Cost-effective hybrid filter for remediation of water from fluoride. *IOP Conference Series: Materials Science and Engineering,* 012038.
[44] Al-Saati N H, Omran I I, Salman A A, Al-Saati Z and Hashim K S 2021. Statistical modeling of monthly streamflow using time series and artificial neural network models: Hindiya Barrage as a case study. *Water Practice and Technology*, **16** 681-91.
[45] Omran I I, Al-Saati N H, Hashim K S, Al-Saati Z N, Patryk K, Khaddar R A, Al-Jumeily D, Shaw A, Ruddock F and Aljefery M 2019. Assessment of heavy metal pollution in the Great Al-Mussaib irrigation channel. *Desalination and Water Treatment*, **168** 165-74.
[46] Abdulredha M, Abdulridha A, Shubbar A, Alkhaddar R, Kot P and Jordan D 2020. Estimating municipal solid waste generation from service processions during the Ashura religious event. *IOP Conference Series: Materials Science and Engineering,* 012075.
[47] Abdulredha M, AlKhaddar R, Kot P, Jordan D and Abdulridha A 2018. Benchmarking of the Current Solid Waste Management System in Karbala, Iraq, Using Wasteaware Benchmark Indicators. *World Environmental and Water Resources Congress 2018: Groundwater, Sustainability, and Hydro-Climate/Climate Change,* 40-8.
[48] Alnaimi H, Idan I J, Al-Janabi A, Hashim K, Gkantou M, Zubaidi S L, Kot P and Muradov M 2020. Ultrasonic-electrochemical treatment for effluents of concrete plants. **888** 1-9.
[49] Emamjomeh M, Kakavand S, Jamali H, Safdari M, Mousavi S and Mousazade M 2020. The treatment of printing and packaging wastewater by electrocoagulation–flotation: the simultaneous efficacy of critical parameters and economics. *Desalination and water treatment*, **205** 161-74.
[50] Zanki A K, Mohammad F H, Hashim K S, Muradov M, Kot P, Kareem M M and Abdulhadi B 2020. Removal of organic matter from water using ultrasonic-assisted electrocoagulation method. *IOP Conference Series: Materials Science and Engineering,* 012033.
[51] Hashim K, Kot P, Zubaid S, Alwash R, Al Khaddar R, Shaw A, Al-Jumeily D and Aljefery M 2020. Energy efficient electrocoagulation using baffle-plates electrodes for efficient Escherichia Coli removal from Wastewater. *Journal of Water Process Engineering*, **33** 101079-86.

[52] Obaid M, Nasr M, Ali I M, Ali A and Hashim K 2021. Performance of green mortar made from locally available waste tiles and silica fume. *Journal of Engineering Science and Technology*, **16** 136-51.
[53] Jasim I, Farhan S, Maliki L and AL-Mamoori S 2021. Climatic Treatments for Housing in the Traditional Holy Cities: A Comparison between Najaf and Yazd Cities. *IOP Conference Series: Earth and Environmental Science,* 012017.
[54] Ali A, Jafer H, Dulaimi A, Atherton W and Sadique M 2018. The development of a low carbon binder produced from the ternary blending of cement, ground granulated blast furnace slag and high calcium fly ash: An experimental and statistical approach. *Construction and Building Materials*, **187** 1051-60.
[55] Abdulredha M, Muhsin A, Al-Janabi A, Alajmi B, Gkantou M, Amoako J, Al-Jumeily D, Mustafina J and AlKhayyat A 2021. Using SF and CKD as cement replacement materials for producing cement mortar. *IOP Materials Science and Engineering,* 012007.
[56] Shubbar A, Jafer H, Abdulredha M, Al-Khafaji Z, Nasr M, Al Masoodi Z and Sadique M 2020. Properties of cement mortar incorporated high volume fraction of GGBFS and CKD from 1 day to 550 days. **30** 101327.
[57] Farhan S, Antón D, Akef V, Zubaidi S and Hashim K 2021. Factors influencing the transformation of Iraqi holy cities: the case of Al-Najaf. *Scientific Review Engineering and Environmental Sciences*, **30** 365-75.
[58] Gkantou M, Muradov M, Kamaris G, Hashim K, Atherton W and Kot P 2019. Novel Electromagnetic Sensors Embedded in Reinforced Concrete Beams for Crack Detection. *Sensors*, **19** 5175-89.
[59] Kot P, Muradov M, Gkantou M, Kamaris G S and Yeboah D 2021. Recent Advancements in Non-Destructive Testing Techniques for Structural Health Monitoring. *Applied Sciences*, **11** 1-28.
[60] Omer G, Kot P, Atherton W, Muradov M, Gkantou M, Shaw A, Riley M, Hashim K and Al-Shamma'a A 2021. A Non-Destructive Electromagnetic Sensing Technique to Determine Chloride Level in Maritime Concrete. *Karbala International Journal of Modern Science*, **7** 61-71.
[61] Ryecroft S, Shaw A, Fergus P, Kot P, Hashim K, Tang A, Moody A and Conway L 2021. An Implementation of a Multi-Hop Underwater Wireless Sensor Network using Bowtie Antenna. *Karbala International Journal of Modern Science*, **7** 113-29.
[62] Ryecroft S, Shaw A, Fergus P, Kot P, Hashim K, Moody A and Conway L 2019. A First Implementation of Underwater Communications in Raw Water Using the 433 MHz Frequency Combined with a Bowtie Antenna. *Sensors*, **19** 1813-23.
[63] Ryecroft S P, shaw A, Fergus P, Kot P, Hashim K and Conway L 2019. A Novel Gesomin Detection Method Based on Microwave Spectroscopy. *12th International Conference on Developments in eSystems Engineering (DeSE),* Kazan, Russia 429-33.
[64] Al-Mansori N J H, Alfatlawi T J M, Hashim K S and Al-Zubaidi L S 2020. The Effects of Different Shaped Baffle Blocks on the Energy Dissipation. *Civil Engineering Journal*, **6** 961-73.
[65] Hashim K, Shaw A and Phipps D, 2019. *Treatment reactor and method of treating a liquid.* WIPO, PCT/GB2019/052493, LJM University, United Kingdom.

IOP Conf. Series: Earth and Environmental Science **877** (2021) 012050 doi:10.1088/1755-1315/877/1/012050

Utilizing industrial by-products as eco-friendly adsorbent for phosphate removal: An experimental approach

S. Mohammed[1]
[1] Faculty of Engineering Department, University of Baghdad, Iraq
* Corresponding author; Email: salam.m@yahoo.com

Abstract. One of the widely present elements in the groundwater and surface water is phosphate due to two reasons; firstly, it is available at high concentrations in the soil, and secondly, it is widely available in wastewaters (industrial, agricultural and domestic wastewaters). Although phosphate causes many problems to the aquatic environment, eutrophication is the most severe problem due to its effects on water quality, economy, and health. Therefore, a number of studies have been made to evaluate the ability of different remedies to eliminate phosphates from wastewaters. Recently, phosphorus extraction may be achieved by filtering the contaminated solution. However, the cost of filtration materials is still high. Due to this reason, research to date has focused on employing inexpensive materials to reduce the cost of the filtering process. In this research, a by-product of steel manufacturing, kiln bottom ashes, was used to extract phosphates from polluted wastewater, considering the impacts of a number of operating parameters, such as to achieve the best possible extraction efficiency for the lowest possible cost. The findings of this study proved the excellent ability of the bottom ash in the extraction of phosphate from wastewater, where it removed more than 90% of 5 mg/L of phosphate after 40 minutes of treatment using 530 mg/L of bottom ash.

Keywords: By-products, phosphate; adsorption; wastewater.

1. Introduction

The increase in the global population led to a rapid expansion in urbanization and the global agricultural and industrial activity, which in turn drove a vast global demand for clean water that exhausts the available freshwater rapidly [1, 2]. This was badly reflected in the availability of freshwater [3-5], global warming [6, 7], and precipitations [8, 9]. The growing population increases the rate of sewage released into waterways, which has severe effects on consumers' health [10-12]. Furthermore, the increase in the population increased the dumping of solid wastes in landfills that pollute groundwater and then surface water [13-15]. Toxins must not be dumped directly into waterways, and all effluents, whether household or manufacturing, must be treated adequately to eliminate or limit the negative impacts of the contaminants. One of the types of impact of pollutants is algae flourishing which is due to phosphorous penetration into the water, which is harmful to the environment [16]. All types of sewage water contain phosphorus [17, 18] and nitrate [19, 20] beside other pollutants such as fluoride [21], heavy metals [22, 23], bacteria [24, 25], coloring pollutants [26-28] and organic matter [29-31]. Additionally, industrial and domestic wastewater [32-35] and atmospheric pollutants [36-38] are playing a role in increasing such pollutants in water bodies. The effluents of the construction industry [39-42], production plants [43-45] and food industry [32, 46].

Surface waters are depleted of oxygen due to algal blooms induced by phosphorus and nitrate contamination [12]. A large proportion of aquatic creatures and plants might die in a short period of time. As just a result of a substantial rise in organic matter content, the taste, smell, and structure of the water will be changed significantly [47-49]. A rise in algal development might have severe effects on health, industries, seafood, and agriculture, and farmers [20, 50]. Many studies revealed that microalgae types could generate poisonous chemicals that cause a range of severe illnesses and, sometimes, fatalities. In drinkable water, the allowable concentration of phosphate is specified at 0.1 mg/L, whereas in streams and ponds, it is 0.05 mg/L [17, 18]. As far as the upper limit pollutant concentration in water is concerning, the Environmental Protection Agency has strict standards. It is therefore important to remove phosphates from contaminated water, depending on what has been said previously. When phosphate is removed via traditional methods, it generally requires a long and complicated procedure. As an instance, the normal filtering process [21], with some modifications, provides an effective approach for phosphate removal, according to the authors. However, the usage of expensive screens makes this approach uneconomical [46, 51]. It is possible to remove phosphate from the water bodies by converting it to a solid-state, which would be subsequently removed from the water via flotation or some other means. Thermal processing, lakes, anaerobic digestion, absorption, and other techniques are used to transform phosphorus [18]. As just a result of the typical operations of species of bacteria and microalgae, phosphate could be biologically processed. Depends on the kind of microbe, anaerobic or aerobic techniques could be used in these situations. The pH, starting dosage, etc. are all factors that affect the efficacy of the treatments. In regards to employing biological intervention to reduce the existence of phosphate, several disadvantages are discovered [17]. Soluble carbon, the fast development of microorganisms, and the pH level of the fluid are all factors to be considered. Phosphate particles can be converted by adding metallic compounds to water polluted with Phosphorus particles. Usually, sewer water encompassing phosphate is treated using chemical techniques. Calcium ion is amongst the most frequently utilized chemical additions for phosphorus removal for its availability, affordability, and convenience of usage. Especially in an acidic medium, aluminium hydroxide is believed to be highly successful in eliminating phosphate from wastewater. As a consequence of the usage of chemical substances, chemical treatment methods frequently lead to toxic sludge [20]. Due to the utilization use reagents, the technique is much more costly than many other procedures. Physically, phosphate extraction methods do not change the chemical composition of the phosphate, which is resulting in a change. After the removal of phosphate pollution using this approach, it is possible to separate these contaminants (phosphate contaminants) from the removal medium. For phosphate removal, filtering and screening technologies are frequently employed [17, 18]. Substances sensitive to pollutants are used in filtering. Water is then separated from pollutants using a variety of processes such as coagulation [10, 27], absorption [11, 52], and straining hybrid methods [19, 29]. To remove phosphates from water, several permeable mediums were employed. The general rule, the expense of filtering is determined based on the cost of the adsorbent materials employed. The phosphate removal capability of membranes has been demonstrated in previous studies. Several methods extract phosphates from water at low levels, but reverse osmosis is among the most popular. Polluted water is forced through a barrier with really limited porosity under high pressure. Assist in removing pollutants in water by means of high-pressured water. Membrane systems are rarely utilized due to their expensive running costs, difficulties in discarding them, and the requirement for specific equipment and expertise. Recently, studies have centred on the use of waste products or natural resources as economical and useful replacements for traditional materials. Fly ash is a standard manufacturing by-product used to remove phosphate from wastewater and is widely used in the construction industry [39, 53]. Water with an acidity level of 7 can be effectively treated with this substance. Due to the above, our research focused primarily on the possibility of using furnaces bottom ashes as an absorbent to extract the phosphates from polluted effluent.

2. Methods

According to the studies, furnace bottom ashes may be utilized as a convenient and efficient adsorbent for the removal of phosphate from the wastewater. Studying the grain size, permeability, porous structure, and relative density of the bottom ashes is essential. The particle size is highly essential because it is directly related to the adsorption surface. When sieve review is performed, sieve sizes of 4,6,10,16,30 and 50 are utilized, and the quantity of mass kept on each sieve is calculated by analyzing the distribution of particle size of the waste. This formula is used to determine the specific gravity test using the gas jar method:

$$\text{Specific Gravity} = \frac{L_2-L_1}{(L_4-L_1)-(L_3-L_2)} \quad (1)$$

Where

L_1: Weight of jar + weight of plates

L_2: The total weight of the sample, plate, and jar, including the plate and the jar itself.

L_3: Water, dry sample, plate, and jar weight together.

L_4: Weight of the jar plus water

Using an X-ray fluorescence analyzer, the chemical test is used to determine the structure of the ashes of the furnaces. It is performed to assess the effectiveness of extracting phosphates from the bottom ashes of the furnace since materials contain calcium, iron, and aluminum can be utilized for phosphate removal. To extract the phosphate in contaminated water, the bottom ashes were employed as a filter. The Box-Behnken design was utilized to improve the technique in terms of burner bottom ash dosage, duration of treatment, and temp. A potassium diphosphate solution in deionized was used to modify the amounts of phosphate in synthetic water.

An experiment was conducted in which polluted water was combined for various lengths of time with the furnace's bottom ashes to determine the effect of duration. In such tests, the retention duration ranging from 10 to 70 min, while the burner ash dosages varied from 100 to 1000 mg/L. Calculations of phosphate levels are made with the use of a Hach Lange spectrophotometer. As a result of this study, the Box-Behnken Design was utilized to create an estimation model in order to obtain the highest water treatment. The response variable was phosphorus removal in this research. The determinants were the phosphorus content level, bottom ash dose, and the duration of treatment. In order to construct the estimation matrix, Minitab 20 was chosen. The ranges of predictor factors considered are shown in Table 1.

Table 1. Varying ranges of independent and dependent variables were examined.

Factors	Code	Level		
		Low	Moderate	High
Concentration of phosphate in water	CoP	5	10	15
The dose of bottom ashes	DoBA	100	550	1000
Time of the treatment	ToT	10	40	70

A powerful determinant of linearity and relevance is the α value, which may be calculated using the formula below:

$$\alpha = (2X)^{1/4} \quad (2)$$

Where X is the number of variables included in the model.

3. Results and Discussion

3.1. Properties of ashes from furnace bottoms

Based on empirical evidence, the majority of the bottom ash grains are maintained on sieves of lower diameters of less than 16. The results of the sieving tests show that the majority of the bottom ash samples consisted of small particles within 1.180 mm and 4.750mm in size. It was computed using the formulas 3 and 4:

$$\text{Uniformity coefficient} = \frac{x_2}{x_1} \quad (3)$$

$$\text{The gradation coefficient} = \frac{x_3^2}{x_1 * x_2} \quad (4)$$

When plotting the passing rate with particle diameter, x_2, x_3, and x_1 show the percentages of particles that pass through each size. Analysis indicated that x_2, x_3, and x_1 the quantity of the first term were 3.06, 1.36, and 0.48, correspondingly. Furthermore, the uniformity coefficient is 6.21 the gradation coefficient is 1.32. The specimens of bottom ash are judged to be of good quality. Because of this, it is unlikely that the bottom ashes would float, as their relative density is greater than those of the liquid. Permeability was 0.61 m^2/g, while the total area was 6.81 m^2/g. A wide surface, good permeability, and high relative density make the bottoms of the furnace excellent for phosphorus extraction. Chemists determined that the bottom ashes included aluminum, iron, and, calcium oxide, magnesium in roughly 25% of the furnace's ash. As a result, it may be argued that the bottom ash of a burner is suitable for removing phosphates.

3.2. *Phosphate Removal*

To investigate the effectiveness of burner ash in removing phosphorus compounds from wastewater, batch experiments were run. For each of the 15 tests, the experimental variables were modified, in order to identify the optimal experimental set for achieving high removal efficiency at low costs. The findings are shown in Table 2 below. As can be seen in Table 2, the Box-Behnken Design and the removal efficiency obtained are provided.

Table 2. Phosphate elimination in batch flow experiments using a Box-Behnken design.

test	DoBA	CoP	ToT	RE%
01	1	0	0	90.1
02	1	0	1	63.1
03	1	-1	0	77.3
04	0	0	0	83.5
05	0	1	1	79.1
06	-1	0	-1	77.2
07	0	0	0	66.4
08	0	-1	-1	58.2
09	0	1	-1	59.9
10	0	-1	1	65.7
11	-1	1	0	80.1
12	1	1	0	79.2
13	-1	0	1	77.8
14	-1	-1	0	79.2
15	0	0	0	70.1

As a result of this investigation, it was shown that the removal efficiencies are proportional to the dose of furnace bottom ashes and the length of time intervals. Phosphorus removal, on the other hand, has an adverse correlation with starting phosphorus concentrations. This is because the accessible surface area of the adsorption is insufficient to collect the phosphorus from the liquid. As a result, there is more surface area accessible for phosphate adsorption from the contaminated water in the furnace bottom ash. It also enhances the interaction of phosphate and the adsorbent. 90.1% of the phosphate concertation in the contaminated water could be removed by using 530 mg/L of bottom ashes for 40 minutes, as shown in Figure 1.

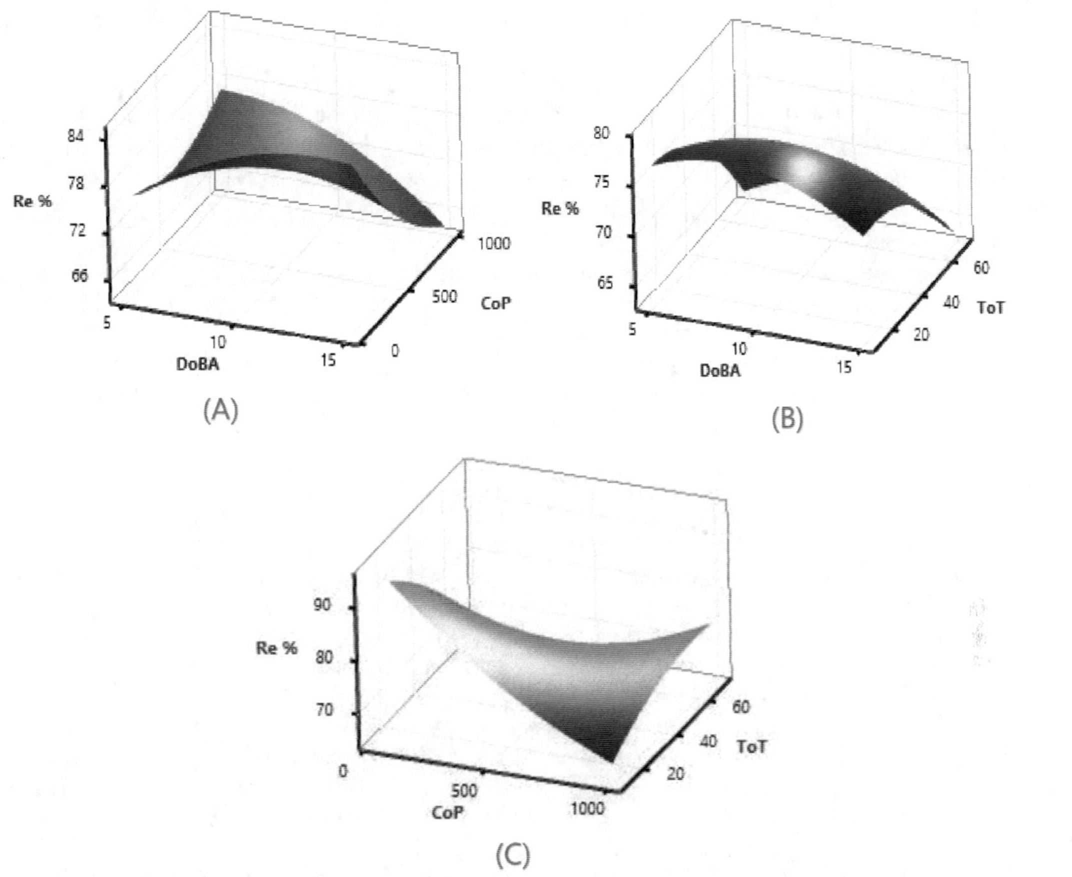

Figure 1: Effects of the studied parameters on phosphate removal by the bottom ash, A) DoBA vs CoP, B) DoBA vs ToT, and C) ToT vs CoP.

To estimate the removal efficiency based on the employed parameters, Box-Behnken Design was used to construct Equation 5.

$$RE\% = 73.33 - 0.44\,DoBA + 2.24\,CoF - 0.10\,ToT + 8.61\,DoBA * DoBA - 2.99\,CoF * CoF - 4.62\,ToT * ToT + 0.25\,DoBA * CoF - 7.17\,DoBA * ToT + 2.93\,CoF * ToT \quad (5)$$

It can be seen from the presented studies about the adsorption method, the depletion of the active sites results in a serious reduction in the adsorption process, therefore; the application of successful sensors (in the literature) helps the operators of the treatment plant to monitor the depletion process, two good examples of the sensors that widely used in the civil and water engineering are the microwave [54-56] and electromagnetic [57-60] sensors.

4. Conclusion

The present study experimentally explored the prospect of using furnace bottom ashes as a treatment method to remediate contaminated water by phosphate. Therefore, the chemical and physical properties of furnace bottom ashes were examined first to check the useability of this by-product for phosphate removal, and then it was experimentally applied for phosphate removal from solutions. It was found the bottom ashes removed more than 90 % of phosphate. Also, it was found that the dose of the bottom ashes and the detention time positively affect the percentage of phosphate removal from the polluted wastewater. With respect to phosphate

adsorption capacity, on the other hand, it is inversely correlated with the initial concentration of the pollutant.
Modelling the relationship between phosphate removal and essential parameters, such as furnace bottom ash dose, phosphate content, and treatment duration is recommended for future studies. To achieve this goal, many software and methods can be used, such as Box–Behnken design.

References

[1] Zubaidi S, Al-Bugharbee H, Muhsen Y, Al-Jumeily D and Aljaaf A J 2019. The Prediction of Municipal Water Demand in Iraq: A Case Study of Baghdad Governorate. *12th International Conference on Developments in eSystems Engineering (DeSE),* Kazan, Russia 274-7.
[2] Salah Z, Abdulkareem I H, Hashim K S, Al-Bugharbee H, Ridha H M, Gharghan S K, Al-Qaim F F, Muradov M, Kot P and Alkhaddar R 2020. Hybridised Artificial Neural Network model with Slime Mould Algorithm: A novel methodology for prediction urban stochastic water demand. *Water*, **12** 1-18.
[3] Zubaidi S, Al-Bugharbee H, Muhsin Y R, Hashim K and Alkhaddar R 2020. Forecasting of monthly stochastic signal of urban water demand: Baghdad as a case study. *IOP Conference Series: Materials Science and Engineering,* 012018.
[4] Al-Saati N H, Omran I I, Salman A A, Al-Saati Z and Hashim K S 2021. Statistical modeling of monthly streamflow using time series and artificial neural network models: Hindiya Barrage as a case study. *Water Practice and Technology*, **16** 681-91.
[5] Salah Z, Ortega-Martorell S, Kot P, Alkhaddar R M, Abdellatif M, Gharghan S K, Ahmed M S and Hashim K 2020. A Method for Predicting Long-Term Municipal Water Demands Under Climate Change. *Water Resources Management*, **34** 1265-79.
[6] Salah Z, Hashim K, Ethaib S, Al-Bdairi N S S, Al-Bugharbee H and Gharghan S K 2020. A novel methodology to predict monthly municipal water demand based on weather variables scenario. *Journal of King Saud University-Engineering Sciences*, **32** 1-18.
[7] Zubaidi S, Ortega-Martorell S, Al-Bugharbee H, Olier I, Hashim K S, Gharghan S K, Kot P and Al-Khaddar R 2020. Urban Water Demand Prediction for a City that Suffers from Climate Change and Population Growth: Gauteng Province case study. *Water*, **12** 1-18.
[8] Zubaidi S L, Kot P, Hashim K, Alkhaddar R, Abdellatif M and Muhsin Y R 2019. Using LARS–WG model for prediction of temperature in Columbia City, USA. *IOP Conference Series: Materials Science and Engineering,* 012026.
[9] Zubaidi Salah L, Al-Bugharbee H, Ortega Martorell S, Gharghan S, Olier I, Hashim K, Al-Bdairi N and Kot P 2020. A Novel Methodology for Prediction Urban Water Demand by Wavelet Denoising and Adaptive Neuro-Fuzzy Inference System Approach. *Water*, **12** 1-17.
[10] Hashim K, Al-Saati N, Alquzweeni S, Kraidi L, Hussein A, Alkhaddar R A and Alwash R 2019. Decolourization of dye solutions by electrocoagulation: an investigation of the effect of operational parameters. *First International Conference on Civil and Environmental Engineering Technologies (ICCEET),* University of Kufa, Iraq 25-32.
[11] Hashim K S, Hussein A H, Kot P, Kraidi L, Alkhaddar R, Shaw A and Alwash R 2019. Effect of initial pH value on the removal of reactive black dye from water by electrocoagulation (EC) method. *2nd International Scientific Conference,* Al-Qadisiyah University, Iraq 12-22.
[12] Hashim K S, Khaddar R A, Jasim N, Shaw A, Phipps D, Kot P, Pedrola M O, Alattabi A W, Abdulredha M and Alawsh R 2019. Electrocoagulation as a green technology for phosphate removal from River water. *Separation and Purification Technology*, **210** 135-44.
[13] Abdulredha M, Abdulridha A, Shubbar A, Alkhaddar R, Kot P and Jordan D 2020. Estimating municipal solid waste generation from service processions during the Ashura religious event. *IOP Conference Series: Materials Science and Engineering,* 012075.
[14] Abdulredha M, Kot P, Al Khaddar R, Jordan D and Abdulridha A 2020. Investigating municipal solid waste management system performance during the Arba'een event in the city of Kerbala, Iraq. *Environment, Development and Sustainability*, **22** 1431-54.
[15] Abdulredha M, Rafid A and Alattabi A 2017. Facing up to waste: how can hotel managers in Kerbala, Iraq, help the city deal with its waste problem? *Procedia engineering*, **196** 771-8.
[16] Hashim K, Al-Saati N, Hussein A and Al-Saati Z 2018. An investigation into the level of heavy metals leaching from canal-dreged sediment: a case study metals leaching from dreged

sediment. *First International Conference on Materials Engineering & Science,* Istanbul Aydın University (IAU), Turkey 12-22.

[17] Hashim K S, Ewadh H M, Muhsin A A, Zubaidi S L, Kot P, Muradov M, Aljefery M and Al-Khaddar R 2020. Phosphate removal from water using bottom ash: Adsorption performance, coexisting anions and modelling studies. *Water Science and Technology*, **83** 1-17.

[18] Alenezi A K, Hasan H A, Hashim K S, Amoako-Attah J, Gkantou M, Muradov M, Kot P and Abdulhadi B 2020. Zeolite-assisted electrocoagulation for remediation of phosphate from calcium-phosphate solution. *IOP: Materials Science and Engineering,* 012031.

[19] Al-Marri S, AlQuzweeni S S, Hashim K S, AlKhaddar R, Kot P, AlKizwini R S, Zubaidi S L and Al-Khafaji Z S 2020. Ultrasonic-Electrocoagulation method for nitrate removal from water. *IOP Conference Series: Materials Science and Engineering,* 012073.

[20] Mohammed A, Hussein A, Yeboah D, Abdulhadi B and Hashim K S 2020. Electrochemical removal of nitrate from wastewater. *IOP Materials Science and Engineering,* 012037.

[21] Alhendal M, Nasir M J, Hashim K S, Amoako-Attah J, Al-Faluji D, Muradov M, Kot P and Abdulhadi B 2020. Cost-effective hybrid filter for remediation of water from fluoride. *IOP Conference Series: Materials Science and Engineering,* 012038.

[22] Abdulraheem F, Al-Khafaji Z, Muradov M and Shubbar A A 2020. Natural filtration unit for removal of heavy metals from water. *IOP Materials Science and Engineering,* 012034.

[23] Omran I I, Al-Saati N H, Hashim K S, Al-Saati Z N, Patryk K, Khaddar R A, Al-Jumeily D, Shaw A, Ruddock F and Aljefery M 2019. Assessment of heavy metal pollution in the Great Al-Mussaib irrigation channel. *Desalination and Water Treatment*, **168** 165-74.

[24] Hashim K, Kot P, Zubaid S, Alwash R, Al-Jumeily D and Aljefery M 2020. Energy efficient electrocoagulation using baffle-plates electrodes for efficient Escherichia Coli removal from Wastewater. *Journal of Water Process Engineering*, **33** 101079-86.

[25] Hashim K S, Ali S S M, AlRifaie J K, Kot P, Shaw A, Al Khaddar R, Idowu I and Gkantou M 2020. Escherichia coli inactivation using a hybrid ultrasonic–electrocoagulation reactor. *Chemosphere*, **247** 125868-75.

[26] Alenazi M, Hashim K S, Hassan A A, Muradov M, Kot P and Abdulhadi B 2020. Turbidity removal using natural coagulants derived from the seeds of strychnos potatorum: statistical and experimental approach. *IOP Conference Series: Materials Science and Engineering,* 012064.

[27] Al-Saati N, Hussein T K, Abbas M, Hashim K, Al-Saati Z, Kot P, Sadique M, Aljefery M H and Carnacina I 2019. Statistical modelling of turbidity removal applied to non-toxic natural coagulants in water treatment: a case study. *Desalination and Water Treatment*, **150** 406-12.

[28] Aqeel K, Mubarak H A, Amoako-Attah J, Abdul-Rahaim L A, Al Khaddar R, Abdellatif M, Al-Janabi A and Hashim K S 2020. Electrochemical removal of brilliant green dye from wastewater. *IOP Conference Series: Materials Science and Engineering,* 012036.

[29] Alnaimi H, Idan I J, Al-Janabi A, Hashim K, Gkantou M, Zubaidi S L, Kot P and Muradov M 2020. Ultrasonic-electrochemical treatment for effluents of concrete plants. **888** 1-9.

[30] Alyafei A, AlKizwini R, Hashim K, Yeboah D, Gkantou M, Al Khaddar R, Al-Faluji D and Zubaidi S L 2020. Treatment of effluents of construction industry using a combined filtration-electrocoagulation method. *IOP Materials Science and Engineering,* 012032.

[31] Zanki A K, Mohammad F H, Hashim K S, Muradov M, Kot P, Kareem M M and Abdulhadi B 2020. Removal of organic matter from water using ultrasonic-assisted electrocoagulation method. *IOP Conference Series: Materials Science and Engineering,* 012033.

[32] Emamjomeh M, Kakavand S, Jamali H, Alizadeh S, Safdari M, Mousavi S, Hashim K and Mousazade M 2020. The treatment of printing and packaging wastewater by electrocoagulation–flotation: the simultaneous efficacy of critical parameters and economics. *Desalination and water treatment*, **205** 161-74.

[33] Abdulhadi B, Kot P, Hashim K, Shaw A, Muradov M and Al-Khaddar R 2021. Continuous-flow electrocoagulation (EC) process for iron removal from water: Experimental, statistical and economic study. *Science of The Total Environment*, **760** 1-16.

[34] Hashim K, AlKhaddar R, Kot P and Al-Shamma'a A 2021. Water purification from metal ions in the presence of organic matter using electromagnetic radiation-assisted treatment. *Journal of Cleaner Production*, **280** 1-17.

[35] Omran I, Al-Saati N, Al-Saati H, Hashim K and Al-Saati Z 2021. Sustainability assessment of wastewater treatment techniques in urban areas of Iraq using multi-criteria decision analysis (MCDA). *Water Practice and Technology*, **16** 648-60.

[36] Grmasha R, Al-sareji O, Salman J, Hashim K and Jasim I 2020. Polycyclic Aromatic Hydrocarbons (PAHs) in Urban Street Dust Within Three Land-Uses of Babylon Governorate, Iraq: Distribution, Sources, and Health Risk Assessment. *Journal of King Saud University - Engineering Sciences*, **33** 1-18.

[37] Al-Sareji O, Grmasha R and Khalid S 2021. Street dust contamination by heavy metals in Babylon governorate, Iraq. *Journal of Engineering Science and Technology*, **16** 28 - 46.

[38] Chabuk A, Hammood Z, Abed S, Kadhim M, Hashim K, Al-Ansari N and Laue J 2021. Noise Level in Textile Industries: Case Study Al-Hillah Textile Factory-Company for Textile Industries, Al-Hillah-Babylon-Iraq. *IOP Earth and Environmental Science,* 012048.

[39] Shubbar A, Al-Shaer A, AlKizwini R, Hashim K, Hawesah H and Sadique M 2019. Investigating the influence of cement replacement by high volume of GGBS and PFA on the mechanical performance of cement mortar. *First International Conference on Civil and Environmental Engineering Technologies (ICCEET),* University of Kufa, Iraq 31-8.

[40] Kadhim A, Sadique M, Al-Mufti R and Hashim K 2020. Developing One-Part Alkali-Activated metakaolin/natural pozzolan Binders using Lime Waste as activation Agent. *Advances in Cement Research*, **32** 1-38.

[41] Majdi H, Shubbar A, Nasr M S, Al-Khafaji Z S, Jafer H, Abdulredha M, Masoodi Z A, Sadique M and Hashim K 2020. Experimental data on compressive strength and ultrasonic pulse velocity properties of sustainable mortar made with high content of GGBFS and CKD combinations. *Data in Brief*, **31** 105961-72.

[42] Al-Jumeily D, Al-Tufaily M and Lunn J 2019. Sustainable and Environmental Friendly Ancient Reed Houses (Inspired by the Past to Motivate the Future). *11th International Conference on Developments in eSystems Engineering (DeSE),* Cambridge, UK 214-9.

[43] Shubbar A A, Sadique M, Nasr M S, Al-Khafaji Z S and Hashim K S 2020. The impact of grinding time on properties of cement mortar incorporated high volume waste paper sludge ash. *Karbala International Journal of Modern Science*, **6** 1-23.

[44] Shubbar A A, Sadique M, Shanbara H K and Hashim K 2020 *The Development of a New Low Carbon Binder for Construction as an Alternative to Cement. In Advances in Sustainable Construction Materials and Geotechnical Engineering*, Berlin: Springer.

[45] Obaid M, Nasr M, Ali I, Ali A and Hashim K 2021. Performance of green mortar made from locally available waste tiles and silica fume. *Journal of Engineering Science and Technology*, **16** 136-51.

[46] Emamjomeh M M, Mousazadeh M, Mokhtari N, Jamali H A, Makkiabadi M, Naghdali Z, Hashim K S and Ghanbari R 2020. Simultaneous removal of phenol and linear alkylbenzene sulfonate from automotive service station wastewater: Optimization of coupled electrochemical and physical processes. *Separation Science and Technology*, **55** 3184-94.

[47] Al-Maliki L, Farhan S, Jasim I, Al-Mamoori S and Al-Ansari N 2021. Perceptions about water pollution among university students: A case study from Iraq. *Cogent Engineering*, **8** 1895473.

[48] Farhan S L, Antón D, Akef V S, Zubaidi S L and Hashim K S 2021. Factors influencing the transformation of Iraqi holy cities: the case of Al-Najaf. *Scientific Review Engineering and Environmental Sciences*, **30** 365-75.

[49] Jasim I A, Farhan S, Maliki L A and AL-Mamoori S K 2021. Climatic Treatments for Housing in the Traditional Holy Cities: A Comparison between Najaf and Yazd Cities. *IOP Conference Series: Earth and Environmental Science,* 012017.

[50] Abdulla G, Kareem M, Muradov M, Abdellatif M and Abdulhadi B 2020. Removal of iron from wastewater using a hybrid filter. *IOP Materials Science and Engineering,* 012035.

[51] Hashim K S, AlKhaddar R, Shaw A, Kot P, Al-Jumeily D, Alwash R and Aljefery M H 2020 *Electrocoagulation as an eco-friendly River water treatment method. In Advances in Water Resources Engineering and Management*, Berline: Springer.

[52] Abdulhadi B A, Kot P, Hashim K S, Shaw A and Khaddar R A 2019. Influence of current density and electrodes spacing on reactive red 120 dye removal from dyed water using electrocoagulation/electroflotation (EC/EF) process. *First International Conference on Civil and Environmental Engineering Technologies (ICCEET),* University of Kufa, Iraq 12-22.

[53] Kadhim A, Sadique M, Al-Mufti R and Hashim K 2020. Long-term performance of novel high-calcium one-part alkali-activated cement developed from thermally activated lime kiln dust. *Journal of Building Engineering*, **32** 1-17.

[54] Gkantou M, Muradov M, Kamaris G, Atherton W and Kot P 2019. Novel Electromagnetic Sensors Embedded in Reinforced Concrete Beams for Crack Detection. *Sensors*, **19** 5175-89.
[55] Ryecroft S P, shaw A, Fergus P, Kot P, Hashim K and Conway L 2019. A Novel Gesomin Detection Method Based on Microwave Spectroscopy. *12th International Conference on Developments in eSystems Engineering (DeSE),* Kazan, Russia 429-33.
[56] Kot P, Hashim K, Muradov M and Al-Khaddar R 2021 *How can sensors be used for sustainability improvement?. In Methods in Sustainability Science*, Elsevier, United Kingdom: Joe Hayton, p 426.
[57] Ryecroft S, Fergus P, Kot P, Hashim K, Moody A and Conway L 2019. A First Implementation of Underwater Communications in Raw Water Using the 433 MHz Frequency Combined with a Bowtie Antenna. *Sensors*, **19** 1813-23.
[58] Kot P, Muradov M, Gkantou M, Kamaris G, Hashim K and Yeboah D 2021. Recent Advancements in Non-Destructive Testing Techniques for Structural Health Monitoring. *Applied Sciences*, **11** 1-28.
[59] Omer G, Kot P, Atherton W, Muradov M, Gkantou M, Riley M, Hashim K and Al-Shamma'a A 2021. A Non-Destructive Electromagnetic Sensing Technique to Determine Chloride Level in Maritime Concrete. *Karbala International Journal of Modern Science*, **7** 61-71.
[60] Ryecroft S, Shaw A, Fergus P, Kot P, Hashim K, Tang A, Moody A and Conway L 2021. An Implementation of a Multi-Hop Underwater Wireless Sensor Network using Bowtie Antenna. *Karbala International Journal of Modern Science*, **7** 113-29.

Electrochemical remediation of wastewater contaminated by phenol

Muhammed A. Shallal[1], Saif S. Radhi[2], Ghusoon J. Shabaa[3], M. Abdulredha[4,*], Mohanad M. Kareem[4], Hayfaa A. Mubarak[5]

[1] MS.c Chemistry, Educational Directorate Thi-Qar, Thi-Qar, Iraq
[2] Al-Mustaqbal University College, Babylon, Iraq
[3] Ministry of Education, School of Gifted Students, Najaf, Iraq
[4] Department of Chemistry, College of Science, University of Babylon, Babylon, Iraq
[5] Chemical Engineering Department, University of Babylon, Babel, Iraq
*Corresponding author; Email: m.abdulredha@yahoo.com

Abstract. One of the most significant contributors to water contamination is the petroleum sector. Large volumes of refinery effluent contaminated with numerous sorts of contaminants are discharged into water sources, causing substantial environmental harm. As a result, researchers looked at the use of a variety of treatment techniques to mitigate the impacts of refinery effluent. Utilising hybrid electrodes (iron as cathodes, and aluminium as anodes) electrodes, this investigation intends to use the electrocoagulation method to minimised phenol contaminants from refinery effluent. In addition, the influence of experimental parameters such as electrical current density, electrode spacing, and duration of treatment on the elimination of phenols was investigated in this study. To eliminate the phenols from the effluent, batch flow investigations were employed. According to the findings, the electrocoagulation technique decreased the number of phenols in petroleum effluent. The hybrid electrocoagulation unit was able to decrease the phenol content by around 45%. With a current density of 4 mA/cm^2 and electrode separation of 2cm, the highest removal efficiency was reached after 110 minutes of treatment. Other experiment factors, such as the original amount of the phenols, must be examined.

Keywords: Petroleum refinery wastewater, phenols, electrocoagulation.

Introduction

The petrol industry has become one of the world's largest, most significant and vital sectors because it provides the primary energy source in several industrialised countries [1, 2]. Furthermore, for several countries throughout the world, the petroleum industry represents the main pile in their national revenue [3]. Nonetheless, this business is regarded as one of the most significant producers of contamination, significantly influencing the environment. The petroleum sector uses a lot of water and generates a lot of effluent from different systems, such as cooling and desalination systems. Researchers have found that to create one barrel of crude oil, more than 250L of water are required. According to the available studies, the petroleum industry generates millions of barrels of polluted wastewater every day that is discharged (some times without proper treatment) to the sources of freshwater [1], adding to this the elevated level of energy consumption, which results in severe air pollution [1, 4]. The main problem in the treatment of petroleum wastewater is the variety of pollutants in this type of wastewater, for instance, the literature showed the wastewaters of the refineries usually contain high concentrations of organic compounds [5, 6], phenols [7], hydrocarbons [8, 9], heavy metals [10-12], greases [13, 14], nitrates [15, 16], phosphate [17, 18], dyes [19, 20], turbidity [21, 22], and biological pollutants [23, 24] other contaminants may be present in wastewater [3, 25].

Although there is a wide range of pollutants in petroleum wastewaters, phenol-based pollutants are highly expected in nearly all forms of petroleum wastewaters [7]. The reports and published studies confirmed that the concentrations of phenol-based pollutants in the wastewaters of the petroleum wastewater could be more than 180 mg/L, noting that the maximum allowable concentration of phenols in water is 1 mg/L, which cause a series of serious environmental and health problems. has been found to be more than 180mg/L. There is a wide body of evidence about petroleum wastewater's direct and indirect effects; for instance, the latter was strongly related to a wide range of diseases and poisoning events [1, 2]. The effects of the petroleum wastewaters are not, unfortunately, limited to water pollution, but also it could cause long-term soil pollution, especially near the refineries or oil production sites [1], and also air pollution [26], that was reflected in the global warming and climate change [27, 28]. As it is well-known the climate change causes serious problems, such as the increase in the global temperature [29, 30], changes in precipitations [31, 32], and water scarcity [33, 34].

The mentioned effects of the petroleum wastewaters highlighted the need to treat refinery wastewaters' effluents properly before the final discharge to the rivers or any other forms of freshwater bodies [35-37]. Therefore, an increasing effort was noticed over the last years to develop proper (efficient and affordable) techniques to remediate the refineries effluents from the mentioned pollutants, which helps maintain the environment and the public health. For instance, physical [38, 39], biological [40, 41], and chemical [42-44] techniques were practised to remove all or some of the above-listed pollutants.

Membrane processes, electrocatalytic, oxidation, absorption, and filtration are some of the technology that has been utilised to determine the effect of one or more contaminants detected in petroleum effluents [38, 39]. Phosphates, nitrates, microbes, organic materials, and heavy metals are among the contaminants eliminated by these methods [40, 41]. Nonetheless, due to changes in the quantity or composition of contaminants in wastewater generated by the petroleum sector, many of these approaches are either too expensive or inefficient to entirely remove or satisfy the disposal requirements limitations. Furthermore, it must be admitted that the refinery industry is not the only source of water and air pollution, but other industry like the cement manufacturing [45-47] and concrete industry [48-51] contribute to the water and air pollution. Also, the rapid expansion in urbanisation contributes to water pollution [52-55]. Moreover, an increasing number of studies are involved in the electrocoagulation technique, which has been shown to eliminate a variety of contaminants from different types of wastewater, such as heavy metals, phosphates, organic matter, colour, and other contaminants [38, 39].

Furthermore, scientists found that electrocoagulation is inexpensive, simple to use, does not need additives, and takes up little space [40, 41]. Additionally, the electrocoagulation sludge includes a high concentration of heavy metals, making it ideal for recycling, particularly in civil engineering applications [47, 56, 57]. This might result in a considerable reduction in trash transfer to landfills and the preservation of landfill sites for other waste kinds [58-60].

This study aims to see if the electrocoagulation technique can be used to remove phenols from refinery effluents. The electrocoagulation unit will be provided with a hybrid electrode, iron (cathodes) and aluminium (anodes)

1. Material and methods

An electrocoagulation system (ES) was used to water purification effluent from a crude oil refining facility. An electrocoagulation system is rectangular, with a wide of 100 mm, a height of 150mm, and a length of 250 mm, with four electrodes; two iron cathodes and two aluminium anodes. Every electrode is 160 mm in length and a width of 100 mm and contains 38 openings with a diameter of 5mm. The openings in electrodes mix the effluent throughout treatment and improve the removal of the contaminants.

110 mg/L of phenols was used in this study to investigate the effectiveness of the ES in the removal of phenol. The effects of electric current density (ECD), treatment period and electrodes displacement (ED) on the performance of the ES regarding phenols elimination from refinery effluent. Table 1 lists the characteristics of the wastewater utilised in this study.

Table 1. The characteristics of the industrial effluents taken from a petroleum refinery.

Parameter	Value
Colour	Brown
PH	7.3
Conductivity (Ms/cm)	9.60
Total suspended solid (mg/l)	114
Phenol (mg/l)	110

By constructing a DC rectifier, three desirable ECDs of 4 mA/cm^2, 8 mA/cm^2, and 12 mA/cm^2 were given EC to eliminate the phenols out from refinery effluent. Furthermore, three-electrode distances of 1.5cm, 2.5cm, and 3.5cm were employed in this study to investigate the influence of the ED on the ES's removal performance. The distances were carefully adjusted.

The elimination of phenols from oil refinery effluent was accomplished by continuously running the effluent through the ES and electrodes. The treatment duration for the phenols in the effluent varied from 1minute to 2 hours. Throughout that time, specimens of 20 mL of the treated wastewater were taken every 10 minutes to investigate how well the ES was removed. A spectrometry (Hach Lange DR 2800) and the standardised cuvette testing LCK 345, which were chosen based on the residual phenol content, were used to measure the residual phenols in the effluent. The degree of the effluent utilised in this study was kept at 25±5°C, which is room temperature. The percentage removal of phenols from refinery effluent may be estimated using the following techniques by using formula 1.

$$\text{Removal Percentage\%} = \frac{L_1 - L_f}{L_1} \times 100\% \qquad (1)$$

L1 and *Lf*, respectively, represent the original and ultimate concentrations of phenols in refinery effluent in mg/L.

2. Results and discussion

2.1. The effect of electrical current density

In the electrocoagulation approach, the current density is a key element because it influences the frequency of coagulant production, which significantly impacts ES removal efficacy. Depending on this, the latter investigation looked at the influence of the ECD impact on the removal performance of phenol from refinery effluent. This was accomplished by employing several ECDs to treat effluent specimens containing 110 mg/l of the contaminant. ECDs of 4 mA/cm^2, 8 mA/cm^2, and 12 mA/cm^2 were utilised in this study.

Furthermore, the ED was maintained at 20 mm, and the treatment time was 110 minutes. The findings (figure 1) revealed that increasing the ECD improves the phenol percentage removal from refinery effluent. After 110 minutes of treatments, the phenol elimination effectiveness rose from 22% to more than 70% when the current density was raised from 4 to 12 mA/cm. This improvement in phenol elimination effectiveness can be attributed to the higher in coagulant production rete.

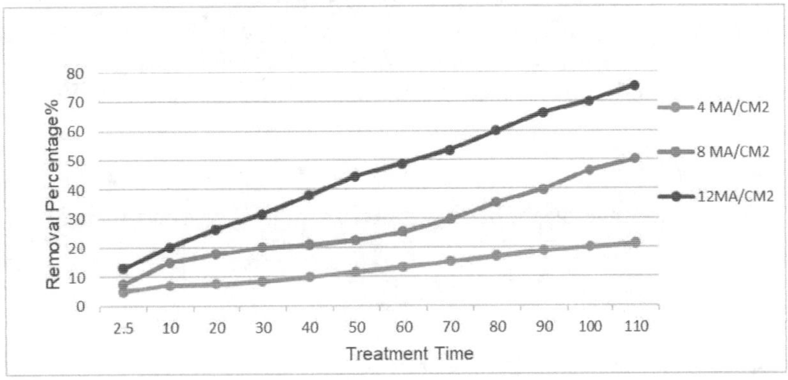

Figure 1: The influence of the ECD on the phenol percentage removal utilising ES.

2.2. *The effect of electrodes displacement*

Another factor that has been observed to have an impact on the electrocoagulation technique's efficacy in terms of different contaminants removal is the distance between electrodes. As a result, the influence of the electrode spacing on the efficiency of electrocoagulation in terms of phenols elimination from refinery effluent has been investigated in this study. The ED was manually adjusted with three separate spacings of 2, 3, and 4 cm. ECD, starting phenol content, and treatment period were retained at 4 mA/cm^2, 120 ppm, and 110 minutes, respectively, for the other experimental factors. Figure 2 shows that when the ED decreases, the percentage removal of phenols using ES rises. Whenever the ED was raised from 2cm to 4cm, the percentage removal of the phenols fell from 40.5% to 12%, as shown in figure 2. Thus, the variation in the electrostatic force of attraction between the electrodes can lead to variation in removal efficiency.

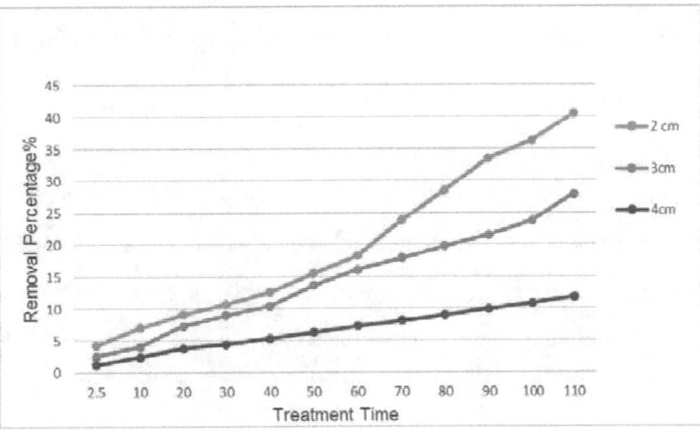

Figure 2: The impact of ED on the effectiveness of phenol removal with ES

2.3. *Influence of treatment time*

Treatment time has been demonstrated to substantially impact the effectiveness of different treatment techniques and applications. In electrocoagulation, Khalid *et al.*, [23] found a favourable relationship between treatment time and coagulant production. The influence of treatment time on the elimination of phenols from petroleum effluent using the suggested ES have been investigated depending on this. To accomplish so, specimens with phenolic concentrations of 110 ppm were treated in the ES for 110 minutes with an ECD of 6 mA/cm^2 and an ED of 2 cm. As seen in Figure 3, the proportion of phenols removed increased continuously with the treatment period. When the treatment period was prolonged from 20 minutes to 110 minutes, the elimination effectiveness of phenols rose from 15% to 59%.

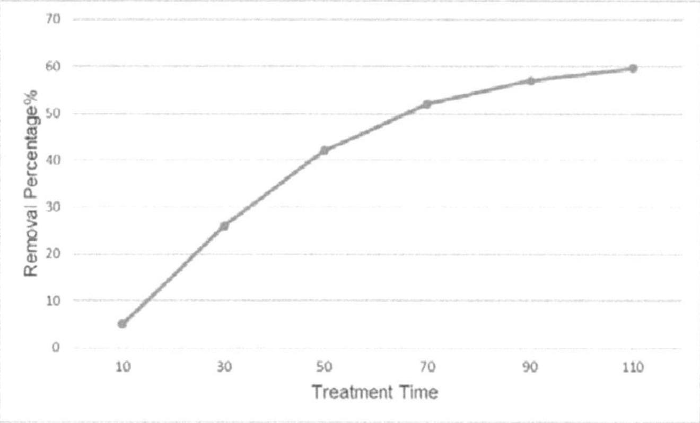

Figure 3: The influence of treatment time on the phenol removal efficiency utilising ES.

Finally, the positive effects of current density on the removal of pollutants from water, unfortunately, is accompanied by power consumption; therefore, it is important to optimise the current density. This can be done using a wide range of technologies, such as wired [61, 62]and wireless [63-65] sensors.

3. Conclusion

Depending on the results observed from this experimental study, the investigation might be summarised in the following:

- The electrocoagulation technique may considerably minimise the amount of phenol contaminants in refinery effluent.
- The elimination of phenols decreased when the current intensity was minimised.
- The elimination of phenols rose when the distance between the electrodes was reduced.

More studies are needed to determine the impact of the original contamination levels on the coagulation/flocculation technique's removal performance.

References

[1] Diya'uddeen B H, Daud W M A W and Aziz A A 2011. Treatment technologies for petroleum refinery effluents: A review. *Process safety and environmental protection*, **89** 95-105.

[2] Abdulredha M, Kadhim N, Almutairi M, Yeboah D and Hashim K 2021. Zeolite as a natural adsorbent for nitrogenous compounds being removed from water. *IOP Materials Science and Engineering,* 012082.
[3] Aljuboury D, Palaniandy P, Abdul Aziz H and Feroz S 2017. Treatment of petroleum wastewater by conventional and new technologies-A review. *Glob. Nest J*, **19** 439-52.
[4] Abdulredha M, Rafid A, Jordan D and Alattabi A J P e 2017. Facing up to waste: how can hotel managers in Kerbala, Iraq, help the city deal with its waste problem? **196** 771-8.
[5] Zanki A K, Mohammad F H, Hashim K S, Muradov M, Kot P, Kareem M M and Abdulhadi B 2020. Removal of organic matter from water using ultrasonic-assisted electrocoagulation method. *IOP Conference Series: Materials Science and Engineering,* 012033.
[6] Emamjomeh M, Kakavand S, Jamali H, Alizadeh S M, Safdari M, Mousavi S and Mousazade M 2020. The treatment of printing and packaging wastewater by electrocoagulation–flotation: the simultaneous efficacy of critical parameters and economics. *Desalination and water treatment*, **205** 161-74.
[7] Emamjomeh M, Mousazadeh M, Mokhtari N, Jamali H A, Makkiabadi M, Naghdali Z, Hashim K and Ghanbari R 2020. Simultaneous removal of phenol and linear alkylbenzene sulfonate from automotive service station wastewater: Optimisation of coupled electrochemical and physical processes. *Separation Science and Technology*, **55** 3184-94.
[8] Grmasha R, Al-sareji O, Salman J and Jasim I A 2020. Polycyclic Aromatic Hydrocarbons (PAHs) in Urban Street Dust Within Three Land-Uses of Babylon Governorate, Iraq: Distribution, Sources, and Health Risk Assessment. *Journal of King Saud University - Engineering Sciences*, **33** 1-18.
[9] Hashim K, AlKhaddar R, Kot P, Al-Jumeily D, Alwash R and Aljefery M 2020 *Electrocoagulation as an eco-friendly River water treatment method. In Advances in Water Resources Engineering and Management*, Berline: Springer.
[10] Abdulla G, Kareem M, Hashim K, Muradov M, Mubarak H, Abdellatif M and Abdulhadi B 2020. Removal of iron from wastewater using a hybrid filter. *IOP Materials Science and Engineering,* 012035.
[11] Abdulraheem F, Al-Khafaji Z, Muradov M, Kot P and Shubbar A 2020. Natural filtration unit for removal of heavy metals from water. *IOP Materials Science and Engineering,* 012034.
[12] Omran I I, Al-Saati N H, Hashim K S, Al-Saati Z N, Patryk K, Khaddar R A, Al-Jumeily D, Shaw A, Ruddock F and Aljefery M 2019. Assessment of heavy metal pollution in the Great Al-Mussaib irrigation channel. *Desalination and Water Treatment*, **168** 165-74.
[13] Alhendal M, Nasir M, Amoako-Attah J, Muradov M, Kot P and Abdulhadi B 2020. Cost-effective hybrid filter for remediation of water from fluoride. *IOP Materials Science and Engineering,* 012038.
[14] Alyafei A, Yeboah D, Al-Faluji D and Zubaidi S 2020. Treatment of effluents of construction industry using a combined filtration-electrocoagulation method. *IOP Materials Science and Engineering,* 88.
[15] Al-Marri S, AlKizwini R S, Zubaidi S L and Al-Khafaji Z S 2020. Ultrasonic-Electrocoagulation method for nitrate removal from water. *IOP Materials Science and Engineering,* 012073.
[16] Mohammed A, Hussein A, Yeboah D , Abdulhadi B, Shubbar A and Hashim K 2020. Electrochemical removal of nitrate from wastewater. *IOP Materials Science and Engineering,* 012037.
[17] Hashim K S, Ewadh H M, Muhsin A A, Zubaidi S L, Kot P, Muradov M, Aljefery M and Al-Khaddar R 2020. Phosphate removal from water using bottom ash: Adsorption performance, coexisting anions and modelling studies. *Water Science and Technology*, **83** 1-17.
[18] Alenezi A K, Hasan H, Amoako-Attah J, Gkantou M, Muradov M, Kot P and Abdulhadi B 2020. Zeolite-assisted electrocoagulation for remediation of phosphate from calcium-phosphate solution. *IOP Conference Series: Materials Science and Engineering,* 012031.
[19] Aqeel K, Mubarak H, Amoako J, Abdellatif M, Al-Janabi A and Hashim K 2020. Electrochemical removal of brilliant green dye from wastewater. *IOP Materials Science and Engineering,* 012036.
[20] Hashim K, Alquzweeni S, Kraidi L, Hussein A H, Andy K and Alwash R 2019. Decolourization of dye solutions by electrocoagulation: an investigation of the effect of operational parameters. *1st Int. Conference on Civil and Environmental Engineering Technologies,* University of Kufa, Iraq.

[21] Alenazi M, Hassan A A, Muradov M, Kot P and Abdulhadi B 2020. Turbidity removal using natural coagulants derived from the seeds of strychnos potatorum: statistical and experimental approach. *IOP Conference Series: Materials Science and Engineering,* 012064.
[22] Al-Saati N H, Hussein T K, Abbas M H, Hashim K, Al-Saati Z N, Kot P, Sadique M, Aljefery M H and Carnacina I 2019. Statistical modelling of turbidity removal applied to non-toxic natural coagulants in water treatment: a case study. *Desalination and Water Treatment*, **150** 406-12.
[23] Hashim K, Kot P, Zubaid S, Alwash R, Al Khaddar R, Shaw A, Al-Jumeily D and Aljefery M 2020. Energy efficient electrocoagulation using baffle-plates electrodes for efficient Escherichia Coli removal from wastewater. *Journal of Water Process Engineering*, **33** 101079-86.
[24] Hashim K S, Ali S, AlRifaie J K, Kot P, Andy A, Al Khaddar R, Idowu I and Gkantou M 2020. Escherichia coli inactivation using a hybrid ultrasonic–electrocoagulation reactor. *Chemosphere*, **247** 125868-75.
[25] Abdulredha M, Muhsin A, Al-Janabi A, Alajmi B, Gkantou M, Amoako-Attah J, Al-Jumeily D, Mustafina J and AlKhayyat A 2021. Using SF and CKD as cement replacement materials for producing cement mortar. *IOP Conference Series: Materials Science and Engineering,* 012007.
[26] Al-Sareji O, Grmasha R, Salman J M, Idowu I and Hashim K S 2021. Street dust contamination by heavy metals in Babylon governorate, Iraq. *Journal of Engineering Science and Technology*, **16** 3528 - 46.
[27] Zubaidi Salah L, Al-Bugharbee H, Ortega Martorell S, Gharghan S, Olier I, Hashim K, Al-Bdairi N and Kot P 2020. A Novel Methodology for Prediction Urban Water Demand by Wavelet Denoising and Adaptive Neuro-Fuzzy Inference System Approach. *Water*, **12** 1-17.
[28] Zubaidi S L, Al-Bugharbee H, Muhsen Y R, Hashim K, Alkhaddar R M, Al-Jumeily D and Aljaaf A J 2019. The Prediction of Municipal Water Demand in Iraq: A Case Study of Baghdad Governorate. *12th International Conference on Developments in eSystems Engineering (DeSE),* Kazan, Russia 274-7.
[29] Zubaidi S, Ortega-Martorell S, Al-Bugharbee H, Olier I, Hashim K S, Gharghan S K, Kot P and Al-Khaddar R 2020. Urban Water Demand Prediction for a City that Suffers from Climate Change and Population Growth: Gauteng Province case study. *Water*, **12** 1-18.
[30] Zubaidi S, Kot P, Hashim K, Alkhaddar R, Abdellatif M and Muhsin Y 2019. Using LARS–WG model for prediction of temperature in Columbia City, USA. *IOP Materials Science and Engineering,* 012026.
[31] Salah Z, Ortega S, Kot P, Alkhaddar R, Abdellatif M, Gharghan S, Ahmed M and Hashim K 2020. A Method for Predicting Long-Term Municipal Water Demands Under Climate Change. *Water Resources Management*, **34** 1265-79.
[32] Zubaidi S, Al-Bugharbee H, Muhsin Y and Alkhaddar R 2020. Forecasting of monthly stochastic signal of urban water demand: Baghdad as a case study. *IOP Materials Science and Engineering,* 012018.
[33] Salah Z, Abdulkareem I, Al-Bugharbee H, Ridha H, Gharghan S, Al-Qaim F and Alkhaddar R 2020. Hybridised Artificial Neural Network model with Slime Mould Algorithm: A novel methodology for prediction urban stochastic water demand. *Water*, **12** 1-18.
[34] Salah Z, Hashim K, Ethaib S, Al-Bdairi S, Al-Bugharbee H and Gharghan S 2020. A novel methodology to predict monthly municipal water demand based on weather variables scenario. *Journal of King Saud University-Engineering Sciences*, **32** 1-18.
[35] Alattabi A, Harris C, Alkhaddar R, Alzeyadi A and Abdulredha M 2017. Online Monitoring of a sequencing batch reactor treating domestic wastewater. *Procedia engineering*, **196** 800-7.
[36] Alwan H, Saleh L and Abdulredha M 2020. Experimental prediction of the discharge coefficients for rectangular weir with bottom orifices. *Journal of Engineering Science and Technology*, **15** 3265-80.
[37] Abdulredha M, Rafid A, Jordan D and Alattabi A 2017. Facing up to waste: how can hotel managers in Kerbala, Iraq, help the city deal with its waste problem? *Procedia engineering*, **196** 771-8.
[38] Hashim K, Hussein A, Zubaidi S, Kraidi L and Alwash R 2019. Effect of initial pH value on the removal of reactive black dye from water by electrocoagulation (EC) method. *2nd International Scientific Conference,* Al-Qadisiyah University, Iraq 12-22.

[39] Hashim K, AlKhaddar R, Kot P and Al-Shamma'a A 2021. Water purification from metal ions in the presence of organic matter using electromagnetic radiation-assisted treatment. *Journal of Cleaner Production*, **280** 1-17.
[40] Al-Saati N H, Omran I I, Salman A A, Al-Saati Z and Hashim K S 2021. Statistical modeling of monthly streamflow using time series and artificial neural network models: Hindiya Barrage as a case study. *Water Practice and Technology*, **16** 681-91.
[41] Hashim K, Al-Saati N, Hussein A and Al-Saati Z 2018. An investigation into the level of heavy metals leaching from canal-dreged sediment: a case study metals leaching from dreged sediment. *First International Conference on Materials Engineering & Science,* Istanbul Aydın University, Turkey 12-22.
[42] Abdulhadi B, Hashim K, Muradov M and Rafid M 2021. Continuous-flow electrocoagulation (EC) process for iron removal from water: Experimental, statistical and economic study. *Science of The Total Environment*, **760** 1-16.
[43] Abdulhadi B, Andy S and Khaddar R 2019. Influence of current density and electrodes spacing on reactive red 120 dye removal from dyed water using electrocoagulation/electroflotation (EC/EF) process. *1st Int. Conference on Civil and Environmental Engineering Technologies,* University of Kufa, Iraq 12-22.
[44] Alnaimi H, Idan I, Al-Janabi A, Hashim K, Gkantou M, Zubaidi S and Muradov M 2020. Ultrasonic-electrochemical treatment for effluents of concrete plants. **888** 1-9.
[45] Shubbar A, Al-Shaer A, AlKizwini R, Hawesah H and Sadique M 2019. Investigating the influence of cement replacement by high volume of GGBS and PFA on the mechanical performance of cement mortar. *1st Int. Conference on Civil and Environmental Engineering Technologies,* University of Kufa, Iraq 31-8.
[46] Shubbar A, Sadique M, Nasr M, Al-Khafaji Z and Hashim K 2020. The impact of grinding time on properties of cement mortar incorporated high volume waste paper sludge ash. *Karbala International Journal of Modern Science*, **6** 1-23.
[47] Shubbar A, Sadique M, Shanbara H and Hashim K 2020 *The Development of a New Low Carbon Binder for Construction as an Alternative to Cement. In Advances in Sustainable Construction Materials and Geotechnical Engineering*, Berlin: Springer.
[48] Kadhim A, Sadique M, Al-Mufti R and Hashim K 2020. Long-term performance of novel high-calcium one-part alkali-activated cement developed from thermally activated lime kiln dust. *Journal of Building Engineering*, **32** 1-17.
[49] Kadhim A, Sadique M and Hashim K 2020. Developing One-Part Alkali-Activated metakaolin/natural pozzolan Binders using Lime Waste as activation Agent. *Advances in Cement Research*, **32** 1-38.
[50] Majdi H, Shubbar A, Nasr M, Al-Khafaji Z, Jafer H, Abdulredha M, Masoodi Z, Sadique M and Hashim K 2020. Experimental data on compressive strength and ultrasonic pulse velocity properties of sustainable mortar made with high content of GGBFS and CKD combinations. *Data in Brief*, **31** 105961-72.
[51] Idowu I, Hashim K, Andy S and Nunes L 2021. Enhancing the fuel properties of beverage wastes as non-edible feedstock for biofuel production. *Biofuels*, **14** 1-8.
[52] Al-Jumeily D, Hashim K, Alkaddar R, Al-Tufaily M and Lunn J 2019. Sustainable and Environmental Friendly Ancient Reed Houses (Inspired by the Past to Motivate the Future). *11th International Conference on Developments in eSystems Engineering (DeSE),* Cambridge, UK 214-9.
[53] Farhan S, Antón D, Akef V, Zubaidi S and Hashim K 2021. Factors influencing the transformation of Iraqi holy cities: the case of Al-Najaf. *Scientific Review Engineering and Environmental Sciences*, **30** 365-75.
[54] Al-Maliki L A, Farhan S L, Jasim I A, Al-Mamoori S K and Al-Ansari N 2021. Perceptions about water pollution among university students: A case study from Iraq. *Cogent Engineering*, **8** 1895473.
[55] Jasim I, Farhan S and AL-Mamoori S 2021. Climatic Treatments for Housing in the Traditional Holy Cities: A Comparison between Najaf and Yazd Cities. *IOP Earth and Environmental Science,* 012017.
[56] Shubbar A, Jafer H, Dulaimi A, Atherton W and Sadique M 2018. The development of a low carbon binder produced from the ternary blending of cement, ground granulated blast furnace slag and high calcium fly ash: An experimental and statistical approach. *Construction and Building Materials*, **187** 1051-60.

[57] Shubbar A, Sadique M, Nasr M, Al-Khafaji Z and Hashim K 2020. The impact of grinding time on properties of cement mortar incorporated high volume waste paper sludge ash. *Karbala International Journal of Modern Science*, **6** 1-23.
[58] Abdulredha M, Abdulridha A, Alkhaddar R, Kot P and Jordan D 2020. Estimating municipal solid waste generation from service processions during the Ashura religious event. *IOP Conference Series: Materials Science and Engineering,* 012075.
[59] Abdulredha M, Jordan D and Abdulridha A 2018. Benchmarking of the Current Solid Waste Management System in Karbala, Iraq, Using Wasteaware Benchmark Indicators. *World Environmental and Water Resources Congress 2018: Groundwater, Sustainability, and Hydro-Climate/Climate Change,* 40-8.
[60] Abdulredha M, Jordan D and Abdulridha A 2020. Investigating municipal solid waste management system performance during the Arba'een event in the city of Kerbala, Iraq. *Environment, Development and Sustainability*, **22** 1431-54.
[61] Gkantou M, Muradov M, Kamaris G, Atherton W and Kot P 2019. Novel Electromagnetic Sensors Embedded in Reinforced Concrete Beams for Crack Detection. *Sensors*, **19** 5175-89.
[62] Kot P, Muradov M, Gkantou M, Kamaris G S, Hashim K and Yeboah D 2021. Recent Advancements in Non-Destructive Testing Techniques for Structural Health Monitoring. *Applied Sciences*, **11** 1-28.
[63] Ryecroft S, Shaw A, Fergus P, Kot P, Moody A and Conway L 2019. A First Implementation of Underwater Communications in Raw Water Using the 433 MHz Frequency Combined with a Bowtie Antenna. *Sensors*, **19** 1813-23.
[64] Kot P, Muradov M and Al-Khaddar R 2021 *How can sensors be used for sustainability improvement?. In Methods in Sustainability Science*, Elsevier, United Kingdom: Joe Hayton, p 426.
[65] Omer G, Kot P, Atherton W, Muradov M, Gkantou M, Shaw A, Riley M, Hashim K and Al-Shamma'a A 2021. A Non-Destructive Electromagnetic Sensing Technique to Determine Chloride Level in Maritime Concrete. *Karbala International Journal of Modern Science*, **7** 61-71.

IOP Conf. Series: Earth and Environmental Science **877** (2021) 012052 doi:10.1088/1755-1315/877/1/012052

The utilization of lignocellular bio-mass as green building thermal insulation material

Muhammad Abdulredha[1,*], Sadiq J. Baqir[2], Sarah M. Ali[3]

[1] Department of Civil Engineering, College of Engineering, University of Kerbala, Kerbala, Iraq
[2] Al-Mustaqbal University College, Babylon, Iraq
[3] Department of pharmacy, Al-Mustaqbal University College, Babylon, Iraq
*Corresponding author; Email: muhammed.r@uokerbala.edu.iq

Abstract. In new residential structures and green architecture, it is necessary to maintain the heat of the internal environment to an appropriate level throughout winter conditions with low electricity usage. This work is thus intended to produce environmentally acceptable isolation substances (organic material). Lignocellular biomass, which is also referred to as Poaceae common reed and Phragmites australis and straw, were used as organic material in this study. During testing of its performance under controlled settings, the insulating effectiveness of these organic compounds was assessed. The exploratory project comprises three forms of isolation: organic made from straw and reed, industrial isolation (fibreglass), and brickwork without insulation. An infrared sensor was used to calculate the quality of isolation. For each isolation situation, the temperature characteristic was produced. The findings show that fibreglass was equivalent to the effectiveness of the organic isolation. Furthermore, the efficiency difference was 0. 84 percent comparing the industrial and organic isolation substances, which shows that Lignocellusic Biomass is a viable environmental-friendly replacement to industrial isolation substances.

1. Introduction

Several nations worldwide have seen great and constant growth in different industries, agriculture, commerce, and urban areas [1, 2]. In this expansion, the number of people, particularly in the petroleum and commercial states, is increasing steadily. Numerous towns and villages have been paved to link them and improve movement among cities and nations [3, 4]. In addition, as a result of persistent population expansion, towns and cities have substantially enlarged their boundaries. Owing to such an expansion, the use of different sorts of products from and to nations and continents, and numerous activities that directly influence the environment as well as people safety and welfare were naturally included. Due to the misuse of unsustainable resources, environmental concerns, including solid wastes accumulation [5-10], climate change [11-13], and global warming [14-16], have become apparent in recent years. For example, worldwide heat is growing rapidly [17-19], freshwater pollution is increasing significantly [20-22], and water use also expanded to an extreme extent [23-25]. Sustainable recycling techniques have thus become a necessity to minimize the health risks associated with contemporary society. For example, some research focused on using by-products of wastewater treatment plants in the construction process, such as the by-products of the electrocoagulation unit [26, 27], filtration units [28-30], and coagulation plants [31-36]. Also, the by-products of furnaces [37-39] and other industries [40-43] are recycled to minimize their environmental effects. However, the by-products of wastewater

treatment plants and any industrial by-products contain harmful pollutants [44-47], which could leach into the indoor environment causing many health problems [4, 48-51].
Sustainable cycles may be only achieved if more natural energy like lignocellulosic materials is recycled and reused. A variety of research is trying to address these problems via the use of new techniques for sewage treatment , freshwater processing , waste management [52-54], and diverse industrial sectors . The concrete industry, which produces large volumes of greenhouse emissions, is among the primary worry for the creation of effective societies . The construction sector has indeed been demonstrated to account for approximately 8 percent of the total atmospheric carbon quantity released to the Earth's atmosphere . Furthermore, the construction sector is rapidly diminishing mineral resources . Hence, for civil engineers and organizations, the building sector has become the most important problem. There have been major attempts to ensure replacements for traditional building materials (development of green buildings). Green building is an active topic of research since this design strategy is expected to significantly decrease the adverse environmental impacts of the construction sector by using very energy-efficient technologies [10].
Massive developments in the construction business worldwide contribute to an increase in energy usage, particularly in the household sector, due to insulation demands. High energy consumption may lead to various difficulties, for example, the depletion of crude energy supplies and environmental concerns, leading to substantial volumes of greenhouse gases in the atmosphere. This is why studies and the need to enhance the energy performance of the buildings throughout the planet arise from such challenges. Therefore, the utilization of alternative products and clean energy becomes a major aspect of future construction [10]. The heat protection of structures plays a key role in reducing energy demand and inconveniencing an interior atmosphere. The use of efficient isolation substances helps achieve apartment complexes' heating with acceptable temperature limitations. The Engineers rely on numerous remedies to avoid high summer temperatures, from applying heat insulation components in built-in walls, flooring, roofing, and even wintertime. To safeguard against extreme heat exposure, sunshades, curtains, and aborigines are extensively utilized. The principle of sustainability can furthermore be promoted by using and renewable resources as insulators. The use of non-toxic, sustainable lignocellulosic biomass is one of the earliest construction products used for protecting humans from external threats and protecting them from harm. In several countries of the globe, for example, timber and strokes are still being used. Such resources can readily be used with different kinds of construction components. Thus, a concept of the environment in planning, building, and maintenance will assess the possible construction techniques involving green, natural materials in a low-energy approach. This would help provide low-energy sustainable structures, mainly for buildings. The use of building Isolation can improve current or future structures' energy efficiency. The use of a suitable thermal insulation system might significantly enhance construction sector development [55, 56].
The objective of these works has been to examine a sustainable way of building through a simulated hollow wall with a sustainable isolating element that employs biotechnological resources, such as lignocellulosic biomass, and is compressed by a simple shaft during insolation. In future studies, the performance of the suggested insulation material can be monitored and evaluated using sensors such as those used in the monitoring of cracks in concrete bodies [57-59] or pollution monitoring [60-63].

2. Methodology

2.1 Experimental procedure

The main substance adopted in this study is straw, which is typically used as isolation in the form of a bale. However, this approach takes a great deal of floor area. The objective of the current work is thus to efficiently employ insulation material to decrease land usage. The straw was placed over the subject in a dense layer supported by a reed frame. An electrical heater was provided on just one side of the cavity and the temperatures were measured using an infrared sensor at the opposite side of the cavity. The use of materials required in building constructions was minimized using various hollow wall techniques, such as the rat-trap brick bonding. The combined cavity-wall construction and renewable, insulating material can protect valuable natural resources to enhance the overall efficiency of new infrastructure projects. It is worth noting that the technique given above is relatively simple; issues like the specific density and quality of the raw materials and the humidity are not considered. The current study approach seeks to show the possibility for future construction projects to use recycled isolation materials. A simple strategy was chosen to assess the effectiveness of straw and reed as thermal insulation materials. As mentioned before, hollow walls may be used to introduce the insulating materials into them (like Rat-Trap Bond) [64]. A portion of the hollow walls was used to simulate the process. The hollow wall was made using conventional clay and cement mortars.

Compressing straw in a layer of about 40 mm thick, back strengthened by reeds, which function vertically as supports, was incorporated for natural insulating components. The straw substrate was coated with a protective carton layer which had a minimal influence on heat output to ease the assembly and movement of samples. Infrared imagery after supplying a heat source was used to detect fluctuations in temperature over time. At the commencement of the non-isolation test, the impact of the once utilized is tested. Similarly, the typical fibreglass insulating material was supplied and then used in the cavity wall, and the performance of this kind of insulator was then tested. Temperature measurements were carried out directly in line with the heat supply at the same position on the walls. This method for temperature measuring at the same site reduced the variation in warming of outside lab impacts during experimentation. In addition, a thermometer from both sides of the wall cavity for the accurate description of the temperature differential between the 2 sides of the cavity wall was given with the portion of the simulated wall cavity. Figure 1 provides visual help on the arrangement of the thermocouples. A was directly in touch with the source of heat; B was always in the contrary direction

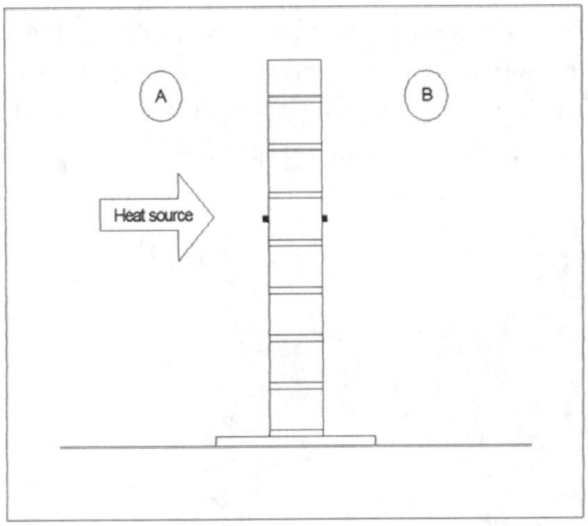

Figure 1: Cavity wall section simulation.

2.2 Data collection

There have been three experiments for every form of isolation: organic, fibreglass, and insulation-free. The temperatures were calibrated in the background (a base value of 19°C), but every type of individual isolation was checked multiple times to guarantee uniformity in the data. Therefore, that heat capacity measured showed insignificant variations; the margin of error was acceptable. The first temperature record for a surface wall was approximately similar to the ambient temperature in the room (19°C). There have been three experiments for every form of isolation: organic, fibreglass, and insulation-free. The temperatures were calibrated in the background (a base value of 19°C), but every type of individual isolation was checked multiple times to guarantee uniformity in the data. Therefore, that heat capacity measured showed insignificant variations; the margin of sampling error was allowed. The infrared camera captured the temperature values and transmitted them to the laptop. Around 30 min from the first measurement, the ultimate average temperature record was completed.

3. Results and Discussion

During the investigation, the surface temperature of the walls was monitored after 25 minutes. A thermal profile has been established with a unique key for each heat profile enables comprehension of the expected temperature throughout the wall section for each insulating scenario. With each isolation situation, Figure 2 illustrates the measured temperature variations. The heat pattern of the surface recorded after 25 minutes without isolation conditions is illustrated in Figure 2a. The results indicated that the greatest temperature was 21.4 ˚C and the lowest temperature being 19.8˚C. Figure 3b illustrates the heat profile observed on the wall surface after 25 minutes once a natural insulator was installed. The results indicated that the greatest temperature was 20.5 ˚C and the lowest temperature being 18.6˚C. Figure 3c shows the wall's thermal profile when industrial isolation (fibreglass) is employed. The temperatures were 20.3°C and 18.9 °C, respectively. In the light of the above, no insulating case is likely to transmit the heat through the hollow wall section, which causes the greatest temperature measurements. This indicates that the greatest recorded temperature and the lowest observed temperature are the best of the three test conditions. No insulation was used during the first test, and the testing of both other insulating materials was based on that. Fibreglass isolation has been determined to have the least heat transfer of the surface; it appears reasonable that the highest performance under these test circumstances should be a uniform industrial standard. Fibreglass can thus be viewed as the best thermal insulating quality insulator. The highest recorded fibreglass temperature, being just 0.2 °C higher than the maximal sensing temperature, was also shown by natural materials as high isolation rates, allow them to be used in future applications.

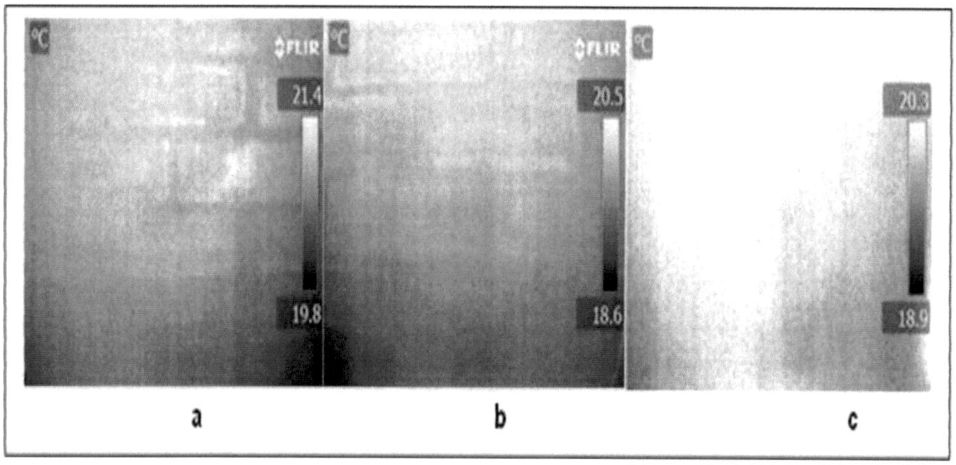

Figure 2: The three insulation scenarios' thermal profile

When looking at Figure 3, one can see that if no insulating was applied, the temperature differential between 2 sides of the wall faces was the smallest of any insulating scenario (62 ˚C). Such results were predicted and suggested that the system transferred more heat than when no insolating elements were used. The temperature variation recorded between the wall sides was 81.5 °C when the wall was insulated with fibreglass. This would be the biggest temperature difference measured and showed the fibreglass stops the most heat loss. The added organic isolation to the wall also exhibited a temperature differential of 79.5 °C on each side. The use of the natural combined materials can significantly enhance the maximum insulation characteristics of the wall construction in comparison with the basic case, which does not include isolation materials. It is noteworthy that variations in the observed temperatures

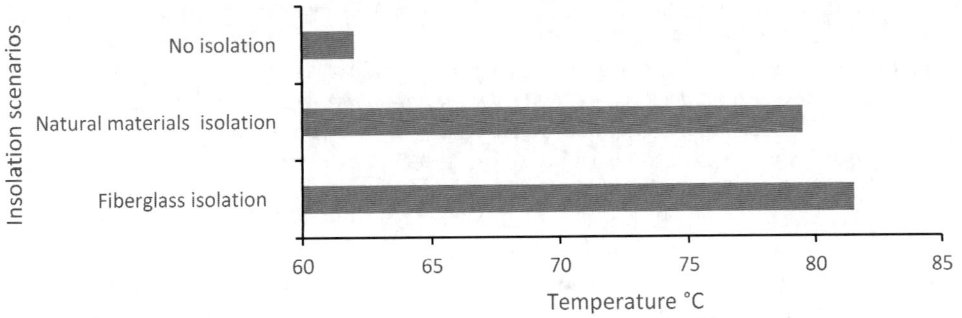

are about the same as when fibreglass is used as conventional in the industry.

Figure 3: Thermal variations in thermocouples

The temperature data from the temperature sensors are shown in Figure 4 in regards to isolation quality. The same overall model is seen in Figure 3. Nevertheless, it is useful to recognize the quality similarities between fibreglass and natural insulators. Only a 0.84% distinction between the 2 insulators was found in terms of proportion effectiveness. Both substances have significantly enhanced the insulating structural performance, as demonstrated by the less efficient use of no insulation.

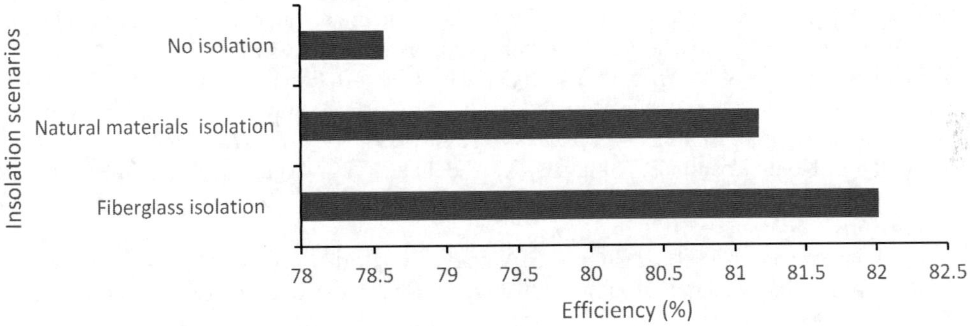

Figure 4: Efficiency of different insolation scenarios.

4. Conclusion

A variety of environmentally friendly techniques were successfully implemented using the mixture of sustainable materials in cavity wall constructions as a natural insulator. The experimental data indicated that natural isolation offers promising results as a heat barrier as contrasted to an industrial insulator such as fibreglass. The results enable the possible use of these organic compounds as a natural source to encourage movement toward green construction. Further research must nevertheless be done to ensure success and extendability for the sustainable construction sector and the healthy home environment with a range of alternative construction techniques.

References

[1] Al-Jumeily D, Hashim K, Alkaddar R and Lunn J 2019. Sustainable and Environmental Friendly Ancient Reed Houses (Inspired by the Past to Motivate the Future). *11th International Conference on Developments in eSystems Engineering (DeSE),* Cambridge, UK 214-9.

[2] Farhan S L, Antón D, Akef V S, Zubaidi S L and Hashim K S 2021. Factors influencing the transformation of Iraqi holy cities: the case of Al-Najaf. *Scientific Review Engineering and Environmental Sciences*, **30** 365-75.
[3] Farhan S L and Nasar A Z 2021. Urban Identity in the Pilgrimage Cities of Iraq: Analysis Trends of Architectural Designers in the City of Karbala. *J. Urban Regen. Renew*, **14** 2-14.
[4] Al-Saati N H, Hussein T K, Abbas M H, Hashim K, Al-Saati Z N, Kot P, Sadique M, Aljefery M H and Carnacina I 2019. Statistical modelling of turbidity removal applied to non-toxic natural coagulants in water treatment: a case study. *Desalination and Water Treatment*, **150** 406-12.
[5] Abdulredha M, Abdulridha A, Shubbar A, Alkhaddar R, Kot P and Jordan D 2020. Estimating municipal solid waste generation from service processions during the Ashura religious event. *IOP Conference Series: Materials Science and Engineering,* 012075.
[6] Abdulredha M, al-Khaddar R, Kot P, Jordan D and Abdulridha A 2018. Benchmarking of the Current Solid Waste Management System in Karbala, Iraq, Using Wasteaware Benchmark Indicators. *World Environmental and Water Resources Congress 2018: Groundwater, Sustainability, and Hydro-Climate/Climate Change,* 40-8.
[7] Abdulredha M, Kot P, Al Khaddar R, Jordan D and Abdulridha A 2020. Investigating municipal solid waste management system performance during the Arba'een event in the city of Kerbala, Iraq. *Environment, Development and Sustainability*, **22** 1431-54.
[8] Abdulredha M, Rafid A, Jordan D and Alattabi A 2017. Facing up to waste: how can hotel managers in Kerbala, Iraq, help the city deal with its waste problem? *Procedia engineering*, **196** 771-8.
[9] Alattabi A W, Harris C, Alkhaddar R, Alzeyadi A and Abdulredha M 2017. Online Monitoring of a sequencing batch reactor treating domestic wastewater. *Procedia engineering*, **196** 800-7.
[10] Alwan H H, Saleh L A, Al-Mohammed F M and Abdulredha M A 2020. Experimental prediction of the discharge coefficients for rectangular weir with bottom orifices. *Journal of Engineering Science and Technology*, **15** 3265-80.
[11] Zubaidi Salah L, Al-Bugharbee H, Ortega Martorell S, Gharghan S, Olier I, Hashim K, Al-Bdairi N and Kot P 2020. A Novel Methodology for Prediction Urban Water Demand by Wavelet Denoising and Adaptive Neuro-Fuzzy Inference System Approach. *Water*, **12** 1-17.
[12] Zubaidi S, Muhsen Y R, Hashim K, Alkhaddar R M, Al-Jumeily D and Aljaaf A J 2019. The Prediction of Municipal Water Demand in Iraq: A Case Study of Baghdad Governorate. *12th International Conference on Developments in eSystems Engineering (DeSE),* Kazan, Russia 274-7.
[13] Zubaidi S L, Kot P, Hashim K, Alkhaddar R, Abdellatif M and Muhsin Y R 2019. Using LARS–WG model for prediction of temperature in Columbia City, USA. *IOP Conference Series: Materials Science and Engineering,* 012026.
[14] Zanki A K, Mohammad F H, Hashim K S, Muradov M, Kot P, Kareem M M and Abdulhadi B 2020. Removal of organic matter from water using ultrasonic-assisted electrocoagulation method. *IOP Conference Series: Materials Science and Engineering,* 012033.
[15] Zubaidi S, Al-Bugharbee H, Muhsin Y R, Hashim K and Alkhaddar R 2020. Forecasting of monthly stochastic signal of urban water demand: Baghdad as a case study. *IOP Conference Series: Materials Science and Engineering,* 012018.
[16] Zubaidi S, Ortega-Martorell S, Al-Bugharbee H, Olier I, Hashim K S, Gharghan S K, Kot P and Al-Khaddar R 2020. Urban Water Demand Prediction for a City that Suffers from Climate Change and Population Growth: Gauteng Province case study. *Water*, **12** 1-18.
[17] Salah Z, Hashim K, Ethaib S, Al-Bdairi N S S, Al-Bugharbee H and Gharghan S K 2020. A novel methodology to predict monthly municipal water demand based on weather variables scenario. *Journal of King Saud University-Engineering Sciences*, **32** 1-18.
[18] Salah Z, Ortega-Martorell S, Kot P, Alkhaddar R M, Abdellatif M, Gharghan S K, Ahmed M S and Hashim K 2020. A Method for Predicting Long-Term Municipal Water Demands Under Climate Change. *Water Resources Management*, **34** 1265-79.
[19] Salah Z, Abdulkareem I, Al-Bugharbee H, Ridha H, Gharghan S K, Al-Qaim F F, Muradov M, Kot P and Alkhaddar R 2020. Hybridised Artificial Neural Network model with Slime Mould Algorithm: A novel methodology for prediction urban stochastic water demand. *Water*, **12** 1-18.

[20] Hashim K S, AlKhaddar R, Shaw A, Kot P, Al-Jumeily D, Alwash R and Aljefery M H 2020 *Electrocoagulation as an eco-friendly River water treatment method. In Advances in Water Resources Engineering and Management*, Berline: Springer.

[21] Hashim K S, Al-Saati N H, Alquzweeni S S, Zubaidi S L, Kot P, Kraidi L, Hussein A H, Alkhaddar R, Shaw A and Alwash R 2019. Decolourization of dye solutions by electrocoagulation: an investigation of the effect of operational parameters. *First International Conference on Civil and Environmental Engineering Technologies (ICCEET),* University of Kufa, Iraq 25-32.

[22] Hashim K, Al-Saati N, Hussein A H and Al-Saati Z 2018. An investigation into the level of heavy metals leaching from canal-dreged sediment: a case study metals leaching from dreged sediment. *1st Int. Conference on Materials Engineering & Science,* Istanbul Aydın University, Turkey 12-22.

[23] Hashim K S, Ewadh H M, Muhsin A A, Zubaidi S L, Kot P, Muradov M, Aljefery M and Al-Khaddar R 2020. Phosphate removal from water using bottom ash: Adsorption performance, coexisting anions and modelling studies. *Water Science and Technology*, **83** 1-17.

[24] Hashim K S, Hussein A H, Zubaidi S L, Kot P, Kraidi L, Alkhaddar R, Shaw A and Alwash R 2019. Effect of initial pH value on the removal of reactive black dye from water by electrocoagulation (EC) method. *2nd International Scientific Conference,* Al-Qadisiyah University, Iraq 12-22.

[25] Hashim K S, Shaw A, AlKhaddar R, Kot P and Al-Shamma'a A 2021. Water purification from metal ions in the presence of organic matter using electromagnetic radiation-assisted treatment. *Journal of Cleaner Production*, **280** 1-17.

[26] Hashim K, Ali S, AlRifaie J, Kot P, Idowu I and Gkantou M 2020. Escherichia coli inactivation using a hybrid ultrasonic–electrocoagulation reactor. *Chemosphere*, **247** 125868-75.

[27] Mohammed A, Hussein A, Yeboah D, Abdulhadi B, Ali A and Hashim K 2020. Electrochemical removal of nitrate from wastewater. *IOP Materials Science and Engineering,* 012037.

[28] Abdulraheem F S, Al-Khafaji Z, Muradov M, Kot P and Shubbar A A 2020. Natural filtration unit for removal of heavy metals from water. *IOP Materials Science and Engineering,* 012034.

[29] Alenezi A K, Hasan H A, Hashim K S, Amoako-Attah J, Gkantou M, Muradov M, Kot P and Abdulhadi B 2020. Zeolite-assisted electrocoagulation for remediation of phosphate from calcium-phosphate solution. *IOP Conference Series: Materials Science and Engineering,* 012031.

[30] Alhendal M, Nasir M J, Hashim K S, Amoako-Attah J, Al-Faluji D, Muradov M, Kot P and Abdulhadi B 2020. Cost-effective hybrid filter for remediation of water from fluoride. *IOP Conference Series: Materials Science and Engineering,* 012038.

[31] Al-Saati N H, Omran I I, Salman A A, Al-Saati Z and Hashim K S 2021. Statistical modeling of monthly streamflow using time series and artificial neural network models: Hindiya Barrage as a case study. *Water Practice and Technology*, **16** 681-91.

[32] Emamjomeh M, Kakavand S, Jamali H A, Mousavi S, Hashim K and Mousazade M 2020. The treatment of printing and packaging wastewater by electrocoagulation–flotation: the simultaneous efficacy of critical parameters and economics. *Desalination and water treatment*, **205** 161-74.

[33] Emamjomeh M M, Mousazadeh M, Mokhtari N, Jamali H A, Makkiabadi M, Naghdali Z, Hashim K S and Ghanbari R 2020. Simultaneous removal of phenol and linear alkylbenzene sulfonate from automotive service station wastewater: Optimization of coupled electrochemical and physical processes. *Separation Science and Technology*, **55** 3184-94.

[34] Hashim K, Kot P, Zubaid S, Alwash R, Al Khaddar R, Shaw A, Al-Jumeily D and Aljefery M 2020. Energy efficient electrocoagulation using baffle-plates electrodes for efficient Escherichia Coli removal from Wastewater. *Journal of Water Process Engineering*, **33** 101079-86.

[35] Omran I, Al-Saati N, Al-Saati H, Hashim K and Al-Saati Z 2021. Sustainability assessment of wastewater treatment techniques in urban areas of Iraq using multi-criteria decision analysis (MCDA). *Water Practice and Technology*, **16** 648-60.

[36] Omran I, Al-Saati N, Al-Saati Z, Patryk K, Khaddar R A, Al-Jumeily D, Ruddock F and Aljefery M 2019. Assessment of heavy metal pollution in the Great Al-Mussaib irrigation channel. *Desalination and Water Treatment*, **168** 165-74.

[37] Shubbar A, Al-Shaer A, Hawesah H A and Sadique M 2019. Investigating the influence of cement replacement by high volume of GGBS and PFA on the mechanical performance of cement mortar.

First International Conference on Civil and Environmental Engineering Technologies (ICCEET), University of Kufa, Iraq 31-8.
[38] Ali A, Sadique M, Nasr M S, Al-Khafaji Z S and Hashim K S 2020. The impact of grinding time on properties of cement mortar incorporated high volume waste paper sludge ash. *Karbala International Journal of Modern Science*, **6** 1-23.
[39] Ali A, Sadique M, Shanbara H K and Hashim K 2020 *The Development of a New Low Carbon Binder for Construction as an Alternative to Cement. In Advances in Sustainable Construction Materials and Geotechnical Engineering*, Berlin: Springer.
[40] Kadhim A, Sadique M, Al-Mufti R and Hashim K 2020. Long-term performance of novel high-calcium one-part alkali-activated cement developed from thermally activated lime kiln dust. *Journal of Building Engineering*, **32** 1-17.
[41] Kadhim A, Sadique M, Mufti R and Hashim K 2020. Developing One-Part Alkali-Activated metakaolin/natural pozzolan Binders using Lime Waste as activation Agent. *Advances in Cement Research*, **32** 1-38.
[42] Majdi H, Shubbar A, Nasr M, Jafer H, Abdulredha M, Sadique M and Hashim K 2020. Experimental data on compressive strength and ultrasonic pulse velocity properties of sustainable mortar made with high content of GGBFS and CKD combinations. *Data in Brief*, **31** 105961-72.
[43] Obaid M, Nasr M, Ali I, Ali A and Hashim K 2021. Performance of green mortar made from locally available waste tiles and silica fume. *Journal of Engineering Science and Technology*, **16** 136-51.
[44] Abdulhadi B, Kot P, Hashim K, Shaw A, Muradov M and Al-Khaddar R 2021. Continuous-flow electrocoagulation (EC) process for iron removal from water: Experimental, statistical and economic study. *Science of The Total Environment*, **760** 1-16.
[45] Abdulhadi B A, Kot P, Hashim K S, Shaw A and Khaddar R A 2019. Influence of current density and electrodes spacing on reactive red 120 dye removal from dyed water using electrocoagulation/electroflotation (EC/EF) process. *First International Conference on Civil and Environmental Engineering Technologies (ICCEET),* University of Kufa, Iraq 12-22.
[46] Abdulla G, Kareem M, Muradov M, Mubarak H, Abdellatif M and Abdulhadi B 2020. Removal of iron from wastewater using a hybrid filter. *IOP Materials Science and Engineering,* 012035.
[47] Alenazi M, Hashim K S, Hassan A A, Muradov M, Kot P and Abdulhadi B 2020. Turbidity removal using natural coagulants derived from the seeds of strychnos potatorum: statistical and experimental approach. *IOP Conference Series: Materials Science and Engineering,* 012064.
[48] Alyafei A, AlKizwini R S, Hashim K S, Yeboah D, Gkantou M, Al Khaddar R, Al-Faluji D and Zubaidi S L 2020. Treatment of effluents of construction industry using a combined filtration-electrocoagulation method. *IOP Conference Series: Materials Science and Engineering,* 012032.
[49] Aqeel K, Mubarak H A, Amoako-Attah J, Abdul-Rahaim L A, Al Khaddar R, Abdellatif M, Al-Janabi A and Hashim K S 2020. Electrochemical removal of brilliant green dye from wastewater. *IOP Conference Series: Materials Science and Engineering,* 012036.
[50] Alnaimi H, Idan I J, Al-Janabi A, Hashim K, Gkantou M, Zubaidi S L, Kot P and Muradov M 2020. Ultrasonic-electrochemical treatment for effluents of concrete plants. **888** 1-9.
[51] Al-Marri S, AlQuzweeni S, AlKhaddar R and Al-Khafaji Z S 2020. Ultrasonic-Electrocoagulation method for nitrate removal from water. *IOP Materials Science and Engineering,* 012073.
[52] Al-Sareji O, Grmasha R, Salman J, Idowu I and Hashim K 2021. Street dust contamination by heavy metals in Babylon governorate, Iraq. *Journal of Engineering Science and Technology*, **16** 3528 - 46.
[53] Grmasha R, Al-sareji O, Salman J, and Jasim I A 2020. Polycyclic Aromatic Hydrocarbons (PAHs) in Urban Street Dust Within Three Land-Uses of Babylon Governorate, Iraq: Distribution, Sources, and Health Risk Assessment. *Journal of King Saud University - Engineering Sciences*, **33** 1-18.
[54] Idowu I A, Hashim K and Nunes L J 2021. Enhancing the fuel properties of beverage wastes as non-edible feedstock for biofuel production. *Biofuels*, **14** 1-8.
[55] Abdulredha M, Kadhim N, Hussein A, Almutairi M, Alkhaddar R, Yeboah D and Hashim K 2021. Zeolite as a natural adsorbent for nitrogenous compounds being removed from water. *IOP Conference Series: Materials Science and Engineering,* 012082.
[56] Shubbar A, Abdulredha M, Nasr M, Al Masoodi Z and Sadique M J2020. Properties of cement mortar incorporated high volume fraction of GGBFS and CKD from 1 day to 550 days. **30** 101327.

[57] Gkantou M, Muradov M, Kamaris G, Atherton W and Kot P 2019. Novel Electromagnetic Sensors Embedded in Reinforced Concrete Beams for Crack Detection. *Sensors*, **19** 5175-89.

[58] Omer G, Kot P, Atherton W, Muradov M, Gkantou M, Shaw A, Riley M, Hashim K and Al-Shamma'a A 2021. A Non-Destructive Electromagnetic Sensing Technique to Determine Chloride Level in Maritime Concrete. *Karbala International Journal of Modern Science*, **7** 61-71.

[59] Kot P, , Muradov M and Rafid R 2021 *How can sensors be used for sustainability improvement?. In Methods in Sustainability Science*, Elsevier, United Kingdom: Joe Hayton, p 426.

[60] Kot P, Muradov M, Gkantou M, Kamaris G and Yeboah D 2021. Recent Advancements in Non-Destructive Testing Techniques for Structural Health Monitoring. *Applied Sciences*, **11** 1-28.

[61] Ryecroft S, Shaw A, Fergus P, Kot P, Hashim K, Moody A and Conway L 2019. A First Implementation of Underwater Communications in Raw Water Using the 433 MHz Frequency Combined with a Bowtie Antenna. *Sensors*, **19** 1813-23.

[62] Ryecroft S, Fergus P, Kot P, Tang A, Moody A and Conway L 2021. An Implementation of a Multi-Hop Underwater Wireless Sensor Network using Bowtie Antenna. *Karbala International Journal of Modern Science*, **7** 113-29.

[63] Ryecroft S P, shaw A, Fergus P, Kot P, Hashim K and Conway L 2019. A Novel Gesomin Detection Method Based on Microwave Spectroscopy. *12th International Conference on Developments in eSystems Engineering (DeSE),* Kazan, Russia 429-33.

[64] Gupta A 2017. Building a Green Home Using Local Resources and Sustainable Technology in Jammu Region–A Case Study. *Energy Procedia*, **115** 59-69.

Elimination of phenol from refineries effluents using electrocoagulation method

M Abdulredha[1,*], Amal H. Khalil[2], Sarah A. Ali[3], Ibijoke Idowu[4], J. Amoako-Attah[4]

[1] Department of Civil Engineering, College of Engineering, University of Kerbala, Kerbala, Iraq
[2] Department of Environmental Engineering, College of Engineering, University of Babylon, Iraq
[3] Al-Mustaqbal University College, Babylon, Iraq
[4] School of Civil Engineering and Built Environment, LJM University, UK
*Corresponding author; Email: muhammed.r@uokerbala.edu.iq

Abstract. The crude oil industry is a major source of water pollution because of huge volumes of refining effluents discharged into the aquatic environment. This effluent consequently consists of substances that causes harm to the aquatic environment and depletes the aquatic population due to depleted oxygen. This study investigated the application of various treatment procedures and materials to reduce the effects of refining process effluent on water. The current study proposes to employ the electrocoagulation (EC) method in the removal of e phenol contamination from refining effluent utilising aluminium electrodes. Continuous flow studies have been carried out in order to remove phenolic chemicals from refinery effluent effects of experimental factors such as electrical current density (ECD), distances between electrodes (DE), and treatment durations (TD) while phenols were eliminated were examined. The results show that the EC method reduced the phenol level in petroleum refinery discharge. The EC unit decreased the phenol level by 57% using aluminium as electrodes. The-optimal removal efficiency was found at 120 TD with an ECD of 6 mA/cm^2 and a DE of 20 mm.

Keywords: Refinery effluent, phenols, EC unit.

1. Introduction

The oil business is among the world's largest and most significant sectors as it represents the major source of energy in several industrialized and developing countries [1]. Moreover, for several nations worldwide, the oil sector is the primary source of revenue and the price of oil influences the economy. However, such a sector is among the primary polluters with a significant environmental effect [2]. This sector employs vast quantities of water and creates massive volumes of polluted effluent, comprising sewage and cooling equipment. Studies determined that around 300 litres of water were necessary to obtain 1 barrel of crude oil. The fact that even more than 33 million barrels of polluted effluent are created every day by the oil sector throughout the globe is based on Daud and Aziz [3] prediction further illustrates the quantities of effluents. In addition, the latest developments in energy use show that contamination caused by the oil sector is increasing [3]. Contaminants in wastewater refineries rely greatly on their origin (Oil

IOP Conf. Series: Earth and Environmental Science **877** (2021) 012053 doi:10.1088/1755-1315/877/1/012053

purification processes that implement water) [4-6]. The contaminated water may include several contaminants, including biological materials, phenols [7], oils [8], toxic substances [9], organics [10-12], phosphate [13, 14], nitrates [15] and metals [16-18]. The phenolic compounds are widespread contaminants detected in virtually all kinds of effluent from refineries [7]. Studies indicate that phenolic chemicals may be concentrated up to 200 mg/L in refinery effluent. The permissible phenol content limitation is 1 mg/L as exceeding concentration might have a major influence on environmental, ecological, and public health due to the high phenolic content in refinery effluent. In addition, soils near the refinery plants are typically severely contaminated and require very costly and complicated remediation methods. Refinery wastewater pollution causes severe diseases and negatively impacts ~~on~~ people's health [1, 19]. In order to satisfy the growing demands for freshwater caused by climate change [20-22], humans urgently need to remove pollutants from large quantities of refining effluent. Investigation on the purification of refining polluted water has risen considerably owing to global warming to identify the technologies to eliminate contaminants from refining effluent [23-26]. Systems of refinery wastewater treatment ranging from simple to very sophisticated techniques for eliminating ~~e~~ contaminants combining physical, chemical and biological approaches. These methods include ~~e~~ coagulation flocculation [27-29], sono-oxidation [30-33], adsorption [32-35] and electrocoagulation [36-38] methods. Phosphates, nitrates, microorganisms, organic compounds, and heavy metals, are common pollutants treated through utilizing these methods. Invariably, most of these approaches are costly or inefficient for removing or fulfilling entirely the elimination criteria due to changes in the concentration or structure of pollutants generated by the oil and gas sector. The problem of water pollution by industrial wastewater continues to increase because of the continuous increase in production and expansion in the industry [39-46], such as refinery and construction industries [47-56].

In addition to this, an increasing amount of studies have investigated EC and have shown that this approach may eliminate numerous contaminants, like toxic substances, hydrocarbons, organic matter, dyes, etc. in the different wastewater [57, 58]. Scientists have shown that the method of EC is inexpensive, simple to use, doesn't really utilize solvents and needs little room. The effluent produced in the EC procedure also includes numerous heavy metals, which make it recyclable, especially in construction applications. The volume of waste disposed into sites and waste conservation areas may be substantially reduced, which enables the use of these sites for other waste categories [59, 60].

The study explores the suitability of utilizing the EC technique to remove phenolic compounds from discharge wastewater process from the petrochemical industry. The choice of the EC technology depends on its own appeal and on the benefits previously mentioned, such as little or no chemical, low cost, etc. Phenol is also regarded as one of the most prevalent contaminants in the effluents discharged from oil refining.

2. Methods

Discharge from an oil refinery plant has been treated with an EC reactor. The EC reactor is a rectangular unit with a width of 90 mm, a depth of 150 mm and a length of 250 mm with four electrodes of aluminium. Electrodes sizes are 150 mm high and 90 mm wide with 42 holes of 2.5mm in radius. Electrode layout and the holes of each electrode mix polluted water and improve the elimination effectiveness during treatment procedures. The wastewater utilized in this investigation recorded 120 mg/l of phenols.

In order to study the influence of the experimental parameters on the EC unit performance in terms of phenol removal of oil refinery effluent, both the current density (ECD) as well as the distance between aluminium electrodes (DE) have been adjusted. The effluent characteristics utilized in this investigation are shown in Table 1.

Table 1. The characteristics of the refinery effluents used in this research.

characteristic	Unit	Value
Salinity	%	0.14
Colour	---	Brown
pH scale	---	6.7
Conductivity	ms/cm	9.64
Total Suspended Solids	mg/l	115
Phenols	mg/l	120

The reactor was provided to remove phenols from the refinery effluent by means of a current rectifier with three desirable currents, 4mA/cm^2, 5mA/cm^2 and 6mA/cm^2. Furthermore, in this investigation, the influence of DE on the reactor removal performance was investigated with three different distances 20mm, 23mm, and 26mm. The distances have been manually modified.

The wastewater is pumped continuously through the unit, and the aluminium electrodes removed the phenols from petroleum refinery effluent. The duration of the purification of phenols increased up to 2 hours. The elimination performance of the reactor was done by collecting 15 mL specimens of treated effluents every five minutes intervals. The residual of the phenolic compounds in the treated effluent has been identified on the basis of the phenolic residual concentrations, with a spectrometer (Hach Lange DR 2800) and the standard cuvette LCK 346 or LCK 345. The effluent condition utilized in this study has been kept equal to 20 ±10°C in the room. The effectiveness of phenols removal from refinery effluent is estimated based on the aforementioned procedures and computed by using equation 1. The S1 and S2 indicate the starting and final concentrations of phenolic compounds in mg/L in the refinery effluent after treatment.

$$\text{Removal }\% = \frac{S1-S2}{S1} \times 100\% \qquad \text{equation} \qquad (1)$$

3. Outcomes

3.1. ECD effects

The ECD is regarded as an essential component in the EC procedure, as it impacts the coagulant generation rates that substantially influence the reactor removing effectiveness. This investigation studied the influence of the ECD on the phenol removal performance from refining effluent. This was done utilizing different ECDs to treat samples of wastewater with 120 mg/L contaminants. The ECDs utilized were 4 to 6 mA/ cm^2. In addition, DE was maintained at 20 mm, and TD was maintained for 45 minutes. The outcomes presented in Figure 1 show that the ECD increment enhances the effectiveness of the elimination of phenol from the refinery effluent. The removal effectiveness of phenol increased from 26% to more than 32% when the ECD increased from 4 to 6 mA/cm after 45 min treatment. This improvement in the removal could be linked to the rise in the coagulant production rate.

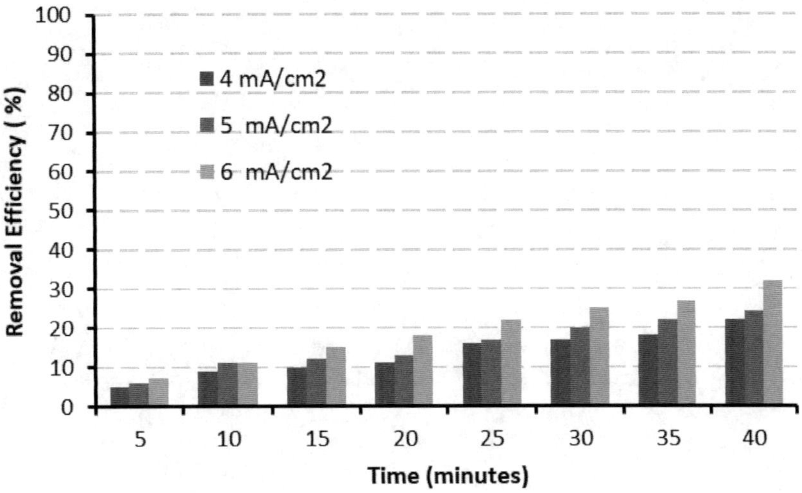

Figure 1: ECD effects on phenol removal

3.2. DE effects

The effect of the EC technique on different pollutants treatment is also highly affected by DE. Consequently, the significance of the phenol treatment and removal from the effluent of the oil refinery using aluminium electrodes was investigated in this research in an EC reactor. Three varied distances of 20, 23 and 26 mm have been used to manually alter the DE. The other experimental aspects, including phenol concentration, ECD, and TD were maintained on 120 mg/L, 4 mA/cm^2 and 45 min, correspondingly. The findings presented in Figure 2 indicated that the effectiveness of phenol extraction using the EC reactor improved when DE was reduced. The extraction effectiveness of phenols with respect to DE from 20 mm to 26 mm decreased from 27% to 17.8% as depicted in figure 2. The decrease in the extraction efficiency can be attributed to the changes in the electrostatic attraction force.

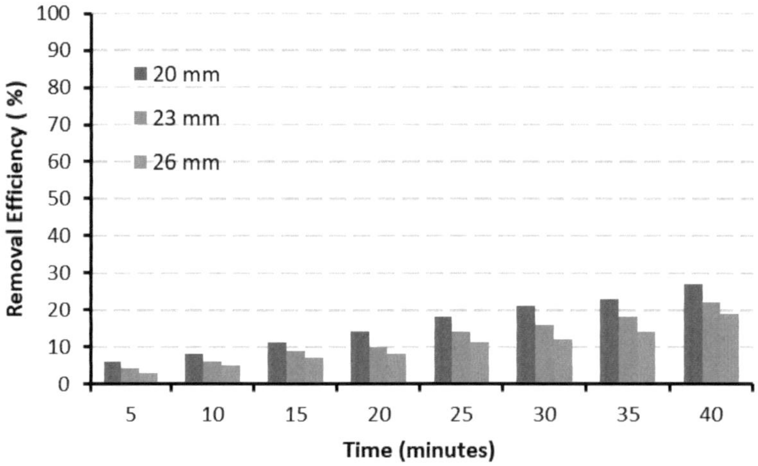

Figure 2: DE effects on phenol removal

3.3. *TD effects*

The performance of different EC technologies and methods was proven to be greatly influenced by the TD. The TD and production of coagulants have a strong correlation. This study examined the impact of the TD on the removal of the phenolic compounds by the EC reactor from the effluent of a refining plant. To achieve this, a ECD of 4 mA/cm^2 and DE of 20 mm were used for 120 minutes for samples containing phenolic contaminants of 120 mg/L in the EC. The proportion of the phenols removed increased with time, as shown in Figure 3. The effectiveness of removal rose from 18 to 57% as the TD extended from 20 to 120 minutes.

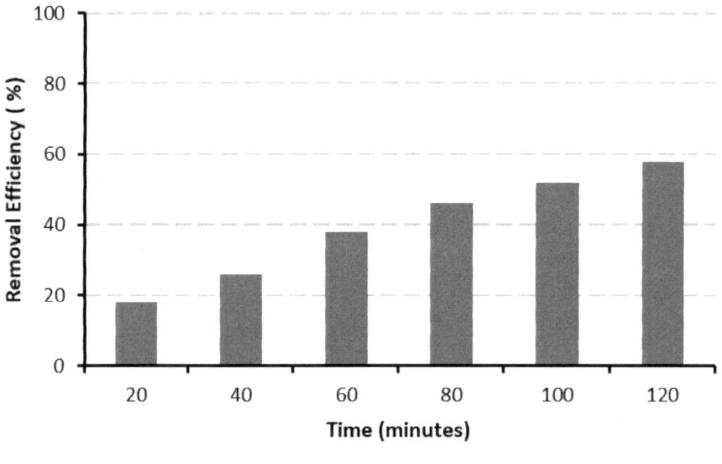

Figure 3: TD effects on phenol removal.

The method studied here could be enhanced by adding on or integrating sensing technologies [54, 55, 61, 62].

4. Conclusion

Water resource continues to dwindle due to the impact of climate change. Compounding the already limited water resource, is the discharged of effluents into our waters by polluters such as the oil industry. This leads to an urgent reversal of this problem through methods that are inexpensive but effective in the reclaiming of water resources from contaminants. Removal of phenols from the wastewaters of crude oil industry using electrocoagulation method was investigated in the present work. In order to remove contaminants, continuous flow tests have been carried. The results of this study have demonstrated that phenol contaminants may be substantially decreased in refinery effluent through the utilisation of the EC method. It was also shown that experimental conditions impacted the removal effectiveness of the phenol contaminants. The elimination of the phenol compounds increases as the DE reduces and the ECD increases.

The effects of other parameters, such as the initial concentration of phenol and solution temperatures, on the removal of phenols is recommended to be investigated in future studies, these parameters were not investigated in this study due to the lack of time.

References

[1] Keramati M and Ayati B 2019. Petroleum wastewater treatment using a combination of electrocoagulation and photocatalytic process with immobilized ZnO nanoparticles on concrete surface. *Process Safety and Environmental Protection*, **126** 356-65.

[2] El-Naas M H, Al-Zuhair S, Al-Lobaney A and Makhlouf S 2009. Assessment of electrocoagulation for the treatment of petroleum refinery wastewater. *J Environ Manage*, **91** 180-5.

[3] Diya'uddeen B H, Daud W M A W and Aziz A A 2011. Treatment technologies for petroleum refinery effluents: A review. *Process safety and environmental protection*, **89** 95-105.

[4] Abdulhadi B A, Kot P, Hashim K S, Shaw A and Khaddar R A 2019. Influence of current density and electrodes spacing on reactive red 120 dye removal from dyed water using electrocoagulation/electroflotation (EC/EF) process. *First International Conference on Civil and Environmental Engineering Technologies (ICCEET),* University of Kufa, Iraq 12-22.

[5] Abdulraheem F S, Al-Khafaji Z S, Muradov M, Kot P and Shubbar A A 2020. Natural filtration unit for removal of heavy metals from water. *IOP Conference Series: Materials Science and Engineering,* 012034.

[6] Alhendal M, Nasir M J, Hashim K S, Amoako-Attah J, Al-Faluji D, Muradov M, Kot P and Abdulhadi B 2020. Cost-effective hybrid filter for remediation of water from fluoride. *IOP Conference Series: Materials Science and Engineering,* 012038.

[7] Emamjomeh M, Mousazadeh M, Mokhtari N, Jamali H, Naghdali Z and Ghanbari R 2020. Simultaneous removal of phenol and linear alkylbenzene sulfonate from automotive service station wastewater: Optimization of coupled electrochemical and physical processes. *Separation Science and Technology*, **55** 3184-94.

[8] Hashim K S, Al-Jumeily D, Alwash R and Aljefery M H 2020 *Electrocoagulation as an eco-friendly River water treatment method. In Advances in Water Resources Engineering and Management*, Berline: Springer.

[9] Hashim K S, Al-Saati N H, Hussein A H and Al-Saati Z N 2018. An investigation into the level of heavy metals leaching from canal-dreged sediment: a case study metals leaching from dreged sediment. *First International Conference on Materials Engineering & Science,* Istanbul Aydın University (IAU), Turkey 12-22.

[10] Zanki A K, Mohammad F H, Muradov M, Kareem M M and Abdulhadi B 2020. Removal of organic matter from water using ultrasonic-assisted electrocoagulation method. *IOP Materials Science and Engineering,* 012033.

[11] Emamjomeh M, Kakavand S, Jamali H A, Alizadeh S M, Safdari M, Mousavi S and Mousazade M 2020. The treatment of printing and packaging wastewater by electrocoagulation–flotation: the simultaneous efficacy of critical parameters and economics. *Desalination and water treatment*, **205** 161-74.

[12] Khalid H, Zubaid S, Alwash R, Al-Jumeily D and Aljefery M 2020. Energy efficient electrocoagulation using baffle-plates electrodes for efficient Escherichia Coli removal from Wastewater. *Journal of Water Process Engineering*, **33** 101079-86.
[13] Hashim K, Ewadh H, Muhsin A, Muradov M, Aljefery M and Al-Khaddar R 2020. Phosphate removal from water using bottom ash: Adsorption performance, coexisting anions and modelling studies. *Water Science and Technology*, **83** 1-17.
[14] Alenezi A, Hasan H, Amoako-Attah J, Gkantou M and Abdulhadi B 2020. Zeolite-assisted electrocoagulation for remediation of phosphate from calcium-phosphate solution. *IOP Materials Science and Engineering,* 012031.
[15] Al-Marri S, AlQuzweeni S, AlKizwini R and Al-Khafaji Z S 2020. Ultrasonic-Electrocoagulation method for nitrate removal from water. *IOP Materials Science and Engineering,* 012073.
[16] Hashim K, Rafid R, Patrik K and Al-Shamma'a A 2021. Water purification from metal ions in the presence of organic matter using electromagnetic radiation-assisted treatment. *Journal of Cleaner Production*, **280** 1-17.
[17] Omran I I, Al-Saati N H, Al-Saati Z N, Patryk K, Khaddar R , Al-Jumeily D, Shaw A, Ruddock F and Aljefery M 2019. Assessment of heavy metal pollution in the Great Al-Mussaib irrigation channel. *Desalination and Water Treatment*, **168** 165-74.
[18] Abdulla G, Kareem M, Muradov M, Mubarak H A, Abdellatif M and Abdulhadi B 2020. Removal of iron from wastewater using a hybrid filter. *IOP Materials Science and Engineering,* 012035.
[19] Alwan H, Saleh L, Mohammed F and Abdulredha M 2020. Experimental prediction of the discharge coefficients for rectangular weir with bottom orifices. *Journal of Engineering Science and Technology*, **15** 3265-80.
[20] Kadhim A, Sadique, Mufti R and Khalid S 2020. Long-term performance of novel high-calcium one-part alkali-activated cement developed from thermally activated lime kiln dust. *Journal of Building Engineering*, **32** 1-17.
[21] Salah Z, Abdulkareem I H, Al-Bugharbee H, Ridha H M, Gharghan S K, Al-Qaim F F, Muradov M, Patrik P and Rafid R 2020. Hybridised Artificial Neural Network model with Slime Mould Algorithm: A novel methodology for prediction urban stochastic water demand. *Water*, **12** 1-18.
[22] Salah Z, Ethaib S, Al-Bdairi N, Al-Bugharbee H and Gharghan S K 2020. A novel methodology to predict monthly municipal water demand based on weather variables scenario. *Journal of King Saud University-Engineering Sciences*, **32** 1-18.
[23] Zubaidi S, Al-Bugharbee H, Muhsin Y, Khalid S and Rafid M 2020. Forecasting of monthly stochastic signal of urban water demand: Baghdad as a case study. *IOP Materials Science and Engineering,* 012018.
[24] Zubaidi S, Ortega S, Al-Bugharbee H, Olier I and Gharghan S 2020. Urban Water Demand Prediction for a City that Suffers from Climate Change and Population Growth: Gauteng Province case study. *Water*, **12** 1-18.
[25] Zubaidi Salah L, Al-Bugharbee H, Ortega Martorell S, Gharghan S, Olier I, Hashim K, Al-Bdairi N and Kot P 2020. A Novel Methodology for Prediction Urban Water Demand by Wavelet Denoising and Adaptive Neuro-Fuzzy Inference System Approach. *Water*, **12** 1-17.
[26] Salah Z, Ortega, Kot P, Alkhaddar R, Abdellatif M, Gharghan S and Ahmed M 2020. A Method for Predicting Long-Term Municipal Water Demands Under Climate Change. *Water Resources Management*, **34** 1265-79.
[27] Alenazi M, Safaa K, Hassan A, Muradov M and Abdulhadi B 2020. Turbidity removal using natural coagulants derived from the seeds of strychnos potatorum: statistical and experimental approach. *IOP Conference Series: Materials Science and Engineering,* 012064.
[28] Alyafei A, AlKizwini R, Yeboah D and Zubaidi S L 2020. Treatment of effluents of construction industry using a combined filtration-electrocoagulation method. *IOP Materials Science and Engineering,* 012032.
[29] Abdulredha M, Jordan D and Abdulridha A 2018. Benchmarking of the Current Solid Waste Management System in Karbala, Iraq, Using Wasteaware Benchmark Indicators. *World Environmental and Water Resources Congress 2018: Groundwater, Sustainability, and Hydro-Climate/Climate Change,* 40-8.
[30] Alnaimi H, Idan I, Al-Janabi A, Gkantou M, AlZubaidi Sand Muradov M 2020. Ultrasonic-electrochemical treatment for effluents of concrete plants. **888** 1-9.
[31] Majdi H, Nasr M, Al-Khafaji Z, Jafer H, Abdulredha M, Masoodi Z and Sadique M 2020. Experimental data on compressive strength and ultrasonic pulse velocity properties of sustainable mortar made with high content of GGBFS and CKD combinations. *Data in Brief*, **31** 105961-72.
[32] Abdulredha M, Rafid A, Jordan D and Alattabi A 2017. Facing up to waste: how can hotel managers in Kerbala, Iraq, help the city deal with its waste problem? *Procedia engineering*, **196** 771-8.

[33] Abdulredha M, Abdulridha A, Shubbar A, Alkhaddar R, Kot P and Jordan D 2020. Estimating municipal solid waste generation from service processions during the Ashura religious event. *IOP Conference Series: Materials Science and Engineering,* 012075.

[34] Abdulredha M, Kot P, Al Khaddar R, Jordan D and Abdulridha A 2020. Investigating municipal solid waste management system performance during the Arba'een event in the city of Kerbala, Iraq. *Environment, Development and Sustainability*, **22** 1431-54.

[35] Alattabi A W, Harris C, Alkhaddar R, Alzeyadi A and Abdulredha M 2017. Online Monitoring of a sequencing batch reactor treating domestic wastewater. *Procedia engineering*, **196** 800-7.

[36] Hashim K S, Al-Saati N H, Alquzweeni S S, Zubaidi S L, Kot P, Kraidi L, Hussein A H, Alkhaddar R, Shaw A and Alwash R 2019. Decolourization of dye solutions by electrocoagulation: an investigation of the effect of operational parameters. *First International Conference on Civil and Environmental Engineering Technologies (ICCEET),* University of Kufa, Iraq 25-32.

[37] Hashim K S, Hussein A H, Zubaidi S L, Kot P, Kraidi L, Alkhaddar R, Shaw A and Alwash R 2019. Effect of initial pH value on the removal of reactive black dye from water by electrocoagulation (EC) method. *2nd International Scientific Conference,* Al-Qadisiyah University, Iraq 12-22.

[38] Aqeel K, Mubarak H A, Amoako-Attah J, Abdul-Rahaim L A, Al Khaddar R, Abdellatif M, Al-Janabi A and Hashim K S 2020. Electrochemical removal of brilliant green dye from wastewater. *IOP Conference Series: Materials Science and Engineering,* 012036.

[39] Ryecroft S P, shaw A, Fergus P, Kot P, Hashim K and Conway L 2019. A Novel Gesomin Detection Method Based on Microwave Spectroscopy. *12th International Conference on Developments in eSystems Engineering (DeSE),* Kazan, Russia 429-33.

[40] Zubaidi S L, Al-Bugharbee H, Muhsen Y R, Hashim K, Alkhaddar R M, Al-Jumeily D and Aljaaf A J 2019. The Prediction of Municipal Water Demand in Iraq: A Case Study of Baghdad Governorate. *12th International Conference on Developments in eSystems Engineering (DeSE),* Kazan, Russia 274-7.

[41] Grmasha R A, Al-sareji O J, Salman J M, Hashim K S and Jasim I A 2020. Polycyclic Aromatic Hydrocarbons (PAHs) in Urban Street Dust Within Three Land-Uses of Babylon Governorate, Iraq: Distribution, Sources, and Health Risk Assessment. *Journal of King Saud University - Engineering Sciences*, **33** 1-18.

[42] Al-Sareji O J, Grmasha R A, Salman J M, Idowu I and Hashim K S 2021. Street dust contamination by heavy metals in Babylon governorate, Iraq. *Journal of Engineering Science and Technology*, **16** 3528 - 46.

[43] Al-Saati N H, Hussein T K, Abbas M H, Hashim K, Al-Saati Z N, Kot P, Sadique M, Aljefery M H and Carnacina I 2019. Statistical modelling of turbidity removal applied to non-toxic natural coagulants in water treatment: a case study. *Desalination and Water Treatment*, **150** 406-12.

[44] Mousazadeh M, Paital B, Naghdali Z, Mortezania Z, Hashemi M, Karamati Niaragh E, Aghababaei M, Ghorbankhani M, Lichtfouse E, Sillanpää M and Emamjomeh M M 2021. Positive environmental effects of the coronavirus 2020 episode: a review. *Environment, Development and Sustainability*, **21** 1-23.

[45] Fergus P, Hussain A, Al-Jumeily D, Huang D-S and Bouguila N 2017. Classification of caesarean section and normal vaginal deliveries using foetal heart rate signals and advanced machine learning algorithms. *Biomedical engineering online*, **16** 1-26.

[46] Ghazali R, Hussain A, Jumeily D and Merabti M 2007. Dynamic ridge polynomial neural networks in exchange rates time series forecasting. *International Conference on Adaptive and Natural Computing Algorithms,* 123-32.

[47] Shubbar A A, Sadique M, Shanbara H K and Hashim K 2020 *The Development of a New Low Carbon Binder for Construction as an Alternative to Cement. In Advances in Sustainable Construction Materials and Geotechnical Engineering,* Berlin: Springer.

[48] Al-Jumeily D, Hashem K, Alkaddar R, Al-Tufaily M and Lunn J 2019. Sustainable and Environmental Friendly Ancient Reed Houses (Inspired by the Past to Motivate the Future). *11th International Conference on Developments in eSystems Engineering (DeSE),* Cambridge, UK 214-9.

[49] Shubbar A A, Sadique M, Nasr M S, Al-Khafaji Z S and Hashim K S 2020. The impact of grinding time on properties of cement mortar incorporated high volume waste paper sludge ash. *Karbala International Journal of Modern Science*, **6** 1-23.

[50] Obaid M K, Nasr M S, Ali I M, Shubbar A A and Hashim K S 2021. Performance of green mortar made from locally available waste tiles and silica fume. *Journal of Engineering Science and Technology*, **16** 136-51.

[51] Farhan S L, Antón D, Akef V S, Zubaidi S L and Hashim K S 2021. Factors influencing the transformation of Iraqi holy cities: the case of Al-Najaf. *Scientific Review Engineering and Environmental Sciences*, **30** 365-75.
[52] Kadhim A, Sadique M, Mufti R and Hashim K 2020. Developing One-Part Alkali-Activated metakaolin/natural pozzolan Binders using Lime Waste as activation Agent. *Advances in Cement Research*, **32** 1-38.
[53] Omer G, Kot P, Atherton W, Muradov M, Gkantou M, Riley M and Al-Shamma'a A 2020. A Non-Destructive Electromagnetic Sensing Technique to Determine Chloride Level in Maritime Concrete. *Karbala International Journal of Modern Science*, **6** 1-14.
[54] Omer G, Atherton W, Muradov M, Gkantou M, Riley M and Al-Shamma'a A 2021. A Non-Destructive Electromagnetic Sensing Technique to Determine Chloride Level in Maritime Concrete. *Karbala International Journal of Modern Science*, **7** 61-71.
[55] Kot P, Muradov M, Gkantou M, Kamaris G and Yeboah D 2021. Recent Advancements in Non-Destructive Testing Techniques for Structural Health Monitoring. *Applied Sciences*, **11** 1-28.
[56] Shubbar A, Al-Shaer A, AlKizwini R, Hawesah H and Sadique M 2019. Investigating the influence of cement replacement by high volume of GGBS and PFA on the mechanical performance of cement mortar. *First International Conference on Civil and Environmental Engineering Technologies,* University of Kufa, Iraq 31-8.
[57] Hashim K S, Ali S S M, AlRifaie J K, Kot P, Shaw A, Al Khaddar R, Idowu I and Gkantou M 2020. Escherichia coli inactivation using a hybrid ultrasonic–electrocoagulation reactor. *Chemosphere*, **247** 125868-75.
[58] Abdulhadi B, Kot P, Hashim K, Shaw A, Muradov M and Al-Khaddar R 2021. Continuous-flow electrocoagulation (EC) process for iron removal from water: Experimental, statistical and economic study. *Science of The Total Environment*, **760** 1-16.
[59] Omran I, Al-Saati N and Al-Saati Z 2021. Sustainability assessment of wastewater treatment techniques in urban areas of Iraq using multi-criteria decision analysis (MCDA). *Water Practice and Technology*, **16** 648-60.
[60] Mohammed A, Hussein A, Yeboah D, Abdulhadi B and Safaa K S 2020. Electrochemical removal of nitrate from wastewater. *IOP Materials Science and Engineering,* 012037.
[61] Ryecroft S, Fergus P, Moody A and Conway L 2019. A First Implementation of Underwater Communications in Raw Water Using the 433 MHz Frequency Combined with a Bowtie Antenna. *Sensors*, **19** 1813-23.
[62] Ryecroft S, Fergus P, Tang A, Moody A and Conway L 2021. An Implementation of a Multi-Hop Underwater Wireless Sensor Network using Bowtie Antenna. *Karbala International Journal of Modern Science*, **7** 113-29.

Reduction of phosphates in sewage using kiln ash as an adsorbent

R A Abdelhadi[1]
[1] Faculaty of Engineering, Kerbala University, Kerbala, Iraq.
* Corresponding author; Email: r.abdulhadi@gmail.com

Abstract. Phosphate is a naturally occurring chemical found in large quantities on the Earth's surface that causes eutrophication when deposited into rivers. A number of studies have been conducted to assess the ability of various treatments to remove phosphates from sewage. Lately, it has been discovered that phosphorus recovery may be accomplished by filtering the water. The cost of the filtering materials, on the other hand, is prohibitively expensive. As a result, current research has concentrated on utilising low-cost ones to minimise the expense of filtering. Steel production residues, such as kiln bottom ashes, are being utilised in this study to recover phosphates from contaminated wastewater. A variety of operational settings have been investigated in order to obtain the highest possible extraction efficiency at the lowest feasible cost. Bottom ashes were demonstrated to be an effective substitute for phosphorus extraction. The highest phosphorus extraction was 90.1 percent after 40 minutes, with a starting concentration of 5 mg/L and an ash dose of 530 mg/L. The findings were used to create a prediction model with a high degree of reliability.

Keywords: Bottom ashes, phosphate; sewage; removal.

1. Introduction

Although phosphate is available at high concentrations in soil and freshwaters, urban spread and industrial expansion drive huge global demand for drinkable water that exhausts the available freshwater rapidly, and in return, increasing the volumes of discharged wastewater that contains high concentrations of phosphate [1, 2]. In addition, the urban spread and industrial expansion badly reflected in the availability of freshwater [3-5], global warming [6, 7], and precipitations [8, 9]. The growing population increases the rate of sewage released into waterways, which has severe effects on the health of consumers [10-12]. Furthermore, the increase in the population increased the dumping of solid wastes in landfills, which resulted from the expansion in the residential areas [13, 14], that pollute groundwater and then surface water [15-18]. Toxins must not be dumped directly into waterways, and all effluents, whether household or manufacturing, must be treated adequately to eliminate or limit the negative impacts of the contaminants. One of the types of impact of pollutants is algae flourishing which is due to phosphorous penetration into the water, which is harmful to the environment [19, 20]. All types of sewage water contain phosphorus [21, 22] and nitrate [23, 24] beside other pollutants such as fluoride [25], heavy metals [26, 27], bacteria [28, 29], coloring pollutants [30-32] and organic matter [33-35]. Additionally, industrial and domestic wastewater [36-39] and atmospheric pollutants [40-42] are playing a role in the increasing of such pollutants in water bodies. The effluents of the construction industry [43-46], production plants [47-49] and food industry [36, 50].

Surface waters are depleted of oxygen due to algal blooms induced by phosphorus and nitrate contamination [12, 51]. A large proportion of aquatic creatures and plants might die in a short period of time. As just a result of a substantial rise in organic matter content, the taste, smell, and structure of the water will be changed significantly [21, 22]. A rise in algal development might have severe effects on health, industries, seafood, and agriculture, and farmers [24, 52]. Many studies revealed that microalgae types can generate poisonous chemicals that cause a range of severe illnesses and, sometimes, fatality. In drinkable water, the allowable concentration of phosphate is specified at 0.1 mg/L, whereas in streams and ponds it is 0.05 mg/L [21, 22]. As far as the upper limit pollutant concentration in water is concerning, the Environmental Protection Agency has strict standards. It is therefore important to remove phosphates from contaminated water, depending on what has been said previously. When phosphate is removed via traditional methods, it generally requires a long and complicated procedure. As an instance, the normal filtering process [25], with some modifications, provides an effective approach for phosphate removal, according to the authors. However, the usage of expensive screens makes this approach uneconomical [50, 53].
It is possible to remove phosphate from the water bodies by converting it to a solid state, which would be subsequently removed from the water via flotation or some other means. Thermal processing, lakes, anaerobic digestion, absorption, and other techniques are used to transform phosphorus [22, 54].
As just a result of the typical operations of species of bacteria and microalgae, phosphate could be biologically processed. Depends on the kind of microbe, anaerobic or aerobic techniques could be used in these situations. The pH, starting dosage, etc. are all factors that affect the efficacy of the treatments. In regards to employing biological intervention to reduce the existence of phosphate, several disadvantages are discovered [21]. Soluble carbon, fast development of microorganisms, and pH level of the fluid are all factors to be considered. Phosphate particles can be converted by adding metallic compounds to water polluted with Phosphorus particles. Usually, sewer water encompassing phosphate is treated using chemical techniques. Calcium ion is amongst the most frequently utilised chemical additions for phosphorus removal for its availability, affordability, and convenience of usage. Especially in an acidic medium, aluminum hydroxide is believed to be highly successful in eliminating phosphate from wastewater. As a consequence of the usage of chemical substances, chemical treatment methods frequently lead to toxic sludge [24]. Due to the utilisation use reagents, the technique is much more costly than many other procedures. Physically, phosphate extraction methods do not change the chemical composition of the phosphate, which is resulting in a change. After the removal of phosphate pollution using this approach, it is possible to separate these contaminants (phosphate contaminants) from the removal medium. For phosphate removal, filtering and screening technologies are frequently employed [21, 22].
Substances sensitive to pollutants are used in filtering. Water is then separated from pollutants using a variety of processes such as coagulation [10, 31], absorption [11, 55], and straining hybrid methods [23, 33]. To remove phosphates from water, several permeable mediums were employed. The general rule, the expense of filtering is determined based on the cost of the adsorbent materials employed. The phosphate removal capability of membranes has been demonstrated in previous studies. There are several methods to extract phosphates from water at low levels, but reverse osmosis is among the most popular. Polluted water is forced through a barrier with really limited porosity under high pressure. Assist in removing pollutants in water by means of high-pressured water. Membrane systems are rarely utilised due to their expensive running costs difficulties in discarding them, and the requirement for specific equipment and expertise. Recently, studies have centered on the use of waste products or natural resources as economical and useful replacements for traditional materials. Fly ash is a standard manufacturing byproduct that would be used to remove phosphate from wastewater, and also it widely used in construction industry [43, 56]. Water with an acidity level of 7 can be effectively treated with this substance. Due to the above, our research focused primarily on the possibility of using furnaces bottom ashes as an absorbent to extract the phosphates from polluted effluent.

2. Methods

According to the studies, furnace bottom ashes may be utilised as a convenient and efficient adsorbent for the removal of phosphate from the wastewater. Studying the grain size, permeability, porous structure and relative density of the bottom ashes is essential. The particle size is highly essential because it is directly related to the adsorption surface. When sieve review is performed, sieve sizes of 4, 6, 10, 16, 30 and 50 are utilised, and the quantity of mass kept on each sieve is calculated by analysing the distribution of particle size of the waste. This formula is used to determine the specific gravity test using the gas jar method:

$$\text{Specific Gravity} = \frac{L_2 - L_1}{(L_4 - L_1) - (L_3 - L_2)} \quad (1)$$

Where

L_1: Weight of jar + weight of plates

L_2: The total weight of the sample, plate, and jar, including the plate and the jar itself.

L_3: Water, dry sample, plate, and jar weight together.

L_4: Weight of the jar plus water

Using an X-ray fluorescence analyser, the chemical test is used to determine the structure of the ashes of the furnaces. It is performed to assess the effectiveness of extracting phosphates from the bottom ashes of the furnace since materials contain calcium, iron, and aluminum can be utilised for phosphate removal. To extract the phosphate in contaminated water, the bottom ashes were employed as a filter. The Box-Behnken design was utilised to improve the technique in terms of burner bottom ash dosage, duration of treatment, and temp. A potassium diphosphate solution in deionised was used to modify the amounts of phosphate in synthetic water.

An experiment was conducted in which polluted water was combined for various lengths of time with the furnace's bottom ashes to determine the effect of duration. In such tests, the retention duration ranging from 10 to 70 min, while the burner ash dosages varied from 100 to 1000 mg/L. Calculations of phosphate levels are made with the use of a Hach Lange spectrophotometer. As a result of this study, the Box-Behnken Design was utilised to create an estimation model in order to obtain the highest water treatment. The response variable was phosphorus removal in this research. The determinants were the phosphorus content level, bottom ash dose, and the duration of treatment. In order to construct the estimation matrix, Minitab 20 was chosen. The ranges of predictor factors considered are shown in Table 1.

Table 1. Varying ranges of independent and dependent variables were examined.

Factors	Code	Level		
		Low	Moderate	High
Concentration of phosphate in water	CoF	5	10	15
The dose of bottom ashes	DoBA	100	550	1000
Time of the treatment	ToT	10	40	70

A powerful determinant of linearity and relevance is the α value, which may be calculated using the formula below:

$$\alpha = (2X)^{1/4} \quad (2)$$

Where X is the number of variables included in the model.

3. Results and Discussion

3.1. Properties of ashes from furnace bottoms

Based on empirical evidence, the majority of the bottom ash grains are maintained on sieves of lower diameters of less than 16. The results of the sieving tests show that the majority of the bottom ash samples consisted of small particles within 1.180 mm and 4.750mm in size. It was computed using the formulas 3 and 4:

$$\text{Uniformity coefficient} = \frac{x_2}{x_1} \quad (3)$$

$$\text{The gradation coefficient} = \frac{x_3^2}{x_1 * x_2} \quad (4)$$

When plotting the passing rate with particle diameter, x_2, x_3, and x_1 show the percentages of particles that pass through each size. Analysis indicated that x_2, x_3, and x_1 the quantity of the first term were 3.06, 1.36, and 0.48, correspondingly. Furthermore, the uniformity coefficient is 6.21 the gradation coefficient is 1.32. The specimens of bottom ash are judged to be of good quality. Because of this, it is unlikely that the bottom ashes would float, as their relative density is greater than those of the liquid. Permeability was 0.61 m^2/g, while the total area was 6.81 m^2/g. A wide surface, good permeability, and high relative density make the bottoms of the furnace excellent for phosphorus extraction. Chemists determined that the bottom ashes included aluminium, iron, and, calcium oxide, magnesium in roughly 25% of the furnace's ash. As a result, it may be argued that the bottom ash of a burner is suitable for removing phosphates.

3.2. Phosphate Removal

To investigate the effectiveness of burner ash in removing phosphorus compounds from wastewater, batch experiments were run. For each of the 15 tests, the experimental variables were modified in order to identify the optimal experimental set for achieving high removal efficiency at low costs. The findings are shown in Table 2 below. As can be seen in Table 2, the Box-Behnken Design and the removal efficiency obtained are provided.

Table 2. Phosphate elimination in batch flow experiments using a Box-Behnken design.

test	DoBA	CoF	ToT	RE%
01	1	0	0	90.1
02	1	0	1	63.1
03	1	-1	0	77.3
04	0	0	0	83.5
05	0	1	1	79.1
06	-1	0	-1	77.2
07	0	0	0	66.4
08	0	-1	-1	58.2
09	0	1	-1	59.9
10	0	-1	1	65.7
11	-1	1	0	80.1
12	1	1	0	79.2
13	-1	0	1	77.8
14	-1	-1	0	79.2
15	0	0	0	70.1

As a result of this investigation, it was shown that the removal efficiencies are proportional to the dose of furnace bottom ashes and the length of time intervals. Phosphorus removal, on the other hand, has an adverse correlation with starting phosphorus concentrations. This is because the accessible surface area of the adsorption is insufficient to collect the phosphorus from the liquid. As a result, there is more surface area accessible for phosphate adsorption from the contaminated water in the furnace bottom ash. It also enhances the interaction of phosphate and the adsorbent. 90.1% of the phosphate concertation in the contaminated water could be removed by using 530 mg/L of bottom ashes for 40 minutes, as shown in Figures 1-3.

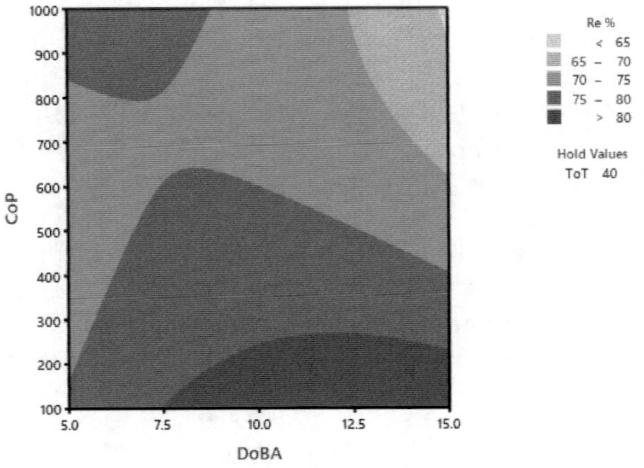

Figure 1: Contour plot of the effects of DoBA and CoP on the removal of phosphate by the bottom ash.

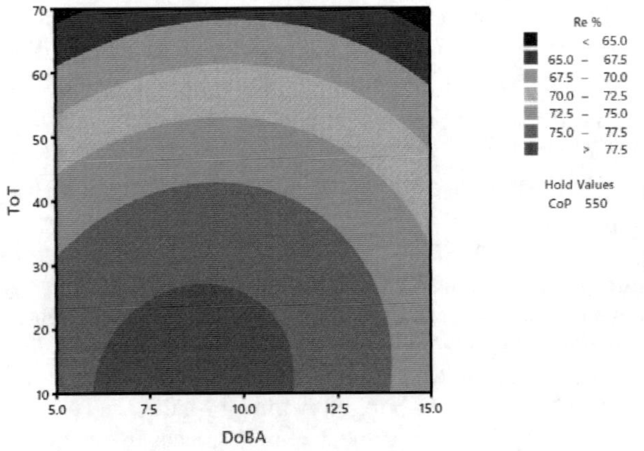

Figure 2: Contour plot of the effects of DoBA and ToT on the removal of phosphate by the bottom ash.

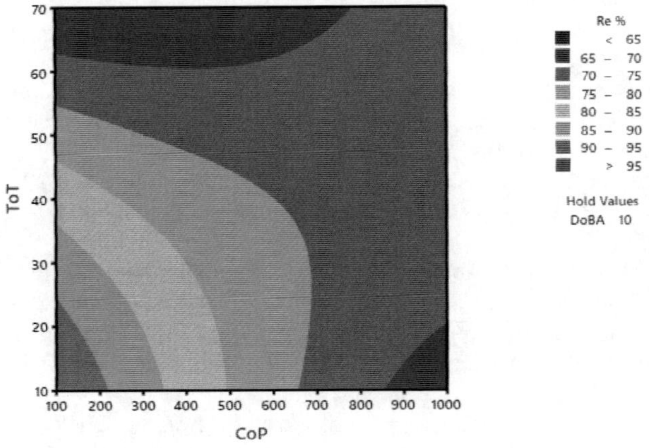

Figure 3: Contour plot of the effects of ToT and CoP on the removal of phosphate by the bottom ash.

To estimate the removal efficiency based on the employed parameters, Box-Behnken Design was used to construct Equation 5.

$$RE\% = 73.33 - 0.44\,DoBA + 2.24\,CoF - 0.10\,ToT + 8.61\,DoBA * DoBA - 2.99\,CoF * CoF - 4.62\,ToT * ToT + 0.25\,DoBA * CoF - 7.17\,DoBA * ToT + 2.93\,CoF * ToT \quad (5)$$

It can be seen from the presented studies about the adsorption method, the depletion of the active sites results in a serious reduction in the adsorption process, therefore; the application of successful sensors (in the literature) helps the operators of the treatment plant to monitor the depletion process, two good examples of the sensors that widely used in the civil and water engineering are the microwave [57-59] and electromagnetic [60-63] sensors.

4. Conclusion

To identify the optimum processing factors, the prospect of using furnace bottom ashes to treat contaminated water for phosphate was studied. The chemical and physical properties of furnace bottom ashes are essential to remove phosphorus from contaminated water, according to research. This is in line with earlier research findings of about 90 % elimination using the bottom ashes. The dose of the bottom ashes as well as the detention time is positively related to the percentage of phosphorus removal from the polluted wastewater. Relative to phosphorus adsorption capacity, on the other hand, it is inversely correlated with the initial concentration of the pollutant. On the basis of furnace bottom ash dosage, phosphate content, and treatment duration, Box-model Behnken's may also forecast how the removal efficiency will vary.

References

[1] Zubaidi S, Al-Bugharbee H, Muhsen Y, Hashim K, Alkhaddar R, Al-Jumeily D and Aljaaf A 2019. The Prediction of Municipal Water Demand in Iraq: A Case Study of Baghdad Governorate. *12th International Conference on Developments in eSystems Engineering (DeSE),* Kazan, Russia 274-7.
[2] Salah Z, Abdulkareem I, Hashim K, Al-Bugharbee H, Ridha H, Gharghan S, Al-Qaim F, Muradov M, Kot P and Alkhaddar R 2020. Hybridised Artificial Neural Network model with Slime Mould Algorithm: A novel methodology for prediction urban stochastic water demand. *Water*, **12** 1-18.
[3] Al-Zubaidi S, Muhsin Y and Alkhaddar R 2020. Forecasting of monthly stochastic signal of urban water demand: Baghdad as a case study. *IOP Materials Science and Engineering,* 012018.
[4] Al-Saati N, Omran I, Salman A, Al-Saati Z and Hashim K 2021. Statistical modeling of monthly streamflow using time series and artificial neural network models: Hindiya Barrage as a case study. *Water Practice and Technology*, **16** 681-91.
[5] Lafta Z, Ortega S, Gharghan S, Ahmed M and Safaa K 2020. A Method for Predicting Long-Term Municipal Water Demands Under Climate Change. *Water Resources Management*, **34** 1265-79.
[6] Lafta Z, Hashim K, Ethaib S, Al-Bdairi N, Al-Bugharbee H and Gharghan S 2020. A novel methodology to predict monthly municipal water demand based on weather variables scenario. *Journal of King Saud University-Engineering Sciences*, **32** 1-18.
[7] Zubaidi S, Ortega, Al-Bugharbee H, Olier I, Hashim K and Gharghan S 2020. Urban Water Demand Prediction for a City that Suffers from Climate Change and Population Growth: Gauteng Province case study. *Water*, **12** 1-18.
[8] Salah L, Abdellatif M and Muhsin Y R 2019. Using LARS–WG model for prediction of temperature in Columbia City, USA. *IOP Materials Science and Engineering,* 012026.
[9] Zubaidi L, Al-Bugharbee H, Ortega S, Gharghan S, Olier I, Al-Bdairi N and Kot P 2020. A Novel Methodology for Prediction Urban Water Demand by Wavelet Denoising and Adaptive Neuro-Fuzzy Inference System Approach. *Water*, **12** 1-17.
[10] Hashim K, Alquzweeni S, Kraidi L, Hussein A and Alwash R 2019. Decolourization of dye solutions by electrocoagulation: an investigation of the effect of operational parameters. *1st Int. Conference on Civil and Environmental Engineering Technologies,* University of Kufa, Iraq 25-32.
[11] Safaa K, Hussein A, Kraidi L and Alwash R 2019. Effect of initial pH value on the removal of reactive black dye from water by electrocoagulation (EC) method. *2nd International Scientific Conference,* Al-Qadisiyah University, Iraq 12-22.

[12] Safaa K, Jasim N, Phipps D, Pedrola M, Alattabi A, Abdulredha M and Alawsh R 2019. Electrocoagulation as a green technology for phosphate removal from River water. *Separation and Purification Technology*, **210** 135-44.
[13] Farhan S, Akef V and Nasar Z 2020. The transformation of the inherited historical urban and architectural characteristics of Al-Najaf's Old City and possible preservation insights. *Frontiers of Architectural Research*,
[14] Farhan S, Jasim I and Al-Mamoori S 2019. The transformation of the city of Najaf, Iraq: Analysis, reality and future prospects. *Journal of Urban Regeneration & Renewal*, **13** 160-71.
[15] Abdulredha M, Rafid A, Jordan D and Alattabi A 2017. Facing up to waste: how can hotel managers in Kerbala, Iraq, help the city deal with its waste problem? *Procedia engineering*, **196** 771-8.
[16] Abdulredha M, Abdulridha A and Jordan D 2020. Estimating municipal solid waste generation from service processions during the Ashura religious event. *IOP Materials Science and Engineering,* 012075.
[17] Abdulredha M, Jordan D and Abdulridha A 2020. Investigating municipal solid waste management system performance during the Arba'een event in the city of Kerbala, Iraq. *Environment, Development and Sustainability*, **22** 1431-54.
[18] Farhan S, Antón D, Akef V and Safaa K 2021. Factors influencing the transformation of Iraqi holy cities: the case of Al-Najaf. *Scientific Review Engineering and Environmental Sciences*, **30** 365-75.
[19] Hashim K, Al-Saati N, Hussein A and Al-Saati Z 2018. An investigation into the level of heavy metals leaching from canal-dreged sediment: a case study metals leaching from dreged sediment. *1st Int. Conference on Materials Engineering & Science,* Istanbul Aydın University, Turkey 12-22.
[20] Alattabi A, Harris C, Alzeyadi A and Abdulredha M 2017. Online Monitoring of a sequencing batch reactor treating domestic wastewater. *Procedia engineering*, **196** 800-7.
[21] Hashim K, Ewadh H, Muhsin A, Muradov M, Aljefery M and Rafid M 2020. Phosphate removal from water using bottom ash: Adsorption performance, coexisting anions and modelling studies. *Water Science and Technology*, **83** 1-17.
[22] Alenezi A, Hasan H, Amoako-Attah J, Gkantou M, Muradov M and Abdulhadi B 2020. Zeolite-assisted electrocoagulation for remediation of phosphate from calcium-phosphate solution. *IOP Materials Science and Engineering,* 012031.
[23] Al-Marri S, AlQuzweeni S, AlKizwini R and Al-Khafaji Z S 2020. Ultrasonic-Electrocoagulation method for nitrate removal from water. *IOP Materials Science and Engineering,* 012073.
[24] Mohammed A-H, Hussein A, Yeboah D, Abdulhadi B and Safaa K 2020. Electrochemical removal of nitrate from wastewater. *IOP Conference Series: Materials Science and Engineering,* 012037.
[25] Alhendal M, Nasir M, Amoako J, Al-Faluji D and Abdulhadi B 2020. Cost-effective hybrid filter for remediation of water from fluoride. *IOP Materials Science and Engineering,* 012038.
[26] Abdulraheem F, Al-Khafaji Z, Hashim K, Muradov M and Shubbar A 2020. Natural filtration unit for removal of heavy metals from water. *IOP Materials Science and Engineering,* 012034.
[27] Omran I, Al-Saati N, Al-Saati Z, Patryk K, Khaddar R A, Al-Jumeily D and Aljefery M 2019. Assessment of heavy metal pollution in the Great Al-Mussaib irrigation channel. *Desalination and Water Treatment*, **168** 165-74.
[28] Hashim K, Zubaid S, Alwash R, Al-Jumeily D and Aljefery M 2020. Energy efficient electrocoagulation using baffle-plates electrodes for efficient Escherichia Coli removal from Wastewater. *Journal of Water Process Engineering*, **33** 101079-86.
[29] Khalid S, Ali S, AlRifaie J, Idowu I and Gkantou M 2020. Escherichia coli inactivation using a hybrid ultrasonic–electrocoagulation reactor. *Chemosphere*, **247** 125868-75.
[30] Alenazi M, Hassan A, Muradov M and Abdulhadi B 2020. Turbidity removal using natural coagulants derived from the seeds of strychnos potatorum: statistical and experimental approach. *IOP Materials Science and Engineering,* 012064.
[31] Al-Saati N, Hussein T, Abbas M, Al-Saati Z, Kot P, Sadique M, Aljefery M H and Carnacina I 2019. Statistical modelling of turbidity removal applied to non-toxic natural coagulants in water treatment: a case study. *Desalination and Water Treatment*, **150** 406-12.
[32] Aqeel K, Mubarak H, Amoako, Abdellatif M, Al-Janabi A and Khalid K S 2020. Electrochemical removal of brilliant green dye from wastewater. *IOP Materials Science and Engineering,* 012036.
[33] Alnaimi H, Idan I J, Al-Janabi A, Hashim K, Gkantou M, Zubaidi S L, Kot P and Muradov M 2020. Ultrasonic-electrochemical treatment for effluents of concrete plants. **888** 1-9.
[34] Alyafei A, AlKizwini R, Yeboah D, Gkantou M, Al-Faluji D and Zubaidi S 2020. Treatment of effluents of construction industry using a combined filtration-electrocoagulation method. *IOP Conference Series: Materials Science and Engineering,* 012032.

[35] Zanki A, Mohammad F, Muradov M, Kareem M and Abdulhadi B 2020. Removal of organic matter from water using ultrasonic-assisted electrocoagulation method. *IOP Materials Science and Engineering,* 012033.

[36] Emamjomeh M, Kakavand S, Jamali H, Safdari M, Mousavi S and Mousazade M 2020. The treatment of printing and packaging wastewater by electrocoagulation–flotation: the simultaneous efficacy of critical parameters and economics. *Desalination and water treatment*, **205** 161-74.

[37] Abdulhadi B, Safaa K, Andy A, Muradov M and Al-Khaddar R 2021. Continuous-flow electrocoagulation (EC) process for iron removal from water: Experimental, statistical and economic study. *Science of The Total Environment*, **760** 1-16.

[38] Hashim K, AlKhaddar R and Al-Shamma'a A 2021. Water purification from metal ions in the presence of organic matter using electromagnetic radiation-assisted treatment. *Journal of Cleaner Production*, **280** 1-17.

[39] Omran I, Al-Saati N, Al-Saati H, Hashim K and Al-Saati Z N 2021. Sustainability assessment of wastewater treatment techniques in urban areas of Iraq using multi-criteria decision analysis (MCDA). *Water Practice and Technology*, **16** 648-60.

[40] Grmasha R, Al-sareji O, Saffa and Jasim I 2020. Polycyclic Aromatic Hydrocarbons (PAHs) in Urban Street Dust Within Three Land-Uses of Babylon Governorate, Iraq: Distribution, Sources, and Health Risk Assessment. *Journal of King Saud University - Engineering Sciences*, **33** 1-18.

[41] Al-Sareji O, Grmasha R, Idowu I and Safaa K 2021. Street dust contamination by heavy metals in Babylon governorate, Iraq. *Journal of Engineering Science and Technology*, **16** 3528 - 46.

[42] Chabuk A, Hammood Z, Abed S, Kadhim M, Al-Ansari N and Laue J 2021. Noise Level in Textile Industries: Case Study Al-Hillah Textile Factory-Company for Textile Industries, Al-Hillah-Babylon-Iraq. *IOP Earth and Environmental Science,* 012048.

[43] Ali A, Al-Shaer A and Sadique M 2019. Investigating the influence of cement replacement by high volume of GGBS and PFA on the mechanical performance of cement mortar. *1st Int. Conference on Civil and Environmental Engineering Technologies,* University of Kufa, Iraq 31-8.

[44] Kadhim A, Sadique M, Al-Mufti R and Khalid S 2020. Developing One-Part Alkali-Activated metakaolin/natural pozzolan Binders using Lime Waste as activation Agent. *Advances in Cement Research*, **32** 1-38.

[45] Majdi H, Ali A, Nasr M, Jafer H, Abdulredha M, Masoodi Z A, Sadique M and Hashim K 2020. Experimental data on compressive strength and ultrasonic pulse velocity properties of sustainable mortar made with high content of GGBFS and CKD combinations. *Data in Brief*, **31** 105961-72.

[46] Al-Jumeily D, Alkaddar R, Al-Tufaily M and Lunn J 2019. Sustainable and Environmental Friendly Ancient Reed Houses (Inspired by the Past to Motivate the Future). *11th International Conference on Developments in eSystems Engineering (DeSE),* Cambridge, UK 214-9.

[47] Shubbar A, Sadique M, Nasr M, Al-Khafaji Z and Safaa K 2020. The impact of grinding time on properties of cement mortar incorporated high volume waste paper sludge ash. *Karbala International Journal of Modern Science*, **6** 1-23.

[48] Ali A, Sadique M, Shanbara H and Khalid K 2020 *The Development of a New Low Carbon Binder for Construction as an Alternative to Cement. In Advances in Sustainable Construction Materials and Geotechnical Engineering*, Berlin: Springer.

[49] Obaid M, Nasr M, Ali I, Shubbar A and Khalid S 2021. Performance of green mortar made from locally available waste tiles and silica fume. *J. of Engineering Science and Technology*, **16** 136-51.

[50] Emamjomeh M, Mousazadeh M, Mokhtari N, Jamali H, Makkiabadi M, Naghdali Z and Ghanbari R 2020. Simultaneous removal of phenol and linear alkylbenzene sulfonate from automotive service station wastewater: Optimization of coupled electrochemical and physical processes. *Separation Science and Technology*, **55** 3184-94.

[51] Abdulredha M, AlKhaddar R, Kot P, Jordan D and Abdulridha A 2018. Benchmarking of the Current Solid Waste Management System in Karbala, Iraq, Using Wasteaware Benchmark Indicators. *World Environmental and Water Resources Congress 2018: Groundwater, Sustainability, and Hydro-Climate/Climate Change,* 40-8.

[52] Abdulla G, Kareem M, Muradov M, Mubarak H, Abdellatif M and Abdulhadi B 2020. Removal of iron from wastewater using a hybrid filter. *IOP Materials Science and Engineering,* 012035.

[53] Khalid S, AlKhaddar R, Al-Jumeily D, Alwash R and Aljefery M H 2020 *Electrocoagulation as an eco-friendly River water treatment method. In Advances in Water Resources Engineering and Management*, Berline: Springer.

[54] Alwan H, Saleh L, Al-Mohammed F and Abdulredha M 2020. Experimental prediction of the discharge coefficients for rectangular weir with bottom orifices. *Journal of Engineering Science and Technology*, **15** 3265-80.

[55] Bareq B and Rafid R 2019. Influence of current density and electrodes spacing on reactive red 120 dye removal from dyed water using electrocoagulation/electroflotation (EC/EF) process. *1st Int. Conference on Civil and Environmental Engineering Technologies,* University of Kufa, Iraq.

[56] Kadhim A, Sadique M, Al-Mufti R and Hashim K 2020. Long-term performance of novel high-calcium one-part alkali-activated cement developed from thermally activated lime kiln dust. *Journal of Building Engineering*, **32** 1-17.

[57] Gkantou M, Muradov M, Kamaris G, Atherton W and Kot P 2019. Novel Electromagnetic Sensors Embedded in Reinforced Concrete Beams for Crack Detection. *Sensors*, **19** 5175-89.

[58] Ryecroft S, Andy S, Fergus P, Safaa K and Conway L 2019. A Novel Gesomin Detection Method Based on Microwave Spectroscopy. *12th International Conference on Developments in eSystems Engineering (DeSE),* Kazan, Russia 429-33.

[59] Omer G, Atherton W, Muradov M, Gkantou M, Riley M and Al-Shamma'a A 2020. A Non-Destructive Electromagnetic Sensing Technique to Determine Chloride Level in Maritime Concrete. *Karbala International Journal of Modern Science*, **6** 1-14.

[60] Ryecroft S, Andy A, Fergus P, Safa K, Moody A and Conway L 2019. A First Implementation of Underwater Communications in Raw Water Using the 433 MHz Frequency Combined with a Bowtie Antenna. *Sensors*, **19** 1813-23.

[61] Kot P, Muradov M, Gkantou M, Kamaris G and Yeboah D 2021. Recent Advancements in Non-Destructive Testing Techniques for Structural Health Monitoring. *Applied Sciences*, **11** 1-28.

[62] Omer G, Kot P, Atherton W, Muradov M, Gkantou M, Riley M and Al-Shamma'a A 2021. A Non-Destructive Electromagnetic Sensing Technique to Determine Chloride Level in Maritime Concrete. *Karbala International Journal of Modern Science*, **7** 61-71.

[63] Ryecroft S, Andy S, Fergus P, Tang A, Moody A and Conway L 2021. An Implementation of a Multi-Hop Underwater Wireless Sensor Network using Bowtie Antenna. *Karbala International Journal of Modern Science*, **7** 113-29.

Functionally Graded Antibacterial Biomaterial Coatings for Titanium Base Alloy

Sura B. Mohammed[1,*], Abdulraheem K. AbidAli[2]

[1] MSc Student, Dept. of Metallurgical Engineering, University of Babylon, Babylon, Iraq.
[2] Department of Metallurgical Engineering, University of Babylon, Babylon, Iraq.

Email: sura.muhammad@student.uobabylon.edu.iq

Abstract. Metal implants made of titanium and its alloys need surface modification operations because they are weak in terms of osseointegration and to protect them from corrosive body fluids and the growth of harmful bacteria that prevent them from performing their function and cause their failure. Hydroxyapatite and titanium dioxide have proven high efficiency by using them as biological and anti-bacterial coating. The basis of the current study is to deposit a functionally graded ceramic coating of titanium dioxide and hydroxyapatite on a Ti6Al4V substrate by electrophoretic deposition (EPD) method at room temperature with optimal conditions of 20 volts and a deposition time of 1 minute, and then sintering in an atmosphere of argon at a temperature of 950°C for one hour. In addition, polarisation, open circuit, anti-bacterial activity, XRD, SEM, EDS was conducted. The laboratory results showed that the corrosion rate decreases for the coated samples compared to the uncoated. As for the SEM examination, the cross-section images showed that the gradient coating is dense, cohesive and free from microscopic cracks. Also, the surface images showed that the surface was uniform. In terms of XRD, the results showed the presence of rutile, anatase, and hydroxyapatite phases in addition to α_CaPO and β_TCP as a result of HAp decomposition. For the anti-bacterial examination, the findings revealed that the coating is anti-bacterial.

1. Introduction

Titanium alloys are broadly used nowadays as metallic biomaterials. They are characterised by good biocompatibility, high corrosion resistance, favourable fatigue strength and high strength to weight ratio. In particular, they are applied in orthopaedic and dental implants because their Young's modules are relatively low among metallic materials [1,2]. Nevertheless, the development of passive oxides layers on titanium's surface makes it inert in the human body but has weak osteoinductive properties [3,4]. In addition, biofilm formation on titanium implants, which has a decisive influence on the persistence of bacterial infections, is a significant problem [5]. Modifying the surface of titanium biomaterials by coatings allows their anti-bacterial properties, biocompatibility and electrochemical corrosion resistance to be improved [6,7].

Electrophoretic Deposition (EPD) could be defined as the deposition of charged colloidal particles, in stable solutions, on the substrates having opposite charges using DC electric fields [8].
EPD is deemed a simple, quick, and multi-purpose coating method that allows changing molecules or charged particles in solution and sequent combination on the electrode to create different bulk materials and coatings with highly homogenous microstructure and suitable thickness [9].

IOP Conf. Series: Earth and Environmental Science **877** (2021) 012055 doi:10.1088/1755-1315/877/1/012055

Nevertheless, high temperatures for the sintering process are required for this type of coating (EPD) to attend mechanical integrity and binding to the substrate. These temperatures on a metallic substrate, which are usually above ≈950°C, cause the undesired breakdown of HA into the $Ca_3(PO_4)_2$ phases [10]. The deposits of HAp nano-particles could be sintered at noticeably low temperatures in comparison with the counterparts of micro-particles. The layers of HAp microparticles could be placed from the stable (non-agglomerated) solution, which is synthesised utilising the micro-size starting powders. Owing to the difference between the thermal expansion coefficient of the sintering process and the substrate, the cracks of HAp coating could happen. The considerable structural and chemical properties variance leads to the difficulty of forming the chemical binding between the HAp coating and Ti-alloy substrate at the interface. Titania (TiO_2) is considered a bio-compatible and bio-active material as it is very efficient against bacterial infections, can resist corrosion and has good mechanical properties. Using the titania into the HAp coating could result in strong binding and high mechanical strength of the composite coating and improve the growth and efficiency of the cells against bacterial infections [10,11].

Functionally graded materials could be defined as a particular class of composite materials, as their changing composition and/or microstructure are controllable in the space [12]. In traditional (not graded) composites, the constituent phases are distributed in average uniform in space, compromising their special properties. Instead, the graded structure combines various materials and does not cause serious interior stress in addition to combining diverse properties in one material system. In other words, the desirable properties of the materials could be totally utilised without forming a sudden interface between macro-domains of heterogeneous composition [12,13], particularly the feasibility of a TiO_2-HAp graded coating (as shown in the literature, [14]) by utilising the plasma spraying technology.

In this study, HAp-titania functionally graded coating was synthesised by EPD in solutions with various contents of HAp and titania microparticles. The optimal amounts of powders were calculated by trial and error.

2. Materials and method

2.1. Coating process

In this investigation, Ti-6Al-4V substrate was used as a cathode and commercially available 316L stainless steel as an anode were soaked in the solution with a constant distance of 15 mm. Before the deposition, the grinding process of Ti-6Al-4V substrates was done by 180 to 3000 grit SiC papers; after that, the polishing process for the specimens was achieved using polish cloth and diamond paste 0.5µm. Then the specimens washing was done using distilled water and ultrasonic with acetone to remove the oxide layer; finally, the specimens were dried and kept in a dissector over a pad of silica gel and utilised for microstructure evolution and electrochemical examination.

The EPD process was performed in a glass beaker; magnetic stirring was applied to keep the dispersion of the homogeneous particles during the whole EPD process, as described in Figure 1.
The system is equipped with a DC power supply that delivers potential ranging from zero to 60 volts. The voltage could be adjusted as needed to attend the optimal deposition. In order to obtain the best electrophoretic parameters, the deposition was carried out at various voltages (10, 20, 30,40, 60 V) and several deposition periods (1, 2, 3, 4, 5, 6, 30 and 60 min) from a suspension containing (50 g/L) % TiO2 and 6 g/L PEI.

Gradient coating was done in five consecutive steps. Each step lasted 12 seconds for each layer. The total time was 1 min. The graduated coating was carried out from solution No. 1, 2, 3, 4 and 5 consecutively, as shown in Table 1. Firstly, the deposition of the internal TiO_2 layer was carried out on the Ti-6Al-4V substrate from solution No.1, then three intermediate layers, TiO_2\HAp composite layers, were utilised from their relatively stable solutions (No. 2 and No. 3 and No. 4) respectively. Finally, the upper coating of HAp was placed on the four prior layers from solution No.5; the amounts of HAp vary progressively from 0 % (weight percentage) near the substrates to 100 % (weight percentage) on the upper coating layer.

IOP Conf. Series: Earth and Environmental Science **877** (2021) 012055 doi:10.1088/1755-1315/877/1/012055

Careful removal for the coated samples from the EPD cell was done, followed by a drying process in the air at room temperature for 24 hours, and finally sintering process using argon furnace at 950 C° for 1 hour. The adopted method in this research is presented in Figure 2, based on previous researches and trail mixes to produce a satisfactory coating.

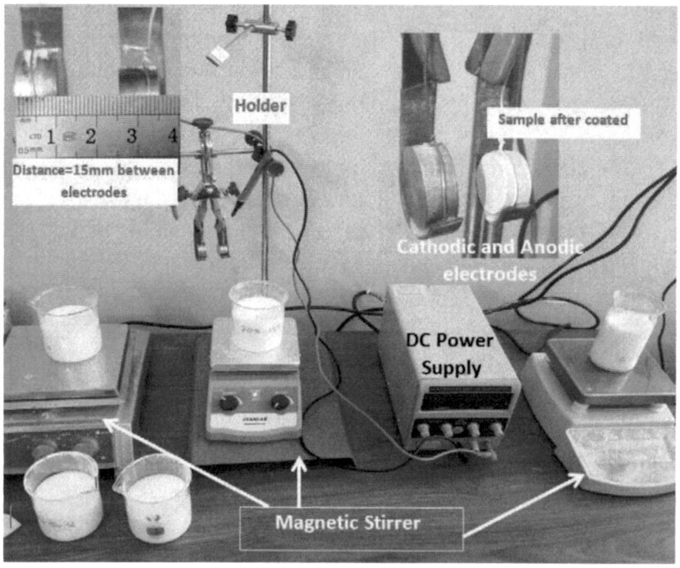

Figure 1: Garmenting processes of FG coating samples.

Table1: Chemical compositions of synthesised specimens.

Solution No.	Layers	TiO_2 (g)	HAp (g)	Absolute Ethanol (ml)	PEI (g/L)
No.1	1st	10	0	200	6
No.2	2nd	7	3		
No.3	3rd	5	5		
No.4	4th	3	7		
No.5	5th	0	10		

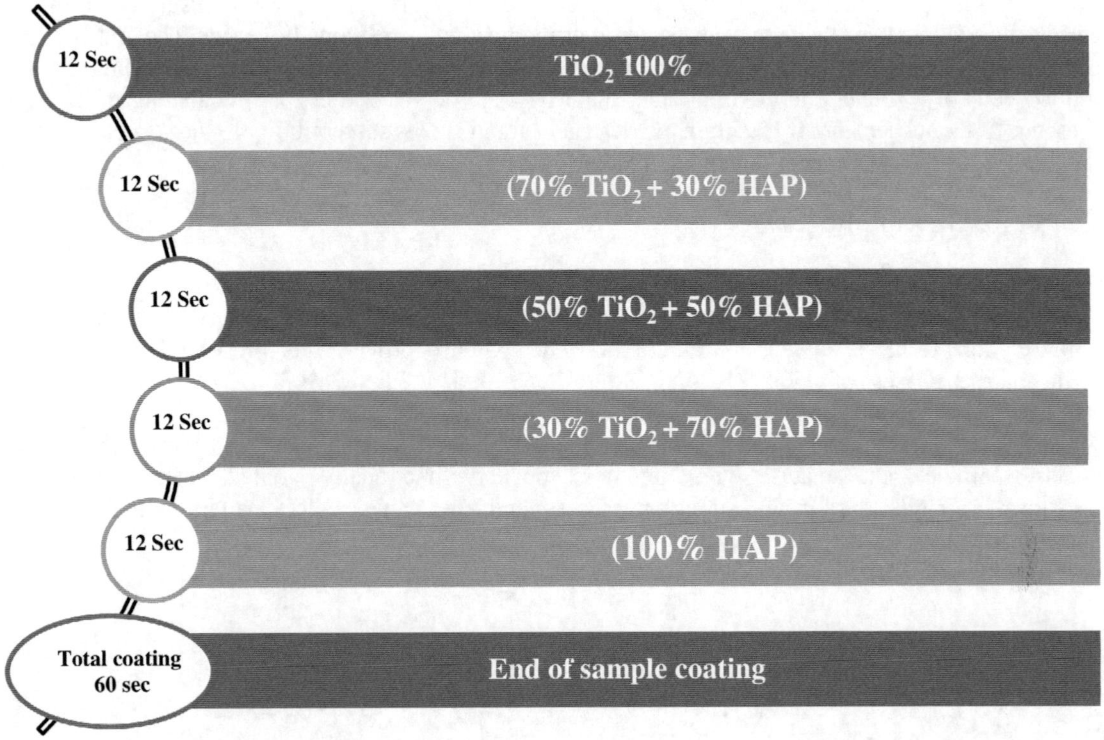

Figure 2: Sample coated time process.

2.2. *Characterisation of the coatings*

A scanning electron microscope (VEGA\\TESCAN) equipped with an energy dispersive spectrometer (EDS) was employed to analyse the coating's microstructural characterisation and elemental composition. In addition, X-ray diffraction (XRD, Philips PW 1480) was also used to analyse Phase compositions of the coated surface.

A potentiodynamic polarisation test was performed in a potentiostat (EG&G 263 A) according to the ASTM G44–99 standard [15] to assess the corrosions resistance of a control (uncoated) and coated specimens, and the samples were immersed into the simulated body fluid Ringer solution electrolyte. A saturated silver/silver chloride electrode was employed as a reference, with a platinum counter electrode; specimens were inserted as a working electrode. The chemical compositions of the ringer solution are illustrated in Table 2. The pH Ringer solution at 37±2°C was 5. The electrochemical test was conducted at a temperature of 37±2°C for all specimens.

Table 2: Chemical compositions of ringer's solution.

Components	Concentrations ($g.L^{-1}$)	PH
Sodium chloride (NaCl)	8.6	5
Potassium chloride (KCl)	0.3	
Calcium chloride dehydrate ($CaCl_2.2H_2O$)	0.33	

In this study, Gram-positive Escherichia coli (E.coli, ATCC8739) was considered as model cariogenic bacteria to evaluate the anti-bacterial properties of the developed surfaces. The anti-bacterial activity was studied by using the inhibition zone method. This method involves spreading about 1 million cells of a single strain over an agar plate by a sterile swab after that, incubating them in the presence of the suspension. If the strain of bacteria or fungi shows susceptibility to the suspension, then the inhibition zone will appear in the agar plate.

3. Test Results and Discussion

3.1. X-Ray Diffraction Test Result

The XRD pattern of sintered HAp/TiO_2 functionally graded coating in argon conditions at 950 °C for 1 hr is given in Figure 3. The pattern proves the presence of HAp and rutile/anatase phases of TiO2 (JCPDS No. 4-0551 and JCPDS No. 21-1272) in the coating. As the phase conversion of anatase to rutile happened at the temperatures range between 400-1000°C (which relays on the microstructural properties, impurity and particles size of anatase particle), the anatase TiO_2 in the deposited functionally graded coating begins to transform to rutile during the sintering process and the gained XRD patterns after sintering show the identical reflections of rutile phase.

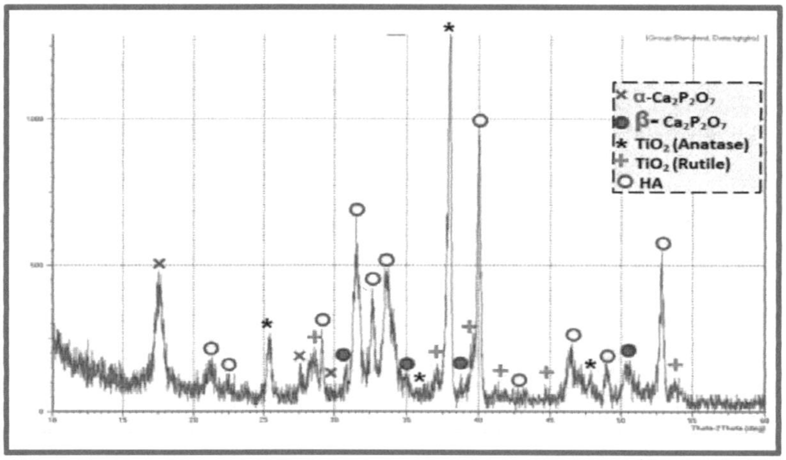

Figure 3: XRD pattern for the functionally graded HAp/TiO_2 coating specimen.

3.2. Scanning Electron Microscopy Analysis

The cross-sectional image of the HAp/TiO_2 functionally graded coating, Figure 4, shows its five layers as TiO_2 buried layer is close to the substrate, HAp upper layer and three HAp\\TiO_2 graded layers in the middle. As could be noticed, the introduced intermediate layer of TiO_2 and the graded compositions contributes to improving the adherence and connection of the coatings. The total thickness of the FGC HAp/TiO2 was calculated at about 173.0 μm. The TiO_2 layer is a portion of the total coating thickness. As the major purpose of the TiO_2 layer is to enhance HAp adherence. Figure 5 shows that the morphology of functionally graded coating, including different layers of HAp layer, TiO_2 layer, HAp/TiO_2 intermediate graded layers that were deposited at the deposition 20V for 1 minute.

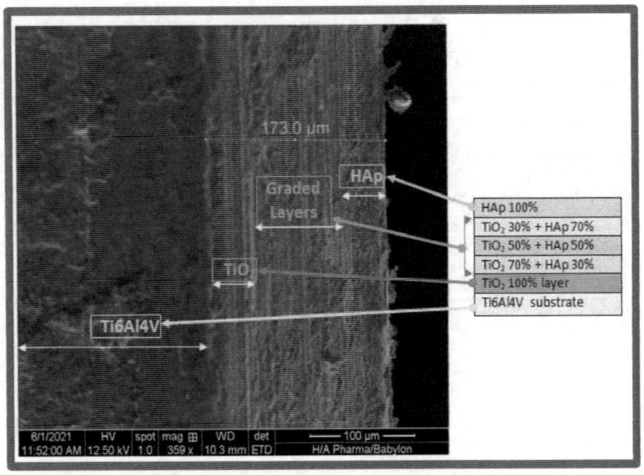

Figure 4: Schematic of SEM cross-section image for functionally graded sample.

A smooth, dense (microstructures with low void ratio), good homogenous surface with fewer pores in structure, no cracks and high uniformity in the coating were attended. The roughness and high porosity of the coating surface make nucleations and osseointegration of bone accumulated easily in forming the bone-producing cells.

Figure 5: SEM micrograph of the surface morphology of functionally graded coating on Ti6Al4V alloy.

3.3. Electrochemical Corrosion

The potentiodynamic polarisation curves of base Ti6Al4V, HAp, TiO_2 and FGC samples in the SBF (Ringer's suspension) are described in Figure 6. The metal's corrosion properties are mainly reported in the polarisation curve; this curve is usually utilised in studies of electrochemical corrosion. Higher corrosion resistance can be attended as a higher possibility of free corrosion, and a smaller current passivation yield was achieved.

The corrosion parameters gained from the polarisation curve applying the Tafel least square fitting method are in Table 3.

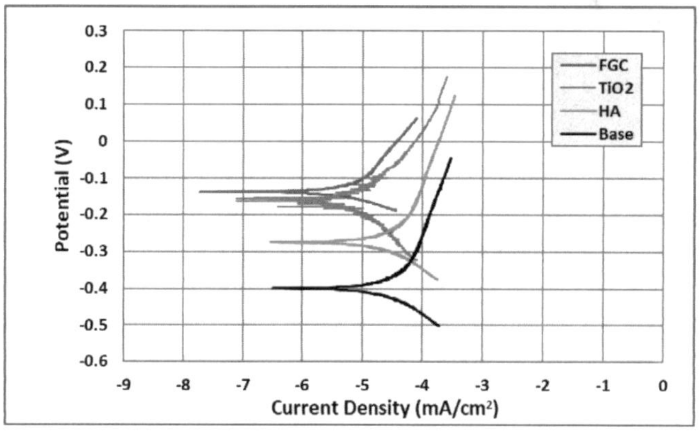

Figure 6: Potentiodynamic polarisation curves of coated and uncoated samples in the ringer's solution.

Table 3: Electrochemical parameters of Ti6Al4V substrate and coated samples in different media.

Media	Sample	E_{corr} (V)	I_{corr}(μA/cm^2)	CR (mpy)	$R_p \times (10^{-3})$
Ringer's Solution	Base	-0.84	20.09	0.176	1.29
	HA	-0.29	4.00	0.035	6.50
	TiO_2	-0.26	3.18	0.027	8.17
	FGC	-0.23	2.52	0.022	9.90

3.4. Anti-bacterial Study

The anti-bacterial effect of FGC HAp/TiO_2 and uncoated specimens were investigated against E. coli culture, as shown in Figure 7. The formation of a clear region around the disc refers to the bacterial inhibition zone. It can be observed that each case exhibited a good anti-bacterial effect towards three types of bacteria after 24 h of the incubation period, but the uncoated specimen has a small inhibition zone compared with the HAp/TiO_2 one. This is attributed to the anti-bacterial activity of the anatase TiO_2 itself [16], and therefore a strong anti-bacterial effect of the composite particles is obtained. Therefore, the presence of TiO_2 particles in the HAp coating increased its effectiveness in preventing the adhesion of bacteria on the surface, which prevented bacterial growth and would improve anti-bacterial activity.

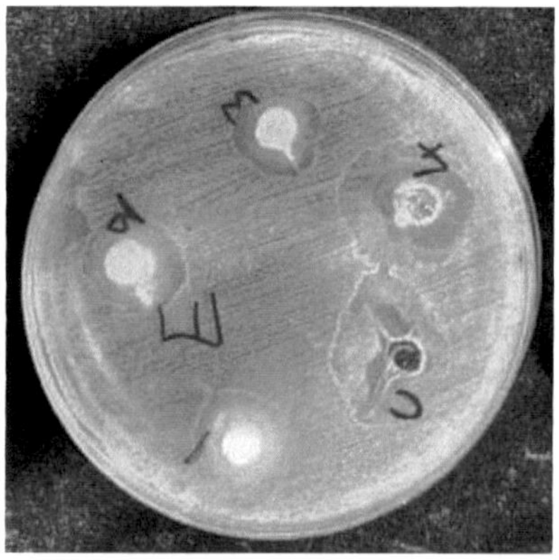

Figure 7: Images of the anti-bacterial tests of HAp/TiO_2 against E. coli.

4. Conclusions

The following points were obtained from this study:

1. The XRD analysis revealed the desired phases (hydroxyapatite and TiO2 (rutile and anatase)) needed in the coating applying synthesised suspension and heating remediation at 950°C. In addition to some bio-phases resulting from the decomposition of hydroxyapatite at sintering temperature.
2. The added amount of PEI to the HAp/TiO2 EPD bath led to more stability in suspension and deposition of uniform cracks free coating.
3. SEM images approved good adherence and cohesion of gradient coatings.
4. HAp/TiO_2 functionally graded coating offers protective effects and could inhibit releasing the ions of toxic metals from Ti6Al4V alloy in physiological environments. The levels of released ions were increased by decreasing the pH value of the solution.
5. The obtained coating could be effectively utilised to alter the surface of Ti6Al4V implants to enhance the biological properties and anti-bacterial activity.

References

[1] Jugowiec, D.; Kot, M.; Moskalewicz, T. Electrophoretic deposition and characterisation of chitosan coatings on near-_ titanium alloy. Arch. Metall. Mater. 2016, 61, 657–664.

[2] Khan, M.A.;Williams, R.L.;Williams, D.F. The corrosion behaviour of Ti-6Al-4V, Ti-6Al-7Nb and Ti-13Nb-3Zr in protein solutions. Biomaterials 1999, 20, 631–637.

[3] Elias, C.N.; Lima, J.H.C.; Valiev, R.; Meyers, M.A. Biomedical applications of titanium and its alloys. JOM 2008, 60, 46–49.

[4] Tamaddon, M.; Samizadeh, S.; Wang, L.; Blunn, G.; Liu, C. Intrinsic Osteoinductivity of Porous Titanium Scaffold for Bone Tissue. Engineering. Int. J. Biomater. 2017, 2017, 1–11.

[5] Secinti, K.D.; Özalp, H.; Attar, A.; Sargon, M.F. Nanoparticle Silver Ion Coatings Inhibit Biofilm Formation on Titanium Implants. J. Clin. Neurosci. 2011, 18, 391–395.

[6] Hea, D.; Zheng, S.; Pub, J.; Zhang, G.; Hub, L. Improving tribological properties of titanium alloys by combining laser surface texturing and diamond-like carbon film. Tribol. Int. 2014, 82, 20–27.

[7] Kaya, S.; Boccaccini, A.R. Electrophoretic deposition of zein coatings. J. Coat. Technol. Res. 2017, 14, 683–689.

[8] Partho Sarkar, Debnath De, Tetsuo Uchikochi and Laxmidhar Besra Electrophoretic Deposition (EPD): Fundamentals and Novel Applications in Fabrication of Advanced Ceramic Microstructures 2012.

[9] Corni I, Ryan MP, Boccaccini AR. Journal of the European Ceramic Society. 2008;28:1353.

[10] Ducheyne, S. Radin, J.C. Heughebaert, M. Heughebaert; "Calcium phosphate ceramics coatings on porous titanium: effect of structure and composition on electrophoretic deposition, vacuum sintering and in vitro dissolution"; Biomaterials 11 (1990) 244-254.

[11] H. Farnoush, J. Aghazadeh Mohandesi, D. Haghshenas Fatmehsari, F. Moztarzadeh; "A kinetic study on the electrophoretic deposition of hydroxyapatite– titania nanocomposite based on a statistical approach"; Ceram. Int.; 38 (2012) 6753–6767.

[12] Y. Miyamoto, W. A. Kaysser, B. H. Rabin, A. Kawasaki, and R. G. Ford, Functionally Graded Materials. Design, Processing and Applications, Kluwer Academic Publishers, Dordrecht, 1999.

[13] K. A. Khor, Y. W. Gu, C. H. Quek, and P. Cheang, ''Plasma Spraying of Functionally Graded Hydroxyapatite/Ti–6Al–4V Coatings,'' Surf. Coat. Technol., 168 195–201 (2003).

[14] V. Cannillo, L. Lusvarghi, and A. Sola, ''Production and Characterisation of Plasma Sprayed TiO2–Hydroxyapatite Functionally Graded Coatings,'' J. Eur. Ceram. Soc., 28 2161–2169 (2008).

[15] ASTM G44-99, 'Standard Practice for Exposure of Metals and Alloys by Alternate Immersion in Neutral 3.5 % Sodium Chloride Solution, January 2000.
[16] A. Kocijan, M. Conradi and M. Hocevar, The influence of surfacewettability and topography on the bioactivity of TiO2/Epoxy coatings on AISI 316L stainless steel, Materials, Vol. 12, No. 11, 2019.

AUTHOR INDEX